	Design or Purpose	Measurement Variables	Ranked Variables	Attributes
1 variable 2 samples only	Independent variates	Back-to-back stem-and-leaf display, Section 2.5 Single-classification anova; unequal sample sizes, Box 9.1; equal sample sizes, Box 9.5 *t*-test comparing two means or other statistics of location, Box 9.6 *t*-test comparing a single observation with a sample mean, Box 9.7 Test of equality of two variances, Box 8.1 Testing difference between two means with unequal variances, Box 13.4	Mann–Whitney *U*-test, Box 13.7 Kolmogorov–Smirnov two-sample test, Box 13.9	Testing difference between two percentages, Box 17.6 Testing difference between two counts, Section 13.8 Sample size required to detect a given difference in percentages, Box 17.13 Log-odds ratios, Section 17.6
	Paired variates	Paired comparisons test, Box 11.3	Wilcoxon's signed ranks test, Box 13.11	Sign test, Section 13.11 Repeated testing of the same individuals, Box 17.16
2 variables 1 sample	Prediction or functional relationship	Regression statistics: computation, Box 14.1; standard errors, Box 14.2; confidence limits, Box 14.3; outliers and residuals, Section 14.11; regression through the origin, Section 14.4 Estimation of *X* from *Y*, Box 14.6 Polynomial regression, Box 16.5 Model II regression, Box 14.11	Kendall's robust line fit method, Section 14.12 Ordering test, Section 14.12	Runs-up-and-down test for trends, Box 18.3
	Association	Computation of correlation coefficient: from unordered sample, Box 15.2; from two-way frequency distribution, Box 15.3 Significance tests, confidence limits, and required sample size for correlation coefficients, Box 15.4; by jackknife, Box 18.6 Principal axes and confidence regions for bivariate scattergram, Box 15.6 Olmstead and Tukey's corner test for association, Box 15.8 Mantel test, Box 18.5	Coefficient of rank correlation: Kendall's, Box 15.7; Spearman's, Section 15.8	2×2 test for independence: by *G*-test, Box 17.6; by chi-square test, Section 17.4; by Fisher's exact test, Box 17.7 The phi coefficient, Section 17.4 $R \times C$ test for independence, Box 17.8 Unplanned tests of homogeneity of $R \times C$ tables, Section 17.4
2 variables ≥ 2 samples	Prediction or functional relationship	Comparison of regression coefficients, Box 14.7; unplanned comparison by T′ nd GT2, Box 14.8 Analysis of covariance, Box 14.9		
	Association	Tests of homogeneity of correlation coefficients, Box 15.5		The Mantel–Haenszel procedure, B[illegible]
2 variables ≥ 3 samples	Prediction or functional relationship	Each sample represents a different value of *X*—regression with more than one value of *Y* per value of *X*, Box 14.4; standard errors, Box 14.2; confidence limits, Box 14.5, also Section 14.10 Test for curvilinear regression by orthogonal polynominals, Box 16.6 Isotonic regression, Box 18.4	Page's *L*-test for ordered alternatives, Section 14.12	Logistic regression, Box 17.15; wi[illegible] multistate categorical variable, B[illegible]
≥ 3 variables 1 sample	Prediction or functional relationship	Multiple regression: computation, Box 16.1; standard errors, Box 16.3; confidence limits, Box 16.2; path coefficients, Table 16.2		
	Association	Coefficient of multiple correlation, Box 16.1; coefficient of partial correlation, Section 16.4	Kendall's coefficient of concordance, Section 15.8	Log-linear model analysis of a thr[illegible] table, Box 17.10, Box 17.11

BIOMETRY

THIRD EDITION

BIOMETRY

THE PRINCIPLES AND PRACTICE OF STATISTICS IN BIOLOGICAL RESEARCH

THIRD EDITION

Robert R. SOKAL and F. James ROHLF
State University of New York at Stony Brook

W. H. FREEMAN AND COMPANY
New York

Library of Congress Cataloging-in-Publication Data

Sokal, Robert R.
Biometry : the principles and practice of statistics in biological research / Robert R. Sokal and F. James Rohlf.—3d ed.
p. cm.
Includes bibliographical references (p. 850) and index.
ISBN 0-7167-2411-1
1. Biometry. I. Rohlf, F. James, 1936– . II. Title.
QH323.5.S63 1995
574′.01′5195—dc20 94-11120
CIP

Printed in the United States of America

2 3 4 5 6 7 8 9 VB 9 9 8 7 6 5

To our parents of blessed memory
Klara and Siegfried Sokal
Harriet and Gilbert Rohlf

CONTENTS

PREFACE xiii

NOTES ON THE THIRD EDITION xvii

1 INTRODUCTION 1

1.1 Some Definitions 1
1.2 The Development of Biometry 3
1.3 The Statistical Frame of Mind 5

2 DATA IN BIOLOGY 8

2.1 Samples and Populations 8
2.2 Variables in Biology 10
2.3 Accuracy and Precision of Data 13
2.4 Derived Variables 16
2.5 Frequency Distributions 19

3 THE HANDLING OF DATA 33

3.1 Computers 34
3.2 Software 35
3.3 Efficiency and Economy in Data Processing 37

4 DESCRIPTIVE STATISTICS 39

4.1 The Arithmetic Mean 40
4.2 Other Means 43
4.3 The Median 44
4.4 The Mode 47
4.5 The Range 48
4.6 The Standard Deviation 49

4.7 Sample Statistics and Parameters 52
4.8 Coding Data Before Computation 53
4.9 Computing Means and Standard Deviations 54
4.10 The Coefficient of Variation 57

5 INTRODUCTION TO PROBABILITY DISTRIBUTION: BINOMIAL AND POISSON 61

5.1 Probability, Random Sampling, and Hypothesis Testing 62
5.2 The Binomial Distribution 71
5.3 The Poisson Distribution 81
5.4 Other Discrete Probability Distributions 93

6 THE NORMAL PROBABILITY DISTRIBUTION 98

6.1 Frequency Distributions of Continuous Variables 98
6.2 Properties of the Normal Distribution 101
6.3 A Model for the Normal Distribution 106
6.4 Applications of the Normal Distribution 109
6.5 Fitting a Normal Distribution to Observed Data 111
6.6 Skewness and Kurtosis 111
6.7 Graphic Methods 116
6.8 Other Continuous Distributions 123

7 ESTIMATION AND HYPOTHESIS TESTING 127

7.1 Distribution and Variance of Means 128
7.2 Distribution and Variance of Other Statistics 136
7.3 Introduction to Confidence Limits 139
7.4 The *t*-Distribution 143
7.5 Confidence Limits Based on Sample Statistics 146
7.6 The Chi-Square Distribution 152
7.7 Confidence Limits for Variances 154
7.8 Introduction to Hypothesis Testing 157
7.9 Tests of Simple Hypotheses Using the Normal and *t*-Distributions 169
7.10 Testing the Hypothesis H_0: $\sigma^2 = \sigma_0^2$ 175

8 INTRODUCTION TO THE ANALYSIS OF VARIANCE 179

8.1 Variances of Samples and Their Means 180
8.2 The *F*-Distribution 184
8.3 The Hypothesis H_0: $\sigma_1^2 = \sigma_2^2$ 189

8.4 Heterogeneity Among Sample Means 190
8.5 Partitioning the Total Sum of Squares and Degrees of Freedom 197
8.6 Model I Anova 201
8.7 Model II Anova 203

9 SINGLE-CLASSIFICATION ANALYSIS OF VARIANCE 207

9.1 Computational Formulas 208
9.2 General Case: Unequal *n* 208
9.3 Special Case: Equal *n* 217
9.4 Special Case: Two Groups 219
9.5 Special Case: A Single Specimen Compared With a Sample 227
9.6 Comparisons Among Means: Planned Comparisons 229
9.7 Comparisons Among Means: Unplanned Comparisons 240
9.8 Finding the Sample Size Required for a Test 260

10 NESTED ANALYSIS OF VARIANCE 272

10.1 Nested Anova: Design 272
10.2 Nested Anova: Computation 275
10.3 Nested Anovas With Unequal Sample Sizes 292
10.4 The Optimal Allocation of Resources 309

11 TWO-WAY ANALYSIS OF VARIANCE 321

11.1 Two-Way Anova: Design 321
11.2 Two-Way Anova With Equal Replication: Computation 323
11.3 Two-Way Anova: Significance Testing 331
11.4 Two-Way Anova Without Replication 342
11.5 Paired Comparisons 352
11.6 Unequal Subclass Sizes 357
11.7 Missing Values in a Randomized-Blocks Design 363

12 MULTIWAY ANALYSIS OF VARIANCE 369

12.1 The Factorial Design 369
12.2 A Three-Way Factorial Anova 370
12.3 Higher-Order Factorial Anovas 381
12.4 Other Designs 385
12.5 Anovas by Computer 387

13 ASSUMPTIONS OF ANALYSIS OF VARIANCE 392

13.1 A Fundamental Assumption 393
13.2 Independence 393
13.3 Homogeneity of Variances 396
13.4 Normality 406
13.5 Additivity 407
13.6 Transformations 409
13.7 The Logarithmic Transformation 413
13.8 The Square-Root Transformation 415
13.9 The Box–Cox Transformation 417
13.10 The Arcsine Transformation 419
13.11 Nonparametric Methods in Lieu of Single-Classification Anovas 423
13.12 Nonparametric Methods in Lieu of Two-Way Anova 440

14 LINEAR REGRESSION 451

14.1 Introduction to Regression 452
14.2 Models in Regression 455
14.3 The Linear Regression Equation 457
14.4 Tests of Significance in Regression 466
14.5 More Than One Value of *Y* for Each Value of *X* 476
14.6 The Uses of Regression 486
14.7 Estimating *X* from *Y* 491
14.8 Comparing Regression Lines 493
14.9 Analysis of Covariance 499
14.10 Linear Comparisons in Anovas 521
14.11 Examining Residuals and Transformations in Regression 531
14.12 Nonparametric Tests for Regression 539
14.13 Model II Regression 541

15 CORRELATION 555

15.1 Correlation and Regression 556
15.2 The Product–Moment Correlation Coefficient 559
15.3 The Variance of Sums and Differences 567
15.4 Computing the Product–Moment Correlation Coefficient 569
15.5 Significance Tests in Correlation 574
15.6 Applications of Correlation 583
15.7 Principal Axes and Confidence Regions 586
15.8 Nonparametric Tests for Association 593

16 MULTIPLE AND CURVILINEAR REGRESSION 609

16.1 Multiple Regression: Computation 610
16.2 Multiple Regression: Significance Tests 623
16.3 Path Analysis 634
16.4 Partial and Multiple Correlation 649
16.5 Choosing Predictor Variables 654
16.6 Curvilinear Regression 665
16.7 Advanced Topics in Regression and Correlation 678

17 ANALYSIS OF FREQUENCIES 685

17.1 Introduction to Tests for Goodness of Fit 686
17.2 Single-Classification Tests for Goodness of Fit 697
17.3 Replicated Tests of Goodness of Fit 715
17.4 Tests of Independence: Two-Way Tables 724
17.5 Analysis of Three-Way and Multiway Tables 743
17.6 Analysis of Proportions 760
17.7 Randomized Blocks for Frequency Data 778

18 MISCELLANEOUS METHODS 794

18.1 Combining Probabilities From Tests of Significance 794
18.2 Tests for Randomness of Nominal Data: Runs Tests 797
18.3 Randomization Tests 803
18.4 The Jackknife and the Bootstrap 820
18.5 The Future of Biometry: Data Analysis 825

APPENDIX: MATHEMATICAL PROOFS 833

BIBLIOGRAPHY 850

AUTHOR INDEX 865

SUBJECT INDEX 871

PREFACE

The success of the first two editions of *Biometry* among teachers, students, and others who use biological statistics has encouraged us to prepare this extensively revised Third Edition.

We wrote and have revised this book because we feel that there is a need for an up-to-date text aimed primarily at the academic biologist—a text that develops the subject from an elementary introduction up to the advanced methods necessary nowadays for biological research and for an understanding of the published literature. Many available texts represent the outlook and interests of agricultural experiment stations. This is quite proper in view of the great application of statistics in this field; in fact, modern statistics originated at such institutions. However, personal inclination and the nature of the institution at which we teach cause us to address ourselves to general biologists, ecologists, geneticists, physiologists, and other biologists working largely on nonapplied subjects in universities, research institutes, and museums. Considerable overlap exists between the needs of these two somewhat artificially contrasted groups and, of necessity, some of our presentation will deal with agricultural experimentation. More broadly, the statistical methods treated here are useful in many applied fields, including medicine and allied health sciences. Since it is a well-known pedagogical dictum that people learn best by familiar examples, we have endeavored to make our examples as pertinent as possible for our readers.

Much agricultural and biological work is by its very nature experimental. This book, while furnishing ample directions for the analysis of experimental work, also stresses descriptive and analytical statistical study of biological phenomena. These powerful methods are often overlooked and sometimes, by implication, the validity of nonexperimental biological work is put in doubt. We think that descriptive, analytical, and experimental approaches are all of value, and we have tried to strike a balance among them. This approach will be appreciated by workers in applied fields such as the health sciences where ethical, financial, or other considerations may prevent free use of direct experiment.

The readers we hope to interest are graduate students in biological departments who require a knowledge of biometry as part of their profesional training and professional biologists in universities, museums, and research positions who for one reason or another did not obtain biometric training during their student careers—that is, people who need to acquire knowledge largely in an autodidactic way. This book has been designed to serve not only as a text to accompany a lecture course but also as a complete volume for self-study.

We aim to instill in our students an ability to think through biological research problems in such a way as to grasp the essentials of the experimental or analytical setup, to know which types of statistical tests to apply in a given case, and to carry out the computations required. A modicum of statistical theory has to be presented to introduce the subject matter, but once this is done, further knowledge of statistical theory is acquired as part of the learning process in solving statistical problems, rather than as a preamble to their solution. Because of limitations of time and the minimal mathematical preparation of most students, discussion of theory must be curtailed and some methods are presented almost entirely in "cookbook" style.

Over the years we have found that students with very limited mathematical backgrounds can handle biometry excellently. There is remarkably little correlation between innate mathematical ability and capacity to understand the ordinary biometric methodologies. By contrast, in our experience a very high correlation exists between success in the biometry course and success in the chosen field of biological specialization. Teachers should encourage students who feel they have no mathematical aptitude, since it does not apply to the study of applied statistics.

We must make our apologies to the mathematical statisticians who may chance on this volume. Compromises had to be made on occasion between precise definitions as required by mathematical statistics and formulations of theory that teaching experience has shown to be most easily understood by beginning classes. In judging our effort, allowance should be made for such changes and circumscriptions.

To make our book more useful to the autodidactic reader, we have employed a more direct and personal style than is customary in textbooks of this sort. No book can replace a dedicated and able teacher, but we have tried to make our written approach as similar as we can to our classroom manner.

ARRANGEMENT OF THIS BOOK

Since this book deviates in several important ways from the customary statistics textbook, we would like to point out the differences so that the prospective reader may derive the maximum benefit from its use. Subject matter is arranged in chapters and sections, numbered by the conventional decimal numbering system: The number before the decimal point refers to the chapter, the number after the point to the section. Topics can be referred to in the table of contents as well as in the index. The tables are also numbered with the decimal system: The first number denotes the chapter, the second the number of the table within the chapter.

Certain special tables are designated "boxes" and are numbered as such, again using the decimal system. The boxes serve a double function. First, they illustrate computational methods for solving various types of biometric problems and can therefore be used as convenient patterns for computation. The boxes usually contain all the steps necessary—from the initial setup of the problem to the final result; hence to students familiar with the book, the boxes can serve as quick summary reminders of the technique. A second important use of the boxes relates to their origin as mimeographed sheets handed out in a biometry course at the University of Kansas. In the allowed lecture time the instructor would not have been able to convey even as much as half the subject matter covered in the course (or in this book) if the material contained in the sheets had to be put on the chalkboard. Asking the students to refer in class to their biometry sheets (now the boxes in this book) saved much time, and, incidentally, students were able to devote their attention to understanding the content of the sheets rather than to copying them. Instructors who employ this book as a text may wish to use the boxes in a similar manner.

Figures are numbered by the same decimal system as tables and boxes. A similar numbering system is also used for (sampling) experiments, (mathematical) expressions, and (homework) exercises.

Since the emphasis in this book is on practical applications of statistics to biology, discussions of statistical theory are deliberately kept to a minimum. However, we feel that some exposure to statistical theory is a good idea for any biologist, so we have provided some in various places. Derivations are given for a number of formulas but these are consigned to the appendix, where they should be studied and reworked by the student.

We present what we consider the minimum statistical knowledge required for a Ph.D. in the biological sciences at the present time. This material is of equal importance for basic research and for applications in areas such as medicine and allied health sciences. The most important topics treated are the simple distributions (binomial, Poisson, and normal), simple statistical tests (including nonparametric tests), analysis of variance through factorial analysis, analysis of frequencies, simple and multiple regression, and correlation. This is more material than is covered in most elementary texts, yet even this much is usually not sufficient to treat even the common problems of everyday biological research. To enable the reader to become acquainted with some advanced methods, we have therefore introduced at appropriate places throughout the book so-called signpost paragraphs, set in italics and bordered by a grey bar to the left. Signpost paragraphs explain some of the advanced methods and their application and refer the reader to books and papers that we have found to be most useful in explaining a given technique in simple and instructive fashion. We hope that these signposted discussions will open up the wider field of statistics for the readers of this book and permit them at the very least to consult professional statisticians intelligently.

Homework exercises for students reading this book in connection with a biometry course (or studying on their own) are given at the end of each chapter. In keeping with our own convictions, these are largely real research problems. Some of them, therefore, require a fair amount of computation for their solution.

We have been dissatisfied with the customary placement of statistical tables at the end of textbooks of biometry and statistics. Users of these books and tables are constantly inconvenienced by having to turn back and forth between the text material on a certain method and the table necessary for the test of significance or for some other computational step. Occasionally, the tables are interspersed throughout a textbook at sites of their initial application; they are then difficult to locate, and turning back and forth in the book is not avoided. Constant users of statistics, therefore, generally use one or more sets of statistical tables, not only because such handbooks usually contain more complete and diverse statistical tables than do the textbooks, but also to avoid the constant turning of pages in the latter.

When we first planned this textbook, we thought we would eliminate statistical tables altogether, asking readers to furnish their own from those available. However, for pedagogical reasons, it was found desirable to refer to a standard set of tables, and we consequently undertook to furnish them, bound separately from the text. The *Statistical Tables,* Third Edition (Rohlf and Sokal, W. H. Freeman and Company, 1995), are identified by boldface letters (for example, Table **A**).

* * * * * * * *

In preparing the First and Second Editions of this book, we had the benefit of constructive criticism from two colleagues. Professor K. R. Gabriel (University of Rochester) and Professor R. C. Lewontin (Harvard University) read the entire manuscript and commented in detail on all aspects of the book. Many of their suggestions were incorporated and have contributed significantly to any merit the book possesses. An equally thorough editorial job was carried out by Dr. Michel Kabay (then of the Université de Moncton in New Brunswick, Canada). Professors F. J. Sonleitner (University of Oklahoma), Theodore J. Crovello (then of the University of Notre Dame), and Albert J. Rowell (University of Kansas) also furnished many detailed comments on the text of the First Edition. Many students in various biometry classes also contributed comments on several versions of the text. Since not all suggestions were accepted, the reviewers named here cannot be held responsible for errors and other failings in the book.

We are obligated to the many generations of biometry students who have enabled us to try out our ideas for teaching biometry and whose responses and criticism have guided us in developing our course. It has been a never-ending source of satisfaction to see these students appropriate biometric techniques as their own and employ them actively in their researches.

Preparation of the Third Edition was greatly facilitated by Barbara Thomson, who word processed all the changes and new materials for the manuscript and helped with checking the proofs. Her familiarity with the book made her assistance invaluable. Donna DiGiovanni assisted with various tasks during the final stages of proofing and indexing.

To Julie we owe a continuing debt of gratitude for her forebearance during the long periods of gestation of three editions of this book.

Robert R. Sokal
F. James Rohlf
July 1994

NOTES ON THE THIRD EDITION

We trust that the changes in the Third Edition of *Biometry,* outlined here, will substantially enhance the book's utility.

When the First Edition was published in 1969, most calculations in biometry were carried out on mechanical desk calculators. Electronic desk calculators were just beginning to make an appearance. The availability of handheld pocket calculators was still several years away. Heavy, repetitive calculations were carried out on large mainframe computers. By now, most statistical computations, except for very large tasks, are carried out on personal computers or work stations, making obsolete the desk calculator formulas we furnished in the text and in the boxes of earlier editions of the book. We have replaced these formulas with simpler structural formulas that readers will find easier to understand because they furnish direct expressions of the nature of the computation. Furthermore, these structural formulas can usually be programmed directly for digital computers and are typically not subject to the same magnitude of rounding error as desk calculator equations.

Removing the desk calculator formulas featured in the previous editions freed up a substantial amount of space, which we have used to introduce a number of new topics that persons embarking on a study or practice of biometry in the 1990s need to learn. Here we list the new topics in the sequence of their appearance in the book. Additional homework exercises have been included as practice in all new methods and as further practice in earlier methods.

In Section 2.5 we now feature back-to-back stem-and-leaf displays for comparing two samples. Chapter 3, on the handling of data, has been rewritten to keep abreast of the developments in computing hardware and software. Section 5.1, on probability, has been revised extensively and now includes Bayes' theorem. We now feature boxplots and quantile–quantile plots in Sections 6.5 and 6.7, respectively. The Bonferroni method of multiple testing is now given in Section 9.6. In Chapter 13 we feature a test of independence of errors in a sequence of continuous variates (Section 13.2), a test for the difference betwen two counts (Section 13.8), and a nonparametric test for replicated two-way analysis of variance (Section 13.12). New features of the chapter on regression, Chapter 14, are regression through the origin (Section 14.4), the logistic equation and logits (Section 14.11), and Kendall's robust line-fit method

(Section 14.13). In Section 16.5 we feature Kruskal's method for determining the relative importance of predictor variables in multiple regression. Chapter 17, "Analysis of Frequencies," contains many changes. In Section 17.2 we present a revised Kolmogorov–Smirnov test using Khamis's adjustment; in Section 17.4 we have added three methods, a nonparametric test of equality of medians, multiple comparisons in $R \times C$ frequency tables, and the phi coefficient of association for a 2×2 table. An entirely new Section 17.6, on the analysis of proportions, covers the log–odds ratio, the Mantel–Haenszel procedure, and logistic regression and mentions Berkson's fallacy, Neyman's example, and Simpson's paradox. Finally, a revised McNemar test, in which the order of presentation of stimuli is tested, is featured in Section 17.7. Extensive changes have also been made in Chapter 18, "Miscellaneous Methods." In addition to mentioning meta-analysis in Section 18.1, we feature the Mantel test for comparing two dissimilarity matrices and isotonic regression in Section 18.3 and the widely used bootstrap procedure in Section 18.4.

Recent statistical literature has presented many improvements of standard statistical tests by transformations of data and corrections to the test statistics. In mentioning some of these and omitting others, we have tried to strike a balance between adding every suggestion in the literature and ignoring them all. The danger in preparing a new edition of a text such as this is adding too many minute changes, making it difficult for the reader to see the main outlines of the methods and to decide among the various alternatives offered. Too many methods are confusing to the novice. Inevitably, however, the book has become more diverse and the number of topics has increased.

The tabular guides to types of problems encountered frequently in biological research, found in the endpapers, have been updated to include the new methods introduced in the Third Edition.

Accompanying the publication of the Third Edition is a new version (3.0) of the BIOM-pc programs. These programs are based on the Third Edition and feature the major methods the new edition lists. The programs have been made easier to use and more powerful. Written by F. James Rohlf, the programs are distributed by Exeter Software, 100 North Country Road, Building B, Setauket, New York 11733.

The Third Edition continues to reflect our many years of experience teaching biometry to classes of graduate students in various departments in the life, health, and earth sciences at the University of Kansas, the University of California, Santa Barbara, the State University of New York at Stony Brook, and various universities abroad. In these courses we have emphasized "doing." For the last twenty years or so our students have been taught to use computer programs. By a series of assignments they are then taught how to solve biological problems statistically, both by thinking through the nature of the problem and by carrying out the computations. Restricting homework exercises to those that are simple to do (hence not time-consuming) was an approach we rejected even during the era of the mechanical calculator. With the present general availability of computers, there is no further reason to shield students from real-life data. Students have to be taught the solution of real biological problems involving complicated and copious data, similar to those that would be obtained in an actual research project.

The fear expressed by some teachers that students would not learn statistics well if they were permitted to use canned computer programs has not been realized in our experience. A careful monitoring of achievement levels before and after the introduction of computers in the teaching of our course revealed no appreciable change in students' performance.

Robert R. Sokal
F. James Rohlf
July 1994

1 INTRODUCTION

Dear Reader:

If you are starting your study of this book without looking at the preface, we respectfully but firmly urge you to turn back and read it. We realize that prefaces are often uninspiring reading. However, the preface in this book contains important information for the user. In it we explain our reasons for writing this book and describe the audience we address. More important, however, we explain our philosophy of teaching statistics and how we expect the arrangement of this book to meet our objectives. Unless the system of the book is understood, we fear that you are not likely to receive full benefit from reading it. Thank you.

The Authors

This introductory chapter is concerned with setting the stage for your study of biometry. First we define the field itself (Section 1.1). Then we cast a necessarily brief glance at its historical development (Section 1.2). Section 1.3 concludes the chapter with a discussion of the attitudes someone trained in statistics brings to biological research. The application of statistics to biology has revolutionized not only the methodology of research but the very interpretation of the phenomena under study. The effects this philosophical attitude has had and is having on biology are outlined briefly. We hope that the reader of this book will absorb and appropriate some of these points of view.

1.1 SOME DEFINITIONS

Scientific etiquette demands that a field be defined before its study is begun. The Greek roots of ''biometry'' are *bios* (''life'') and *metron* (''measure''); hence **biometry** is literally the measurement of life. For the purposes of this book biometry is conceived in a very broad sense. We define it as *the application of statistical methods to the solution of biological problems.* As seen in Section 1.2, the original meaning of biometry was much narrower and implied a special field related to the study of evolution and natural selection; however, the wider definition is customary now. Biometry is also called *biological statistics* or simply *biostatistics.*

Our definition tells us that biometry is the application of statistics to biological problems, but the definition leaves us somewhat up in the air, since "statistics" has not been defined. Statistics is a science that by name at least is well known even to the layperson, who quite often unjustly abuses the term. The number of definitions you can find is limited only by the number of books you wish to consult. In its modern sense we might define **statistics** as *the scientific study of data describing natural variation.* All parts of this definition are important and deserve emphasis.

Scientific study: We are concerned with the commonly accepted criteria of validity of scientific evidence. Objectivity in presenting and evaluating data and the general ethical code of scientific methodology must constantly be in evidence if the old canard that "figures never lie, only statisticians do" is not to be revived.

Data: Statistics generally deals with populations or groups of individuals; hence it deals with *quantities* of information, not with a single *datum.* Thus the measurement of a single animal or the response from a single biochemical test will generally not be of interest; unless a sample of animals is measured or several such tests are performed, statistics ordinarily can play no role. (We are old-fashioned enough to insist that "data" be used only in its plural sense, despite the latest dictionaries.) Unless the data of a study can be quantified in one way or another, they are not amenable to statistical analysis. The data can be measurements (the length or width of a structure or the amount of a chemical in a body fluid) or counts (the number of bristles or teeth, or frequencies of different qualitative states such as mutant eye colors). Types of variables will be discussed in greater detail in Chapter 2.

Natural variation: We use this term in a wide sense to include all those events that happen in animate and inanimate nature not under the direct control of the investigator, plus those that are evoked by the scientist and are partly under his or her control, as in an experiment. Different biologists concern themselves with different levels of variation; other kinds of scientists with yet different levels; but all scientists agree that variation in the chirping of crickets, the number of peas in a pod, and the age at maturity in a chicken is natural. The heartbeat of rats in response to adrenalin or the mutation rate in maize after irradiation may still be considered natural, even though the researcher has manipulated the phenomenon in an experiment. The average biologist, however, would not consider variation in the number of color television sets bought by persons in different states in a given year to be natural, although sociologists or human ecologists might, thus deeming it worthy of study. The qualification "variation in nature" is included in the definition of statistics largely to make certain that the phenomena studied are not arbitrary ones that are entirely under the will and control of the researcher, such as the number of animals employed by the experimenter.

The word statistics is also used in another, related sense: as the plural of the noun *statistic,* which refers to any one of many computed or estimated statistical quantities, such as the mean, the standard deviation, or the correlation coefficient. Each one of these is a statistic. (By extension, the term statistic has also been used to indicate figures or pieces of data—that is, the individuals or items of data to be defined. This is the way the word is used in the familiar warning to the careless driver not to become a statistic. This is not, however, a recommended use of the word.)

1.2 The Development of Biometry

The nature of this book and the space at our disposal do not permit us to indulge in an extended historical review. The origin of modern statistics can be traced back to the seventeenth century, when it derived from two sources. The first of these related to political science and developed as a quantitative description of the various aspects of the affairs of a government or state (hence the term statistics). This subject also became known as political arithmetic. Taxes and insurance caused people to become interested in problems of censuses, longevity, and mortality. Such considerations assumed increasing importance, especially in England, as the country prospered during the development of its empire. John Graunt (1620–1674) and William Petty (1623–1687) were early students of vital statistics, and others followed in their footsteps.

At about the same time came the development of the second root of modern statistics: the mathematical theory of probability engendered by the interest in games of chance among the leisure classes of the time. Important contributions to this theory were made by Blaise Pascal (1623–1662) and Pierre de Fermat (1601–1665), both Frenchmen. Jacques Bernoulli (1654–1705), a Swiss, laid the foundation of modern probability theory as *Ars Conjectandi,* published posthumously. Abraham de Moivre (1667–1754), a Frenchman living in England, was the first to combine the statistics of his day with probability theory in working out annuity values. De Moivre also was the first to approximate the important normal distribution through the expansion of the binomial.

Subsequent stimulus for the development of statistics came from the science of astronomy, in which many individual observations had to be digested into a coherent theory. Many of the famous astronomers and mathematicians of the eighteenth century, such as Pierre-Simon de Laplace (1749–1827) in France and Carl Friedrich Gauss (1777–1855) in Germany were among the leaders in this field. Gauss's lasting contribution to statistics is the development of the method of least squares, which will be encountered later in this book in a variety of forms. Gauss already realized the importance of his ideas to any situation in which errors in observations occur.

Perhaps the earliest important figure in biometric thought was Adolphe Quetelet (1796–1874), a Belgian astronomer and mathematician, who combined the theory and practical methods of statistics and applied them to problems of biology, medicine, and sociology. He introduced the concept of the ''average man'' and was the first to recognize the significance of the constancy of large numbers—an idea of prime importance in modern statistical work. He also developed the notion of statistical variation and distribution.

Much progress was made in the theory of statistics by mathematicians in the nineteenth century, but this development is of less interest to us than is the work of Francis Galton (1822–1911), a cousin of Charles Darwin. Galton has been called the father of biometry and eugenics (a branch of genetics), two subjects that he studied interrelatedly. The publication of Darwin's work and the inadequacy of Darwin's genetic theories stimulated Galton to try to solve the problems of heredity. Galton's major contribution to biology is his application of statistical methodology to the analysis of biological variation, such as the analysis of variability and his study of regression and correlation in biological measurements. His hope of unraveling the laws of genetics through these procedures was in vain. He started with the most difficult material and with the wrong assumptions. His methodology however, has become the foundation for the application of statistics to biology.

A contemporary of Galton's, Florence Nightingale (1820–1910), was the first distinguished female statistician. In addition to being the founder of modern nursing, for which she is universally famous, she was an excellent mathematician who pioneered the compilation and graphic presentation of vital and medical statistics.

Karl Pearson (1857–1936) at University College, London, became interested in the application of statistical methods to biology, particularly in the demonstration of natural selection, through the influence of W. F. R. Weldon (1860–1906), a zoologist at the same institution. Weldon, incidentally, is credited with coining the term biometry for the type of studies he pursued. Pearson continued in the tradition of Galton and laid the foundation for much of descriptive and correlational statistics. The dominant figure in statistics and biometry in the twentieth century has been Ronald A. Fisher (1890–1962). His many contributions to statistical theory will become obvious even to the casual reader of this book. We will acquaint you with the contributions of many of this century's other eminent statisticians as we discuss the topics they developed.

At present, statistics is a broad and extremely active field, whose applications touch almost every science and even the humanities. New applications for statistics are constantly being found, and no one can predict from what branch of statistics new applications to biology will be made. The prospective student of biological statistics need not be overwhelmed by the formidable size of the field, but should be aware that statistics is a constantly changing and growing science and in this respect is quite different from a subject such as elementary algebra or trigonometry. Methods of teaching these subjects may have changed in the last fifty years, but not the basic principles and theorems. Some of the methods presented in this book were published only recently in the technical literature.

Time-honored methods such as tests for normality and statistics of skewness and kurtosis, after a period of disuse, are again commonly employed, because high-speed computers make their calculation feasible. Other techniques, such as the chi-square test and the *t*-test, may become obsolete for a variety of reasons. The entire field of data analysis has become possible only recently, since the development of inexpensive, high-speed computation.

Anyone would be rash to try to predict the scope and practice of biometry 20 years from now. Whatever its nature, however, biometry is more than likely to occupy an even more commanding position in biology in the next century than it does now.

1.3 The Statistical Frame of Mind

The ever-increasing importance and application of statistics to biological data is evident even on cursory inspection. Figure 1.1 shows the results of a survey of eleven decennial volumes of *The American Naturalist;* because of its wide coverage, this journal is a general indicator of trends in biological research. Papers in the decennial volumes were catalogued into those containing no numerical results, those containing numbers but no statistical computations, those containing simple statistics, and those having a major emphasis on mathematics or statistics. The dramatic increase of the percentage of papers involving quantitative

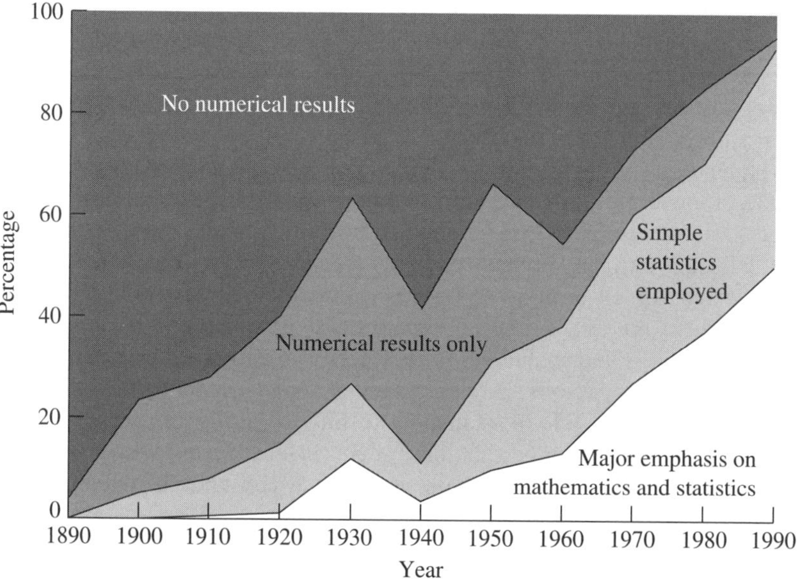

FIGURE 1.1 *Proportions of articles involving numerical and statistical work in decennial issues of* The American Naturalist.

approaches is shown clearly in the graph, in spite of the fluctuation to be expected in a sample of this sort. By 1990, 96% of the papers published during the year employed some statistical tests or were more heavily statistical or mathematical. This trend can be observed in most biological journals.

Why has there been such a marked increase in the use of statistics in biology? It is apparently a result of the realization that in biology the interplay of causal and response variables obeys laws that are not in the classic mold of nineteenth-century physical science. In that century, biologists such as Robert Mayer, Hermann von Helmholtz, and others, in trying to demonstrate that biological processes were nothing but physicochemical phenomena, helped create the impression that the experimental methods and natural philosophy that had led to such dramatic progress in the physical sciences should be imitated fully in biology. Regrettably, opposition to this point of view was confounded with the vitalistic movement, which led to unproductive theorizing.

Thus, many biologists even to this day have retained the tradition of strictly mechanistic and deterministic concepts of thinking, while physicists, as their science became more refined and came to deal with ever more "elementary" particles, began to resort to statistical approaches. In biology most phenomena are affected by many causal factors, uncontrollable in their variation and often unidentifiable. Statistics is needed to measure such variable phenomena with a predictable error and to ascertain the reality of minute but important differences. Whether biological phenomena are in fact fundamentally deterministic and only the variety of causal variables and our inability to control these make these phenomena appear probabilistic, or whether biological processes are truly probabilistic, as postulated in quantum mechanics for elementary particles, is a deep philosophical question beyond the scope of this book and the competence of the authors. The fact remains that most biological phenomena can be discussed only within a probabilistic framework.

A misunderstanding of these principles and relationships has given rise to the attitude of some biologists that if differences induced by an experiment or observed in nature are not clear on plain inspection (and therefore not in need of statistical analysis), they are not worth investigating. This attitude is also related to the search for all-or-none responses, common in fields such as molecular biology. There are still a few legitimate fields of inquiry in which, from the nature of the phenomena studied, statistical investigation is unnecessary. When all animals given injections of a pathogenic organism contract the disease, while none of the controls fall ill, statistical testing is hardly needed. Much more frequent, however, are investigations, such as those determining the relation between smoking and heart disease, where the variability of the outcomes necessitates statistical analysis.

We should stress that statistical thinking is not really different from ordinary disciplined scientific thinking with which we try to quantify our observations. In statistics we express our degree of belief or disbelief as a probability rather than as a vague, general statement. For example, biologists often make statements such as "species A is larger than species B" or "females are more often found sitting on tree M rather than on tree N." Such statements can and should be expressed more precisely in quantitative form. The human mind is a remarkable statistical machine, absorbing many facts from the outside world, digesting these, and regurgitating them in simple summary form. From our experience we know that certain events occur frequently, others rarely. "Someone smoking cigarette" is a frequently observed event; "someone slipping on banana peel" is rare. We know from experience that Japanese are, on the average, shorter than Englishmen and that Egyptians are, on the average, darker than Swedes. We associate thunder with lightning almost always, flies with garbage cans frequently in the summer, but snow with the southern Californian desert extremely rarely. All such knowledge comes to us as a result of our lifetime experience, both directly and through that of others by direct communication or through reading. All these facts have been processed by that remarkable computer, the human brain, which furnishes an abstract. This summary is constantly under revision, and though occasionally faulty and biased, it is on the whole astonishingly sound; it is our knowledge of the moment.

Although statistics arose to satisfy the needs of scientific research, the development of its methodology in turn has affected the sciences in which statistics is applied. Thus, in a positive feedback loop, statistics, created to serve the needs of natural science, has itself affected the philosophy of the biological sciences. Consider the following two examples. First, analysis of variance has had a tremendous effect in influencing the types of experiments researchers carry out; the whole field of quantitative genetics, one of whose problems is the separation of environmental from genetic effects, depends upon the analysis of variance for its realization, and many of the concepts of quantitative genetics have been built directly around the analysis of variance. Second, the techniques of multivariate analysis have given rise to numerous studies designed especially to exploit their ability to analyze many variables simultaneously.

2 DATA IN BIOLOGY

In Section 2.1 we explain the statistical meaning of "sample" and "population," terms used throughout this book. Then we come to the types of observations obtained from biological research material, with which we shall perform the computations in the rest of this book (Section 2.2). In obtaining data we shall run into the problem of the degree of accuracy necessary for recording the data. This problem and the procedure for rounding off figures are discussed in Section 2.3, after which we will be ready to consider in Section 2.4 certain kinds of derived data, such as ratios and indices, frequently used in biological science, which present peculiar problems with respect to their accuracy and distribution. Knowing how to arrange data as frequency distributions is important, because such arrangements permit us to get an overall impression of the shape of the variation present in a sample. Frequency distributions, as well as the presentation of numerical data, are discussed in the last section (2.5) of this chapter.

2.1 SAMPLES AND POPULATIONS

We shall now define a number of important terms necessary for an understanding of biological data. The data in a biometric study are generally based on **individual observations,** which are *observations or measurements taken on the smallest sampling unit.* These smallest sampling units frequently, but not necessarily, are also individuals in the ordinary biological sense. If we measure weight in 100 rats, then the weight of each rat is an individual observation; the hundred rat weights together represent the **sample of observations,** defined as *a collection of individual observations selected by a specified procedure.* In this instance, one individual observation is based on one individual in a biological sense—that is, one rat. However, if we had studied weight in a single rat over a period of time, the sample of individual observations would be all the weights recorded on one rat at successive times. In a study of temperature in ant colonies, where each colony is a basic sampling unit, each temperature reading for one colony is an individual observation, and the sample of observations is the temperatures for all the colonies considered. An estimate of the DNA content of a single mammalian sperm cell is an individual observation, and the corresponding sample of observations is the estimates of DNA concent of all other sperm cells studied in one individual mammal. A synonym for individual observation is "item."

Up to now we have carefully avoided specifying the particular variable being studied because "individual observation" and "sample of observations" as we just used them define only the structure but not the nature of the data in a study. *The actual property measured by the individual observations* is the **variable,** or *character.* The more common term employed in general statistics is *variable.* In evolutionary and systematic biology however, *character* is frequently used synonymously. More than one variable can be measured on each smallest sampling unit. Thus, in a group of 25 mice we might measure the blood pH and the erythrocyte count. The mouse (a biological individual) would be the smallest sampling unit; blood pH and cell count would be the two variables studied. In this example the pH readings and cell counts are individual observations, and two samples of 25 observations on pH and erythrocyte count would result. Alternatively, we may call this example a **bivariate sample** of 25 observations, each referring to a pH reading paired with an erythrocyte count.

Next we define **population.** The biological definition of this term is well known: It refers to all the individuals of a given species (perhaps of a given life history stage or sex) found in a circumscribed area at a given time. In statistics, population always means the *totality of individual observations about which inferences are to be made, existing anywhere in the world or at least within a definitely specified sampling area limited in space and time.* If you take five humans and study the number of leucocytes in their peripheral blood and you are prepared to draw conclusions about all humans from this sample of five, then the population from which the sample has been drawn represents the leucocyte counts of all humankind—that is, all extant members of the species *Homo sapiens.* If, on the other hand, you restrict yourself to a more narrowly specified sample, such as five male Chinese, aged 20, and you are restricting your conclusions to this particular group, then the population from which you are sampling will be leucocyte numbers of all Chinese males of age 20. The population in this statistical sense is sometimes referred to as the *universe.* A population may refer to variables of a concrete collection of objects or creatures—such as the tail lengths of all the white mice in the world, the leucocyte counts of all the Chinese men in the world of age 20, or the DNA contents of all the hamster sperm cells in existence—or it may refer to the outcomes of experiments—such as all the heartbeat frequencies produced in guinea pigs by injections of adrenalin. In the first three cases the population is finite. Although in practice it would be impossible to collect, count, and examine all white mice, all Chinese men of age 20, or all hamster sperm cells in the world, these populations are finite. Certain smaller populations, such as all the whooping cranes in North America or all the pocket gophers in a given colony, may lie within reach of a total census. By contrast, an experiment can be repeated an infinite number of times (at least in theory). An experiment such as the administration of adrenalin to guinea pigs could be repeated as long as the experimenter could obtain material and his or her health and patience held out. The sample of experiments performed is a sample from an infinite number that *could* be performed. Some of the statistical methods to be developed later make a distinction between sampling from finite and from infinite populations. However, although populations are theoretically finite in most applications in biology, they are generally so much larger than samples drawn from them that they can be considered as de facto infinitely sized populations.

2.2 Variables in Biology

Enumerating all the possible kinds of variables that can be studied in biological research would be a hopeless task. Each discipline has its own set of variables, which may include conventional morphological measurements, concentrations of chemicals in body fluids, rates of certain biological processes, frequencies of certain events (as in genetics and radiation biology), physical readings of optical or electronic machinery used in biological research, and many more. We assume that persons reading this book already have special interests and have become acquainted with the methodologies of research in their areas of interest, so that the variables commonly studied in their fields are at least somewhat familiar. In any case, the problems for measurement and enumeration must suggest themselves to the researcher; statistics will not, in general, contribute to the discovery and definition of such variables.

Some exception must be made to this statement. Once a variable has been chosen, statistical analysis may demonstrate it to be unreliable. If several variables are studied, certain elaborate procedures of multivariate analysis can assign weights to them, indicating their value for a given procedure. For example, in taxonomy and various other applications, the method of discriminant functions can identify the combination of a series of variables that best distinguishes between two groups (see Section 16.7). Other multivariate techniques, such as canonical variates analysis, principal components analysis, or factor analysis, can specify characters that best represent or summarize certain patterns of variation (Krzanowski, 1988; Jackson, 1991). As a general rule, however, and particularly within the framework of this book, choosing a variable as well as defining the problem to be solved is primarily the responsibility of the biological researcher.

A more precise definition of variable than the one given earlier is desirable. It is *a property with respect to which individuals in a sample differ in some ascertainable way.* If the property does not differ within a sample at hand or at least among the samples being studied, it cannot be of statistical interest. Being entirely uniform, such a property would also not be a variable from the etymological point of view and should not be so called. Length, height, weight, number of teeth, vitamin C content, and genotypes are examples of variables in ordinary, genetically and phenotypically diverse groups of organisms. Warm-bloodedness in a group of mammals is not a variable because they are all alike in this regard, although body temperature of individual mammals would, of course, be a variable. Also, if we had a heterogeneous group of animals, of which some were homeothermic and others were not, then body temperature regulation (with its two states or forms of expression, "warm-blooded" and "cold-blooded") would be a variable.

We can classify variables as follows:

- Variables
  - Measurement variables
    - Continuous variables
    - Discontinuous variables
  - Ranked variables
  - Attributes

Measurement variables are those whose *differing states can be expressed in a numerically ordered fashion.* There are two types of measurement variables: continuous and discontinuous. **Continuous variables** at least theoretically can assume an infinite number of values between any two fixed points. For example, between the two length measurements 1.5 cm and 1.6 cm an infinite number of lengths could be measured if one were so inclined and had a measuring instrument with sufficiently precise calibration. Any given reading of a continuous variable, such as a length of 1.57 cm, is an approximation to the exact reading, which in practice cannot be known. For purposes of computation, however, these approximations are usually sufficient and, as will be seen below, may be made even more approximate by rounding. Many of the variables studied in biology are continuous. Examples are length, area, volume, weight, angle, temperature, period of time, percentage, and rate.

Discontinuous variables, also known as **meristic variables** (the term we use in this book) or *discrete variables,* are variables that have only certain fixed numerical values, with no intermediate values possible. The number of segments in a certain insect appendage, for instance, may be 4 to 5 or 6 but never $5\frac{1}{2}$ or 4.3. Examples of discontinuous variables are number of a certain structure (such as segments, bristles, teeth, or glands), number of offspring, number of colonies of microorganisms or animals, or number of plants in a given quadrat.

A word of caution: not all variables restricted to integral numerical values are meristic. An example will illustrate this point. If an animal behaviorist were to code the reactions of animals in a series of experiments as (1) very aggressive, (2) aggressive, (3) neutral, (4) submissive, and (5) very submissive, we might be tempted to believe that these five different states of the variable were meristic because they assume integral values. However, they are clearly only arbitrary points (class marks, see Section 2.5) along a continuum of aggressiveness; the only reason that no values such as 1.5 occur is that the experimenter did not wish to subdivide the behavior classes too finely, either for convenience or because of an inability to determine more than five subdivisions of this spectrum of behavior with precision. Thus, this variable is clearly continuous rather than meristic, as it might have appeared at first sight.

Some variables cannot be measured but at least can be ordered or ranked by their magnitude. Thus, in an experiment one might record the rank order of emergence of ten pupae without specifying the exact time at which each pupa emerged. In such a case we would code the data as a **ranked variable,** the order of emergence. Special methods for dealing with ranked variables have been developed, and several are described in this book. By expressing a variable as a series or ranks, such as 1, 2, 3, 4, 5, we do not imply that the difference in magnitude between, say, ranks 1 and 2 is identical to or even proportional to the difference between 2 and 3. Such an assumption is made, however, for measurement variables, discussed above.

Variables that cannot be measured but must be expressed qualitatively are called **attributes,** also known as *categorical* or *nominal variables.* Attributes are properties such as black or white, pregnant or not pregnant, dead or alive, male or female. When such attributes are combined with frequencies, they can be treated statistically. Of 80 mice, for instance, we may state that four are black and the rest gray. When attributes are combined with frequencies into tables suitable for statistical analysis, they are referred to as **enumeration data.** The enumeration data on color in mice just mentioned can be arranged as follows:

Color	Frequency
Black	4
Gray	76
Total number of mice	80

In some cases, if desired, attributes can be changed into variables. Thus colors can be changed into wavelengths or color chart values, which are measurement variables. Certain other attributes can be ranked or ordered as ranked variables. For example, three attributes referring to a structure as "poorly developed," "well developed," and "hypertrophied" could conveniently be coded 1, 2, and 3. These values imply the rank order of development, but not the relative magnitudes of these attribute states.

A term that has not yet been explained is **variate.** In this book we shall use it to mean *a single reading, score, or observation of a given variable.* Thus, if we have measurements of the length of the tails of five mice, tail length will be a continuous variable, and each of the five readings of length will be a variate. In this text we identify variables by capital letters, the most common symbol being

Y. Thus Y may stand for tail length of mice. A variate will refer to a given length measurement; Y_i is the measurement of tail length of the ith mouse, and Y_4 is the measurement of tail length of the fourth mouse in our sample. The use of the terms variable and variate differs somewhat from author to author; frequently the two terms are used synonymously.

2.3 Accuracy and Precision of Data

Scientists are generally aware of the importance of accuracy and precision in their work, and it is self-evident that accuracy must extend to the numerical results of their work and to the processing of these data. Because of the great diversity of approaches employed in different disciplines of biology, it seems futile to attempt to furnish specific rules for obtaining data accurately. We should make certain, however, that whatever the method of obtaining data, it is consistent, so that given the same observational or experimental setup, different numerical readings of the same structure or event would be identical or within acceptable and predictable limits of each other.

Accuracy and precision are used synonymously in everyday speech, but in statistics we distinguish between them. **Accuracy** is *the closeness of a measured or computed value to its true value;* **precision** is *the closeness of repeated measurements of the same quantity* to each other. A biased but sensitive scale might yield inaccurate but precise weight readings. By chance an insensitive scale might result in an accurate reading, but the reading would be imprecise, since another weighing of the same object would be unlikely to yield an equally accurate weight. Unless there is bias in a measuring instrument, precision will lead to accuracy. We need therefore mainly be concerned with the former.

Precise variates are usually but not necessarily whole numbers. Thus, when we count four eggs in a nest, there is no doubt about the exact number of eggs in the nest if we have counted correctly; it is four, not three or five, and clearly it could not be four plus or minus a fractional part. Meristic variables are generally measured as exact numbers. Seemingly, continuous variables derived from meristic ones can, under certain conditions, also be exact numbers. For instance, ratios of exact numbers are themselves also exact. If in a colony of animals there are 18 females and 12 males, the ratio of females to males is 1.5, a continuous variate but also an exact number.

Most continuous variables, however, are approximate: The exact value of the single measurement, the variate, is unknown and probably unknowable. The last digit of the measurement stated should imply precision, that is, the limits on the measurement scale between which we believe the true measurement to lie. Thus a length measurement of 12.3 mm implies that the true length of the structure lies somewhere between 12.25 and 12.35 mm. Exactly where between these **implied limits** the real length is we do not know. Some might object to defining the limits of the number as 12.25 to 12.35 mm because the adjacent measurement of 12.2

would imply limits of 12.15 to 12.25. Where then, they would ask, would a true measurement of 12.25 fall? Would it not equally likely fall in either of the two classes 12.2 and 12.3, clearly an unsatisfactory state of affairs? For this reason some authors define the implied limits of 12.3 as 12.25 to 12.34999. . . . Such an argument is correct, but when we record a number as either 12.2 or 12.3 we imply that the decision whether to put it into the higher or lower class has already been made. This decision was not made arbitrarily, but presumably was based on the best available measurement. If the scale of measurement is so precise that a value of 12.25 would clearly have been recognized, then the measurement should have been recorded originally to four significant figures. Implied limits therefore always carry one figure beyond the last significant one measured by the observer.

Hence it follows that if we record the measurement as 12.32, we are implying that the true value lies between 12.315 and 12.325. If this is what we mean, there is no need to add another decimal figure to our original measurements. If we add another figure, we imply an increase in precision. We see therefore that accuracy and precision in numbers is not an absolute concept, but is relative. Assuming there is no bias, a number becomes increasingly accurate as we are able to write more significant figures for it (that is, increase its precision). To illustrate the relativity of accuracy, consider the following three numbers:

	Implied limits
193	192.5–193.5
192.8	192.75–192.85
192.76	192.755–192.765

Imagine these numbers to be the recorded measurements of the same structure. If we assume that we know the true length of the given structure to be 192.758 units, then the three measurements increased in accuracy from the top down, as the interval between their implied limits decreases.

Meristic variates, though ordinarily exact, may be recorded approximately when large numbers are involved. Thus when counts are reported to the nearest thousand, a count of 36,000 insects in a cubic meter of soil implies that the true number lies between 35,500 and 36,500 insects.

To how many significant figures should we record measurements? If we array the sample by order of magnitude from the smallest individual to the largest one, an easy rule to remember is that *the number of unit steps from the smallest to the largest measurement in an array should usually be between 30 and 300.* For example, if we are measuring a series of shells to the nearest millimeter and the largest is 8 mm and the smallest 4 mm wide, there are only four unit steps between the largest and the smallest measurement. Hence we should have measured our shells to one more significant decimal place. Then the two extreme measurements might have been 8.2 mm and 4.1 mm, with 41 unit steps between them (counting the last significant digit as the unit); this would have been an adequate number of unit steps. The reason for such a role is that an error of 1 in the last significant digit of a reading of 4 mm would constitute an inadmissible error of 25%, but an error of 1 in the last digit of 4.1 is less than 2.5%. By contrast, if we had measured the height of the tallest of a series of plants as 173.2 cm and that of the shortest of these plants as 26.6 cm, the difference between these limits would comprise 1466 unit steps (of 0.1 cm each), which are far too many. We should therefore have recorded the heights to the nearest centimeter, as follows: 173 cm for the tallest and 27 cm for the shortest, yielding 146 unit steps between.

Using the rule stated above, we will record two or three digits for most measurements. On occasion, however, more digits will be necessary—when the leading digits are constant, as in the following example. In a representative series of pH readings measured with great precision, the highest value might be 7.456 and the lowest value 7.434. The first two digits of such measurements are constant for pH readings of the material under study. It is therefore necessary to use three decimal places to obtain an adequate amount of variation for analysis. The perceptive reader may have noticed that the number of unit steps from lowest to highest variate in the pH example was only 22, or less than 30 given as the lower limit for the desired number of steps. This rule, as are all such rules of thumb, is to be taken with a grain of statistical salt. These rules are general guidelines but not laws whose mild infraction would immediately invalidate all further work. With some experience, an appropriate adherence to these rules becomes second nature to the biologist engaging in statistical analysis.

The last digit of an approximate number should always be significant; that is, it should imply a range for the true measurement of from half a unit step below to half a unit step above the recorded score, as illustrated earlier. This rule applies to all digits, zero included. Zeros should therefore not be written at the end of approximate numbers to the right of the decimal point unless they are meant to be significant digits. The measurement 7.80 implies the limits 7.795 to 7.805. If 7.75 to 7.85 is meant to be implied, the measurement should be recorded as 7.8.

When the number of significant digits is reduced, we carry out the process of **rounding** numbers. The rules for rounding are very simple. A digit to be rounded is not changed if it is followed by a digit less than 5. If the digit to be rounded is followed by a digit greater than 5 or by 5 followed by other nonzero digits, it is increased by one. When the digit to be rounded is followed by a 5 standing alone or followed by zeros, it is unchanged if it is even but increased by one if it is odd. The reason for this last rule is that when such numbers are summed in a long series, we should have as many digits raised as are being lowered on the average; these changes would therefore balance out. Practice these rules by rounding the following numbers to the indicated number of significant digits:

Number	Significant digits desired	Answer
26.58	2	27
133.7137	5	133.71
0.03725	3	0.0372
0.03715	3	0.0372
18,316	2	18,000
17.3476	3	17.3

Most pocket calculators or larger computers round their displays using a different rule: They increase the preceding digit when the following digit is a 5 standing alone or with trailing zeros. However, since most of the machines that can be used for statistics also retain eight or ten significant figures internally, the accumulation of rounding errors is minimized. Incidentally, if two calculators give answers with slight differences in the final (least significant) digits, suspect a different number of significant digits in memory as a cause of the disagreement.

2.4 Derived Variables

Most variables in biometric work are observations recorded as direct measurements or counts of biological material or as readings that are the output of various types of instruments. However, there is an important class of variables in biological research, which we may call the **derived** (or computed) **variables,** that are generally based on two or more independently measured variables whose relations are expressed in a certain way. We are referring to ratios, percentages, indices, rates, and the like.

A **ratio** expresses as a single value the relation between two variables. In its simplest form, a ratio is expressed, for example, as 64 : 24, which may represent the number of wild-type versus mutant individuals, the number of males versus females, a count of parasitized individuals versus those not parasitized, and so on. These examples imply ratios based on counts; a ratio based on a continuous variable might be similarly expressed as 1.2 : 1.8, which could represent the ratio of width to length in a sclerite of an insect or the ratio between the concentrations of two minerals contained in water or soil. Ratios may also be expressed as fractions; thus the two ratios above could be expressed as $\frac{64}{24}$ and $\frac{1.2}{1.8}$. For computational purposes, however, it is most useful to express the ratio as a quotient. The two ratios cited here would therefore be 2.666 . . . and 0.666 . . . , respectively. These are pure numbers, not expressed in measurement units of any kind. It is this form for ratios that we will consider further. *Percentages* are also a type of ratio. Ratios and percentages are basic quantities in much biological research. They are widely used and generally familiar.

Not all derived variables are ratios or percentages. The term **index** is used in a general sense for derived variables, although some would limit it to the ratio of one anatomic variable divided by a larger, so-called standard variable. For instance, in a study of the cranial dimensions of cats, Haltenorth (1937) divided all measurements by the basal length of the skull. Thus each measurement was a proportion of the basal length of the skull of the cat being measured. Another example of an index in this sense is the well-known cephalic index in physical anthropology. Conceived in the wide sense, an index could be the average of two measurements—either simply, such as $\frac{1}{2}$(length of A + length of B), or in weighted fashion, such as $\frac{1}{3}$[(2 × length of A) + length of B]. An index may refer to the summation of a series of numerically scored properties. Thus, if an animal is given six behavioral tests in which its score can range from 0 to 4, an index of its behavior might be the sum of the scores of the six tests. Similar indices have been described for determining the degree of hybridity in organisms (the hybrid indices of various authors).

Rates are important in many experimental fields of biology. The amount of a substance liberated per unit weight or volume of biological material, weight gain per unit time, reproductive rates per unit population size and time (birth rates), and death rates would fall in this category. Many counts are really ratios or rates—the number of pulse beats observed in one minute or the number of birds of a given species found in some quadrat. In general, counts are ratios if the unit over which they are counted is not natural, as, for example, an arbitrary time or space interval.

As we shall see, there are some serious drawbacks to the use of ratios and percentages in statistical work. In spite of these disadvantages, the use of these variables is deeply ingrained in scientific thought processes and is not likely to be abandoned. Furthermore, ratios may be the only meaningful way to interpret and understand certain types of biological problems. If the biological process being investigated operates on the ratio of the variables studied, one must examine this ratio to understand the process. Thus, Sinnott and Hammond (1935) found that inheritance of the shapes of the squash *Cucurbita pepo* could be interpreted by a form index based on a length–width ratio, but not in terms of the independent dimensions of shape. Similarly, the evolution of shape in many burrowing animals is a function of the cross-sectional profile of the animal rather than of a single dimension. Selection affecting body proportions will be found to exist in the evolution of almost any organism when properly investigated.

The disadvantages of using ratios are several: First is their relative inaccuracy. Let us return to the ratio $\frac{1.2}{1.8}$ and recall from the previous section that a measurement of 1.2 indicates a true range of measurement of the variable from 1.15 to 1.25; similarly, a measurement of 1.8 implies a range from 1.75 to 1.85. We realize therefore that the true ratio may vary anywhere from $\frac{1.15}{1.85}$ to $\frac{1.25}{1.75}$, or 0.622 and 0.714, respectively. We note a possible maximal error of 4.2% if 1.2 were an original measurement: (1.25 − 1.2)/1.2; the corresponding maximal error for the ratio is 7.0%: (0.714 − 0.667)/0.667. Furthermore, the best estimate of a ratio is not usually the midpoint between its possible ranges. Thus in our example, the midpoint between the implied limits is 0.668 and the true ratio is 0.666 . . . , a difference that here is only slight, but that in other instances could be greater. In many cases, therefore, ratios are not as accurate as measurements obtained directly. This liability can be overcome by empirical determinations of the variability of ratios through measures of their variance (described later in this book).

A second drawback to ratios and percentages is that they may not be approximately normally distributed (see Chapter 6), as required by many statistical tests. This difficulty frequently can be overcome by transformation of the variable (see Chapter 13). Another disadvantage of ratios is that they do not provide information on *how* the two variables whose ratio is being taken are related. Often more may be learned by studying the variables singly first and then examining their relation to each other (bivariate and multivariate analysis).

Table 2.1 POSSIBLE PERCENTAGES OF PERIPHERAL PUPATION IN VIALS SEEDED WITH 10 *DROSOPHILA* EGGS.

Experimental results show number of peripheral pupae in the numerator, total number of pupae in the denominator.

Percent	Possible experimental results yielding this percentage	Percent	Possible experimental results yielding this percentage
00	$\frac{0}{10}, \frac{0}{9}, \frac{0}{8}, \frac{0}{7}, \frac{0}{6}, \frac{0}{5}, \frac{0}{4}, \frac{0}{3}, \frac{0}{2}, \frac{0}{1}$	56	$\frac{5}{9}$
		57	$\frac{4}{7}$
10	$\frac{1}{10}$	60	$\frac{6}{10}, \frac{3}{5}$
11	$\frac{1}{9}$	62	$\frac{5}{8}$
12	$\frac{1}{8}$	67	$\frac{6}{9}, \frac{4}{6}, \frac{2}{3}$
14	$\frac{1}{7}$	70	$\frac{7}{10}$
17	$\frac{1}{6}$	71	$\frac{5}{7}$
20	$\frac{2}{10}, \frac{1}{5}$	75	$\frac{6}{8}, \frac{3}{4}$
22	$\frac{2}{9}$	78	$\frac{7}{9}$
25	$\frac{2}{8}, \frac{1}{4}$	80	$\frac{8}{10}, \frac{4}{5}$
29	$\frac{2}{7}$	83	$\frac{5}{6}$
30	$\frac{3}{10}$	86	$\frac{6}{7}$
33	$\frac{3}{9}, \frac{2}{6}, \frac{1}{3}$	88	$\frac{7}{8}$
38	$\frac{3}{8}$	89	$\frac{8}{9}$
40	$\frac{4}{10}, \frac{2}{5}$	90	$\frac{9}{10}$
43	$\frac{3}{7}$	100	$\frac{10}{10}, \frac{9}{9}, \frac{8}{8}, \frac{7}{7}, \frac{6}{6}, \frac{5}{5}, \frac{4}{4}, \frac{3}{3}, \frac{2}{2}, \frac{1}{1}$
44	$\frac{4}{9}$		
50	$\frac{5}{10}, \frac{4}{8}, \frac{3}{6}, \frac{2}{4}, \frac{1}{2}$		

SOURCE: From a study by Sokal (1966).

Finally, when ratios involve enumeration data or meristic variables, occasionally they give rise to curious distributions. As an example, we cite the percentages obtained from an experiment performed many years ago by Sokal. In this experiment 10 *Drosophila* eggs were put in a vial and the positions of the pupae noted after pupation. Some pupae pupated at the margin or the wall of the vials; these were called peripheral. Others, which pupated away from the wall, were called central. Ideally there should have been 10 pupae—if all the eggs had hatched and there had been no larval mortality whatsoever. In fact, however, because of natural mortality and because of errors in preparing the vials, 10 pupae were not always found. The logical minimum number of survivors on which a result could be reported was one pupa. The proportion of peripheral pupae was then calculated by dividing the number of such pupae in a vial by the total number of pupae found in that vial. Although ratios thus obtained are continuous variables in appearance, they are not so in fact, because certain values can never be obtained. For instance, by limiting the maximum number of pupae to 10, we cannot obtain percentages of peripheral pupation between 0 and 10% or between 90 and 100%, as Table 2.1 shows. Also, we see that some percentages are given by several ratios, others by only one: 33% is obtained by $\frac{3}{9}$ or $\frac{2}{6}$ or $\frac{1}{3}$, but 57% can be obtained only by having 4 out of 7 pupae peripheral. Table 2.1 shows which values are more likely to be encountered if we make the simplest assumption—that there is an equal probability for any given experimental outcome. In using ratios of meristic variates or of counts, such discontinuities and peculiarities of distribution must be taken into account. The problem in this instance was solved in transforming the data to probits (see Section 14.11).

2.5 Frequency Distributions

If we were to sample a population of birth weights of infants, we could represent each measurement by a point along an axis denoting magnitude of birth weights. This example is illustrated in Figure 2.1A for a sample of 25 birth weights. If we sample repeatedly from the population and obtain 100 birth weights, we probably will have to place some of these points on top of other points in order to record them all correctly (Figure 2.1B). As we continue sampling additional hundreds and thousands of birth weights (Figure 2.1C, D), the assemblage of points will continue to increase in size and will assume a fairly definite shape. The curve tracing the outline of the mound of points approximates the distribution of the variable. Remember that a continuous variable such as birth weight can assume an infinity of values between any two points on the abscissa. The refinement of our measurements determines how fine the number of recorded divisions between any two points along the axis are.

The distribution of a variable is of considerable biological interest. A distribution that is asymmetrical and drawn out in one direction tells us that there is,

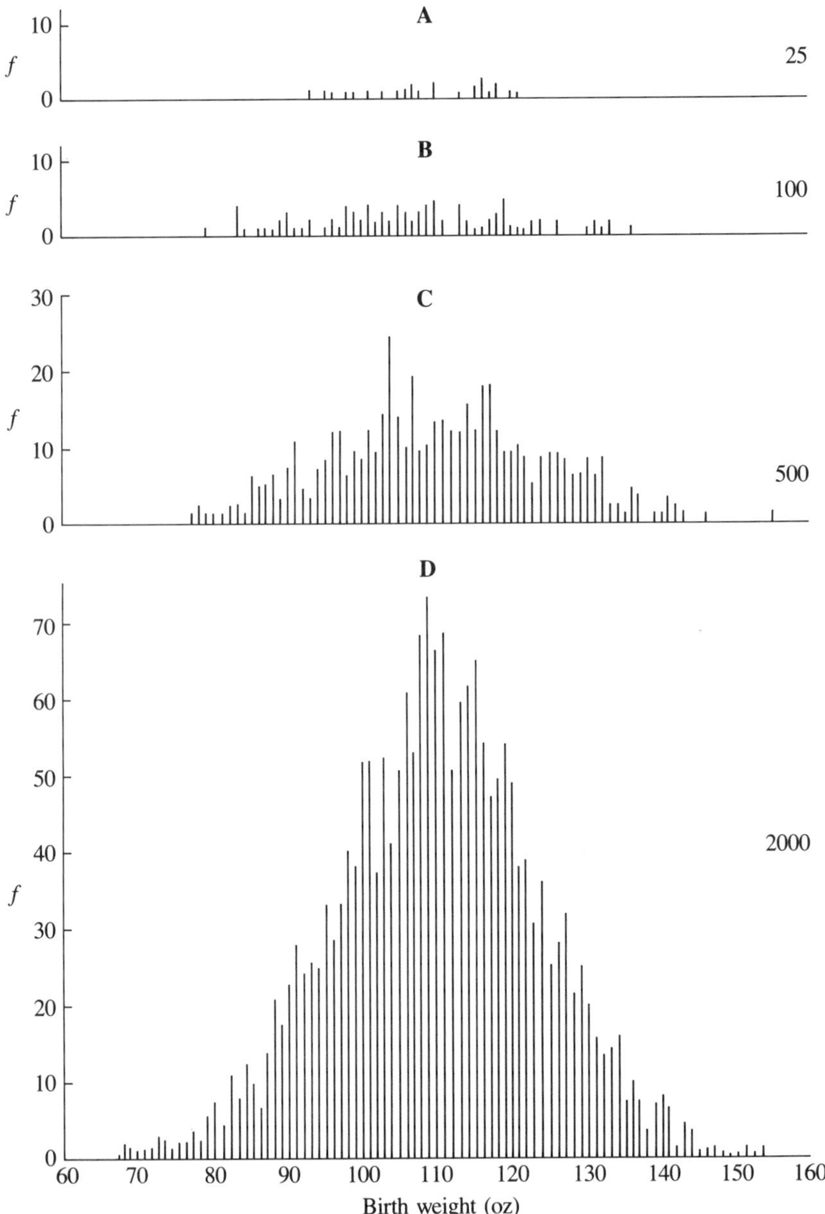

FIGURE 2.1 *Sampling from a population of birth weights of infants (a continuous variable).* **A.** *A sample of 25.* **B.** *A sample of 100.* **C.** *A sample of 500.* **D.** *A sample of 2000.*

perhaps, selection for or against organisms falling in one of the tails of the distribution, or that the scale of measurement chosen may be bringing about a distortion of the distribution. If, in a sample of immature insects, the measurements are bimodally distributed (with two peaks), this would indicate that the population is dimorphic; different species or races may have become intermingled in our sample, or the dimorphism could have arisen from the presence of both sexes or of different instars. There are several characteristic shapes of frequency distributions, the most common of which is the symmetrical bell-shaped distribution (approximated by the graph in Figure 2.1D), the normal frequency distribution discussed in Chapter 6. There are also skewed distributions (drawn out more at one tail than at the other), L-shaped distributions as in Figure 2.2, U-shaped distributions, and others, which impart significant information about certain types of relationships. We will have more to say about the implications of various types of distributions later.

After data have been obtained in a given study, they must be arranged in a form suitable for computation and interpretation. We may assume that variates are randomly ordered initially or are in the order in which the measurements were taken. Researchers should be aware that the sequence of observations may contain valuable information, accessible with techniques discussed in Chapter 18. Records of original observations and their sequence should not be discarded; they may be important in tracing and solving problems of sampling or of measurement.

A simple arrangement would be an *array* of the data by order of magnitude. Thus, for example, the variates 7, 6, 5, 7, 8, 9, 6, 7, 4, 6, 7 could be arrayed in order of decreasing magnitude as follows: 9, 8, 7, 7, 7, 7, 6, 6, 6, 5, 4. Where

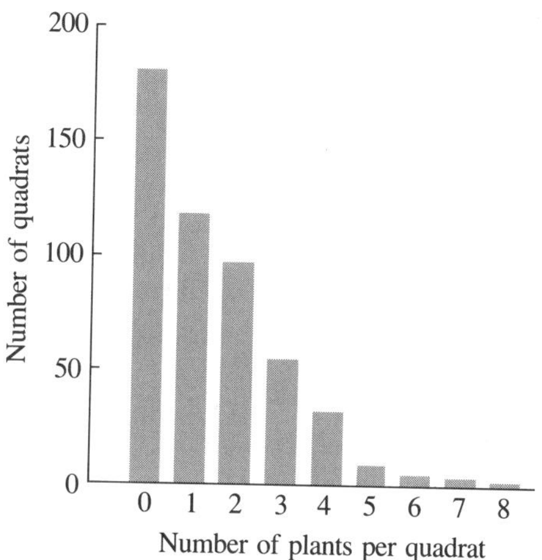

FIGURE 2.2 *Bar diagram. Frequency of the sedge* Carex flacca *in 500 quadrats. Data from Table 2.3.*

there are some variates of the same value, such as the 6's and 7's in this fictitious example, a time-saving device might immediately have occurred to you—namely, to list a frequency for each of the recurring variates: 9, 8, 7(4×), 6(3×), 5, 4. Such shorthand notation is one way to represent a **frequency distribution,** which is simply *an arrangement of the classes of variates with the frequencies of each class indicated.* Conventionally, a frequency distribution is stated in tabular form, as follows for the above example:

Variable Y	Frequency f
9	1
8	1
7	4
6	3
5	1
4	1

This case is an example of a *quantitative* frequency distribution, since Y is clearly a measurement variable. However, arrays and frequency distributions need not be limited to such variables. We can make *qualitative* frequency distributions of attributes. In these, the various classes are listed in some logical or arbitrary order. For example, in genetics we might have a qualitative frequency distribution as follows:

	f
$A-$	86
aa	32

This distribution tells us that there are two classes of individuals, those identified by the $A-$ phenotype, of which 86 were found, and the homozygote recessive aa, of which there were 32 in the sample. In ecology it is quite common to have a species list of inhabitants of a sampled ecological area. The arrangement of such tables is usually alphabetical. An example of an ecological list constituting a qualitative frequency distribution of families is shown in Table 2.2. In this instance the families of insects are listed alphabetically; in other cases the sequence may be by convention, as for the families of flowering plants in botany.

A quantitative frequency distribution based on meristic variates is shown in Table 2.3. This example comes from plant ecology; the number of plants per quadrat sampled are listed at the left in the variable column; the observed frequency is shown at the right.

Table 2.2 A QUALITATIVE FREQUENCY DISTRIBUTION.

Number of individuals of Heteroptera tabulated by family in total samples from a summer foliage insect community.

Family	Observed frequency f
Alydidae	2
Anthocoridae	37
Coreidae	2
Lygaeidae	318
Miridae	373
Nabidae	3
Neididae	5
Pentatomidae	25
Piesmidae	1
Reduviidae	3
Rhopalidae	2
Saldidae	1
Thyreocoridae	10
Tingidae	69
Total Heteroptera	851

SOURCE: Data from Whittaker (1952).

Quantitative frequency distributions based on a continuous variable are the most commonly employed frequency distributions, and you should become thoroughly familiar with them. An example is shown in Box 2.1. It is based on 25 femur lengths of stem mothers (one of the life history stages) of a species of aphids. The 25 readings (measured in coded micrometer units) are shown at the top of the box in the order (reading down columns) in which they were obtained as measurements. The readings could have been arrayed according to their magnitude, but this intermediate step is not necessary in order to proceed to a frequency distribution. We have therefore not shown an array in Box 2.1, but have immediately set up the data in a frequency distribution. The variates increase in magnitude by unit steps of 0.1. The frequency distribution is prepared by entering each variate in turn on the scale and indicating a count by a conventional tally mark. When all of the items have been tallied in the corresponding class, the tallies are converted into numerals indicating frequencies in the next column. Their sum is indicated by Σf.

Let us look at what we have achieved in summarizing our data. The original 25 variates are now represented by only 15 classes. We find that variates 3.6, 3.8,

Table 2.3 A QUANTITATIVE FREQUENCY DISTRIBUTION BASED ON MERISTIC VARIATES.

Number of plants of the sedge Carex flacca *found in 500 quadrats.*

No. of plants per quadrat Y	Observed frequency f
0	181
1	118
2	97
3	54
4	32
5	9
6	5
7	3
8	1
Total	500

SOURCE: Data from Archibald (1950).

and 4.3 have the highest frequencies. However, we also note that there are several classes, such as 3.4 or 3.7, which are not represented by a single aphid. This gives the frequency distribution a drawn-out and scattered appearance. The reason is that we have only 25 aphids, too few to put into a frequency distribution with 15 classes. To obtain a more cohesive and smooth-looking distribution, we have to condense our data into fewer classes. This process is known as **grouping of classes** of frequency distributions; it is illustrated in Box 2.1 and described in the following paragraphs.

What we are doing when we group individual variates into classes of wider range is only an extension of the same process that took place when we obtained the initial measurement. Thus, as we saw in Section 2.3, when we measure an aphid and record its femur length as 3.3 units, we imply that the true measurement lies between the range from 3.25 to 3.35 units but that we were unable to measure to the second decimal place. In recording the measurement initially as 3.3 units we estimated that it fell within this range. Had we estimated that it exceeded the value of 3.35, for example, we would have given it the next higher score, 3.4. Therefore all the measurements between 3.25 and 3.35 were in fact grouped into the class identified by the **class mark** or class midpoint 3.3. Our **class interval** was 0.1 units. If we now make wider class intervals, we are doing nothing but extending the range within which measurements are placed into one class.

Referring to Box 2.1 will clarify this process. We group the data twice to emphasize the flexibility of the process. In the first example of grouping, the class interval has been doubled in width; that is, it was made to equal 0.2 units. If we start at the lower end, the implied class limits will now be from 3.25 to 3.45, the limits for the next class from 3.45 to 3.65, and so forth.

Our next task is to find the class marks. This task was quite simple in the frequency distribution shown at the left side of Box 2.1, in which the original measurements had been used as class marks. However, now we are using a class interval twice as wide, and the class marks are calculated by taking the midpoint of the new class intervals. Thus to find the class mark of the first class we take the midpoint between 3.25 and 3.45, which is 3.35. Note that the class mark has one more decimal place than do the original measurements. Do not believe that we have suddenly achieved greater precision. Whenever we designate a class interval whose last *significant* digit is even (0.2 in this case), the class mark will carry one more decimal place than the original measurements do. On the right side of the table in Box 2.1 the data are grouped once again, using a class interval of 0.3. Because its last significant digit is odd, the class mark now shows as many decimal places as the original variates do, the midpoint between 3.25 and 3.55 being 3.4.

Once the implied class limits and the class mark for the first class have been determined, the others can be written down by inspection without any special computations. Simply add the class interval repeatedly to each of the values. Thus, starting with the lower limit 3.25, by adding 0.2 we obtain 3.45, 3.65, 3.85, and so forth; similarly for the class marks, we obtain 3.35, 3.55, 3.75, and so forth. It should be obvious that the wider the class intervals, the more compact, but also the less precise, the data become. However, looking at the frequency distribution of aphid femur lengths in Box 2.1, we notice that the initial chaotic structure is simplified by grouping. When we group the frequency distribution into five classes with a class interval of 0.3 units, it becomes notably bimodal (that is, it possesses two peaks of frequencies).

In setting up frequency distributions, from 12 to 20 classes should be established. This rule need not be adhered to slavishly but should be employed with some of the common sense that comes from experience in handling statistical data. The number of classes depends largely on the size of the sample studied. Samples of less than 40 to 50 should rarely be given as many as 12 classes, since that would provide too few frequencies per class. On the other hand, samples of several thousand may profitably be grouped into more than 20 classes. If the aphid data of Box 2.1 need to be grouped, they should probably not be grouped into more than 6 classes.

When the sample size is small or the class intervals are wide, the general impression one obtains of a frequency distribution may depend in part on exactly how the class limits are defined. Shifting the class limits may remove the appearance of bimodality or change the apparent location of the mode.

Box 2.1 PREPARATION OF FREQUENCY DISTRIBUTION AND GROUPING INTO FEWER CLASSES WITH WIDER CLASS INTERVALS.

Twenty-five femur lengths of stem mothers of the aphid *Pemphigus populitransversus*. Measurements are in mm $\times 10^{-1}$.

Original measurements				
3.8	3.6	4.3	3.5	4.3
3.3	4.3	3.9	4.3	3.8
3.9	4.4	3.8	4.7	3.6
4.1	4.4	4.5	3.6	3.8
4.4	4.1	3.6	4.2	3.9

Original frequency distribution

Implied limits	Y	Tally marks	f
3.25–3.35	3.3	𝍷	1
3.35–3.45	3.4		0
3.45–3.55	3.5	𝍷	1
3.55–3.65	3.6	𝍷𝍷𝍷𝍷	4
3.65–3.75	3.7		0
3.75–3.85	3.8	𝍷𝍷𝍷𝍷	4
3.85–3.95	3.9	𝍷𝍷𝍷	3
3.95–4.05	4.0		0

Grouping into 8 classes of interval 0.2

Implied limits	Class mark	Tally marks	f
3.25–3.45	3.35	𝍷	1
3.45–3.65	3.55	𝍸	5
3.65–3.85	3.75	𝍷𝍷𝍷𝍷	4
3.85–4.05	3.95	𝍷𝍷𝍷	3

Grouping into 5 classes of interval 0.3

Implied limits	Class mark	Tally marks	f
3.25–3.55	3.4	𝍷𝍷	2
3.55–3.85	3.7	𝍸𝍷𝍷𝍷	8
3.85–4.15	4.0	𝍸	5

<table>
<tr><td>4.05–4.15</td><td>4.1</td><td>||</td><td>2</td><td>4.05–4.25</td><td>4.15</td><td>|||</td><td>3</td><td></td><td></td><td></td><td></td></tr>
<tr><td>4.15–4.25</td><td>4.2</td><td>|</td><td>1</td><td></td><td></td><td></td><td></td><td>4.15–4.45</td><td>4.3</td><td>卌 |||</td><td>8</td></tr>
<tr><td>4.25–4.35</td><td>4.3</td><td>||||</td><td>4</td><td>4.25–4.45</td><td>4.35</td><td>卌 ||</td><td>7</td><td></td><td></td><td></td><td></td></tr>
<tr><td>4.35–4.45</td><td>4.4</td><td>|||</td><td>3</td><td></td><td></td><td></td><td></td><td></td><td></td><td></td><td></td></tr>
<tr><td>4.45–4.55</td><td>4.5</td><td>|</td><td>1</td><td>4.45–4.65</td><td>4.55</td><td>|</td><td>1</td><td>4.45–4.75</td><td>4.6</td><td>||</td><td>2</td></tr>
<tr><td>4.55–4.65</td><td>4.6</td><td></td><td>0</td><td></td><td></td><td></td><td></td><td></td><td></td><td></td><td></td></tr>
<tr><td>4.65–4.75</td><td>4.7</td><td>|</td><td>1</td><td>4.65–4.85</td><td>4.75</td><td>|</td><td>1</td><td></td><td></td><td></td><td></td></tr>
<tr><td>$\sum f$</td><td></td><td></td><td>25</td><td></td><td></td><td></td><td>25</td><td></td><td></td><td></td><td>25</td></tr>
</table>

SOURCE: Data from R. R. Sokal.

Histogram of the original frequency distribution shown above and of the grouped distribution with 5 classes. Line below abscissa shows class marks for the grouped frequency distribution. Hatched bars represent original frequency distribution; hollow bars represent grouped distribution.

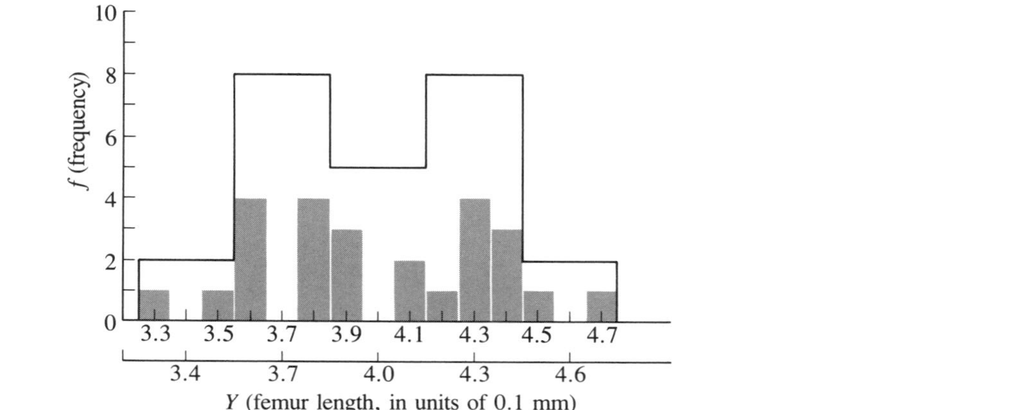

If the original data provide us with fewer classes than we think we should have, then nothing can be done if the variable is meristic, since this is the nature of the data in question. With a continuous variable, however, a scarcity of classes indicates that we probably have not made our measurements with sufficient precision. If we had followed the rule on number of significant figures stated in Section 2.3, this could not have happened.

Whenever there are more than the desired number of classes, grouping should be undertaken. When the data are meristic, the implied limits of continuous variables are meaningless. Yet with many meristic variables, such as a bristle number varying from 13 to 81, it would probably be wise to group them into classes, each containing several counts. This grouping can best be done by using an odd number as a class interval so that the class mark representing the data is a whole rather than a fractional number. Thus if we were to group the bristle numbers 13, 14, 15, and 16 into one class, the class mark would have to be 14.5, a meaningless value in terms of bristle number. It would therefore be better to use a class ranging over 3 bristles or 5 bristles, giving the integral values 14 or 15 as class marks.

Grouping data into frequency distributions was necessary when computations were done by pencil and paper or with mechanical calculators. Nowadays even thousands of variates can be processed efficiently by computer without prior grouping. However, frequency distributions are still extremely useful as a tool for data analysis, especially in an age where it is all too easy for a researcher to obtain a numerical result from a computer program without ever really examining the data for outliers (extreme values) or for other ways in which the sample may not conform to the assumptions of the statistical methods. For this reason most modern statistical computer programs furnish some graphic output of the frequency distribution of the observations.

An alternative to setting up a frequency distribution with tally marks as shown in Box 2.1 is the so-called **stem-and-leaf display** suggested by Tukey (1977). The advantage of this technique is that it not only results in a frequency distribution of the variates of a sample, but also presents them in a form that makes a ranked (ordered) array very easy to construct. It also effectively creates a list of these values—and is very easy to check, unlike the tallies, which can be checked only by repeating the procedure. This technique is therefore useful in computing the median of a sample (see Section 4.3) and in computing various nonparametric statistics that require ordered arrays of the sample variates (see Sections 13.11 and 13.12).

Let us first learn how to construct the stem-and-leaf display. In Box 15.6 we feature measurements of total length recorded for a sample of 15 aphid stem mothers. The unordered measurements are reproduced here: 8.7, 8.5, 9.4, 10.0, 6.3, 7.8, 11.9, 6.5, 6.6, 10.6, 10.2, 7.2, 8.6, 11.1, 11.6. To prepare a stem-and-leaf display we write down the leading digit or digits of the variates in the sample to the left of a vertical line (the ''stem'') as shown below; we then put the next digit of the first variate (a ''leaf'') at that level of the stem corresponding to its leading digit(s):

Step **1**	Step **2** ···	Step **7** ···	Completed Array (step **15**)	Ordered Array
6 |	6 |	6 | 3	6 | 356	6 | 356
7 |	7 |	7 | 8	7 | 82	7 | 28
8 | 7	8 | 75	8 | 75	8 | 756	8 | 567
9 |	9 |	9 | 4	9 | 4	9 | 4
10 |	10 |	10 | 0	10 | 062	10 | 026
11 |	11 |	11 | 9	11 | 916	11 | 169

The first observation in our sample is 8.7. We therefore place a 7 next to the 8. The next variate is 8.5. It is entered by finding the stem level for the leading digit 8 and recording a 5 next to the 7 that is already there. Similarly for the third variate, 9.4, we record a 4 next to the 9, and so on until all 15 variates have been entered (as ''leaves'') in sequence along the appropriate leading digits of the stem. Finally, we order the leaves from smallest (0) to largest (9).

The ordered array is equivalent to a frequency distribution and has the appearance of a histogram or bar diagram (see below), but it also is an efficient ordering of the variates. Thus from the ordered array it becomes obvious that the appropriate ordering of the 15 variates is: 6.3, 6.5, 6.6, 7.2, 7.8, 8.5, 8.6, 8.7, 9.4, 10.0, 10.2, 10.6, 11.1, 11.6, 11.9. The median, the observation having an equal number of variates on either side, can easily be read off the stem-and-leaf display. It is 8.7.

The computational advantage of this procedure is that it orders the variates by their leading digits in a single pass through the data set, relying on further ordering of the trailing digits in each leaf by eye, whereas a direct ordering technique could require up to n passes through the data before all items are correctly arrayed. A FORTRAN program for preparing stem-and-leaf displays is given by McNeil (1977). If most variates are expressed in three digits, the leading digits (to the left of the stem) can be expressed as two-digit numbers (as was done for the last two lines in the example above), or two-digit values can be displayed as leaves to the right of the stem. In the latter case, however, the two-digit leaves would have to be enclosed in parentheses or separated by commas to prevent confusion. Thus if the observations ranged from 1.17 to 8.53, for example, the leading digits to the left of the stem could range from 1 to 8, and the leaves of one class, say 7, might read 26, 31, 47, corresponding to 7.26, 7.31, and 7.47.

In biometric work we frequently wish to compare two samples to see if they differ. In such cases, we may employ **back-to-back stem-and-leaf displays,** illustrated here:

Sample A		Sample B
94	10	05778
988642	11	16
88866531	12	013

This example is taken from Box 13.7 and describes a morphological measurement obtained from two samples of chiggers. By setting up the stem in such a way that it may serve for both samples, we can easily compare the frequencies. Even though these data furnish only three classes for the stem, we can readily see that the samples differ in their distributions. Sample A has by far the higher readings.

When the shape of a frequency distribution is of particular interest, we may often wish to present the distribution in graphic form when discussing the results. This is generally done with frequency diagrams, of which there are two common types. For a distribution of meristic data we use a **bar diagram,** as shown in Figure 2.2 for the sedge data of Table 2.3. The abscissa represents the variable (in our case the number of plants per quadrat), and the ordinate represents the frequencies. What is important about such a diagram is that the bars do not touch each other, which indicates that the variable is not continuous.

By contrast, continuous variables, such as the frequency distribution of the femur lengths of aphid stem mothers, are graphed as a **histogram,** in which the width of each bar along the abscissa represents a class interval of the frequency distribution and the bars touch each other to show that the actual limits of the classes are contiguous. The midpoint of the bar corresponds to the class mark. At the bottom of Box 2.1 are histograms of the frequency distribution of the aphid data, ungrouped and grouped. The height of the bars represents the frequency of each class. To illustrate that histograms are appropriate approximations to the continuous distributions found in nature, we may take a histogram and make the class intervals more narrow, producing more classes. The histogram would then clearly fit more closely to a continuous distribution. We can continue this process until the class intervals approach the limit of infinitesimal width. At this point the histogram becomes the continuous distribution of the variable. Occasionally the class intervals of a grouped continuous frequency distribution are unequal. For instance, in a frequency distribution of ages, we might have more detail on the different stages of young individuals and less accurate identification of the ages of old individuals. In such cases, the class intervals for the older age groups would be wider, those for the younger age groups, narrower. In representations of such data, the bars of the histogram are drawn with different widths.

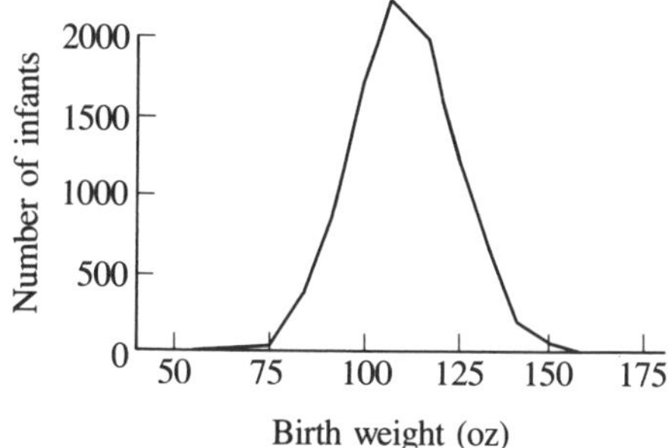

FIGURE 2.3 *Frequency polygon. Birth weights of 9465 male infants. Chinese third-class patients in Singapore, 1950 and 1951. (Data from Millis and Seng, 1954.)*

Figure 2.3 shows another graphic mode of representing a frequency distribution of a continuous variable—birth weight in infants. This is a *frequency polygon,* in which the heights of the class marks in a histogram are connected by straight lines. As we shall see later the shapes of frequency distributions as seen in these various graphs can reveal much about the biological situations affecting a given variable.

EXERCISES 2

2.1 Differentiate between the following pairs of terms and give an example of each. (a) Statistical and biological populations. (b) Variate and individual. (c) Accuracy and precision (repeatability). (d) Class interval and class mark. (e) Bar diagram and histogram. (f) Abscissa and ordinate.

2.2 Round the following numbers to three significant figures: 106.55, 0.06819, 3.0495, 7815.01, 2.9149, and 20.1500. What are the implied limits before and after rounding? Round these same numbers to one decimal place.

2.3 Given 200 measurements ranging from 1.32 to 2.95 mm, how would you group them into a frequency distribution? Give class limits as well as class marks.

2.4 Group the following 40 measurements of interorbital width of a sample of domestic pigeons into a frequency distribution and draw its histogram (data from Olson and Miller, 1958). Measurements are in millimeters.

12.2	12.9	11.8	11.9	11.6	11.1	12.3	12.2	11.8	11.8
10.7	11.5	11.3	11.2	11.6	11.9	13.3	11.2	10.5	11.1
12.1	11.9	10.4	10.7	10.8	11.0	11.9	10.2	10.9	11.6
10.8	11.6	10.4	10.7	12.0	12.4	11.7	11.8	11.3	11.1

2.5 How precisely should you measure the wing length of a species of mosquitoes in a study of geographic variation, if the smallest specimen has a length of about 2.8 mm and the largest a length of about 3.5 mm?

2.6 Transform the 40 measurements in Exercise 2.4 into common logarithms with your calculator and make a frequency distribution of these transformed variates. Comment on the resulting change in the pattern of the frequency distribution from that found before.

2.7 In Exercise 4.3, we feature 120 percentages of butterfat from Ayrshire cows. Make a frequency distribution of these values using a stem-and-leaf display. Prepare an ordered array of these variates from the display. Save the display for use later in Exercise 4.3(b).

3 THE HANDLING OF DATA

We have already stressed in the preface that the skillful and expeditious handling of data is essential to the successful practice of statistics. Since the emphasis of this book is to a large extent on practical statistics, it is necessary to acquaint readers with the various techniques available for carrying out statistical computations. We discuss computer hardware in Section 3.1, software in Section 3.2. In Section 3.3 we focus on the important question of which computational devices and types of software are most appropriate in terms of efficiency and economy.

Lacking mechanical computational aids, we would be reduced to carrying out statistical computations by the so-called pencil-and-paper methods. Textbooks of statistics used to contain extensive sections dealing with clever ways to make computation by hand feasible. At present, however, the use of shortcut or approximate computations by pencil-and-paper methods is a very inefficient use of time and energy. Even handheld calculators are now used much less than they were before, although the more advanced models are capable of performing many of the simpler analyses presented in this text. With microcomputers on most desktops and battery-powered portable computers, computation by hand or with calculators is used mostly to verify the initial results produced by computer.

The benefit from these developments is that researchers can concentrate on the more-important problems of the proper design of an experiment and the interpretation of the statistical results rather than spending time on computational details and special tricks to make computations less tedious. We have eliminated explanation of all the special methods for hand or calculator computation and expect that this will make the material in this text easier to learn.

Another benefit of the routine use of computers is that more-powerful, complex analyses now are usually feasible. Some of these methods are able to give more exact probability values for standard tests of significance. Other methods (see Chapter 18) enable one to make significance tests for nonstandard statistics or for situations in which the usual distributional assumptions are not likely to be true.

3.1 COMPUTERS

Since the publication of the first edition of *Biometry,* a revolution in the types of equipment available for performing statistical computations has occurred. The once standard, electrically driven, mechanical desk calculators that were used for routine statistical calculations in biology have completely disappeared. They were replaced first by a wide variety of electronic calculators (ranging from pocket to desktop models). Except for very simple computations, their use has now been largely replaced by computers. Although some calculators are capable of performing *t*-tests, regression analysis, and similar calculations, they are now used most often just for making rough checks or for manipulating results from a computer.

Calculators range from devices that can only add, subtract, multiply, and divide to scientific calculators with many mathematical and statistical functions built in. Their principal limitation is that they usually can store only limited amounts of data. This means that one must reenter the data in order to try an alternative analysis (e.g., with various transformations, with outliers removed, and so on). Reentry is very tedious and greatly increases the chance for error.

It is difficult to know where to draw the line between the more sophisticated electronic calculators and digital computers. There is a continuous range of capabilities from programmable calculators through palm-top microcomputers, notebook and laptop computers, desktop microcomputers, workstations, and minicomputers, to large-scale mainframe and supercomputers. The largest are usually located in a central computation center of a university or research laboratory.

Programmable calculators are usually programmed in languages unique to the manufacturer of the device. This means that considerable effort may be required to implement an analysis on a new device, which limits the development of software. Even so, collections of statistical programs (often donated by other users) are usually available. These collections do not, however, compare to the depth of libraries of programs available for most computers. Although computers can be programmed in ''machine language'' specific to each line of computers, programs are now usually written using standard high-level languages such as BASIC, C, FORTRAN, or Pascal, enabling the accumulation of large libraries of programs.

Computers consist of three main components: the central processing unit (which performs the calculations and controls the other components), the memory (which stores both data and instructions), and peripheral devices (which handle the input of data and instructions, the output of results, and perhaps intermediate storage of data and instructions). Different devices vary considerably in the capabilities of these three components. In a simple calculator, the processor is the arithmetic unit that adds, subtracts, multiplies, and divides and the memory may consist of only a few registers, each capable of storing a

10-digit number. The only peripheral devices may be the keyboard for entering the data and a light-emitting diode (LED) or liquid crystal (LCD) display.

At present, standard microcomputers usually have 4×10^6 to 8×10^6 eight-bit numbers or characters (bytes) of main memory, 300×10^6 bytes of disk storage, and can perform about 1.8×10^6 average floating-point operations (arithmetic operations on numbers with a decimal place) per second. Large mainframe computers now have capabilities such as 500×10^6 bytes of main memory and 2×10^9 bytes of secondary storage. They can process 800×10^6 floating-point operations per second. There are also specialized computers consisting of many processors tied together to work on a single problem that can achieve even faster combined processing speeds. Of course, with the rapid developments in computing technology over the past few years, we expect these numbers to be out of date by the time this book is in print and to seem very small by the time this book is next revised. These figures are given just for comparison. Because of aggressive marketing, most users are quite impressed with recent developments in microcomputers and workstations. The equally impressive developments in supercomputers needed for the solution of many large-scale problems are not as visible.

The ever-increasing speed of computation is important for many applications. Computers perform no operations that could not, in principle, be done by hand. However, tasks that large-scale computers can perform in a matter of seconds might take thousands of years for humans to complete. Computer hardware is also very reliable. It is unlikely that a person working on a very long calculation would not make an error somewhere along the way unless time were allowed for independent checking of all results. It is now possible to solve problems that one would scarcely even have contemplated before computers were developed.

Generally speaking, the largest and fastest computers are the most efficient and economical for carrying out a given computation, but other factors often influence the decision of what computing hardware is used. Availability is, of course, very important. Because of their cost, very large-scale computers are located only at major universities or laboratories, and their use is shared among hundreds or thousands of authorized users. These computers can, however, be accessed easily from anywhere in the world via terminals or microcomputers attached to high-speed networks. Most statistical analyses require relatively little computation, so the capabilities of a large-scale computer are not needed. (There may be more overhead associated with making the connection to such a computer than with the actual arithmetic.) Often the most important factor is the availability of appropriate software, which is discussed in the next section.

3.2 SOFTWARE

The instructions to a computer are known as the **program,** and the persons who specialize in the writing of computer programs are known as programmers.

Writing instructions in the generally unique language for each model of computer is a very tedious task and is now seldom done by programmers concerned with software written for statistical applications. Fortunately, programs called **compilers** have been developed to translate instructions written in simple, general purpose, problem-oriented languages into the machine language for a particular computer. Some of the best-known compilers are BASIC, C, FORTRAN, and Pascal. FORTRAN is one of the oldest computer languages and is still very popular for numerical calculations, including large-scale vector and parallel computations on many supercomputers. There are also programs called **interpreters** that compile and execute each instruction as it is entered, which can be very convenient for developing software that is changed often and adapted to different applications. The best-known interpreter is for the BASIC language.

Many new computer languages which simplify the development of software for particular types of applications are now available. There are, for example, several powerful systems with built-in commands for statistical, matrix, and graphical operations. Recently systems have been developed that allow the user to write a program by pointing to objects (icons representing input devices or files, particular mathematical operations, statistical analyses, various types of graphs, etc.) on the computer screen and then drawing connections between them to indicate the flow of information between objects (e.g., from an input file, to a particular statistical analysis, to a graph of the results). Menus and dialogue boxes are used to specify various options for the properties of the objects. This **object-oriented programming** approach means that a user no longer has to be a programmer in order to create software appropriate for a certain specialized task. It does, however, make it essential for a researcher to be computer literate.

The material presented in this book consists of relatively standard statistical computations, most of which have been programmed as part of many different statistical packages. The BIOM-pc software for IBM PC compatible computers was developed by one of us (FJR) to carry out the computations for most of the topics covered in this book. An older version, written in FORTRAN, can be adapted to run on other computers, including mini and mainframe computers. (These programs are available from Exeter Software, 100 North Country Road, Setauket, New York 11733.)

The specific steps used during computations on the computer depend on the particular computer and software used. At one time, when using centralized computer centers, one had to employ a **batch mode** of computations. In such an environment, one left card decks of programs and data to be run when computer time was available; printed output was returned hours later or even the next day. Now large computer centers can be accessed via computer terminals (simple devices that consist of just a keyboard and a CRT screen and operate only under the control of the remote computer) or microcomputers that use software that emulates the properties of a computer terminal.

This new environment permits an **interactive mode** of computation in which the computer responds rapidly to each user input. The connections can be via directly connected lines, dial-up phone lines, or various types of network connections allowing convenient access to the resources of a large computer no matter where the computer is physically located. An important limitation of this mode of computing is the speed of communication between the user and the remote computer. Speed limits applications such as interactive graphics (real time rotation of three-dimensional objects), as well as the practicality of applications such as word processing, in which the screen is reformatted as each character is typed and in which very little actual computation is performed.

Although communication speeds are increasing rapidly, these operations are more efficient if performed locally. This is one of the reasons for the popularity of powerful microcomputers and personal workstations where it is practical to have convenient, easy-to-use, and elegant graphical user interfaces. Other reasons for the popularity of small, powerful computers are more psychological. The user generally feels more free to experiment because there are no usage charges, nor the formality of passwords, computer allocations, and logging on to a computer at a remote center. The most important factor in choosing a system should be simply the availability of software to perform the desired tasks. An enormous quantity of software for personal computers has been developed in recent years.

3.3 Efficiency and Economy in Data Processing

From the foregoing description it may seem to the reader that a computer should always be preferred over a calculator. This is usually the case, but simpler calculations can often be done more quickly on a calculator. (One may be done before most computers finish booting.) On the other hand, even simple computations often need to be checked or repeated in a slightly different way. It is very convenient not to have to reenter the input more than once and to have a printed record of all input and operations performed leading to a given result.

The difficult decision is not whether to use a computer but which computer and software to use. Software for most of the computations described in this book is available on most computers. Thus, the choice of hardware usually is determined by factors such as the type of computer already present in one's lab. The computations described in this book can be performed on but do not require large-scale computing systems (except, possibly, for very large randomization tests, Mantel tests, and other complex computations). It is no longer necessary or even efficient for most users to write their own software, since there is usually an overwhelming array of choices of software that can be used for statistical computations. Not all software, however, is of the same quality. There can be different degrees of rounding error, or the software may sometimes produce erroneous results because of bugs in the program. It is a good idea to check a software package by using standard data sets (such as the examples in the boxes in this text) before running one's own data. One should also read published reviews of the software before purchasing it.

Most statistical software can be classified into one of the following major groups (some major packages combine the features of two or more groups): specialized single-operation programs, command language-driven programs, menu-driven programs, and spreadsheets. The earliest statistical software was developed for the batch-computing environment. The program would read a previously prepared file of data, perform a single type of analysis, and then output the results (which usually would include many possible analyses in case they might interest the user).

Alternatively, large statistical packages were developed that combined many types of analysis. With such a package, the user selects the particular operations desired using statements in an artificial command language unique to that software. This type of software requires time to set up the data and to learn the commands and various options needed for a particular analysis, but it is usually a very efficient mode for processing large datasets. These programs also usually offer the widest selection of options. User-friendly, menu-driven programs permit rapid data input and feature easy options for performing standard statistical analyses. Options can be conveniently selected from lists in menus or in dialogue boxes. Some packages combine the advantages of these last two modes by letting the user employ a menu system to prepare the command file needed to carry out the desired analysis.

Although spreadsheet programs are not often used for statistical calculations in biology, their very different mode for computation is often useful. They might be viewed as the computer generalization of the calculator. They simulate a large sheet of paper that has been divided into cells. Each cell can contain a value entered by the user or a result of a computation based on other cells in the spreadsheet. The operations used to produce a particular numerical result or graph are remembered by the program. If an input cell is changed by the user, the consequent changes in the results are immediately recomputed and displayed. This feature makes it very easy to experiment with data and see the results of alternative analyses.

A danger inherent in computer processing of data is that the user may simply obtain the final statistical results without observing the distribution of the variates (in fact without even seeing the data if they are collected automatically by various data acquisition devices). One should take advantage of any options available to display the data that could lead to interesting new insights into their nature, or to the rejection of some outlying observations, or to suggestions that the data do not conform to the assumptions of a particular statistical test. Most statistical packages are capable of providing such graphics. We strongly urge research workers to make use of such operations.

Another danger is that it is too easy to blindly use whatever tests are provided with a particular program without understanding their meanings and assumptions (Searle, 1989). The availability of computers relieves the tedium of computation, but not the necessity to understand the methods being employed.

4 DESCRIPTIVE STATISTICS

An early and fundamental stage in any science is the descriptive stage. Until the facts can be described accurately, analysis of their causes is premature. The question *what* must come before *how*. The application of statistics to biology has followed these general trends. Before Francis Galton could begin to think about the relations between the heights of fathers and those of their sons, he had to have adequate tools for measuring and describing heights in a population. Similarly, unless we know something about the usual distribution of the sugar content of blood in a population of guinea pigs, as well as its fluctuations from day to day and within days, we cannot ascertain the effect of a given dose of a drug upon this variable.

In a sizable sample, obtaining knowledge of the material by contemplating all the individual observations would be tedious. We need some form of summary to deal with the data in manageable form, as well as to share our findings with others in scientific talks and publications. A histogram or bar diagram of the frequency distribution is one type of summary. For most purposes, however, a numerical summary is needed to describe the properties of the observed frequency distribution concisely and accurately. Quantities providing such a summary are called **descriptive statistics.** This chapter will introduce you to some of them and show how they are computed.

Two kinds of descriptive statistics will be discussed in this chapter: statistics of location and statistics of dispersion. **Statistics of location** describe the position of a sample along a given dimension representing a variable. For example, we might like to know whether the sample variates measuring the length of certain animals lie in the vicinity of 2 cm or 20 cm. A statistic of location must yield a representative value for the sample of observations. However, such a statistic (sometimes also known as a measure of central tendency) does not describe the shape of a frequency distribution. This distribution may be long or very narrow; it may be humped or U-shaped; it may contain two humps, or it may be markedly asymmetrical. Quantitative measures of such aspects of frequency distributions are required. To this end we need to define and study **statistics of dispersion.**

The arithmetic mean described in Section 4.1 is undoubtedly the most important single statistic of location, but others (the geometric mean, the harmonic mean, the median, and the mode) are mentioned briefly in Sections 4.2, 4.3, and 4.4. A simple statistic of dispersion, the range, is briefly noted in Section 4.5, and the standard deviation, the most common statistic for describing dispersion, is explained in Section 4.6. Our first encounter with contrasts between sample statistics and population parameters occurs in Section 4.7, in connection with statistics of location and dispersion. Section 4.8 contains a description of methods of coding data to simplify the computation of the mean and standard deviation, which is discussed in Section 4.9. The coefficient of variation (a statistic that permits us to compare the relative amount of dispersion in different samples) is explained in the last section (4.10).

The techniques that will be at your disposal after you have mastered this chapter are not very powerful in solving biological problems, but they are indispensable tools for any further work in biometry.

IMPORTANT NOTE: We will encounter the use of logarithms in this chapter for the first time. To avoid confusion here and in subsequent chapters, common logarithms have been abbreviated consistently as log, and natural logarithms as ln. Thus, $\log x$ means $\log_{10} x$ and $\ln x$ means $\log_e x$.

4.1 The Arithmetic Mean

The most common statistic of location should be familiar. It is the **arithmetic mean,** commonly called the *mean* or *average.* The mean is calculated by summing all the individual observations or items of a sample and dividing this sum by the number of items in the sample. For instance, as the result of a gas analysis in a respirometer, an investigator obtains the following four readings of oxygen percentages:

$$\begin{array}{rr} & 14.9 \\ & 10.8 \\ & 12.3 \\ & \underline{23.3} \\ \text{Sum} = & 61.3 \end{array}$$

The investigator calculates the mean oxygen percentage as the sum of the four items divided by the number of items—here, by 4. Thus the average oxygen percentage is

$$\text{Mean} = \frac{61.3}{4} = 15.325\%$$

Calculating a mean presents the opportunity for learning statistical symbolism. We have already seen that an individual observation is symbolized by Y_i, which stands for the *i*th observation in the sample (see Section 2.2). Four observations could be written symbolically as follows:

$$Y_1, Y_2, Y_3, Y_4$$

We shall define *n*, the **sample size,** as the number of items in a sample. In this particular instance, the sample size *n* is 4. In a large sample, we can symbolize the array from the first to the *n*th item as follows:

$$Y_1, Y_2, \ldots, Y_n$$

When we wish to sum items, we use the following notation:

$$\sum_{i=1}^{i=n} Y_i = Y_1 + Y_2 + \cdots + Y_n$$

The capital Greek sigma, Σ, means sum the items indicated. The expression below Σ (the subscript) gives the first item to be included in the sum; the expression above Σ (the superscript) gives the last item. Thus, in this example, $i = 1$ tells us to begin summing with the first item, Y_1, and $i = n$ tells us to end with the *n*th item, Y_n. The subscript and superscript are necessary to indicate how many items should be summed. The "$i =$ " in the superscript is usually omitted as superfluous. For instance, if we had wished to sum only the first three items, we could have written $\Sigma_{i=1}^{3} Y_i$. On the other hand, had we wished to sum all of them except the first one, we could have written $\Sigma_{i=2}^{n} Y_i$.

We call Σ an operator symbol. It indicates what to do with the variables that follow it. Summation signs with subscripts and superscripts give an explicit mathematical formulation of the summation operation. However, with some exceptions (which will appear in later chapters), it is desirable to omit subscripts and superscripts, especially for the nonmathematical readers of this book. Subscripts generally add to the apparent complexity of the formula and, when unnecessary, distract the student's attention from the important relations expressed by the formula. Below are seen increasing simplifications of the complete summation notation shown at the extreme left:

$$\sum_{i=1}^{i=n} Y_i = \sum_{i=1}^{n} Y_i = \sum_{i} Y_i = \sum^{n} Y = \sum Y$$

The third of the symbols might be interpreted as meaning: sum the Y_i's over all available values of *i*. This is a frequently used notation, although we shall not employ it in this book. The next notation, with *n* as a superscript, tells us to sum *n* items of *Y*; note that the *i* subscript of the *Y* has been dropped as unnecessary. Finally, the simplest notation is shown at the right. It says simply, sum the *Y*'s. We will use this form most frequently, and if a summation sign precedes a variable the summation will be understood to be over *n* items (all the items in the sample) unless subscripts or superscripts specifically tell us otherwise.

We will use the symbol $\bar{Y}$ for the arithmetic mean of the variable *Y*. Some textbooks use $\bar{y}$. Many other textbooks use the symbol *X* for a variable and, consequently, use $\overline{X}$ or $\bar{x}$ for the mean. In agreement with several modern texts we have adopted Y, for reasons that will be explained in Chapter 14. Statistical symbolism is, regrettably, still far from uniform, although laudable progress toward uniformity has been made in recent years. We will try to point out from time to time common alternative symbolisms to those presented in the text to prepare you for encounters with the statistical writings of others.

Our newly acquired knowledge of symbols enables us to write the formula for the arithmetic mean:

$$\overline{Y} = \frac{\sum Y}{n} \tag{4.1}$$

This formula tells us to sum all (n) items and divide the sum by n.

The mean of a sample represents the center of the observations in the sample. If you were to draw a histogram of an observed frequency distribution on a sheet of cardboard, then cut out the histogram and lay it flat against a blackboard, supporting it with a pencil beneath, chances are that it would be out of balance, toppling left or right. If you moved the supporting pencil point to a position about which the histogram would exactly balance, this point of balance would be the arithmetic mean. In fact, this would be an empirical method of finding the arithmetic mean of a frequency distribution.

It is often necessary to average means or other statistics that may differ in their reliabilities because they are based on different sample sizes or for other reasons. In such cases a **weighted average** needs to be computed. A general formula for calculating the weighted average of a set of values Y_i is as follows:

$$\overline{Y}_w = \frac{\sum^{n} w_i Y_i}{\sum^{n} w_i} \tag{4.2}$$

where n variates, each weighted by factor w_i, are being averaged. The values of Y_i in such cases are unlikely to represent variates. They are more likely to be sample means, $\overline{Y}_i$, or other statistics of different reliabilities.

The simplest case in which statistics need to be averaged is when the Y_i are not individual variates but are means. Thus if the following three means are based on differing sample sizes:

$\overline{Y}_i$	n_i
3.85	12
5.21	25
4.70	8

their weighted average using sample sizes n_i as weights w_i will be

$$\overline{Y}_w = \frac{(12)(3.85) + (25)(5.21) + (8)(4.70)}{12 + 25 + 8} = \frac{214.05}{45} = 4.76$$

In this example, computation of the weighted mean is exactly equivalent to adding up all the original measurements and dividing the sum by the total number of the measurements. Thus the sample with 25 observations, will influence the weighted average in proportion to the size of the sample. With data such as those displayed in Table 2.1, we might wish to weight each percentage by its sample size. Clearly, a value of 33% is more reliable when based on nine pupae than it is when based on three pupae.

Occasionally it is desirable to employ an **unweighted average,** which in this case would be $(3.85 + 5.21 + 4.70)/3 = 4.59$. This type of average would be appropriate if we considered each sample mean equally important and wished to represent the entire range of variation of the variable without weighting each mean by the size of its sample.

4.2 Other Means

We will see in Chapters 13 and 14 that variables are sometimes transformed into their logarithms or reciprocals. If we calculate the mean of such a transformed variable and then change the mean back into the original scale, this mean will not be the same as if we had computed the arithmetic mean of the original variable. The resulting means have special names in statistics. The back-transformed mean of a logarithmically transformed variable is called the **geometric mean** (GM). It is computed as

$$GM_Y = \text{antilog}\,\frac{1}{n}\sum \log Y \tag{4.3}$$

which indicates that the geometric mean GM_Y is the antilogarithm of the mean of the logarithms of variable Y. Since adding logarithms is equivalent to multiplying their antilogarithms, another way of representing this quantity is

$$GM_Y = \sqrt[n]{Y_1 Y_2 Y_3 \cdots Y_n} \tag{4.4}$$

The geometric mean permits us to become familiar with the product operator symbol: capital pi, Π. Just as Σ symbolizes *summation* of the items that follow it, so does Π symbolize the *multiplication* of the items that follow it. Subscripts and superscripts of the product operator have exactly the same meaning as in the summation case. Thus Expression (4.4) for the geometric mean can be rewritten more compactly as follows:

$$GM_Y = \sqrt[n]{\prod_{i=1}^{n} Y_i} \tag{4.4a}$$

For any but very small samples, computing the geometric mean by Expression (4.4a) is impractical; thus this computation is carried out by transforming the variates into logarithms.

The reciprocal of the arithmetic mean of reciprocals is called the **harmonic mean.** If we symbolize it by H_Y, the formula for the harmonic mean can be written in concise form (without subscripts and superscripts) as

$$\frac{1}{H_Y} = \frac{1}{n} \sum \frac{1}{Y} \tag{4.5}$$

You may wish to convince yourself that the geometric mean and the harmonic mean of the four oxygen percentages at the beginning of Section 4.1 are 14.65% and 14.09%, respectively. Unless the individual items do not vary, the geometric mean is always less than the arithmetic mean, and the harmonic mean is always less than the geometric mean.

Some beginners in statistics have difficulty accepting the fact that measures of location or central tendency other than the arithmetic mean are permissible or even desirable. They feel that the arithmetic mean is the "logical" average and that any other mean would distort the data. This attitude raises the question of the proper scale of measurement for representing data; this scale is not always the linear scale familiar to everyone, but is sometimes by preference a logarithmic or reciprocal scale. If you have doubts about this question, they may be allayed in Chapter 13, where we discuss the reasons for transforming variables.

4.3 The Median

The **median** (M), a statistic of location occasionally useful in biological research, is defined as the value of the variable (in an ordered array) that has an equal number of items on either side of it. Thus, the median divides a frequency distribution into two halves. In the following sample of five measurements:

14, 15, 16, 19, 23

$M = 16$, since the third observation has an equal number of observations on both sides of it. We can visualize the median easily if we think of an array from largest to smallest—for example, a row of men lined up by their heights. The median individual is the person having an equal number of men on his right and left sides. His height will be the median height of the sample considered. This quantity is evaluated easily from a sample array with an odd number of individuals. When the number in the sample is even, the median is conventionally calculated as the midpoint between two variates—the $(n/2)$th and the $[(n/2) + 1]$th variate. Thus, for the sample of four measurements:

14, 15, 16, 19

the median would be the midpoint between the second and third items, or 15.5.

Whenever one value of a variate occurs more than once, problems may develop in locating the median. Computing the median becomes more involved because all the members of a given class in which the median is located have the same class mark. Computation of the median of a frequency distribution, either with the class marks of the original variate or with variates grouped into wider classes, is illustrated and explained in Box. 4.1.

Box 4.1 COMPUTATION OF THE MEDIAN FROM A FREQUENCY DISTRIBUTION.

Birth weights of male Chinese in ounces.

(1) Y	(2) f	(3) Cumulative frequency F	(4) Reverse cumulative frequency F
59.5	2	2	9465
67.5	6	8	9463
75.5	39	47	9457
83.5	385	432	9418
91.5	888	1320	9033
99.5	1729	3049	8145
107.5	2240	5289	6416
115.5	2007	7296	4176
123.5	1233	8529	2169
131.5	641	9170	936
139.5	201	9371	295
147.5	74	9445	94
155.5	14	9459	20
163.5	5	9464	6
171.5	1	9465	1
	9465		

SOURCE: Millis and Seng (1954).

The data are in the form of a frequency distribution because of the large number of observations in the sample. Column (3) shows a new feature, a *cumulative* frequency distribution, which is produced by successive addition of the frequencies for each class from the column to the left of it, accumulating the values either as the variable increases or as it decreases (reverse cumulative frequencies). The first 2 in column (3) is simply a copy of the 2 to its left, the 8 is the sum of 2 and 6, the 47 is 8 + 39, the 432 is 47 + 385, and so forth. The cumulative frequencies represent the total number of items below the upper class limit of their class [or above the lower class limit in the case of the reverse cumulative frequency distribution shown in column (4)]. Thus there are 432 individuals below 87.5 oz, the upper limit of the 83.5-oz class. The final value in the cumulative frequency column is the sum of all the frequencies in the sample.

The median for ranked data in an array is the $(n + 1)/2$th variate. In a frequency distribution such as the birth weights in this box, however, the median is obtained by interpolation as the $n/2$th variate

$$\frac{n}{2} = \frac{9465}{2} = 4732.5$$

The 4732.5th item is located in the weight class 107.5 oz, somewhere between the implied limits of 103.5 and 111.5. This class contains 2240 items (5289 − 3049 in the

BOX 4.1 CONTINUED

cumulative frequency column), and item 4732.5 is the 4732.5 − 3049 = 1683.5th item in it. Assuming equal distribution of items in the class, we consider item 4732.5 to be

$$\frac{1683.5}{2240} = 0.7516 \text{ of the total class interval,}$$

or 75.16% of the distance from the lower class limit to the upper class limit. Since each class interval is 8.0 oz, the median item is 0.7516 × 8.0 = 6.013 oz above the lower class limit (103.5 oz); that is, the median birth weight is located at 103.5 + 6.013 = 109.513 oz. The median could also have been computed from the other direction. For this approach we would use the reverse cumulative frequency column to compute 4732.5 − 4176 = 556.5. Thus the median would be

$$\frac{556.5}{2240} = 0.2484$$

of a class interval, or 1.987 oz (0.2484 × 8.0), below the *upper* class limit (111.5 oz). The end result, 111.5 − 1.987 = 109.513 oz, is the same.

The median is just one of a family of statistics dividing a frequency distribution into equal proportions. It divides the distribution into two halves. **Quartiles,** on the other hand, cut the distribution at the 25%, 50%, and 75% points—that is, at points dividing the distribution into first, second, third, and fourth quarters by area (and frequencies). The second quartile is, of course, the median. (There are also quintiles, deciles, and percentiles, dividing the distribution into 5, 10, and 100 equal portions, respectively.) A general term for such quantities is **quantiles.**

Medians are most often used for distributions that do not conform to the standard probability models and that thus require nonparametric methods (see Chapter 13). Sometimes the median is a more representative measure of location than is the arithmetic mean. Such instances almost always involve asymmetric distributions. An often-quoted example from economics is a suitable measure of location for the ''typical'' salary of an employee of a corporation. The very high salaries of the few senior executives shift the arithmetic mean, the center of gravity, toward a completely unrepresentative value. The median, on the other hand, is little affected by a few high salaries; it gives the particular point on the salary scale above which and below which lie 50% of the salaries in the corporation.

In biology, an example in which the application of a median is preferred over the arithmetic mean is in populations showing skewed distribution, such as weights. Thus the median weight of American males 50 years old may be a more meaningful statistic than the average weight. The median is also important in cases where it may be difficult or impossible to obtain and measure all the items of a sample necessary to calculate a mean. Consider the following example: An animal behaviorist is studying the time it takes for a sample of animals to perform a certain behavioral step. The variable measured is the time from the beginning of the experiment until each individual has performed. What the behaviorist wants to obtain is an average time of performance. Such an average time, however, could be calculated only after records have been obtained on all the individuals. It may take a long time for the slowest animals to complete their performance, longer than the observer wishes to spend looking at them. Moreover, some of them may never respond appropriately, making the computation of a mean impossible. Therefore, a convenient statistic of location to describe these animals may be the median time of performance, or a related statistic, such as the 75th or 90th percentile. Thus, as long as the observer knows the total sample size, measurements for the right-hand tail of the distribution are not needed. Similar examples are the responses to a drug or poison in a group of individuals (the median lethal or effective dose, LD_{50} or ED_{50}) or the median time for a mutation to appear in a number of genetic lines.

4.4 The Mode

The **mode** refers to the most "fashionable" value of the variable in a frequency distribution, or the value represented by the greatest number of individuals. On a frequency distribution, the mode is the value of the variable at which the curve peaks. In grouped frequency distributions the mode as a point does not have much meaning. Identifying the modal class is usually sufficient. In biology, the mode does not have many applications.

Distributions with two peaks (equal or unequal in height) are called **bimodal;** those with more than two peaks are **multimodal.** In those rare distributions that are U-shaped, we refer to the low point at the middle of the distribution as an **antimode.**

In evaluating the relative merits of the arithmetic mean, the median, and the mode, we must keep a number of considerations in mind. The mean is generally preferred in statistics, since it has a smaller standard error than do other statistics of location (see Section 7.2), is easier to work with mathematically, and has an additional desirable property (see Section 7.1): It tends to be distributed normally even if the original data are not. A disadvantage of the mean is that it is markedly affected by outlying observations, whereas the median and mode are not. The mean is generally more sensitive to changes in the shape of a frequency distribution, so if a statistic reflecting such changes is desired, the mean may be preferred.

In unimodal, symmetrical distributions the mean, the median, and the mode are all identical. A prime example of this is the well-known normal distribution of Chapter 6. In a typical asymmetrical distribution such as the one in Figure 4.1, the relative positions of the mode, median, and mean are generally these: The mean is closest to the drawn-out tail of the distribution, the mode is farthest away from the tail, and the median is in between these. An easy way to remember this sequence is to recall that they occur in alphabetical order from the longer tail of the distribution.

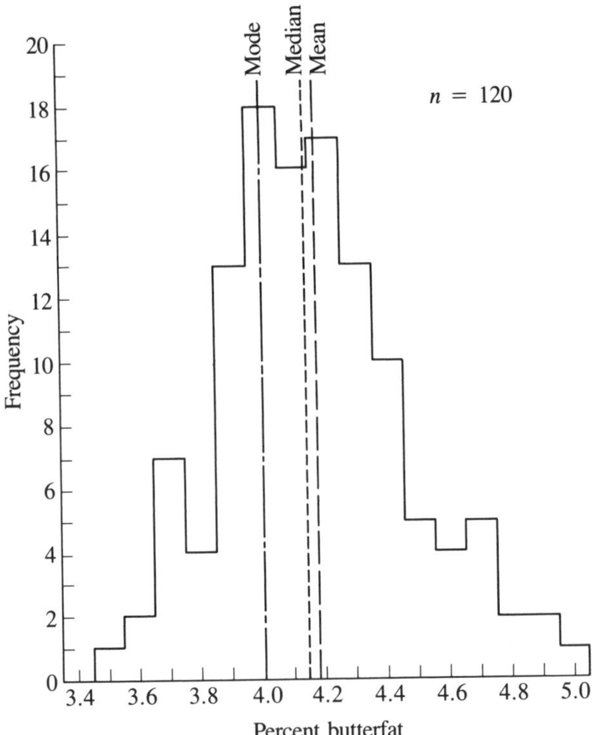

FIGURE 4.1 *An asymmetrical frequency distribution (skewed to the right) showing location of the mean, median, and mode. Percent butterfat in 120 samples of milk (from a Canadian cattle breeder's record book).*

4.5 THE RANGE

We now turn to measures of dispersion or spread. Figure 4.2 demonstrates that distributions that look radically different may possess the identical arithmetic mean. Other ways of characterizing distributions must thus be found.

A simple measure of dispersion is the **range,** the difference between the largest and the smallest items in a sample. The range of the four oxygen percentages from Section 4.1 is

$$\text{Range} = 23.3 - 10.8 = 12.5\%$$

and the range of the birth weights in Box 4.1 is

$$\text{Range} = 171.5 - 59.5 = 112.0 \text{ oz}$$

Since the range is a measure of the span of the variates along the scale of the variable, it is expressed in the same units as the original measurements were. The

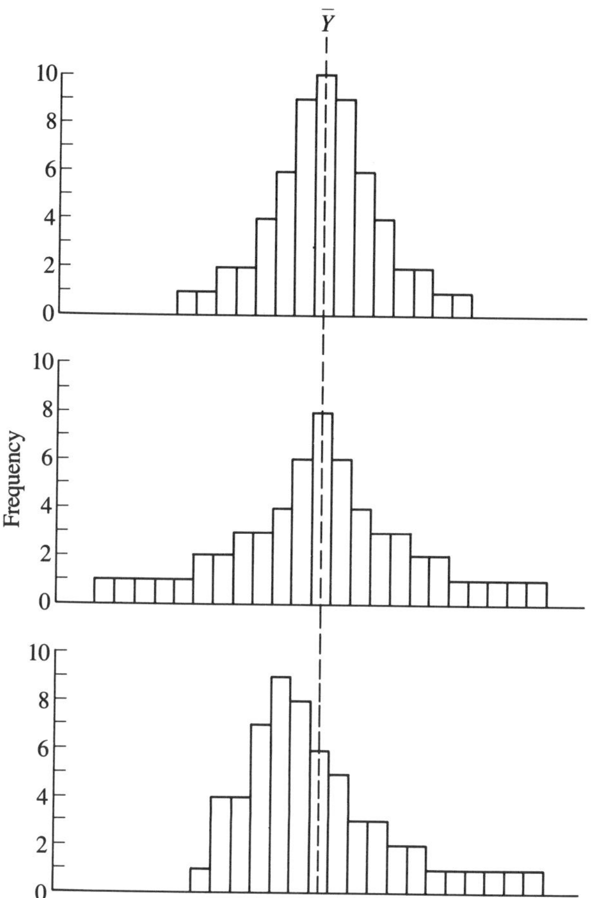

FIGURE 4.2 *Three frequency distributions with identical means and sample sizes but different dispersion patterns.*

range is clearly affected by even a single outlying value; for this reason it is only a rough estimate of the dispersion of all the items in the sample. The range is also affected by sample size: The greater the sample, the wider the range of observations, on the average.

4.6 The Standard Deviation

A desirable measure of dispersion would take all items of a distribution into consideration, weighting each item by its distance from the center of the distribution. We will now try to construct such a statistic. Table 4.1 shows the grouped

Table 4.1 DEVIATIONS FROM THE MEAN.

Grouped femur lengths of aphids (from Box 2.1).

(1) Y	(2) f	(3) fY	(4) $y = Y - \bar{Y}$	(5) fy	(6) $\lvert y \rvert$	(7) $f\lvert y \rvert$	(8) y^2	(9) fy^2
3.4	2	6.8	−0.6	−1.2	0.6	1.2	0.36	0.72
3.7	8	29.6	−0.3	−2.4	0.3	2.4	0.09	0.72
4.0	5	20.0	0.0	0.0	0.0	0.0	0.00	0.00
4.3	8	34.4	0.3	2.4	0.3	2.4	0.09	0.72
4.6	2	9.2	0.6	1.2	0.6	1.2	0.36	0.72
Total	25	100.0		0.0		7.2		2.88

$$\bar{Y} = \frac{\sum fY}{\sum f} = \frac{100.0}{25} = 4.0 \qquad AD = \frac{\sum f\lvert y \rvert}{\sum f} = \frac{7.2}{25} = 0.29 \qquad \sum fy^2 = 2.88$$

$$\text{Variance} = \frac{\sum fy^2}{n} = \frac{2.88}{25} = 0.1152 \qquad \text{Standard deviation} = \sqrt{0.1152} = 0.3394$$

frequency distribution of the femur lengths of the aphid stem mothers from Box 2.1. The first two columns show the class marks and the frequencies. The third column, needed for computing the mean of the frequency distribution (see Section 4.9), is the class mark Y multiplied by the frequency f. Computation of the mean is shown below the table. The mean femur length is 4.0 units.

The distance of each class mark from the mean is computed as the following deviation:

$$y = Y - \bar{Y}$$

Each individual deviation, or **deviate,** is by convention computed as the individual observation minus the mean, $Y - \bar{Y}$, rather than the reverse, $\bar{Y} - Y$. Deviates are symbolized by lowercase letters corresponding to the capital letters of the variables. Column (4) in Table 4.1 gives the deviates computed in this manner. The deviates now have to be multiplied by their respective frequencies f. The results of these computations are shown in column (5).

We now propose to calculate an average deviation that would sum all the deviates and divide them by the number of deviates in the sample. When we sum our deviates, however, we note that they add up to zero, as shown at the bottom of column (5). This is always true of the sum of deviations from the arithmetic mean because the mean is the center of gravity. Consequently an average based on the sum of deviations would also always equal zero. Study Section A.1 in the appendix, which demonstrates that the sum of deviations around the mean of a sample is always equal to zero.

What we are really interested in, however, is the distance from a given variate to the mean. To this end we compute these distances as the absolute values (also known as values with sign neglected) of the deviates. Such values are considered to be positive whether their actual signs are positive or negative. An absolute quantity is symbolized by placement between two vertical lines, as shown at the head of column (6) in Table 4.1. In column (7) these deviates are again multiplied by their frequencies. The sum of these deviates, 7.2, is shown at the foot of the column. The *mean deviation,* or **average deviation** (abbreviated *AD*), is then computed as shown at the bottom of Table 4.1. Its value is 0.29 in the original units of measurement. This statistic, once very popular, has now been almost entirely abandoned in favor of the standard deviation, discussed below.

An alternative way of measuring deviations from the mean is in terms of squared distances from the mean. The sums of such squared deviations have desirable mathematical properties, which we will discuss later. Column (8) lists the squared deviates, all of which are, of course, positive. Finally, column (9) shows the squared deviates multiplied by their frequencies—that is, column (8) multiplied by column (2). The sum of these squared deviates is 2.88. This sum is a very important quantity in statistics, known for short as the **sum of squares** and identified symbolically as Σy^2. In Table 4.1 the sum of squares is symbolized as $\Sigma f y^2$, but the f is usually omitted, since the Σ indicates summation over all the items possible. Another common symbol for the sum of squares is *SS*.

The next step is to obtain the average of the n squared deviations. The resulting quantity is known as the **variance** or the *mean square:*

$$\text{Variance} = \frac{\sum y^2}{n} = \frac{2.88}{25} = 0.1152$$

The variance is a measure of fundamental importance in statistics, and we will employ it throughout this book. At the moment we need only remember that because of the squaring of the deviations, the variance is expressed in squared units. To undo the effect of the squaring we now take the positive square root of the variance and obtain the **standard deviation:**

$$\text{Standard deviation} = +\sqrt{\frac{\sum y^2}{n}} = 0.3394$$

A standard deviation is expressed in the original units of measurement, since it is a square root of the squared units of the variance.

IMPORTANT NOTE: The formula you have just learned is not the one generally employed for computing a standard deviation. A correction for bias must still be applied; it is presented in the next section.

You may have noticed that we have avoided assigning any symbol to either the variance or the standard deviation. We now explain why.

4.7 SAMPLE STATISTICS AND PARAMETERS

Up to now we have calculated statistics from samples without giving too much thought to what these statistics represent. When correctly calculated, a mean and standard deviation are always absolutely true measures of location and dispersion for the samples on which they are based. Thus the true mean of the four oxygen percentage readings in Section 4.1 is 15.325%. The standard deviation of the 25 femur lengths in Table 4.1 is 0.3394 units when the items are grouped as shown. However, rarely in biology (or in statistics in general for that matter) are we interested in measures of location and dispersion only as descriptive summaries of the samples we have studied. Almost always we are interested in the *populations* from which the samples were taken. We therefore would like to know, for example, not the mean of the particular four oxygen percentages, but the true oxygen percentage of the universe of readings from which the four readings were sampled. Similarly, we would like to know the true mean femur length of the population of aphid stem mothers, not merely the mean of the 25 individuals we measured. When studying dispersion we generally wish to learn the true standard deviations of the populations, not those of the samples. These population statistics, however, are unknown and (generally speaking) are unknowable. Who would be able to collect all the stem mothers of this particular aphid population and measure them? Thus we must use **sample statistics** as estimators of *population statistics,* or **parameters.**

It is conventional in statistics to use Greek letters for population parameters and Roman letters for sample statistics. Thus the sample mean, $\overline{Y}$ estimates μ, the parametric mean of the population. Similarly, a sample variance, symbolized by s^2, estimates a parametric variance, symbolized by σ^2. Such estimators should be **unbiased.** By this we mean that samples (regardless of the sample size), taken from a population with a known parameter, should give sample statistics which, when averaged, will give the parametric value. An estimator that does not do so is called **biased.**

The sample mean $\overline{Y}$ is an unbiased estimator of the parametric mean μ. The sample variance as computed in Section 4.6, however, is not unbiased. On the average it will underestimate the magnitude of the population variance σ^2. To overcome this bias, mathematical statisticians have shown that when sums of squares are divided by $n - 1$ rather than by n, the resulting sample variances are unbiased estimators of the population variance. For this reason variances are usually computed by dividing the sum of squares by $n - 1$. The formula for the standard deviation is therefore customarily given as follows:

$$s = +\sqrt{\frac{\sum y^2}{n-1}} \tag{4.6}$$

In the case of the aphid stem mother data (see Table 4.1), the standard deviation would thus be computed as

$$s = \sqrt{\frac{2.88}{24}} = 0.3464$$

This value is only slightly larger than our previous estimate of 0.3394. Of course, the greater the sample size, the less difference there will be between division by n and division by $n - 1$. Regardless of sample size, however, it is good practice to divide a sum of squares by $n - 1$ when computing a variance or standard deviation. It should be assumed that the symbol s^2 refers to a variance obtained by division of the sum of squares by the **degrees of freedom,** as the quantity $n - 1$ is generally called. The only time when division of the sum of squares by n is appropriate is when the interest of the investigator is limited to the sample at hand and to its variance and standard deviation as descriptive statistics of the sample, in contrast to using these as estimates of the population parameters. In the rare cases in which the investigator possesses data on the entire population division by n is justified because then the investigator is not estimating a parameter, but is evaluating it. Thus, for example, the variance of the wing lengths of all adult whooping cranes would be a parametric value; similarly, if the IQs of all winners of the Nobel Prize in physics had been measured, their variance would be a parameter, since it is based on the entire population.

Unfortunately, even though s^2 is an unbiased estimate of σ^2, its square root, s, consistently underestimates σ (Gurland and Tripathi, 1971). This problem can be corrected for by multiplying s by the correction factor C_n:

$$C_n = \frac{[(n - 1)/2]^{1/2}\Gamma[(n - 1)/2]}{\Gamma(n/2)}$$

$\Gamma(n)$ is the gamma function, defined as $\int_0^1 [\ln (1/x)]^{n-1}\, dx$. It can be found in mathematical tables or as subroutines for computer programs. This correction factor is given in Statistical Table **II** for sample sizes up to $n = 30$. For small n, the correction can be important ($C_5 = 1.0638$). With larger n, the correction becomes much smaller ($C_{30} = 1.00866$) and unimportant. For $n > 30$, the approximation $C_n^* \approx 1 + [4(n - 1)]^{-1}$ is sufficiently accurate. In the case of the aphid stem mother data, the unbiased estimate of σ would be 0.3500.

4.8 Coding Data Before Computation

Coding the original data is a far less important subject at the present time, when efficient computational devices are readily available, than it was in earlier days, when it was essential for carrying out most computations. By **coding** we mean the addition or subtraction of a constant number to the original data and/or the multiplication or division of these data by a constant. Data may need to be coded because they were originally expressed in too many digits or are very large numbers that may cause difficulties and errors during data handling. Coding can therefore simplify computation appreciably, and for certain techniques, such as polynomial regression (see Section 16.6), it can be very useful to reduce rounding error (Bradley and Srivastava, 1979). The types of coding shown here are linear transformations of the variables. Persons using statistics should know the effects of such transformations on means, standard deviations, and any other statistics they intend to employ.

Additive coding is the addition or subtraction of a constant (since subtraction is addition of a negative number). Similarly, **multiplicative coding** is the multiplication or division by a constant (since division is multiplication by the reciprocal of the divisor). **Combination coding** is the application of both additive and multiplicative coding to the same set of data. In Section A.2 of the appendix we examine the consequences of the three types of coding for computing means, variances, and standard deviations.

For the case of means, the formula for combination coding and decoding is the most generally applicable one. If the coded variable is $Y_c = D(Y + C)$, then

$$\bar{Y} = \frac{\bar{Y}_c}{D} - C$$

where C is an additive code and D is a multiplicative code. Additive codes have no effect, however, on the sums of squares, variances, or standard deviations. The mathematical proof is given in Section A.2, but this can be seen intuitively because an additive code has no effect on the distance of an item from its mean. For example, the distance from an item of 15 to its mean of 10 would be 5. If we were to code the variates by subtracting a constant of 10, the item would now be 5 and the mean zero, but the difference between them would still be 5. Thus if only additive coding is employed, the only statistic in need of decoding is the mean. Multiplicative coding, on the other hand, does have an effect on sums of squares, variances, and standard deviations. The standard deviations have to be divided by the multiplicative code, just as had to be done for the mean; the sums of squares or variances have to be divided by the multiplicative codes squared because they are squared terms, and the multiplicative factor became squared during the operations. In combination coding the additive code can be ignored.

An example of coding and decoding data is shown in Box 4.3.

4.9 Computing Means and Standard Deviations

Three steps are necessary for computing the standard deviation: (1) finding Σy^2, the sum of squares, (2) dividing by $n - 1$ to give the variance, and (3) taking the square root of the variance to obtain the standard deviation. The procedure used to compute the sum of squares in Section 4.6 can be expressed by the following formula:

$$\sum y^2 = \sum (Y - \bar{Y})^2 \tag{4.7}$$

When the data are unordered, the computation proceeds as in Box 4.2, which is based on the unordered aphid femur length data shown at the head of Box 2.1.

Occasionally original data are already in the form of a frequency distribution, or the person computing the statistics may want to avoid manual entry of large numbers of individual variates, in which case setting up a frequency distribution is also advisable. Data already arrayed in a frequency distribution speed up the computations considerably. An example is shown in Box 4.3. Hand calculations are simplified and data entry into computers is less tedious (and hence there will be less chance for input errors) by coding to remove the awkward class marks. We coded each class mark in Box 4.3 by subtracting 59.5, the lowest class mark of the array. The resulting class marks are the values 0, 8, 16, 24, 32, and so on. Dividing these values by 8 changes them to 0, 1, 2, 3, 4, and so on, which is the desired format, shown in column (3). Details of the computation are given in Box 4.3.

Computer programs that compute the basic statistics $\bar{Y}$, s^2, s, and others that we have not yet discussed are furnished in many commercially available programs. The BIOM-pc program version 3 accepts raw, unordered observations as input, as well as data in the form of a frequency distribution.

An approximate method for estimating statistics is useful when checking the results of calculations because it enables the detection of gross errors in computation. A simple method for estimating the mean is to average the largest and smallest observations to obtain the **midrange.** For the aphid stem mother data of Box 2.1, this value is $(4.7 + 3.3)/2 = 4.0$, which happens to fall almost exactly on the computed sample mean (but, of course, this will not be true of other data sets). Standard deviations can be estimated from ranges by appropriate division of the range:

For samples of	divide the range by
10	3
30	4
100	5
500	6
1000	$6\frac{1}{2}$

The range of the aphid data is 1.4. When this value is divided by 4 we get an estimate of the standard deviation of 0.35, which compares not too badly with the calculated value of 0.3657 in Box 4.2.

A more accurate procedure for estimating statistics is to use Statistical Table **I,** which furnishes the mean range for different sample sizes of a normal distribution (see Chapter 6) with a variance of one. When we divide the range of a sample by a mean range from Table **I,** we obtain an estimate of the standard deviation of the population from which the sample was taken. Thus, for

Box 4.2 CALCULATION OF $\bar{Y}$ AND s FROM UNORDERED DATA.

Based on aphid femur length data, unordered, as shown at the head of Box 2.1.

Computation

$n = 25$; $\sum Y = 100.1$; $\bar{Y} = 4.004$

1. $\sum y^2 = \sum (Y - \bar{Y})^2 = 3.2096$

2. $s^2 = \dfrac{\sum y^2}{n - 1} = \dfrac{3.2096}{24} = 0.1337$

3. $s = \sqrt{0.1337} = 0.3657$

Box 4.3 CALCULATION OF $\bar{Y}$, s, AND V FROM A FREQUENCY DISTRIBUTION.

Birth weights of male Chinese in ounces (from Box 4.1).

(1) Class mark Y	(2) f	(3) Coded class mark Y_c
59.5	2	0
67.5	6	1
75.5	39	2
83.5	385	3
91.5	888	4
99.5	1729	5
107.5	2240	6
115.5	2007	7
123.5	1233	8
131.5	641	9
139.5	201	10
147.5	74	11
155.5	14	12
163.5	5	13
171.5	1	14
	9465 = n	

SOURCE: Millis and Seng (1954).

BOX 4.3 CONTINUED

Coding and decoding

To code:

$$Y_c = \frac{Y - 59.5}{8}$$

To decode $\bar{Y}_c$:

$$\begin{aligned}\bar{Y} &= 8\bar{Y}_c + 59.5\\ &= 50.4 + 59.5\\ &= 109.9 \text{ oz}\end{aligned}$$

To decode s_c:

$$s = 8s_c = 13.594 \text{ oz}$$

Computation

Without coding	With coding
$\sum fY = 1{,}040{,}199.5$	$\sum fY_c = 59{,}629$
$\bar{Y} = 109.8996$	$\bar{Y}_c = 6.300$
$\sum fy^2 = \sum f(Y - \bar{Y})^2$	$\sum fy_c^2 = \sum f(Y_c - \bar{Y}_c)^2$
$= 1{,}748{,}956.7983$	$= 27{,}327.450$
$s^2 = \dfrac{\sum fy^2}{n-1} = 184.8010$	$s_c^2 = \dfrac{\sum fy_c^2}{n-1} = 2.888$
$s = 13.5942$	$s_c = 1.6993$

$$V = \frac{s}{\bar{Y}} \times 100 = \frac{13.594}{109.9} \times 100 = 12.370\%$$

the aphid data we look up $n = 25$ in Table **I** and obtain 3.931. We estimate $s = 1.4/3.931 = 0.356$, a value closer to the sample standard deviation than that obtained by the rougher method discussed above (which, however, is based on the same principle and assumptions).

4.10 THE COEFFICIENT OF VARIATION

Having obtained the standard deviation as a measure of the amount of variation in the data, you may legitimately ask, What can I do with it? At this stage in our comprehension of statistical theory, nothing really useful comes of the computations we have carried out, although the skills learned are basic to all statistical work. So far, the only use that we might have for the standard deviation is as an estimate of the amount of variation in a population. Thus we may wish to compare the magnitudes of the standard deviations of similar populations to see whether population A is more or less variable than population B. When populations differ appreciably in their means, however, the direct comparison of their variances or standard deviations is less useful, since larger organisms usually vary more than smaller ones. For instance, the standard deviation of the tail lengths of elephants is obviously much greater than the entire tail length of a mouse. To compare the relative amounts of variation in populations having different means, the **coefficient of variation,** symbolized by V (or occasionally CV), has been developed. This coefficient is simply the standard deviation expressed as a percentage of the mean. Its formula is

$$V = \frac{s \times 100}{\bar{Y}} \tag{4.8}$$

For example, the coefficient of variation of the birth weights in Box 4.3 is 12.37%, as shown at the bottom of that box. The coefficient of variation is independent of the unit of measurement and is expressed as a percentage. The coefficient of variation as computed above is a biased estimator of the population V. The following estimate V^* is corrected for bias:

$$V^* = \left(1 + \frac{1}{4n}\right) V \tag{4.9}$$

In small samples this correction can make an appreciable difference. Note that when using Expression (4.9), the standard deviation used to compute V should not be corrected using C_n because this would result in an overcorrection. Note also that the correction factor approximates C_n, the correction factor used in Section 4.7.

Coefficients of variation are used extensively when comparing the variation of two populations independent of the magnitude of their means. Whether the birth weights of the Chinese children (see Box 4.1) are more or less variable than the femur lengths of the aphid stem mothers (see Box 2.1) is probably of little interest, but we can calculate the latter as $0.3656 \times 100/4.004 = 9.13\%$, which would suggest that the birth weights are more variable. More commonly, we might wish to test whether a given biological sample is more variable for one character than for another. For example, for a sample of rats, is body weight more variable than blood sugar content? Another frequent type of comparison, especially in systematics, is among different populations for the same character. If, for instance, we had measured wing length in samples of birds from several localities, we might wish to know whether any one of these populations is more variable than the others. An answer to this question can be obtained by examining the coefficients of variation of wing length in these samples.

Employing the coefficient of variation in a comparison between two variables or two populations assumes that the variable in the second sample is proportional to that in the first. Thus we could write $Y_1 = kY_2$, where k is a constant of proportionality. If two variables are related in this manner, their coefficients of variation should be identical. This should be obvious if we remember that k is a multiplicative code. Thus $\bar{Y}_1 = k\bar{Y}_2$ and $s_1 = ks_2$, and consequently

$$V_1 = \frac{100s_1}{\bar{Y}_1} = \frac{100ks_2}{k\bar{Y}_2} = \frac{100s_2}{\bar{Y}_2} = V_2$$

If the variables are transformed to logarithms (see Section 13.7), the relationship between them can be written as $\ln Y_1 = \ln k + \ln Y_2$, and since $\ln k$ is a constant, the variances of $\ln Y_1$ and $\ln Y_2$ are identical. This relation can lead to a test of equality of coefficients of variation (see Lewontin, 1966; Sokal and Braumann, 1980).

At one time systematists put great stock in coefficients of variation and even based some classification decisions on the magnitude of these coefficients. However, there is little, if any, foundation for such actions. More extensive discussion of the coefficient of variation, especially as it relates to systematics, can be found in Simpson et al. (1960), Lewontin (1966), Lande (1977), and Sokal and Braumann (1980).

EXERCISES 4

4.1 Find the mean, standard deviation, and coefficient of variation for the pigeon data given in Exercise 2.4. Group the data into ten classes, recompute $\bar{Y}$ and s, and compare them with the results obtained from the ungrouped data. Compute the median for the grouped data. *Answer:* For ungrouped data, $\bar{Y} = 11.48$ and $s = 0.69178$.

4.2 Find $\bar{Y}$, s, V, and the median for the following data (milligrams of glycine per milligram of creatinine in the urine of 37 chimpanzees; from Gartler et al., 1956).

.008	.018	.056	.055	.135	.052	.077	.026	.440	.300
.025	.036	.043	.100	.120	.110	.100	.350	.100	.300
.011	.060	.070	.050	.080	.110	.110	.120	.133	.100
.100	.155	.370	.019	.100	.100	.116			

4.3 The following are percentages of butterfat from 120 registered three-year-old Ayrshire cows selected at random from a Canadian stock record book.

4.32	4.24	4.29	4.00	3.96	4.48	3.89	4.02	3.74	4.42
4.20	3.87	4.10	4.00	4.33	3.81	4.33	4.16	3.88	4.81
4.23	4.67	3.74	4.25	4.28	4.03	4.42	4.09	4.15	4.29
4.27	4.38	4.49	4.05	3.97	4.32	4.67	4.11	4.24	5.00
4.60	4.38	3.72	3.99	4.00	4.46	4.82	3.91	4.71	3.96
3.66	4.10	4.38	4.16	3.77	4.40	4.06	4.08	3.66	4.70
3.97	3.97	4.20	4.41	4.31	3.70	3.83	4.24	4.30	4.17
3.97	4.20	4.51	3.86	4.36	4.18	4.24	4.05	4.05	3.56
3.94	3.89	4.58	3.99	4.17	3.82	3.70	4.33	4.06	3.89
4.07	3.58	3.93	4.20	3.89	4.60	4.38	4.14	4.66	3.97
4.22	3.47	3.92	4.91	3.95	4.38	4.12	4.52	4.35	3.91
4.10	4.09	4.09	4.34	4.09	4.88	4.28	3.98	3.86	4.58

(a) Calcuate $\bar{Y}$, s, and V directly from the data. (b) Group the data in a frequency distribution and again calculate $\bar{Y}$, s, and V. Compare the results with those of (a). How have the results been affected by grouping? Also calculate the median. *Answer:* For ungrouped data $\bar{Y} = 4.16608$, $s = 0.30238$, $V = 7.25815$.

4.4 What effect would adding the constant 5.2 to all observations have upon the numerical values of the following statistics: $\bar{Y}$, s, V, average deviation, median, mode, and range? What would be the effect of adding 5.2 and then multiplying the sums by 8.0? Would it make any difference in the above statistics if we multiplied by 8.0 first and then added 5.2?

4.5 Estimate μ and σ using the midrange and the range (see Section 4.9) for the data in Exercises 4.1, 4.2, and 4.3. How well do these estimates agree with the estimates given by $\bar{Y}$ and s? *Answer:* Estimates of μ and σ for Exercise 4.2 are 0.224 and 0.1014, respectively.

4.6 Show that the equation for variance can also be written as

$$s^2 = \frac{\sum Y^2 - n\bar{Y}^2}{n-1} \quad \text{or} \quad \frac{\sum Y^2 - \frac{\left(\sum Y\right)^2}{n}}{n-1}$$

4.7 Apply the C_n correction to the estimated standard deviation of the data in Exercise 2.4. Also compute the unbiased estimate of the coefficient of variation, V^*. Comment on the importance of these corrections. *Answer:* $s/C_{40} = 0.69621$, $V^* = 6.06365$.

INTRODUCTION TO PROBABILITY DISTRIBUTIONS: BINOMIAL AND POISSON

Section 2.5 was our first discussion of frequency distributions. For example, Table 2.3 shows a distribution for a meristic, or discrete (discontinuous) variable, the number of sedge plants per quadrat. Examples of distributions for continuous variables are the femur lengths of aphid stem mothers in Box 2.1 or the human birth weights in Box 4.3. Each of these distributions informs us about the absolute frequency of any given class and permits computation of the relative frequencies of any class of variable. Thus, most of the quadrats contained either no sedges or just one or two plants. In the 139.5-oz class of birth weights, we find only 201 out of the 9465 babies recorded; that is, approximately only 2.1% of the infants are in that birth-weight class.

Of course, these frequency distributions are only samples from given populations. The birth weights represent a population of male Chinese infants from a given geographical area. If we knew our sample to be representative of that population, however, we could make all sorts of predictions based on the frequency distribution of the sample. For instance, we could say that approximately 2.1% of male Chinese babies born in this population weigh between 135.5 and 143.5 oz at birth. Similarly, we might say that the probability of the weight at birth of any one baby in this population being in the 139.5-oz birth class is quite low. If each of the 9465 weights were mixed up in a hat, and we pulled one out, the probability that we would pull out one of the 201 in the 139.5-oz class would be very low indeed—only 2.1%. It would be much more probable that we would sample an infant of 107.5 or 115.5 oz, since the infants in these classes are represented by the frequencies 2240 and 2007, respectively.

Finally, if we were to sample from an unknown population of babies and find that the very first individual sampled had a birth weight of 170 oz, we would probably reject any hypothesis that the unknown population was the same as that sampled in Box 4.3. We would arrive at this conclusion because in the distribution in Box 4.3, only one out of almost 10,000 infants had a birth weight that high. Although it is possible to have sampled from the population of male Chinese babies and obtained a birth weight of 170 oz, the probability that the first individual sampled would have such a value is very low indeed. It is much more reasonable to suppose that the unknown population from which we are sampling is different in mean and possibly in variance from the one in Box 4.3.

We have used this empirical frequency distribution to make certain predictions (with what frequency a given event will occur) or make judgments and decisions (whether it is likely that an infant of a given birth weight belongs to this population). In many cases in biology, however, we will make such predictions not from empirical distributions, but on the basis of theoretical considerations that in our judgment are pertinent. We may feel that the data should be distributed in a certain way because of basic assumptions about the nature of the forces acting on the example at hand. If our observed data do not sufficiently conform to the values expected on the basis of these assumptions, we will have serious doubts about our assumptions. This is a common use of frequency distributions in biology. The assumptions being tested generally lead to a theoretical frequency distribution, known also as a **probability distribution.**

A probability distribution may be a simple two-valued distribution such as the 3 : 1 ratio in a Mendelian cross, or it may be a more complicated function that is intended to predict the number of plants in a quadrat. If we find that the observed data do not fit the expectations on the basis of theory, we are often led to the discovery of some biological mechanism causing this deviation from expectation. The phenomena of linkage in genetics, of preferential mating between different phenotypes in animal behavior, of congregation of animals at certain favored places or, conversely, their territorial dispersion are cases in point. We will thus make use of probability theory to test our assumptions about the laws of occurrence of certain biological phenomena. Probability theory underlies the entire structure of statistics, a fact which, because of the nonmathematical orientation of this book, may not be entirely obvious.

In Section 5.1 we present an elementary discussion of probability, limited to what is necessary for comprehension of the sections that follow. The binomial frequency distribution, which not only is important in certain types of studies, such as genetics, but is fundamental to an understanding of the kinds of probability distributions discussed in this book, is covered in Section 5.2. The Poisson distribution, which follows in Section 5.3, is widely applicable in biology, especially for tests of randomness of the occurrence of certain events. Both the binomial and Poisson distributions are discrete probability distributions. Some other discrete distributions are mentioned briefly in Section 5.4. The entire chapter therefore deals with discrete probability distributions. The most common continuous probability distribution is the normal frequency distribution, discussed in Chapter 6.

5.1 PROBABILITY, RANDOM SAMPLING, AND HYPOTHESIS TESTING

We will start this discussion with an example that is not biometrical or biological in the strict sense. We have often found it pedagogically effective to introduce new concepts through situations thoroughly familiar to the student, even if the example is not relevant to the general subject matter of biometry.

Let us imagine ourselves at Matchless University, a state institution somewhere between the Appalachians and the Rockies. Its enrollment figures yield the following breakdown of the student body: 70% of the students are American undergraduates (AU), 26% are American graduate students (AG), and the remaining 4% are from abroad. Of these, 1% are foreign undergraduates (FU) and 3% are foreign graduate students (FG). In much of our work we use proportions rather than percentages as a useful convention. Thus the enrollment consists of 0.70 AUs, 0.26 AGs, 0.01 FUs, and 0.03 FGs. The total student body, corresponding to 100%, is represented by the figure 1.0.

If we sample 100 students at random, we expect that, on the average, 3 will be foreign graduate students. The actual outcome might vary. There might not be a single FG student among the 100 sampled or there may be quite a few more than 3, and the estimate of the proportion of FGs may therefore range from 0 to greater than 0.03. If we increase our sample size to 500 or 1000, it is less likely that the ratio will fluctuate widely around 0.03. The larger the sample, the closer to 0.03 the ratio of FG students sampled to total students sampled will be. In fact, the **probability** of sampling a foreign student can be defined as the limit reached by the ratio of foreign students to the total number of students sampled, as sample size increases. Thus, we may formally summarize by stating that the probability of a student at Matchless University being a foreign graduate student is $P[\text{FG}] = 0.03$. Similarly, the probability of sampling a foreign undergraduate is $P[\text{FU}] = 0.01$, that of sampling an American undergraduate is $P[\text{AU}] = 0.70$, and that for American graduate students, $P[AG] = 0.26$.

Now imagine the following experiment: We try to sample a student at random from the student body at Matchless University. This task is not as easy as might be imagined. If we wanted to do the operation physically, we would have to set up a collection or trapping station somewhere on campus. And to make certain that the sample is truly random with respect to the entire student population, we would have to know the ecology of students on campus very thoroughly so that we could locate our trap at a station where each student had an equal probability of passing. Few, if any, such places can be found in a university. The student union facilities are likely to be frequented more by independent and foreign students, less by those living in organized houses and dormitories. Fewer foreign and graduate students might be found along fraternity row. Clearly we would not wish to place our trap near the International Club or House because our probability of sampling a foreign student would be greatly enhanced. In front of the bursar's window we might sample students paying tuition. The time of sampling is equally important, in the seasonal as well as the diurnal cycle. There seems no easy solution in sight.

Those of you who are interested in sampling organisms from nature will already have perceived parallel problems in your work. If we were to sample only students wearing turbans or saris, their probability of being foreign students would be almost 1. We could no longer speak of a random sample. In the familiar ecosystem of the university these violations of proper sampling procedure are obvious to all of us, but they are not nearly so obvious in real biological instances that are less well known.

How should we proceed to obtain a random sample of leaves from a tree, of insects from a field, or of mutations in a culture? In sampling at random we are attempting to permit the frequencies of various events occurring in nature to be reproduced unalteredly in our records; that is, we hope that on the average the frequencies of these events in our sample will be the same as they are in the natural situation. Another way of stating this goal is that in a random sample we want every individual in the population being sampled to have an equal probability of being included in the sample.

We might go about obtaining a random sample by using records representing the student body, such as the student directory, selecting a page from it at random and a name at random from the page. Or we could assign an arbitrary number to each student, write each on a chip or disk, put these in a large container, stir well, and then pull out a number.

Imagine now that we sample a single student by the best random procedure we can devise. What are the possible outcomes? Clearly the student could be either an AU, AG, FU, or FG. This *set* of four possible outcomes exhausts the possibilities of the experiment. This set, which we can represent as {AU, AG, FU, FG}, is called the **sample space.** Any single experiment would result in only one of the four possible outcomes (elements) in the set. Such an element in a sample space is called a **simple event.** It is distinguished from an **event,** which is any subset of the sample space. Thus in the sample space defined above {AU}, {AG}, {FU}, or {FG} are each simple events. The following sampling results are some of the possible events: {AU, AG, FU}, {AU, AG, FG}, {AG, FG}, {AU, FG}. The meaning of these events, in order, implies being an American, an undergraduate, or both; being an American, a graduate student, or both; being a graduate student; being an American undergraduate or a foreign graduate student. By the definition of event, simple events, as well as the entire sample space, are also events.

Given the sampling space described above, the event **A** = {AU, AG} encompasses all possible outcomes in the space yielding an American student. Similarly, the event **B** = {AG, FG} summarizes the possibilities for obtaining a graduate student. The **intersection** of events **A** and **B,** written **A** ∩ **B,** describes only the events that are shared by **A** and **B.** Clearly only AG qualifies, as can be seen here:

$$\mathbf{A} = \{AU, AG\}$$
$$\mathbf{B} = \qquad \{AG, FG\}$$

Thus **A** ∩ **B** is the event in the sample space that gives rise to the sampling of an American graduate student. When the intersection of two events is empty, as in

$\mathbf{B} \cap \mathbf{C}$, where $\mathbf{C} = \{\text{AU, FU}\}$, the events **B** and **C** are mutually exclusive. These two events have no common element in the sampling space.

We may also define events that are **unions** of two other events in the sample space. Thus $\mathbf{A} \cup \mathbf{B}$ indicates that **A** or **B** or both **A** and **B** occur. As defined above, $\mathbf{A} \cup \mathbf{B}$ describes all students who are Americans, graduate students, or are both (American graduate students).

This discussion of events makes the following relations almost self-evident. Probabilities are necessarily bounded by 0 and 1. Therefore,

$$0 \leq P[\mathbf{A}] \leq 1 \tag{5.1}$$

The probability of an entire sample space is

$$P[\mathbf{S}] = 1 \tag{5.2}$$

where **S** is the sample space of all possible events. Also, for any event **A,** the probability of it not occurring is $1 - P[\mathbf{A}]$. This is known as the complement rule:

$$P[\mathbf{A}^{\text{C}}] = 1 - P[\mathbf{A}] \tag{5.3}$$

where $\mathbf{A}^{\text{C}}$ stands for all events that are not **A.**

Why are we concerned with defining sample spaces and events? Because these concepts lead us to useful definitions and operations regarding the probability of various outcomes. If we can assign a number $0 \leq p \leq 1$ to each simple event in a sample space such that the sum of these p's over all simple events in the space equals unity, then the space becomes a (finite) **probability space.** In our example, the following numbers were associated with the appropriate simple events in the sample space:

$$\{\text{AU, AG, FU, FG}\}$$

$$\{0.70, 0.26, 0.01, 0.03\}$$

Given this probability space, we are now able to make statements regarding the probability of given events. For example, what is the probability of a student sampled at random being an American graduate student? Clearly it is $P[\{\text{AG}\}] = 0.26$. What is the probability that a student is either American or a graduate student? The general formula for the union of the probabilities of events **A** and **B** is

$$P[\mathbf{A} \cup \mathbf{B}] = P[\mathbf{A}] + P[\mathbf{B}] - P[\mathbf{A} \cap \mathbf{B}] \tag{5.4}$$

In terms of the events defined earlier, the answer is

$$\begin{aligned} P[\mathbf{A} \cup \mathbf{B}] &= P[\{\text{AU, AG}\}] + P[\{\text{AG, FG}\}] - P[\{\text{AG}\}] \\ &= 0.96 + 0.29 - 0.26 \\ &= 0.99 \end{aligned}$$

We can see that we must subtract $P[\mathbf{A} \cap \mathbf{B}] = P[\{AG\}]$, because if we did not do so it would be included twice, once in $P[\mathbf{A}]$ and once in $P[\mathbf{B}]$, and would lead to the absurd result of a probability greater than 1.

Now let us assume that we have sampled our single student from the student body of Matchless University. He turns out to be a foreign graduate student. What can we conclude? By chance alone, this result has a probability of 0.03, or would happen 3% of the time, that is, not very frequently. The assumption that we have sampled at random should probably be rejected, since if we accept the hypothesis of random sampling, the outcome of the experiment is improbable. Please note that we said *improbable,* not *impossible.* We could have chanced upon an FG as the first one to be sampled, but it is not very likely. The probability is 0.97 that a single student sampled would be a non-FG. If we could be certain that our sampling method was random (as when drawing student numbers out of a container), we would of course have to reach the conclusion that an improbable event has occurred. The decisions of this paragraph are all based on our definite knowledge that the proportion of students at Matchless University is as specified by the probability space. If we were uncertain about this, we would be led to assume a higher proportion of foreign graduate students as a consequence of the outcome of our sampling experiment.

Now we will sample two students rather than just one. What are the possible outcomes of this sampling experiment? The new sample space can best be depicted by a diagram that shows the set of the sixteen possible simple events as points in a lattice (Figure 5.1). The possible combinations, ignoring which student was sampled first, are {AU, AU}, {AU, AG}, {AU, FU}, {AU, FG}, {AG, AG}, {AG, FU}, {AG, FG}, {FU, FU}, {FU, FG}, and {FG, FG}.

What would be the expected probabilities of these new outcomes? Now the nature of the sampling procedure becomes quite important. We may sample with

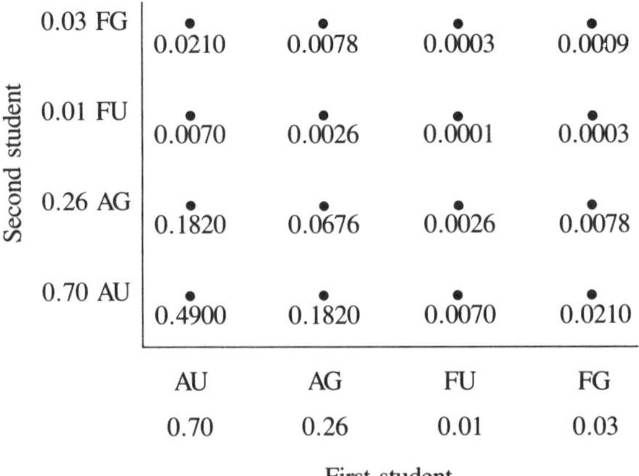

FIGURE 5.1 *Sample space for sampling two students from Matchless University.*

or without **replacement;** that is, we may return the first student sampled to the population or may keep him out of the pool of individuals to be sampled. If we do not replace the first individual sampled, the probability of sampling a foreign graduate student will no longer be exactly 0.03. We can visualize this easily. Assume that Matchless University has 10,000 students. Since 3% are foreign graduate students, there must be 300 FG students at the university. After obtaining a foreign graduate student in the first sample, this number is reduced to 299 out of 9999 students. Consequently, the probability of sampling an FG student now becomes 299/9999 = 0.0299, a slightly lower probability than the value of 0.03 for sampling the first FG student. If, on the other hand, we return the original foreign student to the student population and make certain that the population is thoroughly randomized before being sampled again (that is, give him a chance to lose himself among the campus crowd or, in drawing student numbers out of a container, mix up the disks with the numbers on them), the probability of sampling a second FG student is the same as before—0.03. In fact, if we continue to replace the sampled individuals in the original population, we can sample from it as though it were an infinitely sized population.

Biological populations are, of course, finite, but they are frequently so large that for purposes of sampling experiments we can consider them effectively infinite, whether we replace sampled individuals or not. After all, even in this relatively small population of 10,000 students, the probability of sampling a second foreign graduate student (without replacement) is only minutely different from 0.03. For the rest of this section, we will assume that sampling is with replacement, so the probability of obtaining a foreign student will not change.

There is a second potential source of difficulty in this design. We not only have to assume that the probability of sampling a second foreign student is equal to that of the first, but also that these events are **independent**. By independence of events we mean that the probability of one event is not affected by whether or not another event has occurred. In the case of the students, having sampled one foreign student, is it more or less likely that the second student we sampled is also a foreign student? Independence of the events may depend on where we sample the students or on the method of sampling. If we sampled students on campus, it is quite likely that the events are not independent; that is, having sampled one foreign student, the probability that the second student we sample is foreign increases, since foreign students tend to congregate. Thus, at Matchless University the probability that a student walking with a foreign graduate student is also an FG would be greater than 0.03.

Events **D** and **E** in a sample space are defined as independent whenever

$$P[\mathbf{D} \cap \mathbf{E}] = P[\mathbf{D}]P[\mathbf{E}] \tag{5.5}$$

The probability values assigned to the sixteen points in the sample space of Figure 5.1 have been computed to satisfy this condition. Thus letting $P[\mathbf{D}]$ equal the probability that the first student sampled is an AU—that is, $P[\{AU_1AU_2, AU_1AG_2, AU_1FU_2, AU_1FG_2\}]$—and $P[\mathbf{E}]$ equal the probability that the second student is an FG—that is, $P[\{AU_1FG_2, AG_1FG_2, FU_1FG_2, FG_1FG_2\}]$—we note that the intersection $\mathbf{D} \cap \mathbf{E}$ is $\{AU_1FG_2\}$. This event has a value of 0.0210 in the probability space of Figure 5.1. We find that this value is the product of $P[\{AU]P[\{FG\}] = 0.70 \times 0.03 = 0.0210$. These mutually independent relations have been deliberately imposed upon all points in the probability space. Therefore, if the sampling probabilities for the second student are independent of the type of student sampled first, we can compute the probabilities of the outcomes simply as the product of the independent probabilities. Thus the probability of obtaining two FG students is $P[\{FG\}]P[\{FG\}] = 0.03 \times 0.03 = 0.0009$.

The probability of obtaining one AU and one FG student in the sample might seem to be the product 0.70×0.03. However, it is twice that probability. It is easy to see why. There is only one way of obtaining two FG students, namely by sampling first one FG and then again another FG. Similarly, there is only one way to sample two AU students. However, sampling one of each type of student can be done by first sampling an AU followed by an FG or by first sampling an FG followed by an AU. Thus the probability is $2P[\{AU\}]P[\{FG\}] = 2 \times 0.70 \times 0.03 = 0.0420$.

If we conducted such an experiment and obtained a sample of two FG students, we would be led to the following conclusions: Since only 0.0009 of the samples (9/100th of 1%, or 9 out of 10,000 cases) would be expected to consist of two foreign graduate students, obtaining such a result by chance alone is quite improbable. Given that $P[\{FG\}] = 0.03$ is true, we would suspect that sampling was not random or that the events were not independent (or that both assumptions—random sampling and independence of events—were incorrect).

Random sampling is sometimes confused with randomness in nature. The former is the faithful representation in the sample of the distribution of the events in nature; the latter is the *independence* of the events in nature. Random sampling generally is or should be under the control of the experimenter and is related to the strategy of good sampling. Randomness in nature generally describes an innate property of the objects being sampled and thus is of greater biological interest. The confusion between random sampling and independence of events arises because lack of either can yield observed frequencies of events differing from expectation. We have already seen how lack of independence in samples of foreign students can be interpreted from both points of view in our example from Matchless University.

Expression (5.5) can be used to test whether two events are independent. It is well known that foreign students at American universities are predominantly graduate students, quite unlike the situation for American students. Can we demonstrate this for Matchless University? The proportion of foreign students at the university $P[F] = P[FU] + P[FG] = 0.04$ and of graduate students is $P[G] = P[AG] + P[FG] = 0.29$. If the properties of being a foreign student and of being a graduate student were independent, then $P[FG]$ would equal $P[F]P[G]$. In fact, it does not, since $0.04 \times 0.29 = 0.0116$, not 0.03, the actual proportion of foreign graduate students. Thus nationality and academic rank are not independent. The probability of being a graduate student differs depending on the nationality: Among American students the proportion of graduate students is 0.26/0.96 = 0.27, but among foreign students it is 0.03/0.04 = 0.75.

This dependence leads us directly to the notion of **conditional probability.** We speak of the conditional probability of event **A**, given event **B**, symbolized as $P[\mathbf{A} \mid \mathbf{B}]$. This quantity can be computed as follows:

$$P[\mathbf{A} \mid \mathbf{B}] = \frac{P[\mathbf{A} \cap \mathbf{B}]}{P[\mathbf{B}]} \tag{5.6}$$

Let us apply this formula to our Matchless University example. What is the conditional probability of being a graduate student, given that the student is foreign? We calculate $P[\mathrm{G} \mid \mathrm{F}] = P[\mathrm{G} \cap \mathrm{F}]/P[\mathrm{F}] = 0.03/0.04 = 0.75$. Thus, 75% of the foreign students are graduate students. Similarly, the conditional probability of being a graduate student, given that the student is American, is 0.26/0.96 = 0.2708, a much lower value. Only 27.08% of the American students are graduate students.

Here is another example that requires Expression (5.6) for conditional probability: Let **C** be the event that a person has cancer and $P[\mathbf{C}]$ denote the probability that a particular population has cancer (and $P[\mathbf{C}^C] = 1 - P[\mathbf{C}]$, the probability of not having cancer). In epidemiology $P[\mathbf{C}]$ is known as the **prevalence** of the disease. Let **T** be the event that a certain diagnostic test is positive, i.e., that it indicates that a person has cancer. Upon receiving a positive test result for a patient, what the physician would like to know is $P[\mathbf{C} \mid \mathbf{T}]$, the probability that the patient has cancer, given that the results were positive. Applying Expression (5.6) yields

$$P[\mathbf{C} \mid \mathbf{T}] = \frac{P[\mathbf{C} \cap \mathbf{T}]}{P[\mathbf{T}]} \tag{5.7}$$

However, the information that one collects to determine the reliability of such tests is not usually in the form required by the above equation. Results are collected from tests on individuals known to have cancer and on those known to be cancer-free. Then the proportion of positive tests results among the known cancer cases—$P[\mathbf{T} \mid \mathbf{C}]$—is determined. Epidemiologists call this proportion the **sensitivity** of the test. This value ideally is close to 1. The proportion of negative test results for cancer-free persons—$P[\mathbf{T}^C \mid \mathbf{C}^C]$—is determined next. This value is known as the **specificity** of the test. Again, this value should be as close to 1 as possible. It follows that $P[\mathbf{T} \mid \mathbf{C}^C]$, the probability of a positive result in cancer-free patients equals $1 - P[\mathbf{T}^C \mid \mathbf{C}^C]$, the one-complement of the specificity.

From Expression (5.7), by interchanging **C** and **T** we can show that $P[\mathbf{C} \cap \mathbf{T}] = P[\mathbf{T} \mid \mathbf{C}]P[\mathbf{C}]$. The probability of an event such as **T** can be written as

$$P[\mathbf{T}] = P[\mathbf{T} \mid \mathbf{C}]P[\mathbf{C}] + P[\mathbf{T} \mid \mathbf{C}^C]P[\mathbf{C}^C]$$

which is the sum of the probabilities of having a positive test among those who have cancer and among those who do not have cancer—each weighted by the frequencies of the two populations. Substituting these two results into Expression (5.7) yields

$$P[\mathbf{C} \mid \mathbf{T}] = \frac{P[\mathbf{T} \mid \mathbf{C}]P[\mathbf{C}]}{P[\mathbf{T} \mid \mathbf{C}]P[\mathbf{C}] + P[\mathbf{T} \mid \mathbf{C}^C]P[\mathbf{C}^C]} \tag{5.8}$$

This expression is known as **Bayes' theorem** and can be generalized to allow for an event **C** having more than just two states (the denominator is summed over all events $\mathbf{C}_i$ rather than just **C** and its complement). This famous formula, published posthumously by the eighteenth-century English clergyman Thomas Bayes, hs led to much controversy over the interpretation of the quantity $P[\mathbf{C} \mid \mathbf{T}]$.

Earlier we defined ''probability'' as the proportion that an event occurs out of a large number of trials. In the current example we have only a single patient, who either does or does not have cancer. The patient does not have cancer some proportion of the time. Thus the meaning of $P[\mathbf{C} \mid \mathbf{T}]$ in this case is the degree of one's belief, or the likelihood that this patient has cancer. It is this alternative interpretation of probability and the question of how it should be applied to statistics that is controversial. Kotz and Stroup (1983) give a good introduction to the idea that probability refers to uncertainty of knowledge rather than of events.

Consider the following example, in which Bayes' theorem was applied to a diagnostic test. The figures are based on Watson and Tang (1980). The sensitivity of the radioimmunoassay for prostatic acid phosphatase (RIA-PAP) as a test for prostatic cancer is 0.70. Its specificity is 0.94. The prevalence of prostatic cancer in the white male population is 35 per 100,000, or 0.00035. Applying these values to Expression (5.8), we find

$$P[\mathbf{C} \mid \mathbf{T}] = \frac{P[\mathbf{T} \mid \mathbf{C}]P[\mathbf{C}]}{P[\mathbf{T} \mid \mathbf{C}]P[\mathbf{C}] + P[\mathbf{T} \mid \mathbf{C}^C]P[\mathbf{C}^C]}$$

$$= \frac{0.70 \times 0.00035}{(0.70 \times 0.00035) + [(1 - 0.94)(1 - 0.00035)]} = 0.0041$$

The rather surprising result is that the likelihood that a white male who tests positive for the RIA-PAP test actually has prostate cancer is only 0.41%. This probability is known in epidemiology as the **positive predictive value.** Even if the test had been much more sensitive, say, 0.95 rather than 0.70, the positive predictive value would have been low—0.55 percent. Only for a perfect test (i.e., sensitivity and specificity both = 1) would a positive test imply with certainty that the patient had prostate cancer.

The paradoxically low positive predictive value is a consequence of its dependence on the prevalence of the disease. Only if the prevalence of prostatic cancer were 7895 per 100,000 would there be a 50 : 50 chance that a patient with a positive test result has cancer. This is more than 127 times the highest prevalence ever reported from a population in the United States. Watson and Tang (1980) use these findings (erroneously reported as 1440 per 100,000) and further analyses to make the point that using the RIA-PAP test as a routine screening procedure for prostate cancer is not worthwhile.

Readers interested in extending their knowledge of probability should refer to general texts such as Galambo (1984) or Kotz and Stroup (1983) for a simple introduction.

5.2 The Binomial Distribution

For the discussion to follow, we will simplify our sample space to consist of only two elements, foreign and American students, represented by {F, A}, and ignore whether they are undergraduates or graduates. Let us symbolize the probability space by $\{p, q\}$, where $p = P[\text{F}]$, the probability of being a foreign student, and $q = P[\text{A}]$, the probability of being an American student. As before, we can compute the probability space of samples of two students as follows:

$$\{\text{FF},\ \text{FA},\ \text{AA}\}$$

$$\{p^2,\ 2pq,\ q^2\}$$

If we were to sample three students independently, the probability space of the sample would be

$$\{\text{FFF},\ \text{FFA},\ \text{FAA},\ \text{AAA}\}$$

$$\{p^3,\ 3p^2q,\ 3pq^2,\ q^3\}$$

Samples of three foreign or three American students can be obtained in only one way, and their probabilities are p^3 and q^3, respectively. In samples of three, however, there are three ways of obtaining two students of one kind and one student of the other. As before, if A stands for American and F stands for foreign, then the sampling sequence could be AFF, FAF, and FFA for two foreign students and one American. Thus the probability of this outcome will be $3p^2q$. Similarly, the probability for two Americans and one foreign student is $3pq^2$.

A convenient way to summarize these results is by the binomial expansion, which is applicable to samples of any size from populations in which objects occur independently in only two classes—students who may be foreign or American, individuals who may be dead or alive, male or female, black or white, rough or smooth, and so forth. This summary is accomplished by expanding the binomial term $(p + q)^k$, where k equals sample size, p equals the probability of occurrence of the first class, and q equals the probability of occurrence of the second class. By definition, $p + q = 1$; hence q is a function of p: $q = 1 - p$. We will expand the expression for samples of k from 1 to 3:

For samples of 1 $\quad (p + q)^1 = p + q$

For samples of 2 $\quad (p + q)^2 = p^2 + 2pq + q^2$

For samples of 3 $\quad (p + q)^3 = p^3 + 3p^2q + 3pq^2 + q^3$

These expressions yield the same outcomes discussed previously. The coefficients (the numbers before the powers of p and q) express the number of ways a particular outcome is obtained.

A general formula that gives both the powers of p and q, as well as the binomial coefficients, is

$$\binom{k}{Y} p^Y q^{k-Y} = \frac{k!}{Y!(k-Y)!} p^Y (1-p)^{k-Y} \tag{5.9}$$

In this formula k, p, and q retain their earlier meaning, while Y stands for the number or count of "successes," the items that interest us and whose probability of occurrence is symbolized by p. In our example, Y designates the number of foreign students. The expression $\binom{k}{Y}$ stands for the number of combinations that can be formed from k items taken Y at a time. This expression can be evaluated as $k!/[Y!(k - Y)!]$, where ! means factorial. In mathematics, k factorial is the product of all the integers from 1 up to and including k. Thus: $5! = 1 \times 2 \times 3 \times 4 \times 5 = 120$. By convention, $0! = 1$. In working out fractions containing factorials, note that a factorial always cancels against a higher factorial. Thus $5!/3! = (5 \times 4 \times 3!)/3! = 5 \times 4$. For example, the binomial coefficient for the expected frequency of samples of 5 students containing 2 foreign students is $\binom{5}{2} = 5!/2!3! = (5 \times 4)/2 = 10$.

Now let us turn to a biological example. Suppose we have a population of insects, exactly 40% of which are infected with a given virus X. If we take samples of $k = 5$ insects each and examine each insect separately for the presence of virus, what distribution of samples could we expect if the probability of infection of each insect in a sample were independent from that of other insects in the sample? In this case $p = 0.4$, the proportion infected, and $q = 0.6$, the proportion not infected. The population is assumed to be so large that the question of whether sampling is with or without replacement is irrelevant for practical purposes. The expected frequencies would be the expansion of the binomial:

$$(p + q)^k = (0.4 + 0.6)^5$$

With the aid of Expression (5.9) this expansion is

$$p^5 + 5p^4q + 10p^3q^2 + 10p^2q^3 + 5pq^4 + q^5$$

or

$$(0.4)^5 + 5(0.4)^4(0.6) + 10(0.4)^3(0.6)^2 + 10(0.4)^2(0.6)^3 + 5(0.4)(0.6)^4 + (0.6)^5$$

representing the expected proportions of samples of five infected insects, four infected and one noninfected insects, three infected and two noninfected insects, and so on.

By now you have probably realized that the terms of the binomial expansion yield a type of frequency distribution for these different outcomes. Associated with each outcome, such as "five infected insects," is a probability of occurrence—in this case $(0.4)^5 = 0.01024$. This is a theoretical frequency distribution, or **probability distribution,** of events that can occur in two classes. It describes the expected distribution of outcomes in random samples of five insects, 40% of which are infected. The probability distribution described here is known as the **binomial distribution;** the binomial expansion yields the expected frequencies of the classes of the binomial distribution.

A convenient layout for presentation and computation of a binomial distribution is shown in Table 5.1, based on Expression (5.9). In the first column, which lists the number of infected insects per sample, note that we have revised the order of the terms to indicate a progression from $Y = 0$ successes (infected insects) to $Y = k$ successes. The second column features the binomial coefficient as given by the combinatorial portion of Expression (5.9). Column 3 shows

Table 5.1 EXPECTED FREQUENCIES OF INFECTED INSECTS IN SAMPLES OF 5 INSECTS SAMPLED FROM AN INFINITELY LARGE POPULATION WITH AN ASSUMED INFECTION RATE OF 40%.

(1) Number of infected insects per sample Y	(2) Binomial coefficients $\binom{k}{Y}$	(3) Powers of $p = 0.4$	(4) Powers of $q = 0.6$	(5) Relative expected frequencies $\hat{f}_{rel}$	(6) Absolute expected frequencies $\hat{f}$	(7) Observed frequencies f
0	1	1.00000	0.07776	0.07776	188.4	202
1	5	0.40000	0.12960	0.25920	628.0	643
2	10	0.16000	0.21600	0.34560	837.4	817
3	10	0.06400	0.36000	0.23040	558.3	535
4	5	0.02560	0.60000	0.07680	186.1	197
5	1	0.01024	1.00000	0.01024	24.8	29
			$\sum \hat{f}$ or $\sum f(=n)$	1.00000	2423.0	2423
			Mean	2.00000	2.00004	1.98721
			Standard deviation	1.09545	1.09543	1.11934

increasing powers of p from p^0 to p^5, and column (4) shows decreasing powers of q from q^5 to q^0. The **relative expected frequencies,** which are the probabilities of the various outcomes, are shown in column (5). We label such expected frequencies $\hat{f}_{rel}$. They are the product of columns (2), (3), and (4), and their sum is equal to 1.0, since the events in column (1) exhaust the possible outcomes. We see from column (5) that only about 1% of the samples are expected to consist of 5 infected insects, and 25.9% are expected to contain 1 infected and 4 noninfected insects. We will now test whether these predictions hold in an actual experiment.

EXPERIMENT 5.1. Simulate the case of the infected insects by using a table of random numbers such as Statistical Table **FF**. These are randomly chosen one-digit numbers in which each digit 0 through 9 has an equal probability of appearing. The numbers are grouped in blocks of 25 for convenience. Such numbers can also be obtained from random number keys on some pocket calculators and by pseudorandom number-generating algorithms in computer programs. Since there is an equal probability for any one digit to appear, you can let any four digits (say 0, 1, 2, 3) stand for the infected insects and the remaining digits (4, 5, 6, 7, 8, 9) stand for the noninfected insects. The probability that any one digit selected from the table represents an infected insect (that is, a 0, 1, 2, or 3) is therefore 40% or 0.4, since these are four of the ten possible digits. Also, successive digits are assumed to be independent of the values of previous digits. Thus the assumptions of the binomial distribution should be met in this experiment. Enter the table of random numbers at an arbitrary point (not always at the beginning!) and look at successive groups of five digits, noting in each group how many of the digits are 0, 1, 2, or 3. Take as many groups of five as you can find time to do, but no fewer than 100 groups. (Persons with computer experience can easily generate the data required by this exercise without using Table **FF**. There are also some programs that specialize in simulating sampling experiments.)

Column (7) in Table 5.1 shows the results of such an experiment by a biometry class. A total of 2423 samples of five numbers were obtained from Statistical Table **FF**, and the distribution of the four digits simulating the percentage of infection is shown in this column. The observed frequencies are labeled f. To calculate the expected frequencies for this example, we multiplied the relative expected frequencies, $\hat{f}_{rel}$, of column (5) by $n = 2423$, the number of samples taken. These calculations resulted in **absolute expected frequencies,** $\hat{f}$, shown in column (6). When we compare the observed frequencies in column (7) with the expected frequencies in column (6), we note general agreement between the two columns of figures. The two distributions are illustrated in Figure 5.2. If the observed frequencies did not fit expected frequencies, we might believe that the lack of fit was due to chance alone. Or we might be led to reject one or more of the following hypotheses: (1) that the true proportion of digits 0, 1, 2, and 3 is 0.4 (rejection of this hypothesis would normally not be reasonable, for we may rely on the fact that the proportion of digits 0, 1, 2, and 3 in a table of random

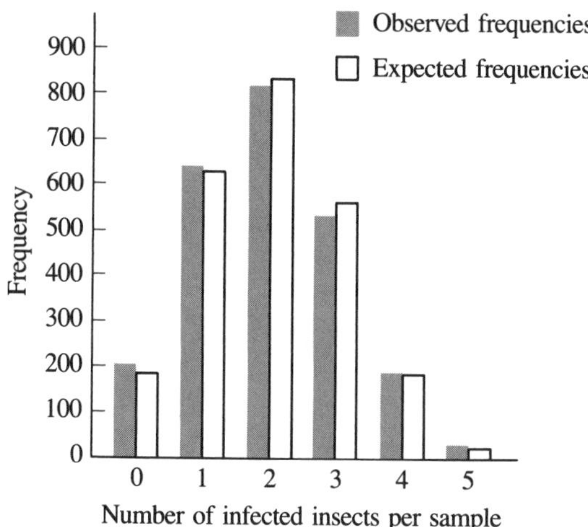

FIGURE 5.2 *Bar diagram of observed and expected frequencies given in Table 5.1.*

numbers *is* 0.4 or very close to it); (2) that sampling was random; and (3) that events are independent.

These statements can be reinterpreted in terms of the original infection model with which we started this discussion. If, instead of a sampling experiment of digits by a biometry class, this had been a real sampling experiment of insects, we would conclude that the insects had indeed been randomly sampled and that we had no evidence to reject the hypothesis that the proportion of infected insects was 40%. If the observed frequencies had not fit the expected frequencies, the lack of fit might be attributed to chance or to the conclusion that the true proportion of infection is not 0.4, or we would have to reject one or both the following assumptions: (1) that sampling was at random, and (2) that the occurrence of infected insects in these samples was independent.

Experiment 5.1 was designed to yield random samples and independent events. How could we simulate a sampling procedure in which the occurrences of the digits 0, 1, 2, and 3 were not independent? We could, for example, instruct the sampler to sample as indicated previously, but every time he found a 3, to search through the succeeding digits until he found another one of the four digits standing for infected individuals and to incorporate this in the sample. Thus, once a 3 was found, the probability would be 1.0 that another one of the indicated digits would be included in the sample. After repeated samples, this procedure would result in higher frequencies of classes of two or more indicated digits and in lower frequencies than expected (on the basis of the binomial distribution) of classes of one event. Many such sampling schemes could be devised. It should be clear that the probability of the second event occurring would be different from and dependent on that of the first.

How would we interpret a large departure of the observed frequencies from expected frequencies in another example? We have not yet learned techniques for testing whether observed frequencies differ from those expected by more than can be attributed to chance alone. This topic will be taken up in Chapter 17. Assume that such a test has been carried out and that it has shown that our observed frequencies are significantly different from the expected frequencies. Two main types of departure from expectation are likely: (1) **clumping** and (2) **repulsion,** shown in fictitious examples in Table 5.2. In real examples we would have no a priori notions about the magnitude of p, the probability of one of the two possible outcomes. In such cases it is customary to obtain p from the observed sample and to calculate the expected frequencies using the sample p. This would mean that the hypothesis that p is a given value cannot be tested, since by design the expected frequencies will have the same p value as the observed frequencies. Therefore, the hypotheses tested are whether the samples are random and the events independent.

The clumped frequencies in Table 5.2 have an excess of observations at the tails of the frequency distribution and consequently a shortage of observations at the center. Such a distribution is also called **contagious.** Remember that the total number of items must be the same in both observed and expected frequencies in order to make them comparable. In the repulsed frequency distribution there are more observations than expected at the center of the distribution and fewer at the tails. These discrepancies are most obvious in columns (4) and (6) of Table 5.2,

Table 5.2 ARTIFICIAL DISTRIBUTIONS TO ILLUSTRATE CLUMPING AND REPULSION.

Expected frequencies from Table 5.1.

(1) Number of infected insects per sample Y	(2) Absolute expected frequencies $\hat{f}$	(3) Clumped (contagious) frequencies f	(4) Deviation from expectation	(5) Repulsed frequencies f	(6) Deviation from expectation
0	188.4	225	+	143	−
1	628.0	703	+	618	−
2	837.4	663	−	943	+
3	558.3	558	0	548	−
4	186.1	227	+	157	−
5	24.8	47	+	14	−
$\sum \hat{f}$ or n	2423.0	2423		2423	
Mean	2.00004	2.00000		2.00000	
Standard deviation	1.09543	1.20074		1.01435	

where the deviations of observed from expected frequencies are shown as plus or minus signs. (These two types of distributions are also called overdispersed and underdispersed, but there has been some confusion in the literature about the meaning of these terms, so we will not use them here.)

What do these phenomena imply? In the clumped frequencies more samples were entirely infected (or largely infected) and similarly more samples were entirely noninfected (or largely noninfected) than you would expect if probabilities of infection were independent. This result could be due to poor sampling design. If, for example, the investigator, in collecting samples of five insects, always tended to pick out like ones—that is, infected ones or noninfected ones—then such a result would likely appear. If the sampling design is sound, however, the results become more interesting. Clumping would then mean that the samples of five are in some way related—that is, if one insect is infected, others in the same sample are more likely to be infected. This relation could be true if the insects came from adjacent locations in a situation in which neighbors are easily infected. Or the insects could be siblings exposed simultaneously to a source of infection. Or the infection could spread among members of a sample between the time the insects are sampled and the time they are examined.

The opposite phenomenon, repulsion, is more difficult to interpret biologically. There are fewer homogeneous groups and more mixed groups in such a distribution, which implies a compensatory phenomenon: If some of the insects in a sample are infected, the others in the sample are less likely to be. If the infected insects in the sample could transmit immunity to their associates in the sample, such a situation could arise logically, but it is biologically improbable. A more reasonable interpretation of such a finding is that for each sampling unit there are a limited number of pathogens available and that once several insects have become infected, the others go free of infection simply because there is no more infectious agent. This situation is unlikely in microbial infections, but in situations in which a limited number of parasites enter the body of the host, repulsion is more reasonable.

From the expected and observed frequencies in Table 5.1, we may calculate the mean and standard deviation of the number of infected insects per sample. These values are given at the bottom of columns (5), (6), and (7) in Table 5.1. We note that the means and standard deviations in columns (5) and (6) are almost identical and differ only trivially because of rounding errors. Column (7), however, being a sample from a population whose parameters are the same as those of the expected frequency distribution in columns (5) or (6), differs. The mean is slightly smaller and the standard deviation is slightly greater than in the expected frequencies. If we wish to know the mean and standard deviation of expected binomial frequency distributions, we need not go through the computations shown in Table 5.1. The mean and standard deviation of a binomial frequency distribution are, respectively,

$$\mu = kp \qquad \sigma = \sqrt{kpq}$$

Substituting the values $k = 5, p = 0.4$, and $q = 0.6$ from the example above, we obtain $\mu = 2.0$ and $\sigma = 1.09545$, which are identical to the values computed from column (5) in Table 5.1. Note that we use the Greek parametric notation here because μ and σ are parameters of an expected frequency distribution, not sample statistics, as are the mean and standard deviation in column (7). The proportions p and q are parametric values also and strictly speaking should be distinguished from sample proportions. In fact, in later chapters we resort to $\hat{p}$ and $\hat{q}$ for parametric proportions (rather than π, which conventionally is used as the ratio of the circumference to the diameter of a circle). Here, however, we prefer to keep our notation simple.

It is interesting to look at the standard deviations of the clumped and repulsed frequency distributions of Table 5.2. We note that the clumped distribution has a standard deviation greater than expected, and that of the repulsed one is less than expected. Comparison of sample standard deviations with their expected values is a useful measure of dispersion in such instances. If we wish to express our variable as a proportion rather than as a count—that is, to indicate mean incidence of infection in the insects as 0.4, rather than as 2 per sample of 5—we can use other formulas for the mean and standard deviation in a binomial distribution:

$$\mu = p \qquad \sigma = \sqrt{pq/k}$$

We will now use the binomial distribution to solve a biological problem. On the basis of our knowledge of the cytology and biology of species A, we expect the sex ratio among its offspring to be 1 : 1. The study of a litter in nature reveals that of 17 offspring, 3 were males and 14 were females. What conclusions can we draw from this evidence? Assuming that $p_{♂}$ (the probability of being a male offspring) $= 0.5$ and that this probability is independent among the members of the sample, the pertinent probability distribution is the binomial for sample size $k = 17$. Expanding the binomial to the power 17 is a nontrivial task, which, as we shall see, fortunately need not be done in its entirety.

The setup of this example is shown in Table 5.3. For the purposes of our problem, we need not proceed beyond the term for 4 males and 13 females. Calculating the relative expected frequencies in column (3), we note that the probability of 3 males and 14 females is 0.005,188,40, a very small value. If we add to this value all "worse" outcomes—that is, all outcomes that are even more unlikely than 14 females and 3 males on the assumption of a 1 : 1 hypothesis—we obtain a probability of 0.006,363,42, still a very small value. In statistics one often needs to calculate the probability of observing a deviation as large or larger than a given value.

On the basis of these findings one or more of the following assumptions is unlikely: (1) that the true sex ratio in species A is 1 : 1; (2) that we have sampled at random in the sense of obtaining an unbiased sample; or (3) that the sexes of the offspring are independent of one another. Lack of independence of events may mean that although the average sex ratio is 1 : 1, the individual sibships, or

Table 5.3 SOME EXPECTED FREQUENCIES OF MALES AND FEMALES FOR SAMPLES OF 17 OFFSPRING ON THE ASSUMPTION THAT THE SEX RATIO IS 1 : 1 [$p_{♂} = 0.5$, $q_{♀} = 0.5$; $(p_{♂} + q_{♀})^k = (0.5 + 0.5)^{17}$].

(1) Y ♂♂	(2) $k - Y$ ♀♀	(3) Relative expected frequencies $\hat{f}_{rel}$	
0	17	0.000,007,63	
1	16	0.000,129,71	
2	15	0.001,037,68	0.006,363,42 (sum of rows 0–3)
3	14	0.005,188,40	
4	13	0.018,157,91	

litters, are largely unisexual—that is, the offspring from a given mating tend to be all (or largely) males or all (or largely) females. To confirm this hypothesis we would need to have more samples and then examine the distribution of samples for clumping, which would indicate a tendency for unisexual sibships.

We must be very precise about the questions we ask of our data. There are really two questions we can ask about the sex ratio: (1) Are the sexes unequal in frequency so that females appear more often than males? and (2) Are the sexes unequal in frequency? We may be concerned with only the first of these questions, since we know from past experience that in this particular group of organisms the males are never more frequent than females. In such a case the reasoning followed above is appropriate. However, if we know very little about this group of organisms and if our question is simply whether the sexes among the offspring are unequal in frequency, then we have to consider both tails of the binomial frequency distribution; departures from the 1 : 1 ratio could occur in either direction. We should then consider not only the probabilities of samples with 3 males and 14 females (and all worse cases) but also the probability of samples of 14 males and 3 females (and all worse cases in that direction). Since this probability distribution is symmetrical (because $p_{♂} = q_{♀} = 0.5$), we simply double the cumulative probability of 0.006,363,42 obtained previously, which results in 0.012,726,84. This new value is still very small, making it quite unlikely that the true sex ratio is 1 : 1.

This is your first experience with one of the most important applications of statistics—hypothesis testing. A formal introduction to this field will be deferred until Section 7.8. We simply point out here that the two approaches just described are known as **one-tailed tests** and **two-tailed tests,** respectively. Students sometimes have difficulty knowing which of the two tests to apply. In future examples, we will try to explain why a one-tailed or a two-tailed test is being used.

We have said that a tendency for unisexual sibships would result in a clumped distribution of observed frequencies. An actual case of this phenomenon in nature is a classic in the literature, the sex ratio data obtained by Geissler (1889) from hospital records in Saxony. Table 5.4 shows the sex ratios of 6115 sibships of 12 children, each from the more extensive study by Geissler. All columns of the table should by now be familiar. To keep you on your toes, and to conform to the layout of the original publication, the meaning of p and q have been reversed from that in the earlier sex ratio example. Now $p_{♀}$ is the proportion of females and $q_{♂}$ that of males. In a binomial, which of the two outcomes is p and which is q is simply a matter of convenience.

The expected frequencies in this example were not calculated on the basis of a 1 : 1 hypothesis, since it is known that in human populations the sex ratio at birth is not 1 : 1. Because the sex ratio varies in different human populations, the best estimate of it for the population in Saxony was obtained simply by using the mean proportion of males in these data. This value can be obtained by

Table 5.4 SEX RATIOS IN 6115 SIBSHIPS OF 12 IN SAXONY.

(1) Y ♀♀	(2) $k - Y$ ♂♂	(3) Relative expected frequencies $\hat{f}_{rel}$	(4) Absolute expected frequencies $\hat{f}$	(5) Observed frequencies f	(6) Deviation from expectation $f - \hat{f}$
0	12	0.000384	2.3	7	+
1	11	0.004264	26.1	45	+
2	10	0.021725	132.8	181	+
3	9	0.067041	410.0	478	+
4	8	0.139703	854.3	829	−
5	7	0.206973	1265.6	1112	−
6	6	0.223590	1367.3	1343	−
7	5	0.177459	1085.2	1033	−
8	4	0.102708	628.1	670	+
9	3	0.042280	258.5	286	+
10	2	0.011743	71.8	104	+
11	1	0.001975	12.1	24	+
12	0	0.000153	0.9	3	+
Total		0.999998	6115.0	6115	

$\bar{Y} = 5.76942 \qquad s^2 = 3.48985$

SOURCE: Data from Geissler (1889).

calculating the average number of females per sibship ($\bar{Y} = 5.76942$) for the 6115 sibships and converting this into a proportion. This value is 0.480,785. Consequently, the proportion of males = 0.519,215. In the deviations of the observed frequencies from the absolute expected frequencies shown in column (6) of Table 5.4, we notice considerable clumping. There are many more instances of families with all female or all male children (or nearly so) than independent probabilities would indicate. The genetic basis for this is not clear, but it is evident that there are some families that ''run to girls'' and similarly others that ''run to boys.'' Other evidence of clumping is the fact that s^2 is much larger than we would expect on the basis of the binomial distribution [$\sigma^2 = kpq = 12(0.480785)0.519215 = 2.99557$].

There is a distinct contrast between the data in Table 5.1 and those in Table 5.4. In the insect infection data of Table 5.1, we had a hypothetical proportion of infection based on outside knowledge. In the sex ratio data of Table 5.4 we had no such knowledge; we used an *empirical value of p obtained from the data,* rather than a *hypothetical value external to the data.* The importance of this distinction will become apparent later. In the sex ratio data of Table 5.3, as in much work in Mendelian genetics, a hypothetical value of p is used.

An alternative, efficient method for calculating expected binomial frequencies is given in Box 5.1. The instructions are self-explanatory. Program BIOM-pc includes an option for computing expected binomial frequencies.

5.3 The Poisson Distribution

In the typical application of the binomial, we had relatively small samples (2 students, 5 insects, 17 offspring, 12 siblings), in which two alternative states occurred at varying frequencies (American and foreign, infected and noninfected, male and female). Quite frequently, however, we study cases in which sample size k is very large, and one of the events (represented by probability q) is much more frequent than the other (represented by p). We have seen that the expansion of the binomial $(p + q)^k$ is quite tiresome when k is large. Suppose you had to expand the expression $(0.001 + 0.999)^{1000}$. In such cases we are generally interested in one tail of the distribution only. This is the tail represented by the terms

$$p^0q^k, \binom{k}{1} p^1q^{k-1}, \binom{k}{2} p^2q^{k-2}, \binom{k}{3} p^3q^{k-3}, \ldots$$

The first term represents no rare events and k frequent events in a sample of k events, the second term represents 1 rare event and $k - 1$ frequent events, the third term 2 rare events and $k - 2$ frequent events, and so forth. The expressions of the form $\binom{k}{i}$ are the binomial coefficients, discussed in the previous section.

Box 5.1 COMPUTING EXPECTED BINOMIAL FREQUENCIES (MODIFIED FROM BLISS AND CALHOUN, 1954).

Insect data of Table 5.1: p (probability of being infected) = 0.4; q (probability of being noninfected) = 0.6; k (sample size) = 5.

(1) Number of infected insects per sample of 5 insects Y	(2) Number of noninfected insects per sample of 5 insects $k - Y$	(3) Relative expected frequencies $\hat{f}_{rel}$
0	5	0.07776
1	4	0.25920
2	3	0.34560
3	2	0.23040
4	1	0.07680
5	0	0.01024

Computation

1. Calculate a quotient $Q = p/q$ to five significant decimal places, $Q = 0.4/0.6 = 2/3 = 0.66667$.
2. Calculate the first relative expected frequency for $Y = 0$, $\hat{f}_{rel_1} = q^k = (0.6)^5 = 0.07776$.
3. Multiply $\hat{f}_{rel_1}$ by n (number of samples in the study) if you want absolute expected frequencies and enter the product as $\hat{f}_1$, the first value in the $\hat{f}$ column. (In our case, since we want relative expected frequencies, we multiply by 1 and enter 0.07776 in the $\hat{f}_{rel}$ column.)
4. The subsequent expected frequencies $\hat{f}_i$ from $\hat{f}_2$ to $\hat{f}_{k+1}$ can be computed as

$$\hat{f}_i = \hat{f}_{i-1} Q \left(\frac{k - i + 2}{i - 1} \right)$$

Thus, in this case

$$\hat{f}_2 = \hat{f}_1 Q(5/1) = (0.07776)\ (2/3)\ (5) \quad = 0.25920$$

$$\hat{f}_3 = \hat{f}_2 Q(4/2) = (0.25920)\ (2/3)\ (2) \quad = 0.34560$$

$$\hat{f}_4 = \hat{f}_3 Q(3/3) = (0.34560)\ (2/3)\ (1) \quad = 0.23040$$

$$\hat{f}_5 = \hat{f}_4 Q(2/4) = (0.23040)\ (2/3)\ (0.5) = 0.07680$$

$$\hat{f}_6 = \hat{f}_5 Q(1/5) = (0.07680)\ (2/3)\ (0.2) = 0.01024$$

Note the regular progression of the third factor from $k/1$ to $1/k$, the numerator and denominator being decreased and increased, respectively, by unity at each step. Each expected frequency is computed from the previous one, so accuracy is important.

5. As a final check, sum the expected frequencies to obtain 1 or n.

Although the desired tail of the curve could be computed by the method of Box 5.1, as long as sufficient decimal accuracy is maintained, it is far easier in such cases to compute another distribution, the Poisson distribution, which closely approximates the desired results. A suggestion for the use of this approximation is that the Poisson may be used to approximate the binomial when the probability of the rare event $p < 0.1$ and the product of sample size and probability $kp < 5$.

The **Poisson distribution** is also a discrete frequency distribution of the number of times a rare event occurs. In contrast to the binomial distribution, however, the number of times that an event does not occur is infinitely large. For our purposes, a Poisson variable will be studied in either a spatial or temporal sample. Examples of the former are the number of moss plants in a sampling quadrat on a hillside or the number of parasites on an individual host; examples of the latter are the number of mutations occurring in a genetic strain in the time interval of one month or the reported cases of influenza in one town during one week.

The Poisson variable, Y, will be the number of events per sample. It can assume discrete values from 0 on up. To be distributed in Poisson fashion, the variable must have two properties: (1) Its mean must be small relative to the maximum possible number of events per sampling unit. Thus the event should be "rare." This requires that the sampling unit should be of sufficient size. For example, a quadrat in which moss plants are counted must be large enough that a substantial number of moss plants would occur there physically if the biological conditions favored the development of moss plants in the quadrat. A quadrat of 1 cm^2 would be far too small for mosses to be distributed in Poisson fashion. Similarly, a time span of 1 minute would be unrealistic for reporting new influenza cases in a town, but within 1 week a great many such cases could occur. (2) An occurrence of the event must be independent of prior occurrences within the sampling unit. Thus, the presence of one moss plant in a quadrat must not enhance or diminish the probability of other moss plants developing in the quadrat. Similarly, the fact that one influenza case has been reported must not affect the probability of subsequent influenza cases being reported. Events that meet these conditions (rare and random events) should be distributed in Poisson fashion.

The purpose of fitting a Poisson distribution to numbers of rare events in nature is to test whether the events occur independently of each other. If they do, they will follow the Poisson distribution. If the occurrence of one event enhances the probability of a second such event, we obtain a clumped or contagious distribution. If the occurrence of one event impedes that of a second such event in the sampling unit, we obtain a repulsed, spatially or temporally uniform distribution. The Poisson distribution can be used as a test for randomness or independence of distribution not only spatially but also in time, as the examples that follow show.

This distribution, named after the French mathematician Siméon-Denis Poisson, who described it in 1837, is an infinite series whose terms add to one (as must be true for any probability distribution). The series can be represented as

$$\frac{1}{e^\mu}, \frac{\mu}{1!e^\mu}, \frac{\mu^2}{2!e^\mu}, \frac{\mu^3}{3!e^\mu}, \frac{\mu^4}{4!e^\mu}, \ldots, \frac{\mu^r}{r!e^\mu}, \ldots \tag{5.10}$$

which are the relative expected frequencies corresponding to the following counts of the rare event Y:

$$0, 1, 2, 3, 4, \ldots, r, \ldots$$

Thus, the first term represents the relative expected frequency of samples containing no rare event; the second term, one rare event; the third term, two rare events; and so on. The denominator of each term contains e^μ, where e is the base of the natural, or Napierian, logarithm, a constant whose value, accurate to five decimal places, is 2.71828. We recognize μ as the parametric mean of the distribution; it is a constant for any given problem. The exclamation mark after the coefficient in the denominator means "factorial," as explained in the previous section.

One way to learn more about the Poisson distribution is to apply it to an actual case. At the top of Box 5.2 is a well-known result from early statistical literature

Box 5.2 CALCULATION OF EXPECTED POISSON FREQUENCIES.

Yeast cells in 400 squares of a hemacytometer: $\bar{Y} = 1.8$ cells per square; $n = 400$ squares sampled.

(1) Number of cells per square Y	(2) Observed frequencies f		(3) Absolute expected frequencies $\hat{f}$		(4) Deviation from expectation $f - \hat{f}$	
0	75		66.1		+	
1	103		119.0		−	
2	121		107.1		+	
3	54		64.3		−	
4	30		28.9		+	
5	13		10.4		+	
6	2		3.1		−	
7	1	17	0.8	14.5	+	+
8	0		0.2		−	
9+	1		0.0		+	
	400		399.9			

SOURCE: Student (1907).

BOX 5.2 (CONTINUED)

Computation—Flow of computation based on Expression (5.11) multiplied by n, since we wish to obtain absolute expected frequencies, $\hat{f}$.

1. Obtain $e^{\bar{Y}}$ by using an exponential key on your calculator or an exponential function in your computer program.

$$e^{\bar{Y}} = e^{1.8} = 6.0496$$

2. $\hat{f}_0 = \dfrac{n}{e^{\bar{Y}}} = \dfrac{400}{6.0496} = 66.12$

3. $\hat{f}_1 = \hat{f}_0 \bar{Y} = 66.12(1.8) = 119.02$

4. $\hat{f}_2 = \hat{f}_1 \dfrac{\bar{Y}}{2} = 119.02\left(\dfrac{1.8}{2}\right) = 107.11$

5. $\hat{f}_3 = \hat{f}_2 \dfrac{\bar{Y}}{3} = 107.11\left(\dfrac{1.8}{3}\right) = 64.27$

6. $\hat{f}_4 = \hat{f}_3 \dfrac{\bar{Y}}{4} = 64.27\left(\dfrac{1.8}{4}\right) = 28.92$

7. $\hat{f}_5 = \hat{f}_4 \dfrac{\bar{Y}}{5} = 28.92\left(\dfrac{1.8}{5}\right) = 10.41$

8. $\hat{f}_6 = \hat{f}_5 \dfrac{\bar{Y}}{6} = 10.41\left(\dfrac{1.8}{6}\right) = 3.12$

9. $\hat{f}_7 = \hat{f}_6 \dfrac{\bar{Y}}{7} = 3.12\left(\dfrac{1.8}{7}\right) = 0.80$

10. $\hat{f}_8 = \hat{f}_7 \dfrac{\bar{Y}}{8} = 0.80\left(\dfrac{1.8}{8}\right) = 0.18$

Total 399.95

$\hat{f}_9$ and beyond 0.05

Note that each term, starting with step **3**, consists of multiplying the previous term by the mean and dividing by i, the index of f_i.

that tests the distribution of yeast cells in 400 squares of a hemacytometer, a counting chamber used in counting blood cells and other microscopic structures suspended in liquid. Column (1) lists the number of yeast cells observed in each hemacytometer square and column (2) gives the observed frequency—the number of squares containing a given number of yeast cells. Note that 75 squares contained no yeast cells, but that most squares held either 1 or 2 cells. Only 17 squares contained 5 or more yeast cells.

Why would we expect this frequency distribution to be distributed in Poisson fashion? We have here a relatively rare event, the frequency of yeast cells per hemacytometer square, the mean of which has been determined as 1.8. That is, on the average there are 1.8 cells per square. Relative to the amount of space in each square and the number of cells that could have come to rest in any one square, the number found is low indeed. We might also expect that the occurrence of individual yeast cells in a square is independent of the occurrence of other yeast cells. This is a commonly encountered class of results to which the Poisson distribution is applied.

The mean of the rare event is the only quantity that we need to know to calculate the relative expected frequencies of a Poisson distribution. Since we do not know the parametric mean of the yeast cells in this problem, we employ an estimate (the sample mean) and calculate expected frequencies of a Poisson distribution whose μ equals the sample mean of the observed frequency distribution of Box 5.2. It is convenient for the purpose of computation to rewrite Expression (5.10) as the following recursion formula:

$$\hat{f}_i = \hat{f}_{i-1}\left(\frac{\bar{Y}}{i}\right) \quad \text{for } i = 1, 2, \ldots, \text{ where } \hat{f}_0 = e^{-\bar{Y}} \tag{5.11}$$

Note that the parametric mean μ has been replaced by the sample mean $\bar{Y}$. Each term developed by this recursion formula is mathematically exactly the same as its corresponding term in Expression (5.10). It is important to make no computational error, since in such a chain multiplication the correctness of each term depends on the accuracy of the term before it. Expression (5.11) yields relative expected frequencies. If, as is more usual, absolute expected frequencies are desired, simply set the first term $\hat{f}_0$ to $n/e^{\bar{Y}}$, where n is the number of samples, and then proceed with the computation as before. The computation is illustrated in Box 5.2, and the expected frequencies so obtained are listed in column (3) of the frequency distribution.

What have we learned from this computation? When we compare the observed frequencies with the expected frequencies, we notice quite a good fit of our observed frequencies to a Poisson distribution of mean 1.8. A statistical test for goodness of fit will be given in Chapter 17. No clear pattern of deviations from expectation is shown. We cannot test a hypothesis about the mean because the mean of the expected distribution was taken from the sample mean of the observed variates. As in the binomial distribution, clumping or aggregation would mean that the probability that a second yeast cell will be found in a square is not independent of the presence of the first one, but is higher. This dependence would result in a clumping of the items in the classes at the tails of the distribution, so there would be some squares with many cells.

The biological interpretation of the dispersion pattern varies with the problem. In this example the yeast cells seem to be randomly distributed in the counting chamber, indicating thorough mixing of the suspension. Unless the proper suspension fluid is used, however, red blood cells often will stick together because of an electrical charge. This so-called rouleaux effect would be indicated by clumping of the observed frequencies.

Note that in Box 5.2, as in the subsequent tables giving examples of the application of the Poisson distribution, we group the low frequencies at one tail of the curve, uniting them by means of a bracket. This grouping tends to simplify the patterns of distribution, but the main reason for it is related to the tests for goodness of fit (of observed to expected frequencies), discussed in Section 17.2. For these tests, no expected frequency $\hat{f}$ should be less than 5.

Before we turn to other examples we need to learn a few more facts about the Poisson distribution. You probably noticed that in computing expected frequencies we needed to know only one parameter—the mean of the distribution. By comparison, in the binomial we needed two parameters, p and k. Thus the mean completely defines the shape of a given Poisson distribution. It follows, then, that the variance is a function of the mean; in a Poisson distribution we have a simple relationship between the two: $\mu = \sigma^2$, the variance being equal to the mean. The variance of the number of yeast cells per square based on the observed frequencies in Box 5.2 equals 1.965, not much larger than the mean of 1.80, indicating again that the yeast cells are distributed approximately in Poisson fashion, that is, randomly. This relationship between variance and mean suggests a rapid test of whether an observed frequency distribution is distributed in Poisson fashion even without fitting expected frequencies to the data. We simply compute a **coefficient of dispersion:**

$$CD = \frac{s^2}{\overline{Y}}$$

This value will be near 1 in distributions that are essentially Poisson, >1 in clumped samples, and <1 in cases of repulsion. In the yeast cell example, $CD = 1.092$. Computer program BIOM-pc includes an option for expected Poisson frequencies.

Figure 5.3 will give you an idea of the shapes of the Poisson distributions of different means. It shows frequency polygons (the lines connecting the mid-

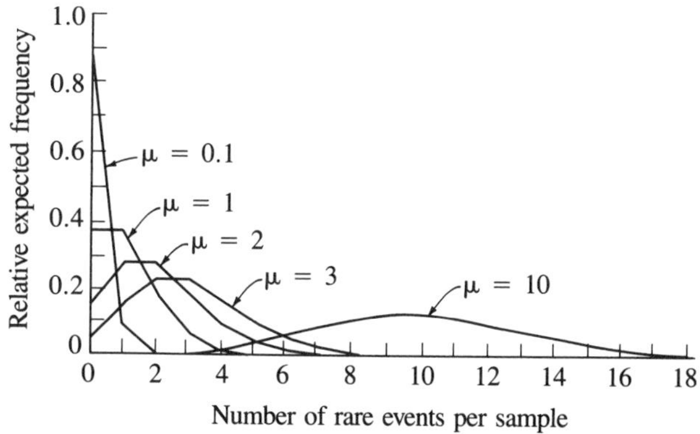

FIGURE 5.3 *Frequency polygons of the Poisson distributions for various values of the mean.*

points of a bar diagram) for five Poisson distributions. For the low value of $\mu = 0.1$, the frequency polygon is extremely L-shaped, but with an increase in the value of μ, the distributions become humped and eventually nearly symmetrical.

We conclude our study of the Poisson distribution by considering several examples from diverse fields. The first example, which is analogous to that of the yeast cells, is from an ecological study of mosses of the species *Hypnum schreberi* invading mica residues of china clay (Table 5.5). These residues occur on deposited "dams" (often 5000 yd^2 in area), on which the ecologists laid out 126 quadrats. In each quadrat they counted the number of moss shoots. Expected frequencies are again computed, using the mean number of moss shoots, $\bar{Y} = 0.4841$, as an estimate of μ. There are many more quadrats than expected at the two tails of the distribution than at its center. Thus, although we would expect only approximately 78 quadrats without a moss plant, we find 100. Similarly, while there are 11 quadrats containing 3 or more moss shoots, the Poisson distribution predicted only 1.7 quadrats. By way of compensation the central classes are shy of expectation. Instead of the nearly 38 expected quadrats with one moss plant each, only 9 were found, and there is a slight deficiency also in the 2-mosses-per-quadrat class. This case is another illustration of clumping, which was encountered first in the binomial distribution. The sample variance $s^2 = 1.308$, much larger than $\bar{Y} = 0.4841$, yields a coefficient of dispersion $CD = 2.702$.

Searching for a biological explanation of the clumping, the investigators found that the protonemata, or spores, of the moss were carried in by water and

Table 5.5 NUMBER OF MOSS SHOOTS (*HYPNUM SCHREBERI*) PER QUADRAT ON CHINA CLAY RESIDUES (MICA).

(1) Number of moss shoots per quadrat Y	(2) Observed frequencies f		(3) Absolute expected frequencies $\hat{f}$		(4) Deviation from expectation $f - \hat{f}$	
0	100		77.7		+	
1	9		37.6		−	
2	6	17	9.1	10.8	−	+
3	8		1.5		+	
4	1		0.2		+	
5	0		0.0		0	
6+	2		0.0		+	
Total	126		126.1			

$\bar{Y} = 0.4841 \quad s^2 = 1.308 \quad CD = 2.702$

SOURCE: Data from Barnes and Stanbury (1951).

Table 5.6 MITES (*ARRENURUS* SP.) INFESTING 589 CHIRONOMID FLIES (*CALOPSECTRA AKRINA*).

(1) Number of mites per fly Y	(2) Observed frequencies f		(3) Poisson expected frequencies $\hat{f}$		(4) Deviation from expectation $f - \hat{f}$	
0	442		380.7		+	
1	91		166.1		−	
2	29		36.2		−	
3	14	27	5.3	5.7	+	+
4	4		0.6		+	
5	6		0.1		+	
6	2		0.0		+	
7	0		0.0		0	
8+	1		0.0		+	
Total	589		589.0			

$\bar{Y} = 0.4363 \qquad s^2 = 0.9709 \qquad CD = 2.225$

SOURCE: Data from F. J. Rohlf.

deposited at random, but that each protonema gave rise to a number of upright shoots, so counts of the latter indicated a clumped distribution. In fact, when the clumps instead of individual shoots were used as variates, the investigators found that the clumps followed a Poisson distribution—that is, were randomly distributed. There are problems in directly applying this approach, since the size of the sampling units can influence the results. This would happen, for example, if the plants produced substances that prevented other plants from growing very close to them (repulsed distribution), but on a larger scale the plants were clumped in favorable regions (clumped distribution). Organisms moving under their own power, when distributed in clumped fashion, may have responded to various social instincts or may have accumulated in clumps, as a result of, or in response to, environmental forces.

The second example features the distribution of a water mite on adults of a chironomid fly (Table 5.6). This example is similar to the case of the mosses except that here the sampling unit is a fly instead of a quadrat. The rare event is a mite's infesting the fly. The coefficient of dispersion, 2.225, reflects the pattern of observed frequencies, which are greater than expected in the tails and less than expected in the center. This relationship is made clear in the last column of the frequency distribution, which shows the signs of the deviations (observed frequencies minus expected frequencies) and shows a characteristic clumped pattern. A possible explanation is that the densities of mites in the several ponds from which the chironomid flies emerged differed considerably. Chironomids that emerged from ponds with many mites would be parasitized by more than one mite, but those from ponds where mites were scarce would show little or no infestation.

The third example tests the randomness of the distribution of weed seeds in samples of grass seed (Table 5.7). Here the total number of seeds in a quarter-ounce sample could be counted, so we could estimate k (which is several thousand), and q, which represents the large proportion of grass seed, as contrasted with p, the small proportion of weed seed. The data are therefore structured as in a binomial distribution with alternative states, weed seed and grass seed. For purposes of computation, however, only the number of weed seeds must be considered. This is an example of a case mentioned at the beginning of this section, a binomial in which the frequency of one outcome is very much smaller than that of the other and the sample size is large. We can use the Poisson distribution as a useful approximation of the expected binomial frequencies for the tail of the distribution. We use the average number of weed seeds per sample of seeds as our estimate of the mean and calculate expected Poisson frequencies for this mean. Although the pattern of the deviations and the coefficient of dispersion again indicate clumping, this tendency is not pronounced, and the methods of Chapter 17 would indicate insufficient evidence for the conclusion that the distribution is not a Poisson. We conclude that the weed seeds are randomly distributed through the samples. If clumping had been significant it

Table 5.7 ***POTENTILLA*** **(WEED) SEEDS IN 98 QUARTER-OUNCE SAMPLES OF GRASS SEEDS (*PHLEUM PRATENSE*).**

(1) Number of weed seeds per sample of seeds Y	(2) Observed frequencies f		(3) Poisson expected frequencies $\hat{f}$		(4) Deviation from expectation $f - \hat{f}$	
0	37		31.3		+	
1	32		35.7		−	
2	16		20.4		−	
3	9	13	7.8	10.6	+	+
4	2		2.2		−	
5	0		0.5		−	
6	1		0.1		+	
7+	1		0.0		+	
Total	98		98.0			

$\bar{Y} = 1.1429$ $\quad s^2 = 1.711$ $\quad CD = 1.497$

SOURCE: Modified from Leggatt (1935).

Table 5.8 EXPECTED FREQUENCIES COMPARED FOR BINOMIAL AND POISSON DISTRIBUTIONS.

(1) Y	(2) Expected binomial frequencies $p = 0.001$ $k = 1000$ $\hat{f}_{rel}$	(3) Expected frequencies approximated by Poisson $\mu = 1$ $\hat{f}_{rel}$
0	0.367695	0.367879
1	0.368063	0.367879
2	0.184032	0.183940
3	0.061283	0.061313
4	0.015290	0.015328
5	0.003049	0.003066
6	0.000506	0.000511
7	0.000072	0.000073
8	0.000009	0.000009
9	0.000001	0.000001
Total	1.000000	0.999999

might have been because the weed seeds stuck together, possibly because of interlocking hairs, sticky envelopes, or the like.

Perhaps you need to be convinced that the binomial under these conditions does approach the Poisson distribution. The mathematical proof is too involved for this text, but we can give an empirical example (Table 5.8). Here the expected binomial frequencies for the expression $(0.001 + 0.999)^{1000}$ are given in column (2), and in column (3) these frequencies are approximated by a Poisson distribution with mean equal to 1, since the expected value for p is 0.001, which is one event for a sample size of $k = 1000$. The two columns of expected frequencies are extremely similar.

The next distribution is extracted from an experimental study of the effects of different densities of parents of the azuki bean weevil (Table 5.9). Larvae of these weevils enter the beans, feed and pupate inside them, and then emerge through a hole. Thus the number of emergence holes per bean is a good measure of the number of adults that have emerged. The rare event in this case is the weevil present in the bean. We note that the distribution is strongly repulsed, a far rarer occurrence than a clumped distribution. There are many more beans containing one weevil than the Poisson distribution would predict. A statistical finding of this sort leads us to investigate the biology of the phenomenon. In this case it was found that the adult female weevils tended to deposit their eggs evenly rather than randomly over the available beans. This prevented too many

Table 5.9 AZUKI BEAN WEEVILS (*CALLOSOBRUCHUS CHINENSIS*) EMERGING FROM 112 AZUKI BEANS (*PHASEOLUS RADIATUS*).

(1) Number of weevils emerging per bean Y	(2) Observed frequencies f	(3) Poisson expected frequencies $\hat{f}$	(4) Deviation from expectation $f - \hat{f}$
0	61	70.4	−
1	50	32.7	+
2	1 }	7.6 }	− }
3	0 } 1	1.2 } 8.9	− } −
4+	0 }	0.1 }	− }
Total	112	112.0	

$\bar{Y} = 0.4643$ $s^2 = 0.269$ $CD = 0.579$

SOURCE: Data from Utida (1943).

eggs being placed on any one bean and precluded heavy competition among the developing larvae. A contributing factor was competition between larvae feeding on the same bean, generally resulting in all but one being killed or driven out. Thus it is easy to understand how these biological phenomena would give rise to a repulsed distribution.

We will end this discussion with a classic, tragicomic application of the Poisson distribution. Table 5.10 is a frequency distribution of men killed by being kicked by a horse in 10 Prussian army corps over 20 years. The basic sampling unit is temporal in this case, one army corps per year. We are not certain exactly how many men are involved, but most likely a considerable number. The 0.610 men killed per army corps per year is the rare event. If we knew the number of men in each army corps, we could calculate the probability of not being killed by a horse in one year. This calculation would give us a binomial that could be approximated by the Poisson distribution. Knowing that the sample size (the number of men in an army corps) is large, however, we need not concern ourselves with values of p and k and can consider the example simply from the Poisson model, using the observed mean number of men killed per army corps per year as an estimate of μ.

This example is an almost perfect fit of expected frequencies to observed ones. What would clumping have meant in such an example? An unusual number of deaths in a certain army corps in a given year could have been due to poor discipline in the particular corps or to a particularly vicious horse that killed

Table 5.10 MEN KILLED BY BEING KICKED BY A HORSE IN 10 PRUSSIAN ARMY CORPS IN THE COURSE OF 20 YEARS.

(1) Number of men killed per year per army corps	(2) Observed frequencies f		(3) Poisson expected frequencies $\hat{f}$		(4) Deviation from expectation $f - \hat{f}$	
0	109		108.7		+	
1	65		66.3		−	
2	22		20.2		+	
3	3	} 4	4.1	} 4.8	−	} −
4	1		0.6		+	
5+	0		0.1		−	
Total	200		200.0			

$\bar{Y} = 0.610$ $s^2 = 0.611$ $CD = 1.002$

SOURCE: Data from Bortkiewicz (1898).

several men before the corps got rid of it. Repulsion might mean that one death per year per corps occurred more frequently than expected. This might have been so if men in each corps were careless each year until someone had been killed, after which time they all became more careful for a while.

Occasionally samples of items that could be distributed in Poisson fashion are taken in such a way that the zero class is missing. For example, an entomologist studying the number of mites per leaf of a tree may not sample leaves at random from the tree, but collect only leaves containing mites, excluding those without mites. For these so-called **truncated Poisson distributions,** *special methods can be used to estimate parameters and to calculate expected frequencies. Cohen (1960) gives tables and graphs for rapid estimation.*

5.4 OTHER DISCRETE PROBABILITY DISTRIBUTIONS

The binomial and Poisson distributions are both examples of discrete probability distributions. For the moss example of the previous section (Table 5.5), this means that there is either no moss plant per quadrat or one moss plant, two plants, three plants, and so forth, but no values in between. Other discrete distributions are known in probability theory, and in recent years many have been suggested as suitable for one application or another. As a beginner, you need not concern yourself with these, but it might be well for you to be familiar by name at least with those we will discuss here briefly.

The **hypergeometric distribution** is the distribution equivalent to the binomial case but sampled from a *finite population without replacement.* In Section 5.1 we estimated the probability of finding a second foreign student in Matchless University. We pointed out that if there are 10,000 students at Matchless University, 400 of whom are foreign, then the probability of sampling one foreigner is indeed 0.04. Once a foreign student has been sampled, however, the probability of sampling another foreign student is no longer 0.04 (even when independent) but is 399/9999, or 0.0399. This probability would be 0.04 only *with replacement*—that is, if we returned the first foreign student sampled to the university population before we sampled again.

The binomial distribution is entirely correct only in cases of sampling with replacement or with infinite population size (which amounts to the same thing). For practical purposes, when small samples are taken from large populations, as in the case of Matchless University, these populations can be considered infinite. But if you have a population of 100 animals, 4% of which carry a mutation, a sample of one mutant reduces the population to 3 out of 99, or from 4% to 3.03%. Thus repeated samples of 5 from this population would follow not the binomial distribution but a different distribution, the hypergeometric distribution. The individual terms of the hypergeometric distribution are given by the expression

$$\frac{\binom{pN}{r}\binom{qN}{k-r}}{\binom{N}{k}}$$

which gives the probability of sampling r items of the type represented by probability p out of a sample of k items from a population of size N. The mean and variance of a hypergeometric distribution are kp and $kpq(N - k)/(N - 1)$.

Note that the mean is the same as that of the binomial distribution, and the variance is that of the binomial multiplied by $(N - k)/(N - 1)$. When N is very large as compared with k, this term is approximately 1, expected since the hypergeometric distribution approximates the binomial. The expected frequencies for any sizable distribution are tedious to compute. We suggest using a digital computer with facilities for double-precision arithmetic to evaluate expected hypergeometric frequencies for a sizable distribution.

In biology, sampling of small samples from a finite distribution occurs in certain problems of evolutionary genetics. Another application is in mark–recapture studies, in which a certain proportion of a population is caught, marked, released, and subsequently recaptured, leading to estimates of the number of the entire population.

A number of probability distributions have been employed as underlying mathematical models for cases of contagious distribution. The difficulty of fitting these varies with the distribution. Greig-Smith (1964) gives a nontechnical account of the application of these distributions to ecology and provides references that may lead the interested reader deeper into the subject. More-quantitative accounts are given by Pielou (1977) and Krebs (1989). Many ecological examples of the application of contagious distributions are given in Williams (1964).

We have used the binomial and Poisson distributions to test whether given data are random or whether they show marked departure from random expectations—either clumping or repulsion. In some cases it may seem unreasonable to assume random occurrences. In social organisms, for example, aggregation is a given. Statisticians and biologists have developed models for such cases which lead to the so-called contagious distributions. Although we cannot discuss these in detail in this volume, we mention two here briefly.

The **negative binomial distribution** has been used probably more frequently than any other contagious distribution. The theoretical conditions that would give rise to a negative binomial distribution are discussed by Bliss and Calhoun (1954), who also give methods of calculating expected frequencies. The account in Bliss and Fisher (1953) is somewhat more rigorous. Krebs (1989) describes methods for estimation and significance testing and provides computer software for the calculations.

The **logarithmic distribution** (or logarithmic series) has been used extensively in studying the distribution of taxonomic units in faunal samples. This distribution has been employed frequently by C. B. Williams (see Williams, 1964). Johnson and Kotz (1969) give general information on many types of discrete distributions.

EXERCISES 5

5.1 In humans the sex ratio of newborn infants is about 100♀♀ : 105♂♂. If we were to take 10,000 random samples of 6 newborn infants each from the total population of such infants for one year, what would be the expected frequency of groups of 6 males, 5 males, 4 males, and so on? *Answer:* For 4 males $\hat{f} = 2456.5$.

5.2 Show algebraically why the computational method of Box 5.1 works.

5.3 The two columns below give fertility of eggs of the CP strain of *Drosophila melanogaster* raised in 100 vials of 10 eggs each (data from R. R. Sokal). Find the expected frequencies, assuming that the mortality for each egg in a vial is independent. Use the observed mean. Calculate the expected variance and compare it with the observed variance. Interpret the results, knowing that the eggs of each vial are siblings and that the different vials contain descendants from different parent pairs. *Answer:* $\sigma^2 = 2.417$, $s^2 = 6.628$.

Number of Eggs Hatched Y	Number of Vials f
0	1
1	3
2	8
3	10
4	6
5	15
6	14
7	12
8	13
9	9
10	9

5.4 Calculate Poisson expected frequencies for the frequency distribution given in Table 2.3 (number of plants of the sedge *Carex flacca* found in 500 quadrats).

5.5 The Army Medical Corps is concerned over the intestinal disease *X*. From previous experience they know that soliders suffering from the disease invariably harbor the pathogenic organism in their feces and that for all practical purposes every diseased stool specimen contains these organisms. The organisms are never abundant, however, and thus only 20% of all slides prepared by the standard procedure contain some of them (we assume that if an organism is present on a slide, it will be seen). How many slides per stool specimen should the laboratory technicians prepare and examine to ensure that if a specimen is positive, it will be erroneously diagnosed negative in less than 1% of the cases (on the average)? On the basis of your answer would you recommend that the Corps attempt to improve their diagnostic methods? *Answer:* 21 slides.

5.6 A cross is made in a genetic experiment in drosophila in which it is expected that one-fourth of the progeny will have white eyes and one-half will have the trait called singed bristles. Assume that the two gene loci segregate independently. (a) What proportion of the progeny should exhibit both traits simultaneously? (b) If 4 flies are sampled at random, what is the probability that they will all have white eyes? (c) What is the probability that none of the 4 flies will have either white eyes or singed bristles? (d) If 2 flies are sampled, what is the probability that at least one of the flies will have white eyes or singed bristles or both traits?

5.7 Readers who have had a semester or two of calculus may wish to try to prove that Expression 5.9 tends to Expression 5.10 as k becomes indefinitely large (and p becomes infinitesimal, so that $\mu = kp$ remains constant).

HINT: $$\left(1 - \frac{x}{n}\right)^n \to e^{-x} \text{ as } n \to \infty$$

5.8 Summarize and compare the assumptions and parameters on which the binomial and Poisson distributions are based.

5.9 The two columns below present an accident record of 647 women working in a munitions plant during a five-week period reported by Greenwood and Yule (1920). The sampling unit is one woman during this period. The model assumes that the accidents are not fatal or very serious and thus do not remove the individual from further exposure. Compute the expected frequency distribution for number of accidents per woman. Do the observed frequencies fit these expectations? Can you come up with an explanation for your findings? Calculate and interpret the coefficient of dispersion. *Answer:* $CD = 1.488$.

Number of accidents per woman Y	Observed frequencies f	
0	447	
1	132	
2	42	
3	21	26 (3, 4, 5+)
4	3	
5+	2	
Total	647	

5.10 Population geneticists work with a variable called gene frequency, which is the proportion of an allele found in the gametes of a population. Thus if a genetic locus has two alternative alleles, A and a, p is the gene frequency or proportion of A, and $q = 1 - p$ is the gene frequency of a. In an insect population, the gene frequency of A is 0.6. If mating is random in the population, what will be the frequencies (that is, the proportions) of the three possible genotypes: AA, Aa, and aa?
HINT: Random mating can be considered as sampling two gametes at random from the gametic pool. The formula that will solve this problem describes the most fundamental relation in population genetics, known as the Hardy–Weinberg Law.

5.11 In Exercise 5.10, allele A is dominant over allele a, so genotypes AA and Aa are phenotypically indistinguishable. When a geneticist examines a single dominant specimen from the population, what is the probability that the specimen is a heterozygote? *Answer:* $P[\mathbf{Aa} \mid \mathbf{D}] = 0.5714$.

6 THE NORMAL PROBABILITY DISTRIBUTION

The theoretical frequency distributions in Chapter 5 were all discrete. Their variables assumed values that changed in integral steps (meristic variables). Thus the number of infected insects per sample was 0 or 1 or 2 but could not assume a value between these. Similarly, the number of yeast cells per hemacytometer square is a meristic variable and requires a discrete probability function to describe it. Most variables encountered in biology, however, are continuous (such as the infant birth weights or the aphid femur lengths used as examples in Chapters 2 and 4). This chapter deals more extensively with the distributions of continuous variables.

Section 6.1 introduces frequency distributions of continuous variables. In Section 6.2 we examine the properties of the most common of such distributions, the normal probability distribution, and we present a model for generating this distribution in Section 6.3. A few applications of the normal distribution are illustrated in Section 6.4. The technique of fitting a normal distribution to observed data is described in Section 6.5, and Section 6.6 discusses some of the reasons for departure from normality in observed frequency distributions and describes techniques for measuring the nature and amount of such departures. Graphic techniques for identifying departures from normality and for estimating the mean and the standard deviation in approximately normal distributions are given in Section 6.7. Section 6.8 comments briefly on other types of theoretical frequency distributions of continuous variables.

6.1 FREQUENCY DISTRIBUTIONS OF CONTINUOUS VARIABLES

For continuous variables the theoretical probability distribution, or **probability density function,** can be represented by a continuous curve (Figure 6.1A). The

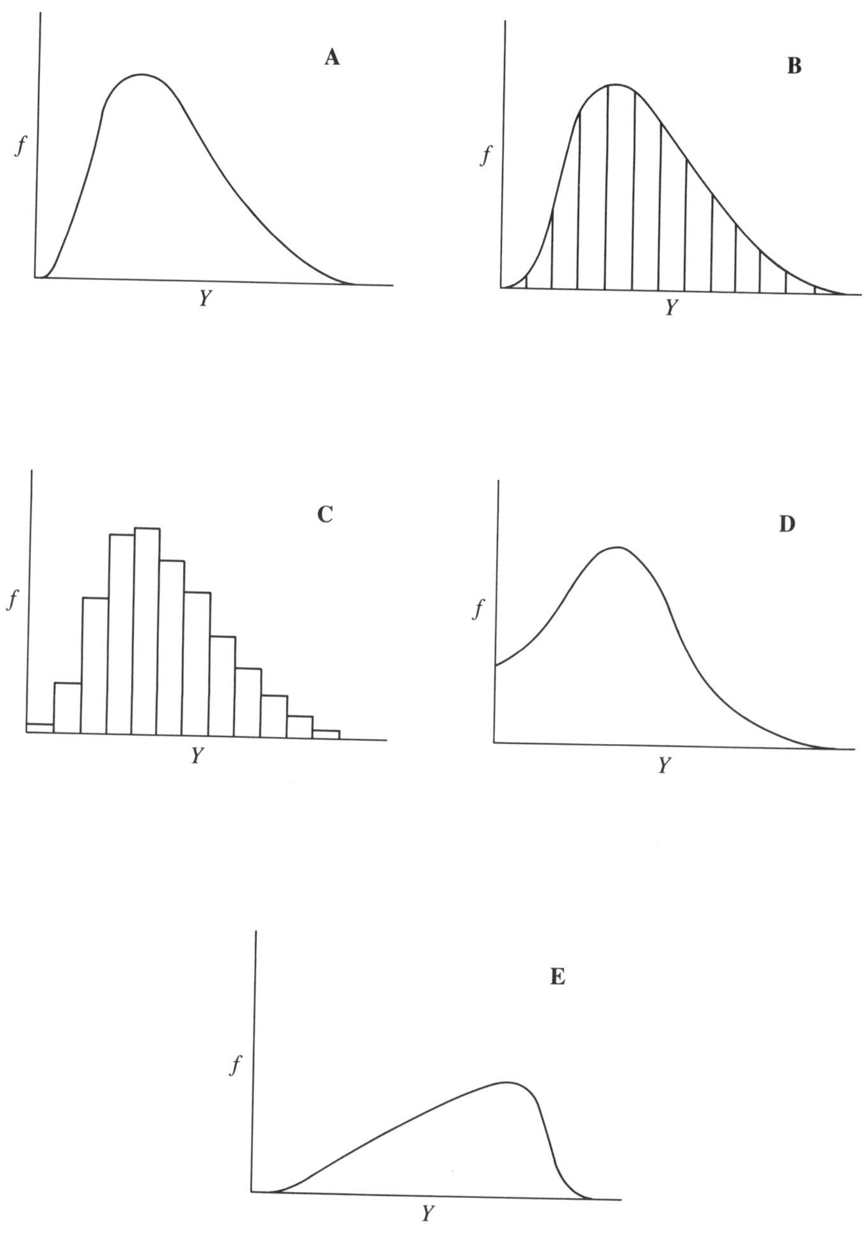

FIGURE 6.1 *Examples of some probability distributions of continuous variables. (See text for explanation.)*

height of the curve gives the density for a given value of the variable. By **density** we mean the relative concentration of variates along the Y-axis (as indicated in Figure 2.1). Note that the Y-axis is the abscissa (horizontal axis) in the figure, and the frequency f, or the density, is the ordinate (vertical axis). To compare the theoretical with the observed frequency distribution, the two must be divided into corresponding classes, as shown by the vertical lines in Figure 6.1B. Probability density functions are defined so that the expected frequency of observations between two class limits (vertical lines) is represented by the area between these limits under the curve. The total area under the curve is therefore equal to the sum of the expected frequencies (1.0 or n, depending on whether relative or absolute expected frequencies were calculated).

When forming a frequency distribution of observations of a continuous variable, the choice of class limits is arbitrary because all values of a variable are theoretically possible. In a continuous distribution one cannot evaluate the probability of the variable being exactly equal to a given value such as 3 or 3.5. One can only estimate the frequency of observations falling between two limits because the area of the curve corresponding to any point along the curve is an infinitestimal.

The usual way to graph continuous distributions is in the form of a histogram —that is, to make each of the segments in Figure 6.1B into rectangles of the same width, as shown in Figure 6.1C. As long as each class interval is the same width, the height of each rectangle is directly proportional to its area, and the frequency of each class may be read directly off the ordinate.

Thus in order to calculate expected frequencies for a continuous distribution, we must calculate the proportions of the area under the curve between the class limits. In Sections 6.2, 6.4, and 6.5 we will see how this is done for normal frequency distributions.

Continuous frequency distributions may start and terminate at finite points along the Y-axis, as in Figure 6.1A, or one or both ends of the curve may extend indefinitely, as in Figures 6.1D and 6.1E. The idea of an area under a curve, when one or both ends go to infinity, may trouble those of you not acquainted with the calculus. Fortunately, however, this is not a great conceptual stumbling block, since in all the cases that we will encounter the tail of the curve approaches the Y-axis rapidly enough that the portion of the area beyond a certain point is, for all practical purposes, zero, and the frequencies it represents are infinitesimal.

We may fit continuous frequency distributions to some sets of meristic data, such as, for example, the number of teeth in an organism. In such cases we have reason to believe that underlying biological variables causing the differences in numbers of the structure are continuous, even though expressed as a discrete variable.

We will now discuss the most important probability density function in statistics, the normal frequency distribution.

6.2 Properties of the Normal Distribution

Formally, the **normal probability density function** can be represented by the expression

$$Z = \frac{1}{\sigma\sqrt{2\pi}} e^{-\frac{1}{2}[(Y-\mu)/\sigma]^2} \tag{6.1}$$

where Z indicates the height of the ordinate of the curve, which represents the density of the items. Z is the dependent variable in the expression because it is a function of the variable Y. There are two constants in the equation: π, well known to be approximately 3.14159, making $1/\sqrt{2\pi}$ equal 0.39894; and e, the base of the Napierian, or natural, logarithms, whose value approximates 2.71828. A normal probability density function has two parameters: the parametric mean μ and the parametric standard deviation σ, which determine the location and shape of the distribution.

Thus there is not just one normal distribution, as one might naively believe when encountering the same bell-shaped image in elementary textbooks. Rather, the number of such curves is infinite, since these parameters can assume an infinite number of values. This variation is illustrated by the three normal curves in Figure 6.2, each of which represents the same total frequencies. Curves A and B, which differ in location but not in shape, represent populations with different means. Curves B and C represent populations that have identical means but different standard deviations. Since the standard deviation of curve C is only half that of curve B, curve C is much narrower.

In theory a normal frequency distribution extends from negative infinity to positive infinity along the axis of the variable (labeled Y, although it is frequently

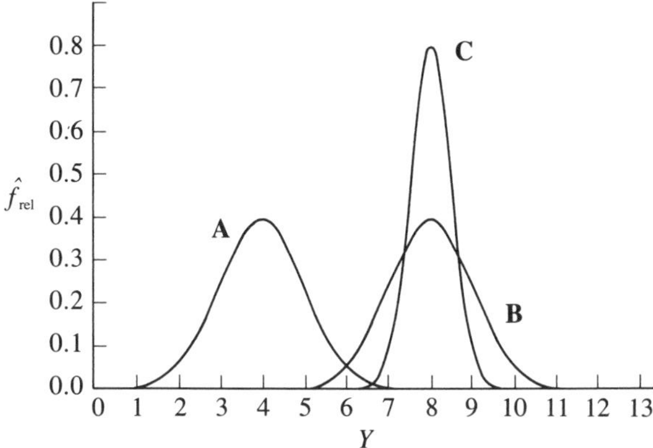

FIGURE 6.2 *Changes in the two parameters of the normal distribution affect the shape and location of the normal probability density function.* **A.** $\mu = 4$, $\sigma = 1$; **B.** $\mu = 8$, $\sigma = 1$; **C.** $\mu = 8$, $\sigma = 0.5$.

the abscissa). A normally distributed variable thus can assume any value, however large or small, although values farther from the mean than plus or minus three standard deviations are quite rare, since their relative expected frequencies are very small. This tendency can be seen from Expression (6.1). When Y is very large or very small, the term $\frac{1}{2}[(Y - \mu)/\sigma]^2$ necessarily becomes very large. Hence e raised to the negative power of that term is very small and Z is therefore very small. The ordinate Z for a normal distribution is tabulated in many books as a function of distance from the mean in standard deviation units (see Pearson and Hartley, 1958, table 1).

The curve is symmetrical around the mean. Therefore the mean, median, and mode of the normal distribution are all at the same point. The following percentages of items in a normal frequency distribution lie within the indicated limits.

$$\mu \pm \sigma \text{ contains } 68.27\% \text{ of the items}$$

$$\mu \pm 2\sigma \text{ contains } 95.45\% \text{ of the items}$$

$$\mu \pm 3\sigma \text{ contains } 99.73\% \text{ of the items}$$

Conversely,

$$50\% \text{ of the items fall between } \mu \pm 0.674\sigma$$

$$95\% \text{ of the items fall between } \mu \pm 1.960\sigma$$

$$99\% \text{ of the items fall between } \mu \pm 2.576\sigma$$

These relations are shown in Figure 6.3. How have these percentages been calculated? The direct calculation of any portion of the area under the normal curve requires an integration of the function shown as Expression (6.1). For Figure 6.3, this integral has been evaluated and is presented in an alternative form: the **normal distribution function** (the theoretical *cumulative* distribution function of the normal probability density function). This function gives the total frequency from negative infinity up to any point along the abscissa. We can therefore directly look up the probability that an observation will be less than a specified value of Y. For example, Figure 6.3 shows that the total frequency up to the mean is 50.00% and the frequency up to a point one standard deviation below the mean is 15.87%. These frequencies are found, graphically, by raising a vertical line from a point, such as $-\sigma$, until it intersects the cumulative distribution curve, and then reading the frequency off the ordinate. The probability of an observation falling between two arbitrary points can be found by subtracting the probability of an observation falling below the lower point from the probability of an observation falling below the upper point. For example, we can see from Figure 6.3 that the probability of an observation falling between the mean and a point one standard deviation below the mean is $0.5000 - 0.1587 = 0.3413$.

The normal distribution function is tabulated in Statistical Table **A,** which gives areas of the normal curve for the normal frequency distribution with the following parameters: mean $= 0$ and standard deviation $= 1$. For convenience

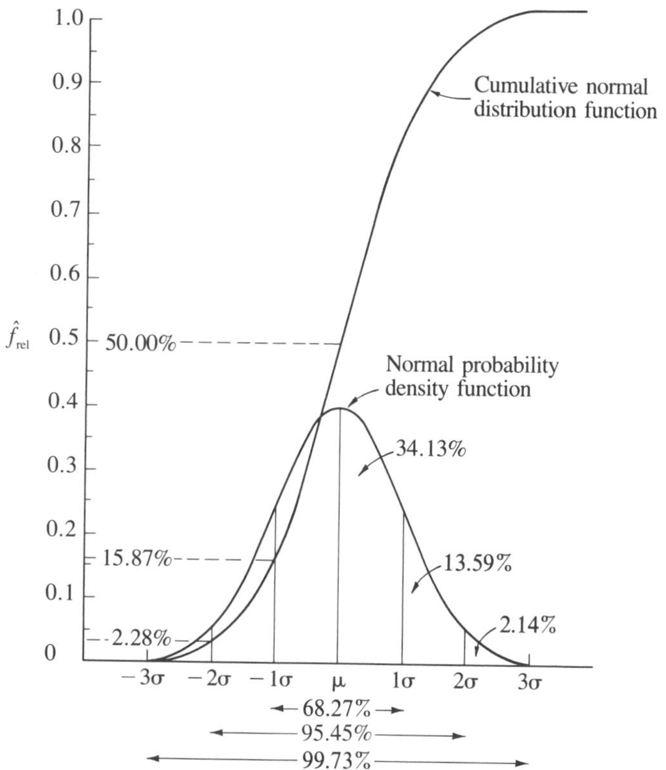

FIGURE 6.3 *Areas under the normal probability density function and the cumulative normal distribution function.*

in later calculations, 0.5 has been subtracted from all of the entries. This table therefore lists the proportion of the area between the mean and any point a given number of standard deviations above it. Thus, for example, the area between the mean and the point 0.50 standard deviations above the mean is 0.1915 of the total area of the curve. Similarly, the area between the mean and the point 2.64 standard deviations above the mean is 0.4959 of the curve. A point 4.0 standard deviations from the mean includes 0.499968 of the area between it and the mean. Since the normal distribution extends from negative to positive infinity, however, one needs to go an infinite distance from the mean to reach fully half of the area under the curve. Section 6.4 will explain how to use the table of areas of the normal curve.

The following sampling experiment will give you a feel for the distribution of items sampled from a normal distribution.

EXPERIMENT 6.1. You are asked to sample from two populations. The first is an approximately normal frequency distribution of 100 wing lengths of houseflies. The

Table 6.1 POPULATIONS OF WING LENGTHS AND MILK YIELDS.

Column 1. *Rank number.* Column 2. *Lengths (in mm × 10^{-1}) of 100 wings of houseflies arrayed in order of magnitude;* $\mu = 45.5$, $\sigma^2 = 15.21$, $\sigma = 3.90$; *distribution is approximately normal.* Column 3. *Total annual milk yield (in hundreds of pounds) of 100 two-year-old registered Jersey cows arrayed in order of magnitude;* $\mu = 66.61$, $\sigma^2 = 124.4779$, $\sigma = 11.1597$; *distribution departs strongly from normality.*

(1)	(2)	(3)	(1)	(2)	(3)	(1)	(2)	(3)	(1)	(2)	(3)	(1)	(2)	(3)
01	36	51	21	42	58	41	45	61	61	47	67	81	49	76
02	37	51	22	42	58	42	45	61	62	47	67	82	49	76
03	38	51	23	42	58	43	45	61	63	47	68	83	49	79
04	38	53	24	43	58	44	45	61	64	47	68	84	49	80
05	39	53	25	43	58	45	45	61	65	47	69	85	50	80
06	39	53	26	43	58	46	45	62	66	47	69	86	50	81
07	40	54	27	43	58	47	45	62	67	47	69	87	50	82
08	40	55	28	43	58	48	45	62	68	47	69	88	50	82
09	40	55	29	43	58	49	45	62	69	47	69	89	50	82
10	40	56	30	43	58	50	45	63	70	48	69	90	50	82
11	41	56	31	43	58	51	46	63	71	48	70	91	51	83
12	41	56	32	44	59	52	46	63	72	48	72	92	51	85
13	41	57	33	44	59	53	46	64	73	48	73	93	51	87
14	41	57	34	44	59	54	46	65	74	48	73	94	51	88
15	41	57	35	44	60	55	46	65	75	48	74	95	52	88
16	41	57	36	44	60	56	46	65	76	48	74	96	52	89
17	42	57	37	44	60	57	46	65	77	48	74	97	53	93
18	42	57	38	44	60	58	46	65	78	49	74	98	53	94
19	42	57	39	44	60	59	46	67	79	49	75	99	54	96
20	42	57	40	44	61	60	46	67	80	49	76	100	55	98

SOURCE: Column 2—Data adapted from Sokal and Hunter (1955). Column 3—Data from Canadian government records.

second population deviates strongly from normality. It is a frequency distribution of the total annual milk yield of 100 Jersey cows. Both populations are shown in Table 6.1. You are to sample from them with replacement in order to simulate sampling from an infinite population. Obtain samples of 35 items from each of the two populations by obtaining two sets of 35 two-digit random numbers from the table of random numbers (Statistical Table **FF**), with which you became familiar in Experiment 5.1. Write down the random numbers in blocks of 5, and copy next to them the value of Y (for either wing length or milk yield) corresponding to the random number. An example of such a block of 5 numbers and the computations required for it, based on the housefly wing lengths, is shown here:

Random number	Wing length Y
16	41
59	46
99	54
36	44
21	42
$\bar{Y} = 45.4$	
$s = 5.1769$	

By calculating $\bar{Y}$ and s separately for each sample of 5 and then jointly for each sample of 35, we obtain quantities that will be useful to us in the next chapter. Therefore, preserve your data carefully! Sampling experiments such as this one can be performed more efficiently by computer. The BIOMLAB program, for example, can draw samples from various theoretical distributions and can calculate means, variances, and so on for each sample.

In this experiment consider the 35 variates for each variable as a single sample, without breaking them down into groups of 5. Since the true mean and standard deviation (μ and σ) of the two distributions are known, you can calculate the expression $(Y_i - \mu)/\sigma$ for each variate Y_i. Thus for the first housefly wing length sampled above, you compute

$$\frac{41 - 45.5}{3.90} = -1.1538$$

This result means that the first wing length is 1.1538 standard deviations below the true mean of the population. The deviation from the mean measured in standard deviation units is called a *standardized deviate,* or **standard deviate.** The arguments of Statistical Table **A,** expressing distance from the mean in units of σ, are called **standard normal deviates.** Group all 35 variates in a frequency distribution; then do the same for milk yields. Since you know the parametric mean and standard deviation, you need not compute each deviate separately, but can simply write down class limits in terms of the actual variable, as well as in standard deviation form. The class limits for such a frequency distribution are shown in Table 6.2. Combine the results of your sampling with those of your classmates and study the percentage of the items in the distribution one, two, and three standard deviations to each side of the mean. Note the marked differences in distribution between the housefly wing lengths and the milk yields.

Table 6.2 SETUP FOR RECORDING FREQUENCY DISTRIBUTIONS OF STANDARD DEVIATES, $(Y_i - \mu)/\sigma$, FOR SAMPLES OF EXPERIMENT 6.1.

Wing lengths			Milk yields		
Class limit	Included variates	f	Class limit	Included variates	f
$-\infty$			$-\infty$		
-3σ			-3σ		
$-2\frac{1}{2}\sigma$			$-2\frac{1}{2}\sigma$		
-2σ	36, 37		-2σ		
$-1\frac{1}{2}\sigma$	38, 39		$-1\frac{1}{2}\sigma$		
$-\sigma$	40, 41		$-\sigma$	51–55	
$-\frac{1}{2}\sigma$	42, 43		$-\frac{1}{2}\sigma$	56–61	
$\mu = 45.5$	44, 45		$\mu = 66.61$	62–66	
$\frac{1}{2}\sigma$	46, 47		$\frac{1}{2}\sigma$	67–72	
σ	48, 49		σ	73–77	
$1\frac{1}{2}\sigma$	50, 51		$1\frac{1}{2}\sigma$	78–83	
2σ	52, 53		2σ	84–88	
$2\frac{1}{2}\sigma$	54, 55		$2\frac{1}{2}\sigma$	89–94	
3σ			3σ	95–98	
$+\infty$			$+\infty$		

6.3 A Model for the Normal Distribution

When we study continuous frequency distributions in nature, we find that a large proportion of them approximate the normal distribution. Why? There are several ways of generating the normal frequency distribution from models based on elementary assumptions. Many of these require more mathematics than we expect of our readers. We therefore use a largely intuitive approach, which we have found of heuristic value.

Let us consider a binomial distribution of the familiar form $(p + q)^k$, in which k becomes indefinitely large. What type of biological situation could give rise to such a binomial distribution? An example is one in which many factors cooperate additively to produce a biological result. The following hypothetical case is perhaps not too far removed from reality. The intensity of skin pigmentation in an animal is due to the summation of many factors, some genetic, others environmental. To simplify this example let us assume that every factor can occur in two states only: present or absent. When the factor is present it contributes one unit of pigmentation to skin color, but it contributes nothing to pigmentation when it is absent. Each factor, regardless of its nature or origin, has the same effect, and the effects are additive; if three out of five possible factors are present in an individual, the pigmentation intensity would be three units (the sum of three contributions of one unit each). Let us also assume that each factor has an equal probability of being present or absent in a given individual. Thus, $p_F = 0.5$, the probability of the factor being present, while $q_f = 0.5$, the probability of the factor being absent.

With only one factor ($k = 1$), expansion of the binomial $(p_F + q_f)^1$ would yield the following two pigmentation classes among the animals:

F	*f*	pigmentation classes
0.5	0.5	expected frequency
1	0	pigmentation intensity

Half the animals would have intensity 1, the other half 0. With $k = 2$ factors present in the population (the factors are assumed to occur independently of each other), the distribution of pigmentation intensities would be represented by the expansion of the binomial $(p_F + q_f)^2$:

FF	*Ff*	*ff*	pigmentation classes
0.25	0.50	0.25	expected frequency
2	1	0	pigmentation intensity

One-fourth of the individuals would have pigmentation intensity 2, one-half 1, and the remaining one-fourth 0.

The expected frequencies for four, six, and ten factors are graphed as histograms in Figure 6.4 (rather than as bar diagrams, as they should be drawn). They approach the familiar bell-shaped outline of the normal frequency distribution (see Figures 6.2 and 6.3). Were we to expand the expression for $k = 20$, our histogram would be so close to a normal frequency distribution that we could not show the difference between them on a graph the size of this page. Of course, simply stating that the expected frequencies of the binomial approach the shape of a normal probability distribution is not acceptable mathematical proof. The

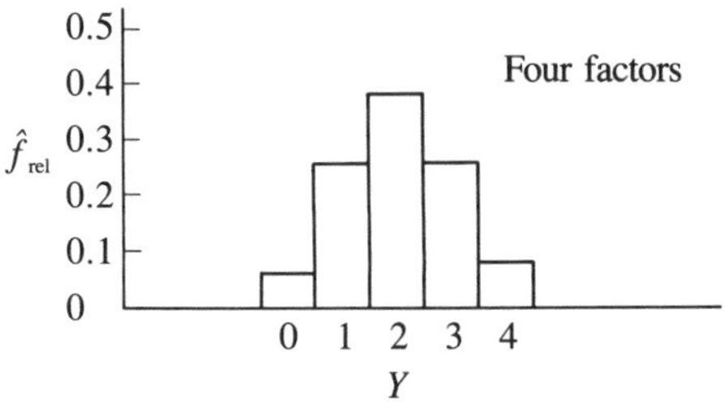

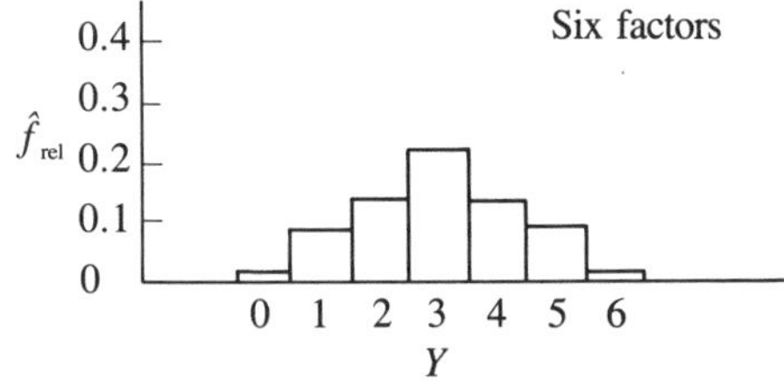

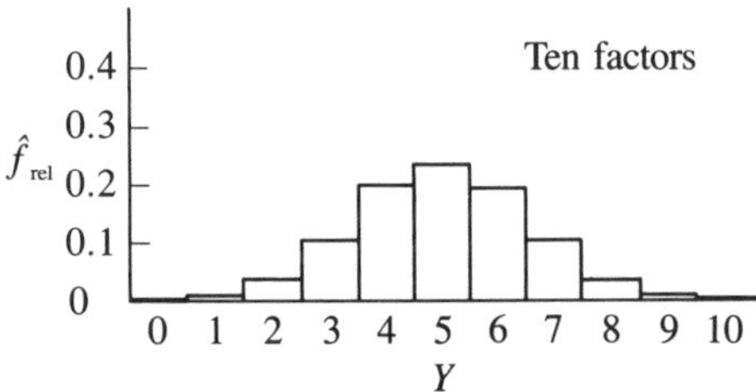

FIGURE 6.4 *Histograms showing relative expected frequencies of* Y, the pigmentation intensity for $k = 4$, 6, and 10 factors. These frequencies are based on the expansion of the binomial $(0.5 + 0.5)^k$.

mathematically inclined reader will find a relatively simple proof of this proposition in Section 2.7 of Ross (1985). Readers not so inclined must take it on faith that the model postulated above, the result of many factors acting independently and additively, will approach normality. At the beginning of this procedure, however, we made some severely limiting assumptions for the sake of simplicity. Let us now return to these and see if we can remove them.

1. The two states of the factors occur in equal frequency—that is, $p = q$. When $p \neq q$, the distribution also approaches normality as k approaches infinity. This is intuitively difficult to see because when $p \neq q$ the histogram is asymmetrical at first. It can be shown mathematically, however, that when k, p, and q are such that $kpq \geq 3$, the normal distribution will be closely approximated.

2. Factors occur in two states only—present or absent. In a more realistic situation factors would occur in several states, one state making a large

contribution, a second state making a smaller contribution, and so forth. It can also be shown, however, that the multinomial $(p + q + r + \cdots + z)^k$ approaches the normal frequency distribution as k approaches infinity.

3. Each factor has the same effect. In reality, different factors may be present in different frequencies and also may have different quantitative effects. As long as these are additive and independent, normality is still approached when k approaches infinity.

Lifting these restrictions makes the assumptions leading to a normal distribution compatible with innumerable biological situations. It is therefore not surprising that many biological variables are approximately normally distributed.

Let us summarize the conditions that tend to produce normal frequency distributions: (1) if there are many factors that are single or composite; (2) if these factors are independent in occurrence; (3) if the factors are independent in effect—that is, if their effects are additive; and (4) if they make equal contributions to the variance. The fourth condition we are not yet in a position to discuss and mention here only for completeness; it will be discussed in Chapter 8.

6.4 Applications of the Normal Distribution

The normal frequency distribution is the most widely used distribution in statistics, and we will have recourse to it in a variety of situations. For the moment we may subdivide its applications as follows:

1. We sometimes have to know whether a given sample is distributed normally before we can apply a certain test to it. To test a sample for normal distribution, we have to calculate expected frequencies for a normal curve of the same mean and standard deviation. The next section will show you how to do this, and Section 6.7 provides some shortcuts.

2. Knowing whether a sample is distributed normally may confirm or reject certain underlying hypotheses about the nature of the factors affecting the phenomenon studied. Refer to the conditions for normality in a frequency distribution (see Section 6.3). Thus if we find a given variable to be distributed normally, we have no reason for rejecting the hypothesis that the causal factors affecting the variable are additive and independent and of equal variance. On the other hand, departures from normality may indicate that certain forces, such as selection, are affecting the variable under study. Skewness of milk-yield data may reflect the fact that these were records of selected cows and that substandard milk cows were not included in the records. Bimodality (see Section 4.4) may indicate a mixture of observations from two populations. Even though measurement variables from a sample are distributed normally, areas and volumes from such samples will not be distributed normally. This is because products and nonlinear transformations of normally distributed variables will not themselves be distributed normally. In many cases transformations of nonnormal variables (see Chapter 13) change the distribution of the transformed variable to normality.

3. If we assume a given distribution to be normal, we may make predictions and tests of given hypotheses based upon this assumption. The following is an example of such an application:

Recall the birth weights of male Chinese children, last illustrated in Box 4.3. The mean of this sample of 9465 birth weights is 109.9 oz, and its standard deviation is 13.593 oz. Sampling at random from the birth records of this population, what is the chance of obtaining a birth weight of 151 oz or heavier? Such a birth weight is considerably above the mean of our sample (the difference is $151 - 109.9 = 41.1$ oz). With a difference in ounces, however, we cannot consult the table of areas of the normal curve. We must **standardize** the difference—that is, divide it by the standard deviation to convert it into a standard deviate. When we divide the difference by the standard deviation we obtain $41.1/13.593 = 3.02$. Thus a birth weight of 151 oz is 3.02 standard deviation units greater than the mean.

Assuming that the birth weights are distributed normally, we may consult the table of areas of the normal curve (Table **A**), where we find a value of 0.4987 for 3.02 standard deviations. This value means that 49.87% of the area of the curve lies between the mean and a point 3.02 standard deviations from it. Conversely, 0.0013 or 0.13% of the area lies beyond 3.02 standard deviation units above the mean. Thus, assuming that we have a normal distribution of birth weights and that $\sigma = 13.593$, only 0.13%, or 13 out of 10,000 of the infants, would have a birth weight of 151 oz *or* a weight *greater than* the mean. That a single sampled item from that population would deviate by so much from the mean is quite improbable. If a random sample of one such weight was obtained from the records of an unspecified population, we might therefore be justified in doubting that the observation came from the population known to us.

The probability in this example was calculated from one tail of the distribution. We found the probability that an individual is 3.02 standard deviations *greater* than the mean. If we have no prior knowledge that the individual will be either heavier or lighter than the mean but are merely concerned with how different it is from the population mean, an appropriate question would be, Assuming that the individual belongs to the population, what is the probability of observing a birth weight of an individual as deviant from the mean in either direction? This probability, which must be computed by using both tails of the distribution, can be obtained simply by doubling the previous probability, since the normal curve is symmetrical. Thus, $2 \times 0.0013 = 0.0026$. This value, too, is so small that we would conclude that a birth weight as deviant as 151 oz is unlikely to have come from the population represented by our sample of male Chinese children.

We can learn one more important point from this example. Our assumption has been that the birth weights are distributed normally. Inspection of the frequency distribution in Box 4.3, however, shows that the distribution is assymmetrical, tapering to the right. Although there are eight classes above the mean class, there are only six classes below it. In view of this asymmetry, conclusions about one tail of the distribution would not necessarily pertain to the second tail. We calculated that 0.13% of the items would be found beyond 3.02 standard deviations above the mean, which corresponds to 151 oz. Our sample, however, contains 20 items (14 + 5 + 1) beyond the 147.5-oz class, the upper limit of which is 151.5 oz. These 20 items (out of the 9465 total) constitute approximately 0.21% of the sample, more than the 0.13% expected from the normal frequency distribution. Although finding a single birth weight as heavy as 151 oz would still be improbable, conclusions based on the assumption of normality might be in error if the exact probability were critical for a given test. Our statistical conclusions are only as valid as our assumptions about the data.

6.5 Fitting a Normal Distribution to Observed Data

A normal distribution is fitted as a curve superimposed upon an observed frequency distribution in the form of a histogram usually only when graphic facilities (plotters) are available. Ordinates are computed by modifying Expression (6.1) to conform to a frequency distribution:

$$Z = \frac{ni}{s\sqrt{2\pi}} e^{-\frac{1}{2}[(Y-\bar{Y})/s]^2} \tag{6.2}$$

In this expression n is the sample size and i the class interval of the frequency distribution. (We show the normal curve superimposed upon the Chinese birth-weight data in Figure 6.7A.)

More common is the technique of calculating expected frequencies for a normal distribution of the same mean and standard deviation as the observed sample. This is the same approach we used for calculating expected binomial or Poisson frequencies (see Chapter 5). The procedure is shown in Box 6.1 for the frequency distribution of birth weights of male Chinese children (see Box 4.3). As with the binomial and Poisson distributions, we have not yet learned a technique for testing the departures from expectation (see see Chapter 17). General inspection shows that the data do not fit a normal distribution very well, but seem skewed to the right (see Section 6.6). Graphic comparisons of observed and expected frequencies are discussed in Section 6.7.

6.6 Skewness and Kurtosis

The previous section showed how to calculate expected frequencies for a normal distribution. In many cases an observed frequency distribution departs obviously from normality; thus statistics that measure the nature and amount of departure are useful. We will emphasize two types of departure from normality. One is **skewness,** which is another name for asymmetry; skewness means that one tail

Box 6.1 CALCULATION OF EXPECTED NORMAL FREQUENCIES.

Birth weights of male Chinese in ounces (from Box 4.3): $n = 9465$; $\overline{Y} = 109.8996$; $s = 13.5942$.

	(1)	(2)	(3)	(4)	(5)	(6)	(7)	(8)
	Y	Lower implied limit of class Y_L	$\frac{\lvert Y_L - \overline{Y}\rvert}{s}$	Area between (3) and $\overline{Y}$	Relative expected frequencies $\hat{f}_{rel}$	Absolute expected frequencies $\hat{f}$	Observed frequencies f	Deviation from expectation $f - \hat{f}$
	$-\infty$	$-\infty$	∞	0.5000				
	59.5	55.5	4.00	0.49997	0.0003	2.8	2	−
	67.5	63.5	3.41	0.4997	0.0021	19.9	6	−
	75.5	71.5	2.82	0.4976	0.0101	95.6	39	−
	83.5	79.5	2.24	0.4875	0.0370	350.2	385	+
	91.5	87.5	1.65	0.4505	0.0951	900.1	888	−
Mean class	99.5	95.5	1.06	0.3554	0.1746	1652.6	1729	+
Mean class	107.5	103.5	0.47	0.1808	0.2286	2163.7	2240	+
Mean class	115.5	111.5	0.12	0.0478	0.2134	2019.8	2007	−
	123.5	119.5	0.71	0.2612	0.1403	1327.9	1233	−
	131.5	127.5	1.29	0.4015	0.0684	647.4	641	−
	139.5	135.5	1.88	0.4699	0.0233	220.5	201	−
	147.5	143.5	2.47	0.4932	0.0057	54.0	74	+
	155.5	151.5	3.06	0.4989	0.0010	9.5	14	+
	163.5	159.5	3.65	0.4999	0.0001	0.9	5	+
	171.5	167.5	4.24	0.5000	0.0000	0.0	1	+
	179.5	175.5	4.83	0.5000				
	$+\infty$	$+\infty$	∞	0.5000				
					1.0000	9464.9	9465	

Computation

1. In column (1) record class marks in increasing order of magnitude of Y. Add one class mark beyond those observed (in this case the class mark 179.5). Bound the column by $-\infty$ and $+\infty$ as shown.
2. In column (2) record the lower implied limit Y_L of the class represented by the class mark in column (1). Once the first lower noninfinite limit is found correctly (55.5 in this case), the others can be obtained by successive addition of the class interval (in this case 8).
3. Divide the absolute magnitude of $Y_L - \bar{Y}$ (that is, with sign neglected) by the calculated standard deviation s of the sample and record the result in column (3). The deviations of the lower class limits from the mean are now expressed in standard deviation units (they are standard deviates).
4. Consult the table of areas of the normal curve (Table **A**) and enter in column (4) the area corresponding to the argument in column (3). Thus in the first two rows 4.00 standard deviations give you 0.49997 of the area, and 3.41 standard deviations give you 0.4997.
5. To calculate the relative expected frequencies, $\hat{f}_{rel}$, start at the head of column (5) and for the second row compute the difference, *sign neglected*, of the area from column (4) in that row and in the one just beneath it. Thus, to obtain the relative expected frequency 0.0003 in column (5), we compute $0.49997 - 0.4997$. Continue in this manner to the bottom of the table. Thus, the value of $\hat{f}_{rel} \neq (0.0233)$ for the 139.5-oz class results from subtracting 0.4699 from 0.4932. Unless the mean falls exactly on a class limit, the row for the class containing the mean is computed differently. To obtain its relative expected frequency, *sum* the areas in column (4) for the class containing the mean and for the class in the line beneath it. In this case $0.1808 + 0.0478 = 0.2286$. The sum of the relative expected frequencies is exactly 1.0000, which is a good check on our computations. If the sum is not close to 1.0, check for computational errors. If the sum is correct and less than 1.0, you may need to extend the frequency distribution by one or more classes at each tail.
6. To obtain the absolute expected frequencies in column (6), multiply the relative expected frequencies by $n = 9465$. Their sum is very close to n, which again furnishes a check on the procedure.
7. In column (7) we show the observed frequencies and in column (8) the signs of the deviations from expectation. Note that classes above the mean generally show an excess and those below the mean show a shortage of observed frequencies with respect to expectation. This is what you would expect in a distribution skewed to the right.

of the curve is drawn out more than the other. In such curves the mean and the median do not coincide. Curves are called skewed to the right or left, depending upon whether the right or left tails are drawn out.

The other type of departure from normality, **kurtosis,** is a more complicated change in distribution. If a symmetrical distribution is considered to have a center, two shoulders, and two tails, kurtosis describes the proportions found in the center and in the tails in relation to those in the shoulders. A **leptokurtic curve** has more items near the center and at the tails, with fewer items in the shoulders relative to a normal distribution with the same mean and variance. A **platykurtic curve** has fewer items at the center and at the tails than the normal curve but has more items in the shoulders. A bimodal distribution is an extreme platykurtic distribution.

Both skewness and kurtosis are vague concepts (as, incidentally, are location and dispersion, or scale) with many ways of being evaluated. For a review of approaches to kurtosis, see Balanda and MacGillivray (1988). In this book we use the conventional sample statistics for measuring skewness and kurtosis, called g_1 and g_2, to represent population parameters γ_1 and γ_2. Their computation is tedious and should always be done by computer. Program BIOM-pc computes g_1 and g_2 when evaluating basic statistics for both ungrouped and grouped data.

The formulas for g_1 and g_2 involve **moment statistics.** A central moment in statistics, as in physics, is $(1/n)\Sigma^n(Y - \bar{Y})^r$, the average of the deviations of all items from the mean, each raised to the power r. The first central moment, $(1/n)\Sigma(Y - \bar{Y})$, always equals zero, as we saw in Section 4.6. The second moment, $(1/n)\Sigma(Y - \bar{Y})^2$, is the variance. The statistic g_1 is the third central moment divided by the cube of the standard deviation, $(1/ns^3)\Sigma(Y - \bar{Y})^3$, and g_2 is 3 less than the fourth central moment divided by the fourth power of the standard deviation, $(1/ns^4)\Sigma(Y - \bar{Y})^4 - 3$. The computation of these two statistics is outlined in Box 6.2 for the birth weights of Chinese males. Just as the sample variance had to be corrected for bias by dividing $\Sigma(Y - \bar{Y})^2$ by $n - 1$ rather than n, so sample g_1 and g_2 need to be corrected to allow for similar bias. These corrections are complicated and are included in Box 6.2. These statistics can also be computed for ungrouped data, in which case the formulas in Box 6.2 would simply drop the f's.

In a normal frequency distribution both γ_1 and γ_2 are zero. A negative g_1 indicates skewness to the left, a positive g_1 skewness to the right. A negative g_2 indicates platykurtosis, while a positive g_2 shows leptokurtosis. Thus a repulsed distribution would have a positive g_2, but a clumped distribution should have a negative g_2. The absolute magnitudes of g_1 and g_2 do not mean much; these statistics have to be tested for "significance" by methods yet to be learned (see Section 7.9 and Box 7.5). It appears that the birth weights in Box 6.2 are skewed to the right (weights generally are skewed in this manner) and that a very slight tendency to leptokurtosis is also present.

When only small samples are taken from highly skewed populations $|g_1|$ will tend to underestimate $|\gamma_1|$, since in a sample of size n it can be shown that

Box 6.2 COMPUTATION OF g_1 AND g_2 FROM A FREQUENCY DISTRIBUTION.

These computations are based on the birth weights of male Chinese in ounces, arranged as in the first two columns of Box 4.3.

Basic sums

$n = \Sigma f = 9465 \qquad \Sigma Y = \Sigma f Y = 1{,}040{,}199.5$

$\bar{Y} = \frac{1}{n}\Sigma f Y = 109.8996 \qquad \Sigma y^2 = \Sigma f(Y - \bar{Y})^2 = 1{,}748{,}956.7983$

$\Sigma y^3 = \Sigma f(Y - \bar{Y})^3 = 4{,}501{,}097.3779$

$\Sigma y^4 = \Sigma f(Y - \bar{Y})^4 = 998{,}109{,}887.1459$

Computation

1. $s^2 = \frac{\sum y^2}{n-1} = \frac{1{,}748{,}956.7983}{9464} = 184.8010$

$s = 13.5942$

2. $g_1 = \frac{n\sum y^3}{(n-1)(n-2)s^3} = \frac{9465(4{,}501{,}097.3779)}{(9464)(9463)(13.5942)^3} = 0.18936$

3. $g_2 = \frac{(n+1)n\sum y^4}{(n-1)(n-2)(n-3)s^4} - \frac{3(n-1)^2}{(n-2)(n-3)}$

$= \frac{(9466)(9465)(998{,}109{,}887.1459)}{(9464)(9463)(9462)(13.5942)^4} - \frac{3(9464)^2}{(9463)(9462)}$

$= 0.08913$

4. If the variable Y has been coded, g_1 and g_2 do not require decoding, since they are scale-free. In a normal frequency distribution γ_1 and γ_2 are both zero. A negative g_1 indicates skewness to the left, a positive g_1 skewness to the right. A negative g_2 indicates platykurtosis, and a positive g_2 shows leptokurtosis.

Compare these results with the graphic analysis in Box 6.3.

$|g_1| \leq \sqrt{n}$ (Wallis et al., 1974). For the sample of 5 milk yields corresponding to the random numbers in the instructions for Experiment 6.1, $g_1 = 1.944$, which is close to its upper bound, $\sqrt{5} = 2.236$. Similar problems exist with estimates of γ_2 in small samples from highly kurtotic populations. The bounds on g_2 are

$$\frac{-2(n-1)}{n-3} \leq g_2 \leq n$$

Thus large samples must be obtained to estimate adequately the parameters of strongly skewed or kurtotic populations.

6.7 Graphic Methods

Many computer programs furnish graphs of the data as byproducts of their computations of statistics describing the frequency distributions. This feature permits visual inspection of the data and is often useful for suggesting the way in which a sample deviates from expectation (such as bimodality or the presence of outliers).

We begin by describing graphs based on cumulative frequency distributions. We first met cumulative frequencies in Box 4.1, when computing the median. In Figure 6.3 we saw that a normal frequency distribution graphed in cumulative fashion describes an S-shaped curve, called a sigmoid curve. If we drop perpendiculars to the abscissa from the cumulative normal curve, at the level of a given proportion we obtain the corresponding quantile (see Section 4.3). In Figure 6.3 the quantile is scaled as the number of standard deviations above or below the mean equivalent to a given proportion. For a normal distribution, these cumulative proportions, transformed into standard deviation scale, are called **normal**

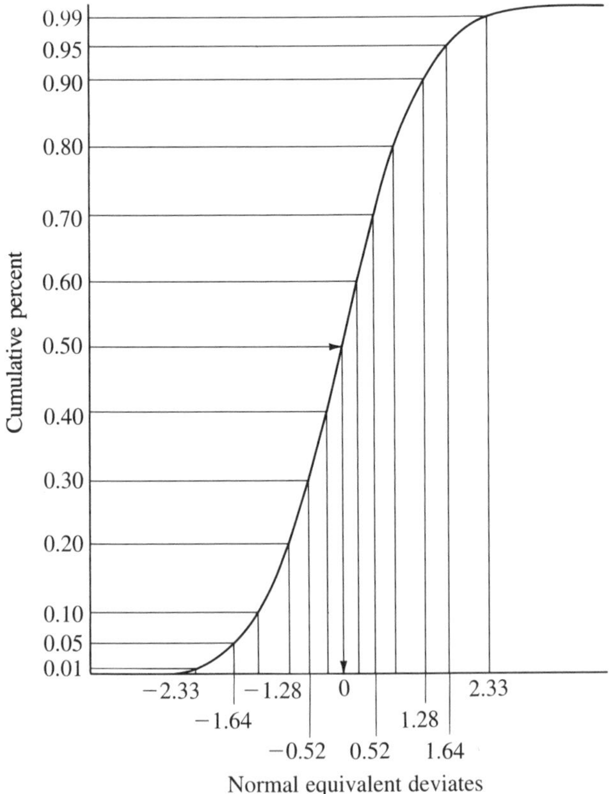

FIGURE 6.5 *Transformation of cumulative percentages into normal equivalent deviates.*

equivalent deviates (NEDs). We show such a transformation in Figure 6.5. Note that there are no 0% or 100% points on the ordinate. These values are not possible since the normal frequency distribution extends from negative to positive infinity, so however long we made our line we would never reach the limiting values of 0% and 100%.

We now construct a graph whose abscissa is in the measurement scale Y of some observed variable, and whose ordinate is in NED units. Assume that we plot a normal frequency distribution above the abscissa. If we measure off suitably spaced quantiles along the abscissa, transform them into NEDs, and plot these points with respect to the measurement and NED scales, the points should fall along a straight line, since the NED transformation is predicated on the area under a normal curve (Figure 6.6A). Therefore, if we take an observed sample

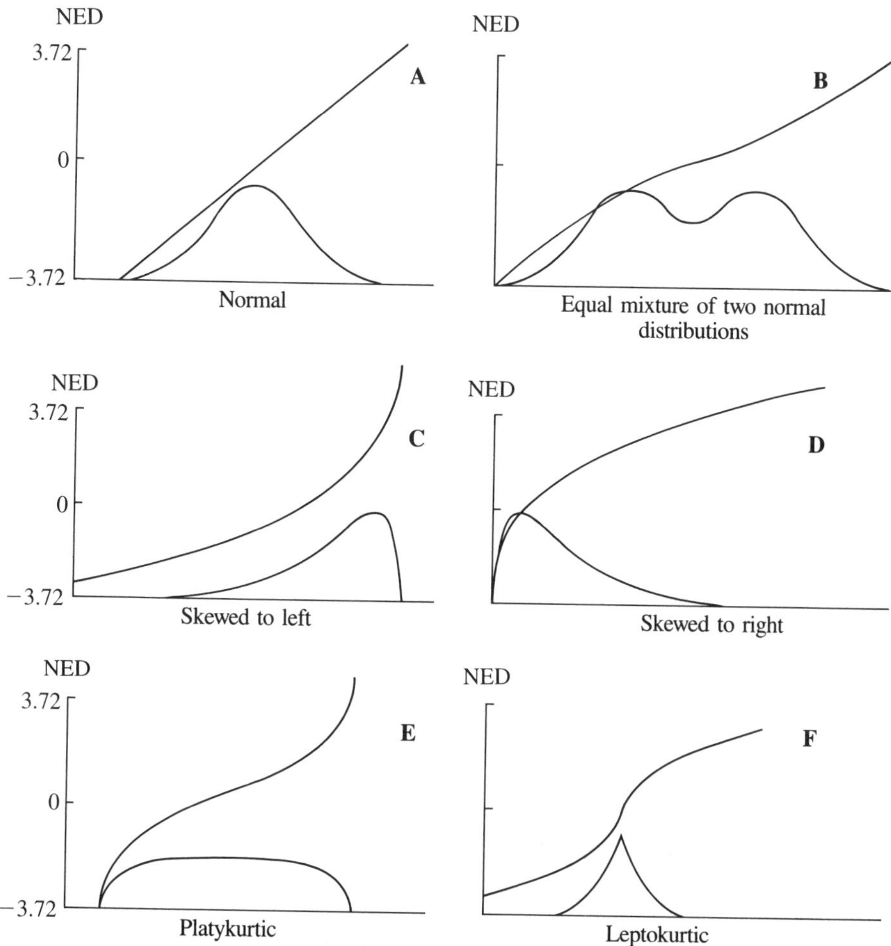

FIGURE 6.6 *Examples of frequency distributions with their cumulative distributions plotted as normal quantile plots. (See Box 6.3 for explanation.)*

with an unknown frequency distribution, plot it in this manner, and find that NEDs against variables fall more or less along a straight line, we may conclude that the sample is consistent with a normal distribution. Such graphs have become known as *quantile–quantile,* or *Q–Q, plots* or as **normal quantile plots** the term we prefer. Nowadays such plots are usually produced by computer programs. Box 6.3, part I, explains what these programs present, and Figure 6.6 illustrates how to interpret departures from normality in normal quantile plots.

Normal quantile plots work best for fairly large ($n > 50$) samples (see Box 6.3, part II). In smaller samples a difference of one item per class would make a substantial difference in the cumulative percentage in the tails. For small samples (< 50) the method of *ranked normal deviates,* or **rankits**, is preferable. With this method, instead of quantiles we use the ranks of each observation in the sample, and instead of NEDs we plot values from a table of rankits, the average positions

Box 6.3 GRAPHIC TEST FOR NORMALITY OF SAMPLES BY MEANS OF NORMAL QUANTILE PLOTS.

I. *Routine computer-processed samples.*

Computation

1. Array samples in order of increasing magnitude of variates.
2. For each variate compute the quantity $p = (i - \frac{1}{2})/n$, where i is the rank order of the ith variate in the array, and n is the sample size. These values will be used for computing NEDs. The correction of $\frac{1}{2}$ prevents the last variate from yielding $p = 1.0$, for which the NED would be positive infinity. For tied values of i, calculate an average p.
3. For each value of p evaluate the corresponding NED. If no computer program is available, the NEDs can be looked up in a table of the inverse cumulative normal distribution (e.g., Table 4 in Pearson and Hartley, 1958, or Table 1.2 in Owen, 1962), or by inverse interpolation in Statistical Table **A.**
4. Plot the NEDs against the original variates and examine the scatterplot visually for linearity. Some programs also plot the straight line expected when the sample is perfectly normally distributed, to serve as a benchmark against which the observed scatterplot can be judged. Alternatively, a straight line is fitted to the points by eye, preferably using a transparent plastic ruler, which permits all the points to be seen as the line is drawn. In drawing the line, most weight should be given to the points between cumulative frequencies of 25% to 75% because a difference of a single item may make appreciable changes in the percentages at the tails. Some programs plot NEDs along the abscissa and the variable Y along the ordinate. We prefer the more common arrangement shown in Figure 6.6 and the figures in this box.

Figure A shows 1400 housefly wing lengths randomly sampled from Table 6.1 as a normal quantile plot. Since these are approximately normal data, it is no surprise that the scatterplot forms a nearly straight line.

BOX 6.3 (CONTINUED)

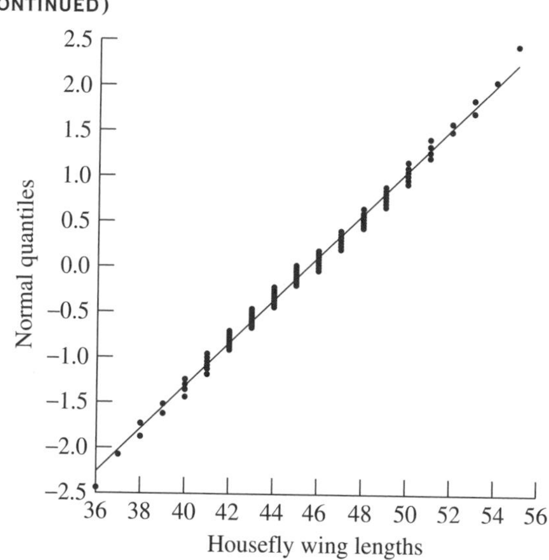

FIGURE A *Normal quantile plot of 1400 random samples of 5 housefly wing lengths from Table 6.1.*

II. *Large frequency distributions*

When the sample is large, plotting every observation may not be worthwhile, and the normal quantile plot can be applied to the frequency distribution as follows. We employ the by now thoroughly familiar Chinese birth weights.

Birth weights of male Chinese in ounces, from Box 4.3.

(1) Class mark Y	(2) Upper class limit	(3) f	(4) Cumulative frequencies F	(5) $p = (F - \frac{1}{2})/n$
59.5	63.5	2	2	0.0002
67.5	71.5	6	8	0.0008
75.5	79.5	39	47	0.0049
83.5	87.5	385	432	0.0456
91.5	95.5	888	1320	0.1394
99.5	103.5	1729	3049	0.3221
107.5	111.5	2240	5289	0.5587
115.5	119.5	2007	7296	0.7708
123.5	127.5	1233	8529	0.9011
131.5	135.5	641	9170	0.9688
139.5	143.5	201	9371	0.9900
147.5	151.5	74	9445	0.9978
155.5	159.5	14	9459	0.9993
163.5	167.5	5	9464	0.9998
171.5	175.5	1	9465	0.9999
		9465		

Box 6.3 (continued)

Computation

1. Prepare a frequency distribution as shown in columns (1), (2), and (3).
2. Form a cumulative frequency distribution as shown in column (4). It is obtained by successive summation of the frequency values. In column (5) express the cumulative frequencies as p-values using the formula in step 2 of part **I.**
3. Graph the NEDs, quantiles, corresponding to these p-values against the upper class limit of each class (Figure B). Notice that the upper frequencies deviate to the right of the straight line. This is typical of data that are skewed to the right (see Figure 6.6D).
4. The graph from step **3** (and those described in the other parts of this box) can also be used for rapid estimation of the mean and standard deviation of a sample. The mean is approximated by a graphic estimation of the median. The more normal the distribution is, the closer the mean will be to the median. The median is estimated by

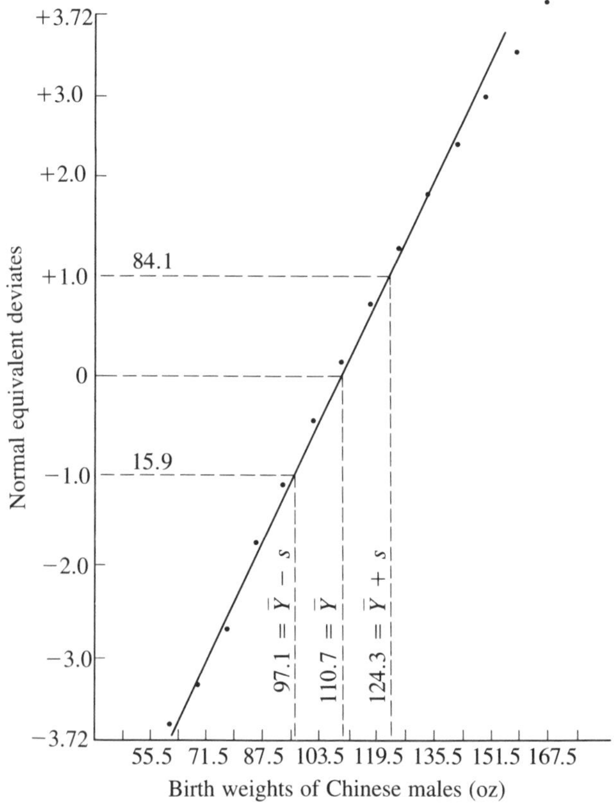

FIGURE B *Graphic analysis of data.*

Box 6.3 (continued)

dropping a perpendicular from the intersection of the 0 point on the ordinate and the cumulative frequency curve to the abscissa (see Figure B). The estimate of the mean of 110.7 oz is quite close to the computed mean of 109.9 oz.

5. The standard deviation is estimated by dropping similar perpendiculars from the intersections of the -1 and the $+1$ points with the cumulative curve, respectively. These points enclose the portion of a normal curve represented by $\mu \pm \sigma$. By measuring the difference between these perpendiculars and dividing this value by 2, we obtain an estimate of one standard deviation. In this instance the estimate is $s = 13.6$, since the difference is 27.2 oz divided by 2. This is a close approximation to the computed value of 13.59 oz.

III. *Small samples ($n \leq 50$)*

Femur lengths of aphid stem mothers, from Box 2.1: $n = 25$.

(1) Y	(2) Rankits from Table **H**	(3) Rankits allowing for ties	(1) Y	(2) Rankits from Table **H**	(3) Rankits allowing for ties
3.3	−1.97	−1.97	4.1	0.10	0.15
3.5	−1.52	−1.52	4.1	0.20	0.15
3.6	−1.26	−1.00	4.2	0.30	0.30
3.6	−1.07	−1.00	4.3	0.41	0.58
3.6	−0.91	−1.00	4.3	0.52	0.58
3.6	−0.76	−1.00	4.3	0.64	0.58
3.8	−0.64	−0.47	4.3	0.76	0.58
3.8	−0.52	−0.47	4.4	0.91	1.08
3.8	−0.41	−0.47	4.4	1.07	1.08
3.8	−0.30	−0.47	4.4	1.26	1.08
3.9	−0.20	−0.10	4.5	1.52	1.52
3.9	−0.10	−0.10	4.7	1.97	1.97
3.9	0.00	−0.10			

Computation

1. Enter the sample arrayed in increasing order of magnitude in column (1). In column (2) put corresponding rankits from Statistical Table **H.** The table gives only the rankits for the half of each distribution greater than the median for any sample size. The other half is the same but is negative in sign. All samples containing an odd number of variates (such as this one) have 0 as the median value. The rankits for this example are looked up under sample size $n = 25$.

2. A special problem illustrated in this example is the case of ties, or variates of identical magnitude. In such a case we sum the rankit values for the corresponding ranks and find their mean. Thus the -1.00 occupying lines 3 to 6 in column (3) is the

BOX 6.3 (CONTINUED)

average of rankits −1.26, −1.07, −0.91, and −0.76 in column (2), which are the third, fourth, fifth, and sixth rankits for a sample of 25, respectively. Care must be taken when ties include the median, because then we must sum positive as well as negative rankits. For example, four rankits of values −0.24, −0.14, −0.05, and +0.05, respectively, would give an average rankit of −0.095.

3. The rankit values (ordinate) are plotted against the variable (abscissa) (see Figure C), and a straight line is fitted by eye. The values do not lie exactly on a straight line and there is some suggestion of bimodality in the data, as is also shown in the figure at the bottom of Box 2.1.

4. Mean and standard deviation can be computed as shown in part **II.**

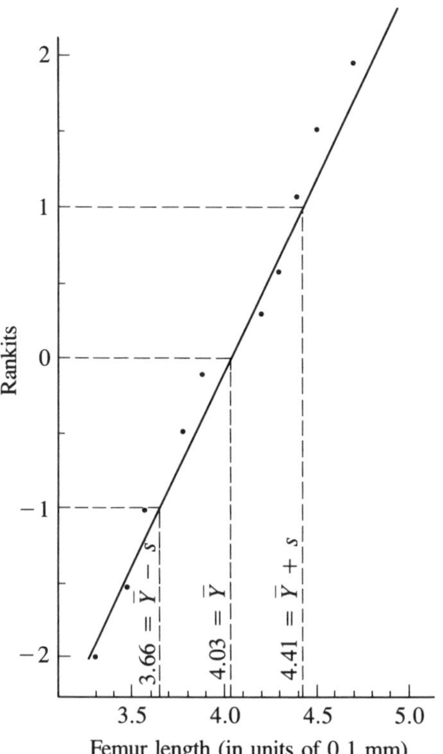

FIGURE C *Graphic analysis of aphid femur length data using rankits.*

in standard deviation units of the ranked items in a normally distributed sample of n items. We could obtain these values empirically by sampling from a standard normal distribution, one whose mean $\mu = 0$, and whose variance (and standard deviation) is $\sigma = \sigma^2 = 1$. If we sample 5 variates from such a distribution, order them from lowest to highest variates, and record their scores, we might get a sample as follows: -1.8, -0.6, -0.1, $+0.9$, $+2.1$. If we repeat this process many times we can calculate the average score of the first item in the ordered sample, of the second, and so forth. Statistical Table **H** shows these average standardized scores, or rankits. We find the rankits for a sample of 5 to be -1.163, -0.495, 0.0, 0.495, 1.163.

These rankits have a variety of uses; one is for testing normality of a frequency distribution, as we will show in a moment. When rankits are plotted against a ranked array of a normally distributed variable, the points will again lie in a straight line. An illustration of the application of rankits to testing for normality and for obtaining estimates of μ and σ is shown in Box 6.3, part III. Figure 6.6 may again be used to interpret the meaning of the resulting curves. Not too much reliance should be placed on single samples, but repeated trends in different samples are likely to be meaningful.

Often we want to compare observed frequency distributions with their expectations without resorting to cumulative frequency distributions. An example is the superimposition of a normal curve on the histogram of an observed frequency distribution as discussed in Section 6.5. In Figure 6.7A we show the frequency distribution of birth weights of male Chinese from Box 6.1 with the ordinates of the normal curve superimposed. An excess of observed frequencies appears at the right tail because of the skewness of the distribution.

Many people find it difficult to compare the heights of bars against the arch of a curve. For this reason John Tukey (see Wainer, 1974) suggested that the bars of the histograms be suspended from the curve. Their departures from expectation are then obvious against the straight abscissa of the graph. Such a **hanging histogram** is shown in Figure 6.7B. The departure from normality is now much clearer.

Because important departures frequently occur in the tails of a curve, it has been suggested that square roots of expected frequencies should be compared with the square roots of observed frequencies. Such a **hanging rootogram** is shown in Figure 6.7C. Note the accentuation of the departure from normality. Finally, one can also use this technique for comparing expected with observed histograms. Figure 6.7D shows the same data plotted in this manner. Square roots of frequencies are again shown. The excess of observed over expected frequencies in the right tail of the distribution is quite evident.

6.8 Other Continuous Distributions

Statisticians have described numerous continuous probability functions and studied their properties. Some observed nonnormal distributions can be made

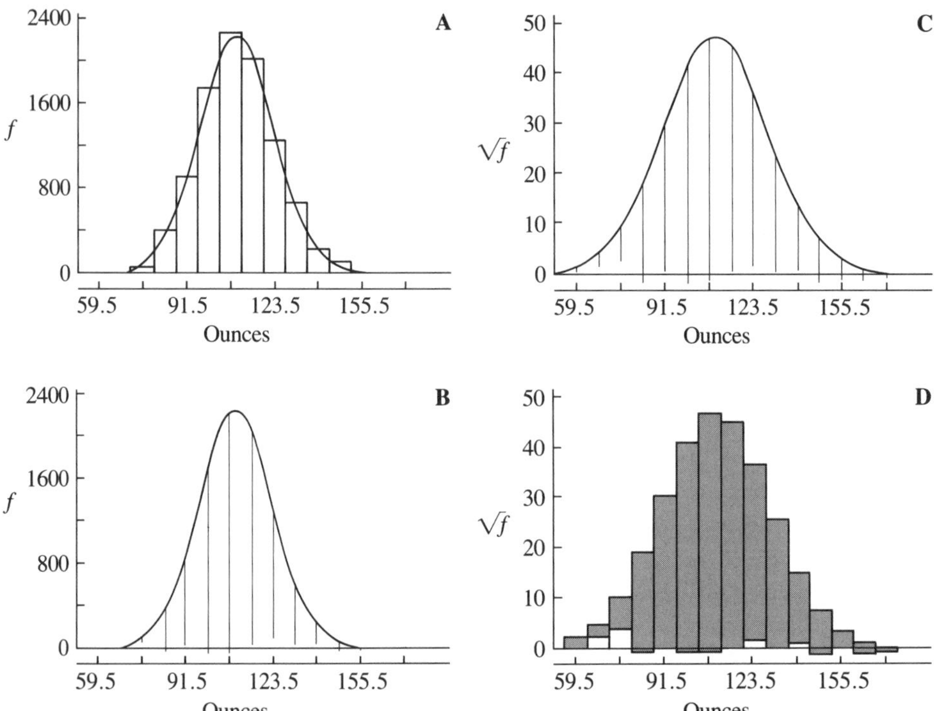

FIGURE 6.7 **A.** Histogram of the observed frequency distribution of birth weights of male Chinese from Box 6.1 with the expected normal curve superimposed. **B.** The data in **A** displayed as a hanging histogram. Frequency bars, narrowed so that they are represented by a single line at the midpoint of each class, are suspended from the expected normal curve. Lines that do not reach the abscissa indicate deficiencies from expectations. Lines extending below the abscissa indicate observed frequencies in excess of expectation. **C.** The data in **B** shown as a hanging rootogram. Observed and expected frequencies are given as square roots of the actual values. Departures from expectation in the tails of the distribution are thus accentuated. **D.** Comparison of the observed and expected frequencies for the birth weight data in **A.** The histogram of the skyline indicates the expected frequencies; that of the "inverted skyline" indicates the observed frequencies. Both frequencies are given in square roots of actual values. Where the inverted skyline does not reach the abscissa, there are fewer observed than expected frequencies. Wherever it reaches below the abscissa, there is an excess of observed frequencies over expected frequencies.

normal by transformation to a suitable scale (see Chapter 13). A common example is the lognormal distribution, which becomes normally distributed when the variates Y_i are transformed to $Y_i' = \ln Y_i$. Quite frequently, morphological variables are lognormally distributed. Even when a variable is normally distributed, various statistics obtained from it (such as the variance) are not normally distributed. We will study several of these distributions and their applications in subsequent chapters.

EXERCISES 6

6.1 Perform the following operations on the data of Exercise 2.4. (a) If you have not already done so, make a frequency distribution from the data and graph the results in the form of a histogram. (b) Compute the expected frequencies for each class based on a normal distribution, with $\mu = \bar{Y}$ and $\sigma = s$. (c) Graph the expected frequencies in the form of a histogram and compare them with the observed frequencies. (d) Comment on the degree of agreement between observed and expected frequencies.

6.2 Carry out the operations in Exercise 6.1 on the transformed data generated in Exercise 2.6.

6.3 Assume that the petal length of a population of plants of species X is normally distributed with a mean of $\mu = 3.2$ cm and a standard deviation of $\sigma = 0.8$. What proportion of the population would be expected to have a petal length (a) greater than 4.5 cm? (b) greater than 1.78 cm? (c) between 2.9 and 3.6 cm? *Answer:* (a) = 0.0521, (b) = 0.9620, and (c) = 0.3373.

6.4 Perform a graphic analysis on the data of Exercises 2.4 and 2.6; examine for normality and estimate the mean and standard deviation. Use a table of rankits (Statistical Table **H**).

6.5 Compute g_1 and g_2 for the data in Exercises 2.4 and 2.6. Do your results agree with what you would expect on the basis of the graphic analyses performed in Exercise 6.4? *Answer:* For Exercise 2.4, $g_1 = 0.2981$ and $g_2 = 0.0512$.

6.6 Perform a graphic analysis (normal quantile plot) of the butterfat data in Exercise 4.3. In addition, plot the data with the abscissa in logarithmic units. Compare the results of the two analyses. Draw hanging rootograms for these two distributions.

6.7 The following data are total lengths (in centimeters) of a collection of bass from a southern lake. Make a histogram of the data, compute g_1 and g_2, and perform the graphic tests. Try to diagnose the departure from normality, if any. *Answer:* Suggestion of bimodality, $g_1 = -0.03691$, $g_2 = 1.42854$.

29.9	40.2	37.8	19.7	30.0	29.7	19.4	39.2	24.7	20.4
19.1	34.7	33.5	18.3	19.4	27.3	38.2	16.2	36.8	33.1
41.4	13.6	32.2	24.3	19.1	37.4	23.8	33.3	31.6	20.1
17.2	13.3	37.7	12.6	39.6	24.6	18.6	18.0	33.7	38.2

6.8 Perform a graphic analysis on the following measurements. Are they consistent with what one would expect in sampling from a normal distribution?

11.44	12.88	11.06	7.02	10.25	6.26	7.92	12.53	6.74
15.81	9.46	21.27	9.72	6.37	5.40	3.21	6.50	3.40
5.60	14.20	6.60	10.42	8.18	11.09	8.74		

6.9 Compute a normal quantile plot for the milk-yield data of Table 6.1. What do we learn from this graph?

7 Estimation and Hypothesis Testing

You are now ready for two very important steps toward learning statistics. In this chapter we provide methods for answering two fundamental statistical questions that all biologists must answer repeatedly in the course of their work: (1) How reliable are the results? (2) How probable is it that the differences between observed results and those expected on the basis of a hypothesis are due to chance alone? The question of reliability is answered by setting confidence limits to sample statistics. The second question involves hypothesis testing. Both subjects belong to the field of statistical inference. The subject matter in this chapter is fundamental to an understanding of all the subsequent chapters. We therefore urge you to study the material until it has been thoroughly mastered.

In Section 7.1 we consider the form of the distribution of means and their variance. In Section 7.2 we examine the distributions and variances of statistics other than the mean. This discussion brings us to the general subject of standard errors—statistics measuring the reliability of an estimate. Confidence limits provide bounds to our estimates of population parameters. We develop the idea of a confidence limit in Section 7.3 and show its application to samples where the true standard deviation is known. However, one usually deals with small, more or less normally distributed samples with unknown standard deviations, in which case the t-distribution must be used. We introduce the t-distribution in Section 7.4. The application of t to the computation of confidence limits for statistics of small samples with unknown population standard deviations is shown in Section 7.5. Another important distribution (chi-square) is explained in Section 7.6 and then applied to setting confidence limits for the variance in Section 7.7. The theory of hypothesis testing is introduced in Section 7.8 and applied to a variety of cases exhibiting the normal or t-distributions in Section 7.9. Finally, Section 7.10 illustrates hypothesis testing for variances by means of the chi-square distribution.

7.1 Distribution and Variance of Means

We commence our study of the distribution and variance of means with a sampling experiment.

EXPERIMENT 7.1. You were asked to retain from Experiment 6.1 the means of the seven samples of 5 housefly wing lengths and the seven similar means of milk yields. We can collect these means from every student in a class, possibly adding them to the

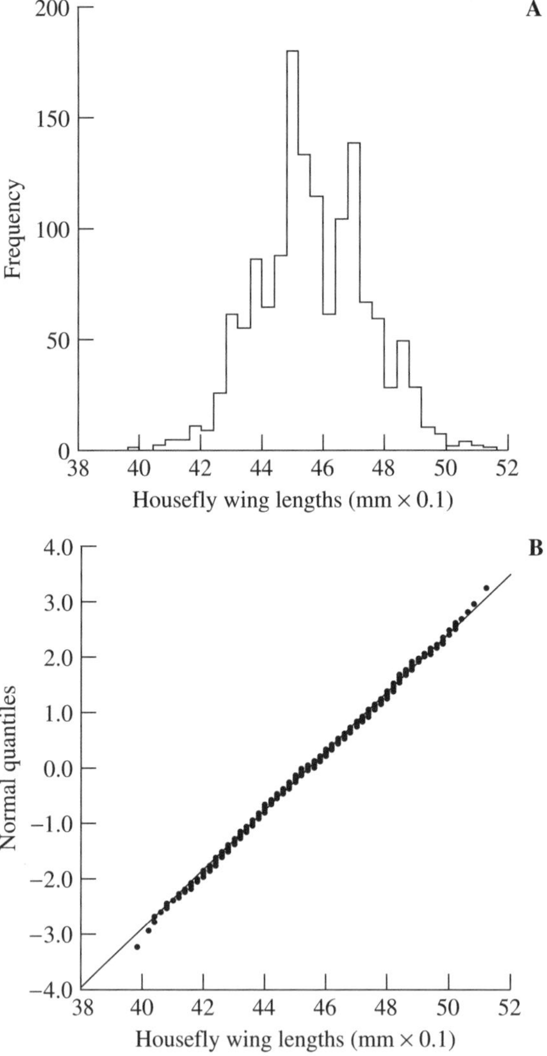

FIGURE 7.1 *Graphs of the means of 1400 random samples of 5 housefly wing lengths.* **A**. *Histogram.* **B**. *Normal quantile plot.* $\overline{Y} = 45.473$, $s = 1.765$, $\sigma_{\overline{Y}} = 1.744$.

sampling results of previous classes, and construct a frequency distribution of these means. Alternatively, each student can generate an arbitrarily large number of samples by computer. For each variable we can also obtain the mean of the seven means, which is a mean of a sample of 35 items. Here again we will make a frequency distribution of these means, although considerably more sampling is required to accumulate a sufficient number of samples of 35 items for a meaningful frequency distribution.

Figure 7.1A shows a histogram of 1400 means of samples of 5 housefly wing lengths obtained by computer. The mean and standard deviation of the 1400

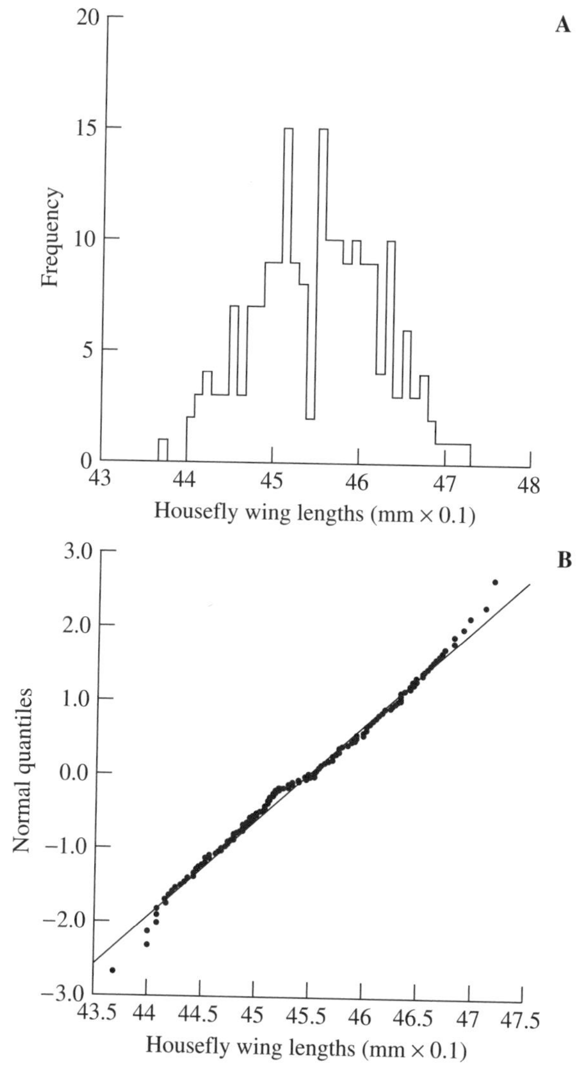

FIGURE 7.2 *Graphs of the means of 200 random samples of 35 housefly wing lengths.* **A**. *Histogram.* **B**. *Normal quantile plot.* $\bar{Y} = 45.473$, $s = 0.718$, $\sigma_{\bar{Y}} = 0.659$.

means are given. Normal quantile plots for these data are shown in Figure 7.1B. Note that the distribution appears quite normal, as does that of the means based on 200 samples of 35 wing lengths (Figure 7.2). This illustrates an important theorem: *The means of samples from a normally distributed population are themselves normally distributed regardless of sample size n*. Thus we note that the means of samples from the normally distributed housefly wing lengths are distributed normally whether they are based on 5 or 35 individual readings.

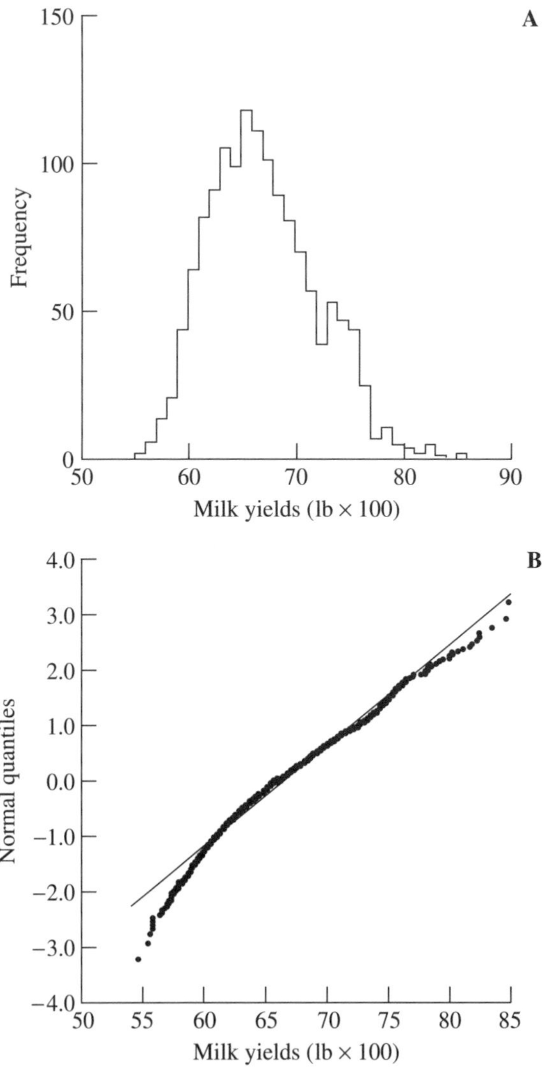

FIGURE 7.3 *Graphs of the means of 1400 random samples of 5 milk yields.* **A**. *Histogram.* **B**. *Normal quantile plot.* $\overline{Y} = 66.542$, $s = 5.066$, $\sigma_{\overline{Y}} = 4.991$.

Both distributions of means of the heavily skewed milk yields (Figures 7.3 and 7.4) appear to be close to normal distributions. The means based on five milk yields, however, do not agree with the normal distribution nearly as well as do the means of 35 items. This illustrates another theorem of fundamental importance in statistics: *As sample size increases, the means of samples drawn from a population of any distribution will approach the normal distribution.* This theorem, when rigorously stated (about sampling from populations with finite

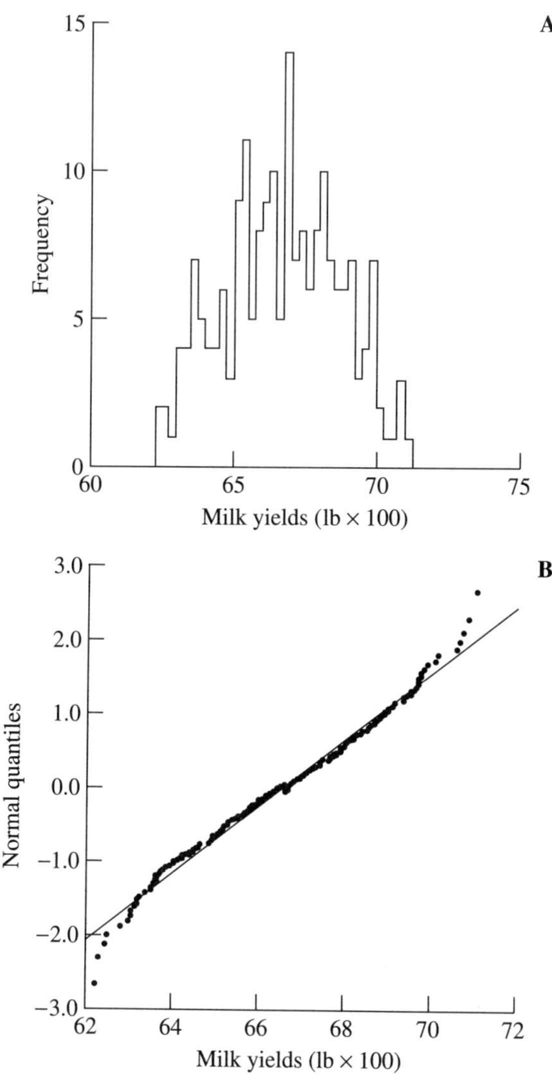

FIGURE 7.4 *Graphs of the means of 200 random samples of 35 milk yields.* **A.** *Histogram.* **B.** *Normal quantile plot.* $\bar{Y} = 66.542$, $s = 2.040$, $\sigma_{\bar{Y}} = 1.886$.

variances), is known as the **central limit theorem** and is illustrated by the near normality of the milk yield means based on 35 items (Figure 7.4) as compared with the considerable skewness of the means based on 5 items (Figure 7.3). The importance of this theorem is that if n is large enough, it permits us to use the normal distribution to make statistical inferences about means of populations even though the items are not at all normally distributed. The necessary size of n depends upon the distribution (skewed populations require larger samples).

Another important fact is that the range of the means is considerably less than that of the original items. Thus the wing length means range from 39.8 to 51.2 in samples of 5 (see Figure 7.1) and from 43.7 to 47.2 in samples of 35 (see Figure 7.2), but the individual wing lengths range from 36 to 55. The milk yield means range from 54.6 to 84.8 in samples of 5 (see Figure 7.3) and from 62.2 to 71.0 in samples of 35 (see Figure 7.4), but the individual milk yields range from 51 to 98. Not only do means show less scatter than the items upon which they are based (an easily understood phenomenon if you give some thought to it), but the range of the distribution of the means diminishes as the sample size upon which the means are based increases.

The differences in ranges are reflected in differences in the standard deviation of these distributions. If we calculate the standard deviations of the means in the four distributions in Figures 7.1 through 7.4, we obtain the following values:

	Observed standard deviations of distributions of means	
	$n = 5$	$n = 35$
Wing lengths	1.765	0.718
Milk yields	5.066	2.040

Note that the standard deviations of the sample means based on 35 items are considerably less than those based on 5 items. This is also intuitively obvious. Means based on large samples should be close to the parametric mean and will not vary as much as will means based on small samples. The variance of means is therefore partly a function of the sample size on which the means are based. It is also a function of the variance of the items in the samples. Thus, in the text table above, the means of milk yields have a much greater standard deviation than means of wing lengths based on comparable sample size simply because the standard deviation of the individual milk yields (11.1597) is greater than that of individual wing lengths (3.90).

That the variance of means decreases as sample size n increases can also be seen in Figures 7.1 through 7.4. The slopes of the normal quantile plots in Figures 7.2B and 7.4B (35 samples each) are steeper than those in Figures 7.1B and 7.3B (5 samples each), respectively, indicating less scatter (hence less variance) in the means based on the larger samples. The increased steepness may not be obvious on casual inspection because both ordinate and abscissa have been rescaled in Figures 7.2B and 7.4B over those of Figure 7.1B and 7.3B, respectively.

The expected value of the variance of sample means can be calculated. By *expected value* we mean the average value to be obtained by infinitely repeated sampling. Thus if we were to take samples of a means of n items repeatedly and were to calculate the variance of these a means each time, the average of these variances would be the expected value. We can visualize the mean as a weighted average of the n independently sampled observations with each weight w_i equal to 1. From Expression (4.2):

$$\bar{Y}_w = \frac{\sum^n w_i Y_i}{\sum^n w_i}$$

we obtain the weighted mean. Section A.2 in the appendix tells us that when variates are multiplied by a factor, say w, their variance will be $w^2 s^2$. We defer to Section 15.3 (and A.12) the proof that the variance of the sum of *independent* variables $Y_1 + Y_2 + \ldots + Y_n$ equals $\sigma_1^2 + \sigma_2^2 + \ldots + \sigma_n^2 = \Sigma^n \sigma_i^2$. By combining these two relations, we can state that the variance of the weighted sum of *independent* items $\Sigma^n w_i Y_i$ is

$$\text{Var}\left(\sum^n w_i Y_i\right) = \sum^n w_i^2 \sigma_i^2 \tag{7.1}$$

where σ_i^2 is the variance of Y_i.

If we change $\Sigma^n w_i Y_i$ to a weighted mean by dividing by $\Sigma^n w_i$, it follows that

$$\sigma^2_{\bar{Y}_w} = \frac{\sum^n w_i^2 \sigma_i^2}{\left(\sum^n w_i\right)^2}$$

If we assume that the variances σ_i^2 are all equal to σ^2, the expected variance of the mean is

$$\sigma^2_{\bar{Y}_w} = \frac{\sigma^2 \sum w_i^2}{\left(\sum w_i\right)^2}$$

Since the Y_i are all individual variates, their weights w_i are equal to one, and

$$\sigma^2_{\bar{Y}} = \frac{\sigma^2 n}{n^2} = \frac{\sigma^2}{n} \tag{7.2}$$

Consequently the expected standard deviation of means is

$$\sigma_{\bar{Y}} = \frac{\sigma}{\sqrt{n}} \tag{7.2a}$$

This formula makes it clear that the standard deviation of means is a function of the standard deviation of items, as well as of the sample size of means. The greater the sample size, the smaller the standard deviation of means. In fact, as sample size increases to a very large number, the standard deviation of means becomes vanishingly small. This makes good sense. Very large sample sizes, averaging many observations, should yield estimates of means less variable than those based on a few items.

When working with samples from a population we do not, of course, know the parametric standard deviation σ of the population, but can only obtain a sample estimate, s, of it. Also, we would be unlikely to have numerous samples of size n from which to compute the standard deviation of means directly. Therefore, we usually have to estimate the standard deviation of means from a single sample by using Expression (7.2a), substituting s for σ:

$$s_{\bar{Y}} = \frac{s}{\sqrt{n}} \tag{7.3}$$

Thus we obtain, from the standard deviation of a single sample, an estimate of the standard deviation of means we would expect were we to obtain a collection of means based on equal-sized samples of n items from the same population. As we shall see, this estimate of the standard deviation of a mean is a very important and frequently used statistic.

Table 7.1 illustrates some estimates of the standard deviations of means that might be obtained from random samples of the two populations that we have been discussing. The means of 5 samples of wing lengths based on 5 individuals ranged from 43.6 to 46.8, their standard deviations from 1.095 to 4.827, and the estimate of standard deviation of the means from 0.490 to 2.159. Ranges for the other categories of samples in Table 7.1 similarly include the parametric values of these statistics. The estimates of the standard deviations of the means of the milk yields cluster around the expected value, since they are not dependent on normality of the variates. However, in a particular sample in which by chance the sample standard deviation is a poor estimate of the population standard deviation (as in the second sample of 5 milk yields), the estimate of the standard deviation of means is equally wide of the mark.

We should emphasize one difference between the standard deviation of items and the standard deviation of sample means: If we estimate a population standard deviation through the standard deviation of a sample, the magnitude of the estimate will not change as we increase the sample size. We may expect that the estimate will improve and will approach the true standard deviation of the population; however, its order of magnitude will be the same, whether the sample is based on 3, 30, or 3000 individuals, as Table 7.1 clearly shows. The values of s are closer to σ in the samples based on $n = 35$ than in samples of $n = 5$. Yet the general magnitude is the same in both instances. The standard deviation of means, however, decreases as sample size increases, as is obvious from

Table 7.1 MEANS, STANDARD DEVIATIONS, AND STANDARD DEVIATIONS OF MEANS (STANDARD ERRORS) OF 5 RANDOM SAMPLES OF 5 AND 35 HOUSEFLY WING LENGTHS AND JERSEY COW MILK YIELDS.

Parametric values for the statistics are given in the sixth line of each category.

	(1) $\bar{Y}$	(2) s	(3) $s_{\bar{Y}}$
	Wing lengths		
	45.8	1.095	0.490
	45.6	3.209	1.435
$n = 5$	43.6	4.827	2.159
	44.8	4.764	2.131
	46.8	1.095	0.490
	$\mu = 45.5$	$\sigma = 3.90$	$\sigma_{\bar{Y}} = 1.744$
	45.37	3.812	0.644
	45.00	3.850	0.651
$n = 35$	45.74	3.576	0.604
	45.29	4.198	0.710
	45.91	3.958	0.669
	$\mu = 45.5$	$\sigma = 3.90$	$\sigma_{\bar{Y}} = 0.659$
	Milk yields		
	66.0	6.205	2.775
	61.6	4.278	1.913
$n = 5$	67.6	16.072	7.188
	65.0	14.195	6.348
	62.2	5.215	2.332
	$\mu = 66.61$	$\sigma = 11.160$	$\sigma_{\bar{Y}} = 4.991$
	65.429	11.003	1.860
	64.971	11.221	1.897
$n = 35$	66.543	9.978	1.687
	64.400	9.001	1.521
	68.914	12.415	2.099
	$\mu = 66.61$	$\sigma = 11.160$	$\sigma_{\bar{Y}} = 1.886$

Sampled from Table 6.1.

Expression (7.3). Thus, means based on 3000 items will have a standard deviation only one-tenth that of means based on 30 items, as illustrated here:

$$\frac{s}{\sqrt{3000}} = \frac{s}{\sqrt{30} \times \sqrt{100}} = \frac{s}{\sqrt{30} \times 10}$$

Since $s = \sqrt{s^2}$ is a biased estimate of σ, $s_{\bar{Y}}$ will also be biased. If an unbiased estimate is desired, $s_{\bar{Y}}$ can be corrected by the same procedure as that described in Section 4.7.

7.2 Distribution and Variance of Other Statistics

Just as we obtained a mean and a standard deviation from each sample of the wing lengths and milk yields, so we could also have obtained other statistics from each sample, such as variance, median, coefficient of variation or g_1. After repeated sampling and computation we would have frequency distributions for these statistics and would be able to compute their standard deviations just as we

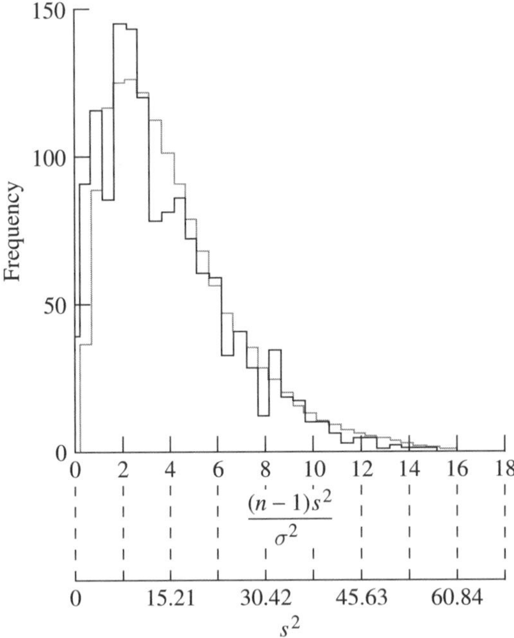

FIGURE 7.5 *Histograms of variances (solid line) based on 1400 samples of housefly wing lengths from Table 6.1 and of expected $\chi^2_{[4]}$ distribution (dotted line). Abscissa is given in terms of s^2 and $(n - 1)s^2/\sigma^2$.*

did for the frequency distribution of means. In many cases the statistics are distributed normally, as was true for the means. In other cases the statistics will be distributed normally only if they are based on samples from a normally distributed population, or if they are based on large samples, or if both these conditions hold. Figure 7.5 shows a frequency distribution of the variances from the 1400 samples of 5 housefly wing lengths (see Figure 7.1). Notice that the distribution is strongly skewed to the right, which is characteristic of the distribution of variances of small samples.

Standard deviations of statistics are generally known as **standard errors.** Beginners sometimes get confused by an imagined distinction between standard deviations and standard errors. The standard error of a statistic such as the mean (or g_1) is the standard deviation of a distribution of means (or g_1's) for samples of a given sample size n. Thus the terms standard error and standard deviation are used synonymously with the following exception: It is not customary to use standard error as a synonym for standard deviation of items in a sample or population. Standard error or standard deviation has to be qualified by referring to a given statistic, such as the standard deviation of g_1, which is the same as the standard error g_1. Used without any qualification, "standard error" conventionally implies the standard error of the mean. "Standard deviation" used without qualification generally means standard deviation of items in a sample or population. Having means, standard deviations, standard errors, and coefficients of variation in a table signifies that arithmetic means, standard deviations of items in samples, standard deviations of their means (= standard errors of means), and coefficients of variation are displayed. The following summary of terms may be helpful:

$$\text{Standard deviation} = s = \sqrt{\sum y^2/(n-1)}$$

$$\text{Standard deviation of a statistic } St = \text{standard error of a statistic } St = s_{St}$$

$$\text{Standard error} = \text{standard error of a mean} = \text{standard deviation of a mean} = s_{\bar{Y}}$$

Standard errors are usually not obtained from a frequency distribution by repeated sampling but are estimated from only a single sample and represent the expected standard deviation of the statistic if a large number of such samples had been obtained. Remember that we estimated the standard error of a distribution of means from a single sample in this manner in the previous section.

Box 7.1 lists the standard errors of common statistics. Column (1) lists the statistic whose standard error is described; column (2) shows the formula for the estimated standard error; column (3) gives the degrees of freedom on which the standard error is based (their use is explained in Section 7.5); column (4) provides comments on the range of application of the standard error. The uses of these standard errors will be illustrated in subsequent sections.

Box 7.1 STANDARD ERRORS FOR COMMON STATISTICS.

	(1) Statistic	(2) Estimate of standard error	(3) df	(4) Comments on applicability
1	$\bar{Y}$	$s_{\bar{Y}} = \dfrac{s}{\sqrt{n}} = \dfrac{s_Y}{\sqrt{n}} = \sqrt{\dfrac{s^2}{n}}$	$n - 1$	True for any population with finite variance
2	Median	$s_{\text{med}} = (1.2533)s_{\bar{Y}}$	$n - 1$	Large samples from normal populations
3	Average deviation (AD)	$s_{\text{AD}} \approx (0.602{,}810{,}3)\dfrac{s}{\sqrt{n}}$	$n - 1$	For large n (>100)
		$= \dfrac{s}{n}\sqrt{\dfrac{2(n-1)}{\pi}\left[\dfrac{\pi}{2} + \sqrt{n(n-1)} - n - \arcsin\left\{\dfrac{1}{(n-1)}\right\}\right]}$	$n - 1$	Samples from normal populations
4	s	$s_s = (0.707{,}106{,}8)\dfrac{s}{\sqrt{n}}$	$n - 1$	Samples from normal populations ($n > 15$)
5	V	$s_V \approx \dfrac{V}{\sqrt{2n}}\sqrt{1 + 2\left(\dfrac{V}{100}\right)^2}$	$n - 1$	Samples from normal populations
		$s_V \approx \dfrac{V}{\sqrt{2n}}$	$n - 1$	Used when $V < 15$
5*	V^*	$s_{V^*} = \left(1 + \dfrac{1}{4n}\right)s_V$	$n - 1$	Samples from normal populations
6	g_1	$s_{g_1} = \sqrt{\dfrac{6n(n-1)}{(n-2)(n+1)(n+3)}} \approx \sqrt{\dfrac{6}{n}}$	∞	Samples from normal populations. The approximate formula is for large n (>100)
7	g_2	$s_{g_2} = \sqrt{\dfrac{24n(n-1)^2}{(n-3)(n-2)(n+3)(n+5)}} \approx \sqrt{\dfrac{24}{n}}$	∞	Samples from normal populations. The approximate formula is for large n (>150)

7.3 Introduction to Confidence Limits

The sample statistics we have been obtaining, such as mean, standard deviation, or g_1, are estimates of the population parameters μ, σ, or γ_1, respectively. So far we have not discussed the reliability of these estimates. First we want to know whether the sample statistics are *unbiased estimators* of the population parameters, as discussed in Section 4.7.

But knowing, for example, that $\bar{Y}$ is an unbiased estimate of μ is not enough. We would like to find out how reliable a measure of μ it is. Of course, what we really wish to know is the magnitude of μ, σ^2, and other parameters. This is an impossible task, however, unless we sample exhaustively. Thus the true values of the parameters almost always remain unknown, and we commonly estimate the reliability of a sample statistic by setting confidence limits to it.

To begin our discussion of this topic, let us start with the unusual case of a population whose parametric mean and standard deviation are known to be μ and σ, respectively. The mean of a sample of n items is symbolized by $\bar{Y}$. The expected standard error of the mean is $\sigma/\sqrt{n}$. As we have seen, the sample means will be normally distributed. Therefore, from Section 6.2, the region from $1.96\sigma/\sqrt{n}$ below μ to $1.96\sigma/\sqrt{n}$ above μ includes 95% of the sample means of size n. Another way of stating this is by the ratio $(\bar{Y} - \mu)/(\sigma/\sqrt{n})$. This ratio is the standard deviate of a sample mean from the parametric mean. Since they are normally distributed, 95% of such standard deviates will lie between -1.96 and $+1.96$. We can express this statement symbolically as follows:

$$P\left\{-1.96 \leq \frac{\bar{Y} - \mu}{\sigma/\sqrt{n}} \leq +1.96\right\} = 0.95$$

This expression means that the probability P that the sample means $\bar{Y}$ will differ by no more than 1.96 standard errors $\sigma/\sqrt{n}$ from the parametric mean μ equals 0.95. The expression in brackets is an inequality, all terms of which can be multiplied by $\sigma/\sqrt{n}$ to yield

$$\left\{-\frac{1.96\sigma}{\sqrt{n}} \leq (\bar{Y} - \mu) \leq +\frac{1.96\sigma}{\sqrt{n}}\right\}$$

We can rewrite this expression as

$$\left\{-\frac{1.96\sigma}{\sqrt{n}} \leq (\mu - \bar{Y}) \leq +\frac{1.96\sigma}{\sqrt{n}}\right\}$$

because $-a \leq b \leq a$ implies $a \geq -b \geq -a$, and this can be written as $-a \leq -b \leq a$. Finally, we can transfer $-\bar{Y}$ across the inequality signs, just as in an equation it could be transferred across the equal sign to yield the final desired expression:

$$P\left\{\bar{Y} - \frac{1.96\sigma}{\sqrt{n}} \leq \mu \leq \bar{Y} + \frac{1.96\sigma}{\sqrt{n}}\right\} = 0.95 \qquad (7.4)$$

or

$$P\{\bar{Y} - 1.96\sigma_{\bar{Y}} \leq \mu \leq \bar{Y} + 1.96\sigma_{\bar{Y}}\} = 0.95 \qquad (7.4a)$$

Thus the probability, P, is 0.95 that the term $\bar{Y} - 1.96\sigma_{\bar{Y}}$ is less than or equal to the parametric mean μ and that the term $\bar{Y} + 1.96\sigma_{Y}$ is greater than or equal to μ. We will call these two terms, $\bar{Y} - 1.96\sigma_{\bar{Y}}$ and $\bar{Y} + 1.96\sigma_{\bar{Y}}$, L_1 and L_2, respectively, representing the lower and upper 95% **confidence limits** of the mean.

Another way of stating the relationship implied by Expression (7.4a) is that if we repeatedly obtained samples of size n from the population and constructed these limits for each, we could expect 95% of the intervals between these limits to contain the true mean, and only 5% of the intervals would miss μ. The interval from L_1 to L_2 is called a **confidence interval.**

If you were not satisfied to have the confidence interval contain the true mean only 95 times out of 100, you might employ 2.576 as a coefficient in place of 1.960. You may remember that 99% of the area of the normal curve lies between $\mu \pm 2.576\sigma$ (see Section 6.2). Thus, to calculate 99% confidence limits, compute the two quantities $L_1 = \bar{Y} - 2.576\sigma/\sqrt{n}$ and $L_2 = \bar{Y} + 2.576\sigma/\sqrt{n}$ as lower and upper confidence limits, respectively. In this case, 99 out of 100 confidence intervals obtained in repeated sampling would contain the true mean. The new confidence interval is wider than the 95% interval (since we have multiplied by a greater coefficient).

If you were still not satisifed with the reliability of the confidence limit you could increase it, multiplying the standard error of the mean by 3.291 in order to obtain 99.9% confidence limits. This value (or 3.891 for 99.99% limits) can be found by inverse interpolation in an extensive table of areas of the normal curve (for example, table 1 in Pearson and Hartley, 1958) or directly in a table of the inverse of the normal probability distribution (for example, table 1.2 in Owen, 1962). The new coefficient would widen the interval further. Notice that you can construct confidence intervals that will be expected to contain μ an increasingly greater percentage of the time. First you would expect to be right 95 times out of 100, then 99 times out of 100, and finally 9999 times out of 10,000. But as your confidence increases, your statement becomes vaguer and vaguer, since the confidence interval lengthens. Consider the following example.

We obtain a sample of 35 housefly wing lengths from the population of Table 6.1 with known mean ($\mu = 45.5$) and standard deviation ($\sigma = 3.90$). Let us assume that the sample mean is 44.8. We can expect the standard deviation of means based on samples of 35 items to be $\sigma_{\bar{Y}} = \sigma/\sqrt{n} = 3.90/\sqrt{35} = 0.6592$. We compute confidence limits as follows:

$$\text{Lower limit is } L_1 = 44.8 - (1.960)(0.6592) = 43.51$$

$$\text{Upper limit is } L_2 = 44.8 + (1.960)(0.6592) = 46.09$$

Remember that this is an unusual case, in which we happen to know the true mean of the population ($\mu = 45.5$), and hence we know that the confidence limits enclose the mean. We expect 95% of such confidence intervals obtained in repeated sampling to include the parametric mean. We could increase the reliability of these limits by going to 99% confidence intervals, replacing 1.960 in the above expression by 2.576 and obtaining $L_1 = 43.10$ and $L_2 = 46.50$. We would have greater confidence that our interval covers the mean, but we would be much less certain about the true value of the mean because of the wider limits. By increasing the degree of confidence still further, say to 99.99%, we would be virtually certain that our confidence limits ($L_1 = 42.24$, $L_2 = 47.36$) contain the population mean, but the bounds enclosing the mean would be so wide as to make our prediction far less useful than previously.

EXPERIMENT 7.2. For the seven samples of 5 housefly wing lengths and the seven similar samples of milk yields last worked with in Experiment 7.1 (Section 7.1), compute 95% confidence limits to the parametric mean for each sample and for the total sample based on 35 items. Base the standard errors of the means on the parametric standard deviations of these populations (housefly wing lengths: $\sigma = 3.90$; milk yields: $\sigma = 11.1597$). Record how many in each of the four classes of confidence limits (wing lengths and milk yields, $n = 5$ and $n = 35$) were correct—that is, contained the parametric mean of the population. Pool your results with those of other class members. (Alternatively, you may obtain 200 samples using a computer program.)

We tried the experiment by computer for 200 samples of 35 wing lengths each, computing confidence limits of the parametric mean by employing the parametric standard error of the mean, $\sigma_{\bar{Y}} = 0.6592$. Figure 7.6 shows these 200 confidence intervals plotted parallel to the ordinate. Of these, 194 (97.0%) cross the parametric mean of the population. This should give you a concrete idea of the reliability of confidence limits.

To reduce the width of the confidence interval, we have to reduce the standard error of the mean, which since $\sigma_{\bar{Y}} = \sigma/\sqrt{n}$, can be done only by reducing either the standard deviation of the items or by increasing the sample size. The first of these alternatives is frequently not available. If we are sampling from a population in nature, ordinarily we have no way of reducing its standard deviation. In many experimental procedures, however, we may be able to reduce the variance of the data. For example, if we are studying heart weight in rats and find that its variance is rather large, we might be able to reduce this variance by taking rats of only one age group, in which the variation of heart weight would be considerably less. Thus, by controlling one of the variables of the experiment, the variance of the response variable, heart weight, is reduced. Similarly, by keeping temperature or other environmental variables constant in a procedure, we can frequently reduce the variance of our response variable and hence obtain more precise estimates of population parameters.

A more common way to reduce the standard error is by increasing sample size. It is obvious from Expression (7.2a) that as n increases, the standard error

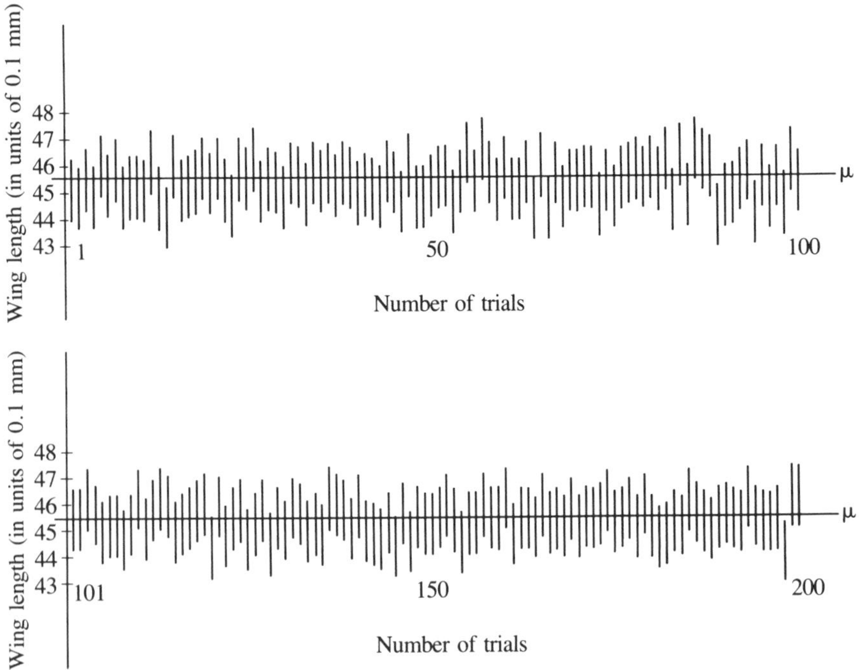

FIGURE 7.6 *Ninety-five percent confidence intervals of means of 200 samples of 35 housefly wing lengths, based on parametric standard errors, $\sigma_{\bar{Y}}$. The heavy horizontal line is the parametric mean μ. The ordinate represents the variable.*

decreases; hence as n approaches infinity, the standard error and the lengths of confidence intervals approach zero. This relationship ties in with what we have already learned: In samples whose size approaches infinity, the sample mean approaches the parametric mean.

We must guard against a common mistake in expressing the meaning of the confidence limits of a statistic. When we have set lower and upper limits (L_1 and L_2, respectively) to a statistic, we imply that the probability of this interval covering the mean is 0.95 or, expressed in another way, that on the average, 95 out of 100 confidence intervals similarly obtained would cover the mean. We *cannot state* that there is a probability of 0.95 that the true mean is contained within any particular observed confidence limits, although this may seem to be saying the same thing. The last statement is incorrect because the true mean is a parameter; hence it is a fixed value and is therefore either inside or outside of the interval. It cannot be inside a particular interval 95% of the time. It is important, therefore, to learn the correct statement and meaning of confidence limits.

Up to now we have considered only means based on normally distributed samples with known parametric standard deviations. We can, however, extend the methods we have learned to samples from populations with unknown standard deviations but where the population is known to be normally distributed and the samples are large, say $n \geq 100$. In such cases, we use the sample standard deviation for computing the standard error of the mean.

When the samples are small ($n < 100$) and we lack knowledge of the parametric standard deviation, however, we must consider the reliability of our sample standard deviation, which requires making use of the so-called t or Student's distribution. We will learn how to set confidence limits using the t-distribution in Section 7.5. First, however, we must become familiar with this distribution.

7.4 THE t-DISTRIBUTION

The deviations $\bar{Y} - \mu$ of sample means from the parametric mean of a normal distribution are themselves distributed normally. If these deviations are divided by the parametric standard deviation, $(\bar{Y} - \mu)/\sigma_{\bar{Y}}$, they are still distributed normally, with $\mu = 0$ and $\sigma = 1$. In fact, they form a standard normal distribution (see Section 6.7). Subtracting the constant μ from every $\bar{Y}_i$ is simply an additive code (see Section 4.8) and will not change the form of the distribution of sample means, which is normal (see Section 7.1). Dividing each deviation by the constant $\sigma_{\bar{Y}}$ reduces the variance to unity, but it does so proportionally for the entire distribution, so that its shape is not altered and a previously normal distribution remains so.

If, on the other hand, we had calculated the variance s_i^2 of each of the samples and calculated the deviation for each mean $\bar{Y}_i$ as $(\bar{Y}_i - \mu)/s_{\bar{Y}_i}$, where $s_{\bar{Y}_i}$ stands for the estimate of the standard error of the mean of the ith sample, we would have found the distribution of the deviations to be wider and flatter than the normal distribution. This distribution is illustrated in Figure 7.7, which shows the ratio $(\bar{Y}_i - \mu)/s_{\bar{Y}_i}$ for the 1400 samples of 5 housefly wing lengths of Figure 7.1. The new distribution is wider than the corresponding normal distribution because the denominator is the sample standard error rather than the parametric standard error and thus will sometimes be smaller, at other times greater, than expected. This increased variation will be reflected in the greater variance of the ratio $(\bar{Y} - \mu)/s_{\bar{Y}}$. The expected distribution of this ratio, called the ***t*-distribution,** is also known as Student's distribution after W. S. Gossett, who first described it publishing under the pseudonym "Student." The t-distribution is a function with a complicated mathematical formula that need not be presented here.

Note that the sample standard deviation s that enters into the denominator of the t-ratio must *not* be corrected for bias by multiplying by C_n in the manner of Section 4.7. The t-distribution itself allows for the bias inherent in the estimate of s.

Like the normal distribution, the t-distribution is symmetric and extends from negative to positive infinity. It differs from the normal distribution, however, in

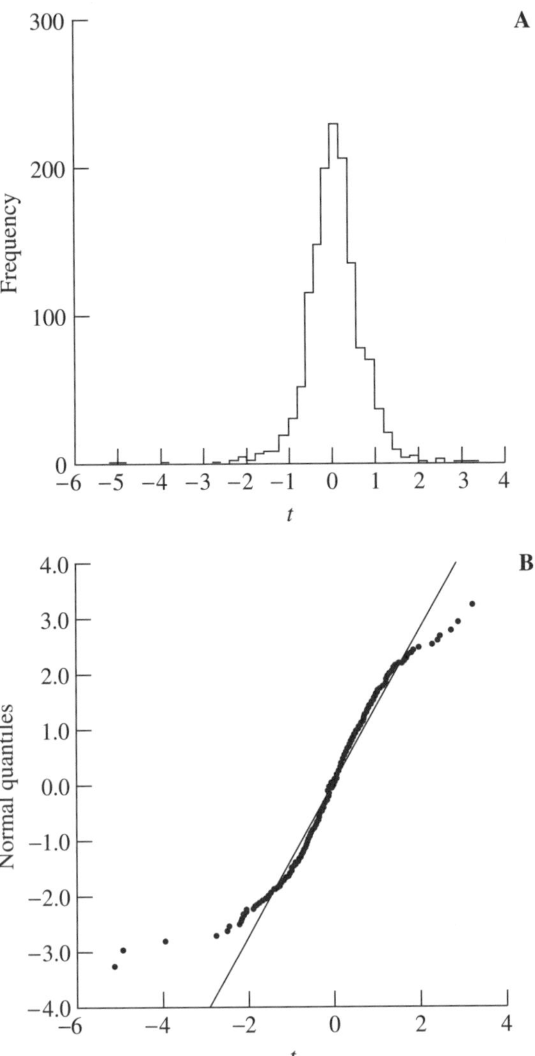

FIGURE 7.7 *Distribution of quantity* $t_s = (\bar{Y} - \mu)/s_{\bar{Y}}$ *along abscissa computed for 1400 samples of 5 housefly wing lengths* **A**. *Histogram.* **B**. *Normal quantile plot.*

that it assumes different shapes depending on the number of degrees of freedom. By degrees of freedom is meant the quantity $n - 1$, where n is the sample size upon which a variance has been based. Remember that the quantity $n - 1$ is the divisor in obtaining an unbiased estimate of the variance from a sum of squares (see Section 4.7). The number of degrees of freedom pertinent to a given t-distribution are the same as the number of degrees of freedom of the standard deviation in the ratio $(\bar{Y} - \mu)/s_{\bar{Y}}$. Degrees of freedom (abbreviated *df* or sometimes ν)

can range from 1 to infinity. A t-distribution for $df = 1$ deviates the most from the normal distribution. As the number of degrees of freedom increases, the t-distribution approaches the shape of the standard normal distribution ($\mu = 0$, $\sigma = 1$) ever more closely, and in a graph the size of this page a t-distribution of $df = 30$ is essentially indistinguishable from a normal distribution. At $df = \infty$ the t-distribution *is* the normal distribution. Thus we can think of the t-distribution as the general case and the normal distribution as the special case of the t-distribution in which $df = \infty$. Figure 7.8 shows t-distributions for 1 and 2 degrees of freedom compared with a normal frequency distribution.

We were able to employ a single table for the areas of the normal curve by coding the argument in standard deviation units (see Section 6.4). However, since the t-distributions differ in shape for differing degrees of freedom, it will be necessary to have a separate t-table, corresponding in structure to the table of the areas of the normal curve, for each value of df. This would make for very cumbersome and elaborate sets of tables. Some values of the cumulative t-distributions for $df = 1$ to 24, 30, 40, 60, 120, and ∞ are given in Table 9 of Pearson and Hartley (1958). If you have access to that source, try looking up a given value of t, say $t = 1.0$, for $df = 5$. The cumulative area includes 0.81839 of the area of the t-distribution for $df = 5$. The corresponding value for the normal distribution, from a table of the areas of the normal curve, is 0.84134. The latter figure can also be found in the table of the cumulative t-distribution under $df = \infty$. Since a greater proportion of the area is drawn out into the tails, less of the area is found between one standard deviation above and below the mean in the t-distribution than in the normal distribution.

Conventional t-tables, however, are arranged differently. Statistical Table **B** shows degrees of freedom and probability as arguments and the corresponding values of t as functions. The probabilities indicate the percent of the area in both tails of the curve (to the right and left of the mean) beyond the indicated value of t. Thus, looking up the *critical value* of t at probability $P = 0.05$ and $df = 5$, we find $t = 2.571$ in Table **B.** Since this is a two-tailed table, the probability of 0.05

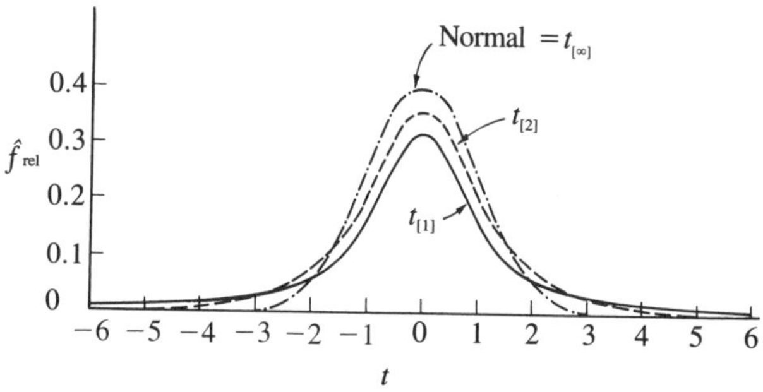

FIGURE 7.8 *Frequency curves of t-distributions for 1 and 2 degrees of freedom compared with the normal distribution.*

means that 0.025 of the area in each tail will be beyond a t-value of 2.571. Recall that the corresponding value for infinite degrees of freedom (for the normal curve) is 1.960. Only those probabilities generally used are shown in Table **B.** You should become very familiar with looking up t-values in this important table. A fairly conventional symbolism is $t_{\alpha[\nu]}$, meaning the tabled t-value for ν degrees of freedom and proportion α in both tails ($\alpha/2$ in each tail), which is equivalent to the t-value for the cumulative probability of $1 - \alpha/2$. Try looking up some of these values to become familiar with the table. For example, convince yourself that $t_{.05[7]}$, $t_{.01[3]}$, $t_{.02[10]}$, and $t_{.05[\infty]}$ correspond to 2.365, 5.841, 2.764, and 1.960, respectively.

We will now use the t-distribution to set confidence limits to means of small samples.

7.5 Confidence Limits Based on Sample Statistics

Armed with knowledge of the t-distribution, we are now able to set confidence limits to the means of samples from a normal frequency distribution whose parametric standard deviation is unknown. The limits are computed as $L_1 = \bar{Y} - t_{\alpha[n-1]}s_{\bar{Y}}$ and $L_2 = \bar{Y} + t_{\alpha[n-1]}s_{\bar{Y}}$ for confidence limits of probability $P = 1 - \alpha$. Thus, for 95% confidence limits we use values of $t_{.05[n-1]}$. We can rewrite Expression (7.4a) as

$$P\{L_1 \leq \mu \leq L_2\} = P\{\bar{Y} - t_{\alpha[n-1]}s_{\bar{Y}} \leq \mu \leq \bar{Y} + t_{\alpha[n-1]}s_{\bar{Y}}\} = 1 - \alpha \quad (7.5)$$

Box 7.2 shows an example of the application of this expression. The following sampling experiment should help convince you of the appropriateness of the t-distribution for setting confidence limits to means of samples from a normally distributed population with unknown σ.

EXPERIMENT 7.3. Repeat the computations and procedures of Experiment 7.2 (in Section 7.3), but base standard errors of the means on the standard deviations computed for each sample and use the appropriate t-value in place of a standard normal deviate.

Figure 7.9 shows 95% confidence limits of 200 sampled means of 35 housefly wing lengths, computed with t and $s_{\bar{Y}}$ rather than with the normal curve and $\sigma_{\bar{Y}}$ as in Figure 7.6. Note that 191 (95.5%) of the 200 confidence intervals cross the parametric mean.

We can use the same technique for setting confidence limits to any statistic, as long as the statistic follows the normal distribution. This technique applies in an approximate way to all the statistics of Box 7.1. Thus, for example, we may set confidence limits to the coefficient of variation of the aphid femur lengths of

Box 7.2 CONFIDENCE LIMITS FOR μ.

Aphid stem mother femur lengths (from Boxes 2.1 and 4.2): $\bar{Y} = 4.004$; $s = 0.366$; $n = 25$.

Values for $t_{\alpha[n-1]}$ from a two-tailed t-table (Statistical Table **B**), where $1 - \alpha$ is the proportion expressing confidence and $n - 1$ is the degrees of freedom:

$$t_{.05[24]} = 2.064 \qquad t_{.01[24]} = 2.797$$

The 95% confidence limits for the population mean, μ, are given by the equations

$$L_1 \text{ (lower limit)} = \bar{Y} - t_{.05[n-1]}\frac{s}{\sqrt{n}}$$

$$= 4.004 - \left(2.064\frac{0.366}{\sqrt{25}}\right) = 4.004 - 0.151$$

$$= 3.853$$

$$L_2 \text{ (upper limit)} = \bar{Y} + t_{.05[n-1]}\frac{s}{\sqrt{n}}$$

$$= 4.004 + 0.151$$
$$= 4.155$$

The 99% confidence limits are

$$L_1 = \bar{Y} - t_{.01[24]}\frac{s}{\sqrt{n}}$$

$$= 4.004 - \left(2.797\frac{0.366}{\sqrt{25}}\right) = 4.004 - 0.205$$

$$= 3.799$$

$$L_2 = \bar{Y} + t_{.01[24]}\frac{s}{\sqrt{n}}$$

$$= 4.004 + 0.205$$
$$= 4.209$$

Boxes 2.1 and 4.2. These are computed as

$$P\{V^* - t_{\alpha[n-1]}s_{V^*} \leq V_P \leq V^* + t_{\alpha[n-1]}s_{V^*}\} = 1 - \alpha$$

where V_P stands for the parametric value of the coefficient of variation. Since the standard error of the coefficient of variation approximately equals $s_V = V/\sqrt{2n}$

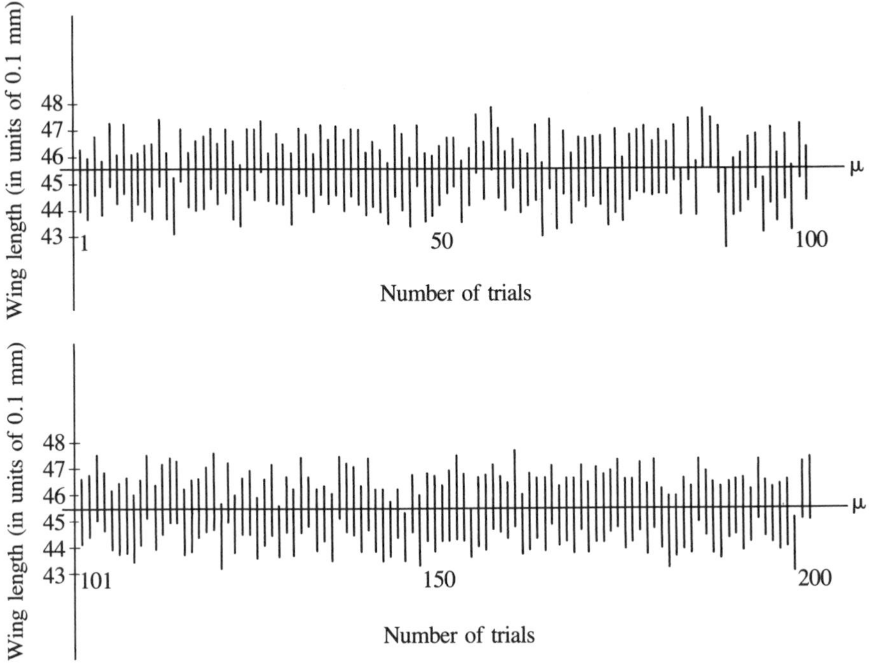

FIGURE 7.9 *Ninety-five percent confidence intervals of means of 200 samples of 35 housefly wing lengths, based on sample standard errors, $s_{\bar{Y}}$. The heavy horizontal line is the parametric mean μ. The ordinate represents the variable.*

and $s_{V*} = (1 + 1/4n)s_V$, we proceed as follows:

$$V = \frac{100s}{\bar{Y}} = \frac{100(0.3656)}{4.004} = 9.1309 \approx 9.13$$

$$V^* \approx \left(1 + \frac{1}{4n}\right) V = \left(1 + \frac{1}{4(25)}\right) 9.1309 = 9.2222 \approx 9.22$$

$$s_V = \frac{9.1309}{\sqrt{2 \times 25}} = \frac{9.1309}{7.0711} = 1.2913$$

$$s_{V*} = \left(1 + \frac{1}{4n}\right) s_V = 1.01\ (1.2913) = 1.3042$$

$$\begin{aligned} L_1 &= V^* - t_{.05[24]} s_{V*} \\ &= 9.2222 - (2.064)(1.3042) \\ &= 9.2222 - 2.6919 \\ &= 6.5303 \approx 6.53 \end{aligned}$$

$$L_2 = V^* + t_{.05[24]}s_{V^*}$$
$$= 9.2222 + 2.6919$$
$$= 11.9141 \approx 11.91$$

The distribution of the coefficient of variation is quite sensitive to departures from normality in the distribution of Y, so this procedure should not be used when Y is markedly nonnormally distributed.

When sample size is very large or when σ is known, the distribution is effectively normal. Rather than turn to the table of areas of the normal curve, however, we usually use $t_{\alpha[\infty]}$, the t-distribution with infinite degrees of freedom.

Although confidence limits are a useful measure of the reliability of a sample statistic, they are not commonly given in scientific publications; instead, the statistic $\pm$ its standard error is cited. Thus, you will frequently see column headings such as "Mean $\pm$ SE," indicating that the reader is free to use the standard error to set confidence limits if so inclined. It should be obvious to you from studying the t-distribution that you cannot set confidence limits to a statistic without knowing the sample size on which it is based, since n is necessary to compute the correct degrees of freedom. Thus, one should not cite means and standard errors without also stating sample size n. (The abbreviation SEM, for standard error of the mean, has become common in computer printouts.)

You must also be careful in interpreting such tables of results. Often they are cited as "Mean and Standard Deviation." The meaning of standard deviation is ambiguous here. Is it the standard deviation of the items or the standard deviation of their mean? Desirable headings for $\bar{Y} \pm s_{\bar{Y}}$ are "Means and Their Standard Errors" or "Means and Their Standard Deviations." If the heading says "Means and Standard Deviations" one would ordinarily assume s, not $s_{\bar{Y}}$. Also, if for any reason you wish to cite confidence limits, do not give them as "Statistic $\pm \frac{1}{2}$(Confidence Interval)" because readers would confuse this with the conventional "Mean $\pm$ Standard Error"; state limits as L_1 and L_2, respectively.

It is important to state a statistic and its standard error to a sufficient number of decimal places. The following rule of thumb helps. *Divide the standard error by three, then note the decimal place of the first nonzero digit of the quotient; give the statistic significant to that decimal place and provide one further decimal for the standard error.* This rule is quite simple, as an example will illustrate. If the mean and standard error of a sample are computed as 2.354 ± 0.363, we divide 0.363 by 3, which yields 0.121. Therefore, the mean should be reported accurate to one decimal place, and the standard error should be reported accurate to two decimal places. Thus we report this result as 2.4 ± 0.36. If, on the other hand, the same mean had a standard error of 0.243, dividing this standard error by 3 would have yielded 0.081, and the first nonzero digit would have been in the second decimal place. Thus the mean should have been reported as 2.35 ± 0.243.

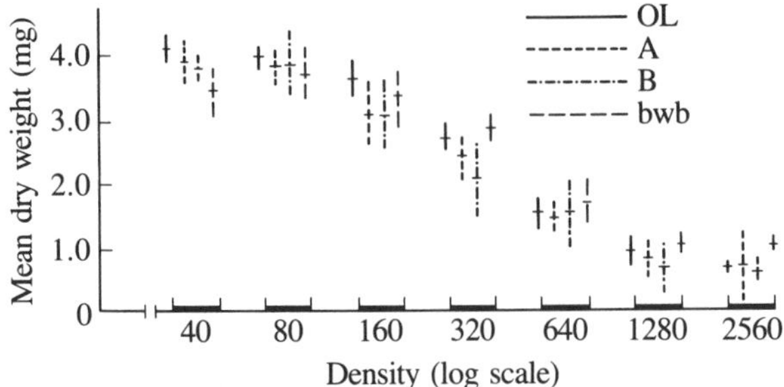

FIGURE 7.10 *Weights of four strains of houseflies in pure culture at varying densities. Means and their 95% confidence limits are shown. The abscissa, density in logarithmic scale, refers to number of eggs per 36 g of medium. The four strains are indicated by different patterns of lines, as shown at upper right. (Data from Sullivan and Sokal, 1965.)*

Graphic representations of the means of samples often indicate confidence limits by a simple method (Figure 7.10). A common alternative is to use wide bars with error bars superimposed on them. We show two versions of such graphs in Figure 7.11. This format is appropriate for frequency data, as in the example we show. Representing means and their standard errors of continuous

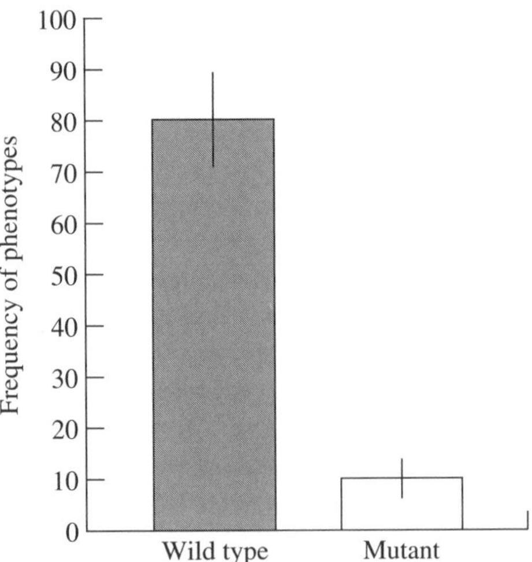

FIGURE 7.11 *Examples of error bars showing one standard error above the mean (left) or one standard error above and below the mean (right) for the observed frequencies given in Table 17.1. We assume that these frequencies (counts) are Poisson distributed and their standard errors (= standard deviations) are the square roots of the frequencies.*

variables in this way, however, tends to mislead the reader into assuming the bars are frequencies (as in Figures 2.2 or 5.2). There are two ways of indicating the error bar or "whisker." The one shown on the left in Figure 7.11 marks off one standard error above the mean; the one on the right marks off two standard errors, one above, the other below the mean. We prefer the right-hand version because most people find it difficult to mentally reflect the error bar below the mean. This recommendation, however, requires that the bar be white, or at least not solid black, so that the whisker beneath the mean is visible. You can see that there is no perfect solution for this scheme.

A more elaborate scheme, which has become quite common recently, is the method of **boxplots** (Tukey, 1977; McGill et al., 1978). A boxplot generally accompanies computer output of basic statistics programs. Such plots for two data sets are shown in Figure 7.12. The sample is represented as a box whose top and bottom are drawn at the lower and upper quartiles. The box is divided at the median. The length of the box is known as the **interquartile range.** A vertical line is drawn from the top of the box to the largest observation within 1.5 interquartile ranges of the top. A comparable line is drawn from the bottom to the smallest observation within 1.5 interquartile ranges of the bottom. All observations beyond these limits are plotted individually. Observations more than 3 interquartile ranges away from the box are given special prominence. The great

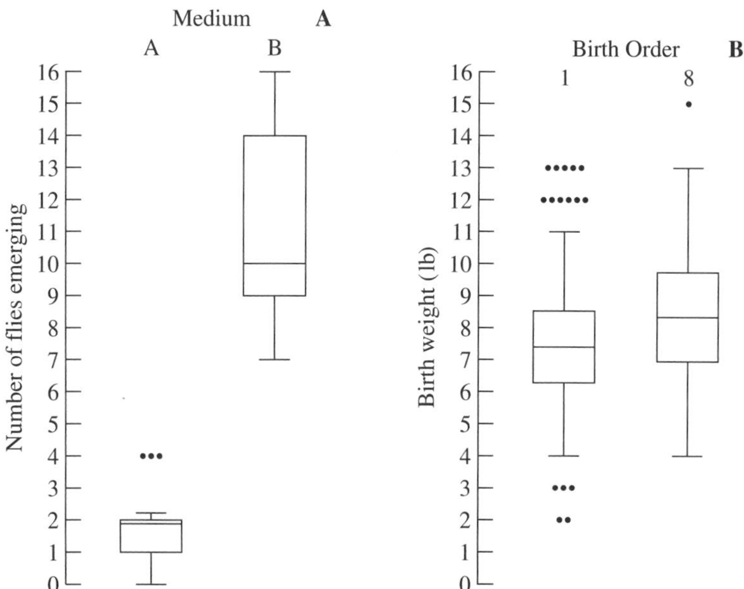

FIGURE 7.12 **A**. *Boxplots of number of flies emerging from different media (Table 13.3).* **B**. *Boxplots of birth weight data from Exercise 9.3.*

utility of boxplots is that they furnish measures of location (the median line), dispersion (the length of the box and the distance between the upper and lower whiskers), skewness (asymmetry of the upper and lower portions of the box, asymmetry of the upper and lower whiskers, or inequalities in the number of extreme individual observations plotted), and long-drawn-out tails (distance between the ends of the whiskers in relation to the length of the box). Boxplots are useful when two or more samples must be compared, as in both Figures 7.12A and 7.12B. In Figure 7.12A we note very clearly that more flies emerged from medium B than from medium A. There is no overlap between the two distributions. In Figure 7.12B we note that the birthweights of firstborns appear to be lighter than those of birth order 8. The lengths of the whiskers and the larger number of observations in the upper tails of the distributions show that the data are skewed to the right. This is to be expected in birth weights. For further elaborations of boxplots see Benjamini (1988).

7.6 The Chi-Square Distribution

Another continuous distribution of great importance in statistics is the distribution of χ^2 (read *chi-square*). We need to learn it now in connection with the distribution and confidence limits of variances.

The **chi-square distribution** is a probability density function whose values range from zero to positive infinity. Thus, unlike the normal distribution or t, the function approaches the χ^2-axis asymptotically only at the right-hand tail of the curve, not at both tails. The function describing the χ^2-distribution is complicated and will not be given here. As in the case of t, there is not merely one χ^2-distribution, but one distribution for each number of degrees of freedom. Therefore, χ^2 is a function of ν, the number of degrees of freedom. Figure 7.13 shows probability density functions for the χ^2-distributions for 1, 2, 3, and 6 degrees of freedom. Notice that the curves are strongly skewed to the right, L-shaped at first, but more or less approaching symmetry for higher degrees of freedom.

We can generate a χ^2-distribution from a population of standard normal deviates. Recall that we standardize a variable Y_i by subjecting it to the operation $(Y_i - \mu)/\sigma$ (see Section 6.2). Let us symbolize a standardized variable as $Y_i' = (Y_i - \mu)/\sigma$. Now imagine repeated samples of n variates Y_i from a normal population with mean μ and standard deviation σ. For each sample we transform every variate Y_i to Y_i', as defined above. The quantities $\sum^n Y_i'^2$ computed for each sample will be distributed as a χ^2-distribution with n degrees of freedom. Using the definition of Y_i', we can rewrite $\sum^n Y_i'^2$ as

$$\sum^n \frac{(Y_i - \mu)^2}{\sigma^2} = \frac{1}{\sigma^2} \sum^n (Y_i - \mu)^2 \tag{7.6}$$

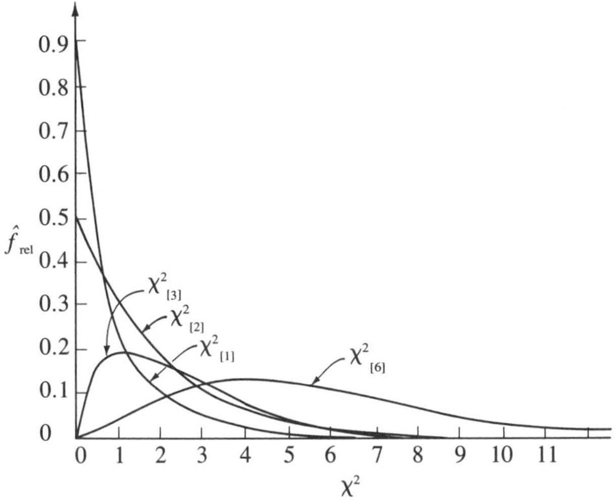

FIGURE 7.13 *Frequency curves of χ^2-distributions for 1, 2, 3, and 6 degrees of freedom.*

When we change the parametric mean μ to a sample mean, this expression becomes

$$\frac{1}{\sigma^2}\sum^{n}(Y_i - \bar{Y})^2 \tag{7.7}$$

which is simply the sum of squares of the variable divided by a constant, the parametric variance. Another common way of stating this expression is

$$\frac{(n-1)s^2}{\sigma^2} \tag{7.8}$$

Here we have replaced the numerator of Expression (7.7) with $n - 1$ times the sample variance, which is, of course, the sum of squares.

If we were to sample repeatedly n items from a normally distributed population, Expression (7.8) computed for each sample would yield a χ^2-distribution with $n - 1$ degrees of freedom. Notice that, although we have samples of n items, we have lost a degree of freedom because we are now employing a sample mean rather than the parametric mean. Figure 7.5, a sample distribution of variances, has a second scale along the abscissa, which is the first scale multiplied by the constant $(n - 1)/\sigma^2$. This scale converts the sample variances s^2 of the first scale into Expression (7.8). Since the second scale is proportional to s^2, the distribution of the sample variances serves to illustrate a sample distribution approximating χ^2. The distribution is strongly skewed to the right, as would be expected in a χ^2-distribution.

Figure 7.5 also shows the absolute expected frequencies of a χ^2-distribution for 4 degrees of freedom. The observed and expected frequencies do not agree well. There are more observations near the low values of χ^2, probably because the housefly wing length data of Table 6.1 are not quite normal. They do not extend from $-\infty$ to $+\infty$ but only from -2.44σ to $+2.44\sigma$. Means sampled from these data apparently were approximately normally distributed, but these ratios seem more sensitive to a departure from normality in the original population. In order to bring the population closer to normality we would have had to furnish a base population $n = 1000$ or more, rather than the mere 100 wing lengths of Table 6.1.

The expected frequencies are obtained from tables (for example, Table 7 in Pearson and Hartley, 1958) or by direct computation of the cumulative frequencies of the χ^2-distribution. Values of χ^2 and ν (degrees of freedom) serve as arguments, and the portion of the area of the curve to the right of a given value of χ^2 and for the given *df* is the function. Extensive tables of this sort are too cumbersome and also do not furnish the exact probability levels customarily required. Therefore, conventional χ^2-tables, such as Statistical Table **D,** give these common probabilities and degrees of freedom as arguments and list the χ^2 corresponding to the probability and the *df* as the functions. Each chi-square in Table **D** is the value of χ^2 beyond which the area under the χ^2-distribution for ν degrees of freedom represents the indicated probability. Just as we used subscripts to indicate the cumulative proportion of the area, as well as the degrees of freedom represented by a given value of t, we subscript χ^2 as follows: $\chi^2_{\alpha[\nu]}$ indicates the χ^2-value to the right of which is found proportion α of the area under a χ^2-distribution for ν degrees of freedom.

Let us learn how to use Table **D.** Looking at the distribution of $\chi^2_{[2]}$, we note that 90% of all values of $\chi^2_{[2]}$ would be to the right of 0.211, but only 5% of all values of $\chi^2_{[2]}$ would be greater than 5.991. Mathematical statisticians have shown that the expected value of $\chi^2_{[\nu]}$ (the mean of a χ^2-distribution) equals its degrees of freedom ν. Thus the expected value of a $\chi^2_{[5]}$-distribution is 5. When we examine 50% values (the medians) in the χ^2-table, we notice that they are generally lower than the expected values (the means). Thus for $\chi^2_{[5]}$ the 50% point is 4.351, which illustrates the asymmetry of the χ^2-distribution: The mean is to the right of the median. Our first application of the χ^2-distribution will be in the next section. Its most extensive use, however, will occur in Chapter 17.

7.7 Confidence Limits for Variances

We saw in the last section that the ratio $(n - 1)s^2/\sigma^2$ is distributed as χ^2 with $n - 1$ degrees of freedom. We take advantage of this fact when we set confidence limits to variances.

We can make the following statement about the ratio $(n - 1)s^2/\sigma^2$:

$$P\left\{\chi^2_{(1-(\alpha/2))[n-1]} \leq \frac{(n-1)s^2}{\sigma^2} \leq \chi^2_{(\alpha/2)[n-1]}\right\} = 1 - \alpha$$

This expression is similar to those in Section 7.3 and implies that the probability P that the indicated boundary values of $\chi^2_{[n-1]}$ cover this ratio is $1 - \alpha$. Simple algebraic manipulation of the quantities in the inequality within brackets yields

$$P\left\{\frac{(n-1)s^2}{\chi^2_{(\alpha/2)[n-1]}} \leq \sigma^2 \leq \frac{(n-1)s^2}{\chi^2_{(1-(\alpha/2))[n-1]}}\right\} = 1 - \alpha \tag{7.9}$$

Since $(n - 1)s^2 = \sum y^2$, we can simplify Expression (7.9) to

$$P\left\{\frac{\sum y^2}{\chi^2_{(\alpha/2)[n-1]}} \leq \sigma^2 \leq \frac{\sum y^2}{\chi^2_{(1-(\alpha/2))[n-1]}}\right\} = 1 - \alpha \tag{7.10}$$

This expression still looks formidable, but it means simply that if we divide the sum of squares $\sum y^2$ by the two values of $\chi^2_{[n-1]}$ bounding $1 - \alpha$ of the area of the $\chi^2_{[n-1]}$-distribution, the two quotients will enclose the true value of the variance σ^2 with a probability of $P = 1 - \alpha$.

A numerical example will make this clear. Suppose we have a sample of 5 housefly wing lengths with a sample variance of $s^2 = 13.52$. If we wish to set 95% confidence limits to the parametric variance, we evaluate Expression (7.10) for the sample variance s^2. We first calculate the sum of squares for this sample: $4 \times 13.52 = 54.08$. Then we look up the values for $\chi^2_{.025[4]}$ and $\chi^2_{.975[4]}$. Since 95% confidence limits are required, α in this case is equal to 0.05. These χ^2-values span between them 95% of the area under the χ^2-curve. They correspond to 11.143 and 0.484, respectively, and the limits in Expression (7.10) then become

$$L_1 = 54.08/11.143 \qquad \text{and} \qquad L_2 = 54.08/0.484$$

or

$$L_1 = 4.85 \qquad \text{and} \qquad L_2 = 111.74$$

This confidence interval is very wide, but we must not forget that the sample variance is, after all, based on only 5 individuals. Note also that the interval is asymmetrical around 13.52, the sample variance, in contrast to the confidence intervals encountered earlier, which were symmetrical around the sample statistic.

This method is called the **equal tails method** because an equal amount of probability is placed in each tail (for example, $2\frac{1}{2}\%$). It can be shown that in view of the skewness of the distribution of variances, this method does not yield the shortest possible confidence intervals. One may wish the confidence interval to be "shortest" in the sense that the ratio L_2/L_1 be as small as possible. Box 7.3 shows how to obtain these **shortest unbiased confidence intervals** for σ^2 using

Box 7.3 CONFIDENCE LIMITS FOR σ^2. METHOD OF SHORTEST UNBIASED CONFIDENCE INTERVALS.

Aphid stem mother femur lengths (from Boxes 2.1 and 4.2): $n = 25$; $s^2 = 0.1337$.

The factors from Statistical Table **O** for $\nu = n - 1 = 24$ *df* and confidence coefficient $1 - \alpha = 0.95$ are

$$f_1 = 0.5943 \qquad f_2 = 1.876$$

and for a confidence coefficient of 0.99 they are

$$f_1 = 0.5139 \qquad f_2 = 2.351$$

The 95% confidence limits for the population variance, σ^2, are given by the equations

$$L_1 = (\text{lower limit}) = f_1 s^2 = 0.5943(0.1337) = 0.07946$$
$$L_2 = (\text{upper limit}) = f_2 s^2 = 1.876(0.1337) = 0.2508$$

The 99% confidence limits are

$$L_1 = f_1 s^2 = 0.5139(0.1337) = 0.06871$$
$$L_2 = f_2 s^2 = 2.351(0.1337) = 0.3143$$

Box 7.4 CONFIDENCE LIMITS FOR p, THE BINOMIAL PARAMETER, AND λ, THE MEAN OF THE POISSON DISTRIBUTION. METHOD OF SHORTEST UNBIASED CONFIDENCE INTERVALS.

Binomial parameter, p

In Section 5.2 we described a sibship consisting of 17 offspring, 3 males and 14 females. The proportion of males is $3/17 = 0.1765$. If we look up the 95% shortest unbiased confidence limits for $k = 17$ and $Y = 3$ in Table **P**, we find $L_1 = 0.0499$ and $L_2 = 0.4165$. The limits are broad, indicating uncertainty in the location of the true proportion (although we can be confident in excluding 0.5).

Poisson parameter, λ

In a bacterial culture, six mutations of a certain type were found in one generation. What are the 99% confidence limits of the mean? According to Table **N**, 99% limits are $L_1 = 1.786$ and $L_2 = 15.813$. Such intervals have a 99% chance of including the true mean number of mutations.

Statistical Table **O,** based on the method of Tate and Klett (1959). With the help of Table **O,** which gives $(n - 1)/\chi^2_{p[n-1]}$, where p is an adjusted value of $\alpha/2$ or $1 - \alpha/2$ designed to yield the shortest unbiased confidence intervals, the computation is very simple and much faster than calculating confidence limits by means of χ^2.

Similar shortest unbiased confidence limits can be found for binomial parameters p and for the mean λ of a Poisson variable. Box 7.4 shows the simple computation by means of tables based on methods described in Crow (1956) and Crow and Gardner (1959).

7.8 Introduction to Hypothesis Testing

The most frequent application of statistics in biological research is to test a hypothesis. Statistical methods are important in biology because results of experiments are usually not clear-cut and therefore need statistical tests to support decisions between alternative hypotheses. A statistical test examines a set of sample data and, on the basis of an expected distribution of the data, leads to a decision about whether to accept the hypothesis underlying the expected distribution or reject that hypothesis and accept an alternative one. The nature of the tests varies with the data and the hypothesis, but the same general philosophy of hypothesis testing is common to all tests. Study the material in this section very carefully because it is fundamental to an understanding of every subsequent chapter in this book!

Let us consider again the sample of 17 animals of species A, 14 of which were females and 3 of which were males, which we discussed in Section 5.2. If our question is whether this litter could have come from a population with a true sex ratio of 1 : 1, we can visualize the following experiment. First we must create an infinitely sized population, half of whose members are females, the other half males. From this we can now repeatedly sample litters of 17 animals and examine how often we obtain a result of 14 females and 3 males. Since assembling an infinite population is impossible, we accomplish the same result with a finite population by sampling with replacement. This approach to hypothesis testing through randomization, almost always carried out by computer, is very common today, and, as we will see in Chapter 18, is the preferred approach in many cases.

In the example with the litter of 17 animals, however, we can simplify our task by relying on probability theory, using what we learned about the binomial distribution in Chapter 5. Assuming that the litter was sampled at random from a binomially distributed population, we concluded from Table 5.3 that if the sex ratio in the population was 1 : 1 ($p_{♀} = q_{♂} = 0.5$), the probability of obtaining a sample with 14 females and 3 males is 0.005188, making it very unlikely that such a result could be obtained by chance alone. We learned that it is conventional to include all ''worse'' outcomes—that is, all those that deviate even more from the outcome expected on the hypothesis $p_{♀} = q_{♂} = 0.5$. Including all worse outcomes, the probability is 0.006363, still a very small value. This computation is based on the one-tailed test, in which we are interested only in departures from the 1 : 1 sex ratio that show a preponderance of females. If we have no preconception about the direction of the departures from expectation, we must calculate the probability of obtaining a sample as deviant as 14 females and 3 males *in either direction* from expectation—that is, the probability either of obtaining a sample of 3 females and 14 males (and all worse samples) or of obtaining 14 females and 3 males (and all worse samples). Such a test is two-tailed, and since the distribution is symmetrical, we can simply double the previously discussed probability to yield 0.012726.

What does this probability mean? Our hypothesis is that $p_♀ = q_♂ = 0.5$. Let us call this hypothesis H_0, the **null hypothesis,** which is the hypothesis under test. It is called the null hypothesis because it states that there is no real difference between the true value of p in the population from which we sampled and the hypothesized value of $\hat{p} = 0.5$. For instance, in the current example we believe that our sample does not exhibit a 1 : 1 sex ratio only because of sampling error. As stated earlier, this probability was computed on the assumptions that the litter was sampled at random, that the population of samples is binomially distributed, and that $\hat{p} = 0.5$. In singling out the last assumption as the null hypothesis and considering it to be improbable in view of the outcome of the experiment, we are implicitly expressing our confidence in the validity of the other two assumptions. If the litter was not sampled at random and/or the data were not binomially distributed, the probability of obtaining the observed outcome would be seriously in error. When carrying out a statistical test, we usually make several assumptions associated with the test. The investigator chooses which of these to make the null hypothesis and which others to consider valid assumptions.

If the null hypothesis $p_♀ = q_♂ = 0.5$ is true, then approximately 13 samples out of 1000 will be as deviant or more deviant than this one in either direction *by chance alone.* Thus, it is quite *possible* to have arrived at a sample of 14 females and 3 males or 14 males and 3 females by chance, but it is not very *probable,* since so deviant an event would occur only about 13 out of 1000 times, or 1.3% of the time. If we actually obtain such a sample, we may make one of two decisions: that the null hypothesis is true (that is, the sex ratio is 1 : 1) and that the sample obtained by us just happened to be one of those in the tail of the distribution, or that so deviant a sample is too improbable an event to justify acceptance of the null hypothesis—that is, that the hypothesis about the sex ratio being 1 : 1 is not true.

Either of these decisions may be correct, depending upon the truth of the matter. If the 1 : 1 hypothesis is correct, then the first decision (to accept the null hypothesis) will be correct. If we decide to reject the hypothesis under these circumstances, we commit an error. *The rejection of a true null hypothesis* is called a **type I error.** On the other hand, if the true sex ratio of the population is other than 1 : 1, the first decision (to accept the 1 : 1 hypothesis) is an error, a so-called **type II error,** which is *the acceptance of a false null hypothesis.* Finally, if the 1 : 1 hypothesis is not true and we do decide to reject it, then again we make the correct decision. Thus, there are two kinds of correct decisions, accepting a true null hypothesis and rejecting a false null hypothesis, and two kinds of errors, type I, rejecting a true null hypothesis, and type II, accepting a false null hypothesis. The relationships between hypotheses and decisions can be summarized as follows:

		Null hypothesis	
		Accepted	Rejected
Null hypothesis	True	Correct decision	Type I error
	False	Type II error	Correct decision

Before we carry out a test, we have to decide what magnitude of type I error (rejection of true hypothesis) we will allow. There will always be some samples that by chance are very deviant. The most deviant of these are likely to mislead us into believing the null hypothesis to be untrue. If we permit 5% of samples to lead us into a type I error, then we will reject 5 out of 100 samples from the population, deciding that these are not samples from the given population. In the distribution under study, this means that we would reject all samples of 17 animals containing 13 of one sex and 4 of the other sex, as you can see by referring to column (3) of Table 7.2, where the expected frequencies of the outcomes on the hypothesis $p_{♀} = q_{♂} = 0.5$ are shown. This table is an extension of Table 5.3, which showed only a tail of the distribution. Actually, you would obtain a type I error slightly less than 5% if you summed relative expected frequencies for both tails starting with the class of 13 of one sex and 4 of the other. Summing the frequencies in Table 7.2, the relative expected frequency in the two tails is $2 \times 0.024{,}520{,}9 = 0.049{,}041{,}8$. In a discrete frequency distribution, such as the binomial, we cannot calculate errors of exactly 5% as we can in a continuous frequency distribution, where we can measure off exactly 5% of the area. If we decide on an approximate 1% error, we reject the hypothesis $p_{♀} = q_{♂}$ for all samples of 17 animals having 14 or more of one sex (from Table 7.2 we find that the $\hat{f}_{rel}$ in the tails sum to $2 \times 0.006{,}362{,}9 = 0.012{,}725{,}8$). Thus, the smaller the type I error we are prepared to accept, the more deviant a sample has to be for us to reject the null hypothesis H_0. Your natural inclination might be to have as little error as possible. You may decide to work with an extremely small type I error, such as 0.1% or even 0.01%, accepting the null hypothesis unless the sample is extremely deviant. The difficulty with such an approach is that although guarding against an error of the first kind, you might be falling into an error of the second kind (type II), accepting the null hypothesis when in fact it is not true and an alternative hypothesis H_1 is true. We will show later in the discussion how this comes about.

Table 7.2 RELATIVE EXPECTED FREQUENCIES FOR SAMPLES OF 17 ANIMALS UNDER TWO HYPOTHESES.

Binomial distribution.

(1) ♀♀	(2) ♂♂	(3) $H_0: p_♀ = q_♂ = \frac{1}{2}$ $\hat{f}_{rel}$	(4) $H_1: p_♀ = 2q_♂ = \frac{2}{3}$ $\hat{f}_{rel}$
17	0	0.000,007,6	0.001,015,0
16	1	0.000,129,7	0.008,627,2
15	2	0.001,037,6	0.034,508,6
14	3	0.005,188,0	0.086,271,5
13	4	0.018,158,0	0.150,975,2
12	5	0.047,210,7	0.196,267,7
11	6	0.094,421,4	0.196,267,7
10	7	0.148,376,5	0.154,210,4
9	8	0.185,470,6	0.096,381,5
8	9	0.185,470,6	0.048,190,7
7	10	0.148,376,5	0.019,276,3
6	11	0.094,421,4	0.006,133,4
5	12	0.047,210,7	0.001,533,3
4	13	0.018,158,0	0.000,294,9
3	14	0.005,188,0	0.000,042,1
2	15	0.001,037,6	0.000,004,2
1	16	0.000,129,7	0.000,000,2
0	17	0.000,007,6	0.000,000,0
Total		1.000,000,2	0.999,999,9

First let us learn some more terminology. A type I error is most frequently expressed as a probability and is symbolized by α. When expressed as a percentage it is known as the **significance level.** Thus a type I error of $\alpha = 0.05$ corresponds to a significance level of 5% for a given test. When we cut off areas proportional to α, the type I error, on a frequency distribution, the portion of the abscissa under the area that has been cut off is called the **rejection region,** or *critical region,* of a test, and the portion of the abscissa that would lead to acceptance of the null hypothesis is called the **acceptance region.** Figure 7.14A is a bar diagram showing the expected distribution of outcomes in the sex ratio example, given H_0. The dashed lines separate approximate 1% rejection regions from the 99% acceptance region.

Now let us take a closer look at the type II error, the probability of accepting the null hypothesis when it is false. If you try to evaluate the probability of a type II error, you immediately run into a problem. If the null hypothesis H_0 is false,

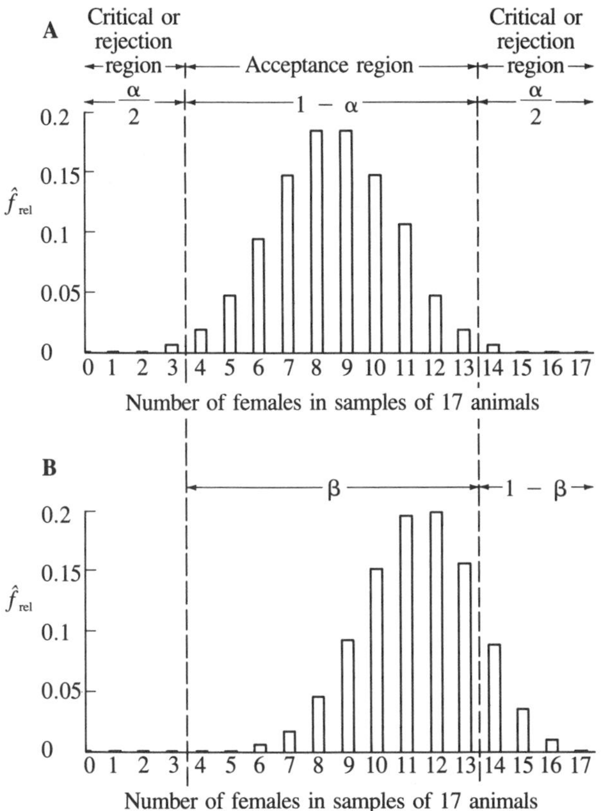

FIGURE 7.14 *Expected distributions of outcomes when sampling 17 animals from two hypothetical populations.* **A.** *$H_0: p♀ = q♂ = \frac{1}{2}$.* **B.** *$H_1: p♀ = 21♂ = \frac{2}{3}$. Dashed lines separate rejection, or critical, regions from the acceptance region of the distribution of* **A**. *Type I error α equals approximately 0.01.*

some other hypothesis H_1 must be true. But unless you can specify H_1, you are not in a position to calculate a type II error. Consider the following example. Suppose in our sex ratio case we have only two reasonable possibilities—(1) our old hypothesis $H_0: p_♀ = q_♂$, or (2) an alternative hypothesis $H_1: p_♀ = 2q_♂$, which states that the sex ratio is 2:1 in favor of females, so $p_♀ = \frac{2}{3}$ and $q_♂ = \frac{1}{3}$. We now have to calculate expected frequencies for the binomial distribution $(p_♀ + q_♂)^k = (\frac{2}{3} + \frac{1}{3})^{17}$ to find the probabilities of the outcomes under this hypothesis. These frequencies are shown graphically in Figure 7.14B and are tabulated and compared with expected frequencies of the earlier distribution in Table 7.2.

Suppose we had decided on a type I error of $\alpha \approx 0.01$, as shown in Figure 7.14A. At this significance level we would accept the H_0 for all samples of 17 having 13 or fewer animals of one sex. Approximately 99% of all samples will fall into this category. However, what if H_0 is not true and H_1 is true? Clearly, from the population represented by hypothesis H_1, we could also obtain outcomes in which one sex was represented 13 or fewer times in samples of 17. We have to calculate what proportion of the curve representing hypothesis H_1 will overlap the acceptance region of the distribution representing hypothesis H_0. In this case we find that 0.8695 of the distribution representing H_1 overlaps the acceptance region of H_0 (see Figure 7.14B). Thus, if H_1 is really true (and H_0 correspondingly false), we would erroneously accept the null hypothesis 86.95% of the time. This percentage corresponds to the proportion of samples from H_1 that fall within the limits of the acceptance regions of H_0. This proportion is called β, the type II error expressed as a proportion. In this example β is quite large. A sample of 17 animals is clearly unsatisfactory to discriminate between the two hypotheses. Although 99% of the samples under H_0 would fall in the acceptance region, 87% would do so under H_1. A single sample that falls in the acceptance region would not enable us to reach a decision between the hypotheses with a high degree of reliability. If the sample had 14 or more females, we would conclude that H_1 was correct. If it had 3 or fewer females we might conclude that neither H_0 nor H_1 was true. As H_1 approaches H_0 (as in $H_1: p_♀ = 0.55$, for example), the two distributions would overlap more and more and the magnitude of β would increase, making discrimination between the hypotheses even less likely. Conversely, if H_1 represented $p_♀ = 0.9$, the distributions would be much farther apart and type II error β would be reduced. Clearly, then, the magnitude of β depends, among other things, on the parameters of the alternative hypothesis H_1 and cannot be specified without knowledge of these parameters.

When the alternative hypothesis is fixed, as in the previous example ($H_1: p_♀ = 2q_♂$), the magnitude of the type I error α we are prepared to tolerate will determine the magnitude of the type II error β. The smaller the rejection region α in the distribution under H_0, the greater the acceptance region $1 - \alpha$ in this distribution. The greater $1 - \alpha$, however, the greater its overlap with the distribution representing H_1, and hence the greater β. Convince yourself of this in Figure 7.14. By moving the dashed lines outward we are reducing the critical regions representing type I error α in diagram A. But as the dashed lines move outward, more of the distribution of H_1 in diagram B will lie in the acceptance region of the null hypothesis. Thus, by decreasing α we are increasing β and in a sense defeating our own purposes.

In most applications, scientists want to keep both of these errors small, since they do not wish to reject a null hypothesis when it is true, nor do they wish to accept it when another hypothesis is correct. We will see what steps can be taken to decrease β while holding α constant at a preset level. Note, however, that there are special applications, often nonscientific, in which one type of error is less serious than the other, and our strategy of testing or method of procedure would obviously take this into account. Thus, if you were a manufacturer producing a certain item according to specifications (which correspond to the null hypothesis in this case), you would wish to maximize your profits and to reject as few as possible, which is equivalent to making α small. That is why in industrial statistics α is known as **producers' risk.** You might not be as concerned with samples that come from a population specified by the alternative hypothesis H_1 but that appear to conform with H_0 because they fall within its acceptance range. Such products could conceivably be marketed as conforming to specification H_0. On the other hand, as a consumer you would not mind so much a large value of α, representing a large proportion of rejects in the manufacturing process. You would, however, be greatly concerned about keeping β as small as possible, since you would not wish to accept items as conforming to H_0, which in reality were samples from the population specified by H_1 that might be of inferior quality. For this reason β is known in industrial statistics as **consumers' risk.**

Let us summarize what we have learned up to this point. When we have to carry out a statistical test, we first specify a null hypothesis H_0 and establish a significance level that corresponds to a probability of α for a type I error. In the case of the sex ratios, we defined $H_0 : p_{♀} = q_{♂}$ and $\alpha \approx 0.01$. Having done this, we take a sample and test whether the sample statistic is within the acceptance region of the null hypothesis. Our sample turned out to be 14 females and 3 males. Since this sample statistic falls beyond the acceptance region, we reject the null hypothesis and conclude that this sample came from a population in which $p_{♀} \neq q_{♂}$.

If we can specify an alternative hypothesis, we can calculate the probability of type II error. In this case $H_1 : p_{♀} = 2q_{♂}$ and $\beta = 0.8634$. This is the probability of accepting the null hypothesis when, in fact, the alternative hypothesis is true. In certain special situations in which the alternative hypotheses can be clearly specified, as in genetics and in this example, one might then test the previous alternative hypothesis, changing it into the null hypothesis. Thus, we might now wish to test whether the true sex ratio is $2_{♀♀} : 1_{♂}$. From Figure 7.14B it is obvious that the probability of 14 males and 3 females is very small and can be ignored. The probability of obtaining 14 or more females under the new null hypothesis is $1 - \beta$ of the old alternative hypothesis, as illustrated in Figure 7.14B. Hence this probability would be 0.1366 and, if we accept $\alpha = 0.05$, we cannot reject the new null hypothesis.

Significance levels can be varied at will by the investigator. The choices are limited, however, because for many tests cumulative probabilities of the appropriate distributions have not been tabulated. One must use published probability levels, which are commonly 0.05, 0.01, and 0.001, although several others are occasionally encountered. When a null hypothesis has been rejected at a specified level of α, we say that the sample is **significantly different** from the parametric or hypothetical population at probability $P \leq \alpha$. Generally, values of α greater than 0.05 are not considered to be **statistically significant.** A significance level of 5% ($P = 0.05$) corresponds to one type I error in 20 trials, a level of 1% ($P = 0.01$) to one error in 100 trials. Significance levels less than 1% ($P \leq 0.01$) are nearly always adjudged significant; those between 5% and 1%

may be considered significant at the discretion of the investigator. Since statistical significance has a special technical meaning (H_0 rejected at $P \leq \alpha$), we shall use the adjective *significant* only in this sense; its use in scientific papers and reports, unless such a technical meaning is clearly implied, should be discouraged. For general descriptive purposes synonyms such as important, meaningful, marked, noticeable, and others can serve to underscore differences and effects.

A brief remark on null hypotheses represented by asymmetrical probability distributions is in order here. Suppose our null hypothesis in the sex ratio case had been $H_0: p_♀ = \frac{2}{3}$, as discussed above. The distribution of samples of 17 offspring from such a population is shown in Figure 7.14B. Because this distribution is clearly asymmetrical, the critical regions have to be defined independently. For a given two-tailed test we can either double the probability P of a deviation in the direction of the closer tail and compare $2P$ with α, the conventional level of significance; or we can compare P with $\alpha/2$, half the conventional level of significance. In the latter case, 0.025 is the maximal value of P conventionally considered significant.

We review now what we have learned by means of a second example, this time involving a continuous frequency distribution—the normally distributed housefly wing lengths from Table 6.1—of parametric mean $\mu = 45.5$ and variance $\sigma^2 = 15.21$. Means based on 5 items sampled from these are also distributed normally (see Figure 7.1). Assume that someone presents you with a single sample of 5 housefly wing lengths and you wish to test whether it could have come from the specified population. Your null hypothesis will be $H_0: \mu = 45.5$ or $H_0: \mu = \mu_0$, where μ is the true mean of the population from which you sampled and μ_0 stands for the hypothetical parametric mean of 45.5. Assume for the moment that we have no evidence that the variance of our sample is very much greater or smaller than the parametric variance of the housefly wing lengths. (If it were, it would be unreasonable to assume that our sample comes from the specified population. There is a critical test of the assumption about the sample variance, which we will discuss later.) The curve at the center of Figure 7.15 represents the expected distribution of means of samples of 5 housefly wing lengths from the specified population. Acceptance and rejection regions for a type I error $\alpha = 0.05$ are delimited along the abscissa. The boundaries of the rejection regions are computed as follows (remember that $t_{[\infty]}$ is equivalent to the normal distribution):

$$L_1 = \mu_0 - t_{.05[\infty]}\sigma_{\bar{Y}} = 45.5 - (1.96)(1.744) = 42.08$$

and

$$L_2 = \mu_0 + t_{.05[\infty]}\sigma_{\bar{Y}} = 45.5 + (1.96)(1.744) = 48.92$$

Thus we would consider it improbable for means less than 42.08 or greater than 48.92 to have been sampled from this population. For such sample means, we would therefore reject the null hypothesis. The test we are proposing is two-tailed because we have no a priori assumption about the possible alternatives to

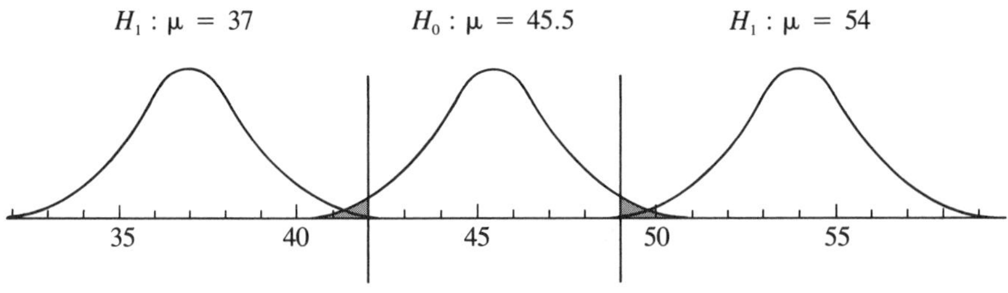

FIGURE 7.15 *Expected distribution of means of samples of 5 housefly wing lengths from normal populations specified by μ as shown above curves and $\sigma_{\bar{Y}} = 1.744$. Center curve represents null hypothesis, $H_0: \mu = 45.5$, curves at sides represent alternative hypotheses, $\mu = 37$ or $\mu = 54$. Vertical lines delimit 5% rejection regions for the null hypothesis ($2\frac{1}{2}$% in each tail, shaded).*

our null hypothesis. If we could assume that the true mean of the population from which the sample was taken could be only equal to or greater than 45.5, the test would be one-tailed.

Now let us examine alternative hypotheses. One alternative hypothesis might be that the true mean of the population from which our sample stems is 54.0, but that the variance is the same as before. We can express this assumption as $H_1: \mu = 54.0$ or $H_1: \mu = \mu_1$, where μ_1 stands for the alternative parametric mean 54.0. From the table of the areas of the normal curve (Table **A**) and our knowledge of the variance of the means, we can calculate the proportion of the distribution implied by H_1 that would overlap the acceptance region implied by H_0. We find that 54.0 is 5.08 measurement units from 48.92, the upper boundary of the acceptance region of H_0. This corresponds to $5.08/1.744 = 2.91\sigma_{\bar{Y}}$ units. From Statistical Table **A** we find that 0.0018 of the area will lie beyond 2.91σ at one tail of the curve. Thus under this alternative hypothesis 0.0018 of the distribution of H_1 will overlap the acceptance region of H_0. This is β, the type II error under this alternative hypothesis. Actually this is not entirely correct. Since the left tail of the H_1 distribution goes all the way to negative infinity, it will leave the acceptance region and cross over into the left-hand rejection region of H_0. However, this represents only an infinitesimal amount of the area of H_1 (the lower critical boundary of H_0, 42.08, is $6.83\sigma_{\bar{Y}}$ units from $\mu_1 = 54.0$; the area is less than 1×10^{-9}) and can be ignored.

Our alternative hypothesis H_1 specified that μ_1 is 8.5 units greater than μ_0. As we said, however, we may have no a priori reason to believe that the true mean of our sample is either greater or less than μ. Therefore we may simply assume that the true mean is 8.5 measurement units away from 45.5. In such a case we must similarly calculate β for the alternative hypothesis: $\mu_1 = \mu_0 - 8.5$. Thus the alternative hypothesis becomes $H_1: \mu = 54.0$ or 37.0, or $H_1: \mu = \mu_1$, where μ_1 represents either 54.0 or 37.0, the alternative parametric means. Since the distributions are symmetrical, β is the same for both alternative hypotheses. Type II error for hypothesis H_1 is therefore 0.0018, regardless of which of the two alternative hypotheses is correct. If H_1 is really true, 18 out of 10,000 samples would lead to an incorrect acceptance of H_0, a very low proportion of error. These relations are shown in Figure 7.15.

You may rightly ask what reason we have to believe that the alternative parametric value for the mean is 8.5 measurement units to either side of $\mu_0 =$ 45.5. Any justification for such a belief would be quite unusual. As a matter of

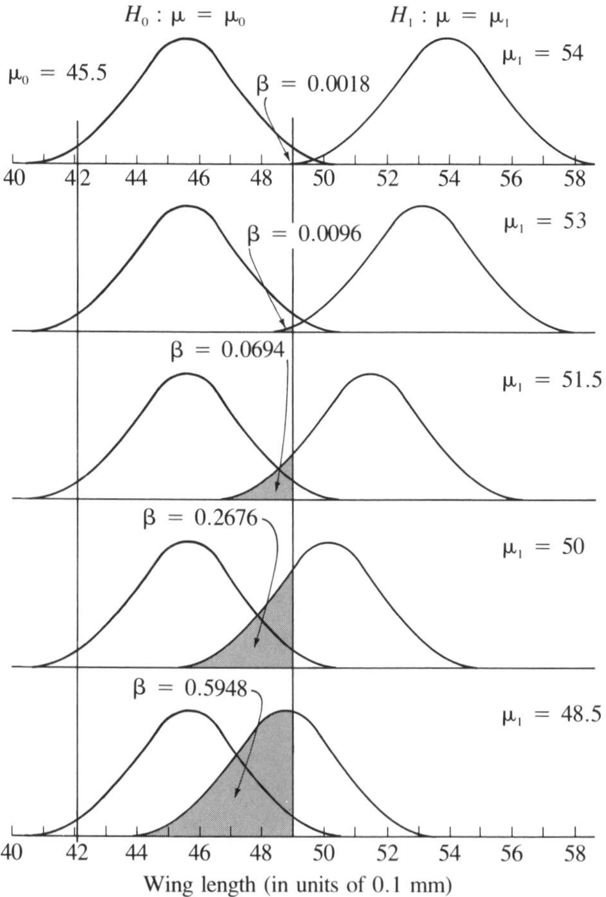

FIGURE 7.16 *Diagram to illustrate increases in type II error, β, as alternative hypothesis, H_1, approaches null hypothesis, H_0—that is, μ_1 approaches μ. Shading represents β. Vertical lines mark off 5% critical regions ($2\frac{1}{2}$% in each tail) for the null hypothesis. To simplify the graph, the alternative distributions are shown for one tail only. Data identical to those in Figure 7.15.*

fact, the true mean may just as well be 7.5 or 6.0 or any number of units to either side of μ_0. If we draw curves for $H_1: \mu = \mu_0 \pm 7.5$, we find that β has increased considerably, since the curves for H_0 and H_1 are now closer together. Thus the magnitude of β depends on how far the alternative parametric mean is from the parametric mean of the null hypothesis. As the alternative mean approaches the parametric mean, β increases to a maximum value of $1 - \alpha$, which is the area of the acceptance region under the null hypothesis. At this maximum the two distributions would be superimposed upon each other. Figure 7.16 illustrates the increase in β as μ_1 approaches μ, starting with the test illustrated in Figure 7.15. To simplify the graph, the alternative distributions are shown for one tail only. Thus we see clearly that β is not a fixed value but varies with the nature of the alternative hypothesis.

An important concept in connection with hypothesis testing is the **power** of a test. The power is $1 - \beta$, the complement of β, and is the probability of rejecting the null hypothesis when it is false and the alternative hypothesis is correct. For any given test, we would like to have the quantity $1 - \beta$ be as large as possible and the quantity β as small as possible. Since we generally cannot specify a given alternative hypothesis, we have to describe β or $1 - \beta$ for a continuum of alternative values. When $1 - \beta$ is graphed in this manner the result is called a **power curve** for the test under consideration. Figure 7.17 shows the power curve for the housefly wing length example just discussed. This figure can be compared with Figure 7.16, from which it is directly derived.

Figure 7.16 emphasizes the type II error β, and Figure 7.17 graphs the complement of this value, $1 - \beta$. Note that the power of the test falls off sharply as the alternative hypothesis approaches the null hypothesis. Common sense confirms these conclusions: We can make clear and firm decisions about whether our sample comes from a population of mean 45.5 or 60.0. The power is essentially 1. But if the alternative hypothesis is that $\mu_1 = 45.6$, differing only by 0.1 from the value assumed under the null hypothesis, deciding which of these hypotheses is true is difficult and the power will be very low.

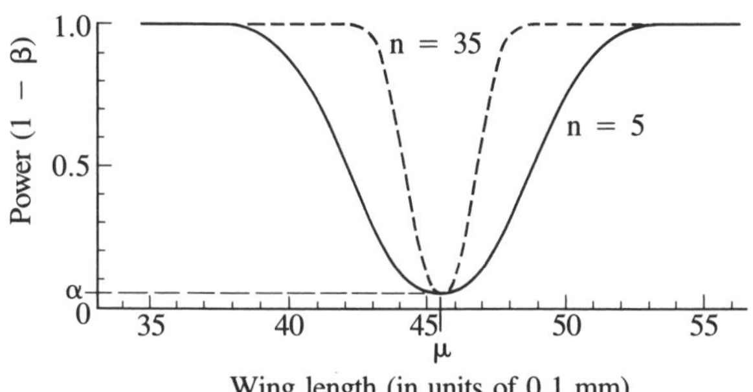

FIGURE 7.17 *Power curves for testing $H_0: \mu = 45.5$. $H_1 \mu \neq 45.5$ for $n = 5$ (as in Figures 7.15 and 7.16) and for $n = 35$.*

To improve the power of a given test (that is, decrease β) while keeping α constant for a stated null hypothesis, we must increase sample size. If, instead of sampling 5 wing lengths we had sampled 35, the distribution of means would be much narrower. Thus rejection regions for the identical type I error would now commence at 44.21 and 46.79. Although the acceptance and rejection regions would remain the same proportionately, the acceptance region would become much narrower in absolute value. Previously we could not, with confidence, reject the null hypothesis for a sample mean of 48.0. Now, when based on 35 individuals, a mean as deviant as 48.0 would occur only 15 times out of 100,000 and the hypothesis would, therefore, be rejected.

What has happened to type II error? Since the distribution curves are not as wide as before, they overlap less. If the alternative hypothesis $H_1: \mu = 54.0$ or 37.0 is true, the probability that the null hypothesis could be accepted by mistake (type II error) is infinitesimally small. If we let μ_1 approach μ_0, β will increase, of course, but it will always be smaller than the corresponding value for sample size $n = 5$. This comparison is shown in Figure 7.17, where the power for the test with $n = 35$ is much higher than that for $n = 5$. If we were to increase our sample size to 100 or 1000, the power would be increased still further. Thus we reach an important conclusion: If a given test is not sensitive enough, we can increase its sensitivity (= power) by increasing sample size.

There is yet another way of increasing the power of a test. If we cannot increase sample size, we may increase the power by changing the nature of the test. Different statistical techniques testing roughly the same hypothesis may differ substantially in both the magnitude and the slopes of their power curves. Tests that maintain higher power levels over substantial ranges of alternative hypotheses are clearly to be preferred. The popularity of the nonparametric tests mentioned in several places in this book has grown not only because of their computational simplicity but in many cases also because their power curves are less affected by failure of assumptions than are those of the parametric methods. Nonparametric tests, however, have lower overall power than do parametric ones, when all the assumptions of the parametric test are met.

While on the subject of different tests we should discuss briefly the **size** of a test. When we follow the protocols of a given statistical test and decide on a specified type I error α, the correctness of that value depends on how sensitive the test is to departures from its assumptions in the sample analyzed. The intended type I error rate may be only nominal, and the actual error rate—termed, somewhat infelicitously, the size of the test—may be greater or smaller. In the former case we speak of a **liberal test** (i.e., we reject the null hypothesis more often than we should), the latter case represents a **conservative test** (i.e., we reject the null hypothesis less often).

Let us look briefly at a one-tailed test. The null hypothesis is $H_0: \mu_0 = 45.5$ as before. The alternative hypothesis, however, assumes that we have reason to believe that the parametric mean of the population from which our sample has been taken cannot be less than $\mu_0 = 45.5$. If it is different from that value, it can be only greater than 45.5. We might have two grounds for such an hypothesis.

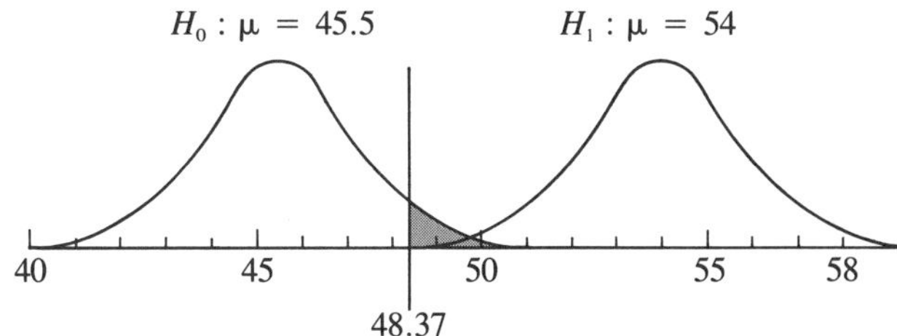

FIGURE 7.18 *One-tailed significance test for the distribution of Figure 7.15. The vertical line now cuts off 5% rejection region from one tail of the distribution (shaded area).*

First, we might have a biological reason for such a belief. For example, our parametric flies might be a dwarf population, making any other population from which our sample could have come necessarily bigger. A second reason might be that we are interested in only one direction of difference. For example, we may be testing the effect of a chemical in the larval food intended to increase the size of the sample of flies. Therefore, we would expect that $\mu_1 \geq \mu_0$, and we are not interested in testing for any μ_1 that is less than μ_0 because such an effect is the opposite of what we anticipate. Similarly, if we are investigating the effect of a certain drug as a cure for cancer, we might wish to compare the untreated population that has a mean fatality rate θ (from cancer) with the treated population, whose rate is θ_1. Our alternative hypotheses would be $H_1 : \theta_1 < \theta$. That is, we would not be interested in any θ_1 that is greater than θ because if our drug increases mortality from cancer, it certainly is not much of a prospect for a cure.

When such a one-tailed test is performed, the rejection region along the abscissa is only under one tail of the curve representing the null hypothesis. Thus, for our housefly data (distribution of means of sample size $n = 5$) the rejection region will be in one tail of the curve only and for a 5% type I error will appear as shown in Figure 7.18. We compute the critical boundary as $45.5 + (1.645)(1.744) = 48.37$. (The 1.645 is $t_{.10[\infty]}$, which corresponds to the 5% value for a one-tailed test.) Compare this rejection region, which rejects the null hypothesis for all means greater than 48.37, with the two rejection regions in Figure 7.16, which reject the null hypothesis for means lower than 42.08 and greater than 48.92. In Figure 7.18 the alternative hypothesis is considered for one tail of the distribution only, and the power curve of the test is not symmetrical but is drawn with respect to only one side of the distribution.

7.9 Tests of Simple Hypotheses Using the Normal and *t*-Distributions

We will now apply our new knowledge of hypothesis testing to some simple examples involving the normal and *t*-distributions.

The first test is fairly uncommon. Given a single variate Y_1 or a sample mean $\bar{Y}$, we wish to know whether it differs significantly from a parametric mean μ—that is, whether it could have been sampled from a population with parameter μ for the character in question. Such a test assumes that we know the parametric mean of a population, which is not the usual case. It is more likely that μ represents a generally accepted standard or one promulgated by a government control or inspection agency. Examples might be the mean percentage of butterfat in a type of milk, or the mean number of microorganisms permitted in a grade of milk or in a vaccine. In systematics, a certain mean may represent a species or subspecies population, and we might test whether a given sample can be considered part of that population.

If we know the variance of the standard or parametric population, and if the variable is distributed normally the test is straightforward. Either

$$\frac{\bar{Y}_1 - \mu_0}{\sigma} \tag{7.11}$$

or

$$\frac{\bar{Y} - \mu_0}{\sigma_{\bar{Y}}} \tag{7.12}$$

would be distributed normally. As an example, we will test whether a single housefly with wing length of 43 units belongs to the population we are familiar with ($\mu = 45.5$, $\sigma = 3.90$). We compute Expression (7.11):

$$\frac{43 - 45.5}{3.90} = \frac{-2.5}{3.90} = -0.641$$

We ignore the negative sign of the numerator, since this is a two-tailed test, so it is immaterial in which direction the item deviates from the mean. This is a two-tailed test because the question posed is whether the sample could belong to the population ($H_0: \mu = 45.5$). The alternative hypothesis is that the sample belongs to a population whose mean is either greater than or less than 45.5 ($H_1: \mu \neq 45.5$). Since the individual reading is 0.641 standard deviations from the mean, the table of areas of the normal curve (Statistical Table **A**, with linear interpolation) informs us that 0.2392 of the items under the normal curve lie between μ_0 and 0.641 standard deviations, and hence 0.2608 of the items lie beyond 0.641 standard deviations on one side of the mean; thus 0.5216 of the items will be 0.641 or more standard deviations from the mean on both sides of the distribution. Therefore, by any of the conventional levels of significance (5% or 1%), we would accept the null hypothesis and conclude that the housefly wing length of 43 units belonged to the specified population. Ordinarily we would simply use the *t*-table (Statistical Table **B**) and look up $df = \infty$. We would note that $t_{.9[\infty]} = 0.126$ and $t_{.5[\infty]} = 0.674$. Therefore the probability of such a deviation is between 0.50 and 0.90, but closer to 0.50, as we found more accurately from the table of the areas of the normal curve.

Similarly, a sample of 100 milk yields that gave a mean of 70.05 could be tested to determine whether it belonged to the population of Table 6.1 ($\mu = 66.61$, $\sigma = 11.1597$). We compute Expression (7.12),

$$\frac{70.05 - 66.61}{11.1597/\sqrt{100}} = \frac{3.44}{1.11597} = 3.08$$

and find that the sample mean is 3.08 standard deviations above the parametric mean. Only 0.0011 of the area of the normal curve (from Table **A**) extends beyond 3.08 standard deviations, and even when doubled for the two-tailed test appropriate in this case (0.0022 or 0.22% of the area of the curve), the percentage is much smaller than permitted at the conventional level of significance. We are led to reject the null hypothesis in this case. The mean based on the sample of 100 items does not seem to belong to the parametric population we know.

Now let us proceed to a test that is similar except that we do not know the parametric standard deviation. Government regulations prescribe that the standard dosage in a certain biological preparation should be 600 activity units per cubic centimeter. We prepare 10 samples of this preparation and test each for potency. We find that the mean number of activity units per sample is 592.5 units per cubic centimeter and the standard deviation of the samples is 11.2. Does our sample conform to the government standard? Stated more precisely, our null hypothesis is $H_0: \mu = \mu_0$. The alternative hypothesis is that the dosage is not equal to 600, or $H_1: \mu \neq \mu_0$. We calculate the significance of the deviation $\bar{Y} - \mu_0$ expressed in standard deviation units. The appropriate standard deviation of means (the standard error of the mean), *not* the standard deviation of items, because the deviation is that of a sample mean around a parametric mean. We therefore calculate $s_{\bar{Y}} = s/\sqrt{n} = 11.2/\sqrt{10} = 3.542$. We next test the deviation $(\bar{Y} - \mu_0)/s_{\bar{Y}}$. We saw in Section 7.4 that a deviation divided by an estimated standard deviation will be distributed according to the *t*-distribution with $n - 1$ degrees of freedom. We therefore write

$$t_s = \frac{\bar{Y} - \mu_0}{s_{\bar{Y}}} \tag{7.13}$$

This expression indicates that we would expect this deviation to be distributed as a *t*-variate. Note that in Expression (7.13) we wrote t_s. In most textbooks you will find this ratio simply identified as *t*, but in fact the *t*-distribution is a parametric and theoretical distribution that generally is only approached, but never equaled, by observed, sampled data. This may seem a minor distinction, but readers should be quite clear that in any hypothesis testing of samples we are only *assuming* that the distributions of the tested variables follow certain theoretical probability distributions. To conform with general statistical practice, the *t*-distribution should really have a Greek letter (such as τ), with *t* serving as the sample statistic. Since this would violate long-standing practice, however, we prefer to use the subscript *s* to indicate the sample value.

The test is very simple. We calculate Expression (7.13),

$$t_s = \frac{592.5 - 600}{3.542} = \frac{-7.5}{3.542} = -2.12 \qquad df = n - 1 = 9$$

and compare it with the expected values for t at 9 degrees of freedom. Since the t-distribution is symmetrical, we ignore the sign of t_s and always look up its positive value in Table **B.** The two values on either side of $t_s = 2.12$ are $t_{.05[9]} = 2.26$ and $t_{.10[9]} = 1.83$. These are t-values for two-tailed tests, appropriate in this instance because the alternative hypothesis is that $\mu \neq 600$; that is μ can be smaller or greater. The significance level of our value of t_s appears to be between 5% and 10%; if the null hypothesis is true, the probability of obtaining a deviation as great or greater than 7.5 is somewhere between 0.05 and 0.10. By conventional levels of significance this probability is insufficient for declaring the sample mean as significantly different from the stnadard. We thus accept the null hypothesis. In conventional language we would report the results of the statistical analysis as follows: "The sample mean is not significantly different from the accepted standard." Such a statement in a scientific report should always be backed up by a probability value, and the proper presentation is $0.10 > P > 0.05$, which means that the probability of such a deviation is between 0.05 and 0.10. Another way of saying this is that the value of t_s is *not significant* (frequently abbreviated *ns*).

Asterisks often appear after the computed value of the significance test—for example, $t_s = 2.86^{**}$. The symbols generally represent the following probability ranges:

$$* = 0.05 \geq P > 0.01 \qquad ** = 0.01 \leq P > 0.001 \qquad *** = P \leq 0.001$$

However, since some authors occasionally imply other ranges by these asterisks, the meaning of the symbols must be specified in each scientific report.

It might be argued that in the case of the biological preparation, the concern of the tester should not be whether the sample differs significantly from a standard, but whether it is significantly *below* the standard. This may be one of those biological preparations in which an excess of the active component is of no harm but a shortage would make the preparation ineffective at the conventional dosage. Then the test becomes one-tailed, performed in exactly the same manner except that the critical values of t for a one-tailed test are at half the probabilities of the two-tailed test. Thus 2.26, the former 0.05-value, becomes $t_{.025[9]}$, and 1.83, the former 0.10-value, becomes $t_{.05[9]}$, making our observed t_s-value of 2.12 "significant at the 5% level" or, more precisely stated, significant at $0.05 > P > 0.025$. If we are prepared to accept a 5% significance level, we would consider the preparation significantly below the standard.

You may be surprised that the same example, employing the same data and significance tests, can lead to two different conclusions, and you may begin to wonder whether some of the things you hear about statistics and statisticians are not, after all, correct. The explanation lies in the fact that the two results are answers to different questions. If we test whether our sample is significantly different from the standard in either direction, we must conclude that it is not different enough for us to reject the null hypothesis. If, on the other hand, we exclude from consideration the fact that the true sample mean μ could be greater than the established standard μ_0, the difference as we find it appears significant. It is obvious from this example that in any statistical test one must clearly state whether a one-tailed or a two-tailed test has been performed, if the nature of the example is such that there could be any doubt about the matter. We should also point out that such a difference in the outcome of the results is not necessarily typical. It is only because the outcome in this case is in a borderline area between clear significance and nonsignificance. Had the difference between sample and standard been 10.5 activity units, the sample would have been unquestionably significantly different from the standard by the one-tailed or the two-tailed test.

The promulgation of a standard mean is generally insufficient for establishing a rigid standard for a product. If the variance among the samples is sufficiently large, establishing a significant difference between the standard and the sample mean will never be possible. This point is important and should be quite clear to you. Remember that the standard error can be increased in two ways—by lowering sample size or by increasing the standard deviation of the replicates. Both of these are undesirable aspects of any experimental setup. Yet, in the relatively rare cases where the purpose of the investigator is to show *no* difference rather than to show a significant difference, this result can be obtained superficially simply by having a large enough error due to small sample size or to poor technique. One of us once had to review a research article, the main point of which was that the unpleasant side effects of drug *B* (newly developed) were no different from those of drug *A* (which was well established). The writers of the article used the appropriate statistical test but used it on only very few individuals. They apparently failed to understand that by using few enough individuals you could never refute the null hypothesis (that effect of drug *B* equals effect of drug *A*). One needs to calculate beforehand the sample size necessary to establish differences of a given magnitude at a certain level of significance (see Section 9.8).

The test described earlier for the biological preparation leads us to a general test for the significance of any statistic—that is, for the significance of a deviation of any statistic from a parameter—which is illustrated in Box 7.5. Such a test applies whenever the statistics are expected to be distributed normally. When the standard error is estimated from the sample, the t-distribution is used. However, since the normal distribution is just a special case, $t_{[\infty]}$, of the t-distribution, most statisticians uniformly apply the t-distribution with the appropriate degrees of freedom from 1 to infinity.

In Box 7.5, as an example, we are testing the significance of g_1; that is, we are testing whether the deviation of the observed value of g_1 is significantly different from the expected value of γ_1 for a normal distribution, which is zero. The significance of a number of coefficients, such as g_1, g_2, regression coefficients,

Box 7.5 TESTING THE SIGNIFICANCE OF A STATISTIC (THE SIGNIFICANCE OF A DEVIATION FROM A PARAMETER). FOR NORMALLY DISTRIBUTED STATISTICS.

Computation

1. Compute t_s as the following ratio

$$t_s = \frac{St - St_p}{s_{St}}$$

where St is a sample statistic, St_p is the parametric value against which the sample statistic is to be tested, and s_{St} is its estimated standard error, obtained from Box 7.1.

2. The pertinent hypotheses are

$$H_0: St = St_p \qquad H_1: St \neq St_p$$

for a two-tailed test, and

$$H_0: St = St_p \qquad H_1: St > St_p$$

or

$$H_0: St = St_p \qquad H_1: St < St_p$$

for a one-tailed test.

3. In the two-tailed test look up the critical value of $t_{\alpha[\nu]}$ in Statistical Table **B,** where α is the type I error agreed upon and ν is the degrees of freedom pertinent to the standard error employed (see Box 7.1). In the one-tailed test look up the critical value of $t_{2\alpha[\nu]}$ for a significance level of α.
4. Accept or reject the appropriate hypothesis in step **2** on the basis of the t_s value in step **1** compared with critical values of t in step **3.**

Example of computation

Test the significance of g_1 from Box 6.2 ($g_1 = 0.18936$, $n = 9465$). Note from Box 7.1 that the standard error of g_1 is $\sqrt{6/n}$ for samples of $n > 150$.

1. $$t_s = \frac{(g_1 - \gamma_1)}{s_{g1}} = \frac{0.18934 - 0}{\sqrt{6/9465}} = \frac{0.18934}{0.025178} = 7.52$$

2. The null hypothesis is that the distribution is not skewed—that is, that $\gamma_1 = 0$. The pertinent test is clearly two-tailed, since g_1 can be either negative (skewed to the left) or positive (skewed to the right). We wish to test whether there is *any* skewness; that is, the alternative hypothesis is that $\gamma_1 \neq 0$. Thus

$$H_0: \gamma_1 = 0 \qquad H_1: \gamma_1 \neq 0$$

BOX 7.5 (CONTINUED)

3. The appropriate number of degrees of freedom employed with s_{g_1} is ∞ (see Box 7.1). We therefore employ critical values of t with degrees of freedom $\nu = \infty$:

$$t_{.05[\infty]} = 1.960 \qquad t_{.01[\infty]} = 2.576 \qquad t_{.001[\infty]} = 3.291$$

4. $t_s = 7.52^{***}$ ($P \ll 0.001$). This expression is a terse symbolism for stating that the observed g_1 is significant at $P < 0.001$. Actually, g_1 is significant at far less than $P < 0.001$, since 7.52 is so much larger than the critical value 3.291. We indicate this by the doubled less than sign. The probability is thus much less than one in a thousand that a g_1 as observed can be sampled from a population in which γ_1 equals zero (no skewness). We accept the alternative hypothesis and conclude that $\gamma_1 \neq 0$. Since g_1 is positive, this indicates that $\gamma_1 > 0$; that is, the birth weights are skewed to the right. Thus the distribution of the birth weights is asymmetrical, drawn out at the right tail.

and correlation coefficients, is tested in this manner, by the appropriate standard error. Values of g_1 and g_2 based on small samples are only approximately normally distributed. For critical tests of such values consult D'Agostino and Tietjen (1973) and D'Agostino and Pearson (1973).

In Box 7.5 the conclusions are stated in statistical shorthand (the conventional way in which such results are reported in the scientific literature) and in terms of the actual detailed implications of the test, which you should understand thoroughly before reading further.

7.10 TESTING THE HYPOTHESIS $H_0: \sigma^2 = \sigma_0^2$

The method of Box 7.5 can be used only if the statistic is distributed normally. In the case of the variance this is not so. As we saw in Section 7.6, sums of squares divided by σ^2 follow the χ^2-distribution. Therefore, testing the hypothesis that a sample variance is different from a parametric variance requires that we employ the χ^2-distribution.

Let us use the biological preparation of the last section as an example. We were told that the standard deviation based on 10 samples was 11.2. Therefore the variance must have been 125.44. Suppose the government postulates that the variance of samples from the preparation should be no greater than 100.0. Is our sample variance significantly above 100.0? Remembering from Expression (7.8) that $(n - 1)s^2/\sigma^2$ is distributed as $\chi^2_{[n-1]}$, we proceed as follows. We first calculate

$$\begin{aligned} X^2 &= (n-1)s^2/\sigma^2 \\ &= (9)125.44/100 \\ &= 11.290 \end{aligned}$$

Note that we compute the quantity X^2 rather than χ^2, again to emphasize that we are obtaining a sample statistic that we shall compare to the parametric distribution. The use of X^2 to denote the sample statistic approximating a χ^2-distribution has become quite widely established. Following the general outline of Box 7.5 we next establish our null and altenative hypotheses, which are $H_0: \sigma^2 \leq \sigma_0^2$ and $H_1: \sigma^2 > \sigma_0^2$; that is, we are to perform a one-tailed test. The critical value of χ^2 is found next as $\chi^2_{\alpha[\nu]}$, where α indicates the type I error and ν the pertinent degrees of freedom. Quantity α represents the proportion of the χ^2-distribution to the right of the given value as described in Section 7.6, and you see now why we used the symbol α for that portion of the area; it corresponds with the type I error. For 9 degrees of freedom we find in statistical Table D that

$$\chi^2_{.05[9]} = 16.919 \qquad \chi^2_{.10[9]} = 14.684 \qquad \chi^2_{.50[9]} = 8.343$$

Notice that the probability of getting a χ^2 as large as 11.290 is therefore higher than 0.10 but less than 0.50, assuming that the null hypothesis is true. Thus X^2 is not significant at the 5% level; we have no basis for rejecting the null hypothesis, and we must conclude that the variance of the 10 samples of the biological preparation may be no greater than the standard permitted by the government. If we had decided to test whether the variance is different from the standard, permitting it to deviate in either direction, the hypotheses for this two-tailed test would have been $H_0: \sigma^2 = \sigma_0^2$ and $H_1: \sigma^2 \neq \sigma_0^2$, and a 5% type I error would have yielded the following critical values for the two-tailed test:

$$\chi^2_{.975[9]} = 2.700 \qquad \chi^2_{.025[9]} = 19.023$$

The values represent chi-squares at points cutting off $2\frac{1}{2}\%$ rejection regions at each tail of the χ^2-distribution. A value of $\chi^2 < 2.700$ or >19.023 would have been evidence that the sample variance did not belong to this population. Our value of $X^2 = 11.290$ would have led again to an acceptance of the null hypothesis.

In the next chapter we will introduce another significance test to test the hypotheses about variances of this section. This test is the mathematically equivalent F-test, which is a more general test, allowing us to test the hypothesis that two sample variances come from populations with equal variances.

Investigators who wish to test the null hypothesis $H_0: \sigma^2 = \sigma_0^2$ may wish to know how large a sample they should use to test this hypothesis with a specified power. An example of how to determine the appropriate sample size is featured in Zar (1984, section 8.12).

EXERCISES 7

7.1 Differentiate between type I and type II errors. What do we mean by the power of a statistical test?

7.2 Since it is possible to test a statistical hypothesis with a sample of any size, why are larger sample sizes preferred?

7.3 The 95% confidence limits for μ as obtained in a given sample were 4.91 and 5.67 g. Is it correct to say that 95 times out of 100 the population mean, μ, falls inside the interval from 4.91 to 6.57 g? If not, what would the correct statement be?

7.4 Set 95% confidence limits to the means in Table 7.1. Are these limits all correct? (That is, do they contain μ?)

7.5 Set 99% confidence limits to the mean, median, coefficient of variation, variance, and g_2 for the birth weight data given in Boxes 4.1, 4.3, and 6.2 *Answer:* The lower limits are 109.540, 109.060, 12.136, 178.698, and −0.0405, respectively.

7.6 In Section 5.3 the coefficient of dispersion was given as an index of whether or not data agree with a Poisson distribution. Since in a true Poisson distribution the mean, μ, equals the parametric variance, σ^2, the coefficient of dispersion is analogous to Expression (7.8). Using the weed seed data from Table 5.7, test the hypothesis that the true variance is equal to the sample mean—in other words, that we have sampled from a Poisson distribution (in which the coefficient of dispersion should equal unity).

7.7 In a study measuring bill length of the dusky flycatcher, Johnson (1966) found that the bill length for the males had a mean of 8.14 ± 0.021 and a coefficient of variation of 4.67%. On the basis of this information, infer how many specimens must have been used. *Answer:* $n = 328$.

7.8 In a study of mating calls in the tree toad *Hyla ewingi,* Littlejohn (1965) found the note duration of the call in a sample of 39 observations from Tasmania to have a mean of 189 ms and a standard deviation of 32 ms. Assign 95% confidence intervals to the mean and to the variance.

7.9 Using the method described in Exercise 7.6, test the agreement of the observed distributions to Poisson distributions by testing the hypothesis that the true coefficient of dispersion equals unity for the data of Tables 5.5, 5.6, 5.9, and 5.10. Note that in these examples the chi-square table is not adequate, so approximate critical values must be computed using the method given with Statistical Table **D.** Section 8.3 presents an alternative significance test that avoids this problem. *Answer:* For the data in Table 5.6, $(n - 1) \times \mathrm{CD} = 1308.30$, $\chi^2_{.025[588]} \approx 656.61$.

7.10 In direct klinokinetic behavior relating to temperature, animals turn more often in the warm end of a gradient and less often in the cold end, but the direction of turning is random. In a computer simulation of such behavior, the following results were found. The mean position along a temperature gradient was found to be −1.352. The standard deviation was 12.267, and n equaled 500 individuals. The gradient was marked off in units: Zero corresponded to the middle of the gradient, the starting point of the animals; minus corresponded to the cold end; and plus corresponded to the warmer end. Test the hypothesis that direct klinokinetic behavior did not result in a tendency toward aggregation in either the warm or the cold end; that is, test the hypothesis that μ, the mean position along the gradient, was zero.

7.11 Distinguish between power and size of a test.

7.12 Box 9.4 shows the effects of 5 treatments on growth in plant tissue cultures. For each sample prepare a boxplot. Line up the boxplots to permit comparison among the treatments.

7.13 We observe 10 weed seeds in a 2-ounce sample of grass seeds. Estimate the confidence limits of the number of weed seeds *per ounce* of grass seed.

7.14 Prepare a boxplot for the birth weight data of Boxes 4.1, 4.3, and 6.2. What do you learn from inspecting this plot?

8 Introduction to Analysis of Variance

This chapter focuses on the analysis of variance, a method, developed by R. A. Fisher, that is fundamental to much of the application of statistics in biology and especially to experimental design. Analysis of variance is used to test for differences among sample means and differences among linear combinations of means. Its name is derived from the fact that variances are used to measure the differences among means. A simple application of the analysis of variance is to test whether two or more sample means could have been obtained from populations with the same parametric mean. Where only two samples are involved, the *t*-test has traditionally been used to test significant differences between means. The analysis of variance, however, is a more general test, which permits testing two samples as well as many, and we are therefore introducing it at this early stage in order to equip the reader with a powerful weapon for his or her statistical arsenal. We will discuss the *t*-test for two samples as a special case in Section 9.4.

A knowledge of analysis of variance is indispensable to any modern biologist, and after you have mastered it, you will undoubtedly use it many times to test scientific hypotheses. The analysis of variance, however, is more than a technique for statistical analysis. Once it is understood, analysis of variance is a tool that can provide an insight into the nature of variation of natural events, into Nature in short, which is possibly of even greater value than the knowledge of the method as such. Like other models in science, however, analysis of variance may create constructions of nature in the mind of the scientist that give rise to misleading or unproductive conclusions. See Lewontin (1974) for an in-depth discussion of this issue.

In Section 8.1 we approach this subject via familiar ground, the sampling experiment of the housefly wing lengths. From these samples we will obtain two independent estimates of the population variance. We digress in Section 8.2 to introduce yet another continuous distribution, the *F*-distribution, needed for the significance test in analysis of variance. Section 8.3 is another digression, in which we show how the *F*-distribution can be used to test whether two samples have the same variance. We are then ready for Section 8.4, in which we examine the effects of subjecting the samples to different treatments. Section 8.5 describes the partitioning of sums of squares of degrees of freedom, the actual *analysis* of variance. Sections 8.6 and 8.7 take up in a more formal way the two scientific models for which the analysis of variance is appropriate, the so-called fixed treatment effects model (Model I) and the variance component model (Model II).

Except for Section 8.3, the entire chapter is largely theoretical. We will postpone the practical details of computation until Chapter 9. A thorough understanding of the material in Chapter 8 is necessary, however, for working out the examples of analysis of variance in Chapter 9.

One final comment: Professor J. W. Tukey of Princeton University has contributed many a well-turned phrase or term (in addition to many important theoretical concepts) to the field of statistics. The abbreviation *anova* for "analysis of variance" is one of them, and it quickly became standard. We will use *anova* interchangeably with *analysis of variance* throughout the text.

8.1 Variances of Samples and Their Means

In this section we discuss analysis of variance in terms of the familiar sampling experiment of housefly wing lengths (see Experiment 6.1 and Table 6.1), in which we combined seven samples of 5 wing lengths to form samples of 35. We have reproduced one such sample in Table 8.1. The seven samples of 5, here called groups, are listed vertically in the upper half of the table. Before we explain Table 8.1 further, we must become familiar with some additional terminology and symbolism for dealing with this kind of problem. We call our samples **groups;** they are sometimes called *classes* or other names that we will learn later. In any analysis of variance problem we will have two or more such samples or groups, and the symbol a will represent the number of groups. Thus in this example $a = 7$. Each group or sample is based on n items as before; in Table 8.1, $n = 5$. The total number of items in the table is a times n, which in this case equals 7×5, or 35.

In an anova, summation signs can no longer be as simple as heretofore. We can sum either the items of one group only or the items of the entire table. We therefore have to use superscripts with the summation symbol. In line with our policy of using the simplest possible notation whenever this is not likely to lead to misunderstanding, we will use $\Sigma^n Y$ to indicate the sum of the items of a group and $\Sigma^{an} Y$ to indicate the sum of all the items in the table. The mean of each group, symbolized by $\bar{Y}_i$, is shown in the row underneath the horizontal dividing line and is computed simply as $\Sigma^n Y_i/n$. The bottom row in Table 8.1 contains $\Sigma^n y_i^2$, the sum of squares of $\bar{Y}$, separately for each group.

From each of these sums of squares we can obtain an estimate of the population variance of housefly wing length. Thus in the first group $\Sigma^n y^2 = 29.2$.

Table 8.1 SEVEN SAMPLES (GROUPS) OF 5 WING LENGTHS OF RANDOMLY SELECTED HOUSEFLIES.

Parametric mean, $\mu = 45.5$; variance, $\sigma^2 = 15.21$.

	a groups ($a = 7$)								
	1	2	3	4	5	6	7	Computation of sum of squares of means	Computation of total sum of squares
n individuals per group ($n = 5$)	41	48	40	40	49	40	41		
	44	49	50	39	41	48	46		
	48	49	44	46	50	51	54		
	43	49	48	46	39	47	44		
	42	45	50	41	42	51	42		
$\bar{Y}_i$	43.6	48.0	46.4	42.4	44.2	47.4	45.4	$\bar{\bar{Y}} = 45.34$	$\bar{Y} = 45.34$
$\sum^{n} y_i^2$	29.2	12.0	75.2	45.2	98.8	81.2	107.2	$\sum^{a} (\bar{Y} - \bar{\bar{Y}})^2 = 25.417$	$\sum^{an} y^2 = 575.886$

Data from Experiment 6.1 and Table 6.1.

Therefore our estimate of the population variance is

$$s^2 = \frac{\sum^n y^2}{n - 1} = \frac{29.2}{4} = 7.3$$

a rather low estimate compared to what we know the true variance to be. Since we have a sum of squares for each group, we could obtain an estimate of the population variance from each of these. It stands to reason, however that we would get a better estimate if we averaged these separate variance estimates in some way. We can obtain this value by computing the weighted average of the variances using Expression (4.2) in Section 4.1. Actually, in this instance a simple average would suffice, since each estimate of the variance is based on samples of the same size. We prefer to give the general formula, however, which works equally well for this case and for instances of unequal sample sizes, where the weighted average is necessary.

In this case each sample variance s_i^2 is weighted by its degrees of freedom, $w_i = n_i - 1$, resulting in a sum of squares (Σy_i^2), since $(n_i - 1)s_i^2 = \Sigma y_i^2$. Thus the numerator of Expression (4.2) is the sum of the sums of squares. The denominator is $\Sigma^a(n_i - 1) = 7 \times 4$, the sum of the degrees of freedom of each group. The average variance therefore is

$$s^2 = \frac{29.2 + 12.0 + 75.2 + 45.2 + 98.8 + 81.2 + 107.2}{28} = \frac{448.8}{28} = 16.029$$

This quantity is an estimate of 15.21, the parametric variance of housefly wing lengths. We might call this estimate, based on 7 independent estimates of variances of groups, the *average variance within groups* or simply **variance within groups.** Note that we use the expression *within* groups, although in previous chapters we used the term variance *of* groups. The reason is that the variance estimates used for computing the average variance have up to now all come from sums of squares measuring the variation within one column. As we will see, one can also compute variances among groups, cutting across group boundaries.

To obtain a second estimate of the population variance we treat the seven group means, $\bar{Y}$, as though they were a sample of seven observations. The resulting statistics are shown in the lower right of Table 8.1, headed "Computation of sum of squares of means." There are seven means in this example; in the general case there will be a means. First we compute $\bar{\bar{Y}}$, the grand mean of the group means, as $\bar{\bar{Y}} = \Sigma^a\bar{Y}/a$. The sum of the seven means is $\Sigma^a\bar{Y} = 317.4$ and the grand mean is $\bar{\bar{Y}} = 45.34$, a fairly close approximation to the parametric mean $\mu = 45.5$. The sum of squares represents the deviations of the group means from the grand mean, $\Sigma^a(\bar{Y} - \bar{\bar{Y}})^2$, which, when evaluated, is 25.417.

From the sum of squares of the means we obtain a **variance among the means** in the conventional way as follows: $\Sigma^a(\bar{Y} - \bar{\bar{Y}})^2/(a - 1)$. We divide by $a - 1$ rather than $n - 1$ because the sum of squares was based on a items (means). Thus variance of the means $s_{\bar{Y}}^2 = 25.417/6 = 4.2362$. We learned in

Expression (7.2) that when randomly sampling from a single population,

$$\sigma_{\bar{Y}}^2 = \frac{\sigma^2}{n}$$

and hence

$$\sigma^2 = n\sigma_{\bar{Y}}^2$$

Thus, we can estimate a variance of items by multiplying the variance of means by the sample size on which the means are based (assuming we have sampled at random from a common population). When we do this for our present example, we obtain $s^2 = 5 \times 4.2362 = 21.181$. This value is another estimate of the parametric variance 15.21. It is not as close to the true value as the previous estimate based on the average variance within groups, but this is to be expected since it is based on only 7 "observations." To distinguish it from the variance of means from which it was computed, as well as from the variance within groups with which it will be compared, we will call this variance the **variance among groups;** its value, n times the variance of means, is an independent estimate of the parametric variance σ^2 of the housefly wing lengths. It may not be clear at this stage why the two estimates of σ^2 that we have obtained, the variance within groups and the variance among groups, are independent. We ask you to take on faith that they are.

Let us review what we have done so far by expressing in a more formal way the data table for an anova and the computations carried out up to now. Table 8.2 is a generalized table for data such as the samples of housefly wing lengths. Do not be put off by the unfamiliar notation! Each individual wing length is represented by Y, subscripted to indicate its position in the data table. The wing length of the jth fly from the ith sample or group is given by Y_{ij}. Thus you will notice

Table 8.2 DATA ARRANGED FOR SIMPLE ANALYSIS OF VARIANCE, SINGLE CLASSIFICATION, COMPLETELY RANDOMIZED.

		a groups						
		1	2	3	⋯	i	⋯	a
n items	1	Y_{11}	Y_{21}	Y_{31}	⋯	Y_{i1}	⋯	Y_{a1}
	2	Y_{12}	Y_{22}	Y_{32}	⋯	Y_{i2}	⋯	Y_{a2}
	3	Y_{13}	Y_{23}	Y_{33}	⋯	Y_{i3}	⋯	Y_{a3}
	⋮	⋮	⋮	⋮	·	⋮	·	⋮
	j	Y_{1j}	Y_{2j}	Y_{3j}	⋯	Y_{ij}	⋯	Y_{aj}
	⋮	⋮	⋮	⋮	·	⋮	·	⋮
	n	Y_{1n}	Y_{2n}	Y_{3n}	⋯	Y_{in}	⋯	Y_{an}
Means	$\bar{\bar{Y}}$	$\bar{Y}_1$	$\bar{Y}_2$	$\bar{Y}_3$	⋯	$\bar{Y}_i$	⋯	$\bar{Y}_a$

that the first subscript changes with each column representing a group in the table, and the second subscript changes with each row representing an individual item. (Those of you familiar with matrix algebra may wonder why the subscript notation here is the reverse of the one in that field. It is simply a well-established usage in anova, and we prefer not to change it.) Using this notation we can compute the variance of sample 1 as

$$\frac{1}{n-1}\sum_{j=1}^{j=n}(Y_{1j}-\bar{Y}_1)^2$$

The variance within groups, which is the average variance of the samples, is computed as

$$\frac{1}{a(n-1)}\sum_{i=1}^{i=a}\sum_{j=1}^{j=n}(Y_{ij}-\bar{Y}_i)^2$$

Note the double summation. It means that we first set $i = 1$ (i being the index of the outer Σ), when we start with the first group. We sum the squared deviations of all items from the mean of the first group, changing index j of the inner Σ from 1 to n in the process. We then return to the outer summation, set $i = 2$, and sum the squared deviations for group 2 from $j = 1$ to $j = n$. This process is continued until i, the index of the outer Σ, is set to a. In other words, we sum all the squared deviations within one group first and add this sum to similar sums from all the other groups. The variance among groups is computed as

$$\frac{n}{a-1}\sum_{i=1}^{i=a}(\bar{Y}_i-\bar{\bar{Y}})^2$$

Now that we have two independent estimates of the population variance, what shall we do with them? We might wish to find out whether they do in fact estimate the same parameter. To test this hypothesis we need a statistical test that will evaluate the probability that the two sample variances are from the same population. Such a test employs the F-distribution, which we discuss in the next section.

8.2 THE *F*-DISTRIBUTION

Let us devise yet another sampling experiment. This one is quite tedious if done by hand, so we won't ask you to carry it out. Assume that you are sampling at random from a normally distributed population, such as the housefly wing lengths with mean μ and variance σ^2. The sampling procedure consists of first sampling n_1 items and calculating their variance s_1^2, followed by sampling n_2 items and calculating their variance s_2^2. The degrees of freedom of the two variances are $\nu_1 = n_1 - 1$ and $\nu_2 = n_2 - 1$. Sample sizes n_1 and n_2 may or may not be equal to each other but are fixed for any one sampling experiment. Thus, for example, we might always sample 8 wing lengths for the first sample (n_1) and

6 wing lengths for the second sample (n_2). After each pair of values (s_1^2 and s_2^2) has been obtained, we calculate

$$F_s = \frac{s_1^2}{s_2^2}$$

This ratio will be close to 1, because these variances are estimates of the same quantity. Its value will depend on the relative magnitudes of variances s_1^2 and s_2^2. If we repeatedly take samples of sizes n_1 and n_2, calculating the ratios F_s of their variances, the average of these ratios will approach $(n_2 - 1)/(n_2 - 3)$, which is close to 1.0 when n_2 is large. Statisticians have worked out the expected distribution of this statistic, which is called the **F-distribution** in honor of R. A. Fisher. This distribution is another described by a complicated mathematical function that need not concern us here.

Unlike the t- and χ^2-distributions, the shape of the F-distribution is determined by *two* values for degrees of freedom, ν_1 and ν_2. Thus for every possible combination of values ν_1, ν_2, each ν ranging from 1 to infinity, there exists a separate F-distribution. Remember that, like the t-distribution and the χ^2-distribution, the F-distribution is a theoretical probability distribution. Variance ratios based on sample variances, s_1^2/s_2^2, are sample statistics that may or may not follow the F-distribution. We have, therefore, distinguished the sample variance ratio by calling it F_s, conforming to our convention of separate symbols for sample statistics as distinct from probability distributions (such as t_s and X^2 for t and χ^2, respectively).

We have discussed the generation of an F-distribution by repeatedly taking two samples from the same normal distribution. We could also have generated the F-distribution by sampling from two separate normal distributions differing in their mean but identical in their parametric variances—that is, with $\mu_1 \neq \mu_2$ but $\sigma_1^2 = \sigma_2^2$. Thus we obtain an F-distribution whether the samples come from the same normal population or from different ones, as long as their variances are identical.

Figure 8.1 shows several representative F-distributions. For very low degrees of freedom the distribution is L-shaped, but it becomes humped and strongly skewed to the right as both degrees of freedom increase. Statistical Table **F** shows the cumulative probability distribution of F for several selected probability values. The values in the table represent $F_{\alpha[\nu_1,\nu_2]}$, where α is the proportion of the F-distribution to the right of the given F-value (in one tail) and ν_1, ν_2 are the degrees of freedom pertaining to the numerator and the denominator of the variance ratio, respectively. The table is arranged with ν_1, the degrees of freedom pertaining to the upper (numerator) variance, across the top and ν_2, the degrees of freedom pertaining to the lower (denominator) variance, along the left margin. At each intersection of degree of freedom values, we list several values of F for decreasing magnitudes of α. The first value is $\alpha = 0.75$. Note that for $\alpha = 0.5$, F is not $(n_2 - 1)/(n_2 - 3)$, as you might expect, because F for $\alpha = 0.5$ is the median and not the mean of the asymmetrical F-distribution. Since in tests of

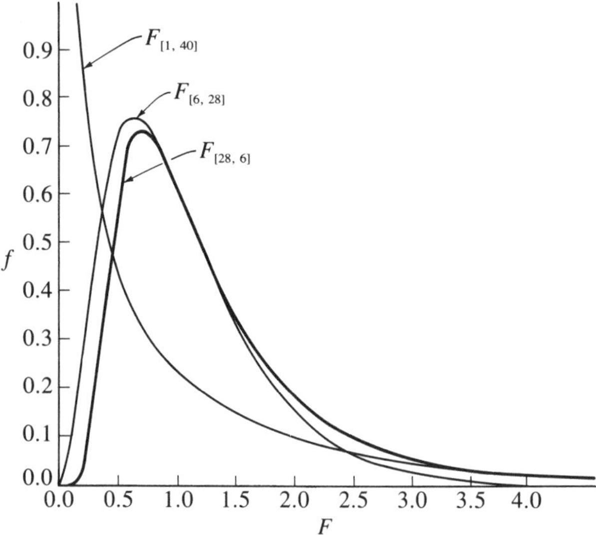

FIGURE 8.1 *Three representative F-distributions.*

hypotheses we are usually interested in small values of α, the right tail of the curve is especially emphasized in most F-tables. For example, an F-distribution with $\nu_1 = 6$, $\nu_2 = 28$ is 2.45 at $\alpha = 0.05$. Thus 0.05 of the area under the curve lies to the right of $F = 2.45$, as Figure 8.2 shows. Only 0.01 of the area under the curve lies to the right of $F = 3.53$. Thus if we have a null hypothesis $H_0: \sigma_1^2 = \sigma_2^2$, with the alternative hypothesis $H_1: \sigma_1^2 > \sigma_2^2$, we would use a one-tailed F-test as illustrated by Figure 8.2.

We can now test the two variances obtained in the sampling experiment of Section 8.1 and Table 8.1. The variance among groups based on 7 means was 21.180 and the variance within 7 groups of 5 individuals was 16.029. Our null hypothesis is that the two variances estimate the same parametric variance; the alternative hypothesis in an anova is always that the parametric variance estimated by the variance among groups is greater than that estimated by the variance within groups. The reason for this restrictive alternative hypothesis, which leads to a one-tailed test, will be explained in Section 8.4.

We calculate the variance ratio: $F_s = s_1^2/s_2^2 = 21.181/16.029 = 1.32$. Before we can inspect the F-table we have to know the appropriate degrees of freedom for this variance ratio. We will learn simple formulas for degrees of freedom in an anova later, but at the moment let us reason it out for ourselves. The upper variance (among groups) was based on the variance of 7 means; hence it should have $a - 1 = 6$ degrees of freedom. The lower variance was based on an average of 7 variances, each of them based on 5 individuals yielding 4 degrees of freedom per variance; $a(n - 1) = 7 \times 4 = 28$ degrees of freedom. Thus the

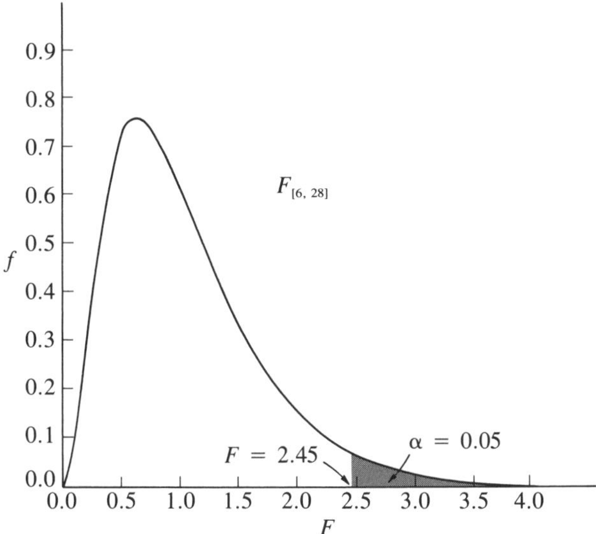

FIGURE 8.2 *Frequency curve of the F-distribution for 6 and 28 degrees of freedom, respectively. A one-tailed 5% rejection region is marked off at F = 2.45.*

upper variance has 6, the lower variance 28 degrees of freedom. If we check Table **F** for $\nu_1 = 6$, $\nu_2 = 28$ we find that $F_{.05[6,28]} = 2.45$. For $F = 1.32$, corresponding to the F_s-value actually obtained, α is between 0.5 and 0.25. Thus, we may expect between half and one-fourth of all variance ratios of samples based on 6 and 28 degrees of freedom, respectively, to have F_s-values greater than 1.32. We have no evidence to reject the null hypothesis and therefore conclude that the two sample variances estimate the same parametric variance. This conclusion corresponds, of course, to what we knew anyway from our sampling experiment. Since the seven samples were taken from the same population, the estimate using the variance of their means is expected to yield another estimate of the parametric variance of housefly wing length.

Whenever the alternative hypothesis is more general, i.e., that the two parametric variances are unequal (rather than the restrictive hypothesis $H_1 : \sigma_1^2 > \sigma_2^2$), the sample variance s_1^2 could be smaller as well as greater than s_2^2. This leads to a two-tailed test, and in such cases a 5% type I error means that rejection regions of $2\frac{1}{2}\%$ will occur at each tail of the curve. Because the F-distribution is asymmetrical, these rejection regions will not look identical, as they do in the t-distribution. Figure 8.3 shows the 5% two-tailed rejection regions for the same F-distribution illustrated in Figure 8.2. Since usually only the right half of the F-distribution is tabulated (Table **F** shows only $\alpha = 0.75$ in the left half of the curve), when faced with a two-tailed test we deliberately divide the greater variance by the lesser variance, and for a type I error of α, we look up the F-value

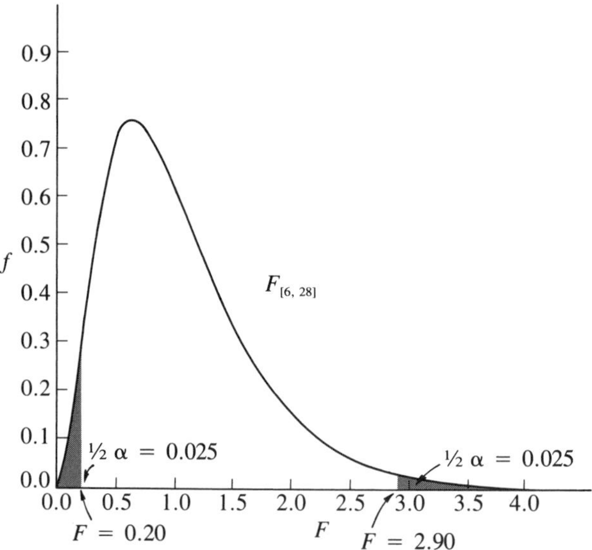

FIGURE 8.3 *Frequency curve of the F-distribution for 6 and 28 degrees of freedom, respectively. A two-tailed 5% rejection region ($2\frac{1}{2}$% in each tail) is marked off at F = 0.20 and F = 2.90, respectively.*

for $\alpha/2$ in the F-table. Computational details of this method are illustrated in Section 8.3.

It is sometimes necessary to obtain F-values for $\alpha > 0.5$ (that is, in the left half of the F-distribution). Since, as we have just learned, these values are rarely tabulated, they can be obtained by using the following simple relationship:

$$F_{\alpha[\nu_1,\,\nu_2]} = \frac{1}{F_{(1-\alpha)[\nu_2,\nu_1]}} \tag{8.1}$$

Note that $F_{.05[5,24]} = 2.62$. If we wish to obtain $F_{.95[5,24]}$ (the F-value to the right of which lies 95% of the area of the F-distribution with 5 and 24 degrees of freedom, respectively), we first have to find $F_{.05[24,5]} = 4.53$. Then $F_{.95[5,24]}$ is the reciprocal of 4.53, which equals 0.221. Thus 95% of an F-distribution with 5 and 24 degrees of freedom lies to the right of 0.221.

There is an important relationship between the F-distribution and the χ^2-distribution. You may remember that the ratio $X^2 = \Sigma y^2/\sigma^2$ was distributed as a χ^2 with $n - 1$ degrees of freedom. If you divide the numerator of this expression by $n - 1$, you obtain the ratio $F_s = s^2/\sigma^2$, which is a variance ratio with an expected distribution of $F_{[n-1,\infty]}$. The upper degrees of freedom are $n - 1$ (the degrees of freedom of the sum of squares or sample variance). The lower degrees of freedom are infinite because only on the basis of an infinite number of items can we obtain the true, parametric variance of a population. Therefore, by dividing a value of X^2 by $n - 1$ degrees of freedom, we obtain an F_s-value with $n - 1$ and ∞ *df*, respectively. In general $\chi^2_{[\nu]}/\nu = F_{[\nu,\infty]}$. We can convince ourselves of this fact by inspecting the F- and χ^2-tables. From the χ^2-table (statistical Table **D**) we find that $\chi^2_{.05[10]} = 18.307$. Dividing this value by 10 *df*, we obtain 1.8307. From the F-table (Table **F**) we find, for $\nu_1 = 10$, $\nu_2 = \infty$, that $F_{.05[10,\infty]} = 1.83$. Thus the two statistics of significance are closely related and, lacking a χ^2-table, we could make do with an F-table alone, using the values of $\nu F_{[\nu,\infty]}$ in place of $\chi^2_{[\nu]}$.

Before we return to analysis of variance, we will quickly apply our new knowledge of the F-distribution to testing a hypothesis about two sample variances.

8.3 THE HYPOTHESIS $H_0: \sigma_1^2 = \sigma_2^2$

A test of the null hypothesis that two normal populations represented by two samples have the same variance is illustrated in Box 8.1. As we will see later, some tests leading to a decision about whether the means of two samples come from the same population assume that the population variances are equal. This test, however, is of interest in its own right. It is often necessary to test whether two samples could have come from populations that have the same variance. In genetics we may need to know whether an offspring generation is more variable for a character than the parent generation. In systematics we might like to find

Box 8.1 TESTING THE SIGNIFICANCE OF DIFFERENCES BETWEEN TWO VARIANCES.

Survival in days of the cockroach *Blattella vaga* when kept without food or water.

Females	$n_1 = 10$	$\bar{Y}_1 = 8.5$ days	$s_1^2 = 3.6$
Males	$n_2 = 10$	$\bar{Y}_2 = 4.8$ days	$s_2^2 = 0.9$

$H_0: \sigma_1^2 = \sigma_2^2$ $\qquad$ $H_1: \sigma_1^2 \neq \sigma_2^2$

SOURCE: Data modified from Willis and Lewis (1957).

The alternative hypothesis is that the two variances are unequal. We have no reason to suppose that one sex should be more variable than the other. In view of the alternative hypothesis this is a two-tailed test. Since only the right tail of the F-distribution is tabulated extensively in Statistical Table **F** and in most other F-tables, we calculate F_s as the ratio of the greater variance over the lesser one:

$$F_s = \frac{s_1^2}{s_2^2} = \frac{3.6}{0.9} = 4.00$$

BOX 8.1 CONTINUED

Because the test is two-tailed, we look up the critical value $F_{\alpha/2[\nu_1,\nu_2]}$, where α is the type I error accepted, $\nu_1 = n_1 - 1$ and $\nu_2 = n_2 - 1$, the degrees of freedom for the upper and lower variance, respectively. Whether we look up $F_{\alpha/2[\nu_1,\nu_2]}$ or $F_{\alpha/2[\nu_2,\nu_1]}$ depends on whether sample 1 or sample 2 has the greater variance and was placed in the numerator.

From Table **F** we find $F_{.025[9,9]} = 4.03$, $F_{.05[9,9]} = 3.18$, and $F_{.10[9,9]} = 2.44$. Because this is a two-tailed test we double these probabilities. Thus, the F-value of 4.03 represents a probability of $\alpha = 0.05$, since the right-hand tail area of $\alpha = 0.025$ is matched by a similar left-hand area to the left of $F_{.975[9,9]} = 1/F_{.025[9,9]} = 0.248$. Therefore, assuming the null hypothesis is true, the probability of observing an F-value greater than 4.00 and smaller than $1/4.00 = 0.25$ is $0.10 > P > 0.05$. Strictly speaking, the two sample variances are not significantly different—the two sexes were equally variable in their duration of survival. The outcome is close enough to the 5% significance level, however, to make us suspicious that the variances were in fact different. It would be desirable to repeat this experiment in the hope that more decisive results would emerge.

out whether two local populations are equally variable. In experimental biology we may wish to demonstrate under which of two experimental setups the readings will be more variable. In general, the less variable setup would be preferred; if both setups were equally variable the experimenter would pursue the one that is simpler or less costly to undertake.

The test employed is usually a two-tailed test, since the alternative hypothesis is generally $H_1: \sigma_1^2 \neq \sigma_2^2$.

8.4 Heterogeneity Among Sample Means

We will now modify the data of Table 8.1 (see Section 8.1). Suppose these seven groups of houseflies did not represent random samples from the same population but resulted instead from the following experiment: Each sample was reared in a separate culture jar and the medium in each of the culture jars was prepared in a different way. Some had more water added, others more sugar, yet others more solid matter. Let us assume that sample 7 represents the standard medium against which we propose to compare the other samples. The changes in the medium affect the sizes of the flies that emerge from it; the sizes of the flies in turn affect the wing lengths we have been measuring.

We shall assume the following effects resulting from treatment of the medium:

Medium 1—decreases average wing length of a sample by 5 units
2—decreases average wing length of a sample by 2 units
3—does not change average wing length of a sample
4—increases average wing length of a sample by 1 unit

5—increases average wing length of a sample by 1 unit
6—increases average wing length of a sample by 5 units
7—(control) does not change average wing length of a sample

We will symbolize the effect of treatment i as α_i. (Please note that this use of α is not related to its use as a symbol for type I error. There are, regrettably, more symbols needed in statistics than there are letters in the Greek and Roman alphabets.) Thus α_i assumes the following values for these treatment effects:

$$\alpha_1 = -5 \qquad \alpha_4 = 1$$
$$\alpha_2 = -2 \qquad \alpha_5 = 1$$
$$\alpha_3 = 0 \qquad \alpha_6 = 5$$
$$\alpha_7 = 0$$

Note that $\Sigma^a \alpha_i = 0$; that is, the sum of the effects cancels out. This cancellation is a convenient property that is generally postulated but it is unnecessary for our argument. We can now modify Table 8.1 by adding the appropriate values of α_i to each sample. In sample 1 the value of α_1 is -5, so the first wing length, 41, now becomes 36; the second wing length, 44, becomes 39; and so on. For the second sample α_2 is -2, changing the first wing length from 48 to 46. Where α_i is 0, the wing lengths do not change; where α_i is positive, they are increased by the magnitude indicated. The changed values can be inspected in Table 8.3, which is arranged identically to Table 8.1.

We now repeat our previous computations. We calculate the sum of squares of the first sample and find it to be 29.2. If you compare this value with the sum of squares of the first sample in Table 8.1, you find the two values to be identical. Similarly, all other values of $\Sigma^n y_i^2$, the sum of squares of each group, are identical to their previous values. Why is this so? The effect of adding α_i to each group is simply that of an additive code, since α_i is constant for any one group. Section A.2 in the appendix shows that additive codes do not affect sums of squares or variances. Therefore, not only is each separate sum of squares the same as before, but the average variance within groups is still 16.029. Now let us compute the variance of the means. It is $100.617/6 = 16.770$, a value much higher than the variance of means found before, 4.236. When we multiply by $n = 5$ to get an estimate of σ^2, we obtain the variance of groups, which now is 83.848, no longer even close to an estimate of σ^2. We then repeat the F-test with the new variances and find that $F_s = 83.848/16.029 = 5.23$, which is much greater than the critical value of $F_{.05[6,28]} = 2.45$. In fact, the observed value of F_s is greater than $F_{.005[6,28]} = 4.02$. Clearly, the upper variance, representing the variance among groups, has become significantly larger. The two variances are most unlikely to represent the same parametric variance.

What has happened? Table 8.4 explains by representing Table 8.3 symbolically in the manner that Table 8.2 represented Table 8.1. Note that each group has a constant α_i added and that this constant changes the means of these groups

Table 8.3 DATA OF TABLE 8.1 WITH FIXED-TREATMENT EFFECTS α_i OR RANDOM EFFECTS A_i ADDED TO EACH SAMPLE.

	a groups ($a = 7$)							Computation of sum of squares of means	Computation of total sum of squares
	1	2	3	4	5	6	7		
α_i or A_i	−5	−2	0	+1	+1	+5	0		
	36	46	40	41	50	45	41		
	39	47	50	40	42	53	46		
n individuals per group ($n = 5$)	43	47	44	47	51	56	54		
	38	47	48	47	40	52	44		
	37	43	50	42	43	56	42		
$\bar{Y}_i$	38.6	46.0	46.4	43.4	45.2	52.4	45.4	$\bar{\bar{Y}} = 45.34$	$\bar{Y} = 45.34$
$\sum^{n} y_i^2$	29.2	12.0	75.2	45.2	98.8	81.2	107.2	$\sum^{a} (\bar{Y} - \bar{\bar{Y}})^2 = 100.617$	$\sum^{an} y^2 = 951.886$

Table 8.4 DATA OF TABLE 8.3 ARRANGED IN THE MANNER OF TABLE 8.2.

		a groups						
		1	2	3	···	i	···	a
n items	1	$Y_{11} + \alpha_1$	$Y_{21} + \alpha_2$	$Y_{31} + \alpha_3$	···	$Y_{i1} + \alpha_i$	···	$Y_{a1} + \alpha_a$
	2	$Y_{12} + \alpha_1$	$Y_{22} + \alpha_2$	$Y_{32} + \alpha_3$	···	$Y_{i2} + \alpha_i$	···	$Y_{a2} + \alpha_a$
	3	$Y_{13} + \alpha_1$	$Y_{23} + \alpha_2$	$Y_{33} + \alpha_3$	···	$Y_{i3} + \alpha_i$	···	$Y_{a3} + \alpha_a$
	⋮	⋮	⋮	⋮	·	⋮	·	⋮
	j	$Y_{1j} + \alpha_1$	$Y_{2j} + \alpha_2$	$Y_{3j} + \alpha_3$	···	$Y_{ij} + \alpha_i$	···	$Y_{aj} + \alpha_a$
	⋮	⋮	⋮	⋮	·	⋮	·	⋮
	n	$Y_{1n} + \alpha_1$	$Y_{2n} + \alpha_2$	$Y_{3n} + \alpha_3$	···	$Y_{in} + \alpha_i$	···	$Y_{an} + \alpha_a$
Means		$\bar{Y}_1 + \alpha_1$	$\bar{Y}_2 + \alpha_2$	$\bar{Y}_3 + \alpha_3$	···	$\bar{Y}_i + \alpha_i$	···	$\bar{Y}_a + \alpha_a$

by α_i. In Section 8.1 we computed the variance within groups as

$$\frac{1}{a(n-1)} \sum_{i=1}^{i=a} \sum_{j=1}^{j=n} (Y_{ij} - \bar{Y}_i)^2$$

If we try to repeat this computation, our formula becomes more complicated because to each Y_{ij} and each $\bar{Y}_i$ there has now been added α_i. We therefore write

$$\frac{1}{a(n-1)} \sum_{i=1}^{i=a} \sum_{j=1}^{j=n} [(Y_{ij} + \alpha_i) - (\bar{Y}_i + \alpha_i)]^2$$

When we open the parentheses inside the square brackets, the second α_i changes sign and the α_i's cancel out, leaving the expression exactly as before, substantiating our earlier observation that the variance within groups does not change, despite the treatment effects.

The variance of means was previously calculated by the formula

$$\frac{1}{a-1} \sum_{i=1}^{i=a} (\bar{Y}_i - \bar{\bar{Y}})^2$$

From Table 8.4, however, we see that the new grand mean equals

$$\frac{1}{a} \sum_{i=1}^{i=a} (\bar{Y}_i + \alpha_i) = \frac{1}{a} \sum_{i=1}^{i=a} \bar{Y}_i + \frac{1}{a} \sum_{i=1}^{i=a} \alpha_i = \bar{\bar{Y}} + \bar{\alpha}$$

When we substitute the new values for the group means and the grand mean, the formula appears as

$$\frac{1}{a-1} \sum_{i=1}^{i=a} [(\bar{Y}_i + \alpha_i) - (\bar{\bar{Y}} + \bar{\alpha})]^2$$

which in turn yields

$$\frac{1}{a-1}\sum_{i=1}^{i=a}[(\bar{Y}_i - \bar{\bar{Y}}) + (\alpha_i - \bar{\alpha})]^2$$

Squaring the expression in the square brackets, we obtain the terms

$$\frac{1}{a-1}\sum_{i=1}^{i=a}(\bar{Y}_i - \bar{\bar{Y}})^2 + \frac{1}{a-1}\sum_{i=1}^{i=a}(\alpha_i - \bar{\alpha})^2 + \frac{2}{a-1}\sum_{i=1}^{i=a}(\bar{Y}_i - \bar{\bar{Y}})(\alpha_i - \bar{\alpha})$$

The first of these terms we recognize immediately as the previous variance of the means, $s_{\bar{Y}}^2$. The second is a new quantity but is familiar by general appearance as a variance or at least a quantity akin to a variance. The third expression is a new type, a so-called covariance, which we have not yet encountered. We will not be concerned with it at this stage except to say that in cases such as this one, where the magnitude of treatment effects α_i is assumed to be independent of the $\bar{Y}_i$ to which they are added, the expected value of this quantity is zero; hence it would not contribute to the new variance of means.

The independence of the treatment effects and the sample means is an important concept. If we had not applied different treatments to the medium jars, but simply had treated all jars as controls, we would still have obtained differences among the wing length means. These are the differences found in Table 8.1 with random sampling from the same population. By chance some of these means are greater, some are smaller. In planning the experiment we had no way of predicting which sample means would be small and which would be large. Therefore, in planning our treatments we had no way of matching up large treatment effect, such as that of medium 6, with the mean that by chance would be the greatest, as that in sample 2. Also, the smallest sample mean (sample 4) is not associated with the smallest treatment effect. Only if the magnitude of the treatment effects were deliberately correlated with the sample means (this would be difficult to do in the experiment designed here) would the third term in the expression, the covariance, have an expected value other than zero.

We have seen that the first quantity is simply the variance of means as described before. The second quantity clearly is added as a result of the treatment effects and is analogous to a variance but cannot be called such since it is not based on a random variable but rather on deliberately chosen treatments largely under our control. By changing the magnitude and nature of the treatments we can more or less alter this variancelike quantity at will. We shall, therefore, call it the **added component due to treatment effects.** Since the α_i's are coded so that $\bar{\alpha} = 0$, we can rewrite the middle term as

$$\frac{1}{a-1}\sum_{i=1}^{i=a}(\alpha_i - \bar{\alpha})^2 = \frac{1}{a-1}\sum_{i=1}^{i=a}\alpha_i^2 = \frac{1}{a-1}\sum^{a}\alpha^2$$

In analysis of variance we multiply the variance of the means by n to estimate the parametric variance of the items. As you know, we call the quantity obtained in this way the variance of groups. When we do this for the case in which treatment effects are present, we obtain

$$n\left(s_{\bar{Y}}^2 + \frac{1}{a-1}\sum^a \alpha^2\right) = s^2 + \frac{n}{a-1}\sum^a \alpha^2$$

Thus we see that the estimate of the parametric variance of the population is increased by the quantity

$$\frac{n}{a-1}\sum^a \alpha^2$$

which is n times the added component due to treatment effects. We found the variance ratio F_s to be significantly greater than could be reconciled with the null hypothesis. It is now obvious why. We were testing the variance ratio expecting to find F approximately equal to $\sigma^2/\sigma^2 = 1$. In fact, however, the expected ratio is

$$\frac{\sigma^2 + \frac{n}{a-1}\sum^a \alpha^2}{\sigma^2}$$

It is clear from this formula, deliberately displayed in this lopsided manner, that the F-test is sensitive to the presence of the added component due to treatment effects.

At this point you have an additional insight into the analysis of variance. The method permits us to test whether there are added treatment effects—that is, to test whether a group of means can be considered random samples from the same population or whether we have sufficient evidence to conclude that the treatments that have affected each group separately have resulted in shifting these means sufficiently so that they can no longer be considered samples from the same population. If the latter is true, an added component due to treatment effects will be present and may be detected by an F-test in the significance test of the analysis of variance.

In such a study we are generally not interested in the magnitude of

$$\frac{n}{a-1}\sum^a \alpha^2$$

but we are interested in the magnitude of the separate values of α_i. In our example these are the effects of different formulations of the medium on wing length. If instead of housefly wing length we were measuring blood pressure in samples of rats, and the different groups had been subjected to different drugs or different doses of the same drug, the quantities α_i would represent the effects of drugs on the blood pressure, the issue of interest to the investigator. We may also be interested in studying differences of the type $\alpha_1 - \alpha_2$, leading us to the question of the significance of the differences between the effects of any two types of medium or any two drugs. But we are a little ahead of our story.

Analysis of variance that involves treatment effects of the type just studied is called a **Model I anova.** In Section 8.6 Model I will be defined precisely. There is another model, called a **Model II anova,** in which the added effects for each group are not fixed treatments but are random effects—that is, we have not deliberately planned or fixed the treatment for any one group, and the effects on each group are random and only partly under our control.

Suppose that the seven samples of houseflies in Table 8.3 represented the offspring of seven randomly selected females from a population reared on a uniform medium. There would be genetic differences among these females, which their seven broods would reflect. The exact nature of these differences is unclear and unpredictable. Before measuring them, we have no way of knowing whether brood 1 would have longer wings than brood 2, nor have we any way of controlling this experiment so that brood 1 would grow longer wings. As far as we can ascertain, the genetic factors for wing length are distributed in an unknown manner in the population of houseflies (we might hope that they are distributed normally), and our sample of seven is a random sample of these factors.

Another example of a Model II anova is that instead of making up our seven cultures from a single batch of medium, we have prepared seven batches separately, one right after the other, and are now analyzing the variation among the batches. We would not be interested in the exact differences from batch to batch. Even if we measured these, we would not be in a position to interpret them. Not having deliberately varied batch 3, we have no idea why it should produce longer wings than batch 2, for example. We would, however, be interested in the magnitude of the variance of the added effects. Thus, if we used seven jars of medium derived from one batch, we could expect the variance of the jar means to be $\sigma^2/5$, since there were 5 flies per jar. When based on different batches of medium, however, the variance could be expected to be greater, because all the imponderable accidents of formulation and environmental differences during medium preparation that make one batch of medium different from another would come into play. Interest would focus on the added variance component arising from differences among batches. Similarly, in the first example we would be interested in the added variance component arising from genetic differences among the females.

We will now take a quick look at the algebraic formulation of the Model II anova. In Table 8.3 the first row at the head of the data columns shows not only α_i but also A_i, which is the symbol we will use for a random group effect. We use a capital letter to indicate that the effect is a variable. As with the treatment effects, we will consider the sum of the random effects to be zero:

$$\frac{1}{a}\sum^{a} A_i = \bar{A} = 0$$

The algebra of calculating the two estimates of the population variance is the same as in Model I, except that in place of α_i we imagine A_i substituted in Table

8.4. The estimate of the variance among means now represents the quantity

$$\frac{1}{a-1}\sum_{i=1}^{i=a}(\bar{Y}_i - \bar{\bar{Y}})^2 + \frac{1}{a-1}\sum_{i=1}^{i=a}(A_i - \bar{A})^2 + \frac{2}{a-1}\sum_{i=1}^{i=a}(\bar{Y}_i - \bar{\bar{Y}})(A_i - \bar{A})$$

The first term is the variance of means $s_{\bar{Y}}^2$ as before, and the last term is the covariance between the group means and the random effects A_i, the expected value of which is zero (as before) because the random effects are independent of the magnitude of the means. The middle term is a true variance, since A_i is a random variable. We symbolize it by s_A^2 and call it the **added variance component among groups.** In the previous two examples, this term would represent the added variance component among females or among medium batches. The existence of this added variance component is tested for by the F-test. If the groups are random samples, we may expect F to approximate $\sigma^2/\sigma^2 = 1$, but with an added variance component the expected ratio is

$$\frac{\sigma^2 + n\sigma_A^2}{\sigma^2}$$

Note that σ_A^2, the parametric value of s_A^2, is multiplied by n, since we have to multiply the variance of means by n to obtain an independent estimate of the variance of the population.

In a Model II anova we are not interested in the magnitude of any A_i or in differences such as $A_1 - A_2$, but we are interested in the magnitude of σ_A^2 and its relative magnitude with respect to σ^2, which is generally expressed as the percentage $100s_A^2/(s^2 + s_A^2)$. Since the variance among groups estimates $\sigma^2 + n\sigma_A^2$, we can calculate s_A^2 as

$$\frac{1}{n}(\text{variance among groups} - \text{variance within groups})$$
$$= \frac{1}{n}[(s^2 + ns_A^2) - s^2] = \frac{1}{n}(ns_A^2) = s_A^2$$

If we interpret the example in Table 8.3 as a Model II anova, $s_A^2 = \frac{1}{5}(83.848 - 16.029) = 13.56$. This added variance component among groups is

$$\frac{100 \times 13.56}{16.029 + 13.56} = \frac{1356}{29.589} = 45.83\%$$

of the sum of the variances among and within groups.

Model II anovas will be discussed formally in Section 8.7; the methods of estimating variance components are treated in detail in the next chapter.

8.5 Partitioning the Total Sum of Squares and Degrees of Freedom

You may be discouraged to find more algebraic formulas ahead in this section, but these describe only simple properties and, given a little good will, can be mastered by anyone. We present these quantities because they are useful in computational methods for analysis of variance.

Until now we have ignored one other variance that can be computed from the data in Table 8.1. If we remove the classification into groups, we can consider the housefly data to be a single sample of $an = 35$ wing lengths and calculate the mean and variance of these items in the conventional manner. The quantities necessary for this computation are shown in the last column at the right in Tables 8.1 and 8.3, headed "Computation of total sum of squares." We obtain a mean of $\bar{Y} = 45.34$ for the sample in Table 8.1, which is, of course, the same as the quantity $\bar{\bar{Y}}$ computed previously from the seven group means. The sum of squares of the 35 items is 575.886, which gives a variance of 16.938 when divided by 34 degrees of freedom. Repeating these computations for the data in Table 8.3 we obtain $\bar{Y} = 45.34$ (the same as in Table 8.1 because $\Sigma^a \alpha_i = 0$) and $s^2 = 27.997$, which is considerably greater than the corresponding variance from Table 8.1. The total variance computed from all an items is another estimate of σ^2. It is a good estimate in the first case, but in the second sample (Table 8.3), where added components due to treatment effects or added variance components are present, it is a poor estimate of the population variance.

However, the purpose of calculating the total variance in an anova is not to use it as yet another estimate of σ^2 but to introduce an important mathematical relationship between it and the other variances. This relationship is most obvious when we arrange our results in a conventional *analysis of variance table,* or **anova table,** (Table 8.5). Such a table is divided into four columns. The first identifies the source of variation as among groups, within groups, and total (groups amalgamated to form a single sample). The column headed *df* gives the degrees of freedom by which the sums of squares pertinent to that source of variation must be divided to yield the variances. The degrees of freedom for variation among groups is $a - 1$, for variation within groups $a(n - 1)$, and for total variation $an - 1$. The next two columns show sums of squares and variances, respectively. Notice that the sums of squares entered in the anova table are the sum of squares among groups, the sum of squares within groups, and the sum of squares of the total sample of an items. In anovas variances are not called

Table 8.5 ANOVA TABLE FOR DATA IN TABLE 8.1.

	(1) Source of variation	(2) *df*	(3) Sums of squares, *SS*	(4) Mean squares, *MS*
$\bar{Y} - \bar{\bar{Y}}$	Among groups	6	127.086	21.181
$Y - \bar{Y}$	Within groups	28	448.800	16.029
$Y - \bar{\bar{Y}}$	Total	34	575.886	16.938

such but are generally called **mean squares,** since, in a Model I anova, they do not estimate a population variance. These quantities are not true *mean* squares because the sums of squares are divided by the degrees of freedom rather than by sample size. The sums of squares and mean squares are frequently abbreviated *SS* and *MS*, respectively.

The sums of squares and mean squares in Table 8.5 are the same as those obtained previously, except for minute rounding errors. Note, however, an important property of the sums of squares: They are obtained independently of each other, but when we add the *SS* among groups to the *SS* within groups we obtain the total *SS*. The sums of squares are additive! Another way of saying this is that we can decompose the total sum of squares into a portion due to variation among groups and another portion due to variation within groups. In Section A.3 of the appendix we demonstrate why the sums of squares within and among groups add up to the total sum of squares. Observe that the degrees of freedom are also additive and that the total of 34 *df* can be decomposed into 6 *df* among groups and 28 *df* within groups. Thus, if we know any two of the sums of squares (and their appropriate degrees of freedom), we can compute the third and complete our analysis of variance. Note that the mean squares, however, are not additive, which should be obvious, since generally $(a + b)/(c + d) \neq a/c + b/d$.

Before we continue, let us review the meaning of the three mean squares in the anova. The total *MS* is a statistic of dispersion of the 35 (an) items around their mean, the grand mean (in this example, 45.34). It describes the variance in the entire sample due to all and sundry causes and estimates σ^2 when there are no added treatment effects or variance components among groups. The within-group *MS*, also known as the *individual* or *intragroup* or **error mean square,** gives the average dispersion of the 5 (n) items in each group around the group means. If the a groups are random samples from a common homogeneous population, the within-group *MS* should estimate σ^2. The *MS* among groups is based on the variance of group means, which describes the dispersion of the 7 (a) group means around the grand mean. If the groups are random samples from a homogeneous population, the expected variance of their mean will be σ^2/n. Therefore, in order to have all three variances of the same order of magnitude, we multiply the variance of means by n to obtain the variance among groups. If there are no added treatment effects or variance components, the *MS* among groups is an estimate of σ^2. Otherwise it is an estimate of

$$\sigma^2 + \frac{n}{a-1}\sum^{a} \alpha^2 \qquad \text{or} \qquad \sigma^2 + n\sigma_A^2$$

depending on whether the anova is Model I or II.

The additivity relations we have just learned are independent of the presence of added treatment or random effects. We could show this algebraically, but Table 8.6 makes this clear simply by summarizing the anova of Table 8.3, in which α_i or A_i is added to each sample. The additivity relation still holds, although the values for group *SS* and total *SS* are different from those in Table 8.5.

Table 8.6 ANOVA TABLE FOR DATA IN TABLE 8.3.

	(1) Source of variation	(2) df	(3) Sums of squares, SS	(4) Mean squares, MS
$\bar{Y} - \bar{\bar{Y}}$	Among groups	6	503.086	83.848
$Y - \bar{Y}$	Within groups	28	448.800	16.029
$Y - \bar{\bar{Y}}$	Total	34	951.886	27.997

Another way of looking at the partitioning of the variation is to study the deviation from means in a particular case. Referring to Table 8.1, we can look at the wing length of the first individual in the seventh group, which is 41. Its deviation from its group mean is

$$Y_{71} - \bar{Y}_7 = 41 - 45.4 = -4.4$$

The deviation of the group mean from the grand mean is

$$\bar{Y}_7 - \bar{\bar{Y}} = 45.4 - 45.34 = 0.06$$

and the deviation of the individual wing length from the grand mean is

$$Y_{71} - \bar{\bar{Y}} = 41 - 45.34 = -4.34$$

Note that these deviations are additive. The deviation of the item from the group mean and that of the group mean from the grand mean add to the total deviation of the item from the grand mean. These deviations are stated algebraically as $(Y - \bar{Y}) + (\bar{Y} - \bar{\bar{Y}}) = (Y - \bar{\bar{Y}})$. Squaring and summing these deviations for an items will result in

$$\sum^{a} \sum^{n} (Y - \bar{Y})^2 + n \sum^{a} (\bar{Y} - \bar{\bar{Y}})^2 = \sum^{an} (Y - \bar{\bar{Y}})^2$$

Before squaring, the deviations were in the relationship $a + b = c$. After squaring we would expect them to be in the form $a^2 + b^2 + 2ab = c^2$. What happened to the cross-product term corresponding to $2ab$? This is

$$2 \sum^{an} (Y - \bar{Y})(\bar{Y} - \bar{\bar{Y}}) = 2 \sum^{a} [(\bar{Y} - \bar{\bar{Y}}) \sum^{n} (Y - \bar{Y})]$$

a covariance-type term that is always zero, since $\Sigma^n(Y - \bar{Y}) = 0$ for each of the a groups (see proof in Section A.1 of the appendix).

We identify the deviations represented by each level of variation at the left margins of the anova tables (see Tables 8.5 and 8.6). Note that the deviations add up correctly: The deviation among groups plus the deviation within groups equals the total deviation of items in the analysis of variance, $(\bar{Y} - \bar{\bar{Y}}) + (Y - \bar{Y}) = (Y - \bar{\bar{Y}})$.

8.6 Model I Anova

We have already mentioned the two different models in analysis of variance. An important point is that the basic setup of data, as well as the computation and significance test, in most cases is the same for both models. The purposes of analysis of variance differ for the two models, as do some of the supplementary tests and computations following the initial significance test. The two models were first defined by Eisenhart (1947). Although analysis of variance dates back to the 1930s, the distinction between the models had not been made in earlier work.

Let us now try to resolve the variation found in an analysis of variance case. This process will not only lead us to a more formal interpretation of anova but will also give us a deeper understanding of the nature of variation itself. For this discussion we will refer to the housefly wing lengths of Table 8.3. We ask the following question: What makes any given housefly wing length assume the value it does? The third wing length of the first sample of flies is 43 units. How can we explain such a reading?

Knowing nothing else about this individual housefly, our best guess as to its wing length is the grand mean of the population, which we know to be $\mu = 45.5$. However, we have additional information about this fly. It is a member of group 1, which has undergone a treatment shifting the mean of the group downward by 5 units. Therefore, $\alpha_1 = -5$, and we would expect our individual Y_{13} (the third individual of group 1) to measure $45.5 - 5 = 40.5$ units. In fact, however, it is 43 units, which is 2.5 units above this latest expectation. To what can we ascribe this deviation? To individual variation of the flies within a group because of the variance of individuals in the population ($\sigma^2 = 15.21$). All the genetic and environmental effects that make one housefly different from another housefly come into play to produce this variance.

By means of carefully designed experiments we might learn something about what causes this variance and attribute it to certain specific genetic or environmental factors. We might also be able to eliminate some of the variance. For instance, by using only full sibs (brothers and sisters) in any one culture jar, we would decrease the genetic variation in individuals, and undoubtedly the variance within groups would be smaller. However, it is hopeless to try to eliminate all variance completely. Even if we could remove all genetic variance, there would still be environmental variance, and even in the most improbable case in which we could remove both, there would remain measurement error; thus we would never obtain exactly the same reading, even on the same individual fly. The within-groups *MS* always remains as a residual, greater or smaller from experiment to experiment—part of the nature of things. This is why the within-groups variance is also called the error variance or error mean square. It is not an error in the sense of someone having made a mistake, but in the sense of providing you with a measure of the variation you have to contend with when trying to estimate significant differences among the groups. The error variance is composed of deviations for each individual, symbolized by ϵ_{ij}, the random component of the jth individual variate in the ith group. In our case, $\epsilon_{13} = 2.5$, since the actual observed value is 2.5 units above its expectation of 40.5.

We will now state this relationship more formally. In a Model I analysis of variance we assume that the differences among group means, if any, are due to the fixed treatment effects determined by the experimenter. The purpose of the analysis of variance is to estimate the true differences among the group means. Any single variate can be decomposed as follows:

$$Y_{ij} = \mu + \alpha_i + \epsilon_{ij} \tag{8.2}$$

where $i = 1, \ldots, a$, $j = 1, \ldots, n$, ϵ_{ij} represents an independent, normally distributed variable with mean $\epsilon_{ij} = 0$, and variance $\sigma_\epsilon^2 = \sigma^2$. Therefore a given reading is composed of the grand mean μ of the population, a fixed deviation α_i of the mean of group i from the grand mean μ, and a random deviation ϵ_{ij} of the jth individual of group i from its expectation, which is $(\mu + \alpha_i)$. Remember that both α_i and ϵ_{ij} can be either positive or negative. The expected value (mean) of the ϵ_{ij}'s is zero, and their variance is the parametric variance of the population, σ^2. For all the assumptions of the analysis of variance to hold, the distribution of ϵ_{ij} must be normal.

In a Model I anova we test for differences of the type $\alpha_1 - \alpha_2$ among the group means by testing for the presence of an added component due to treatments. If we find that such a component is present, we reject the null hypothesis that the groups come from the same population and accept the alternative hypothesis that at least some of the group means are different from each other, which indicates that at least some of the α_i's are unequal in magnitude. Next we generally wish to test which α_i's are different from each other. This is done by significance tests, with alternative hypotheses such as $H_1 : \alpha_1 > \alpha_2$ or $H_1 : \frac{1}{2}(\alpha_1 + \alpha_2) > \alpha_3$. In words, these test whether the mean of group 1 is significantly greater than the mean of group 2, or whether the mean of group 3 is smaller than the average of the means of groups 1 and 2.

Some examples of Model I anovas in biological disciplines follow. An experiment in which we try the effects of different drugs on batches of animals results in a Model I anova. We are interested in the results of the treatments and the differences between them. The treatments are fixed and determined by the experimenter. The same is true when we test the effects of different doses of a given factor—for example, a chemical or the amount of light to which a plant has been exposed, or temperatures at which culture bottles of insects have been reared. The treatment does not have to be entirely understood and manipulated by the experimenter; as long as it is fixed and repeatable, Model I will apply. Comparing the birth weights of the Chinese children in a hospital in Singapore with weights of Chinese children born in a hospital in China would also be a Model I anova. The treatment effects then would be "China versus Singapore," which sums up a whole series of different factors, genetic and environmental—

some known to us but most of them not understood. However, this is a definite treatment that we can describe and also repeat; namely, we can, if we wish, again sample birth weights of infants in Singapore as well as in China.

Another example of a Model I anova is the study of body weights for several age groups of animals. The treatments would be the ages, which are fixed. If we find that there is a significant difference in weight among the ages, we might follow this up with the question of whether there is a difference from age 2 to age 3 or only from age 1 to age 2. To a very large extent Model I anovas are the result of an experiment and of deliberate manipulation of factors by the experimenter. However, the study of differences such as the comparison of birth weights from two countries, although not an experiment proper, also falls into this category.

The null hypothesis in a Model I anova can be written as $H_0 : \alpha_1 = \alpha_2 = \cdots = \alpha_a$. What this means in words is that the different levels of the treatment do not differ in their effect on variable Y. This does not say, however, that the treatment variable itself has no effect on the observed variable, but only that differences in the level or quality of the treatment do not affect the response. Thus we emphasize contrasts of the type $(\alpha_1 - \alpha_2)$ among the groups. Accepting the null hypothesis implies that $(\alpha_1 - \alpha_2) = 0$ and all other contrasts among treatment levels are equally zero, but the factor itself may well have an effect on the variable. Thus different oxygen concentrations may have no effect on the heartbeat of samples of animals, yet the presence of oxygen is essential—in its absence the heartbeat would drop to zero. What Model I anova tests is the *differential* effects of the treatment.

Let us review once more the idea of decomposing an individual reading, using the birth weights as an example. An individual birth weight of an infant in Singapore can be thought of as consisting of the parametric mean μ of the birth weights of all Chinese children plus the treatment effects of being in Singapore α_1, plus a random component ϵ_{1j}, which represents the genetic and environmental differences making this individual infant deviate from its population mean.

8.7 MODEL II ANOVA

The structure of variation in a Model II anova is similar to that in Model I:

$$Y_{ij} = \mu + A_i + \epsilon_{ij} \tag{8.3}$$

where $i = 1, \ldots, a$, $j = 1, \ldots, n$, ϵ_{ij} represents an independent, normally distributed variable with mean $\bar{\epsilon}_{ij} = 0$ and variance $\sigma^2_\epsilon = \sigma^2$, and A_i represents a normally distributed variable, independent of all ϵ's, with mean $\bar{A} = 0$ and variance σ^2_A. The main distinction is that in place of fixed treatment effects α_i, we now consider random effects A_i, which differ from group to group. Since the effects are random, it is futile to estimate the magnitude of these random effects for any one group, or the differences from group to group, but we can estimate their variance, the added variance component among groups σ_A^2. We test for its presence and estimate its magnitude s_A^2, as well as its percentage contribution to the variation in a Model II analysis of variance.

The following examples illustrate applications of Model II anova. Suppose we wish to determine the DNA content of rat liver cells. We take five rats and make three preparations from each of the five livers obtained. The assay readings will be for $a = 5$ groups with $n = 3$ readings per group. The five rats presumably are sampled at random from the colony available to the experimenter. They must be different in various ways, genetically and environmentally, but we have no definite information about the nature of these differences. Thus, if we learn that rat 2 has slightly more DNA in its liver cells than rat 3, we can do little with this information, because we are unlikely to have any basis for following up this problem. We will, however, be interested in estimating the variance of the three replicates within any one liver and the variance among the five rats; that is, we want to know if there is variance σ_A^2 among rats in addition to the variance σ^2 expected on the basis of the three replicates. The variance among the three preparations presumably arises only from differences in technique and possibly from differences in DNA content in different parts of the liver (unlikely in a homogenate). Added variance among rats, if it existed, might be due to differences in ploidy or related phenomena. The relative amounts of variation among rats and "within" rats (= among preparations) would guide us in designing further studies of this sort. If there were little variance among the preparations and relatively more variation among the rats, we would need fewer preparations and more rats. On the other hand, if the variance among rats were proportionally smaller, we would use fewer rats and more preparations per rat. We will discuss the efficient planning of experiments in Section 10.4.

In a study of the amount of variation in skin pigment in human populations we might wish to look at different families within a homogeneous racial group and brothers and sisters within each family. The variance within families would be the error mean square, and we would test for an added variance component among families. We would expect an added variance component σ_A^2 because there are genetic differences among families that determine amount of skin pigmentation. We would be especially interested in the relative proportions of the two variances σ^2 and σ_A^2 because they would provide us with important genetic information. From our knowledge of genetic theory we would expect the variance among families to be greater than the variance among brothers and sisters within a family.

These examples illustrate the two types of problems involving Model II anovas that are most likely to arise in biological work. One type is concerned with the general problem of the design of an experiment and the magnitude of the experimental error at different levels of replication, such as error among replicates within rat livers and among rats, error among batches or experiments, and so forth. The other relates to variation among and within families, among and within females, among and within populations, and so forth, being concerned with the general problem of the relation between genetic and phenotypic variation.

It is sometimes difficult to decide whether a given study is a Model I or Model II anova. The distinction depends not on the nature of the factor differentiating the groups but on whether levels of that factor can be considered a random sample of more such levels or are fixed treatments whose differences the investigator wishes to contrast. Thus the factor "years" might simply be a kind of replication in which it would be a random factor. On the other hand, one might be interested in specific years because of events that occurred in those years, which would make years a fixed factor.

In the next chapter we will practice computing analyses of variance using several different examples. These examples will provide an opportunity to distinguish cases of Model I and Model II anovas and to learn the proper way of analyzing each model and of presenting the results of such analyses.

EXERCISES 8

8.1 In a study comparing the chemical composition of the urine of chimpanzees and gorillas (Gartler et al., 1956), the following results were obtained: For 37 chimpanzees the variance for the amount of glutamic acid in milligrams per milligram of creatinine was 0.01082 (see Exercise 4.2). A similar study based on six gorillas yielded a variance of 0.12442. Is there a significant difference between the variability in chimpanzees and gorillas? *Answer:* $F_s = 11.499$, $F_{.025[5,36]} \approx 2.90$.

8.2 Show that it is possible to represent the value of an individual variate as follows: $Y_{ij} = (\bar{\bar{Y}}) + (\bar{Y}_i - \bar{\bar{Y}}) + (Y_{ij} - \bar{Y}_i)$. What do each of the terms in parentheses estimate in a Model I anova and in a Model II anova?

8.3 The following data are from an experiment by Sewall Wright in which he crossed Polish and Flemish giant rabbits and obtained 27 F_1 rabbits. These were inbred, and 112 F_2 rabbits were obtained. We have extracted the following data on femur length of these rabbits

	n	$\bar{Y}$	s
F_1	27	83.39	1.65
F_2	112	80.50	3.81

Is there a significantly greater amount of variability in femur lengths among the F_2 than among the F_1 rabbits? What well-known genetic phenomenon is illustrated by these data? *Answer:* Yes; segregation and recombination.

8.4 For the data in Table 8.3, make tables to represent partitioning of the value of each variate into its three components: $\bar{\bar{Y}}$, $(\bar{Y}_i - \bar{\bar{Y}})$, and $(Y_{ij} - \bar{Y}_i)$. The first table will consist of 35 values, all equal to the grand mean. In the second table all entries in a given column will be equal to the difference between the mean of that column and the grand mean. The last table will consist of the deviations of each individual variate from its column mean. These tables represent estimates of the individual components of Expression (8.2). Compute the mean and sum of squares for each table.

9 SINGLE-CLASSIFICATION ANALYSIS OF VARIANCE

We are now ready to study applications of the analysis of variance. This chapter deals with the simplest kind of anova, **single-classification analysis of variance,** in which the groups of samples are classified (grouped) by only a single criterion. Both interpretations of the seven samples of housefly wing lengths studied in the last chapter, different medium formulations (Model I) and progenies of different females (Model II), would represent a single criterion for classification. Other examples are different temperatures at which groups of animals were raised or different soils in which samples of plants have been grown.

Section 9.1 restates the basic computational formulas for analysis of variance, based on the topics covered in the previous chapter. Section 9.2 gives an example of the general case with unequal sample sizes, since all groups in the anova need not necessarily have the same sample size. We will illustrate this case using a Model II anova. Some computations unique to a Model II anova—estimating variance components and setting confidence limits to them—are also shown. Since the basic computations for the analysis of variance are the same in either model, it is not necessary to repeat the illustration with a Model I anova. Model I is featured in Section 9.3, which shows the computational simplification resulting from equal sample sizes. Formulas become especially simple for the two-sample case, as we will see in Section 9.4. In this case a t-test can be applied as well, and we shall show that it is mathematically equivalent to an analysis of variance F-test when there are only two groups. Section 9.5 illustrates another special case, the comparison of a single specimen with a sample.

When a Model I analysis of variance has been found to be significant, leading to the conclusion that the means are not from the same population, we wish to test the means in various ways to discover which pairs of means are different from each other and whether the means can be divided into groups that are significantly different from each other. These so-called multiple comparisons tests are covered in Sections 9.6 and 9.7. Section 9.6 deals with planned comparisons designed before the test is run, and Section 9.7 discusses a-posteriori tests that suggest themselves to the experimenter as a result of his or her analysis.

Experimenters frequently ask statisticians what sample sizes are required for a given study. This topic is taken up in Section 9.8.

In contrast to some previous chapters, which dealt largely with theory, much of the discussion in this chapter centers on boxes illustrating practical applications. Computational details are discussed in the boxes; general implications and additional comments are found in the text.

9.1 COMPUTATIONAL FORMULAS

We saw in Section 8.5 that the total sum of squares and degrees of freedom can be partitioned additively into those of variation among groups and those of variation within groups. For the analysis of variance proper, we need only the sum of squares among groups and the sum of squares within groups, not the total sum of squares. Not all computer programs furnish all three sums of squares. When only two are given, the third can be obtained easily from the other two by addition or subtraction, depending on which two are initially available. The three sums of squares are computed as follows:

$$SS_{\text{among}} = n \sum^{a} (\bar{Y} - \bar{\bar{Y}})^2$$

$$SS_{\text{within}} = \sum^{a} \sum^{n} (Y - \bar{Y})^2$$

$$SS_{\text{total}} = \sum^{a} \sum^{n} (Y - \bar{\bar{Y}})^2$$

The formula for the sum of squares among groups (SS_{among}) is based on the simplest model: equal sample sizes and all groups in the analysis being tested. The following more general formula permits any set of groups to be tested, with each group based on its own, different sample size n_i:

$$\sum n_i (\bar{Y}_i - \bar{Y}_w)^2$$

where $\bar{Y}_w$ is the weighted mean of means based on Expression (4.2).

9.2 GENERAL CASE: UNEQUAL n

Our first computational example of a single-classification analysis of variance has unequal sample sizes. We will develop general formulas for carrying out such an analysis, and we will see in the next section that the more usual anova, with equal sample sizes, is only a special case, slightly simpler to compute. In this example we use a Model II anova. Up to and including the F-test for significance, the computations are the same whether the anova is Model I or Model II.

We will identify the stage in the computations at which the operations change depending on the model.

The example, shown in Box 9.1, concerns a series of morphological measurements of the width of the scutum (dorsal shield) of samples of tick larvae obtained from four different host individuals of the cottontail rabbit. These four hosts were obtained at random from one locality. We know nothing about their origins or their genetic constitution. They represent a random sample of the population of host individuals from the given locality. We would not be in a position to interpret differences between larvae from any given host, since we know nothing of the origins of the individual rabbits. Population biologists are nevertheless interested in such analyses because they provide an answer to the following question.

Are the variances of means of larval characters among hosts greater than expected on the basis of variances of the characters within hosts? We can calculate the average variance of width of larval scutum on a host. This will be our "error" term in the analysis of variance. We then test the observed mean square among groups and see if it contains an added component of variance. What would such an added component of variance represent? The mean square within host individuals (that is, of larvae on any one host) represents genetic differences among larvae and differences in environmental experiences of these larvae. Added variance among hosts demonstrates significant differentiation among the larvae possibly due to differences among the hosts affecting the larvae or to genetic differences among the larvae (assuming each host carries a family of ticks, or at least a population whose individuals are more related to each other than they are to tick larvae on other host individuals). The emphasis in this example is on the magnitudes of the variances. In view of the random choice of hosts, this is a clear case of a Model II anova.

Note that in order to compute quantity **2,** we feature the means of the four groups in Box 9.1. Because this is a Model II anova, however, the means hold no interest for us except as an aid to computation. We are not interested in the individual means or possible differences among them. Of course, a reason for ignoring this advice appears at the beginning of the analysis. One might wish to look at the group means to spot outliers, which might represent readings in error.

The computation is illustrated in Box 9.1. After quantities 2 through 4 have been evaluated, they are entered into an anova table, as shown in the box. General formulas for such a table are shown first, followed by a table filled in with actual values for the specific example. We note 3 degrees of freedom among groups, since there are four host individuals, and 33 *df* within groups (the sum of 7, 9, 12, and 5 degrees of freedom for groups 1 through 4, respectively). Note that the mean square among groups (MS_{among}) is considerably greater than the error mean square (MS_{within}), giving rise to suspicion that an added variance component is present. If MS_{among} is equal to or less than MS_{within}, we do not bother going on with the analysis, for we would not have evidence for the presence of an added variance component. You may wonder how MS_{among} can be

Box 9.1 SINGLE-CLASSIFICATION ANOVA WITH UNEQUAL SAMPLE SIZES.

Width (in μm) of scutum (dorsal shield) of larvae of the tick *Haemaphysalis leporispalustris* in samples from 4 cottontail rabbits. This is a Model II anova.

	Hosts (a = 4)				
	1	2	3	4	$\sum^{a}$
	380	350	354	376	
	376	356	360	344	
	360	358	362	342	
	368	376	352	372	
	372	338	366	374	
	366	342	372	360	
	374	366	362		
	382	350	344		
		344	342		
		364	358		
			351		
			348		
			348		
$\sum^{n_i} Y$	2978	3544	4619	2168	13,309
$\bar{Y}_i$	372.25	354.40	355.31	361.33	359.70
n_i	8	10	13	6	37
$\sum^{n_i} y_i^2$	379.50	1278.40	954.77	1165.33	3778.00

SOURCE: Data from P. A. Thomas (unpublished results).

Preliminary computations

1. Grand mean $= \bar{\bar{Y}} = \dfrac{1}{\sum^{a} n_i} \sum_{i=1}^{a} \sum_{j=1}^{n_i} Y_{ij} = \dfrac{1}{37}(2978 + 3544 + \ldots + 2168)$

$= 359.70$

IMPORTANT NOTE: For simplicity we will abbreviate our notation in the rest of this box as follows:

$$\sum^{a} = \sum_{i=1}^{a} \qquad \sum^{n} = \sum_{j=1}^{n_i} \qquad Y = Y_{ij}$$

BOX 9.1 CONTINUED

2. SS_{among}, sum of squares among groups $= \sum^{a} n_i(\bar{Y}_i - \bar{\bar{Y}})^2$

$$= \sum^{a} n_i \left(\frac{\sum^{n} Y_i}{n_i} - \bar{\bar{Y}} \right)^2$$

$$= 8(372.25 - 359.70)^2 + 10(354.40 - 359.70)^2 + \ldots + 6(361.33 - 359.70)^2 = 1807.73$$

3. SS_{within}, sum of squares within groups, $= \sum^{a}\sum^{n} y^2 = \sum^{a}\sum^{n}(Y - \bar{Y}_i)^2 = 3778.00$

4. $SS_{\text{total}} = SS_{\text{among}} + SS_{\text{within}} =$ quantity 2 + quantity 3 $= 1807.73 + 3778.00 = 5585.73$

The anova table of formulas is constructed as follows. (Boldface numbers represent quantities calculated in the corresponding computational steps.)

	Source of variation	df	SS	MS	F_s	Expected MS
$\bar{Y} - \bar{\bar{Y}}$	Among groups	$a - 1$	**2**	$\frac{\mathbf{2}}{a-1}$	$\frac{MS_{\text{among}}}{MS_{\text{within}}}$	$\sigma^2 + n_0\sigma_A^2$
$Y - \bar{Y}$	Within groups	$\sum^{a} n_i - a$	**3**	$\frac{\mathbf{3}}{\sum^{a} n_i - a}$		σ^2
$Y - \bar{\bar{Y}}$	Total	$\sum^{a} n_i - 1$	**4**			

Substituting the computed values, we obtain the following completed anova table

	Source of variation	df	SS	MS	F_s
$\bar{Y} - \bar{\bar{Y}}$	Among groups (among hosts)	3	1807.73	602.58	5.26**
$Y - \bar{Y}$	Within groups (error; among larvae on a host)	33	3778.00	114.48	
$Y - \bar{\bar{Y}}$	Total	36	5585.73		

$F_{.05[3,33]} = 2.89$ $F_{.01[3,33]} = 4.44$ $F_{.001[3,33]} = 6.89$

* = $0.01 < P \leq 0.05$, ** = $0.001 < P \leq 0.01$, *** = $P \leq 0.001$. These conventions will be followed throughout the text and will not be explained in subsequent boxes and tables.

Conclusion

We reject the null hypothesis at $P < 0.01$. There is a significant added variance component among hosts for width of scutum in larval ticks.

less than MS_{within}. Remember that these two values are independent estimates. If there is no added variance component among groups, the estimate of the variance among groups is as likely to be less as it is to be greater than the variance within groups.

Expressions for the expected values of the mean squares are also shown in the first anova table of Box 9.1. These are the expressions you learned in the previous chapter for a Model II anova, with one exception: In place of n is n_0. Since sample size n_i differs among groups in this example, no single value of n is appropriate in the formula. We therefore use an average n; this, however, is not simply $\bar{n}$, the arithmetic mean of the n_i's, but is

$$n_0 = \frac{1}{a-1}\left(\sum^{a} n_i - \frac{\sum^{a} n_i^2}{\sum^{a} n_i}\right) \tag{9.1}$$

which is an average usually close to, but always less than $\bar{n}$, unless sample sizes are equal, in which case $n_0 = \bar{n}$. Mathematical statisticians have determined that the appropriate mean value of n for the expected mean square among groups is n_0 as defined here.

The critical 5%, 1%, and 0.1% values of F are shown below the second anova table in Box 9.1 (2.89, 4.44, and 6.89, respectively). You should confirm them for yourself in Table **F**. Note that the argument $\nu_2 = 33$ is not given. You must therefore interpolate between arguments representing 30 and 40 degrees of freedom, respectively. The values in Box 9.1 we computed using harmonic interpolation (see section on interpolation in the Introduction to *Statistical Tables*). It is frequently not necessary, however, to carry out such an interpolation. One can simply look up the "conservative" value of F, which is the next tabled F with less degrees of freedom—in our case $F_{\alpha[3,30]}$, which is 2.92 for $\alpha = 0.05$ and 4.51 for $\alpha = 0.01$. The observed value of F_s is 5.26, considerably greater than the interpolated as well as the conservative value of $F_{.01}$. We therefore reject the null hypothesis ($H_0: \sigma_A^2 = 0$) that there is no added variance component among groups and that the two mean squares estimate the same variance, allowing a type I error of less than 1%. We accept instead the alternative hypothesis of the existence of an added variance component σ_A^2.

What is the biological meaning of this conclusion? The ticks on different host individuals differ more from each other than do individual ticks on any one host either because of some modifying influence of individual hosts on the ticks (biochemical differences in blood, differences in the skin, differences in the environment of the host individual; all of them are rather unlikely in this case) or because of genetic differences among the ticks. The ticks on each host may represent a sibship (the descendants of a single pair of parents), and thus the differences of the ticks among host individuals could represent genetic differences among families; or selection may have acted differently on the tick populations on each host; or the hosts may have migrated to the collection locality from different geographic areas in which the ticks differ in width of scutum. Of these possibilities, genetic differences among sibships seem most reasonable, in view of the biology of the organism.

The computations up to this point would have been identical in a Model I anova. (If this had been Model I, the conclusion would have been that there is a significant treatment effect rather than an added variance component.) Now, however, we must complete the computations appropriate to a Model II anova. These may include estimating the added variance component, computing its confidence limits, and calculating percentage variation at the two levels. Computer programs often include the computations for variance components, regardless of whether the anova is Model I or Model II, rather than have the user specify which model is intended. The user must then decide on the correct model and should ignore variance components when these are inappropriate. Our computer program for single-classification and nested anovas operates in this way.

The steps for computing the variance components for the example in Box 9.1 are given in Box 9.2. The expressions for the expected mean squares in the anova table show how the variance component among groups σ_A^2 and the error variance σ^2 are obtained. Of course, since the values that we obtain are sample estimates, they are written as s_A^2 and s^2. We are frequently interested not so much in the absolute values of these variance components as in their relative magnitudes. For this purpose we sum these two values and express each as a percentage of this sum. Thus $s^2 + s_A^2 = 114.48 + 54.179 = 168.659$, and s^2 and s_A^2 are 67.9% and 32.1% of this sum, respectively.

Relatively more variation occurs within groups (larvae on a host) than among groups (hosts). The proportion of the variation among groups is also known as r_I, the **coefficient of intraclass correlation.** A high value of this coefficient means that most of the variation in the sample is among groups. A value of unity would indicate that all of the variance in the sample is among groups—i.e., that there is no variance within groups. In such a case, all the individuals within each group would be identical. This coefficient is a measure of the similarity, or *correlation,* among the individuals within a group relative to the amount of differences found among the groups. This meaning is not the sense in which we use correlation nowadays, and the percentage variation among and within groups seems a more expressive way to represent this statistic.

Formulas for computing exact confidence intervals of variance components from an anova with unequal sample sizes have not been developed. We can compute such limits, however, by means of the bootstrap procedure, which is discussed in Section 18.4, where we take up this very problem as an example. In the meantime, for the sake of completeness, we show in Box 9.3 an example of the direct computation of confidence limits for another anova with equal sample sizes. The method used here is the Moriguti–Bulmer procedure recommended by Boardman (1974). Several other procedures have been proposed, but according to Boardman only this method and another one by Williams and Tukey

Box 9.2 ESTIMATION OF VARIANCE COMPONENTS IN A SINGLE-CLASSIFICATION ANOVA WITH UNEQUAL SAMPLE SIZES.

Data from Box 9.1.

Anova table

Source of variation	*df*	*MS*	Expected *MS*
Among groups (among hosts)	3	602.58	$\sigma^2 + n_0\sigma_A^2$
Within groups (error; among larvae on a host)	33	114.48	σ^2

Computation

1. The sizes of the four samples are 8, 10, 13, and 6. Hence from Expression (9.1) the average sample size, n_0, is

$$n_0 = \frac{1}{a-1}\left(\sum^a n_i - \frac{\sum^a n_i^2}{\sum^a n_i}\right)$$

$$= \frac{1}{4-1}\left([8+10+13+6] - \frac{8^2+10^2+13^2+6^2}{8+10+13+6}\right) = 9.009$$

   As a check, we confirm that $n_0 \leq \bar{n} = 9.25$.

2. We may estimate the variance component as follows:

$$s_A^2 = \frac{MS_{\text{among}} - MS_{\text{within}}}{n_0} = \frac{602.58 - 114.48}{9.009} = 54.179$$

3. If interested only in the relative magnitudes of the variance component and the error variance, we often express these as a percentage of their sum, $s^2 + s_A^2 = 114.48 + 54.179 = 168.659$.

   Percent of variation among groups (hosts):

$$\frac{s_A^2}{s^2 + s_A^2} \times 100 = \frac{54.179}{114.48 + 54.179} \times 100 = \frac{54.179}{168.659} \times 100 = 32.1\%$$

   (The coefficient of intraclass correlation r_I = proportion of variation among groups = 0.321).

   Percent of variation within groups (among larvae on a host):

$$\frac{s^2}{s^2 + s_A^2} \times 100 = \frac{114.48}{168.659} \times 100 = 67.9\%$$

Box 9.3 ADDED VARIANCE COMPONENTS AND THEIR APPROXIMATE CONFIDENCE LIMITS IN A MODEL II ANOVA WITH EQUAL SAMPLE SIZES: THE MORIGUTI–BULMER PROCEDURE.

Fertility (number of larvae hatched out of 10 eggs deposited) for the offspring of 7 females of *Drosophila melanogaster,* PP strain ($a = 7$ females). There were $n = 3$ replicates for each female.

Anova table

Source of variation	df	MS	Expected MS	F_s
Among groups (among females)	6	9.111	$\sigma^2 + n\sigma_A^2$	2.20 *ns*
Within groups (replicates)	14	4.142	σ^2	
Total	20			

SOURCE: Data from R. R. Sokal (unpublished results)

Since MS_{within} is an unbiased estimate of σ^2 and MS_{among} is an unbiased estimate of $\sigma^2 + n\sigma_A^2$ (where σ_A^2 is the added variance component due to differences among females), σ_A^2 may be estimated as follows:

$$s_A^2 = \frac{1}{n}(MS_{\text{among}} - MS_{\text{within}})$$

$$= \frac{1}{3}(9.111 - 4.142) = 1.656$$

It is also possible, with the following computational steps, to calculate approximate confidence limits for σ_A^2 (the confidence limits for MS_{within}, σ^2, may be calculated using the methods described in Section 7.7).

1. For degree of freedom $\nu_1 = df_{\text{among}} = 6$, $\nu_2 = df_{\text{within}} = 14$, $\alpha =$ desired level of significance $= .05$ (for 95% confidence intervals), look up the following tabulated F-values in Statistical Table **F:**

$$F_{\alpha/2[\nu_1,\nu_2]} = F_{.025[6,14]} = 3.50$$

$$F_{\alpha/2[\nu_2,\nu_1]} = F_{.025[14,6]} = 5.30 \quad \text{(obtained by interpolation)}$$

$$F_{\alpha/2[\nu_1,\infty]} = F_{.025[6,\infty]} = 2.41$$

$$F_{\alpha/2[\infty,\nu_1]} = F_{.025[\infty,6]} = 4.85$$

Box 9.3 Continued

2. The lower confidence limit is

$$L_1 = \frac{MS_{\text{within}}}{n}\left[\frac{F_s}{F_{\alpha/2[\nu_1,\infty]}} - 1 - \left(\frac{F_{\alpha/2[\nu_1,\nu_2]}}{F_{\alpha/2[\nu_1,\infty]}} - 1\right)\frac{F_{\alpha/2[\nu_1,\nu_2]}}{F_s}\right]$$

$$= \frac{4.142}{3}\left[\frac{2.20}{2.41} - 1 - \left(\frac{3.50}{2.41} - 1\right)\frac{3.50}{2.20}\right]$$

$$= 1.38067\{0.91286 - 1 - [0.45228(1.59091)]\}$$

$$= 1.38067(-0.80668)$$

$$= -1.11376$$

The lower limit is negative because F_s was not significant. Conventionally the lower limit is set equal to zero in such cases, because a negative variance is nonsensical. Thus $L_1 = 0$.

3. The upper confidence limit is

$$L_2 = \frac{MS_{\text{within}}}{n}\left[F_s F_{\alpha/2[\infty,\nu_1]} - 1 + \left(1 - \frac{F_{\alpha/2[\infty,\nu_1]}}{F_{\alpha/2[\nu_2,\nu_1]}}\right)\frac{1}{F_{\alpha/2[\nu_2,\nu_1]}F_s}\right]$$

$$= \frac{4.142}{3}\left[(2.20)(4.85) - 1 + \left(1 - \frac{4.85}{5.30}\right)\frac{1}{5.30(2.20)}\right]$$

$$= 1.38067\left[10.6700 - 1 + \frac{1 - 0.91509}{11.66}\right]$$

$$= 1.38067(9.67728)$$

$$= 13.361$$

consistently yield the desired $100(1 - \alpha)\%$ coverage. In a study using the Monte Carlo method (see Section 18.3 for an explanation of this approach) of the properties of variance components in a variety of examples, we were able to confirm the reliable behavior of the Moriguti–Bulmer method.

The example in Box 9.3 expresses fertility as the number of larvae hatched out of 10 eggs laid by females of *Drosophila melanogaster.* From each of the 7 females in the study, which had been chosen randomly from a certain genetic strain, 3 replicate egg samples (of 10 eggs each) were tested for hatchability. We therefore have a basic data table (not shown) of 7 groups (females) and 3 replicates (hatch counts) within groups. The anova table is shown in Box 9.3. Note that the F-value is not significant; thus we cannot reject the null hypothesis that the variance component is zero. However, we will set confidence limits to the F-value anyway in order to obtain experience with this computation. The lower confidence limit is negative. This result was to be expected, since the variance component among groups is not significant. The computation is straightforward and is laid out in Box 9.3.

9.3 SPECIAL CASE: EQUAL *n*

Equal sample sizes in a single-classification analysis of variance result in slightly simpler computational formulas, although these are basically not different from the general case shown in the previous section. We will illustrate a single-classification anova with equal sample sizes by a Model I example. As we have stated before, the computation up to and including the first test of significance is identical for both models. Thus the computation of Box 9.4 could also apply to a Model II anova with equal sample sizes.

The data in Box 9.4, which are from an experiment in plant physiology, record the length in coded units of pea sections grown in tissue culture with auxin present. The purpose of the experiment was to test the effects of various sugars on growth as measured by length. Four experimental groups, representing three different sugars and one mixture of sugars, were used, plus one control without sugar. Ten observations (replicates) were made for each treatment. (The term "treatment" already implies a Model I anova.) The five groups obviously do not represent random samples from all possible experimental conditions but were deliberately designed to test the effects of certain sugars on the growth rate. We are interested in the effect of the sugars on length, so our null hypothesis is that there is no added component due to treatment effects among the five groups; that is, the population means are all assumed to be equal. In the preliminary computation, the formulas for steps 1 to 3 are simpler, since there is now only a single, constant n, rather than n_i for each group. The anova table is similar to that for the general case (see Box 9.1), except that the expressions for number of degrees of freedom are simpler.

Since ν_2 is relatively large, the critical values of F have been computed by harmonic interpolation in Statistical Table **F** (see the section on interpolation in the Introduction to *Statistical Tables*). These values have been given here only to present a complete record of the analysis. Ordinarily, when confronted with this example, you would not bother working out these values of F. Comparison of the observed variance ratio $F_s = 49.33$ with $F_{.001[4,40]} = 5.70$, the conservative critical value, would convince you to reject the null hypothesis. The probability that the five groups differ as much as they do by chance is almost infinitesimally small. Clearly the sugars produce an added treatment effect, apparently inhibiting growth and consequently reducing the length of the pea sections.

At this stage we are not in a position to say whether each treatment is different from every other treatment, or whether the sugars are different from the control but not different from each other. Such tests are necessary to complete a Model I analysis, but we defer their discussion until Sections 9.6 and 9.7.

Box 9.4 SINGLE-CLASSIFICATION ANOVA WITH EQUAL SAMPLE SIZES.

The effect of different sugars on length, in ocular units ($\times 0.114$ = mm), of pea sections grown in tissue culture with auxin present; $n = 10$ (replications per group). This is a Model I anova.

	Treatments ($a = 5$)				
Observations, i.e., replications	Control	2% glucose added	2% fructose added	1% glucose + 1% fructose added	2% sucrose added
1	75	57	58	58	62
2	67	58	61	59	66
3	70	60	56	58	65
4	75	59	58	61	63
5	65	62	57	57	64
6	71	60	56	56	62
7	67	60	61	58	65
8	67	57	60	57	65
9	76	59	57	57	62
10	68	61	58	59	67
$\sum^{n} Y$	701	593	582	580	641
$\bar{Y}$	70.1	59.3	58.2	58.0	64.1

SOURCE: Data from W. Purves (unpublished results).

Preliminary computations

1. Grand mean $= \bar{\bar{Y}} = \frac{1}{an} \sum^{a} \sum^{n} Y = \frac{1}{50}(701 + 593 + \quad . \; . \; . \quad + 641) = 61.94$

2. SS_{among}, sum of squares among groups, $= n \sum^{a} (\bar{Y} - \bar{\bar{Y}})^2 = 1077.32$

3. SS_{within}, sum of squares within groups, $= \sum^{a} \sum^{n} y^2 = \sum^{a} \sum^{n} (Y - \bar{Y})^2 = 245.50$

4. $SS_{\text{total}} = SS_{\text{among}} + SS_{\text{within}}$ = quantity **2** + quantity **3** = 1077.32 + 245.50 = 1322.82

Box 9.4 Continued

Anova table: formulas

	Source of variation	*df*	*SS*	*MS*	F_s	Expected *MS*
$\bar{Y} - \bar{\bar{Y}}$	Among groups	$a - 1$	**2**	$\frac{\mathbf{2}}{a-1}$	$\frac{MS_{\text{among}}}{MS_{\text{within}}}$	$\sigma^2 + \frac{n}{a-1}\sum^{a}\alpha^2$
$Y - \bar{Y}$	Within groups	$a(n - 1)$	**3**	$\frac{\mathbf{3}}{a(n-1)}$		
$Y - \bar{\bar{Y}}$	Total	$an - 1$	**4**			

Completed anova table

	Source of variation	*df*	*SS*	*MS*	F_s
$\bar{Y} - \bar{\bar{Y}}$	Among groups (among treatments)	4	1077.32	269.33	49.33***
$Y - \bar{Y}$	Within groups (error, replicates)	45	245.50	5.46	
$Y - \bar{\bar{Y}}$	Total	49	1322.82		

$F_{.05[4,45]} = 2.58 \quad F_{.01[4,45]} = 3.77 \quad F_{.001[4,45]} = 5.57$

Conclusions

Since $P \ll 0.001$, there is a highly significant added component due to treatment effects in the mean square among groups (treatments). The different sugar treatments clearly have a significant effect on growth of the pea sections.

See Sections 9.6 and 9.7 and Boxes 9.8 through 9.12 for the completion of a Model I analysis of variance—i.e., to determine which means are significantly different from each other.

9.4 Special Case: Two Groups

A frequent test in statistics is to establish the *significance of the difference between two means,* which can be done easily by means of an *analysis of variance for two groups.* Box 9.5 shows this procedure for a Model I anova, the more frequent model in the two-sample case. If the assumptions of a Model II anova are met, however, it too can be carried out for two groups.

Box 9.5 shows the results of an experiment with water fleas, *Daphnia longispina,* in which the onset of reproductive maturity is measured as the average age

Box 9.5 SINGLE-CLASSIFICATION ANOVA WITH TWO GROUPS WITH EQUAL SAMPLE SIZES.

Average age (in days) at beginning of reproduction in *Daphnia longispina* (each variate is a mean based on approximately similar numbers of females). Two series derived from different genetic crosses and containing 7 clones each are compared; $n = 7$ clones per series. This is a Model I anova.

	Series ($a = 2$)	
	I	II
	7.2	8.8
	7.1	7.5
	9.1	7.7
	7.2	7.6
	7.3	7.4
	7.2	6.7
	7.5	7.2
$\sum^{n} Y$	52.6	52.9
$\bar{Y}$	7.5143	7.5571
s^2	0.5047	0.4095

SOURCE: Data by Ordway from Banta (1939).

Preliminary computation

1. Grand mean $= \bar{\bar{Y}} = \dfrac{1}{2n}\sum^{2}\sum^{n} Y = \dfrac{1}{14}(52.6 + 52.9) = 7.5357$

2. $SS_{\text{among}} = n\sum^{2}(\bar{Y} - \bar{\bar{Y}})^2 = 0.00641$

3. $SS_{\text{within}} = \sum^{2}\sum^{n} y^2 = 5.48571$

4. $SS_{\text{total}} =$ quantity **2** + quantity **3** $= 5.49214$

NOTE: With only two equal-sized groups, the SS_{among} can be computed directly by the following simple formula, avoiding calculation of quantity **2:**

$$SS_{\text{among}} = \frac{n}{2}(\bar{Y}_1 - \bar{Y}_2)^2$$

$$= \frac{7}{2}(7.5143 - 7.5571)^2 = 0.00641$$

BOX 9.5 CONTINUED

Anova table

	Source of variation	df	SS	MS	F_s
$\bar{Y} - \bar{\bar{Y}}$	Between groups (series)	1	0.00641	0.00641	0.0140
$Y - \bar{Y}$	Within groups (error; clones within series)	12	5.48571	0.45714	
$Y - \bar{\bar{Y}}$	Total	13	5.49212		

$F_{.05[1,12]} = 4.75$

Conclusions

Since $F_s \ll F_{.05[1,12]}$, the null hypothesis is accepted. The means of the two series are not significantly different; that is, the two series do not differ in average age at beginning of reproduction.

(in days) at the beginning of reproduction. Since each variate in the table is an average, a possible flaw in the analysis is that these averages may not be based on equal sample sizes. We are not given this information, however, and therefore must proceed on the assumption that each reading in the table is an equally reliable variate. The two series represent different genetic crosses, and the seven replicates in each series are clones derived from the same genetic cross. This example is clearly a Model I, since the question to be answered is whether average age at the beginning of reproduction in series I differs from that in series II. Inspection of the data shows that the mean age at beginning of reproduction is very similar for the two series. We would be surprised therefore to find that they are significantly different, but we will carry out a test anyway. As you realize by now, one cannot tell from the absolute magnitude of a difference whether it is significant. Significance depends on the magniutde of the error mean square, representing the variance within series.

The computations are shown in Box 9.5. Since the sample sizes in this example are equal, we follow the computational outline for Box 9.4. If the sample sizes in the two groups were unequal, we would follow the outline of Box 9.1. With equal sample sizes and only two groups, we can make one more computational shortcut. We can compute quantity **2,** SS_{among}, by the very simple formula shown in the note in Box 9.5. There is only 1 degree of freedom between the two groups. The critical value of $F_{.05[1,12]}$ is given underneath the anova table, but it is really not necessary to consult it. Inspection of the mean squares in the anova shows that MS_{among} is much smaller than MS_{within}; therefore the value of F_s is far below unity, and so we cannot estimate an added component due to treatment effects between the series. In cases where $MS_{\text{among}} \leq MS_{\text{within}}$, we do not usually bother to calculate F_s because the analysis of variance could not possibly be significant.

This example is unusual in that the mean square between groups is very much lower than that within groups. You may wonder whether these two could really be estimates of the same parametric variance, as they are supposed to be according to theory. Let us test whether the two variances are equal. Our null hypothesis is that $H_0: \sigma^2_{\text{among}} = \sigma^2_{\text{within}}$ and the alternative hypothesis is $H_1: \sigma^2_{\text{among}} \neq \sigma^2_{\text{within}}$. In view of the alternative hypothesis, this is a two-tailed test analogous to that in Box 8.1. We calculate $F_s = MS_{\text{within}}/MS_{\text{among}} = 0.45714/0.00643 = 71.09$. Since $F_{.025[12,1]} = 977$, we accept H_0 and conclude that the two mean squares are estimates of the same parametric variance. We should not be surprised that there is such a great discrepancy between the variance estimates. After all, the mean square between groups is based on only 1 degree of freedom.

Some statisticians would object to the procedure just described on the grounds that after a one-tailed test has been made with the alternative hypothesis ($H_1: \sigma^2_{\text{among}} > \sigma^2_{\text{within}}$) and the null hypothesis has been accepted, it is not valid to retest the data with another alternative hypothesis ($H_1: \sigma^2_{\text{among}} \neq \sigma^2_{\text{within}}$). The net effect of such a procedure is that of a two-tailed test, with twice the intended type I error. A second alternative hypothesis should therefore be tested only with great caution.

A mean square among groups significantly lower than the error mean square would mean either that an unusual event has occurred by chance (type I error) or that the group means are less variable than they should be on the basis of the variation of their items. A low mean square can occur whenever items are not sampled at random but are sampled with a type of compensatory bias. A grocer filling baskets of strawberries may not wish to fill them at random but may fill them so that the smaller, less desirable strawberries are at the bottom of each basket, larger, more attractive ones are on top on view to the customer. Filling the strawberry baskets in this way would result in considerable within-basket variance of strawberry size, but remarkable constancy in means among baskets (if the grocer has done a thorough job of biased filling of baskets), so that the mean square among groups (baskets) would be remarkably low. In theory at least, MS_{among} would be zero if the grocer is able to fill each basket with an identical assortment of strawberries.

In scientific work similar poor sampling may result from unrecognized biases that cause the person doing the sampling to compensate unconsciously for obvious deviations from what he or she believes to be the norm. When biased sampling can be ruled out, the underdispersion of sample means may be due to a compensation effect in the natural phenomenon under study. Whenever resources for a process are limited, some individuals may obtain a large part and others may have to be satisfied with only a small part. Thus, if there is a limited space in which to live, several large individuals may occupy most of it, leaving space for only a few small ones. This phenomenon is related to the idea of repulsion encountered in the binomial distribution (see Section 5.2). Situations such as these indicate that the basic assumptions of an anova are not met (see Chapter 13), and ordinarily one would not continue the anova.

The reader should not confuse two distinct sets of null and alternative hypotheses that can be considered in connection with an example such as the two strains of water fleas. We have just discussed the one-sidedness of the alternative hypothesis $H_1 : \sigma^2_{\text{among}} > \sigma^2_{\text{within}}$. We might refer to this hypothesis and its accompanying null hypothesis as **operational hypotheses,** since they are related to the set of necessary computational operations. That is, to carry out an analysis of variance we have to calculate a mean square among groups and test whether it is greater than the mean square within groups. Thus we use an operationally one-tailed hypothesis, and it is for this reason that the F-table (Statistical Table **F**) is presented as a one-tailed table. These operational hypotheses, however, should be distinguished from **fundamental hypotheses,** or *conceptual hypotheses,* which are concerned with the nature of the original experiment or analysis undertaken. In the water flea example, we had no a priori reason for the mean of strain I to be greater than or less than that of strain II, so we applied a two-sided alterative hypothesis $H_1 : \mu_{\text{I}} \neq \mu_{\text{II}}$. Had there been reason to formulate the conceptual alternative hypothesis in a one-sided manner, say $H_1 : \mu_{\text{I}} > \mu_{\text{II}}$, then the critical value of F to be looked up in Table **F** should have been for probability 2α, where the desired type I error is α. Keep the distinction between a conceptual and an operational alternative hypothesis in mind; we will refer to other conceptually one-sided tests from time to time. Note that for analyses of variance with more than a single degree of freedom in the numerator of an F-statistic ($a > 2$ groups), a one-sided conceptual alternative hypothesis cannot be defined.

Another method of solving a Model I two-sample analysis of variance is a ***t*-test** *of the differences between two means.* This t-test is the traditional method of solving such a problem, and may already be familiar to you from previous acquaintance with statistical work. It has no real advantage in ease of either computation or understanding, and as you will see below, it is mathematically equivalent to the anova that was just computed in Box 9.5. The t-test is presented here mainly for completeness. It would seem too much of a break with tradition not to have the t-test in a biometry text.

In Section 7.4 we learned about the t-distribution and saw that a t-distribution of $n - 1$ *df* could be obtained from a distribution of the term $(\bar{Y}_i - \mu)/s_{\bar{Y}i}$, where $s_{\bar{Y}i}$ has $n - 1$ degrees of freedom and $\bar{Y}$ is normally distributed. The numerator of this term represents a deviation of a sample mean from a parametric mean, and the denominator represents a standard error for such a deviation. The expression

$$t_s = \frac{(\bar{Y}_1 - \bar{Y}_2) - (\mu_1 - \mu_2)}{\sqrt{\left[\dfrac{(n_1 - 1)s_1^2 + (n_2 - 1)s_2^2}{n_1 + n_2 - 2}\right]\left(\dfrac{n_1 + n_2}{n_1 n_2}\right)}} \tag{9.2}$$

is also distributed as t. Expression (9.2) looks complicated, but it really has the same structure as the simpler term for t. The numerator is a deviation, this time not between a single sample mean and the parametric mean, but between a single difference between two sample means, $\bar{Y}_1$ and $\bar{Y}_2$, and the true difference between the means of the populations represented by these means. In a test of this sort our null hypothesis is that the two samples come from the same population; that is, they must have the same parametric mean. Thus the difference $\mu_1 - \mu_2$ is assumed to be zero. We therefore test the deviation of the difference $\bar{Y}_1 - \bar{Y}_2$ from zero. The denominator of Expression (9.2) is a standard error, the standard error of the difference between two means, $s_{\bar{Y}1 - \bar{Y}2}$. The left portion of (the denominator in square brackets) is a weighted average of the variances of the two samples, s_1^2 and s_2^2, computed in the manner of Section 8.1. The right term of the standard error is the computationally easier form of $(1/n_1) + (1/n_2)$, which is the factor by which the average variance within groups must be multiplied to convert it into a variance of the difference of means. The analogy of multiplying a sample variance s^2 by $1/n$ to transform it into a variance of a mean $s_{\bar{Y}}^2$ should be obvious.

The test as outlined here assumes equal variances in the two samples. This assumption has been made with all the analyses of variance we have carried out thus far, although we have not stressed this. If the variances of the two samples are significantly different from each other, other methods have to be applied. These methods are discussed in Section 13.3, where we also present the tests for checking whether several variances are equal. With only two variances, equality may be tested by the procedure in Box 8.1.

When sample sizes are equal in a two-sample test, Expression (9.2) simplifies to

$$t_s = \frac{(\bar{Y}_1 - \bar{Y}_2) - (\mu_1 - \mu_2)}{\sqrt{\dfrac{1}{n}(s_1^2 + s_2^2)}} \tag{9.3}$$

which is the formula applied in Box 9.6 to the water flea example of Box 9.5. When the sample sizes are unequal but rather large and hence the differences between n_i and $n_i - 1$ are relatively trivial, Expression (9.2) reduces to the simpler

$$t_s = \frac{(\bar{Y}_1 - \bar{Y}_2) - (\mu_1 - \mu_2)}{\sqrt{\dfrac{s_1^2}{n_2} + \dfrac{s_2^2}{n_1}}} \tag{9.4}$$

The derivation of Expressions (9.3) and (9.4) from Expression (9.2) is shown in Section A.4 of the appendix. The pertinent degrees of freedom for Expressions (9.2) and (9.4) are $n_1 + n_2 - 2$ and for Expression (9.3) df is $2(n - 1)$. When the calculation is done by computer, there is, of course, no reason to use the approximation.

Box 9.6 A t-TEST OF THE HYPOTHESIS THAT TWO SAMPLE MEANS COME FROM A POPULATION WITH EQUAL μ; ALSO CONFIDENCE LIMITS OF THE DIFFERENCE BETWEEN TWO MEANS.

Average age at beginning of reproduction in *Daphnia*. Data from Box 9.5.

This test assumes that the variances in the populations from which the two samples were taken are identical. If in doubt about this hypothesis, test by the method of Box 8.1, and if variances of the two samples are highly significantly different do not employ this test but carry out the test shown in Section 13.3.

The appropriate formula for t_s is one of the following:

Expression (9.2), when sample sizes are unequal and n_1 or n_2 or both are small (<30): $df = n_1 + n_2 - 2$

Expression (9.3), when n_1 and n_2 are identical (regardless of size): $df = 2(n - 1)$

Expression (9.4), when n_1 and n_2 are unequal but large (>30): $df = n_1 + n_2 - 2$

For the present data, since sample sizes are equal, we choose Expression (9.3):

$$t_s = \frac{(\bar{Y}_1 - \bar{Y}_2) - (\mu_1 - \mu_2)}{\sqrt{\frac{1}{n}(s_1^2 + s_2^2)}}$$

Since we are testing the null hypothesis that $\mu_1 - \mu_2 = 0$, we replace $\mu_1 - \mu_2$ in this equation by zero. Thus

$$t_s = \frac{7.5143 - 7.5571}{\sqrt{(0.5047 + 0.4095)/7}} = \frac{-0.0428}{\sqrt{0.9142/7}} = \frac{-0.0428}{0.3614} = -0.1184$$

The degrees of freedom for this example are $2(n - 1) = 2 \times 6 = 12$. The critical value (from Table **B**) of $t_{.05[12]} = 2.179$. Since the absolute value of our observed t_s is less than the critical t-value, the means are found to be not significantly different, which is the same result as was obtained by the anova of Box 9.5.

A comparison of the results of the present t-test with those of the F-test of Box 9.5 shows close correspondence. We have learned that $F_{[1,\nu_2]} = t^2_{[\nu_2]}$. Since $t_s = -0.1184$, $t_s^2 = 0.0140$, which is equal to the F_s-value from Box 9.5.

Confidence limits of the difference between two means

$$L_1 = (\bar{Y}_1 - \bar{Y}_2) - t_{\alpha[\nu]} s_{\bar{Y}_1 - \bar{Y}_2}$$

$$L_2 = (\bar{Y}_1 - \bar{Y}_2) + t_{\alpha[\nu]} s_{\bar{Y}_1 - \bar{Y}_2}$$

In this case $\bar{Y}_1 - \bar{Y}_2 = -0.0428$, $t_{.05[12]} = 2.179$, and $s_{\bar{Y}1 - \bar{Y}2} = .03614$ as computed earlier as the denominator of the t-test. Therefore

$$L_1 = -0.0428 - (2.179)(0.3614) = -0.8303$$

$$L_2 = -0.0428 + (2.179)(0.3614) = 0.7447$$

The 95% confidence limits contain the zero point (no difference), as was to be expected, since the difference $\bar{Y}_1 - \bar{Y}_2$ was found to be not significant.

The test of significance for differences between means using the t-test is shown in Box 9.6. This is a two-tailed test because our alternative hypothesis is $H_1: \mu_1 \neq \mu_2$. The results of this test are identical to those of the anova in Box 9.5: the two means are not significantly different. We mentioned earlier that these two results are in fact mathematically equivalent. We can show this most simply by squaring the obtained value for t_s, which should be identical to the F_s-value of the corresponding analysis of variance. Since $t_s = -0.1184$ in Box 9.6, $t_s^2 = 0.0140$. This value equals the F_s obtained in the anova of Box 9.5. Why? We will show two reasons for this. We learned that $t_{[\nu]} = (\bar{Y} - \mu)/s_{\bar{Y}}$, where ν is the degrees of freedom of the variance of the mean $s_{\bar{Y}}^2$; therefore $t_{[\nu]}^2 = (\bar{Y} - \mu)^2/s_{\bar{Y}}^2$. However, this expression can be regarded as a variance ratio. The denominator is clearly a variance with ν degrees of freedom. The numerator, also a variance, is a single deviation squared, which represents a sum of squares possessing 1 rather than zero degrees of freedom (since it is a deviation from the true mean μ rather than from a sample mean). A sum of squares based on 1 degree of freedom is also a variance. Thus t^2 is a variance ratio; specifically, $t_{[\nu]}^2 = F_{[1,\nu]}$, as we have already seen. In Section A.5 of the appendix we demonstrate algebraically that the t_s^2 obtained in Box 9.6 and the F_s-value obtained in Box 9.5 are identical quantities.

We can also demonstrate the relationship between t and F numerically from the Statistical Tables. In Table **F** we find that $F_{.05[1,10]} = 4.96$. This value should be equal to $t_{.05[10]}^2$. The square root of 4.96 is 2.227 and when we look up $t_{.05[10]}$ in Table **B** we find it to be 2.228, which is within acceptable rounding error of 2.227. Since t approaches the normal distribution as its degrees of freedom approach infinity, $F_{\alpha[1,\nu]}$ approaches the distribution of the square of the normal deviate as $\nu \to \infty$.

The relation between t^2 and F leads to another relation. We have just learned that when $\nu_1 = 1$, $F_{[\nu_1,\nu_2]} = t_{[\nu_2]}^2$. We know from Section 8.2 that $\chi_{[\nu_1]}^2/\nu_1 = F_{[\nu_1,\infty]}$. Therefore, when $\nu_1 = 1$ and $\nu_2 = \infty$, $\chi_{[1]}^2 = F_{[1,\infty]} = t_{[\infty]}^2$. This relation can be demonstrated from Tables **D**, **F**, and **B**:

$$\chi_{.05[1]}^2 = 3.841$$

$$F_{.05[1,\infty]} = 3.84$$

$$t_{.05[\infty]} = 1.960 \qquad t_{.05[\infty]}^2 = 3.8416$$

The interrelationships between the t- and F-distributions (as well as the normal and χ^2-distributions) are evident in a diagrammatic view of an F-table, as shown in Figure 9.1.

The t-test for differences between two means is useful when we wish to set confidence limits to such a difference. Box 9.6 shows how to calculate 95% confidence limits for the difference between the series means in the *Daphnia* example. The appropriate standard error and degrees of freedom depend on whether Expression (9.2), (9.3), or (9.4) is chosen for t_s. We are not surprised to find that the confidence limits of the difference in this case enclose the value of

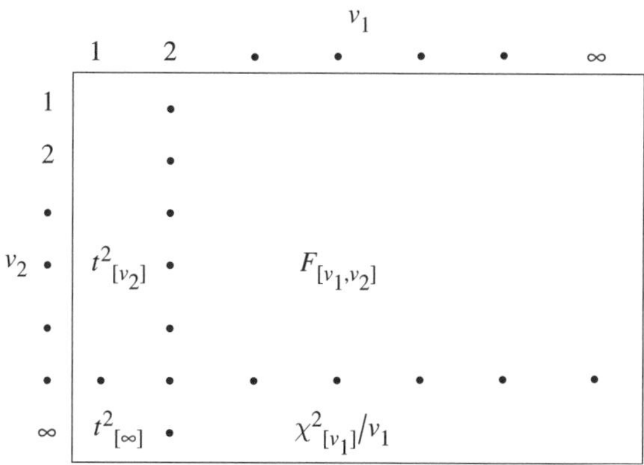

FIGURE 9.1 *The leftmost column of this diagram of an F-table represents* $F_{[1,\nu_2]}$*, which equals* $t^2_{[\nu_2]}$*. The bottom row is* $F_{[\nu_1,\infty]}$*, which equals* $\chi^2_{[\nu_1]}/\nu_1$*. The entry in the lower left hand corner is the intersection of* $F_{[1,\nu_2]}$ *and* $F_{[\nu_1,\infty]}$*, which is* $F_{[1,\infty]}$ *and equals* $t^2_{[\infty]}$*.*

zero, ranging from -0.8303 to $+0.7447$. This condition must be true when a difference is found to be not significantly different from zero. We can interpret from this situation that we cannot exclude zero as the true value of the difference between the means of the two series. In fact, this confidence interval contains all parameter values for the difference between two means, for which hypotheses would not be rejected.

Had the alternative hypothesis been one-sided conceptually, as for example $H_1: \mu_{\text{I}} > \mu_{\text{II}}$, then we should have employed the t-table using probability 2α for an intended type I error of α.

9.5 Special Case: A Single Specimen Compared With a Sample

A type of significance test quite frequent in taxonomy, but also of use in other fields, is whether a single specimen sampled at random could belong to a given population. If the parameters and distribution of the population are known, this test presents no new features. Thus in the case of a normal distribution with mean μ and standard deviation σ, if we sample a single specimen and measure it for the variable in question, recording it as Y_1, we can test whether it belongs to the population by considering $(Y_1 - \mu)/\sigma$ as a standard normal deviate (see Section 6.2) and inspecting its magnitude in the table of areas of the normal curve (Statistical Table **A**).

If the population is known to us only through a sample, however, the way to deal with this problem is to consider it a two-sample case similar to the one discussed in Section 9.4. The new aspect here is that one of the samples is represented by only a single variate, so it does not contribute to the degrees of freedom or to the estimate of the variance within groups. An example of such a test is shown in Box 9.7. The appropriate formula is

$$t_s = \frac{Y_1 - \bar{Y}_2 - (\mu_1 - \mu_2)}{s_2\sqrt{\dfrac{n_2 + 1}{n_2}}} \tag{9.5}$$

where Y_1 is the single specimen, $\bar{Y}_2$ is the sample mean, s_2 is the sample standard deviation, and n_2 is the sample size. Expression (9.5) is easily obtained from

Box 9.7 **COMPARISON OF A SINGLE OBSERVATION WITH THE MEAN OF A SAMPLE.**

Wing lengths (in millimeters) of a sample of mosquitos, *Aedes canadensis*, from Minnesota and a single specimen from Kansas. Data from F. J. Rohlf (unpublished results).

A special case of the method of Box 9.6, in which one of the two samples contains only one observation, so the variance of that sample cannot be estimated.

The modified formula is Expression (9.5):

$$t_s = \frac{Y_1 - \bar{Y}_2 - (\mu_1 - \mu_2)}{s_2\sqrt{\dfrac{n_2 + 1}{n_2}}}$$

$$df = n_2 - 1$$

Sample from Minnesota:

Y	4.02	3.88	3.34	3.87	3.18

$$n = 5 \qquad \bar{Y} = 3.658$$

$$s^2 = 0.1387 \qquad s = 0.3725$$

Single variate from Kansas:

$$Y_1 = 3.02$$

$$t_s = \frac{3.02 - 3.658 - 0}{0.3725\sqrt{(5+1)/5}} = \frac{-0.64}{0.3725 \times 1.09544} = \frac{-0.64}{0.40805} = -1.5684$$

$$df = 5 - 1 = 4$$

$$t_{.05[4]} = 2.776 \qquad t_{.10[4]} = 2.132 \qquad t_{.20[4]} = 1.533$$

The probability of obtaining a variate as extreme as 3.02 mm is between 0.20 and 0.10 under the null hypothesis that both samples were from the same statistical population. We therefore accept the null hypothesis and conclude that the Kansas specimen could well have come from a population whose wing length did not differ from that of the Minnesota population. In view of the small sample size, the conclusion could also be that we do not have sufficient evidence to reject the hypothesis that the Kansas specimen came from a population with the same mean as the Minnesota population.

Expression (9.2) by substituting Y_1 for $\overline{Y}_1$, and 1 for n_1. The pertinent degrees of freedom are as before: $n_1 + n_2 - 2$, which reduces to $n_2 - 1$.

The problem in Box 9.7 is whether a single mosquito collected in Kansas could belong to a population with the same mean wing length as the Minnesota mosquito population. This example is a two-tailed test because our alternative hypothesis is that the mean of the Kansas population from which the single specimen comes could be greater or less than that of the Minnesota population ($H_1 : \mu_1 \neq \mu_2$).

The results from Box 9.7 show that between 10 and 20% of wing lengths sampled from the Minnesota population would be as far from the Minnesota mean in either direction as is the Kansas specimen. We therefore do not have sufficient evidence to reject the null hypothesis, and we conclude that the Kansas population does not differ in wing length from the Minnesota population.

Note that such a test has few, if any, direct taxonomic implications. It does not prove or disprove the assertion that the Kansas specimen is in the same species, subspecies, or in any other taxon with the Minnesota population. The Kansas specimen could well be in another genus and still not be different in wing length. Conversely, although there may be a statistical difference in wing length between the Kansas population and the Minnesota population for a variety of reasons it may not be wise to recognize this specimen in any formal sense as a taxonomically distinct unit. In conventional taxonomy, criteria for specific distinction are usually not statistical but biological (Mayr and Ashlock, 1991). Criteria for subspecies may involve tests for spatial homogeneity with respect to many characters rather than simple tests of significance.

Occasionally the comparison of single individuals with sample means is of individuals that are very deviant from their sample mean. This situation involves considerations of statistical theory that are beyond the scope of this text. If you wish to carry out such tests for outliers, consult Barnett and Lewis (1978), Dixon and Massey (1969), or Gumbel (1954), who have treated this problem in depth.

9.6 Comparison Among Means: Planned Comparisons

After the initial significance test we complete a Model II anova by estimating the added variance components. We usually complete a Model I anova of more than two groups by examining the data in greater detail, testing which means are different from which other means or which groups of means are different from other such groups or from single means. Two-sample cases, such as the example of Box 9.5 examining the average age of water fleas in two series, are of no interest here. Their mean square between groups possesses only one degree of freedom and admits of only one test—i.e., do the two means differ?

When the number of groups, a, is greater than two, however, further tests are possible and usually are of interest. Thus the data in Box 9.4 on length of pea sections show a significant difference among the five treatments (based on 4 degrees of freedom). Although we conclude that the means are not all equal, we do not know which ones differ from which other ones. This leads us to the subject of tests among pairs and groups of means. For example, we might test the control against the average of the 4 experimental treatments representing added sugars. The question to be tested would be, Does the addition of sugars have an effect on length of pea sections? We might also test for differences among the sugar treatments. A reasonable test might be the average of the pure sugars (glucose, fructose, and sucrose) versus the mixed sugar treatment (1% glucose, 1% fructose).

An important point about such tests is that they are designed and chosen *independently* of the results of the experiment. They should be planned *before* the experiment has been carried out and the results obtained. Such comparisons are thus called **planned comparisons,** or *a-priori comparisons.* By contrast, after the experiment has been carried out we might wish to compare certain means that we notice to be markedly different. For instance, sucrose with a mean of 64.1 appears to have had less of a growth-inhibiting effect than fructose with a mean of 58.2 (see Box 9.4). We might therefore wish to test whether there is a significant difference between the effects of fructose and sucrose. Such comparisons, which suggest themselves *as a result of* the completed experiment, are called **unplanned comparisons** (also known as *a posteriori,* or *multiple comparisons*). They include tests of the comparisons between all possible pairs of means. When there are a means, there can, of course, be $a(a - 1)/2$ possible comparisons between the means. The reason we make this distinction between planned and unplanned comparisons is that the tests of significance appropriate for the two are different. A simple analogous example will explain why.

Let us assume we have sampled from an approximately normal population of heights of men. We have computed their mean and standard deviation. If we sample two men at a time from this population, we can predict the difference between them on the basis of ordinary statistical theory. Some men will be very similar, others relatively different. Their differences will be distributed normally with a mean of 0 and an expected variance of $2\sigma^2$, for reasons explained in Section 15.3. Thus, a large difference between a randomly sampled pair of men will have to be a sufficient number of standard deviations greater than zero in order for us to reject our null hypothesis that the two men come from the specified population.

If, on the other hand, we were to look at the heights of the people before sampling them and then take pairs that seem to be very different, we would repeatedly obtain differences between pairs of men that are several standard deviations apart. Such differences would be outliers in the expected frequency distribution of differences, and time and again we would reject our null hypothesis when in fact it is true. The men would have been sampled from the same population, but because they would not have been sampled at random but would have been inspected before being sampled, the probability distribution on which our hypothesis rested would no longer be valid. Obviously the tails in a large sample from a normal distribution will be anywhere from 5 to 7 standard deviations apart, and if we deliberately take individuals from each tail and compare them, they will appear to be highly significantly different from each other, even though they belong to the same population.

When we select means differing greatly from each other in an analysis of variance, we are doing exactly the same thing as taking a tall and a short man from the frequency distribution of heights. If we wish to know whether these differ significantly, we cannot use the ordinary probability distribution on which the analysis of variance rested; we have to use special tests of significance. Unplanned tests will be discussed in the next section. In this section we concentrate on how to carry out comparisons planned before the execution of the experiment.

In carrying out planned comparisons we first must decide how many and which comparisons to carry out. If the sum of the degrees of freedom of the comparisons does not exceed the $a - 1$ degrees of freedom among groups, and if the set of planned comparisons exhibits a property known as orthogonality (explained later in this section), we should use the method of **orthogonal comparisons**. There are two ways of computing orthogonal comparisons: We can subdivide the treatment sum of squares and the treatment degrees of freedom into separate comparisons, carrying these out as part of the analysis of variance, using F-tests as tests of significance, or we can carry out t-tests for comparisons between two groups. The t-tests, however, are mathematically equivalent to corresponding F-tests (for the same reason that F-tests and t-tests were equivalent in Section 9.4), and we prefer the F-tests because of their simplicity and elegance and because they tie in well with the overall anova.

The general rule for making a planned comparison is simple and relates to the rule for obtaining the sum of squares for any set of groups (see Section 9.1). To compare k groups of any size n_i, compute $SS = \Sigma^k n_i (\bar{Y}_i - \bar{\bar{Y}})^2$, where the summation is over the k groups, $\bar{Y}_i$ is the mean of each group, and $\bar{\bar{Y}} = \Sigma n_i \bar{Y}_i / \Sigma n_i$ is the average over the k groups. The following example will illustrate.

Table 9.1 lists the means and sample sizes of the experiment with the pea sections first discussed in Box 9.4. We now wish to test whether the controls are different from the four treatments representing addition of sugar. There will thus be two groups, the control group and the "sugars" group. The mean of the sugars group is 59.9, and its sample size is 40. We therefore compute

$$\begin{aligned} SS_{(\text{control vs. sugars})} &= \sum n_i(\bar{Y}_i - \bar{\bar{Y}})^2 \\ &= 10(70.1 - 61.94)^2 + 40(59.9 - 61.94)^2 \\ &= 832.32 \end{aligned}$$

In this case $\bar{\bar{Y}}$ is the same as for the anova because it involves all the groups of the study. The result is a sum of squares for the comparison between these two groups. Since a comparison between two groups has only 1 degree of freedom, the sum of squares is also a mean square. This mean square is tested over the

Table 9.1 **MEANS, GROUP SUMS, AND SAMPLE SIZES FROM THE DATA IN BOX 9.4.**

Length of pea sections (in ocular units) grown in tissue culture. Coefficients of linear comparisons are also given for two comparisons.

	Control	2% glucose added	2% fructose added	1% glucose + 1% fructose added	2% sucrose added	$\sum$
$\bar{Y}$	70.1	59.3	58.2	58.0	64.1	$(61.94 = \bar{\bar{Y}})$
n	10	10	10	10	10	50
Comparisons						
Control vs. treatments (c_{i1})	+4	−1	−1	−1	−1	
Mixed vs. pure sugars (c_{i2})	0	−1	−1	+3	−1	

error mean square of the anova to give the following comparisons:

$$F_s = \frac{MS_{(\text{control vs. sugars})}}{MS_{\text{within}}} = \frac{832.32}{5.46} = 152.44$$

$$F_{.05[1,45]} = 4.05 \qquad F_{.01[1,45]} = 7.23 \qquad F_{.001[1,45]} = 12.4$$

This comparison is highly significant, showing that the additions of sugars significantly retarded the growth of the pea sections.

Next we test whether the mixture of sugars is significantly different from the pure sugars. We first calculate the mean of the pure sugars

$$\bar{Y}_{\text{pure sugars}} = \frac{59.3 + 58.2 + 64.1}{3} = 60.5333$$

We use a simple average since the sample sizes are equal. Then we evaluate the new grand mean (of all sugars):

$$\bar{\bar{Y}}_{\text{sugars}} = \frac{59.3 + 58.2 + 58.0 + 64.1}{4} = 59.9$$

Thus,

$$\begin{aligned} SS_{(\text{mixed sugar vs. pure sugars})} &= 10(58.0 - 59.9)^2 + 30(60.5333 - 59.9)^2 \\ &= 48.13 \end{aligned}$$

Here the $\overline{\overline{Y}}$ is different because it is based on the sugars only. The appropriate test statistic is

$$F_s = \frac{MS_{(\text{mixed sugars vs. pure sugars})}}{MS_{\text{within}}} = \frac{48.13}{5.46} = 8.82$$

This result is significant in view of the critical values of $F_{\alpha[1,45]}$ given above.

A final test is among the three pure sugars. This mean square has 2 degrees of freedom, since it is based on three means. Thus we compute

$$SS_{(\text{among pure sugars})} = 10(59.3 - 60.5333)^2 + 10(58.2 - 60.5333)^2 + 10(64.1 - 60.5333)^2 = 196.87$$

$$MS_{(\text{among pure sugars})} = \frac{SS_{(\text{among pure sugars})}}{df} = \frac{196.87}{2} = 98.433$$

$$F_s = \frac{MS_{(\text{among pure sugars})}}{MS_{\text{within}}} = \frac{98.433}{5.46} = 18.03$$

This F_s is highly significant, since $F_{.001[2,45]} = 8.09$.

We conclude that the addition of the three sugars retards growth in the pea sections, that mixed sugars affect the sections differently from pure sugars, and that the pure sugars are significantly different among themselves, probably because of the sucrose, which has a far higher mean. We cannot test the sucrose against the other two, because that would be an unplanned test, which suggested itself to us only after we had looked at the results. Such a test requires the methods discussed in the next section.

Our a-priori tests might have been quite different, however, depending entirely on our initial hypotheses. Thus, we alternatively could have planned to test control versus sugars initially, followed by disaccharides (sucrose) versus monosaccharides (glucose, fructose, glucose + fructose), followed by mixed versus pure monosaccharides, and finally by glucose versus fructose.

The pattern and number of planned comparison tests are determined by one's hypotheses about the data. There are certain restrictions, however, if the comparisons are to be orthogonal. As stated earlier, we should not use up more than the $a - 1$ degrees of freedom among the groups, and we should structure the tests in such a way that each comparison tests an independent relationship among the means (as was done in the sugars example). For example, we would prefer not to test if means one, two, and three differed if we had already found that mean one differed from mean three, since significance of the latter implies significance of the former. The following discussion provides rules for establishing the orthogonality of the planned comparisons.

At the bottom of Table 9.1 are two rows of **coefficients of linear comparisons** labeled c_{ij}. The subscript i refers to the group $(1, 2, \ldots, i, \ldots, a)$ in the anova, while j identifies the particular linear comparison $(1, 2, \ldots, j, \ldots, a - 1)$. How are these coefficients determined? For convenience, they are usually expressed as integers, although they may be fractional. For any one comparison, the coefficients must sum to zero when weighted by the sample sizes of the groups; $\Sigma_{i=1}^{a} n_i c_{ij} = 0$. The purpose of the first row of coefficients is to test whether the control is different from the mean of the four treatment groups. Note that to indicate the contrast, the four treatment groups are given a minus sign and the control group is given a plus sign. To obtain the mean of the four treatment groups we have to compute the following: $\frac{1}{4}(\bar{Y}_2 + \bar{Y}_3 + \bar{Y}_4 + \bar{Y}_5)$. Thus, each treatment mean is weighted by a coefficient of $\frac{1}{4}$, as contrasted with the coefficient 1 for the control. To avoid fractions we multiply the coefficients by 4.

An easy way of working out the coefficient to be assigned to each sum in a linear comparison is to assign it the integer representing *the number of samples in the contrasted set.* In this example there is a control set and a sugar set. The control set has one sample; the sugar set has four samples. Therefore we assign a coefficient of 4 to the control set and a coefficient of 1 to each member of the sugar set. The signs of the two sets must be unlike, but computationally it makes no difference whether the plus is assigned to the 1's or to the 4.

To reinforce what we have just learned, let us work out the second row of coefficients in Table 9.1. We now wish to contrast mixed sugars and pure sugars, that is, to test the mean of the three pure sugars against the single mixed sugar, leaving out the control group altogether. Since the control group does not enter in this comparison, we assign it a value $c_{1,2} = 0$. The comparison is between $\frac{1}{3}(\bar{Y}_2 + \bar{Y}_3 + \bar{Y}_5)$ and $\bar{Y}_4$. By the rule of thumb given here, the single mixed sugar is assigned a coefficient of 3 and the pure sugars each a coefficient of 1, since the comparison is between a set of three pure sugars and a set of one mixed sugar. Again, we assign negative signs to the 1's and a plus sign to the 3, but we could just as well switch the signs around.

A table of coefficients permits an easy test for orthogonality of comparisons. On the basis of such a table two comparisons are orthogonal if and only if the sum of the sample sizes times the products of their coefficients equals zero, or expressed as a formula, $\Sigma_{i=1}^{a} n_i c_{ij} c_{ik} = 0$. The subscripts j and k refer to two separate comparisons and i to the various groups. Let us test the two comparisons in Table 9.1, which we have just completed. Since n_i is constant ($= 10$) for all groups, this factor can be ignored.

$$(4 \times 0) + (-1 \times -1) + (-1 \times -1) + (-1 \times 3) + (-1 \times -1) = 0$$

This calculation shows that these two comparisons are orthogonal. The minor complications of specifying the coefficients when sample sizes are unequal are presented in Section 14.10.

A property of an orthogonal set of comparisons is that their sums of squares add up to the sum of squares among groups. The three sums of squares we have obtained thus far, based on 1, 1, and 2 *df*, respectively, together add up to the sum of squares among treatments of the original analysis of variance based on 4 degrees of freedom:

$$
\begin{array}{lrr}
SS_{\text{(control vs. sugars)}} & = \;\; 832.32 & 1 \\
SS_{\text{(mixed vs. pure sugars)}} & = \;\;\; 48.13 & 1 \\
SS_{\text{(among pure sugars)}} & = \;\; \underline{196.87} & \underline{2} \\
SS_{\text{(among treatments)}} & = 1077.32 & 4
\end{array}
$$

This property illustrates once again the elegance of analysis of variance. The treatment sums of squares can be decomposed into separate parts that are sums of squares in their own right, with degrees of freedom pertaining to them. One sum of squares measures the difference between the controls and the sugars, the second that between the mixed sugars and the pure sugars, and the third the remaining variation among the three sugars. We can present all of these results as an anova table, as Table 9.2 shows.

As an example of a *non*orthogonal set of comparisons, we replace $SS_{\text{(mixed sugar vs. pure sugars)}}$ with the SS for the average of glucose and fructose versus glucose and fructose mixed. What would be the coefficients of linear comparisons for this test? Since control and sucrose are not involved in the comparison, they are assigned values of zero. The average of the two pure sugars is represented by 1's for the sugars, the mixed sugar by a 2 to balance the number of samples in the contrasted set. Thus the coefficients are 0, 1, 1, −2, 0. Although this comparison is orthogonal to the first of our tests (control versus treatments), it is not orthogonal to the test of mixed versus pure sugars. To confirm these statements, we may compute $(4 \times 0) + (-1 \times 1) + (-1 \times 1) + (-1 \times -2) + (-1 \times 0) = 0$ and $(0 \times 0) + (-1 \times 1) + (-1 \times 1) + (+3 \times -2) + (-1 \times 0) = -8$. The second sum of products is not equal to zero, confirming the nonorthogonality of mixed versus pure sugars to the average of glucose and fructose versus mixed sugars in the same set.

Table 9.2 ANOVA TABLE FROM BOX 9.4 WITH TREATMENT SUM OF SQUARES DECOMPOSED INTO ORTHOGONAL COMPARISONS.

Source of variation	SS	df	MS	F_s
Treatments	1077.32	4	269.33	49.33***
Control vs. sugars	832.32	1	832.32	152.44***
Mixed vs. pure sugars	48.13	1	48.13	8.82**
Among pure sugars	196.87	2	98.43	18.03***
Within	245.50	45	5.46	
Total	1322.82	49		

To calculate the sum of squares for the comparison, we first compute the mean of glucose and fructose: $(59.3 + 58.2)/2 = 58.75$. Then we obtain the grand mean of glucose, fructose, and mixed sugars: $(59.3 + 58.2 + 58.0)/3 = 58.5$. Finally, SS for the average of glucose and fructose versus glucose and fructose mixed is calculated as follows:

$$20(58.75 - 58.5)^2 + 10(58.0 - 58.5)^2 = 3.75$$

With 3.75 replacing 48.13 (see Table 9.2), the sums of squares obviously will no longer add up to 1077.32.

Now let us apply these techniques to another example. Box 9.8 shows egg-laying records from 25 females of three genetic lines of *Drosophila*, two selected for resistance and susceptibility to DDT, the third a nonselected control line. The data show that average fecundity is considerably higher in the nonselected line. Since we are interested in differences among the lines, each of which was treated in a specific manner, this is clearly a Model I anova. The computations are the same as for Box 9.4, which also was a Model I single-classification anova with equal sample sizes. The steps are shown less fully because we assume that you are by now familiar with the basic computations of the analysis of variance. The anova table shows a highly significant difference among lines. The three critical values are no longer given, but the three asterisks indicate a probability of $P < 0.001$ assuming that the null hypothesis is correct.

The investigator planned to test two hypotheses: Did the selected lines (RS and SS) differ in fecundity from the nonselected line (NS)? Second, did the line selected for resistance differ in fecundity from that selected for susceptibility? These hypotheses form a set of orthogonal comparisons and are tested in the manner you just learned. The results show that the selected lines have a significantly different (lower) fecundity from that of the nonselected line and that the two selected lines do not differ from each other in fecundity. These data are tabulated in Box 9.8 in a new anova table, which includes both one-degree-of-freedom tests. Again you can see that the sums of squares are additive, together yielding the sum of squares among lines. Note also that this anova table is given in more abbreviated form than earlier ones to show you how such tables are published in the literature. The F's are not given in this table, nor is the total sum of squares. As you will see later, sometimes even the sums of squares are not stated, leaving only the essential information: the degrees of freedom and the mean squares.

When the planned comparisons are not orthogonal (note that when they involve more than $a - 1$ degrees of freedom they cannot be orthogonal), we carry out the tests as shown here but adjust the values of the type I error α. If the comparisons are nonorthogonal, they lack independence. Thus if the outcome of a single such comparison is significant, the outcomes of subsequent comparisons might more likely be significant as well, and decisions based on conventional

Box 9.8 COMPARISONS OF MEANS AND GROUPS OF MEANS IN A SINGLE-CLASSIFICATION MODEL I ANOVA.

Per diem fecundity (number of eggs laid per female per day for the first 14 days of life) for 25 females from each of 3 lines of *Drosophila melanogaster*. The RS and SS lines were selected for resistance and for susceptibility to DDT, respectively; and the NS line is a nonselected control strain; $n = 25$ females per line.

	Lines ($a = 3$)					
	Resistant (RS)		Susceptible (SS)		Nonselected (NS)	
	12.8	22.4	38.4	23.1	35.4	22.6
	21.6	27.5	32.9	29.4	27.4	40.4
	14.8	20.3	48.5	16.0	19.3	34.4
	23.1	38.7	20.9	20.1	41.8	30.4
	34.6	26.4	11.6	23.3	20.3	14.9
	19.7	23.7	22.3	22.9	37.6	51.8
	22.6	26.1	30.2	22.5	36.9	33.8
	29.6	29.5	33.4	15.1	37.3	37.9
	16.4	38.6	26.7	31.0	28.2	29.5
	20.3	44.4	39.0	16.9	23.4	42.4
	29.3	23.2	12.8	16.1	33.7	36.6
	14.9	23.6	14.6	10.8	29.2	47.4
	27.3		12.2		41.7	
$\sum^{n} Y$	631.4		590.7		834.3	
$\bar{Y}$	25.256		23.628		33.372	

SOURCE: Data from R. R. Sokal (unpublished results).

I. *Anova*

1. $\bar{\bar{Y}} = \frac{1}{an}\sum^{a}\sum^{n} Y = \frac{1}{75}(631.4 + 590.7 + 834.3) = 27.419$

2. $SS_{\text{among}} = n\sum^{a}(\bar{Y}_i - \bar{\bar{Y}})^2$

$= 25((25.256 - 27.419)^2 + (23.628 - 27.419)^2 + (33.372 - 27.419)^2)$

$= 1362.21$

3. $SS_{\text{within}} = \sum^{a}\sum^{n}(Y - \bar{Y}_i)^2 = 5659.02$

4. $SS_{\text{total}} = SS_{\text{among}} + SS_{\text{within}} = 7021.23$

BOX 9.8 CONTINUED

Anova table

	Source of variation	df	SS	MS	F_s
$\bar{Y} - \bar{\bar{Y}}$	Among groups (among genetic lines)	2	1362.21	681.11	8.67***
$Y - \bar{Y}$	Within groups	72	5659.02	78.60	
$Y - \bar{\bar{Y}}$	Total	74	7021.23		

II. *Planned tests among the means*

There were two interesting a-priori hypotheses. Their rationale is discussed in the text.

1. $H_0: \mu_{(\text{RS and SS})} = \mu_{\text{NS}} \qquad H_1: \mu_{(\text{RS and SS})} \neq \mu_{\text{NS}}$
The mean for RS and SS is (25.256 + 23.628)/2 = 24.442. The grand mean is quantity **1** from part **I.**

$$SS_{(\text{RS}+\text{SS})\text{ vs. NS}} = 50(24.442 - 27.419)^2 + 25(33.372 - 27.419)^2 = 1329.08$$

Because there is only one df for this comparison, $SS = MS$.

The significance test for this hypothesis is

$$F_s = \frac{MS_{(\text{RS}+\text{SS})\text{ vs. NS}}}{MS_{\text{within}}} = \frac{1329.08}{78.60} = 16.91^{***}$$

Since $F_{.001[1,60]} = 12.0$, the comparison is highly significant. We can reject the null hypothesis far below the 0.1% level.

2. $H_0: \mu_{\text{RS}} = \mu_{\text{SS}} \qquad H_1: \mu_{\text{RS}} \neq \mu_{\text{SS}}$

Since, in this case, the two groups to be compared have equal sample sizes, instead of testing them with the equation

$$n_{\text{RS}}(\bar{Y}_{\text{RS}} - \bar{\bar{Y}})^2 + n_{\text{SS}}(\bar{Y}_{\text{SS}} - \bar{\bar{Y}})^2$$

we may use a similar equation for the sum of squares, first learned in Box 9.5:

$$SS_{(\text{RS vs. SS})} = \frac{n}{2}(\bar{Y}_{\text{RS}} - \bar{Y}_{\text{NS}})^2 = \frac{25}{2}(25.256 - 23.628)^2 = 33.13$$

There is one df for this comparison, and again $SS = MS$.

The significance test for this hypothesis is

$$F_s = \frac{MS_{\text{RS vs. SS}}}{MS_{\text{within}}} = \frac{33.13}{78.60} = 0.4215$$

which is not significant.

Box 9.8 Continued

Anova table

Source of variation	*df*	*SS*	*MS*
Among groups (among genetic lines)	2	1362.21	681.11***
Selected vs. nonselected lines	1	1329.08	1329.08***
RS vs. SS	1	33.13	33.13 *ns*
Within groups (error; females within genetic lines)	72	5659.02	78.60

levels of significance might be in doubt. For this reason we employ a conservative approach in which one lowers the type I error of the statistic of significance for each comparison so that the probability of making any type I error at all in the entire series of tests does not exceed α. This probability is called the **experimentwise error rate**. Assuming that an investigator plans to make k independent significance tests, each using a critical value corresponding to a significance level of α', then the probability of making no type I errors in any of the k tests is $(1 - \alpha')^k$. The experimentwise error rate (i.e., the probability of making at least one type I error) is then $\alpha = 1 - (1 - \alpha')^k$. In order to achieve a specified experimentwise error rate, α, this equation can be solved for α' to yield

$$\alpha' = 1 - (1 - \alpha)^{1/k}$$

This method has been called the **Dunn–Šidák method** by Ury (1976), who showed that employing α' for each comparison results in a conservative test (experimentwise error rate $\leq \alpha$) when the individual significance tests are not independent. (It is exact when the tests are independent.)

Applying this approach to the pea section data from Box 9.4 let us assume that the investigator has good reason to test the comparisons between the following treatments, given here in abbreviated form: (C) vs. (G, F, S, G + F); (G, F, S) vs. (G + F); (G) vs. (F) vs. (S); and (G, F) vs. (G + F). The four tests to be undertaken in this series require that each individual test be adjusted to a significance level of

$$\alpha' = 1 - 1(1 - 0.05)^{1/4} = 0.01274$$

for an experimentwise error rate of $\alpha = 0.05$. Thus the critical values for the F_s-ratios of these comparisons are $F_{.01274[1,45]}$ or $F_{.01274[2,45]}$ as appropriate. Such nonstandard values of α are not found in any F-tables. We can circumvent this problem for critical values of $F_{\alpha'[1,\nu_2]}$ by making use of the relationship $F_{\alpha'[1,\nu_2]} = t^2_{\alpha'[\nu_2]}$. These values can be found by turning to Statistical Table **C**, which contains critical values of $t_{\alpha'[\nu_2]}$ for experimentwise error rate α and degrees of freedom ν_2 for k comparisons. For critical values of $F_{\alpha'[\nu_1,\nu_2]}$, where $\nu_1 \geq 2$, there are no exact tables available, so one must either interpolate in the F-table between the bracketing values of α or use a computer program to compute the desired critical value. Alternatively one can use the next lower tabled value of α for a conservative test.

In spite of the change in critical value, the conclusions concerning the first three tests are unchanged. The last test, the average of glucose and fructose versus a mixture of the two, is not significant, since $F_s = 3.75/5.46 = 0.687$. Adjusting the critical value is a conservative procedure; individual comparisons are less likely to be significant using this approach.

An alternative method for limiting the overall experimentwise error rate is to test each comparison using a significance level of $\alpha'' = \alpha/k$. This is called the **Bonferroni method**. It has been shown to be slightly more conservative (its experimentwise error rate is smaller) than the Dunn–Šidák method. The Bonferroni method is still used occasionally, since the calculation of α'' is slightly simpler than that of α'. However, unless the number of comparisons k yields a tabled probability value of α'', interpolation may be messy, and it is far easier to use the Dunn–Šidák method with Table **C**. Note also that the Bonferroni method, the Dunn–Šidák method, and other similar approaches are frequently referred to as Bonferroni techniques, using the term in a more inclusive sense. Some may be tempted to use the Bonferroni method, since α'' values are more likely to correspond to those in standard tables. For example, at the $\alpha = 0.05$ level, if one is fortunate enough to require $k = 5$ comparisons, then $\alpha'' = 0.01$ and one can look up critical values in standard tables. An investigator should not be tempted, however, to use a particular number of comparisons just to avoid the bother of interpolation or using special tables.

The approaches we have just described have the desired property of controlling the probability of making any type I error at all using a set of k significance tests. As a consequence, each individual test is rather conservative if k is not small. The power of the tests is thus low if more than one null hypothesis is false. Holm (1979) developed a sequential test that increases the power of the individual tests when the overall null hypothesis is rejected. Although the original proposal was based on the Bonferroni method, we describe it in Box 9.9 in terms of the Dunn–Šidák method, since the latter method results in slightly less conservative tests.

Using an experimentwise error approach we can also set confidence limits to means and differences between means. We discuss this topic in Section 14.10.

9.7 Comparisons Among Means: Unplanned Comparisons

The development of proper methods for testing hypotheses that are suggested by inspection of the results of analysis of variance has occupied the attention of statisticians since the 1950s. Several different approaches have been proposed,

Box 9.9 SEQUENTIAL BONFERRONI TESTS: k COMPARISONS BY THE DUNN–ŠIDÁK METHOD AT A SPECIFIED EXPERIMENTWISE ERROR RATE α.

We will perform five planned but nonorthogonal tests on the pea section data of Box 9.4.

1. List the tests in the order performed and furnish probabilities for each that are as precise as possible. This can be done with extended t-tables for single degree of freedom tests, with detailed interpolation, or with computer programs that furnish probabilities. All but the last test in our example have been carried out previously in this section. The last test is carried out in an analogous manner.

Contrasts	P_i Probability
Sugars vs. control (C)(G, F, G + F, S)	$<10^{-6}$
Mixed vs. pure sugars (G + F)(G, F, S)	0.004765
Among pure sugars (G)(F)(S)	0.000002
Mixed vs. average of G and F (G + F)(G, F)	0.418749
Monosaccharides vs. disaccharides (G, F)(S)	$<10^{-6}$

2. Order these probability values from smallest to largest.

P_i	Associated α
$<10^{-6}$	0.0102
$<10^{-6}$	0.0127
0.000002	0.0170
0.004765	0.0253
0.418749	0.0500

3. Compare the smallest probability P_1 to $1 - (1 - \alpha)^{1/k}$. For a 5% experimentwise error rate, this comparison yields $1 - (1 - .05)^{1/5} = 0.0102$. If P_1 is greater than this critical value, declare all tests not significant. If $P_1 \leq$ the critical value, as in our example, declare this contrast significant and compare the second smallest probability, P_2, to $1 - (1 - \alpha)^{1/(k-1)} = 0.0127$. If $P_2 > 0.0127$, declare the remaining tests nonsignificant. If $P_2 \leq 0.0127$, as in our example, declare this contrast significant and continue testing the successively larger probabilities in the same manner.

As a result of this procedure, we decide that all the tests except that comparing the mixed glucose and fructose with the average of its constituents are significant. Note that

BOX 9.9 CONTINUED

these results do not differ from what we had already established. Nevertheless this procedure is preferable, because it is the correct test when the tests exhibit nonorthogonality (which must be the case whenever more tests are made than there are degrees of freedom among groups). Furthermore, the use of a blanket Dunn–Šidák probability of $1 - (1 - \alpha)^{1/k}$ for all tests is less powerful than this sequential procedure.

NOTE: Much of the labor involved in this procedure can be eliminated when one can place the tests in order of their probability without having to compute the probabilities. For example, if all the tests are based on pairs of means with equal sample sizes (unlike the present example), one can order the tests by the observed differences in means and then make the series of tests using Table **C** with $k' = k$, $k - 1, \ldots$, stopping as soon as a test is nonsignificant.

and a complete consensus has not yet been achieved on which method to employ. There have been so many diverse approaches because the problem of testing all of the "interesting" differences—those differences among means that loom large in the observed sample—is subject to a variety of interpretations. We will present a few techniques that appear to be reasonable on the basis of current knowledge and that should suffice for most problems.

Most unplanned multiple comparisons tests adopt a conservative approach in that they employ experimentwise error rates for the type I error. The experimentwise error rate of "making any type I error at all" means making such an error in the overall test of significance of the anova or in any of the subsidiary comparisons among means or sets of means needed to complete the analysis of the experiment. As already noted in the previous section, although the probability of any error at all is α, the probability of error for any particular test of some subset, such as a test of the difference among three or between two means, is necessarily less than α. The significance level α' for any one test may be much less than the experimentwise α, and if there are many means in the anova this actual error α' may be $\frac{1}{10}$th, $\frac{1}{100}$th, or even $\frac{1}{1000}$th of the experimentwise α (Gabriel, 1964). For this reason the tests employing experimentwise error rates are conservative. That is, they are not very sensitive to differences of individual means or differences within small subsets. Obviously not many differences are going to be considered significant if α' is minute. This is the price one pays for testing unplanned comparisons.

One way of carrying out unplanned comparisons of means is to test potentially significant differences among all possible pairs of means. In an anova with a means there could be $a(a - 1)/2$ such comparisons. We might reasonably be motivated to adopt such a philosophy if we intended to test the difference between the largest and the smallest mean and between some other pairs of means that appear to differ greatly. Although we may carry out only a few such tests (far fewer than the $a(a - 1)/2$ possible such comparisons), in choosing the most different pair we have implicitly surveyed all possible pairs.

Many methods examine the difference between each pair of means. *If such a test were a planned comparison* one could test it as

$$t_s = \frac{(\bar{Y}_1 - \bar{Y}_2) - (\mu_1 - \mu_2)}{\sqrt{\left[\left(\frac{1}{n_1} + \frac{1}{n_2}\right) MS_{\text{within}}\right]}} \tag{9.6}$$

or for equal sample sizes as

$$t_s = \frac{(\bar{Y}_1 - \bar{Y}_2) - (\mu_1 - \mu_2)}{\sqrt{\frac{2}{n} MS_{\text{within}}}} \tag{9.6a}$$

with the same number of degrees of freedom as possessed by the error mean square. Although in any one test fewer individuals may be involved than in the full anova, the number of degrees of freedom is based on MS_{within}. This is an important difference from previous t-tests.

If we assume that $\mu_1 - \mu_2 = 0$ and set t_s equal to $t_{\alpha[\nu]}$, Expression (9.6a) can be arranged to read

$$|\bar{Y}_1 - \bar{Y}_2| = t_{\alpha[\nu]} \sqrt{\frac{2}{n} MS_{\text{within}}} = LSD \tag{9.7}$$

It should be obvious that the quantity $|\bar{Y}_1 - \bar{Y}_2|$ is the difference of means that will just be significant between any pair of means in a study with equal sample sizes at a probability α. Any two means that differ by at least this amount will, when put into Expression (9.6), give a significant value of t_s. This quantity is called the **least significant difference** (*LSD*) and has much appeal because it is a single value which, once computed, permits the investigator to look at any pair of means and tell whether they are significantly different. In the *Drosophila* example of Box 9.8 we can compute

$$LSD = t_{.01[72]} \sqrt{\frac{2}{n} MS_{\text{within}}} = 2.646 \times 2.507 = 6.634$$

Thus any two means that we planned to compare and that differ by 6.634 or more are significantly different from each other at $P \leq 0.01$. The *LSD* indicates (correctly) that the nonselected line differs from either selected line in fecundity, but that the two selected lines are not different from each other. As we have pointed out already, however, *such an approach is valid for planned comparisons only*.

To generalize the procedure to include unplanned comparisons, we rewrite the formula for the *LSD* as

$$MSD = (\text{critical value}) \times SE \tag{9.8}$$

where *MSD* stands for **minimum significant difference,** the critical value refers to the statistical distribution appropriate to the given test and *SE* is the appropriate standard error. Most of the tests discussed below will be based on this general formula, but the critical values and the standard errors will vary from test to test.

One such critical value is $Q_{\alpha[k,\nu]}$, a special statistic for significance testing, called the **studentized range,** and shown in Statistical Table **J.** This table gives critical values for the distribution of *Range/s* for k normally distributed items, where s is an independent estimate of the standard deviation of the items. The table has two arguments: Across the top is k, the number of items in the range, and along the left margin is ν, the degrees of freedom of the standard deviation. The studentized range could be employed if we wanted to test the largest observed difference in our set of means, since we would in effect be testing the significance of the range of the means. Analogous to Expression (9.8) a **minimum significant range** could be defined as

$$MSR = Q_{\alpha[k,\nu]}\sqrt{\frac{MS_{\text{within}}}{n}} \tag{9.9}$$

Note that the standard error in the equation has changed also. But this difference is only apparent. The missing 2 from inside the square root is incorporated in the value of Q. For the special case $k = 2$ (when there are only 2 means in the entire anova), *MSR* is the same as *LSD* because $Q_{\alpha[2,\nu]} = \sqrt{2}t_{\alpha[\nu]}$.

When there are more than two means in an anova, the maximum difference between these means is the range of the sample means.

For all tests of any pair of means we set $k = a$ and $\nu = a(n - 1)$. Then for the pea sections of Box 9.4,

$$MSR = Q_{.05[5,45]}\sqrt{\frac{5.46}{10}} = 4.018\sqrt{0.546} = 2.969$$

This *MSR* can be used to test the range of all means, as well as differences between any pair of means, for *equal sample sizes (n)* only. This technique has various names. In our first edition we called it the *range STP procedure,* also known as *Tukey's honestly significant difference method.* In the recent literature the term **T-method** has become widely used. In Box 9.10 we show an application of the T-method to the pea section data.

When sample sizes are unequal, the strategy to be pursued is still not settled. Techniques for unequal sample sizes employ modifications of the critical values and of the standard errors. In recent years two techniques for testing all differences between pairs of means have been widely recommended: the **T′-method** (Spjøtvoll and Stoline, 1973), which employs the **studentized augmented range** distribution Q' as a critical value, and the **GT2-method** (Hochberg, 1974), which employs the studentized maximum modulus distribution m. The **studentized maximum modulus** is the maximum over k comparisons of the quantity $|\bar{Y}_i - \mu|/s_{\bar{Y}}$. The $\bar{Y}_i$ are assumed to be normally distributed with a mean

Box 9.10 MULTIPLE COMPARISONS AMONG PAIRS OF MEANS BASED ON EQUAL SAMPLE SIZES (UNPLANNED COMPARISONS): THE T-METHOD.

Lengths of pea sections grown in tissue culture. Data from Box 9.4.

1. Substitute into Expression (9.8), MSD = (critical value) × SE, as follows:

For the critical value use $Q_{\alpha[k,\nu]}$, the studentized range from statistical Table **J**. Set $k = a$, and let ν equal the degrees of freedom of MS_{within}. Consulting Table **J** we find that $Q_{.05[5,45]} = 4.018$. The Q-value was obtained by harmonic interpolation between the entries for $\nu = 40$ and $\nu = 60$ degrees of freedom. The standard error for the T-test is

$$s_{\bar{Y}} = \sqrt{\frac{MS_{\text{within}}}{n}} = \sqrt{\frac{5.46}{10}} = 0.7389$$

2. A pair of means, $\bar{Y}_i$ and $\bar{Y}_j$, is declared significantly different at the experimentwise error rate α if and only if the difference between the sample means equals or exceeds the critical difference, i.e., if $|\bar{Y}_i - \bar{Y}_j| \geq MSD$. In the present example

$$MSD = 4.018(0.7389) = 2.9691$$

We can now test the difference between any pair of means against MSD. Thus the means of the control versus that of sucrose yield a difference of $(70.1 - 64.1) = 6.0$, which is highly significant, since $6.0 \gg 2.9691$.

3. A convenient method for systematically testing all $a(a - 1)/2$ pairs of means is to compute upper and lower comparison limits for each mean such that two means are significantly different if and only if their limits do not overlap. The limits (Gabriel, 1978) are simply

$$l_i = \bar{Y}_i - \frac{1}{2}MSD = \bar{Y}_i - 1.4845$$

$$u_i = \bar{Y}_i + \frac{1}{2}MSD = \bar{Y}_i + 1.4845$$

Since the MSD above was calculated for $\alpha = 0.05$, these comparison limits are for an experimentwise 0.05 error rate. The ordered group means and their limits are as follows:

	Control	2% sucrose added	2% glucose added	2% fructose added	1% glucose + 1% fructose added
$\bar{Y}_i$	70.1	64.1	59.3	58.2	58.0
l_i	68.62	62.62	57.82	56.72	56.52
u_i	71.58	65.58	60.78	59.68	59.48

BOX 9.10 CONTINUED

Group means whose comparison intervals overlap (for example, glucose, fructose, and glucose + fructose) are not significantly different at the experimentwise error rate of $\alpha = 0.05$. Those whose intervals do not overlap are significantly different. We can display this conveniently by plotting the group means and their comparison intervals as shown in the figure below. Clearly control and sucrose means are significantly different from each other and from all other means at the 5% experimentwise error rate. The glucose, fructose, and glucose plus fructose means are not significantly different from each other.

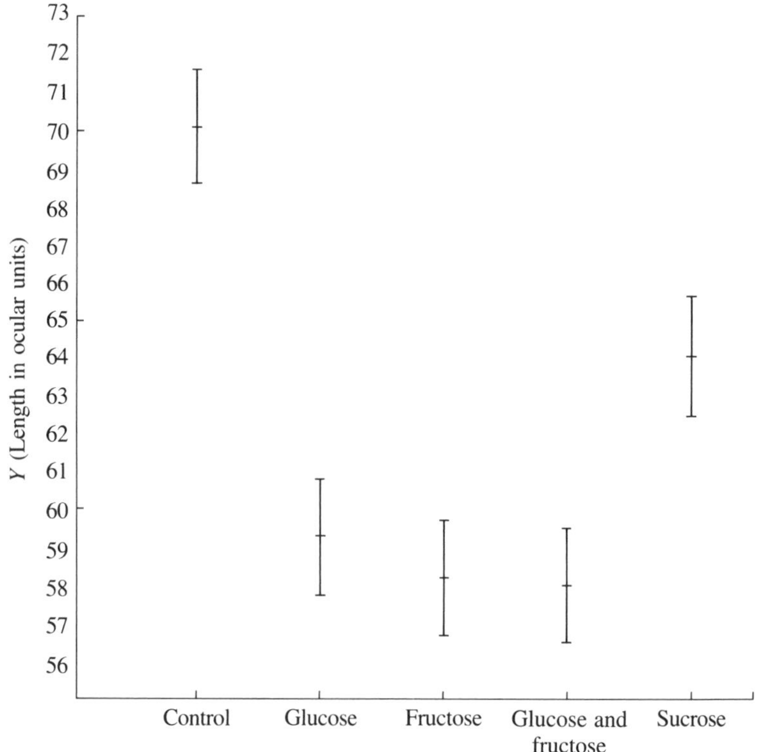

95% comparison intervals by the T-method for the means of the pea section data. Means whose intervals do not overlap are significantly different.

of μ and standard deviation $\sigma_{\bar{Y}}$. Critical values of these distributions are given in Statistical Tables **L** and **M,** respectively. The procedures are explained in detail in Box 9.11. If the sample sizes are nearly equal the T′-method will always be preferred, since it will yield a more powerful test. For unequal sample sizes the choice between the two methods depends upon the heterogeneity of the sample sizes of the mean. When sample sizes are very unequal, the GT2-method is preferred.

Box 9.11 MULTIPLE COMPARISONS AMONG PAIRS OF MEANS BASED ON UNEQUAL SAMPLE SIZES (UNPLANNED COMPARISONS): THE GT2-, T′-, AND TUKEY-KRAMER METHODS.

Head widths (in millimeters) of a species of tiger beetle *(Cicindela circumpicta)* collected from 8 localities.

	Localities (ordered by magnitude of mean)							
	Barnard, Kansas	Roswell, New Mexico	Nebraska	Talmo, Kansas	Okeene, Okla-homa	Kackley, Kansas	Mayfield, Okla-homa	Stafford Co., Kansas
$\bar{Y}$	3.5123	3.6940	3.7223	3.7229	3.7376	3.7479	3.7843	3.8451
n_i	20	20	20	14	10	15	11	20
s^2	0.0357	0.0445	0.0135	0.0284	0.0204	0.0184	0.0338	0.0157

SOURCE: Selected data from H. L. Willis (unpublished results).

Compute the pooled variance = weighted average variance = MS_{within}:

$$MS_{\text{within}} = \frac{\sum^a (n_i - 1)s_i^2}{\sum^a (n_i - 1)}$$

$$= \frac{(20 - 1)0.0357 + (20 - 1)0.0445 + \ldots + (20 - 1)0.0157}{19 + 19 + \ldots + 10 + 19}$$

$$= \frac{3.2270}{122}$$

$$= 0.02645$$

MS_{within} has degrees of freedom, ν, equal to $\Sigma(n_i - 1)$. In this example $\nu = 122$.

The GT2-method

1. Because the n_i's seem quite unequal we employ the GT2-method. Substitute into Expression (9.8), MSD = (critical value) × SE, as follows:

   For the critical value use $m_{\alpha[k^*,\nu]}$, the two-tailed studentized maximum modulus from Statistical Table **M.** Set $k = a$, and $k^* = k(k - 1)/2 = a(a - 1)/2 = 8(8 - 1)/2 = 28$ in this example. As already stated, $\nu = 122$. For a 5% experimentwise error rate use the tabled value $m_{.05[28,120]} = 3.183$ as a conservative approximation of the desired value of $m_{.05[28,122]}$. When the required degrees of freedom are not as close to a tabled value of ν as in this case, it is recommended that the average between the linearly and harmonically interpolated values of m be employed.

Box 9.11 Continued

The standard error for the GT2-test is

$$SE_{ij} = \sqrt{s^2_{\bar{Y}_i} + s^2_{\bar{Y}_j}} = \sqrt{\frac{MS_{\text{within}}}{n_i} + \frac{MS_{\text{within}}}{n_j}}$$

for a comparison of means $\bar{Y}_i$ and $\bar{Y}_j$. For convenience, compute an array of

$$s^2_{\bar{Y}_i} = \frac{MS_{\text{within}}}{n_i}$$

as shown here:

0.001323 0.001323 0.001323 0.001889 0.002645 0.001763 0.002405 0.001323

2. A pair of means $(\bar{Y}_i, \bar{Y}_j)$ is declared significantly different at the experimentwise error rate α if and only if their difference equals or exceeds the critical difference, MSD_{ij}, that is, if $|\bar{Y}_j - \bar{Y}_i| \geq MSD_{ij}$. Note that in computing

$$MSD_{ij} = m_{\alpha[k^*,\nu]}\,(s^2_{\bar{Y}_i} + s^2_{\bar{Y}_j})^{1/2}$$

the critical value $m_{\alpha[k^*,\nu]}$ is a constant for a given problem. For example $MSD_{12} = 3.183(.001323 + .001323)^{1/2} = 3.183(.051439) = 0.16370$. Thus localities 1 and 2 differ significantly in their means, since $(3.6940 - 3.5123) = 0.1817 > 0.16370$.

To test all pairs of means it is convenient to prepare a table of $|\bar{Y}_j - \bar{Y}_i|$ and MSD_{ij} values for all pairs of means i and j. In the table below, $|\bar{Y}_j - \bar{Y}_i|$ is given below the diagonal and MSD_{ij} above the diagonal.

	Ranked localities							
Ranked localities	1	2	3	4	5	6	7	8
1	—	.1637	.1637	.1804	.2005	.1768	.1943	.1637
2	.1817*	—	.1637	.1804	.2005	.1768	.1943	.1637
3	.2100*	.0283	—	.1804	.2005	.1768	.1943	.1637
4	.2106*	.0289	.0006	—	.2143	.1924	.2086	.1804
5	.2253*	.0436	.0153	.0147	—	.2113	.2262	.2005
6	.2356*	.0539	.0256	.0250	.0103	—	.2055	.1768
7	.2720*	.0903	.0620	.0614	.0467	.0364	—	.1943
8	.3328*	.1511	.1228	.1222	.1075	.0972	.0608	—

The differences larger in absolute value than their *MSD* value are significantly different at the 0.05 level and are indicated by an asterisk.

3. A simple approximate method proposed by Gabriel (1978), which avoids the tedium of preparing the $k \times k$ table of differences and *MSD* values, is to compute lower and upper comparison limits (l, u) for each sample mean such that two means are declared significantly different if and only if their intervals do not overlap. For the ith

BOX 9.11 CONTINUED

mean the limits are simply

$$l_i = \bar{Y}_i - \sqrt{\tfrac{1}{2}}\, m_{\alpha[k^*,\nu]} s_{\bar{Y}_i}$$

$$u_i = \bar{Y}_i + \sqrt{\tfrac{1}{2}}\, m_{\alpha[k^*,\nu]} s_{\bar{Y}_i}$$

For the first mean these are

$$l_1 = 3.5123 - \sqrt{\tfrac{1}{2}}\,(3.183)(0.001323)^{1/2}$$
$$= 3.5123 - 0.081866$$
$$= 3.4304$$
$$u_1 = 3.5123 + 0.081866$$
$$= 3.5942$$

The limits for all 8 localities are as follows:

l_i	3.4304	3.6121	3.6404	3.6251	3.6218	3.6534	3.6739	3.7632
u_i	3.5942	3.7759	3.8042	3.8207	3.8534	3.8424	3.8947	3.9270

Only the limits for locality 1 are distinct from those of the other localities—i.e., the comparison interval of this locality does not overlap with those of the other localities, all of which overlap mutually. This is most conveniently seen in the figure below, which shows that the mean for locality 1 is significantly different from the others and that the others do not differ significantly from one another.

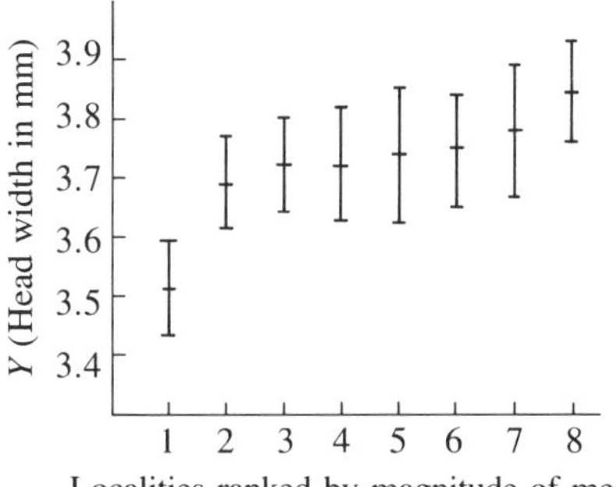

95% comparison intervals by the GT2-method for the means of the tiger beetle data. Means whose intervals do not overlap are significantly different.

The T′-method

1. Had the sample sizes been more nearly equal, the T′-method adjusted for unequal sample sizes would have been the preferred approach. We illustrate below the T′-

BOX 9.11 CONTINUED

method applied to the geographic variation data in the tiger beetles, although the disparity in sample sizes makes it likely that the GT2-method will have greater power.

Substitute into Expression (9.8), MSD = (critical value) × SE, as follows: For the critical value use $Q'_{\alpha[k,\nu]}$, the studentized augmented range from Statistical Table **L.** Set $k = a$, and employ the degrees of freedom of MS_{within} for ν. In this case $k = 8$, and $\nu = 122$ as noted earlier. Consulting Table **L,** we find $Q'_{.05[8,122]} = 4.363$. This value will be constant for all tests.

The standard error for unequal sample sizes is

$$SE_{ij} = \sqrt{\frac{MS_{\text{within}}}{min(n_i,\, n_j)}}$$

Thus for the comparison of the means of localities 3 and 4, min (20, 14) = 14, and we compute

$$SE_{3,4} = \sqrt{\frac{0.02645}{14}} = \sqrt{0.001889} = 0.04347$$

Therefore

$$MSD_{3,4} = 4.363(0.04347) = 0.1896$$

2. A pair of means $(\bar{Y}_1, \bar{Y}_j)$ is declared significantly different at the experimentwise error rate α if and only if their difference equals or exceeds the critical difference MSD_{ij}, i.e., if $|\bar{Y}_j - \bar{Y}_i| \geq MSD_{ij}$. Note that the critical value $Q'_{\alpha[k,\nu]}$ is a constant for a given problem.

Thus, for example, in the comparison between means $\bar{Y}_3$ and $\bar{Y}_4$, the actual difference $|\bar{Y}_4 - \bar{Y}_3| = 0.0006 \ll 0.1896$, and the two means are clearly not significantly different. Note that the MSD for this comparison by the T′-method is greater than that obtained for the same comparison by the GT2-method (for which $MSD_{3,4} = 0.1804$ can be found in the $k \times k$ table featured earlier in this box). The higher value of the MSD confirms that in this case with markedly unequal sample sizes, GT2 is more powerful than the T′-method. In borderline cases all values of MSD_{ij} would have to be computed by both methods (another $k = k$ table should be set up for the T′-method), and the lesser of the two values of MSD_{ij} would be chosen. It is possible that in a single analysis, one test may be more powerful for some comparisons, whereas the other test may be more powerful for other comparisons.

3. As in step 3 of the GT2-method, one can compute approximate comparison limits to the sample means by the T′-method. For the ith mean the limits are

$$l_i = \bar{Y}_i - \frac{1}{2} Q'_{\alpha[k,\nu]} s_{\bar{Y}_i}$$

$$u_i = \bar{Y}_i + \frac{1}{2} Q'_{\alpha[k,\nu]} s_{\bar{Y}_i}$$

If the sample sizes are approximately equal, one can compare the means graphically, as before. In this example, however, the sample sizes are substantially different, so this method should not be used.

BOX 9.11 CONTINUED

The Tukey-Kramer Method

1. Another method that can be applied to these data is the Tukey-Kramer procedure. Substitute into Expression (9.8), MSD = (critical value) × SE, as follows: For the critical value use $Q_{\alpha[k,\nu]}$, the studentized range from Statistical Table **J.** Set $k = a$, and employ the degrees of freedom of MS_{within} for ν. In this case $k = 8$, and $\nu = 122$ as noted before. In Table **J** we find that $Q_{.05[8,122]} = 4.362$. This value will be used for all tests.

   The equation for computing the standard error is

$$SE_{ij} = \sqrt{\frac{MS_{\text{within}}\left(\frac{1}{n_i} + \frac{1}{n_j}\right)}{2}}$$

   Thus for the comparison of the means of localities 3 and 4,

$$SE_{3,4} = \sqrt{\frac{0.02645\left(\frac{1}{20} + \frac{1}{14}\right)}{2}}$$

$$= \sqrt{0.001606} = 0.040074$$

   Therefore

$$MSD_{3,4} = 4.362(0.040074) = 0.17480$$

2. A pair of means $(\bar{Y}_i, \bar{Y}_j)$ is declared significantly different at the experimentwise error rate α if and only if their difference equals or exceeds the critical difference MSD_{ij}, i.e., if $|\bar{Y}_j - \bar{Y}_i| \geq MSD_{ij}$.

   For example, means $\bar{Y}_3$ and $\bar{Y}_4$ are not significantly different, since $|\bar{Y}_4 - \bar{Y}_3| = 0.0006 \ll 0.17480$. Note that the MSD for this comparison is smaller than that obtained for the same comparison by both the GT2- and the T′-methods and thus yields the most powerful test of the three tests examined.

   To test all pairs of means it is most convenient to prepare a table of $|\bar{Y}_j - \bar{Y}_i|$ and MSD_{ij} values, as was shown for the GT2-method:

	Ranked localities							
	1	2	3	4	5	6	7	8
1	—	.1586	.1586	.1748	.1943	.1713	.1883	.1586
2	.1817*	—	.1586	.1748	.1943	.1713	.1883	.1586
3	.2100*	.0283	—	.1748	.1943	.1713	.1883	.1586
4	.2106*	.0289	.0006	—	.2077	.1864	.2021	.1748
5	.2253*	.0436	.0153	.0147	—	.2048	.2192	.1943
6	.2356*	.0539	.0256	.0250	.0103	—	.1991	.1713
7	.2720*	.0903	.0620	.0614	.0467	.0364	—	.1883
8	.3328*	.1511	.1228	.1222	.1075	.0972	.0608	—

Box 9.11 Continued

The $|\bar{Y}_j - \bar{Y}_i|$ values are given below the diagonal, and MSD_{ij} values are given above the diagonal. Differences larger in absolute value than their *MSD* value are significant at the 0.05 level and are marked with an asterisk. In this case the results are the same as those obtained using the GT2-method.

A third, older method for unequal sample sizes was revived by Dunnett (1980), following a simulation study. This method, the **Tukey–Kramer procedure,** employs the studentized range but uses average sample sizes in the standard errors. Dunnett's results suggest that the T′- and GT2-methods are too conservative, whereas the Tukey–Kramer method yields results which, while still conservative, are much closer to the intended significance level α. The Tukey–Kramer method is also given in Box 9.11. When in doubt about which method to use, carry out all three procedures and employ the one that yields the smallest *MSD*. Choice of the method that gives the best result (i.e., the smallest *MSD*) is valid, since the decision is not based upon a consideration of sample statistics but only upon the sample sizes. All three methods of multiple comparisons are carried out by the BIOM-pc computer program.

A fourth technique for testing all differences between pairs of means, the **Welsch step-up procedure** (Welsch, 1977), employs a stepwise approach. The means are arrayed in order of magnitude. Then pairs of adjacent means are tested for significant differences. Only if this test is not significant for a given pair does one proceed to test larger subsets of adjacent means that include the pair. Larger subsets of size j are tested only if smaller subsets contained in them are not significant. The procedure differs from others discussed in this section in that the error rate for each test is adjusted to reflect the size of the subset being tested. The adjustment is computed as $\alpha_j = \alpha(j/k)$, where α_j is the type I error for testing a range of j adjacent means. The test is illustrated in Box 9.12 for equal sample sizes only. Because of the adjustment of the critical value for the subset size, the Welsch procedure has greater power than the T- and GT2-methods discussed above. This method does not, however, permit the construction of simultaneous confidence limits to differences or of graphical tests by comparison limits. Since the T- and GT2-methods are very simple and fast, many workers prefer them to the Welsch test, even though they are less powerful. The BIOM-pc computer program carries out the Welsch test.

In addition to testing pairs of means, the investigator may wish to make certain *unplanned contrasts among means* suggested by the results of the experiment. To carry out such tests we employ the coefficients of linear comparisons introduced in the previous section. In Box 9.13 we feature three approaches—the Scheffé method, the T-method, and the GT2-method. The **Scheffé method** is most powerful when a and ν are both small and most coefficients of linear

Box 9.12 MULTIPLE COMPARISONS AMONG MEANS BASED ON EQUAL SAMPLE SIZES (UNPLANNED TESTS): THE WELSCH STEP-UP PROCEDURE.

Lengths of pea sections grown in tissue culture. Data from Box 9.4.

1. First calculate the standard error of a group mean, using the error mean square (MS_{within}) from the anova as a pooled estimate of the variance among items within a group:

$$s_{\bar{Y}} = \sqrt{\frac{MS_{\text{within}}}{n}} = \sqrt{\frac{5.46}{10}} = \sqrt{0.546} = 0.7389$$

This standard error is based on $a(n-1) = 45$ degrees of freedom (the *df* of MS_{within}).

2. Next calculate the minimum significant ranges, MSR_j, for subranges of $j = 2, 3, \ldots, k$ means. These are ranges of means analogous to minimum significant differences *(MSD)* and are defined as

$$MSR_j \text{ (for } k \text{ groups)} = Q_{\alpha_j[k,\nu]} s_{\bar{Y}}$$

In this expression Q_{α_j} is the studentized range tabled for nonstandard probabilities $\alpha_j = \alpha(\frac{j}{k})$ (except for $j = k - 1$, where $\alpha_{k-1} = \alpha$). Its value can be found in Statistical Table **K.** Harmonic interpolation is necessary for $\nu = 45$.

For an experimentwise error rate $\alpha = 0.05$, $\nu = 45$ degrees of freedom, and k set equal to $a = 5$, the α_j's, Q's, and MSR's are the following:

	j			
	2	3	4	5
α_j	.02	.03	.05	.05
Q_{α_j}	3.41	3.77	3.77	4.07
MSR_j	2.52	2.79	2.79	3.00

The group means are then arrayed in order of magnitude:

Glucose + Fructose	Fructose	Glucose	Sucrose	Control
$\bar{Y}_1$	$\bar{Y}_2$	$\bar{Y}_3$	$\bar{Y}_4$	$\bar{Y}_5$
58.0	58.2	59.3	64.1	70.1

The sequence of steps for testing is as follows:

1. Compare ranges of adjacent pairs of means with MSR for $j = 2$, which equals 2.52.

Box 9.12 Continued

$$\bar{Y}_5 - \bar{Y}_4 = 70.1 - 64.1 = 6.0 > 2.52$$

$$\bar{Y}_4 - \bar{Y}_3 = 64.1 - 59.3 = 4.8 > 2.52$$

$$\bar{Y}_3 - \bar{Y}_2 = 59.3 - 58.2 = 1.1 < 2.52$$

$$\bar{Y}_2 - \bar{Y}_1 = 58.2 - 58.0 = 0.2 < 2.52$$

We do not test any wider range that includes the significant ranges $(\bar{Y}_5 - \bar{Y}_4)$ and $(\bar{Y}_4 - \bar{Y}_3)$, since any range including a significant subrange is declared significant by this test.

2. Compare the range of three means with MSR for $j = 3$, which equals 2.79. For example, $\bar{Y}_3 - \bar{Y}_1 = 59.3 - 58.0 = 1.3 < 2.79$, so this group is considered homogeneous.

   Our conclusions can be summarized as $\mu_1 = \mu_2 = \mu_3 < \mu_4 < \mu_5$. Conclusions are often shown diagrammatically by underlining the means that are not significantly different:

Glucose + Fructose	Fructose	Glucose	Sucrose	Control
$\bar{Y}_1$	$\bar{Y}_2$	$\bar{Y}_3$	$\bar{Y}_4$	$\bar{Y}_5$
58.0	58.2	59.3	64.1	70.1

(The underline spans the first three means: 58.0, 58.2, 59.3.)

comparisons c_i are nonzero. With equal or nearly equal sample sizes, the T-method will be the best method, while in other cases the GT2-method should be employed. Borderline cases may again require trying all three methods to find the most sensitive one. The computations are not difficult and are laid out in detail in Box 9.13.

The Scheffé test, as carried out in Box 9.13, is mathematically equivalent to the sum of squares simultaneous test procedure *(SS-STP)* prominently featured in our first edition as an a-posteriori multiple-comparison method (see discussion below). The relation between the Scheffé test and *SS-STP* is analogous to that between the t-test and the F-test.

More than just pairs of means and contrasts between averages based on two groups of means may be of interest to an investigator examining the results of an analysis of variance. Are certain partitions of the set of means significantly heterogeneous? To return to the pea section example, it seems that control differs from sucrose, which in turn differs from the three other sugar treatments; but these three seem homogeneous—that is, they do not differ. Many hypotheses similar to the one just stated could be tested. A conservative method that tests *all possible sets of comparisons* among means with an experimentwise error rate α is the **sum of squares simultaneous test procedure** *(SS-STP)* proposed by

Box 9.13 CONFIDENCE LIMITS AND TESTS OF UNPLANNED CONTRASTS AMONG MEANS USING THE SCHEFFÉ, T-, AND GT2-METHODS.

Length of pea sections grown in tissue culture. Data from Box 9.4.

1. Set up the desired contrasts. For example, in the pea section data the mean for sucrose is considerably higher than the average of the other sugars. The contrast would be as follows. (For convenience the coefficients c_i have been scaled so that their sum in each direction equals unity, but the coefficients could have been written 0, 1, 1, 1, −3.)

	Control	2% glucose added	2% fructose added	1% glucose + 1% fructose added	2% sucrose added
$\bar{Y}$	70.1	59.3	58.2	58.0	64.1
n_i	10	10	10	10	10
c_i	0	1/3	1/3	1/3	−1

$\sum n_i c_i = 0$

The contrast between the means is C, computed as follows:

$$C = \sum c_i \bar{Y}_i = 0 + \tfrac{1}{3}(59.3) + \tfrac{1}{3}(58.2) + \tfrac{1}{3}(58.0) - 1(64.1) = -5.60$$

2. For simultaneous confidence limits and tests of significance with experimentwise error, the most efficient method depends upon a, the number of means, ν, the degrees of freedom of MS_{within}, and the particular contrast being tested. Generally the Scheffé procedure will be the most powerful (yield the shortest intervals) for small values of a and ν and contrasts with nonzero c_i for most of the means. The T-method will be best if the sample sizes are equal or nearly equal. In other cases the GT2-method (or the Gabriel approximation) should be used. To be certain one has selected the optimal method (with the shortest interval), all three methods should be tried in critical studies.

2a. *Scheffé method*

The simultaneous $100(1 - \alpha)\%$ confidence limits for a contrast $C = \Sigma c_i \bar{Y}_i$ are

$$C \pm [(a - 1)\, F_{\alpha[a-1,\nu_2]}]^{1/2}\, s_c$$

where

$$s_c = \left(MS_{\text{within}} \sum \frac{c_i^2}{n_i} \right)^{1/2}$$

Box 9.13 Continued

In this example

$$s_c = \left\{ 5.46 \left(0 + \frac{(\frac{1}{3})^2}{10} + \frac{(\frac{1}{3})^2}{10} + \frac{(\frac{1}{3})^2}{10} + \frac{(-1)^2}{10} \right) \right\}^{1/2}$$

$$= [5.46(0.13333)]^{1/2} = 0.85324$$

so the limits to $C = -5.6$ are

$$-5.6 \pm [(5 - 1)F_{.05[4,45]}]^{1/2} 0.85324 = -5.6 \pm (4(2.58))^{1/2} 0.85324$$

$$= -5.6 \pm 2.741$$

Since the limits $(-8.341, -2.859)$ exclude zero, C is significantly different from zero.

2b. *The T-method*

The simultaneous $100(1 - \alpha)\%$ confidence limits for a contrast are

$$C \pm Q_{\alpha[a,v]} s_c$$

Note that the critical value is Q, the studentized range from Statistical Table **J,** rather than Q', the studentized augmented range from Table **L.** The standard error is computed as follows: Let c_i^+ and c_i^- denote those c_i's that are positive and negative, respectively. Then

$$s_c = [MS_{\text{within}}]^{1/2} \max \left\{ \sum \frac{c_i^+}{\sqrt{n_i}}, -\sum \frac{c_i^-}{\sqrt{n_i}} \right\}$$

In our example

$$s_c = (5.46)^{1/2} \max \left\{ \frac{\frac{1}{3}}{\sqrt{10}} + \frac{\frac{1}{3}}{\sqrt{10}} + \frac{\frac{1}{3}}{\sqrt{10}}, -\frac{(-1)}{\sqrt{10}} \right\}$$

$$= 2.3367 \max \{0.31623, 0.31623\}$$

$$= 0.73892$$

After obtaining $Q_{\alpha[a,v]}$ by interpolating in Table **J,** the 95% limits are computed as

$$-5.6 \pm 4.018(0.73892) = -5.6 \pm 2.969$$

These limits $(-8.569$ to $-2.631)$ are slightly wider than those found using the Scheffé method. When a and v are not small, one expects the Scheffé limits to be wider and hence to provide a less powerful test.

2c. *The GT2-method*

The simultaneous $100(1 - \alpha)\%$ confidence limits for a contrast, C, are

$$C \pm m_{\alpha[k^*,v_2]} s_C$$

where the standard error s_C is computed as follows. (As in the T-method let c_i^+ and c_i^- denote the positive and negative coefficients, respectively.)

$$s_C = [MS_{\text{within}}]^{1/2} \frac{-\sum (c_i^+ c_j^-) \left(\frac{1}{n_i} + \frac{1}{n_j} \right)^{1/2}}{\frac{1}{2} \sum |c_i|}$$

Box 9.13 Continued

The summation in the numerator is over all possible combinations of a positive c_i with a negative c_j. The summation in the denominator is over all c_i.

$$s_C = (5.46)^{1/2} \frac{[-(\frac{1}{3})(-1)(\frac{1}{10}+\frac{1}{10})^{1/2}] + [-(\frac{1}{3})(-1)(\frac{1}{10}+\frac{1}{10})^{1/2}] + [-(\frac{1}{3})(-1)(\frac{1}{10}+\frac{1}{10})^{1/2}]}{\frac{1}{2}(0+\frac{1}{3}+\frac{1}{3}+\frac{1}{3}+|-1|)}$$

$$= 2.3367 \frac{(\frac{1}{10}+\frac{1}{10})^{1/2}}{\frac{1}{2}(2)} = 2.3367(0.2)^{1/2}$$

$$= 1.0450$$

After $m_{\alpha[k^*,\nu]}$ is obtained from Table M, the 95% limits are computed as

$$-5.6 \pm 2.937(1.0450) = -5.6 \pm 3.069$$

These limits $(-8.669, -2.531)$ are even wider than the limits obtained by the T-method, which is to be expected since the sample sizes are equal.

3. Gabriel (1978) proposed a simple approximation to the T or GT2 confidence limits for a contrast making use of the comparison limits given in Box 9.11. The limits are

$$l = \sum c_i^+ l_i - \sum |c_i^-| u_i$$

and

$$u = \sum c_i^+ u_i - \sum |c_i^-| l_i$$

where c_i^+ and c_i^- are the positive and negative coefficients of linear comparison, respectively, and l_i and u_i are the lower and upper comparison limits as shown in Boxes 9.10 and 9.11. In our example, because the sample sizes are equal, we apply the T-method and make use of the comparison limits from Box 9.10. We obtain

$$l = [\tfrac{1}{3}(57.82) + \tfrac{1}{3}(56.72) + \tfrac{1}{3}(56.52)] - |-1|65.58 = 57.02 - 65.58 = -8.56$$

$$u = [\tfrac{1}{3}(60.78) + \tfrac{1}{3}(59.68) + \tfrac{1}{3}(59.48)] - |-1|62.62 = 59.98 - 62.62 = -2.64$$

The approximate 95% limits are quite close to the exact ones obtained by the T-method in Step **2b.** If sample sizes are unequal, these approximate limits require much less effort when many contrasts are to be tested.

Gabriel (1964), which is an extension of the Scheffé method. ''All possible sets of comparisons'' subsumes all pairs of means, all possible contrasts between groups of means, and all possible partitions of the a means—a very large number of tests indeed.

The rationale of the *SS-STP* method can be developed as follows. In a single-classification anova, the implication of an overall significant result is that

$$\frac{MS_{\text{among}}}{MS_{\text{within}}} \geq F_{\alpha[a-1,a(n-1)]} \qquad (9.10)$$

Since $MS_{\text{among}}/MS_{\text{within}} = SS_{\text{among}}/[(a-1)MS_{\text{within}}]$, we can rewrite Expression (9.10) as

$$SS_{\text{among}} \geq (a-1)MS_{\text{within}}F_{\alpha[a-1,a(n-1)]} \tag{9.11}$$

For example, in Box 9.4, where the anova is significant, $SS_{\text{among}} = 1077.32$. Substituting into Expression (9.11), we obtain

$$1077.32 > (5-1)(5.46)(2.58) = 56.35$$

It is therefore possible to compute a critical *SS* value for a test of significance of an anova.

Another way of calculating overall significance, then, is to see whether SS_{among} is greater than this critical *SS*. It is of interest to investigate why the SS_{among} is as large as it is and to test for the significance of the contributions made to this *SS* by differences among the sample means. This was discussed in the previous section, where separate sums of squares were computed based on comparisons among means planned before the data had been examined. A comparison was called significant if its F_s-ratio was greater than $F_{\alpha[k-1,a(n-1)]}$, where k is the number of means being compared. We can now also state this in terms of sums of squares. An *SS* is significant if it is greater than $(k-1)MS_{\text{within}}F_{\alpha[k-1,a(n-1)]}$.

The tests we just described were unplanned comparisons. Following a procedure applied in other *unplanned* comparisons, we set $k = a$ in the last formula, no matter how many means we compare. Thus the critical value of the *SS* will be larger than in the previous method, making it more difficult to demonstrate the significance of a sample *SS*. We do this to allow for the fact that we choose for testing the differences between group means that appear to be contributing substantially to the significance of the overall anova.

We are now able to test any differences between means and groups of means that attract our attention. For example, are the three sugar treatments other than sucrose different from each other? To test this, we compute the *SS* among these three means by the usual formula:

$$SS = n\sum(\bar{Y}_i - \bar{\bar{Y}})^2 = 10[(59.3-58.5)^2 + (58.2-58.5)^2 + (58.0-58.5)^2] = 10 \times 0.98 = 9.8$$

Note that the grand mean $\bar{\bar{Y}} = 58.5$ is computed only from the three means involved in this contrast, not from all five means in the pea section data. The differences among these means are not significant because this *SS* is less than the critical *SS* (56.35) calculated earlier.

The sucrose mean looks suspiciously different from the means of the other sugars. To test this we compute

$$SS = 10(64.1-59.9)^2 + 30(58.5-59.9)^2 = 235.2$$

Note that the grand mean, 59.9, is calculated from the four means involved in the contrast, and that 58.5 is the mean of the three sugars other than sucrose. The resulting *SS* is greater than the critical *SS*. We conclude, therefore, that sucrose retards growth significantly less than the other sugars do. We may continue in this fashion, testing all the differences that look suspicious or even testing all possible sets of means, considering them 2, 3, 4, and 5 at a time. This approach may require a computer if there are more than 5 means to be compared, since very many tests could be made. Computer program BIOM-pc carries out such tests.

The diagrammatic representation of the results of a multiple comparisons test will not always be as simple as those in Boxes 9.10 and 9.11. Often there are no clear boundaries (gaps) between sets of means not significantly different from each other, and such sets overlap. For example, in the following data, which represent means of the character thorax width for the aphid *Pemphigus populitransversus* studied in 23 localities in eastern North America, the means are arrayed by order of magnitude in coded units and the lines beside them represent nonsignificant sets of means. Sokal and Rinkel (1963) employed this symbolism at the margin of a geographic distribution map to provide an easy visual test of significance of any pair of means. When there are too many overlapping lines, they are difficult to keep apart by eye and a different technique (illustrated here on the right) is preferable (Sokal and Thomas, 1965).

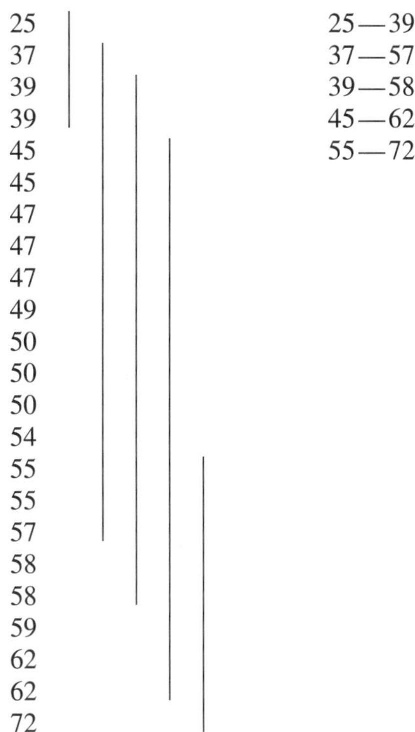

Any pair of means enclosed by the range of any one line is not significantly different. Thus means 47 and 50 are not different from each other because they are enclosed in the range 39 to 58, but means 39 and 62 are different from each other because there is no range that encloses them both.

Although all the procedures described here lead to the same conclusions in the pea section data (see Boxes 9.4, 9.10, 9.12, and 9.13), such agreement need not, in general, occur. When the tiger beetle data in Box 9.11 were analyzed by the *SS-STP,* we found the following pattern of significances:

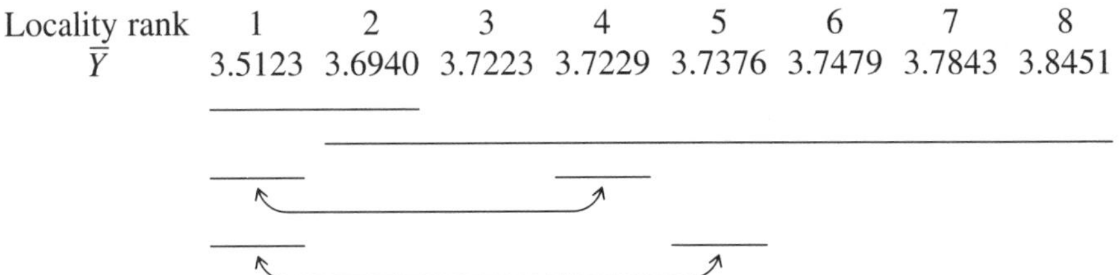

Locality rank	1	2	3	4	5	6	7	8
$\bar{Y}$	3.5123	3.6940	3.7223	3.7229	3.7376	3.7479	3.7843	3.8451

which is not consistent with the results indicated in Box 9.11, even though all these methods are testing the same null hypothesis: equality of means. Since sample sizes are unequal, means in nonsignificant sets are not necessarily adjacent, as indicated by the double-headed arrows. A reason for this difference is that *SS* and simultaneous confidence limit tests differ in their sensitivity to various alternative hypotheses.

To forestall confusion about the various methods of multiple comparisons among means presented in this and the previous section, we summarize current recommendations about their application in Table 9.3.

The subject of multiple comparisons is an active research field, and new methods will undoubtedly be developed in the future. Among the methods not covered in this text, perhaps the best known is the Student–Newman Keuls (SNK) *procedure, which is another multiple-range procedure and which is computationally identical to the Welsch step-up procedure (see Box 9.12), except for the use of another table of critical values. Computational steps for the* SNK *procedures are illustrated in the first edition of this book (Sokal and Rohlf, 1969). Gill (1978) discusses a procedure attributed to Dunnett that compares all treatment means with a control in a Model I anova.*

9.8 Finding the Sample Size Required for a Test

Statisticians are frequently asked how large a sample must be for an experiment. Regrettably, there is usually no simple answer. Without an estimate of the variability of the items, no answer at all can be given. The statistician needs to know what the experimenter wishes to test. Does he wish to establish a significant difference between means or to estimate the variance of the items to assign confidence limits to a single mean? Given a precise statement of the problem, statisticians have provided a number of computational solutions. In the case of

Table 9.3 RECOMMENDATIONS CONCERNING SUITABLE MULTIPLE COMPARISONS TESTS IN ANALYSIS OF VARIANCE.

Type of comparisons	Number k of comparisons	Method[a]	Qualifications	Method described in
Planned orthogonal	$k \leq (a-1)$	Orthogonal decomposition of MS_{among}		Section 9.6
Planned nonorthogonal and unplanned	Less than all pairwise comparisons $k < \frac{a(a-1)}{2}$	Dunn–Šidák method		Section 9.6
		Bonferroni method		Section 9.6
		Sequential Dunn–Šidák method		Box 9.9
	All pairwise comparisons $k = k^* = \frac{a(a-1)}{2}$	T-method	Equal sample sizes	Box 9.10
		T′-method	Nearly equal sample sizes	Box 9.11
		GT2-method	Unequal sample sizes	Box 9.11
		Tukey–Kramer method	Unequal sample sizes	Box 9.11
		Welsch method	Equal sample sizes	Box 9.12
	All possible contrasts $k > \frac{a(a-1)}{2}$	Scheffé method	Small a and ν; most c_i are nonzero	Box 9.13
		T-method	Equal or nearly equal sample sizes	Box. 9.13
		GT2-method	Unequal sample sizes	Box 9.13
	All possible sets of comparisons $k \gg \frac{a(a-1)}{2}$	*SS–STP*		Section 9.7

[a] All methods employ experimentwise error rates except the orthogonal decomposition, which employs comparisonwise error rates.

a-posteriori tests there is again some disagreement about whether to obtain experimentwise error rates or single-test error rates, but we will be concerned only with a-priori tests here.

We will present a simple test that provides an answer to the most common question: How big a sample size must I obtain in order to be able to show that if a true difference δ exists, it is significant at a significance level α, with a probability P that the significance will be found? The formula for this test is shown near the top of Box 9.14, where the meaning of each symbol is explained. To use this formula, we must specify the true standard deviation σ, which can be estimated from a preliminary analysis. Thus it is always necessary to have a prior idea of how variable the data are.

For example, assume that you are about to start a study of the geographic variation of birds, that wing length is one of the characters you will measure, and that you wish to know how many specimens of birds to study per locality. You must have studied at least a few localities previously by analysis of variance to estimate the within-locality variance of wing length for this particular species of bird. Similarly, for an experiment in which you are going to determine the glycogen content of rat livers, you must have several prior determinations to estimate their variance before you can calculate how many replicates you will need.

Next you need to establish δ, the smallest true difference you desire to detect. If you are interested in picking up only a very crude difference (a large value of δ), you will need relatively few specimens; but for a fine difference (a small value of δ), a large sample is required. The significance level at which you are prepared to consider two means significantly different is α; the desired probability that a significant difference will be found (if it exists and is as small as δ) is P. This probability is, of course, the power $1 - \beta$ of the significance test. One property of the formula in Box 9.14 is that we do not need to have actual values for σ and δ but need to know only their ratio. The example in the box makes this clear. We only know the coefficient of variation and the percentage of difference we want to detect. We can therefore write $(6 \times \bar{Y})/(5 \times \bar{Y}) = 6/5 = 1.2$, since this is the ratio of a 6% coefficient of variation over a 5% difference of means. The mean $\bar{Y}$ cancels out and a ratio of 1.2 results.

The example in Box 9.14 is solved in iterative fashion; that is, we enter a value for n into the formula and use it to obtain a better estimate of n. We need a first estimate of n because without it, we would not know the degrees of freedom and could not look up t in Statistical Table **B.** Once we have such an estimate, however far off, we can then get a better estimate by applying the formula, using it to look up a better value of t, and so on. Iterative procedures are frequently employed in mathematics when a direct solution for a term is difficult or impossible.

If you are not familiar with iteration, an analogy from Thomson (1951) may help. This is the problem of Robinson Crusoe wishing to make a wooden lathe. To do so he needs to have wheels and spindles, but he needs a lathe to make

Box 9.14 FINDING SAMPLE SIZE REQUIRED AND POWER FOR AN ANOVA.

I. *Number of replications needed to detect a given "true" difference between two means*

From previous studies you know that the coefficient of variation of wing length of a species of bird is about 6%. You plan to study four populations by analysis of variance. How many measurements need be made from each population to be 80% certain of detecting a 5% difference between two of the four means at the 1% level of significance? Since $a = 4$, the error MS will have $\nu = 4(n - 1)$ degrees of freedom.

The appropriate formula is

$$n \geq 2\left(\frac{\sigma}{\delta}\right)^2 \{t_{\alpha[\nu]} + t_{2(1-P)[\nu]}\}^2$$

where n = number of replications

σ = true standard deviation

δ = The smallest true difference that you want to detect. (*Note:* It is necessary to know only the ratio of σ to δ, not their actual values.)

ν = degrees of freedom of the sample standard deviation ($\sqrt{MS_{\text{within}}}$) with a groups and n replications per group

α = significance level (such as 0.05)

P = desired probability that a difference will be found to be significant (if it is as small as δ). This is the intended power of the test.

$t_{\alpha[\nu]}$ and $t_{2(1-P)[\nu]}$ = values from a two tailed t-table with ν degrees of freedom and corresponding to probabilities of α and $2(1 - P)$, respectively. *Note:* If $P = \frac{1}{2}$ then $t_1 = 0$.

Iterative solution

We try an initial value of $n = 20$ as a reasonable guess. Then $\nu = 4(20 - 1) = 4 \times 19 = 76$. Since $V = 6\%$, $s = 6\bar{Y}/100$. We wish δ to be 5% of the mean; that is, $\delta = 5\bar{Y}/100$. Using s as an estimate of σ, we obtain $\sigma/\delta = (6\bar{Y}/100)/(5\bar{Y}/100) = 6/5$. Thus

$$n \geq 2\left(\frac{6}{5}\right)^2 \{t_{.01[76]} + t_{2(1-0.80)[76]}\}^2$$

$$= 2\left(\frac{6}{5}\right)^2 [2.642 + 0.847]^2 = 2(1.44)12.171 = 35.1$$

Next we try $n = 35$, making $\nu = 4(35 - 1) = 136$:

$$n \geq 2(1.44)[2.612 + 0.845]^2 = 2.88(11.95) = 34.4$$

We round up to $n = 35$ to be conservative. This is the same as the previous solution, indicating that we have iterated to stability. It appears that 35 replications per population sample are necessary.

BOX 9.14 CONTINUED

II. *Finding the required sample size and estimating the power of an anova*

To estimate sample size for the overall anova, we employ a graphic method. To use these graphs (see Statistical Table **GG**) we must evaluate a quantity ϕ, related to the parameter of the noncentral F-distribution. The method for computing this quantity depends on the question asked of the data. First let us imagine that the investigator who carried out the pea section experiments of Box 9.4 wishes to do a new series of $a = 6$ treatments and wants to know how many replicates are needed to achieve a power of $1 - \beta = 0.80$ at the $\alpha = 0.05$ level of significance, assuming a difference δ of at least 10.0 ocular units between the most different means.

The formula for ϕ is $(n\delta^2/2as^2)^{1/2}$. Since the formula includes n, the desired result, we will make a first guess that the previously employed sample size $n = 10$ is a good number to employ. Thus $\phi = [(10)(10.0)^2/(2)(6)(5.46)]^{\frac{1}{2}} = 3.91$. The intended degrees of freedom among treatments is $\nu_1 = a - 1 = 5$, so we employ the graph in Table **GG** for $\nu_1 = 5$. The abscissa features two scales for ϕ: one for $\alpha = 0.05$, the other for $\alpha = 0.01$. We choose the one for $\alpha = 0.05$ and look for the curve that is closest to $\nu_2 = a(n - 1) = 6(9) = 54$ degrees of freedom. Clearly the power is near 100%. We can therefore drastically reduce the sample size n if we are statisfied with a power of 80%.

Let us try $n = 2$. Then $\phi = [(2)(10.0)^2/(2)(6)(5.46)]^{1/2} = 1.75$. In this case $\nu_2 = 6(2 - 1) = 6$, and the power approximates 56%, lower than intended. Raising the sample size to $n = 3$ increases the power to approximately 92%. Thus, to be conservative, the investigator should use 3 replicates each for the 6 treatments.

NOTE: It should be obvious that we can also use this approach to find the power of an intended test when the sample size is known but other parameters are unspecified.

Sometimes the investigator wishes to know the power of an anova that has already been carried out. In such a case, the appropriate formula is

$$\phi = \left[\frac{(a - 1)(MS_{\text{among}} - MS_{\text{within}})}{aMS_{\text{within}}}\right]^{1/2}$$

This formula assumes that the true difference among groups is equal to the observed difference. Suppose we wish to know the power with which the original pea section anova had been carried out. We compute $\phi = [(5 - 1)(269.33 - 5.46)/(5.46)]^{1/2} = 6.22$. Even at $\alpha = 0.01$ this value of ϕ is off the graph. The power of this anova is effectively 100%, largely because of the great replicability (small MS_{within}) of these data.

these. So Robinson Crusoe fashions a crude wheel by whittling with a knife, constructing a crude lathe with the wheel. He now can prepare a slightly better wheel from which he constructs a better lathe, which yields a yet better wheel, and so on. Such iterative procedures are often convergent; that is, they give increasingly better estimates. Even if you make a mistake in the computations along the way, you will only be slowed down in finding the right solution. The correct procedure should always home in on the correct answer eventually.

Note that the left portion of the expression in Box 9.14, $2(6/5)^2 = 2(1.44) = 2.88$, remains constant, while the expression in square brackets changes. For the first iteration we guess a sample size of $n = 20$, and the solution of the equation yields an estimate of $n = 35.1$. Substituting this value into the expression we arrive at $n = 34.4$ after the second iteration. We round up to $n = 35$ to be conservative. Thus we conclude that for a study of these bird wing lengths and for the percentage of difference required, as well as for the other specifications of the problem (α and P), 35 birds per locality sample should be adequate.

A little experimentation with this equation will show that it is not too sensitive to changes in α and P but is very sensitive to change in the ratio of σ/δ, indicating that very large samples are required to detect small differences. We can also see that any refinement of experimental technique that reduces σ decreases the sample sizes required or increases the possibility of detecting smaller differences.

This computational technique is strictly correct only for a test between two means within an anova of a means. If we wish to know the sample size n required for the overall anova for a given α, and $P = 1 - \beta$, or conversely wish to know the power P of an anova, given a fixed sample size n, we have two options. We can use a computer program to carry out the far more complex computations and iterations (a review of such programs is found in Goodstein, 1989), or we can turn to some classic graphs, which provide approximate answers. The second part of Box 9.14 provides instructions on how to use these graphs based on the method of Pearson and Hartley (1951). We show how to estimate both the sample size n and the power P of an anova.

EXERCISES 9

9.1 The following is an example with easy numbers to help you become familiar with the analysis of variance. A plant ecologist wishes to test the hypothesis that the height of plant species X depends on the type of soil it grows in. He measures the height of 3 plants in each of 4 plots representing different soil types, all 4 plots being contained in an area two miles square. His results are tabulated below. (Height is given in centimeters.) Does your analysis support this hypothesis? *Answer:* $F_s = 6.950$, $F_{.05[3,8]} = 4.07$.

Observation number	Plots			
	1	2	3	4
1	15	25	17	10
2	9	21	23	13
3	4	19	20	16

9.2 The following are measurements (in coded micrometer units) of the thorax length of the aphid *Pemphigus populitransversus*. The aphids were collected in 28 galls on the cottonwood, *Populus deltoides*. Four alate (winged) aphids were randomly selected from each gall and measured. The alate aphids of each gall are mostly isogenic (identical twins), being descended parthenogenetically from one stem mother. Thus, any variance within galls should be due to environment only. Variance among different galls may be due to differences in genotype and also to environmental differences among galls. If this character, thorax length, is affected by genetic variation, significant intergall variance must be present. The converse is not necessarily true; significant intergall variance need not indicate genetic variation; it could as well be due to environmental differences (data from Sokal, 1952). Analyze the variance of thorax length. Is there significant intergall variance? Give estimates of the added component of intergall variance, if present, and set confidence limits to it. What percentage of the variance is controlled by intragall and what percentage by intergall factors? Discuss your results.

Gall No.	Thorax length				Gall No.	Thorax length			
1.	6.1,	6.0,	5.7,	6.0	15.	6.3,	6.5,	6.1,	6.3
2.	6.2,	5.1,	6.1,	5.3	16.	5.9,	6.1,	6.1,	6.0
3.	6.2,	6.2,	5.3,	6.3	17.	5.8,	6.0,	5.9,	5.7
4.	5.1,	6.0,	5.8,	5.9	18.	6.5,	6.3,	6.5,	7.0
5.	4.4,	4.9,	4.7,	4.8	19.	5.9,	5.2,	5.7,	5.7
6.	5.7,	5.1,	5.8,	5.5	20.	5.2,	5.3,	5.4,	5.3
7.	6.3,	6.6,	6.4,	6.3	21.	5.4,	5.5,	5.2,	6.3
8.	4.5,	4.5,	4.0,	3.7	22.	4.3,	4.7,	4.5,	4.4
9.	6.3,	6.2,	5.9,	6.2	23.	6.0,	5.8,	5.7,	5.9
10.	5.4,	5.3,	5.0,	5.3	24.	5.5,	6.1,	5.5,	6.1
11.	5.9,	5.8,	6.3,	5.7	25.	4.0,	4.2,	4.3	4.4
12.	5.9,	5.9,	5.5,	5.5	26.	5.8,	5.6,	5.6,	6.1
13.	5.8,	5.9,	5.4,	5.5	27.	4.3,	4.0,	4.4,	4.6
14.	5.6,	6.4,	6.4,	6.1	28.	6.1,	6.0,	5.6,	6.5

9.3 Millis and Seng (1954) published a study on the relation of birth order to the birth weights of infants. The data below on first-born and eighth-born infants are extracted from a table of birth weights of male infants of Chinese third-class patients at the Kandang Kerbau Maternity Hospital in Singapore in 1950 and 1951.

Birth weight (lb : oz)	Birth order 1	8
3:0—3:7	—	—
3:8—3:15	2	—
4:0—4:7	3	—
4:8—4:15	7	4
5:0—5:7	111	5
5:8—5:15	267	19
6:0—6:7	457	52
6:8—6:15	485	55
7:0—7:7	363	61
7:8—7:15	162	48
8:0—8:7	64	39
8:8—8:15	6	19
9:0—9:7	5	4
9:8—9:15	—	—
10:0—10:7	—	1
10:8—10:15	—	—
	1932	307

Which birth order appears to be accompanied by heavier infants? Is this difference significant? Can you conclude that birth order causes differences in birth weight? (Computational note: The variable should be coded as simply as possible.) Re-analyze, using the t-test, and verify that $t_s^2 = F_s$. *Answer:* $t_s = 11.016$.

9.4 The following cytochrome oxidase assessments of male *Periplaneta* roaches in cubic millimeters per ten minutes per milligram were taken from a larger study by Brown and Brown (1956):

	n	$\bar{Y}$	$s_{\bar{Y}}$
24 hours after methoxychlor injection	5	24.8	0.9
Control	3	19.7	1.4

Are the two means significantly different?

9.5 The following data are measurements of five random samples of domestic pigeons collected during the months of January, February, and March in Chicago in 1955. The variable is the length from the anterior end of the narial opening to the tip of the bony beak and is recorded in millimeters. Data from Olson and Miller (1958).

Samples				
1	2	3	4	5
5.4	5.2	5.5	5.1	5.1
5.3	5.1	4.7	4.6	5.5
5.2	4.7	4.8	5.4	5.9
4.5	5.0	4.9	5.5	6.1
5.0	5.9	5.9	5.2	5.2
5.4	5.3	5.2	5.0	5.0
3.8	6.0	4.8	4.8	5.9
5.9	5.2	4.9	5.1	5.0
5.4	6.6	6.4	4.4	4.9
5.1	5.6	5.1	6.5	5.3
5.4	5.1	5.1	4.8	5.3
4.1	5.7	4.5	4.9	5.1
5.2	5.1	5.3	6.0	4.9
4.8	4.7	4.8	4.8	5.8
4.6	6.5	5.3	5.7	5.0
5.7	5.1	5.4	5.5	5.6
5.9	5.4	4.9	5.8	6.1
5.8	5.8	4.7	5.6	5.1
5.0	5.8	4.8	5.5	4.8
5.0	5.9	5.0	5.0	4.9

Are the five samples homogeneous? *Answer:* $F_s = 1.977$, $df = 4, 95$.

9.6 If a study such as that given in Exercise 9.5 on pigeons were to be repeated with only two samples (one a control and the other representing pigeons fed a special diet), what sample size should be used to be 80% certain of observing a true difference between two means as small as a tenth of a millimeter at the 5% level of significance? Assume that the error variance in this new experiment would be the same as in the previous one.

9.7 P. E. Hunter (1959, detailed data unpublished) selected two strains of *Drosophila melanogaster,* one for short larval period (SL) and one for long larval periods (LL). A nonselected control strain (CS) was also maintained. At generation 42 the following data were obtained for the larval period (measured in hours). Analyze and interpret.

	Strain		
	SL	CS	LL
n_i	80	69	33
$\sum^{n_i} Y$	8070	7291	3640

$$\sum^{3}\sum^{n_i} Y^2 = 1{,}994{,}650$$

Note that part of the computation has already been performed for you. Perform planned tests among the three means (short versus long larval periods and each against the control). Are these three tests independent? Set 95% confidence limits to the observed differences of means for which these comparisons are made. *Answer:* $MS_{(\text{SL vs. LL})} = 2076.6697$.

9.8 The following data were taken from a study of the systematics of honeybees (DuPraw, 1965). Of the 15 variables reported in the original study, the means and sample sizes for wing length variable A–O (in mm × 83) are given below for 15 localities. The pooled standard deviation within localities is 9.284. Perform an analysis of variance and compare all pairs of means to determine which are significantly different using an experimentwise error rate of 5%. Graph the comparison limits for the means. Interpret your analysis. You may wish to compare your results with those in the original article based upon all 15 measurements.

	Locality						
	1	2	3	5	6	7	8
n	20	20	20	16	12	11	16
$\bar{Y}$	379.9	384.1	376.8	362.9	374.1	360.4	341.8

	Locality							
	9	10	11	13	14	15	16	17
n	20	20	20	10	10	10	10	10
$\bar{Y}$	371.0	353.3	354.2	330.7	342.6	378.6	368.5	368.8

Locality codes:

1. Europe northwest of the Alps
2. Europe southeast of the Alps
3. Italy
5. Cyprus
6. Caucasus
7. Eastern Mediterranean
8. Egypt
9. North Africa
10. Central Africa
11. Cape of Good Hope
13. India and Pakistan
14. Japan
15. Mt. Kilimanjaro and vicinity
16. Yugoslavia
17. Yellow "oasis bee" from Sahara

9.9 The following data were taken from a study of blood protein variations in deer (Cowan and Johnston, 1962). The variable is the mobility of serum protein fraction *II* expressed as 10^{-5} cm²/V · s.

Samples	$\bar{Y}$	$s_{\bar{Y}}$
1 Sitka	2.8	0.07
2 California blacktail	2.5	0.05
3 Vancouver Island blacktail	2.9	0.05
4 Mule deer	2.5	0.05
5 Whitetail	2.8	0.07

$n = 12$ for each mean. Perform an analysis of variance and a multiple-comparisons test using a variety of procedures. Compare the results of the various unplanned multiple-comparisons tests. Which method is preferred in this example? *Answer:* $MS_{\text{within}} = 0.0416$, maximal nonsignificant sets (at $P = 0.05$) are samples (1, 3, 5) and (2, 4) by the *SS-STP* procedure. The corresponding sets for the T-method (the preferred method) are (1, 5), (2, 4), and (3).

9.10 In a study of the effect of density on dry weight of hybrid houseflies, Bhalla and Sokal (1964) obtained the following results. Density is given as number of individuals per 36 grams of medium; the means $\bar{Y}$ are stated in milligrams and are based on 5 replicates.

Density	$\bar{Y}$	$s_{\bar{Y}}$
40	3.23	0.146
80	2.63	0.153
160	2.24	0.141
320	1.57	0.047
640	1.40	0.201
1280	0.73	0.031
2560	0.74	0.065

Carry out an analysis of variance. Earlier experiments had demonstrated the effect of density on percent emergence of these flies to be the following: density 40—suboptimal; densities 80 to 320—optimal; density 640—some stress; densities 1280 and 2560—severe stress. Design an orthogonal set of comparisons to test whether differences in stress on these houseflies affect their dry weights.

9.11 An investigator testing the effect of two heavy metal ions on the growth of rainbow trout established the following treatment classes:

1. Control
2. Copper, low concentration
3. Copper, high concentration
4. Zinc, low concentration
5. Copper and zinc, both at low concentrations
6. Copper and zinc, both at high concentrations

Give an example of a set of orthogonal single-degree-of-freedom comparisons and a set of nonorthogonal comparisons. *Answer:* One such set of orthogonal comparisons is:

Control	Cu low	Cu high	Zn low	Cu + Zn low	Cu + Zn high
−5	1	1	1	1	1
0	2	2	2	−3	−3
0	1	1	−2	0	0
0	0	0	0	−1	+1
0	1	−1	0	0	0

9.12 Estimate the power of the analysis of variance carried out in Exercise 9.7.

9.13 What sample size should be used in a new density experiment based on Exercise 9.10? The investigator plans to use 5 densities, wishes to establish a significant (0.01) difference of at least 1.0 between the most distant means, and desires a power of 80%. *Answer:* A sample size of $n = 5$ yields power of about 93%.

10 NESTED ANALYSIS OF VARIANCE

In this chapter we discuss an extension of single-classification anova to handle cases where each class or group is divided into two or more randomly chosen subgroups. Section 10.1 explains the reasons for employing such a design. Section 10.2 shows the computation for subsamples of equal size. Some complications ensue when subsamples are of unequal sizes; these are treated in Section 10.3. Section 10.4 introduces the concept of relative efficiency of designs. We consider in this final section how best to plan an experiment to obtain a maximum amount of information for any given investment of effort and money.

10.1 NESTED ANOVA: DESIGN

The simple design of a single-classification anova (see Chapter 9) is frequently insufficient to represent the complexity of a given experiment and to extract all the relevant information from it. This chapter deals with cases in which each major grouping or class is divided into randomly chosen subgroups. Thus for each of the seven batches of medium from Section 8.4 we might prepare five jars, and a sample of houseflies could be reared from each of these jars. Each group in such an analysis would represent a type of medium, each subgroup a jar for one medium, and the items in a subgroup would be the individual flies being measured from each jar.

What other information do we obtain from such a design? In an experiment with a single jar for each different medium formulation, we can never be certain that observed differences among the flies in different jars are due to different formulations. Ecological and other unintentional differences among the jars could result in morphological differences among the flies, which would have occurred even if the same formulation had been used in all of the jars. The only way to separate these two effects is to have two or more jars for each medium formulation. If we do not find any differences among the jars *within* medium formulations—beyond what we should expect on the basis of the amount of variation observed within a jar—we can ascribe differences among the media to their formulations. Even if we find a significant amount of added variance among the jars within a formulation, however, we can still test whether the variation among the media is greater than that to be expected on the basis of the observed variation among the jars that were treated alike. If the tested mean square is significant, we can conclude that differences among medium formulations are present above and beyond differences expected among jars.

We call the analysis of such a design a **nested analysis of variance** because the subordinate classification is nested within the higher level of classification. Thus, in our example the jars are subordinate to the medium formulations. The design is also often called a *hierarchic analysis of variance.* A crucial requirement of this type of analysis is that groups representing a subordinate level of classification be randomly chosen. Thus the replicate jars within one medium formulation must not be deliberately picked to represent certain microenvironments. The subordinate level of a nested anova is always Model II. The highest level of classification in a nested anova may be Model I or Model II. If it is Model II, we speak of a **pure Model II nested anova.** If the highest level is Model I, we called it a **mixed model nested anova.**

There are two general classes of applications of nested anova. The first is similar to the hypothetical example just discussed. This class commonly is used to ascertain the magnitude of error at various stages of an experiment or an industrial process. In the hypothetical housefly example we wish to know whether there is substantial variation among jars as well as among flies within jars, and we need to know the magnitude of this variation to enable us to test the magnitude of the added component due to treatments (= different formulations of medium). Nested anovas are not limited to the two levels discussed so far. We can divide the subgroups into subsubgroups, and even further, as long as these are chosen randomly.

Thus we might design an experiment in which we test the effect of five drugs on the quantity of pigment in the skin of an animal. The five drugs and one control (six groups) are the major classification and clearly are fixed treatment effects (Model I). For each drug we might use five randomly selected rats. These would provide a measure of variance of rats within a drug class or of "rats within drugs," as one would say in statistical jargon. From each rat we might take three small skin samples at random from its ventral side. This presents us with a new subordinate level of variation—skin samples within rats. Each skin sample is macerated and divided into two lots, which are hydrolyzed separately. This level is hydrolysates within skin samples. Finally, the quantity of pigment might be read as an optical density, and two replicate readings might be made of each hydrolysate. The basic error variance would be the variance of the replicate readings per hydrolysate, but we would also have estimates of the variance between hydrolysates within one skin sample, among skin samples within one rat, and among rats within one drug.

These variance estimates are important in designing similar experiments because they indicate at which level of the experiment most of our sampling efforts should be concentrated. The most variable aspect of our experiment will need the greatest replication or will need better experimental control. Thus if we find that the two hydrolysates show the greatest proportion of variance, our method of hydrolysis clearly is not standardized enough and should be improved, or if this is impossible, we should divide each macerated skin sample into more lots to provide several more hydrolysates. We discuss this topic in greater detail in Section 10.4. The drug–rats–skin samples–hydrolysates–optical density example discussed here is clearly a mixed model nested anova, with drugs representing the fixed treatment effect and all other levels representing the random effects.

The second major use of a nested analysis of variance is in cases that are usually pure Model II. Such cases frequently are from the field of quantitative genetics, where we wish to know the magnitude of the variance attributable to various levels of variation in a study. Systematists interested in discovering the sources of variation in natural populations also use this type of approach. To illustrate such an application, we propose an experiment in which separate litters by the same mother have been sired by different fathers. If there are ten females (dams), each with five litters sired by five different males (sires), the variance can be subdivided into that among dams, that among sires within the same dam, and that among offspring within the same sire. Genetic theory (using the statistical consequences of Mendelian genetics) can predict the relative magnitudes of such components on various assumptions. One of the aims of the science of quantitative genetics has been to verify such predictions and analyze departures from them. A similar approach can be used in geographic variation studies in which variances among samples within one locality and among localities can be estimated and may lead to important conclusions about the population distribution pattern of an organism. For further information on the application of analysis of variance to quantitative genetics, refer to the very readable account by Falconer (1981) and the more extended treatment by Weir (1990).

Before we compute an actual nested anova, we should examine the linear model upon which it is based. In Expressions (8.2) and (8.3) we gave equations for decomposing variates in a single-classification anova for Model I and Model II, respectively. For a two-level nested anova (Model II) this equation becomes

$$Y_{ijk} = \mu + A_i + B_{ij} + \epsilon_{ijk} \tag{10.1}$$

where Y_{ijk} is the kth observation in the jth subgroup of the ith group, μ is the parametric mean of the population, A_i is the random contribution for the ith group of the higher level A, B_{ij} is the random contribution for the jth subgroup (level B) of the ith group, and ϵ_{ijk} is the error term of the kth item in the jth subgroup of the ith group. We assume that A_i, B_{ij}, and ϵ_{ijk} are distributed normally, with means of zero and variances of σ_A^2, $\sigma_{B\subset A}^2$, and σ^2 respectively. We use the symbol $\sigma_{B\subset A}^2$, rather than σ_B^2, to indicate that the variance is of level B

within level A. We shall use A, B, C, . . . , as subscripts for levels of variance components. Statistical workers frequently use letters that help them remember the meaning of variance components. For example, σ_G^2 might signify genetic variance, σ_D^2 variance among dogs, and so on. In a textbook, however, a uniform and logical system is necessary.

When the model is mixed (that is, the highest level has fixed treatment effects) we decompose a variate as follows:

$$Y_{ijk} = \mu + \alpha_i + B_{ij} + \epsilon_{ijk} \qquad (10.2)$$

This expression is the same as Expression (10.1) except that α_i is a fixed treatment effect, unlike A_i in the earlier model. Expressions (10.1) and (10.2) can be expanded to further levels, which are identified by C, D, and so forth, with the subscripts extended to l, m, and so on.

We will now compute a nested anova.

10.2 Nested Anova: Computation

The example we have chosen to illustrate the computation of a nested analysis of variance is of a pure Model II anova as described in Section 10.1. This anova studies the percentage of phenotypic variation at two sampling levels. The data, on rearing field-caught mosquito pupae *(Aedes intrudens)* in the laboratory, were obtained by Rohlf. Twelve female pupae were brought into the laboratory and divided at random into three rearing cages, in which four pupae each were reared. The purpose of dividing them among different cages was to see whether the differences among cages would add variance to the overall variation among females that is due to their presumable genetic and ecological differences. The structure of the experiment and of the analysis, therefore, is three cages of four females each. Length of the left wing was measured on each female, and the measurement was repeated once, the mosquitoes being presented to the measurer in a random order so as not to bias the results. The measurements are shown at the top of Box 10.1.

Since this is a pure Model II nested anova, the means of groups and subgroups are not of interest. They are furnished, however, in Box 10.1 to explain both notation and computation. Rather than employing the more precise notation $\bar{Y}_{ij}$ and $\bar{Y}_i$ for subgroup and group means, respectively, we use the simpler $\bar{Y}_B$ and $\bar{Y}_A$ to indicate levels B and A. The sum of squares of groups (cages) is computed as in the single-classification anova of Chapter 9. In step **1** we compute the grand mean of all the observations in the usual way. In step **2** the deviations of the $a = 3$ group means $\bar{Y}_A$ from the grand mean $\bar{Y}$ are squared and summed, and then multiplied by their sample sizes *(nb),* since we need sums of squares of groups rather than of means. Because there are a group means, the degrees of freedom pertaining to the sum of squares of groups is $a - 1$, in this case 2. Next, in step **3,** we compute the *sum of squares of subgroups within groups.* The

Box 10.1 TWO-LEVEL NESTED ANOVA: EQUAL SAMPLE SIZES.

Two independent measurements of the left wings of each of 4 female mosquitoes *(Aedes intrudens)* reared in each of 3 cages; $b = 4$ females. The data are given in micrometers. This is a Model II anova.

		Groups of females ($a = 3$)											
		Cage 1				Cage 2				Cage 3			
		1	2	3	4	1	2	3	4	1	2	3	4
Measurements	1	58.5	77.8	84.0	70.1	69.8	56.0	50.7	63.8	56.6	77.8	69.9	62.1
($n = 2$)	2	59.5	80.9	83.6	68.3	69.8	54.5	49.3	65.8	57.5	79.2	69.2	64.5
Subgroup sums	$\sum^{n} Y$	118.0	158.7	167.6	138.4	139.6	110.5	100.0	129.6	114.1	157.0	139.1	126.6
Subgroup means	$\bar{Y}_B$	59.0	79.35	83.8	69.2	69.8	55.25	50.0	64.8	57.05	78.5	69.55	63.3
Group means	$\bar{Y}_A$	72.8375				59.9625				67.1000			

SOURCE: Data from F. J. Rohlf (unpublished results).

Preliminary computation

1. Grand mean $\bar{\bar{Y}} = \frac{1}{abn}\sum^{a}\sum^{b}\sum^{n} Y = 66.6333$

2. SS_{among} (among groups) $= nb\sum^{a}(\bar{Y}_A - \bar{\bar{Y}})^2 = 665.6758$

3. SS_{subgr} (subgroups within groups) $= n\sum^{a}\sum^{b}(\bar{Y}_B - \bar{Y}_A)^2 = 1720.6775$

4. SS_{within} (within subgroups; error SS) $= \sum^{a}\sum^{b}\sum^{n}(Y - \bar{Y}_B)^2 = 15.6200$

Now fill in the anova table.

Anova table: formulas

	Source of variation	df	SS	MS	F_s	Expected MS
$\bar{Y}_A - \bar{\bar{Y}}$	Among groups	$a - 1$	**2**	$\frac{\mathbf{2}}{a-1}$	$\frac{MS_{\text{among}}}{MS_{\text{subgr}}}$	$\sigma^2 + n\sigma^2_{B \subset A} + nb\sigma^2_A$
$\bar{Y}_B - \bar{Y}_A$	Among subgroups within groups	$a(b - 1)$	**3**	$\frac{\mathbf{3}}{a(b-1)}$	$\frac{MS_{\text{subgr}}}{MS_{\text{within}}}$	$\sigma^2 + n\sigma^2_{B \subset A}$
$Y - \bar{Y}_B$	Within subgroups	$ab(n - 1)$	**4**	$\frac{\mathbf{4}}{ab(n-1)}$		σ^2
$Y - \bar{\bar{Y}}$	Total	$abn - 1$	**2 + 3 + 4**			

Completed anova

	Source of variation	df	SS	MS	F_s
$\bar{Y}_A - \bar{\bar{Y}}$	Among groups (among cages)	2	665.6758	332.8379	1.741 *ns*
$\bar{Y}_B - \bar{Y}_A$	Among subgroups within groups (among females within cages)	9	1720.6775	191.1864	146.88***
$Y - \bar{Y}_B$	Within subgroups (error; between measurements on each female)	12	15.6200	1.3017	
$Y - \bar{\bar{Y}}$	Total	23	2401.9733		

$F_{.05[2,9]} = 4.26 \qquad F_{.001[9,12]} = 7.48$

BOX 10.1 CONTINUED

Conclusions

There is no evidence ($P > 0.05$) for a variance component among cages, but there is a highly significant ($P \ll 0.001$) added variance component among females for wing length in these mosquitoes.

From the components of the expected mean squares shown in the anova table presenting general formulas, we can derive a procedure for estimating the variance components:

Within subgroups (error; between measurements on each female) $= s^2 = 1.3017$

$$\text{Among subgroups within groups (among females within cages)} = s^2_{B \subset A} = \frac{MS_{\text{subgr}} - MS_{\text{within}}}{n} = \frac{191.1864 - 1.3017}{2} = 94.9424$$

$$\text{Among groups (among cages)} = s^2_A = \frac{MS_{\text{among}} - MS_{\text{subgr}}}{nb} = \frac{332.8379 - 191.1864}{8} = 17.7064$$

Since we are frequently interested only in their relative magnitudes, the variance components can be expressed as percentages of the sum of their variances:

$$s^2 + s^2_{B \subset A} + s^2_A = 1.3017 + 94.9424 + 17.7064 = 113.9505$$

$$s^2 \text{ represents } \frac{100 \times 1.3017}{113.9505} = 1.14\%$$

$$s^2_{B \subset A} \text{ represents } \frac{100 \times 94.9424}{113.9505} = 83.32\%$$

$$s^2_A \text{ represents } \frac{100 \times 17.7064}{113.9505} = 15.54\%$$

deviations between subgroup means $\bar{Y}_B$ and group means $\bar{Y}_A$ are squared and summed for the b subgroups within a group and over the a groups. To turn this SS of means into one of groups, we multiply by n, the sample size of each subgroup. The degrees of freedom in each subgroup are $b - 1$, where b is the number of subgroups per group. Since there are a groups, the df for subgroups within groups is $a(b - 1)$, which for the example in Box 10.1 is 9. Finally, in step **4** we compute the error sum of squares, the variation within subgroups. This value is the sum of the squared deviations of the items Y from the subgroup means $\bar{Y}_B$, the summation extending over all subgroups and all groups. Because there are n items per subgroup, there are $n - 1$ degrees of freedom for each of the ab subgroups in the experiment. Thus, the error df is computed as $ab(n - 1)$, which here is 12.

We now arrange the sums of squares in an anova table as shown in Box 10.1. Mean squares are obtained as before by dividing the sums of squares by their respective degrees of freedom. Note that the total SS is merely the sum of the other three sums of squares and that the degrees of freedom are similarly additive. The total SS is the quantity you would obtain if you calculated the sum of squares of the entire dataset of abn items. Its df would necessarily be $abn - 1$, which can easily be shown to be the sum of the three separate degrees of freedom $a - 1$, $a(b - 1)$, and $ab(n - 1)$. As is our practice, general formulas are shown first, followed by the specific example. Notice especially the expected mean squares in a nested anova. The variance for each level above the error contains within it the variances of all levels below. Thus the expected variance of subgroups within groups is $\sigma^2 + n\sigma^2_{B \subset A}$, which resembles the familiar expression from a single-classification anova. The only new aspect is that σ^2_B has changed to $\sigma^2_{B \subset A}$ to symbolize that it is the variance of level B *within* level A. The expected mean square among groups contains the terms below it plus $nb\sigma^2_A$. From these expected mean squares the tests of significance are obvious.

For reasons explained later in this section, we always test lower levels before testing upper levels. Thus we first test $MS_{\text{subgr}}/MS_{\text{within}}$ for the significance of $\sigma^2_{B \subset A}$ and then test $MS_{\text{among}}/MS_{\text{subgr}}$ for the existence of σ^2_A. In the example of Box 10.1 we find an added variance component among females within cages, which is highly significant with an F_s value of 146.88, corresponding to a probability of type I error far less than 1%. When we test the cages over the MS_{subgr} however, we find the F_s-ratio to be only 1.741, which is not significant at $\alpha = 0.05$. Thus we have no reason to reject the null hypothesis $H_0: \sigma^2_A = 0$. Although the females differed from each other for genetic reasons, or possibly because of different environmental experiences before being brought into the laboratory or during development, the fact that they were reared in different cages did not add significant variation to their wing lengths.

It is instructive to rearrange these nested data in two ways, each time creating a different single-classification anova. In part I of Box 10.2 we show the first of these arrangements, in which we ignore the fact that the data are divided into three cages but simply treat each of the 12 females as one group in a

Box 10.2 ANALYSES OF REARRANGED DATA FOR A TWO-LEVEL NESTED ANOVA.

Data from Box 10.1.

I. Rearrangement of data: first example

	a females ($a = 12$)											
	1	2	3	4	5	6	7	8	9	10	11	12
n (measurements)	58.5	77.8	84.0	70.1	69.8	56.0	50.7	63.8	56.6	77.8	69.9	62.1
($n = 2$)	59.5	80.9	83.6	68.3	69.8	54.5	49.3	65.8	57.5	79.2	69.2	64.5
$\sum^{n} Y$	118.0	158.7	167.6	138.4	139.6	110.5	100.0	129.6	114.1	157.0	139.1	126.6
$\bar{Y}$	59.0	79.35	83.8	69.2	69.8	55.25	50.0	64.8	57.05	78.5	69.55	63.3

Preliminary computation

1. $\bar{\bar{Y}} = \frac{1}{an}\sum^{a}\sum^{n} Y = 66.6333$

2. $SS_{\text{among}} = n\sum^{a}(\bar{Y} - \bar{\bar{Y}})^2 = 2386.3533$

3. $SS_{\text{within}} = \sum^{a}\sum^{n}(Y - \bar{Y})^2 = 15.6200$

Anova table

	Source of variation	*df*	*SS*	*MS*	F_s
$\bar{Y} - \bar{\bar{Y}}$	Among groups (females)	11	2386.3533	216.9412	166.66**
$Y - \bar{Y}$	Within groups (between measurements on each female)	12	15.6200	1.3017	

II. Rearrangement of data: second example

	a cages ($a = 3$)		
	1	2	3
n measurements ($n = 8$)	58.5	69.8	56.6
	59.5	69.8	57.5
	77.8	56.0	77.8
	80.9	54.5	79.2
	84.0	50.7	69.9
	83.6	49.3	69.2
	70.1	63.8	62.1
	68.3	65.8	64.5
$\sum^{n} Y$	582.7	479.7	536.8
$\bar{Y}$	72.8375	59.9625	67.1000

Preliminary computation

1. $\bar{\bar{Y}} = \frac{1}{an} \sum^{a} \sum^{n} Y = 66.6333$

2. $SS_{\text{among}} = n \sum^{a} (\bar{Y} - \bar{\bar{Y}})^2 = 665.6758$

3. $SS_{\text{within}} = \sum^{a} \sum^{n} (Y - \bar{Y})^2 = 1736.2975$

Anova table

	Source of variation	*df*	*SS*	*MS*	F_s
$\bar{Y} - \bar{\bar{Y}}$	Among groups (cages)	2	665.6758	332.838	4.026*
$Y - \bar{Y}$	Within groups (among measurements on all females in a cage)	21	1736.2975	82.681	

single-classification anova. Convince yourself that the data at the head of Box 10.2 are the same as those at the head of Box 10.1. Now we simply have $a = 12$ females, each of which was measured $n = 2$ times. The fact that these 12 females come from three cages is hidden by this arrangement of the data. The analysis of variance of this table is straightforward, and the quantities shown beneath it yield the ensuing anova table.

The error mean square in the resulting anova table represents variance within females, which estimates the average variance between the two measurements for each female. The upper mean square represents variance among females, but in this arrangement of the data it confounds two sources of variation: the differences among females within a cage and possible differences from cage to cage. Note that this mean square yields a highly significant $F_s = 166.66$, which would occur with an infinitesimal probability if the null hypothesis were correct. Obviously this mean square should have been broken down into its components, representing variation of females within a cage and that among cages. It is reassuring to find this mean square significant because this tells us that there is evidence of some significant differentiation at either or both of the upper levels of variation in the study. No differences among the females, whether inherent or because of the different cages, would have been reflected in the simplified analysis we have just considered.

The second rearrangement of the data of Box 10.1 is shown in part II of Box 10.2. We can remove the division into subgroups and simply treat data as a single-classification anova of items within the higher level of classification. This procedure yields eight measurements in each of three cages. Sample sizes a and n now assume new values and should not be confused with the same symbols used in part I of Box 10.2. The analysis of variance shows significant variation among cages, but only at the 5% level. The error term in this anova represents a mixture of variances among measurements and among females in a cage. One way of interpreting a nested classification is as a further partitioning of the SS within groups of a single-classification anova. Thus, in this case we separate variation between measurements of one female from variation among females in a cage.

You may wonder why the MS among cages yields significant results in the second anova of Box 10.2 but does not do so in Box 10.1. The reason is that we tested it over a different, smaller error term. The error term in the second anova of Box 10.2 is a mixture of the mean squares from the two lower levels of the analysis of variance in Box 10.1. If you added the two sums of squares and their degrees of freedom, you would obtain 1736.2975, the SS_{within} of the second anova of Box 10.2, and 21 error degrees of freedom. This error term, however, confounds two separate sources of variation: the variation among measurements with 12 degrees of freedom and a sum of squares of 15.6200, and differences among females within cages with 9 degrees of freedom and a sum of squares of 1720.6775. Clearly, if each cage is represented by several females, the mean square among the latter must be included in the error term.

This Model II nested anova is completed by estimating the three variance components as shown in Box 10.1. The operations are simple extensions of the method of estimating variance components presented in Chapter 9. When we divide the overall variation into percentages attributable to each level, the error variance (measurements within one female) represents only 1.14% of the total variation. The measurements evidently did not differ much from each other, and in future work it would probably suffice to measure each wing only once. Most of the variation (83.32%) represents variation among females (within cages), presumably largely genetic but possibly also environmental, as already discussed.

You may question the 15.54% of variance among cages. Having decided earlier that there is no added variance component among cages, why did we estimate it at all? One reason is to provide a general outline of all the computations in the box, so that you can use the example as a model for others you will carry out in your research, but a more important reason is that, even though we were not in a position to reject the null hypothesis, we cannot prove that there is not a small contribution to the variance from cage differences. If we had had more degrees of freedom (more cages), perhaps we could have shown such a difference to be significant. Thus we might quite legitimately estimate the variance component and calculate its percentage contribution to the total variation, since the computation gives the best estimate of this variance component if it does indeed exist.

In the example in Box 10.1 the test for subgroups within groups was significant; hence we used the mean square for subgroups as a divisor to test for differences among groups. What should we have done, however, if we had accepted the null hypothesis that there were no differences among subgroups? From the expected mean squares in Box 10.1 we see that, in that case, the MS_{subgr} estimates σ^2, as does MS_{within}. One might well pool these two estimates of σ^2 to obtain a better estimate of σ^2 based on more degrees of freedom. Statisticians do not agree about the conditions under which mean squares should be pooled and about the desirability of pooling. The experimenter cannot go wrong by not pooling, but increasing the reliability of the error estimate can improve the confidence placed in the conclusions. Such pooling is done by dividing the sum of the sums of squares by the sum of the degrees of freedom of the implicated sources of variation. A conservative set of rules for pooling is illustrated in Box 10.3.

Our second illustration of nested anova makes use of a design that is perhaps the most common in the literature: a two-level mixed model, shown in Box 10.4. The groups represent different strains of houseflies, some known to be DDT-resistant, others not. The investigators wanted to test whether there were differences in various morphological characters among these strains and to see whether these differences could be related to DDT resistance. Since the morphology of flies is generally quite labile, depending substantially on environmental

Box 10.3 RULES FOR POOLING MEAN SQUARES IN ANOVA (WHEN INTERMEDIATE-LEVEL *MS* IS NONSIGNIFICANT).

Notation

Levels of variation	*df*	*SS*	*MS*	Variance estimated
3	df_3	SS_3	MS_3	σ_3^2
2	df_2	SS_2	MS_2	σ_2^2
1	df_1	SS_1	MS_1	σ_1^2

The question often arises in anovas whether to pool two mean squares, in different lines of an anova table, that appear to be estimates of the same quantity. Research workers like to do this, since the pooled mean squares have more degrees of freedom, but because the actual population parameters estimated by the mean squares are unknown, it is difficult to know when one should pool. The rules listed here were proposed by Bancroft (1964).

Dichotomous key to the rules for pooling mean squares

1. **(a)** On the basis of prior knowledge (from related experiments), σ_2^2/σ_1^2 is thought to be small (between 1 and 2) **2**

(b) No prior knowledge of σ_2^2/σ_1^2, or ratio thought to be large (>2). Set $\alpha_1 = 0.75$ (*Note:* If σ_2^2/σ_1^2 is actually small, the significance levels of the subsequent tests will be affected. For this reason many statisticians recommend that one never pool.) **3**

2. **(a)** $df_3 \geq df_2$ and $df_1 \geq 5 \times df_2$. Set $\alpha_1 = 0.50$ **3**

(b) Relations not as stated above. Set $\alpha_1 = 0.25$ **3**

3. **(a)** $F_s = MS_2/MS_1$ is significant at $P \leq \alpha_1$. Do not pool. Test MS_3 over MS_2 or MS_1 as indicated by anova procedure. Use conventional levels of significance.

(b) $F_s = MS_2/MS_1$ is not significant at $P \leq \alpha_1$. Pool; that is, test MS_3 over $(SS_2 + SS_1)/(df_2 + df_1)$ at conventional levels of significance.

The following hypothetical examples illustrate these rules. Assume that the anova tables below represent Model II nested anovas, so MS_3 should be tested over MS_2. It is in such examples that pooling is most frequently applied, when permissible, because df_2 is usually small, and by pooling MS_2 with MS_1 we can increase our degrees of freedom.

Example A

Level	*df*	*SS*	*MS*
3	1	50	50
2	10	250	25
1	20	400	20

BOX 10.3 CONTINUED

Assume no prior knowledge of σ_2^2/σ_1^2. Therefore, according to **1(b)** set $\alpha_1 = 0.75$. $F_s = MS_2/MS_1 = 25/20 = 1.25$ is greater than $F_{.75[10,20]} = 0.656$; hence according to **3(a)** do not pool and test MS_3 over MS_2. $F_s = 50/25 = 2$, which is less than $F_{.05[1,10]} = 4.96$ and therefore not significant.

Example B

Level	*df*	*SS*	*MS*
3	1	50	50
2	10	100	10
1	20	400	20

Assume no prior knowledge of σ_2^2/σ_1^2. According to **1(b)** set $\alpha_1 = 0.75$. $F_s = MS_2/MS_1 = 10/20 = 0.5$, which is less than $F_{.75[10,20]} = 0.656$. Therefore according to **3(b)** one may pool and test MS_3 over $(SS_2 + SS_1)/(df_2 + df_1) = (100 + 400)/(10 + 20) = 16.67$. $F_s = 50/16.67 = 3.00$, which is less than $F_{.05[1,30]} = 4.17$ and therefore not significant. Note that if we had not pooled, we would have found that $F_s = MS_3/MS_2 = 50/10 = 5.00$, which is greater than $F_{.05[1,10]} = 4.96$ and therefore significant. Although this test using a pooled error *MS* is therefore less powerful than the test one would perform without pooling, presumably the pooled error *MS* is a better estimate of the population variance σ^2.

Example C

Level	*df*	*SS*	*MS*
3	1	100	100
2	10	250	25
1	20	400	20

On the basis of prior knowledge we believe σ_2^2/σ_1^2 to be small. We therefore choose option **1(a)** and proceed to decision **2.** Since $df_3 = 1$ is not $\geq df_2 = 10$ and $df_1 = 20$ is not $\geq 5 \times df_2 = 5 \times 10 = 50$, we follow **2(b)** and set $\alpha_1 = 0.25$. $F_s = 25/20 = 1.25$ is not significant compared with $F_{.25[10,20]} = 1.40$. Therefore according to **3(b)** we test MS_3 over the pooled error $MS = (250 + 400)/(10 + 20) = 650/30 = 21.67$. $F_s = MS_3/MS_{\text{pooled}} = 100/21.67 = 4.615 > F_{.05[1,30]} = 4.17$ and is therefore significant. If we had not pooled, we would have found that $F_s = 100/25 = 4.0$, which is less than $F_{.05[1,10]} = 4.96$.

These rules can also be used for pooling in orthogonal analyses of variance (see Chapter 11), in which the treatment mean square has to be tested over the error mean square. In such cases MS_3 is tested over MS_1, whose degrees of freedom might be increased by pooling it with MS_2 (the interaction mean square) if these rules permit.

Box 10.4 TWO-LEVEL NESTED ANOVA: MIXED MODEL.

Eight housefly strains differing significantly in DDT resistance were tested for morphological differences. Three jars of each strain were prepared, and all jars were incubated together. Eight females were taken from each jar after all the adults had emerged. The character analyzed here is setae number on the third abdominal sternum. To conserve space, original variates are not shown. We give only the means for each jar; n(number of items per subgroup) = 8; b(number of subgroups per group) = 3; a(number of groups) = 8; abn(total sample size) = 192.

Strain	Jar	Jar means	Strain means	Strain	Jar	Jar means	Strain means
LDD	1	27.000		LC	1	28.500	
	2	27.750			2	26.875	
	3	26.625	27.1250		3	27.000	27.4583
OL	1	33.375		RH	1	29.500	
	2	38.125			2	30.375	
	3	31.250	34.2500		3	28.250	29.3750
NH	1	27.500		NKS	1	30.125	
	2	26.625			2	29.625	
	3	28.500	27.5417		3	31.750	30.5000
RKS	1	31.750		BS	1	27.875	
	2	31.750			2	25.625	
	3	35.250	32.9167		3	27.500	27.0000

Preliminary computation

1. $\bar{\bar{Y}} = 29.5208$
2. $SS_{\text{among}} = 1323.42$
3. $SS_{\text{subgr}} = 357.25$
4. $SS_{\text{within}} = 4663.25$

Anova table

	Source of variation	*df*	*SS*	*MS*	F_s	Expected *MS*
$\bar{Y}_A - \bar{\bar{Y}}$	Among strains (groups)	7	1323.42	189.06	8.47***	$\sigma^2 + n\sigma^2_{B \subset A} + nb \frac{\sum \alpha^2}{a - 1}$
$\bar{Y}_B - \bar{Y}_A$	Among jars within strains (subgroups within groups)	16	357.25	22.33	*ns*	$\sigma^2 + n\sigma^2_{B \subset A}$
$Y - \bar{Y}_B$	Within jars (error)	168	4663.25	27.76		σ^2
$Y - \bar{\bar{Y}}$	Total	191	6343.92			

SOURCE: Data from Sokal and Hunter (1955).

influence when they are immature, three jars of the same medium were prepared for each strain in order to separate the differences among strains from the variances among separate jars. Clearly, if each strain is reared in only one jar, we cannot separate differences among jars from differences among strains. The character reported here is setae number on the third abdominal sternum.

The data in Box 10.4 are in abbreviated form. Setae on eight females were counted per jar, but only the means of these eight counts are given in the table; the individual variates are omitted to conserve space. Therefore, the sum of squares within jars (quantity **4,** provided) cannot be computed from the information furnished. This is a nested anova because the jars were allocated at random, three to a strain. The quantities necessary for the computation are shown in the box, as is the completed anova. The box shows that there is no evidence for an added variance component among the jars, since the mean square among jars is less than that within jars. It appears therefore that differences in the microclimate of the jars do not affect setae number in these flies. This is not true, however, of other characters. Sokal and Hunter (1955) found that many measurements of lengths of structures, such as wing length, were strongly affected by variation among the jars within strains.

In this way we learn about the differential responses of various characters to environmental conditions. Should we pool MS_{subgr} with MS_{within} before testing the significance of MS_{among}? Consulting Box 10.3, we make no assumption about the ratio of the two variances and set $\alpha_1 = 0.75$. $F_s = MS_{\text{subgr}}/MS_{\text{within}} = 0.804$, which is clearly significant at $\alpha_1 = 0.75$, since even $F_{.75[20,\infty]} = 0.773 < F_s$. Note that to avoid interpolation, we use the F-value for the next higher tabled degrees of freedom in place of $F_{.75[16,168]}$, because F-values below 1 increase as their degrees of freedom increase, while those above 1 show the reverse relationship. In view of the significance of the variance ratio, we do not pool and test $MS_{\text{among}}/MS_{\text{subgr}}$ at the conventional level of significance. We find the mean square among strains highly significant. Since the highest level of classification is Model I, we would ordinarily complete the analysis of these data with multiple comparisons (Sections 9.6 or 9.7), using MS_{subgr} as the error term.

We now take up a three-level nested anova. The example that we have chosen is a mixed model; the highest level of classification is Model I, and the next two levels of classification are randomly chosen subgroups and subsubgroups (Model II). Box 10.5 features data on the glycogen content of rat livers. The measurement is in arbitrary units. Duplicate readings were made on each of three preparations of rat livers from each of two rats for three different treatments, as shown in the box. The setup of the data should by now be familiar. Note that we have a new symbol, c, the number of subsubgroups per subgroup, which in this case equals 3. If you were to have yet lower level classifications, you could introduce symbols d, e, and so on.

The preliminary computations in Box 10.5 follow the same pattern as those in Box 10.1. Note that we find the grand mean in step **1,** followed by the SS of groups, subgroups within groups, subsubgroups within subgroups, and within

Box 10.5 THREE-LEVEL NESTED ANOVA: EQUAL SAMPLE SIZES, MIXED MODEL.

Glycogen content of liver in arbitrary units. Duplicate readings on each of 3 preparations of rat livers from each of 2 rats for each of 3 treatments.

Treatments ($a = 3$)	Control						Compound 217						Compound 217 plus sugar					
Rats ($b = 2$)	1			2			1			2			1			2		
Preparations ($c = 3$)	1	2	3	1	2	3	1	2	3	1	2	3	1	2	3	1	2	3
Readings ($n = 2$)	131	131	136	150	140	160	157	154	147	151	147	162	134	138	135	138	139	134
	130	125	142	148	143	150	145	142	153	155	147	152	125	138	136	140	138	127
Preparation sums	261	256	278	298	283	310	302	296	300	306	294	314	259	276	271	278	277	261
Preparation means	130.5	128.0	139.0	149.0	141.5	155.0	151.0	148.0	150.0	153.0	147.0	157.0	129.5	138.0	135.5	139.0	138.5	130.5
Rat means	132.5000			148.5000			149.6667			152.3333			134.3333			136.0000		
Treatment means	140.5000						151.0000						135.1667					
Grand sum	5120																	

BOX 10.5 CONTINUED

Preliminary computation

1. $\bar{\bar{Y}} = \frac{1}{abcn} \sum^{a} \sum^{b} \sum^{c} \sum^{n} Y = \frac{1}{36} 5120 = 142.2222$

2. SS_{among} (among groups) $= bcn \sum^{a} (\bar{Y}_A - \bar{\bar{Y}})^2 = 1557.55$

3. SS_{subgr} (subgroups within groups) $= cn \sum^{a} \sum^{b} (\bar{Y}_B - \bar{Y}_A)^2 = 797.67$

4. SS_{subsubgr} (subsubgroups within subgroups) $= n \sum^{a} \sum^{b} \sum^{c} (\bar{Y}_C - \bar{Y}_B)^2 = 594.00$

5. SS_{within} (within subsubgroups; error) $= \sum^{a} \sum^{b} \sum^{c} \sum^{n} (Y - \bar{Y}_C)^2 = 381.00$

Rules for working out the degrees of freedom are given in the text.

Anova table

	Source of variation	*df*	*SS*	*MS*	F_s	Expected *MS*
$\bar{Y}_A - \bar{\bar{Y}}$	Among groups (treatments)	2	1557.55	778.78	2.93 *ns*	$\sigma^2 + n\sigma^2_{C\subset B} + nc\sigma^2_{B\subset A} + ncb \frac{\sum \alpha^2}{a-1}$
$\bar{Y}_B - \bar{Y}_A$	Among subgroups within groups (rats within treatments)	3	797.67	265.89	5.37*	$\sigma^2 + n\sigma^2_{C\subset B} + nc\sigma^2_{B\subset A}$

$\bar{Y}_C - \bar{Y}_B$	Among subsubgroups within subgroups (preparations within rats)	12	594.00	49.50	2.34*	$\sigma^2 + n\sigma^2_{C\subset B}$
$Y - \bar{Y}_C$	Within subsubgroups (error; readings within preparations)	18	381.00	21.17		σ^2
$Y - \bar{\bar{Y}}$	Total	35	3330.22			

$F_{.05[12,18]} = 2.34$ $F_{.01[12,18]} = 3.37$ $F_{.05[3,12]} = 3.49$ $F_{.01[3,12]} = 5.95$ $F_{.05[2,3]} = 9.55$

Estimation of variance components

$$s^2_{B\subset A} = (265.89 - 49.50)/6 = 36.06 \text{ (rats within treatments)}$$

$$s^2_{C\subset B} = (49.50 - 21.17)/2 = 14.16 \text{ (preparations within rats)}$$

$$s^2 = 21.17 \text{ (readings within preparations)}$$

These can also be expressed as percentages:

$$\text{Since } s^2_{B\subset A} + s^2_{C\subset B} + s^2 = 71.40$$

$$s^2_{B\subset A} \text{ (rats within treatments)} = 100 \times 36.06/71.40 = 50.5\%$$

$$s^2_{C\subset B} \text{ (preparations within rats)} = 100 \times 14.16/71.40 = 19.8\%$$

$$s^2 \text{ (readings within preparations)} = 100 \times 21.17/71.40 = 29.6\%$$

subsubgroups in steps **2** to **5.** The formulas are given in the box. It is easy to construct additional formulas if yet another level of variation is added.

The number of degrees of freedom could be provided by formula as in Box 10.1. We feel it is more profitable, however, to reason it out for the individual case. There are $a = 3$ treatments and hence $a - 1 = 2$ degrees of freedom among groups (treatments). There are $b = 2$ rats in each treatment, corresponding to $b - 1 = 1$ degree of freedom for rats in each treatment. Since there are $a = 3$ treatments, we have $a(b - 1) = 3$ *df* for subgroups within groups (rats within treatments). From each rat were made $c = 3$ preparations, yielding $c - 1 = 2$ degrees of freedom for preparations within one rat. Since there were $ab = 6$ rats in the study, we have $ab(c - 1) = 12$ *df* among subsubgroups within subgroups (preparations within rats). Finally, $n = 2$ readings per preparation yield $n - 1 = 1$ *df* per preparation, and since there are $abc = 18$ preparations in the study, there are $abc(n - 1) = 18$ *df* within subsubgroups, corresponding to the error term (readings within preparations). These degrees of freedom, shown in the anova table in Box 10.5, total 35, as they should, since there are $abcn = 36$ readings in all.

The anova table is shown in Box 10.5. We proceed to test each mean square over the one immediately beneath it, starting with the subsubgroups. Since the mean squares of subsubgroups and subgroups are significant, the question of pooling does not arise, and the treatment mean square is tested over that of subgroups. The F-ratio for the latter test, however, is not large enough to be significant. We conclude that although there is significant added variance among preparations within rats and among rats within treatments, we cannot establish significant differences among treatments. Note that this test is based on very few degrees of freedom. We might suspect that there really are differences among the treatments but that we cannot pick them up with so few degrees of freedom. We would be tempted to repeat this experiment using more rats per treatment and might well be able to demonstrate significant differences among the treatments. We will return to this example in Section 10.4 to examine the efficiency of its design.

10.3 Nested Anovas With Unequal Sample Sizes

Regrettably, we do not always have equal sample sizes in an experiment or a study that lends itself to a nested anova. The complications arising from unequal replication are sufficiently annoying to urge us to design our work in such a way that we have equal sample sizes whenever possible. The basic principles of a nested anova are the same, even when sample sizes are unequal. However, there are three additional problems: (1) more complex and cumbersome notation is needed, (2) computations become much more tedious, and (3) there are no exact tests of significance (and the approximate tests can be very inexact). The second of these considerations is becoming less important as more and more of the computational work is carried out by sophisticated computer programs.

If we were to provide the fully correct mathematical symbolism, including subscripts for the variates and subscripts and superscripts for the summation signs, the formulas would look forbiddingly complex for what is essentially a simple computational setup. Let us look at an example of a two-level nested anova with unequal sample sizes to see what the notational and computational problems are. Box 10.6 records such a study, carried out on the blood pH of mice. In this experiment 15 female mice (dams) were successively mated over a period of time to either two or three males (sires). These sires were different for the 15 dams; thus a total of 37 sires was employed. The litters resulting from each mating were kept separate and the blood pH of female members of these litters was determined. The number of female offspring in these litters ranged from three to five.

The data are arranged in the data table in Box 10.6, which shows individual readings, as well as litter means (the means of the several readings for each litter), litter sizes, dam means (the means of the readings for the several litters of any one dam), the sample sizes on which these dam means are based (the sum of litter sizes), and finally the grand sum of all readings as well as the total sample size of 160 readings. There are $a = 15$ groups (dams) in this study, each of them containing b_i subgroups (sires), where the i refers to the dam number. Thus b_7 is the number of sires of dam number 7, which equals 3. The number of replicates (individual mice in a litter) must be symbolized with two subscripts, n_{ij}, where i refers to the dam number and j to the sire number of dam i. Thus $n_{7,2}$ is the sample size of the number of female offspring from the mating of dam 7 and sire 2, which equals 5. The number of replicates for the ith dam is given as n_i which is computed as $\sum^{b_i} n_{ij}$. Thus the first dam has $4 + 4 = 8$ offspring. To express symbolically how many sires there are in the study we have to write $\sum^a b_i = 37$, and to find out how many mice were measured in the entire study we have to sum the number of mice in each litter for all sires and over all dams as $\sum^a \sum^{b_i} n_{ij} = 160$.

The computations are straightforward and are given in Box 10.6. The degrees of freedom can be worked out directly from the table. There are 15 dams, and hence 14 degrees of freedom among groups; there are 37 sires within dams, which lose one degree of freedom per dam, so their $df = 37 - 15 = 22$. The within-subgroups degrees of freedom are based on 160 mice belonging to 37 litters, and hence losing 37 degrees of freedom, which yields 123 *df*. By formula, the degrees of freedom could be obtained as $a - 1$, $\sum^a b_i - a$, and $\sum^a \sum^{b_i} n_{ij} - \sum^a b_i$, respectively.

The first significance test is of the MS_{subgr} to test for the presence of the subgroups variance component. Note that $F_s = 1.470 < F_{.05[24,\infty]} = 1.52$; therefore it must be $< F_{.05[22,123]}$, the correct critical value. Should we pool MS_{subgr} with MS_{within}? No assumption is made about the variance ratio (see Box 10.3),

Box 10.6 TWO-LEVEL NESTED ANOVA WITH UNEQUAL SAMPLE SIZES.

Blood pH for 3 to 5 female mice within litters resulting from matings of 2 or 3 different sires to each of 15 dams. The data have been coded by subtracting 7.0 and then multiplying by 100. $a = 15$ dams; $b_i = 2$ or 3 sires per dam where $i = 1, \ldots, a$; $n_{ij} = 3, 4,$ or 5 mice per litter (from one sire) where $i = 1, \ldots, a$, and $j = 1, \ldots, b_i$. This is a Model II anova.

Dam number	Sire number	Blood pH readings of individual mice					Litter size n_{ij}	Litter mean $\bar{Y}_B = \frac{1}{n_{ij}} \sum_{k=1}^{n_{ij}} Y_{ijk}$	Sample size of dam mean $n_i = \sum^{b_i} n_{ij}$	Dam mean $\bar{Y}_A = \frac{1}{n_i} \sum_{j=1}^{b_i} \sum_{k=1}^{n_{ij}} Y_{ijk}$
		1	2	3	4	5				
1	1	48	48	52	54		4	50.50000		
	2	48	53	43	39		4	45.75000	8	48.12500
2	1	45	43	49	40	40	5	43.40000		
	2	50	45	43	36		4	43.50000	9	43.44444
3	1	40	45	42	48		4	43.75000		
	2	45	33	40	46		4	41.00000		
	3	40	47	40	47	47	5	44.20000	13	43.07692
4	1	38	48	46			3	44.00000		
	2	37	31	45	41		4	38.50000	7	40.85714
5	1	44	51	49	51	52	5	49.40000		
	2	49	49	49	50		4	49.25000		
	3	48	59	59			3	55.33333	12	50.83333
6	1	54	36	36	40		4	41.50000		
	2	44	47	48	48		4	46.75000		
	3	43	52	50	46	39	5	46.00000	13	44.84615
7	1	41	42	36	47		4	41.50000		
	2	47	36	43	38	41	5	41.00000		
	3	53	40	44	40	45	5	44.40000	14	42.35714

8	1	52	53	48			3	51.00000		
	2	40	48	50	40	51	5	45.80000	8	47.75000
9	1	40	34	37	45		4	39.00000		
	2	42	37	46	40		4	41.25000	8	40.12500
10	1	39	31	30	41	48	5	37.80000		
	2	50	44	40	45		4	44.75000	9	40.88889
11	1	52	54	52	56	53	5	53.40000		
	2	56	39	52	49	48	5	48.80000	10	51.10000
12	1	50	45	43	44	49	5	46.20000		
	2	52	43	38	33		4	41.50000	9	44.11111
13	1	39	37	33	43	42	5	38.80000		
	2	43	38	44			3	41.66667		
	3	46	44	37	54		4	45.25000	12	41.66667
14	1	50	53	51	43		4	49.25000		
	2	44	45	39	52		4	45.00000		
	3	42	48	45	51	48	5	46.80000	13	47.00000
15	1	47	49	45	43	42	5	45.20000		
	2	45	42	52	51	32	5	44.40000		
	3	51	51	53	45	51	5	50.20000	15	46.60000
Grand sum $\sum_{i=1}^{a}\sum_{j=1}^{b_i}\sum_{k=1}^{n_{ij}} Y_{ijk}$							7197			
Total sample size $\sum_{i=1}^{a}\sum_{j=1}^{b_i} n_{ij}$									160	

The structure of the analysis is within litters, among litters due to different sires (sires within dams), and among dams.

BOX 10.6 CONTINUED

Preliminary Computation

1. Grand mean $\bar{\bar{Y}} = \dfrac{1}{\sum^{a}\sum^{b_i} n_{ij}} \sum^{a}\sum^{b_i}\sum^{n_{ij}} Y = \dfrac{1}{160}\,7197 = 44.98125$

2. SS_{among} (among groups) $= \sum^{a} n_i(\bar{Y}_A - \bar{\bar{Y}})^2 = 1780.174$

3. SS_{subgr} (subgroups within groups) $= \sum^{a}\sum^{b_i} n_{ij}(\bar{Y}_B - \bar{Y}_A)^2 = 800.237$

4. SS_{within} (within subgroups; error SS) $= \sum^{a}\sum^{b_i}\sum^{n_{ij}} (Y - \bar{Y}_B)^2 = 3042.533$

Anova table

	Source of variation	df	SS	MS	F_s or (F'_s)	Expected MS
$\bar{Y}_A - \bar{\bar{Y}}$	Among groups (dams)	14	1780.174	127.155	3.496 (3.473**)[a]	$\sigma^2 + n'_0\sigma^2_{B\subset A} + (nb)_0\sigma^2_A$
$\bar{Y}_B - \bar{Y}_A$	Among subgroups (sires within dams)	22	800.237	36.374	1.470 *ns*	$\sigma^2 + n_0\sigma^2_{B\subset A}$
$Y - \bar{Y}_B$	Within subgroups (error; among mice of one litter—that is, from one sire)	123	3042.533	24.736		σ^2
$Y - \bar{\bar{Y}}$	Total	159	5622.944			

$F_{.05[24,\infty]} = 1.52$, therefore $F_s = MS_{subgr}/MS_{within}$ with 22 and 123 df cannot be significant. By the rules of Box 10.3, we do not pool these two mean squares, but test MS_{among} over MS_{subgr}.

[a] Since the coefficient n_0 is not equal to n'_0 there is no MS over which we can test MS_{among} exactly. See below how this value of F'_s has been obtained.

Tests of significance

As shown at the bottom of the anova table above, MS_{subgr} is tested directly over MS_{within}. It is not significant. The coefficients of the variance components in a two-level nested anova with unequal sample sizes are calculated as follows:

1. $\sum^{a}\sum^{b_i} n_{ij} = 4 + 4 + \ldots + 5 + 5 = 160$

2. $\sum^{a}\sum^{b_i} n_{ij}^2 = 4^2 + 4^2 + \ldots + 5^2 + 5^2 = 708$

3. $\sum^{a}\left(\sum^{b_i} n_{ij}\right)^2 = 8^2 + 9^2 + \ldots + 13^2 + 15^2 = 1800$

4. $\sum^{a}\left(\frac{\sum^{b_i} n_{ij}^2}{\sum^{b_i} n_{ij}}\right) = \frac{4^2 + 4^2}{8} + \frac{5^2 + 4^2}{9} + \ldots + \frac{5^2 + 5^2 + 5^2}{15} = 65.690$

$$n_0' = \frac{\sum^{a}\left(\frac{\sum^{b_i} n_{ij}^2}{\sum^{b_i} n_{ij}}\right) - \frac{\sum^{a}\sum^{b_i} n_{ij}^2}{\sum^{a}\sum^{b_i} n_{ij}}}{df_{\text{among}}} = \frac{\text{quantity } \mathbf{4} - (\text{quantity } \mathbf{2}/\text{quantity } \mathbf{1})}{df_{\text{among}}} = \frac{65.690 - (708/160)}{14} = 4.376$$

$$n_0 = \frac{\sum^{a}\sum^{b_i} n_{ij} - \sum^{a}\left(\frac{\sum^{b_i} n_{ij}^2}{\sum^{b_i} n_{ij}}\right)}{df_{\text{subgr}}} = \frac{\text{quantity } \mathbf{1} - \text{quantity } \mathbf{4}}{df_{\text{subgr}}} = \frac{160 - 65.690}{22} = 4.287$$

$$(nb)_0 = \frac{\sum^{a}\sum^{b_i} n_{ij} - \frac{\sum^{a}\left(\sum^{b_i} n_{ij}\right)^2}{\sum^{a}\sum^{b_i} n_{ij}}}{df_{\text{among}}} = \frac{\text{quantity } \mathbf{1} - (\text{quantity } \mathbf{3}/\text{quantity } \mathbf{1})}{df_{\text{among}}} = \frac{160 - (1800/160)}{14} = 10.625$$

Solving for estimates of the 3 unknown variance components:

$$MS_{\text{among}} = 127.155 = s^2 + 4.376 s_{B\subset A}^2 + 10.625 s_A^2$$

$$MS_{\text{subgr}} = 36.374 = s^2 + 4.287 s_{B\subset A}^2$$

$$MS_{\text{within}} = 24.736 = s^2$$

BOX 10.6 CONTINUED

Therefore

$$s^2 = MS_{\text{within}} = 24.736$$

$$s^2_{B\subset A} = \frac{MS_{\text{subgr}} - MS_{\text{within}}}{4.287} = \frac{36.374 - 24.736}{4.287} = 2.715$$

$$s^2_A = \frac{MS_{\text{among}} - MS_{\text{within}} - 4.376 s^2_{B\subset A}}{10.625} = \frac{127.155 - 24.736 - (4.376)(2.715)}{10.625} = 8.521$$

Before we can perform a test of significance for MS_{among} we must check whether the given data satisfy the Gaylor and Hopper (1969) conditions under which the Satterthwaite approximation can be used. First verify that $df_{\text{subgr}} < 100$ and that $df_{\text{subgr}} < 2df_{\text{within}}$. In this case, since $df_{\text{subgr}} = 22$ and $2df_{\text{within}} = 246$, the conditions are satisfied. Next test that the ratio R is greater than the criterion C.

$$R = \frac{n_0'}{n_0' - n_0} \times \frac{MS_{\text{subgr}}}{MS_{\text{within}}}$$

$$= \frac{4.376}{(4.376 - 4.287)} \times \frac{36.374}{24.736} = 72.302$$

$$C = F_{.025[df_{\text{within}},\, df_{\text{subgr}}]} \times F_{.5[df_{\text{subgr}},\, df_{\text{within}}]}$$

$$= 2.08(0.975) = 2.03$$

Since $R > C$, it is safe to use Satterthwaite's approximation. (If these conditions were not met, only the simple approximate test, $F_s = MS_{\text{among}}/MS_{\text{subgr}}$ could be made.) For Satterthwaite's approximation, one evaluates the following weights

$$w_2 = \frac{n_0'}{n_0} = 1.0208$$

$$w_1 = 1 - w_2 = -0.0208$$

which are used to compute a new, synthetic denominator:

$$\begin{aligned} MS'_{\text{subgr}} &= w_1 MS_{\text{within}} + w_2 MS_{\text{subgr}} \\ &= -0.0208(24.736) + 1.0208(36.374) \\ &= 36.616 \end{aligned}$$

This mean square is equal to

$$s^2 + n'_0 s^2_{B \subset A} = 24.736 + 4.376(2.715) = 36.617$$

(within rounding error). The Satterthwaite formula for the degrees of freedom of the reconstituted mean square MS'_{subgr} is

$$\begin{aligned} df'_{\text{subgr}} &= \frac{(MS'_{\text{subgr}})^2}{\dfrac{(w_1 MS_{\text{within}})^2}{df_{\text{within}}} + \dfrac{(w_2 MS_{\text{subgr}})^2}{df_{\text{subgr}}}} \\ &= \frac{(36.616)^2}{\dfrac{(-0.0208 \times 24.736)^2}{123} + \dfrac{(1.0208 \times 36.374)^2}{22}} \\ &= \frac{1340.7315}{\dfrac{0.2647}{123} + \dfrac{1378.6799}{22}} \\ &= \frac{1340.7315}{0.0022 + 62.6673} = \frac{1340.7315}{62.6694} \\ &= 21.39 \approx 21 \end{aligned}$$

Finally,

$$F'_s = \frac{MS_{\text{among}}}{MS'_{\text{subgr}}} = \frac{127.154}{36.617} = 3.473$$

Since this value is greater than $F_{.01[12,21]} = 3.17$ (the conservative critical value employed to avoid interpolation for $F_{.01[14,21]}$ in Statistical Table **F**), there is a significant variance component among dams, but as we have seen, there is no added variance among sires.

and $F_s = 1.470 \gg F_{.75[24,\infty]} = 0.793$, the approximate critical value. We do not pool and plan to test MS_{among} over MS_{subgr}.

Unequal sample sizes present problems for additional significance tests. Not only must average sample sizes be computed (using a method similar to the special formula learned in Box 9.2), but a glance at the expected mean squares of the anova table in Box 10.6 shows that the average coefficients of the same variance components do not correspond at different levels. Notice that $\sigma^2_{B \subset A}$ has n_0 as a coefficient in MS_{subgr}, but has another quantity, n'_0, as a coefficient in MS_{among}. The formulas for these coefficients are also shown in Box 10.6. In this example, the coefficients have similar values—4.287 for n_0 and 4.376 for n'_0—but in other examples they can differ considerably. Next we estimate the three unknown variance components—within subgroups, among subgroups within groups, and among groups—by simple substitutions as shown in Box 10.6.

Thus a general problem arises from unequal sample sizes: There are no exact tests of significance for testing the upper $k - 1$ levels in a k-level nested anova. Box 10.6 shows that the ratio $F_s = MS_{among}/MS_{subgr}$ does not simply estimate $(nb)_0\sigma^2_A$, since $n_0 \neq n'_0$. We can test significance by synthesizing a new denominator mean square (employing what is commonly referred to as the Satterthwaite approximation) against which to test the mean square of groups. This denominator mean square, MS'_{subgr}, has $s^2_{B \subset A}$ multiplied by $n'_0 = 4.376$ instead of $n_0 = 4.287$ as before. We now obtain a variance ratio $F'_s = 3.473$ whose significance must be evaluated against a critical value of F. The degrees of freedom of the denominator mean square are then calculated. First we rewrite the new mean square as a linear combination of the previous mean squares, as shown in Box 10.6. Then we calculate df'_{subgr} from the coefficients of the old mean squares, which in this case is 21.39. To be conservative we use 21 degrees of freedom for the denominator mean square and compare $F_s = 3.473$ with $F_{.01[14,21]}$. In Statistical Table **F** we find that $F_{.01[12,21]} = 3.17$, showing the added variance component among dams to be highly significant.

The test based on the Satterthwaite approximation can be unreliable when n_0 is different from n'_0 and $\sigma^2_{B \subset A}$ is small in comparison to σ^2 (Gaylor and Hopper, 1969; Boardman, 1974). Gaylor and Hopper (1969) suggest two rules for deciding when the Satterthwaite approximation can be used. These are given in Box 10.6. The conditions implied by these rules are satisfied for the example in Box 10.6. If these conditions are not met, then probably the only test one can do is the ordinary F-test (MS_{among}/MS_{subgr}). Using the Satterthwaite approximation when the required conditions are not satisfied tends to yield excessively conservative tests (in some Monte Carlo simulations we have observed a type I error of 0.4% at an intended 5% level of significance). This subject requires more theoretical work so that at least good approximate tests can be made in all cases.

The biological interpretation of this example is interesting. Since sires within dams did not show an added variance component, significant differentiation does not seem to depend on the sires. Either the sires came from an inbred strain and were identical genetically, or the variance among dams, but not among sires, can be explained through maternal effects. In this way the analysis of variance can be of great help in suggesting further research in problems of genetics.

The next example is a three-level nested anova with unequal sample sizes. This analysis is carried out not for a planned experiment but to analyze a series of data that had accumulated in the laboratory. The data are diameters of pollen grains of a species of tree that were deposited in deep layers at the margin of a marsh. Samples of this pollen were analyzed; the variable measured was mean diameter of 100 pollen grains found on one slide. That the basic variate in this study is a mean need not concern us here. The pollen samples were grouped into three depths—the highest-level category in this study—which clearly are fixed treatment effects; hence this is a Model I anova. For the first two depths two core samples each had been obtained, but three were available for the third depth. Varying numbers of preparations had been made from these core samples and varying numbers of slides examined from these preparations. The structure of the analysis is shown in the table in Box 10.7; to conserve space the data themselves are not given. The core samples within depths and the preparations within these samples can be considered to have been obtained at random; for this reason the analysis becomes a mixed model nested anova.

Although not shown in Box 10.7, the computations (given the data) could easily be derived from a combination of Boxes 10.5 and 10.6. The result of these computations is the anova table in Box 10.7. The purpose of Box 10.7 is to show the analysis of this table because of the complexities of testing significance and estimating variance components in an example with unequal sample sizes. The first significance test is exact, *MS* of subsubgroups over the error *MS*, which in this case is not significant but is large enough according to the rules in Box 10.3 to contraindicate pooling.

Since both MS'_{subsubgr} and MS'_{subgr} satisfy the criteria, the Satterthwaite approximation can be used with MS_{subgr} and MS_{among} to test for corresponding variance components (only the former was found to be significant). Again we find that F and F' are similar. A Monte Carlo study of this design shows that if $\sigma^2_{B \subset A}$ were equal to zero then the F'-test would have been excessively conservative. Out of 1000 replications of this design, only 0.4% were declared significant at the 5% level, and none were significant at the 1% level. The F-test yielded the correct probabilities. Thus the Satterthwaite approximation cannot be used uncritically. The estimation of the variance components is also shown in detail in Box 10.7. No new procedures are introduced; they simply become more involved because of the extra level of variation. Since this is a mixed model, the highest level is Model I, and we do not estimate a variance component for it but simply test the significance of added treatment effects. The box shows computation of a variance component for this level simply to complete the outline of all necessary computations for other examples that might be pure Model II.

The conclusions from the analysis are that there is a significant added variance component among subgroups (core samples), while neither of the other two mean squares yields significant results. We therefore conclude that the

Box 10.7 THREE-LEVEL NESTED ANOVA WITH UNEQUAL SAMPLE SIZES.

Diameter of pollen grains of a species of tree from a paleobotanical study. All core samples were collected from the same locality at the margin of a marsh. Several core samples were obtained at each of three depths; varying numbers of preparations were made from each core sample and varying numbers of slides examined from each preparation. Variates are means of diameters of 100 pollen grains on a slide. Actual data are not shown here. This is a mixed-model anova.

The structure of the analysis and the sample sizes

Depths (a)	1				2					3				
Core samples (b_i)	1		2		1			2		1		2		3
Preparations (c_{ij})	1	2	1	2	1	2	3	1	2	1	2	1	2	1
Number of slides per preparation (n_{ijk})	4	5	4	4	5	4	4	5	5	3	4	2	2	5
Number of slides in core sample (n_{ij})	9		8		13			10		7		4		5
Number of slides per depth (n_i)	17				23					16				
Total sample size (n)	56													

From the original data, the various sums of squares were computed following the procedures in Boxes 10.5 and 10.6. The results of these computations are given below.

Anova table

	Source of variation	df	SS	MS	F_s	F'_s
$\bar{Y}_A - \bar{\bar{Y}}$	Among groups (depths)	2	396.64	198.32	1.239 *ns*	1.131 *ns*
$\bar{Y}_B - \bar{Y}_A$	Among subgroups within groups (core samples within depths)	4	640.44	160.11	4.194*	4.466*
$\bar{Y}_C - \bar{Y}_B$	Among subsubgroups within subgroups (preparations within core samples)	7	249.41	35.63	1.712 *ns*	
$Y - \bar{Y}_C$	Within subsubgroups (error; within preparations)	42	874.02	20.81		
$Y - \bar{\bar{Y}}$	Total	55	2160.51			

Note: The F'_s values are not computed as simple variance ratios of the mean squares in the table because of unequal coefficients in the expected mean squares. Their computation is explained below.

BOX 10.7 CONTINUED

As in Box 10.6, only the next-to-the-lowest $MS(MS_{\text{subsubgr}})$ may be tested exactly. [In the present case it is not significant, and since we have no prior knowledge about the magnitude of the ratio of the variance components, we do not pool because $F_s = 1.712$ is greater than $F_{.75[7,60]} = 0.604$, which is a conservative approximation in Statistical Table **F** for $F_{.75[7,42]}$ (see Box 10.3).]

For the approximate tests of significance involving the other mean squares we must calculate the coefficients of the variance components, estimate the components, and then synthesize new mean squares to use as the denominators of the F_s ratios if the Satterthwaite approximation is used.

The expected mean squares

Source of variation	Expected *MS* for a Model II
Among groups	$\sigma^2 + n_0''\sigma^2_{C\subset B} + (nc)_0'\sigma^2_{B\subset A} + (ncb)_0\sigma^2_A$
Subgroups within groups	$\sigma^2 + n_0'\sigma^2_{C\subset B} + (nc)_0\sigma^2_{B\subset A}$
Subsubgroups within subgroups	$\sigma^2 + n_0\sigma^2_{C\subset B}$
Within subsubgroups (error)	σ^2

In our case we have a mixed model. The highest level is differentiated by fixed treatment effects. The top line of the expected mean squares (MS_{among}) should therefore read

$$\sigma^2 + n_0''\sigma^2_{C\subset B} + (nc)_0'\sigma^2_{B\subset A} + (ncb)_0 \frac{\sum \alpha^2}{a-1}$$

Preliminary computation

The basic quantities needed to compute the coefficients of the variance components are the following:

1. $\sum^{a}\sum^{b_i}\sum^{c_{ij}} n^2_{ijk} = 4^2 + 5^2 + \ldots + 2^2 + 5^2 = 238$

2. $\sum^{a}\sum^{b_i} n^2_{ij} = 9^2 + 8^2 + \ldots + 4^2 + 5^2 = 504$

3. $\sum^{a} n^2_i = 17^2 + 23^2 + 16^2 = 1074$

4. $\sum^{a}\sum^{b_i}\sum^{c_{ij}} n_{ijk} = \sum^{a}\sum^{b_i} n_{ij} = \sum^{a} n_i = 56$

5. $\sum^{a}\sum^{b_i}\left(\frac{\sum^{c_{ij}} n^2_{ijk}}{\sum^{c_{ij}} n_{ijk}}\right) = \frac{4^2+5^2}{9} + \frac{4^2+4^2}{8} + \ldots + \frac{2^2+2^2}{4} + \frac{5^2}{5} = 28.512$

Box 10.7 Continued

6. $\sum^{a}\left(\dfrac{\sum^{b_i}\sum^{c_{ij}} n_{ijk}^2}{\sum^{b_i}\sum^{c_{ij}} n_{ijk}}\right)$

$$= \frac{4^2 + 5^2 + 4^2 + 4^2}{17} + \ldots + \frac{3^2 + 4^2 + 2^2 + 2^2 + 5^2}{16} = 12.571$$

7. $\sum^{a}\left(\dfrac{\sum^{b_i} n_{ij}^2}{\sum^{b_i} n_{ij}}\right) = \dfrac{9^2 + 8^2}{17} + \dfrac{13^2 + 10^2}{23} + \dfrac{7^2 + 4^2 + 5^2}{16} = 25.850$

$$n_0'' = \frac{\sum^{a}\left(\dfrac{\sum^{b_i}\sum^{c_{ij}} n_{ijk}^2}{\sum^{b_i}\sum^{c_{ij}} n_{ijk}}\right) - \left(\dfrac{\sum^{a}\sum^{b_i}\sum^{c_{ij}} n_{ijk}^2}{\sum^{a}\sum^{b_i}\sum^{c_{ij}} n_{ijk}}\right)}{df_{\text{among}}}$$

$$= \frac{\text{quantity } \mathbf{6} - (\text{quantity } \mathbf{1}/\text{quantity } \mathbf{4})}{df_{\text{among}}} = \frac{12.571 - (238/56)}{2} = 4.160$$

$$n_0' = \frac{\sum^{a}\sum^{b_i}\left(\dfrac{\sum^{c_{ij}} n_{ijk}^2}{\sum^{c_{ij}} n_{ijk}}\right) - \sum^{a}\left(\dfrac{\sum^{b_i}\sum^{c_{ij}} n_{ijk}^2}{\sum^{b_i}\sum^{c_{ij}} n_{ijk}}\right)}{df_{\text{subgr}}} = \frac{\text{quantity } \mathbf{5} - \text{quantity } \mathbf{6}}{df_{\text{subgr}}}$$

$$= \frac{(28.512 - 12.571)}{4} = 3.985$$

$$n_0 = \frac{\sum^{a}\sum^{b_i}\sum^{c_{ij}} n_{ijk} - \sum^{a}\sum^{b_i}\left(\dfrac{\sum^{c_{ij}} n_{ijk}^2}{\sum^{c_{ij}} n_{ijk}}\right)}{df_{\text{subsubgr}}} = \frac{\text{quantity } \mathbf{4} - \text{quantity } \mathbf{5}}{df_{\text{subsubgr}}}$$

$$= \frac{(56 - 28.512)}{7} = 3.927$$

$$(nc)_0' = \frac{\sum^{a}\left(\dfrac{\sum^{b_i} n_{ij}^2}{\sum^{b_i} n_{ij}}\right) - \left(\dfrac{\sum^{a}\sum^{b_i} n_{ji}^2}{\sum^{a}\sum^{b_i} n_{ij}}\right)}{df_{\text{among}}}$$

$$= \frac{\text{quantity } \mathbf{7} - (\text{quantity } \mathbf{2}/\text{quantity } \mathbf{4})}{df_{\text{among}}} = \frac{25.850 - (504/56)}{2} = 8.425$$

Box 10.7 Continued

$$(nc)_0 = \frac{\sum^a \sum^{b_i} n_{ij} - \sum^a \left(\frac{\sum^{b_i} n_{ij}^2}{\sum^{b_i} n_{ij}} \right)}{df_{\text{subgr}}} = \frac{\text{quantity } \mathbf{4} - \text{quantity } \mathbf{7}}{df_{\text{subgr}}}$$

$$= \frac{56 - 25.850}{4} = 7.538$$

$$(ncb)_0 = \frac{\sum^a n_i - \left(\frac{\sum^a n_i^2}{\sum^a n_i} \right)}{df_{\text{among}}} = \frac{[\text{quantity } \mathbf{4} - (\text{quantity } \mathbf{3}/\text{quantity } \mathbf{4})]}{df_{\text{among}}}$$

$$= \frac{56 - (1074/56)}{2} = 18.411$$

To solve for estimates of the 4 unknown variance components, we must solve the 4 simultaneous equations in which the 4 variances are the unknowns

$$MS_{\text{among}} = 198.32 = s^2 + 4.160 s^2_{C \subset B} + 8.425 s^2_{B \subset A} + 18.411 s^2_A$$

$$MS_{\text{subgr}} = 160.11 = s^2 + 3.985 s^2_{C \subset B} + 7.538 s^2_{B \subset A}$$

$$MS_{\text{subsubgr}} = 35.63 = s^2 + 3.927 s^2_{C \subset B}$$

$$MS_{\text{within}} = 20.81 = s^2$$

Therefore

$$s^2 = MS_{\text{within}} = 20.81$$

$$s^2_{C \subset B} = \frac{MS_{\text{subsubgr}} - MS_{\text{within}}}{3.927} = \frac{35.63 - 20.81}{3.927} = 3.774$$

$$s^2_{B \subset A} = \frac{MS_{\text{subgr}} - MS_{\text{within}} - 3.985 s^2_{C \subset B}}{7.538}$$

$$= \frac{160.11 - 20.81 - [(3.985)(3.774)]}{7.538} = 16.485$$

Variance components may also be expressed as percentages:

Since $s^2_{B \subset A} + s^2_{C \subset B} + s^2 = 41.069$

$s^2_{B \subset A}$ (core samples within depths) $= 100 \times 16.485/41.069 = 40.1\%$

$s^2_{C \subset B}$ (preparations within core samples) $= 100 \times 3.774/41.069 = 9.2\%$

s^2(slides within preparations) $= 100 \times 20.81/41.069 = 50.7\%$

BOX 10.7 CONTINUED

In this example there is no s_A^2 because the highest-level classification is due to fixed treatment effects. However, we show the computation of s_A^2 here to provide a complete pattern for all computations in a Model II three-level anova.

$$s_A^2 = \frac{MS_{\text{among}} - MS_{\text{within}} - 4.160s_{C\subset B}^2 - 8.425s_{B\subset A}^2}{18.411}$$

$$= \frac{198.32 - 20.81 - [(4.160)(3.774)] - [(8.425)(16.485)]}{18.411} = 1.245$$

Tests of significance

MS_{subsubgr} may be tested directly over MS_{within}:

$$F_s = \frac{35.63}{20.81} = 1.712\ ns$$

Following the rules in Box 10.3 we do not pool.

Before we can test MS_{subgr} we must check whether the Satterthwaite approximation can be used, as was done in Box 10.6.

First we verify that $df_{\text{subsubgr}} = 7 < 2\ df_{\text{within}} = 2(42) = 84$. Then we compute the ratio.

$$R = \left(\frac{n_0'}{n_0' - n_0}\right)\frac{MS_{\text{subsubgr}}}{MS_{\text{within}}} = \left(\frac{3.985}{3.985 - 3.927}\right)\frac{35.63}{20.81} = 117.657$$

This value is to be compared to the criterion

$$C = F_{.025[df_{\text{within}},\ df_{\text{subsubgr}}]} \times F_{.5[df_{\text{subsubgr}},\ df_{\text{within}}]}$$

$$= F_{.025[42,7]} \times F_{.5[7,42]}$$

$$= 4.31(0.922) = 3.974$$

Since $R > C$ it is safe to use the Satterthwaite approximation given below.

If these conditions were not met, then one would have to carry out the approximate test $F_s = MS_{\text{subgr}}/MS_{\text{subsubgr}}$ (which in this case happens to give very similar results).

For the Satterthwaite approximation compute the weights:

$$w_2 = \frac{n_0'}{n_0} = \frac{3.985}{3.927} = 1.0148$$

$$w_1 = 1 - w_2 = -0.0148$$

The new synthesized mean square is

$$MS'_{\text{subsubgr}} = w_1 MS_{\text{within}} + w_2 MS_{\text{subsubgr}}$$

$$= -0.0148(20.81) + 1.0148(35.63)$$

$$= 35.8493$$

BOX 10.7 CONTINUED

Its degrees of freedom are computed as

$$df'_{\text{subsubgr}} = \frac{(MS'_{\text{subsubgr}})^2}{\dfrac{(w_1 MS_{\text{within}})^2}{df_{\text{within}}} + \dfrac{(w_2 MS_{\text{subsubgr}})^2}{df_{\text{subsubgr}}}}$$

$$= \frac{(35.8493)^2}{\dfrac{[-0.0148(20.81)]^2}{42} + \dfrac{[1.0148(35.63)]^2}{7}}$$

$$= \frac{1285.1723}{186.7668} = 6.88 \approx 7$$

$$F'_s = \frac{MS_{\text{subgr}}}{MS'_{\text{subsubgr}}} = \frac{160.11}{35.8493} = 4.466$$

The critical value is $F_{.05[4,7]} = 4.12$, so the subgroup *MS* (among core samples within a depth) is significant.

Similar procedures are used to test MS_{among}. First check that $df_{\text{among}} = 2 < 2\ df_{\text{subgr}} = 2(4) = 8$. Then compute the now much more complex ratio

$$R = \frac{n_0 (nc)'_0 MS_{\text{subgr}}}{[n''_0 (nc)_0 + (nc)'_0 n_0 - n_0 (nc)_0 - (nc)'_0 n'_0] MS_{\text{within}} + [(nc)'_0 n'_0 - n''_0 (nc)_0] MS_{\text{subsubgr}}}$$

$$= \frac{3.927(8.425)160.11}{[4.160(7.538) + 8.425(3.927) - 3.927(7.538) - 8.425(3.985)]20.81 + [8.425(3.985) - 4.160(7.538)]35.63}$$

$$= \frac{5297.2353}{1.2677(20.81) + 2.2155(35.63)} = 50.296$$

The criterion C is now

$$C = F_{.025[df_{(\text{within + subsubgr})},\ df_{\text{subgr}}]} \times F_{.5[df_{\text{subgr}},\ df_{(\text{within + subsubgr})}]}$$

$$= F_{.025[49,4]} \times F_{.5[4,49]}$$

$$= 8.38(0.852) = 7.140$$

Since $R > C$ we may use the Satterthwaite approximation again. Compute the weights

$$w_5 = \frac{(nc)'_0}{(nc)_0} = \frac{8.425}{7.538} = 1.1177$$

$$w_4 = \frac{n''_0}{n_0} - \frac{w_5 n'_0}{n_0} = \frac{4.160}{3.927} - \frac{(1.1177)3.985}{3.927} = -0.0749$$

$$w_3 = 1 - w_5 - w_4 = 1 - 1.1177 - (-0.0749) = -0.0428$$

$$MS'_{\text{subgr}} = w_3 MS_{\text{within}} + w_4 MS_{\text{subsubgr}} + w_5 MS_{\text{subgr}}$$

$$= -0.0428(20.81) - 0.0749(35.63) + 1.1177(160.11)$$

$$= 175.3956$$

BOX 10.7 CONTINUED

The F' value is therefore

$$F'_s = \frac{MS_{\text{among}}}{MS'_{\text{subgr}}} = \frac{198.32}{175.3956} = 1.131$$

It is unlikely that such a small F-value is significant, but we compute the degrees of freedom just to check.

$$df'_{\text{subgr}} = \frac{(MS'_{\text{subgr}})^2}{\frac{(w_3 MS_{\text{within}})^2}{df_{\text{within}}} + \frac{(w_4 MS_{\text{subsubgr}})^2}{df_{\text{subsubgr}}} + \frac{(w_5 MS_{\text{subgr}})^2}{df_{\text{subgr}}}}$$

$$= \frac{(175.3956)^2}{\frac{[-0.0428(20.81)]^2}{42} + \frac{[-0.0749(35.63)]^2}{7} + \frac{[1.1177(160.11)]^2}{4}}$$

$$= \frac{30763.6165}{8007.2546} = 3.842 \approx 4$$

Therefore the critical value of the F'_s given above is $F_{.05[2,4]} = 6.94$, and F'_s is clearly nonsignificant.

diameter of the pollen grains shows added variance among core samples but not among preparations and does not differ for the depths recorded in this study. Thus there must be some local heterogeneity in this character, and different core samples appear to yield different pollen samples. This heterogeneity is reflected in the percentage of variation attributable to the several levels: 40.1% among core samples, 9.2% among preparations within core samples, and 50.7% among slides within preparations. The preparations seem to represent core samples with high replicability, but the techniques for making slides and measuring the pollen grains on them merit further examination and refinement in view of the high percentage of variation of that level. Similarly, these results show substantial variation among core samples. Each depth should be adequately sampled by cores.

The general method of analysis of nested anovas is the same for any number of levels. To enable you to generate coefficients for variance components for cases with unequal samples sizes other than those discussed in this section, general rules are given in Raktoe et al. (1981).

Nested anovas with more than three levels are usually carried out by computer. Program BIOM-pc performs the computations described in Sections 10.2

and 10.3. It computes nested anovas up to 10 levels, and in the case of unequal sample sizes, with and without the Satterthwaite approximation.

10.4 The Optimal Allocation of Resources

In Box 9.14 we learned a method for estimating the sample size necessary to detect a given difference between means. In the designs implied by nested analyses of variance this problem is more complex. Before we can determine how many replications to use, we must decide how these replications should be structured. Take the experiment of Box 10.5 as an example. We wish to test for differences among treatments, and each treatment is represented by one or more rats. But how should these rats be represented? Recall that three preparations were made from each rat liver and two readings taken on each preparation. Would it have been wiser to take four readings per preparation but only one preparation per rat, or would it have been better to take five preparations from one rat liver but only one reading per preparation? Only after making such decisions can we solve for required sample size (number of rats per treatment) in the manner of Box 9.14.

We approach the problem of optimal allocation of resources by determining the expected variance of means of various designs. In a Model I single-classification analysis of variance (as in Box 9.4) the anova is structured as follows:

Treatments (groups)	$\sigma^2 + n\dfrac{\sum \alpha^2}{a-1}$
Replicates (within groups)	σ^2

The estimated error variance is s^2, and the expected variance of a group mean is $s_{\bar{Y}}^2 = s^2/n$, which is simply the familiar formula for the variance of the mean.

In a mixed model two-level nested anova (as in Box 10.4) the expected mean squares are structured as follows:

Treatments (groups)	$\sigma^2 + n\sigma_{B \subset A}^2 + nb\dfrac{\sum \alpha^2}{a-1}$
Subgroups within groups	$\sigma^2 + n\sigma_{B \subset A}^2$
Replicates within subgroups	σ^2

and the variance of group means can be estimated as MS_{subgr}/nb, which yields

$$s_{\bar{Y}}^2 = \frac{s^2}{nb} + \frac{s_{B \subset A}^2}{b} \tag{10.3}$$

where s^2 and $s_{B \subset A}^2$ are estimates of σ^2 and $\sigma_{B \subset A}^2$, respectively. Note that the error variance is now divided by nb, the total number of variates on which a subgroup mean is based, while the variance component of subgroups within groups is divided only by b, the number of subgroups constituting a mean.

In a mixed model three-level nested anova (as in box 10.5), expected mean squares have the following structure:

Treatments (groups)	$\sigma^2 + n\sigma^2_{C\subset B} + nc\sigma^2_{B\subset A} + ncb\dfrac{\sum \alpha^2}{a-1}$
Subgroups within groups	$\sigma^2 + n\sigma^2_{C\subset B} + nc\sigma^2_{B\subset A}$
Subsubgroups within subgroups	$\sigma^2 + n\sigma^2_{C\subset B}$
Replicates within subsubgroups	σ^2

with s^2, $s^2_{C\subset B}$, and $s^2_{B\subset A}$ estimating variance components for increasing hierarchic levels. The expected variance of a group mean based on b subgroups, c subsubgroups, and n replicates within subsubgroups is MS_{subgr}/ncb, which yields

$$s^2_{\bar{Y}} = \frac{s^2}{ncb} + \frac{s^2_{C\subset B}}{cb} + \frac{s^2_{B\subset A}}{b} \tag{10.4}$$

The variances of group means are expectations of the variances among means sampled from identical populations. They do not therefore include added treatment or random effects among groups. Thus we always divide the MS just below MS_{among} by the number of replicates on which a group mean is based, yielding $s^2_{\bar{Y}}$, as shown in Expressions (10.3) and (10.4).

The following example will make these formulas more meaningful to you. In a hypothetical pharmacological experiment testing the effects of various drugs, each drug was tested on several rats. Several replicated readings of the chemical composition of each rat's blood were made. A mixed model two-level nested anova (among treatments, among rats within treatments, readings within rats) resulted in the following variance component estimates:

Among rats	$s^2_{B\subset A} = 25$
Readings within rats	$s^2 = 100$

We will now contrast two of many possible designs for further research with these drugs. In design (1) we propose to employ $b = 10$ rats per treatment, and to make $n = 2$ readings on each rat. The expected variance of treatment means (if no added treatment effects are present) can be obtained from Expression (10.3):

$$s^2_{\bar{Y}}(1) = \frac{100}{2 \times 10} + \frac{25}{10} = 7.50$$

Design (2) might consist of 2 rats per treatment, but with 9 readings on each rat. Its expected variance would be

$$s^2_{\bar{Y}}(2) = \frac{100}{9 \times 2} + \frac{25}{2} = 18.06$$

Thus design (2) would have a considerably greater variance than design (1) and, all other considerations being equal, would be less desirable because it would be far less sensitive than design (1) to differences among treatment means.

To compare two designs we compute the **relative efficiency** (*RE*) of one design with respect to the other. *RE* is a ratio, usually expressed as a percentage, of the variances resulting from the two designs. The design whose variance is in the denominator is the one whose relative efficiency is being evaluated. Thus we measure the relative efficiency of design (2) with respect to design (1) by

$$RE = \frac{s^2_{\bar{Y}}(1)}{s^2_{\bar{Y}}(2)} \times 100 \tag{10.5}$$

In the previous example the relative efficiency of design (1) with respect to design (2) is

$$RE = \frac{100 s^2_{\bar{Y}}(2)}{s^2_{\bar{Y}}(1)} = \frac{100 \times 18.06}{7.50} = 240.8\%$$

The increase in efficiency on changing from design (2) to design (1) is

$$240.8\% - 100\% = 140.8\%$$

The meaning of relative efficiency can be explained as follows. If we take the design whose relative efficiency is being measured and divide its replication by *RE* (expressed as a ratio), the resulting design will be as sensitive as the original design with which it is being compared. Thus in the pharmacological experiment if we take $10/2.408 = 4.153$ rats per treatment (with 2 readings per rat), we will get a denominator *MS* for testing the treatment *MS* approximately equal to that of design (2), 2 rats with 9 readings per rat. The equality is only approximate because the denominator mean squares are based on different degrees of freedom that affect the sensitivity of the design. Also, we obviously cannot apportion 4.153 rats per treatment, but would need to settle for 4 rats.

The relative efficiency of one design with respect to another is not very meaningful, however, unless the relative costs of obtaining the two designs are taken into consideration. Clearly, if one design is twice as efficient as another (that is, has half the other's variance) but at the same time is ten times as expensive to achieve, we might not choose it. To introduce the idea of cost we write a so-called **cost function,** which tells us the cost per treatment class. For a two-level mixed model,

$$C = bc_{B \subset A} + nbc_{R \subset B} \tag{10.6}$$

where C is the cost per treatment class, $c_{B \subset A}$ is the cost of one subgroup unit, $c_{R \subset B}$ is the cost of one replicate reading within a subgroup, and b and n have their familiar meanings.

Let us assume for the pharmacological experiment that the cost of one subgroup unit (the cost of buying and feeding one rat) is $c_{B \subset A} = \$15$. Similarly, suppose the cost of one replicate within the subgroup (the price of one blood test on a rat) is $c_{R \subset B} = \$3$; then the total cost of design (1) by Expression (10.6) will be

$$C(1) = (10 \times \$15) + (10 \times 2 \times \$3) = \$210$$

and of design (2)

$$C(2) = (2 \times \$15) + (2 \times 9 \times \$3) = \$84$$

Although more efficient than design (2), design (1) costs considerably more: It is $210/84 = 2.50$ times more expensive. Thus the cost efficiency of design (1) with respect to design (2) is $2.408/2.50 = 0.9632$—i.e., the first design is slightly less efficient than the second.

We now ask two questions: (a) For a given amount of money per treatment class what is the optimal (most efficient) design? (b) What is the least expensive design for obtaining a given variance of a treatment mean? To answer these questions we need one more formula:

$$n = \sqrt{\frac{c_{B \subset A} s^2}{c_{R \subset B} s^2_{B \subset A}}} \tag{10.7}$$

which yields the optimal number of replicates n per subgroup unit—that is, a number that will result in minimal cost and minimal variance.

Expression (10.7) shows that the optimal number of replicates per subgroup depends on the ratio of costs of the two levels, as well as on the ratio of their variance components. If the costs (= effort) of obtaining a replication at either hierarchic level are the same, Expression (10.7) becomes the ratio of the standard deviations corresponding to the variance components. Expression (10.7) may yield $n < 1$. In such a case we set $n = 1$, since obviously we must have at least one reading per subgroup. A rat will be of no use to us in the pharmacological experiment unless we take at least one blood sample from it! Cases in which the optimal number of replicates per subgroup is <1 are those in which variance among subgroups is great and the error variance is relatively small. In our problem

$$n = \sqrt{\frac{15 \times 100}{3 \times 25}} = \sqrt{\frac{1500}{75}} = \sqrt{20} = 4.472$$

Therefore the optimal number of readings per rat is 4, ignoring the decimal fraction. Thus we answer questions (a) and (b) as follows:

(a) Assume that we wish to spend the same amount of money as in design (1), that is, \$210. What is the optimal design at this cost? We substitute the known values in Expression (10.6) and solve for b:

$$C = bc_{B \subset A} + nbc_{R \subset B}$$

$$210 = (b \times 15) + (4 \times b \times 3) = 27b$$

$$b = 210/27 = 7.778 \approx 8$$

Thus 8 rats with 4 readings per rat should yield the most efficient way of spending approximately \$210. Note that we said "approximately," for obviously we cannot use a fractional number of rats or blood samples, which would be necessary to yield an exact cost of \$210. To check our results we calculate the new cost of the design, which is

$$C = (8 \times \$15) + (4 \times 8 \times \$3) = \$216$$

close to the figure of \$210 we aimed at. We can show that the new design is more efficient than the old one. We calculate the new variance of a treatment mean based on 8 rats and 4 blood tests per rat by substituting in Expression (10.3):

$$s_{\bar{Y}}^2 = \frac{100}{8 \times 4} + \frac{25}{8} = 6.25$$

The new variance is smaller than that of design (1) at nearly the same cost. The relative efficiency,

$$RE = \frac{7.50 \times 100}{6.25} = 120\%$$

has increased by 20%, but the increase in cost is only 2.9%. Clearly the new design is preferable, even though design (1) was relatively not a bad design to start with.

The process may also be shown graphically. The solid line in Figure 10.1 is a graph of the cost equation $\$210 = \$15b + \$3nb$ and represents all combinations of n and b that would yield an experiment costing \$210. The dotted lines represent the equation $s_{\bar{Y}}^2 = 100/nb + 25/b$ for various values of $s_{\bar{Y}}^2$. All intersections between the line for cost = \$210 and equations for the various $s_{\bar{Y}}^2$'s give the $s_{\bar{Y}}^2$, n, and b for all possible experiments costing \$210. What we want in this case is the one with the smallest $s_{\bar{Y}}^2$. One can see that no line representing an experiment with $s_{\bar{Y}}^2 < 6.41$ can intersect the cost equation, so $s_{\bar{Y}}^2 = 6.41$ is the best we can expect. These two lines intersect at $n = 4$ and $b = 7.8$ (≈ 8), which is the best combination of n and b for our requirements.

(b) If we wish to achieve a given variance, as, for example, the value of 18.06 resulting from design (2), what is the least expensive way to do so? We substitute the known and given terms in Expression (10.3) and solve for b:

$$s_{\bar{Y}}^2 = \frac{s^2}{nb} + \frac{s_{B \subset A}^2}{b}$$

$$18.06 = \frac{100}{4b} + \frac{25}{b}$$

$$= \frac{100 + (4 \times 25)}{4b}$$

$$b = \frac{100 + 100}{4 \times 18.06} = 2.8 \approx 3$$

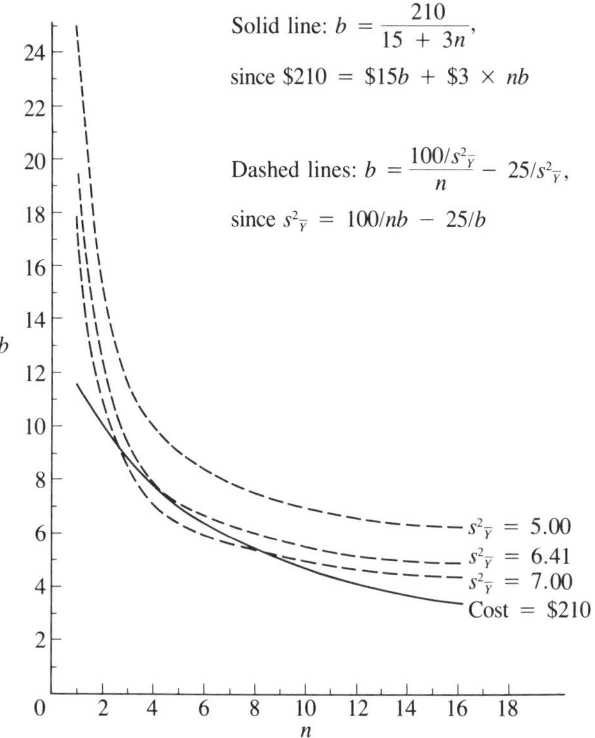

FIGURE 10.1 *Equations for cost and estimated variance of a group mean in a two-level anova.*

Thus 3 rats with 4 readings per rat will give the most economic design for a treatment mean square approximating 18.06. When we check this result we find

$$s^2_{\bar{Y}} = \frac{100}{(3 \times 4)} + \frac{25}{3} = 16.67$$

which is a value less than the aimed-at variance (because we used whole numbers of replicates), but the new cost would be

$$C = (3 \times \$15) + (4 \times 3 \times \$3) = \$81$$

Thus a 6% decrease in cost would be accompanied by an 8% increase in relative efficiency. This result may also be determined graphically by finding the smallest cost equation that intersects the line of the equation for the desired $s^2_{\bar{Y}}$.

Box 10.8 summarizes the computations of optimal allocation of resources for two and three levels of sampling. The formulas in the latter case are similar to those in the former and should be self-explanatory.

Box 10.8 OPTIMAL ALLOCATION OF RESOURCES IN TWO- AND THREE-LEVEL SAMPLING: OUTLINE OF PROCEDURE.

Two levels

In a study of variation in the aphid *Pemphigus populitransversus,* Sokal (1962) measured the forewing length of each of 2 alates for each of 15 galls from each of 23 localities in eastern North America: $n = 2$; $b = 15$. The estimates of the variance components were:

Galls within localities	$s^2_{B \subset A} = 10.507$
Within galls	$s^2 = 6.813$

In this instance the costs of obtaining another gall or another alate in the same gall are roughly equal. Therefore $c_{R \subset B} = c_{B \subset A}$.

Optimum sample size n per subgroup is computed from Expression (10.7):

$$n = \sqrt{\frac{c_{B \subset A} s^2}{c_{R \subset B} s^2_{B \subset A}}} = \sqrt{\frac{6.813}{10.507}} = \sqrt{0.6484} = 0.805$$

Although $n < 1$ we must use at least one alate per gall in further studies. The number of galls to be measured depends on the desired variance of locality means:

$$s^2_{\bar{Y}} = \frac{6.813}{1 \times b} + \frac{10.507}{b}$$

Thus the variance of a locality mean based on 10 galls would be

$$\frac{6.813}{1 \times 10} + \frac{10.507}{10} = 0.6813 + 1.0507 = 1.7320$$

Three levels

In Box 10.5 we studied the effect of 3 treatments on glycogen content in rat livers. The experiment was structured into b rats within treatments ($s^2_{B \subset A} = 36.1$), c preparations within rats ($s^2_{C \subset B} = 14.2$), and n readings within preparations ($s^2 = 21.1$). Let us assume that a rat costs $c_{B \subset A} = \$5$, a preparation $c_{C \subset B} = \$20$, and a reading $c_{R \subset C} = \$2$. Extending Expression (10.6) for two-level samples, the cost C of a treatment class for a given design is

$$C = bc_{B \subset A} + cbc_{C \subset B} + ncbc_{R \subset C}$$

where b, c, and n are the number of replicates at the B, C, and lowest levels, and $c_{B \subset A}$, $c_{C \subset B}$, and $c_{R \subset C}$ are the costs per replicate at these levels, respectively.

Expression (10.7) yields

$$c = \sqrt{\frac{c_{B \subset A} s^2_{C \subset B}}{c_{C \subset B} s^2_{B \subset A}}} \qquad n = \sqrt{\frac{c_{C \subset B} s^2}{c_{R \subset C} s^2_{C \subset B}}}$$

as the optimal number of replicates c and n in the subgroups and subsubgroups, respectively.

Box 10.8 Continued

Solving for our case, we get

$$c = \sqrt{\frac{5 \times 14.2}{20 \times 36.1}} = \sqrt{\frac{71.0}{722}} = \sqrt{0.0983} = 0.3136$$

$$n = \sqrt{\frac{20 \times 21.2}{2 \times 14.2}} = \sqrt{\frac{424.0}{28.4}} = \sqrt{14.930} = 3.864$$

Thus employing 4 readings per preparation and 1 preparation per rat will give us the most efficient design.

If we can spend \$200 per treatment, we substitute this amount in the cost formula above and solve for b:

$$200 = (b \times 5) + (1 \times b \times 20) + (4 \times 1 \times b \times 2) = 5b + 20b + 8b$$

$$b = \frac{200}{33} = 6.06$$

We should use 6 rats per treatment, make 1 preparation from each rat, and carry out 4 readings on each preparation.

If we wish to obtain the most efficient design for a given variance, say 22, we proceed as follows. Substituting in Expression (10.4) and solving for b we get

$$s_{\bar{Y}}^2 = \frac{s^2}{ncb} + \frac{s_{C \subset B}^2}{cb} + \frac{s_{B \subset A}^2}{b}$$

$$22 = \frac{21.2}{4 \times 1 \times b} + \frac{14.2}{1 \times b} + \frac{36.1}{b} = \frac{21.2 + 4(14.2 + 36.1)}{4b}$$

$$b = \frac{222.4}{4 \times 22} = \frac{222.4}{88} = 2.53$$

We decide to take 3 rats per treatment.

Thus our expected variance of a treatment mean based on 3 rats, 1 preparation, and 4 readings is

$$s_{\bar{Y}}^2 = \frac{21.2}{4 \times 1 \times 3} + \frac{14.2}{1 \times 3} + \frac{36.1}{3} = 1.767 + 4.733 + 12.03 = 18.533$$

This value is not exactly 22, since $b = 3$ is fairly far from the exact solution to the equation ($b = 2.53$). Cost for this design would be

$$C = (3 \times 5) + (1 \times 3 \times 20) + (4 \times 1 \times 3 \times 2) = \$99$$

Let us compare this with the original design in Box 10.5, in which we employed 2 rats per treatment, 3 preparations per rat, and 2 readings per preparation. By Expression (10.4) this yields the following expected MS of means

$$\frac{21.2}{2 \times 3 \times 2} + \frac{14.2}{3 \times 2} + \frac{36.1}{2} = 1.767 + 2.367 + 18.050 = 22.183$$

When we multiply by $ncb = 12$ we obtain $MS_{\text{subgr}} = 266.20$, which differs from the actually calculated value of 265.89 because of rounding errors. The relative efficiency

BOX 10.8 CONTINUED

of the optimal design computed above is $RE = 22.183/18.533 = 119.7\%$. The cost of the design in Box 10.5 is thus

$$C = (2 \times 5) + (3 \times 2 \times 20) + (2 \times 3 \times 2 \times 2) = \$154$$

55.6% higher than the optimal design. Clearly 1 preparation per rat and 4 readings per preparation are a more efficient allocation of resources for this study.

We should point out that whenever the cost of an experiment is of little consideration, but the design of the most efficient experiment (lowest variance) for a given total number of observations is important, the most efficient procedure will always be to take a single observation for each of the subgroups; that is, set $n = 1$ in a two-level anova and both n and $c = 1$ in a three-level anova. This design is also often used when the cost of making measurements (such as concentrations of various metals in environmental monitoring programs) is very expensive. But, of course, such studies would yield no information on the magnitude of variation at lower levels, nor information on the statistical distribution of the variable.

EXERCISES 10

10.1 J. A. Weir (unpublished results) took blood pH readings on mice of two strains that had been selected for high and low blood pH and obtained the following data on male litter mates. Only litters with at least 4 males were considered, and 4 males were selected at random whenever more than 4 males were present in a litter. Data are presented on seven litters for each strain. Analyze the data, test whether the pHH (high) strain has a higher average pH value than the pHL (low) strain (a one-sided test), and compute variance components and percent variation among males within litters and litters within strains. *Answer:* Percent variation of litters within strains = 23.40%, $F_s = 2.222$ with $df = 12, 42$.

Strain	Litter code number	pH readings			
pHH	387	7.43	7.38	7.49	7.49
	388	7.39	7.46	7.50	7.55
	389	7.53	7.50	7.63	7.47
	401	7.39	7.39	7.44	7.55
	402	7.48	7.43	7.47	7.44
	404	7.43	7.55	7.44	7.50
	405	7.49	7.49	7.51	7.54

(*Table continued on page 318.*)

Strain	Litter code number	pH readings			
pHL	392	7.40	7.46	7.43	7.42
	408	7.35	7.40	7.46	7.38
	413	7.51	7.39	7.42	7.43
	414	7.46	7.53	7.49	7.45
	415	7.48	7.53	7.52	7.43
	434	7.43	7.40	7.48	7.47
	446	7.53	7.47	7.50	7.53

10.2 Hasel (1938) made a study of sampling practices in timber surveys. The variate studied is board feet of timber in a ponderosa pine forest. He studied 9 sections, each subdivided into 4 quarter sections [each of which can be subdivided into 4 forty-acre tracts, or 16 ten-acre tracts, or 32 five-acre tracts, or 64 basic plots (of two-and-a-half acres)]. If sampling and distribution of timber were entirely at random, no added variance should exist beyond the basic plots.

Interpret the analysis of variance given below and isolate variance components. If it is standard practice among lumbermen to sample 16 basic plots of a forest, how would you (with the limited information at your disposal, and assuming that cost is not important) recommend that the plots be distributed in order to make efficient comparisons in a new experiment to compare differences between forests? Sampling all plots within the same forty-acre tract or sampling only one basic plot per section are the two extremes. Compute their efficiencies relative to each other. To keep the notation uniform, identify variance components as follows: s^2 for basic plots, s_5^2 for fives, s_{10}^2 for tens, s_{40}^2 for forties, s_Q^2 for quarter sections, and s_S^2 for sections.

Source of variation	*df*	*SS*
Between sections	8	109,693.35
Quarter sections within sections	27	57,937.40
Forties within quarter sections	108	66,569.53
Tens within forties	432	161,109.77
Fives within tens	576	121,940.82
Basic plots within fives	1152	133,337.74
Total	2303	650,588.61

After trying several intermediate solutions between the extreme sampling procedures, you will have convinced yourself that sampling one basic plot per section is the most efficient (in terms of obtaining the lowest variance). This method, however, which involves sampling from 16 different sections of land, would require considerable work and travel. Thus, to assume that cost is unimportant is not realistic in this problem, since work and travel must be equated to cost. Design a sampling procedure that would involve less travel but would still result in a small variance of the mean (<50.0).

10.3 The following are unpublished results from a study by R. R. Sokal on geographic variation in the aphid *Pemphigus populitransversus.* Mean length for antennal segment IV of stem mothers was computed for each of 75 localities in eastern North America. The means were grouped and coded 1 through 8. The data were subdivided according to drainage system along which the localities occur. A hierarchic classification was set up as follows: The drainages were divided into three coasts: East Coast, Gulf Coast, and Great Lakes. Each coast was subdivided into a number of areas, one of which, the Mississippi area, was further subdivided into ten river systems. Analyze the data to determine whether differences in the length of antennal segment IV occur among the coasts. At what hierarchic level is there the greatest amount of variation? Interpret your findings. This problem might appear to be a Model I anova throughout. For our purposes, however, the areas and river systems are considered random samples from the available equivalent sampling units, and the emphasis is on the relative amounts of variation at these hierarchic levels. The data are presented in summary form for ease of computation.

Coast	Area	River system	n_{ijk}	$\sum^{n_{ijk}} Y_{ijkl}$
East	New England		3	13
	Hudson		2	12
	Chesapeake Bay		2	11
	North Carolina		1	4
	Florida		1	1
Gulf	East		4	14
	Central		4	12
	Mississippi	Red	2	6
		Arkansas	7	29
		Lower Mississippi	10	33
		Illinois	6	32
		Missouri	3	15
		Upper Mississippi	6	34
		Lower Ohio	5	13
		Tennessee-Cumberland	3	9
		Wabash	2	9
		Middle Ohio	7	27
Great Lakes			7	39

$$SS_{\text{total}} = \sum^{a}\sum^{b_i}\sum^{c_{ij}}\sum^{n_{ijk}} (Y_{ijkl} - \bar{\bar{Y}})^2 = 226.74667$$

Answer: $MS_{\text{Areas} \subset \text{Coasts}} = 4.0133$; *EMS* for this *MS* is $\sigma^2 + 2.533\sigma^2_{R \subset A} + 3.544\sigma^2_{A \subset C}$; $s^2_{A \subset C}$ is negative.

10.4 How could you improve the design of Weir's experiment (Exercise 10.1) using the same number of mice he did? First assume that costs are not important, then rework the problem, assuming that the extra cost of obtaining an additional litter of mice is about three times the cost of measuring an additional mouse. Were your improvements worthwhile? That is, if the difference between the means of the two strains were the same in the new experiment as it was in the previous one, would you now expect to be able to reject the null hypothesis?

10.5 Hanna (1953) studied hair pigment concentration (measured with a spectrophotometer and expressed as optical densities) in 39 pairs of monozygous and 40 pairs of dizygous twins. Two hair samples were taken from each person and hydrolyzed separately. Three readings were taken for each sample. Throughout the study, samples were randomized to avoid bias in reading optical densities.

	Monozygous twins		Dizygous twins	
Source of variation	*df*	*MS*	*df*	*MS*
Among pairs	38	2676.1734	39	2509.9174
Between twins within pairs	39	44.2005	40	307.0163
Between samples within twins	78	3.2133	80	2.5451
Among readings within samples (error)	312	0.0662	320	0.0334

Complete the analysis of variance and estimate the added variance components. Compare and interpret the differences in the estimated variance components for monozygous and dizygous twins. *Answer:* $s^2_{T \subset P}$ equals 6.8312 for the monozygous twins and 50.7452 for the dizygous twins.

11 Two-Way Analysis of Variance

This chapter covers designs in which the effects of two factors are considered simultaneously. Single-factor analyses were discussed in Chapter 9. Although Chapter 10 tested subclasses representing a separate source of variation, these were hierarchic, or nested within the major classification. In this chapter the two sources of variation are of equal rank. Thus we can group the variates by two separate kinds of treatments—for example, different doses of a drug and different temperatures at which these drugs are tested. Section 11.1 gives further details of this design. The computation of such an anova for replicated subclasses (more than one variate per subclass or factor combination) is shown in Section 11.2, which also contains a discussion of the meaning of interaction as used in statistics. Significance testing in a two-way anova is the subject of Section 11.3. Section 11.4 discusses two-way anovas without replication (only a single variate per subclass). The well-known method of paired comparisons is treated in Section 11.5 as a special case of a two-way anova without replication. Section 11.6 deals with complications arising from unequal subclass sizes in a two-way anova, and Section 11.7 presents a method for analyzing the design of Section 11.4 in which values for some subclasses are missing.

11.1 Two-Way Anova: Design

From the single-classification anova of Chapter 9 we progress to the two-way anova of this chapter by a single logical step. Individual items may be grouped into classes representing the different possible combinations of two treatments or factors. The experiment testing five sugar treatments on pea sections (see Box 9.4), for example, could be carried out at two different pH levels. In this case we would want to know not only whether sucrose induced a different growth than the control medium did, but also whether pH treatment 1 affected the pea sections differently from treatment 2. Obviously each combination of factors should be represented by a randomly chosen sample of pea sections. Thus for five sugars and two pH levels we need at least $5 \times 2 = 10$ samples. Such a design is a **two-way analysis of variance** of the effects of sugars and of pH treatments.

This method of anova assumes that a given pH and a given sugar each contribute a certain amount to the growth of a pea section and that these two contributions add their effects without influencing each other. In the next section we will see how departures from this assumption are measured. Consideration of the expression for decomposing variates in a two-way anova is also deferred to Section 11.2.

The two factors in the present design may represent either Model I or Model II effects or one of each, in which case we talk of a **mixed model.**

Beginners are often confused between two-level nested anovas and two-way anovas. For example, if we subject samples of the pea sections not only to different sugar treatments but also to two pH levels, in what way does this differ from a nested design in which each sugar solution is prepared twice, so there are two batches of sugar made up for each of the treatments? We represent the design of the two-way anova as follows. The asterisks represent the observed variates.

Sugar treatments

	1	2	3	4	5
pH level 1	* * * * *	* * * * *	* * * * *	* * * * *	* * * * *
pH level 2	* * * * *	* * * * *	* * * * *	* * * * *	* * * * *

The nested anova, on the other hand, is represented by the following arrangement. Again the asterisks represent the observed variates.

Sugar treatments

	1		2		3		4		5	
Sugar batches	1	2	1	2	1	2	1	2	1	2
	* * * * *	* * * * *	* * * * *	* * * * *	* * * * *	* * * * *	* * * * *	* * * * *	* * * * *	* * * * *

Why can't we rearrange these data into a two-way table with batches at right angles to sugar treatments as in the first table? And why not nest the pH levels inside sugar treatments, as in the second table? The reason is that the first (two-way) arrangement implies that the two pH classes are *common* to the entire study—that is, that pH level 1 is logically the same for all the sugar treatments and so is pH level 2, although their effects may not be the same for all sugars for the variable tested. Arranging the pH levels nested within each sugar treatment would imply that the two pH levels per treatment were random samples from all possible such levels and that pH level 1 in treatment 1 is not the same as pH level 1 in treatment 2, which is, of course, nonsense for these data. Conversely, if we tried to arrange the data from the nested analysis as a two-way anova, we would imply that batches 1 and 2 had the same meaning for all the strains in the experiment. This is not so. Batch 1 for treatment 1 has no closer relation to batch 1 in treatment 2 than it does to batch 2 in that treatment. Batches 1 and 2 are simply arbitrary designations for the two randomly prepared sugar solutions that represent each treatment. By contrast, if all batches labeled 1 were prepared on one day, while all batches labeled 2 were made on the following day, the "1" and "2" would represent common information for the study that should properly be arranged as a two-way anova.

The critical question to be asked is always, *Does the arrangement of the data into a two-way table correctly imply a correspondence across the classes?* If there is a correspondence across classes, the two-way design is appropriate. If there is no correspondence across classes and we recognize that the factor represents only random subdivisions of the classes of another factor (as the batches within treatments), we have a nested anova.

This example also illustrates the difficulty in carrying out the laboratory procedure in a manner compatible with the statistical design. In order to have two levels of pH for each sugar treatment, one must make up two batches. Thus the random variation due to batches within a sugar treatment is confounded with the effect of pH. To be a proper two-way design, each pH–sugar combination would have to be prepared independently (and in a random order to guard against subtle biases).

We will now illustrate the computation of a two-way anova. You will obtain closer insight into the structure of this design as we explain the computations.

11.2 Two-Way Anova With Equal Replication: Computation

We will explain the computation of a two-way anova using the simplest possible example—two factors, each divided into two classes. In a study of the inactivation of the effects of vitamin A by rancid fat it became important to study differences in food consumption when rancid lard was substituted for fresh lard in the diet of rats. Box 11.1 shows the results of an experiment conducted on 12 rats, 6 males and 6 females, in which 3 randomly selected individuals of each sex were fed fresh lard, and the other 3 rancid lard. The following analysis of these

Box 11.1 TWO-WAY ANOVA WITH REPLICATION.

Differences in food consumption when rancid lard was substituted for fresh lard in the diet of rats. Food eaten (in grams) during 73 days by 12 rats aged 30 to 34 days at the start of the experiment. The data are classified in two ways, by fat (fresh vs. rancid lard) and by sex (male vs. female). Number of rats per fat and sex combination, $n = 3$.

	Sex ($r = 2$)	Fat ($c = 2$) Fresh	Rancid	Row means $\bar{R}$
		709	592	
	♂	679	538	
		699	476	
Subgroup means $\bar{Y}$		695.6667	535.3333	615.5000
		657	508	
	♀	594	505	
		677	539	
Subgroup means $\bar{Y}$		642.6667	517.3333	580.0000
Column means $\bar{C}$		669.1667	526.3333	
Grand sum		7173		

Preliminary computation

1. $\bar{\bar{Y}} = \frac{1}{rcn}\sum^{r}\sum^{c}\sum^{n} Y = \frac{1}{12}7173 = 597.7500$

2. $SS_{\text{subgr}} = n\sum^{rc}(\bar{Y} - \bar{\bar{Y}})^2 = 65{,}903.5833$

3. $SS_{\text{within}} = \sum^{rc}\sum^{n}(Y - \bar{Y})^2 = 11{,}666.6667$

4. SS_{total} = quantity **2** + quantity **3** = 77,570.2500

Preliminary anova

	Source of variation	*df*		*SS*	*MS*
$\bar{Y} - \bar{\bar{Y}}$	Among subgroups	3	$rc - 1$	65,903.5833	21,967.85
$Y - \bar{Y}$	Within subgroups (error)	8	$rc(n - 1)$	11,666.6667	1,458.33
$Y - \bar{\bar{Y}}$	Total	11	$rcn - 1$	77,570.2500	

BOX 11.1 CONTINUED

The analysis proceeds as follows:

5. SS_{rows} (SS due to sex) $= \sum^{r} cn(\bar{R} - \bar{\bar{Y}})^2 = 3780.7500$

6. SS_{columns} (SS due to fat) $= \sum^{c} rn(\bar{C} - \bar{\bar{Y}})^2 = 61{,}204.0833$

7. $SS_{\text{interaction}}$ (SS due to sex × fat) = quantity **2** − quantity **5** − quantity **6** = 918.7500

Completed anova

	Source of variation	df		SS	MS
$\bar{R} - \bar{\bar{Y}}$	Between rows (sex)	1	$(r - 1)$	3780.7500	3780.75 *ns*
$\bar{C} - \bar{\bar{Y}}$	Between columns (fat)	1	$(c - 1)$	61,204.0833	61,204.08***
$\bar{Y} - \bar{R} - \bar{C} + \bar{\bar{Y}}$	Interaction (sex × fat)	1	$(r - 1)(c - 1)$	918.7500	918.75 *ns*
$Y - \bar{Y}$	Error	8	$rc(n - 1)$	11,666.6667	1458.33
$Y - \bar{\bar{Y}}$	Total	11	$rcn - 1$	77,570.2500	

SOURCE: Data from Powick (1925).

data is designed to be as instructive as possible. It is not, however, the most elegant method of carrying out the computational steps; an elegant method is illustrated in Box 11.2, which will be considered later in this section. Notice some minor changes in symbolism in Box 11.1. We have called the number of rows in the table r and the number of columns c. This notation will be easy to remember for now, although in Box 11.2 we will adopt a more general symbolism consistent with that of previous and later chapters.

We commence by computing a single-classification anova of the four subgroups or subclasses, each representing readings on three rats. If we had no further classification of these four subgroups by sex or freshness of lard, such an anova would test whether there is variation among the four subgroups over and above the variance within the subgroups. Since we have subdivisions by sex and type of fat, however, our only purpose here is to compute some quantities necessary for further analysis. The quantities **1** through **3** are computed in the familiar manner of Box 9.4, although the symbolism is slightly different, since in place of a groups we now have rc subgroups.

We continue the computation by finding the sums of squares for rows and columns of the table. First we sum the squared differences between row means and the grand mean; then we perform the same operation for column means and the grand mean. The squared differences are weighted by the sample sizes of the row and column means, respectively.

Let us return for a moment to the preliminary analysis of variance, in which we divided the total sum of squares into two parts: the sum of squares among the four subgroups and that within the subgroups, the error sum of squares. The new sums of squares pertaining to row and column effects clearly are not part of the error, but must contribute to the differences that comprise the sum of squares among the four subgroups. We therefore subtract row and column *SS* from the subgroup *SS*, which is 65,903.5833. The row *SS* is 3780.75, and the column *SS* is 61,204.0833. Together they add up to 64,984.8333, almost but not quite the value of the subgroup sum of squares. The difference represents a third sum of squares, called the **interaction sum of squares,** whose value in this case is 918.7500. We will discuss the meaning of this new sum of squares presently. At the moment let us say only that it is almost always present (but not necessarily large enough to be important). The interaction sum need not be independently computed but may be obtained as illustrated above—by the subtraction of the row *SS* and the column *SS* from the subgroup *SS*. This procedure is shown graphically in Figure 11.1, which illustrates the decomposition of the total sum of squares into the subgroup *SS* and error *SS*, the former being subdivided into the row *SS*, column *SS*, and interaction *SS*. The relative magnitudes of these sums of squares differ from experiment to experiment. In Figure 11.1 they are not shown proportional to their actual values in the rat experiment; if they were, the area representing the row *SS* would have to be about $\frac{1}{16}$th of that allotted to the column *SS*.

Before we can intelligently test for the presence of effects due to sex or fat in this anova, we must understand the meaning of interaction. We can best explain interaction in a two-way anova by means of an artificial illustration based on the rat data we have just studied. If we interchange the readings for rancid and fresh

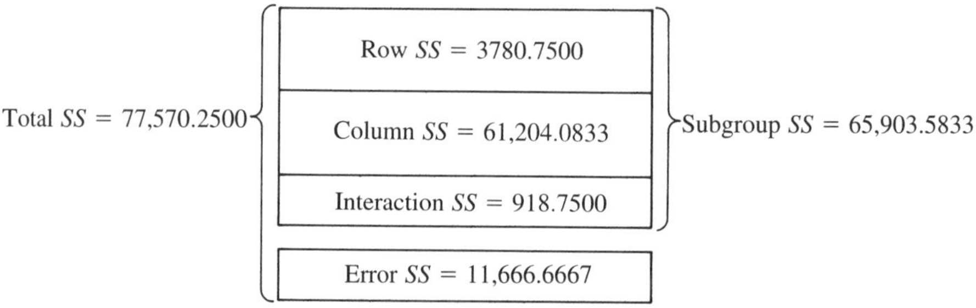

FIGURE 11.1 *Diagram of the partitioning of the total sums of squares* (SS) *in a two-way orthogonal anova. The areas of the subdivisions are not shown proportional to the magnitudes of the sums of squares.*

Table 11.1 AN ARTIFICIAL EXAMPLE TO ILLUSTRATE THE MEANING OF INTERACTION.

The readings for rancid and fresh lard of the male rats in Box 11.1 have been interchanged. Only subgroup and marginal means are given below.

	Fat		
Sex	Fresh	Rancid	$\bar{R}$
♂	535.3333	695.6667	615.5000
♀	642.6667	517.3333	580.0000
$\bar{C}$	589.0000	606.5000	$597.7500 = \bar{\bar{Y}}$

Completed anova

Source of variation	*df*	*SS*	*MS*
Sex	1	3780.7500	3780.75 *ns*
Fat	1	918.7500	918.75 *ns*
S × F	1	61,204.0833	61,204.08***
Error	8	11,666.6667	1458.33
Total	11	77,570.2500	

lard *for the male rats only,* we obtain Table 11.1. Only the means of the subgroups, rows, and columns are shown. We complete the analysis of variance in the manner presented already; note the results at the bottom of Table 11.1. The total, subgroup, and error *SS* are the same as before (see Box 11.1). This should not surprise you, since we are using the same data. All we did was interchange the contents of the upper two cells of the table.

When we partition the subgroup *SS*, we do find some differences. Note that the *SS* between sexes (between rows) is unchanged. Since the change we made was within one row, the total for that row was not altered and consequently the row *SS* did not change. The means of the columns, however, have been altered appreciably as a result of the interchange of the readings for fresh and rancid lard in the males. The mean for fresh lard is now very close to that for rancid lard, and the difference between the fats, previously quite marked, is now no longer so. By contrast, the interaction *SS*, obtained by subtracting the sums of squares of rows and columns from the subgroup *SS*, is now a large quantity. Remember that the subgroup *SS* is the same in the two examples. In the first example we subtracted sums of squares due to the effects of both fat and sex, leaving only a tiny residual representing the interaction. In the second example these two *main effects* (fat and sex) account only for little of the subgroup sum of squares, leaving the interaction sum of squares as a substantial residual. What is the essential difference between these two examples?

If we inspect the subgroup and marginal means for the original data from Box 11.1 and for the "doctored" data in Table 11.1, the original results are quite clear: Males eat slightly more than females, and this is true for both fresh and rancid fats. Note also that fresh lard is eaten in considerably greater quantity than rancid lard, and again this is true of males and females. Thus our statements about differences due to sex or to the freshness of the lard can be made independently of each other. When we interpret the artificial data (see Table 11.1), however, we note that although males still ate more fat than females (since row sums have not changed), this difference depends greatly on the nature of the fat. Males seem to eat considerably less fresh lard than the females, but of the rancid fat they appear to eat considerably more. Thus we can no longer make an unequivocal statement about the amount of fat eaten by the two sexes. We have to qualify our statement by the type of lard that was consumed. For fresh fat, $\bar{Y}_{♂} < \bar{Y}_{♀}$, but for rancid fat, $\bar{Y}_{♂} > \bar{Y}_{♀}$. If we examine the effects of fresh versus rancid fats in the artificial example, we notice that slightly more rancid than fresh lard was consumed. However, again we have to qualify this statement by the sex of the consuming rat; the males ate more rancid fat, the females ate more fresh fat.

This dependence of the effect of one factor on the level of another factor is called **interaction.** It is a common and fundamental scientific idea. The sum of squares for interaction measures the departure of the subgroup means from the values expected on the basis of additive combinations of the row and column means. Any given combination of levels of factors such as fresh lard–male rats may result in a positive or negative deviation from the expected value based on the means for fresh lard and male sex. In common biological terminology a large *positive* deviation of this sort is called **synergism.** When drugs act synergistically, the result of the interaction of the two drugs may be above and beyond the separate effects of each drug. When a combination of levels of two factors *inhibit* each other's effects, we call it **interference.** Synergism and interference both tend to magnify the interaction *SS.*

Testing for interaction is an important procedure in analysis of variance. If the artificial data of Table 11.1 were real, stating that rancid lard was consumed in slightly greater quantities than fresh lard would be of little value. This statement would cover up the important differences in the data, which are that males ate more rancid fat and females preferred fresh fat.

We are now able to write an expression symbolizing the decomposition of a single variate in a two-way analysis of variance in the manner of Expressions (8.2) and (10.2) for single-classification and nested anovas. Expression (11.1) assumes that both factors represent fixed treatment effects, Model I. This assumption seems reasonable, since sex and the freshness of the lard are fixed treatments. Variate Y_{ijk} is the kth item in the subgroup representing the ith group of treatment A and the jth group of treatment B. It is decomposed as follows:

$$Y_{ijk} = \mu + \alpha_i + \beta_j + (\alpha\beta)_{ij} + \epsilon_{ijk} \tag{11.1}$$

where μ equals the parametric mean of the population, α_i is the fixed treatment effect for the ith group of treatment A, β_j is the fixed treatment effect of the jth group of treatment B, $(\alpha\beta)_{ij}$ is the interaction effect in the subgroup representing the ith group of factor A and the jth group of factor B, and ϵ_{ijk} is the error term of the kth item in subgroup ij. We make the usual assumption that ϵ_{ijk} is distributed normally with a mean of 0 and a variance of σ^2. If one or both of the factors are Model II, we replace the α_i and β_j in the formula by A_i and B_j, respectively.

In previous chapters we learned that each sum of squares represents a sum of squared deviations. What deviations does an interaction SS represent? Let us refer back to the anovas of Box 11.1. The variation among subgroups is represented by $\bar{Y} - \bar{\bar{Y}}$, where $\bar{Y}$ stands for the subgroup means and $\bar{\bar{Y}}$ for the grand mean. When we subtract the deviations due to rows $\bar{R} - \bar{\bar{Y}}$ and columns $\bar{C} - \bar{\bar{Y}}$ from those of subgroups, we obtain

$$\begin{aligned}(\bar{Y} - \bar{\bar{Y}}) - (\bar{R} - \bar{\bar{Y}}) - (\bar{C} - \bar{\bar{Y}}) &= \bar{Y} - \bar{\bar{Y}} - \bar{R} + \bar{\bar{Y}} - \bar{C} + \bar{\bar{Y}} \\ &= \bar{Y} - \bar{R} - \bar{C} + \bar{\bar{Y}}\end{aligned}$$

This complicated expression is the deviation due to interaction. When we evaluate one such expression for each subgroup, square it, sum these squares, and multiply the sum by n, we obtain the interaction SS. This partition of the deviations also holds for their squares because the sums of the cross products of the separate terms cancel out.

The deviations leading to the sums of squares are shown in Table 11.2. Calculate some of these deviations to confirm for yourself where they come from. Notice the large deviations due to interaction and their alternating signs, minus in the left upper and right lower quadrants and plus in the right upper and left lower quadrants. A simple method for revealing the nature of the interaction in the data is to inspect the means of the original data table. The original data (see Box 11.1), showing no interaction, would yield the following pattern of relative magnitudes:

	Fresh		Rancid
♂		>	
♀		>	

The relative magnitudes of the means in the artificial example (see Table 11.1) yielding interaction can be summarized as follows:

	Fresh		Rancid
♂		<	
♀		>	

Table 11.2 DEVIATIONS FROM MEANS.

Artificial data of Table 11.1. Differences from values obtained in Box 11.1 and Table 11.1 are due to rounding errors.

Sex	Fat: Fresh	Fat: Rancid	$\bar{R} - \bar{\bar{Y}}$
♂	−62.4167 −71.4167	+97.9167 +71.4167	+17.7500
♀	+44.9167 +71.4167	−80.4167 −71.4167	−17.7500
$\bar{C} - \bar{\bar{Y}}$	−8.7500	+8.7500	

Left upper number in each quadrant: $\bar{Y} - \bar{\bar{Y}}$ (= subgroups). *Right lower number in each quadrant:* $\bar{Y} - \bar{R} - \bar{C} + \bar{\bar{Y}}$ (= interaction). $\bar{R} - \bar{\bar{Y}}$ (= rows); $\bar{C} - \bar{\bar{Y}}$ (= columns).

$$n \sum^{rc} (\bar{Y} - \bar{\bar{Y}})^2 = 65{,}903.5833 \qquad n \sum^{rc} (\bar{Y} - \bar{R} - \bar{C} + \bar{\bar{Y}})^2 = 61{,}204.0833$$

When the pattern of signs expressing relative magnitudes is not uniform, as in the second table here, interaction is indicated. As long as the pattern of means is uniform, as in the first table, interaction may not be present. However, interaction is often present without change in the *direction* of the differences; only the relative magnitudes may be affected. In any case the statistical test needs to be performed to test whether the deviations are larger than can be expected from chance alone.

In summary, when the effect of two treatments applied together cannot be predicted from the average responses of the separate factors, statisticians call this phenomenon interaction and test for the presence of its effect by means of an interaction mean square. This is a very common scientific relationship. If we say that the effect of density on the fecundity or weight of a beetle depends on its genotype, we imply that a genotype × density interaction is present. If the geographic variation of a parasite depends on the nature of the host species it attacks, we speak of a host × locality interaction. If the effect of temperature on a metabolic process is independent of the effect of oxygen concentration, we say that temperature × oxygen interaction is absent.

We will discuss significance testing in a two-way anova in the next section. We now proceed to a formal presentation of an efficient computational method for carrying out such an anova. Before doing so, however, we should point out that computational steps **5** and **6** of Box 11.1 could have been shortened by employing the simplified formula for a sum of squares between *two* groups illustrated in Box 9.5. In an analysis with only two rows and two columns the interaction *SS* can be computed directly as

$$\frac{n}{4}\text{(sum of means of one diagonal} - \text{sum of means on the other diagonal)}^2$$

which, when carried out to a sufficient number of decimal places, would yield the correct value $\frac{3}{4}(695.6665 + 517.3333 - 642.6667 - 535.3333)^2 = 918.75$ for the data of Box 11.1.

The computations for a two-way anova with replication are illustrated in Box 11.2. In this example the oxygen consumption of two species of limpets was measured at three concentrations of seawater. The two species represent one factor, and the three concentrations of seawater are the other. Since both factors are fixed treatment effects, this is a Model I anova. The outline of the computations, shown step by step in Box 11.2, follows the familiar format of previous analyses. We fill in the analysis of variance table and proceed to the tests of significance described in the next section.

11.3 Two-Way Anova: Significance Testing

Before we can test hypotheses about the sources of variation isolated in Box 11.2, we must become familiar with the expected mean squares for this design. In the anova table of Box 11.2 we first show the expected mean squares for Model I, both species differences and seawater concentrations being fixed treatment effects. Incidentally, this model would also be appropriate for the earlier example of fat consumption in rats (sex and freshness of lard are fixed treatments). Note that the within-subgroups or error *MS* again estimates the parametric variance of the items. The most important fact to remember about a Model I anova is that the mean square at each level of variation carries only the added effect due to that level of treatment; except for the parametric variance of the items, it does not contain any term from a lower line. Thus the expected *MS* of factor *A* contains only the parametric variance of the items plus the added term due to factor *A*, but does not also include interaction effects. In Model I the significance test is therefore simple and straightforward. Any source of variation is tested by the variance ratio of the appropriate mean square over the error *MS*. Thus for the appropriate tests we employ variance ratios ***A/Error, B/Error,*** and ***(A × B)/Error,*** where each boldface term signifies a mean square. Thus $\boldsymbol{A} = MS_A$, $\boldsymbol{Error} = MS_{\text{within}}$.

When we apply these tests to the example of Box 11.2, we are led to reject the null hypothesis that factor *B*, salinity, has no effect. However, we cannot similarly reject null hypotheses concerning factor *A* or the interaction. Using the conventional verbal shorthand of research publications, we find "salinity significant, while species and salinity × species interaction are not." We conclude

Box 11.2 TWO-WAY ANOVA WITH REPLICATION.

Oxygen consumption rates of 2 species of limpets, *Acmaea scabra* and *A. digitalis,* at 3 concentrations of seawater. The variable measured is $\mu\ell$ O_2/mg dry body weight/min at 22°C. There are 8 replicates per combination of species and salinity ($n = 8$). This is a Model I anova.

Factor B: Seawater concentration ($b = 3$)	Factor A: Species ($a = 2$) *Acmaea scabra*		*Acmaea digitalis*		$\bar{Y}_B$
	7.16	8.26	6.14	6.14	
	6.78	14.00	3.86	10.00	
100%	13.60	16.10	10.40	11.60	
	8.93	9.66	5.49	5.80	
	$\bar{Y} = 10.5612$		$\bar{Y} = 7.4288$		8.9950
	5.20	13.20	4.47	4.95	
	5.20	8.39	9.90	6.49	
75%	7.18	10.40	5.75	5.44	
	6.37	7.18	11.80	9.90	
	$\bar{Y} = 7.8900$		$\bar{Y} = 7.3375$		7.6138
	11.11	10.50	9.63	14.50	
	9.74	14.60	6.38	10.20	
50%	18.80	11.10	13.40	17.70	
	9.74	11.80	14.50	12.30	
	$\bar{Y} = 12.1738$		$\bar{Y} = 12.3262$		12.2500
$\bar{Y}_A$	10.2083		9.0308		461.74 = Grand sum

SOURCE: Data from F. J. Rohlf (unpublished results).

Preliminary computation

1. Grand mean $\bar{\bar{Y}} = \frac{1}{abn}\sum^{a}\sum^{b}\sum^{n} Y = \frac{1}{48} 461.74 = 9.6196$

2. SS_A (SS of columns) $= nb\sum^{a}(\bar{Y}_A - \bar{\bar{Y}})^2 = 16.6381$

3. SS_B (SS of rows) $= na\sum^{b}(\bar{Y}_B - \bar{\bar{Y}})^2 = 181.3210$

4. $SS_{A\times B}$ (Interaction SS) $= n\sum^{a}\sum^{b}(\bar{Y} - \bar{Y}_A - \bar{Y}_B + \bar{\bar{Y}})^2 = 23.9262$

5. SS_{within} (Within subgroups; error SS) $= \sum^{a}\sum^{b}\sum^{n}(Y - \bar{Y})^2 = 401.5213$

Now fill in the anova table.

	Source of variation	df	SS	MS	Expected MS (Model I)
$\bar{Y}_A - \bar{\bar{Y}}$	A (columns)	$a - 1$	**2**	$\frac{\mathbf{2}}{(a-1)}$	$\sigma^2 + \frac{nb}{a-1}\sum^{a}\alpha^2$
$\bar{Y}_B - \bar{\bar{Y}}$	B (rows)	$b - 1$	**3**	$\frac{\mathbf{3}}{(b-1)}$	$\sigma^2 + \frac{na}{b-1}\sum^{b}\beta^2$
$\bar{Y} - \bar{Y}_A - \bar{Y}_B + \bar{\bar{Y}}$	$A \times B$ (interaction)	$(a-1)(b-1)$	**4**	$\frac{\mathbf{4}}{(a-1)(b-1)}$	$\sigma^2 + \frac{n}{(a-1)(b-1)}\sum^{ab}(\alpha\beta)^2$
$Y - \bar{Y}$	Within subgroups	$ab(n-1)$	**5**	$\frac{\mathbf{5}}{ab(n-1)}$	σ^2
$Y - \bar{\bar{Y}}$	Total	$abn - 1$	**2 + 3 + 4 + 5**		

Since this example is a Model I anova for both factors, the expected *MS* above are correct. The following are the corresponding expressions for other models.

BOX 11.2 CONTINUED

Source of variation	Model II	Mixed model (*A* fixed, *B* random)
A	$\sigma^2 + n\sigma^2_{AB} + nb\sigma^2_A$	$\sigma^2 + n\sigma^2_{AB} + \frac{nb}{a-1}\sum^a \alpha^2$
B	$\sigma^2 + n\sigma^2_{AB} + na\sigma^2_B$	$\sigma^2 + na\sigma^2_B$
$A \times B$	$\sigma^2 + n\sigma^2_{AB}$	$\sigma^2 + n\sigma^2_{AB}$
Within subgroups	σ^2	σ^2

Anova table

Source of variation	*df*	*SS*	*MS*	F_s
A (columns; species)	1	16.6381	16.6381	1.740 *ns*
B (rows; salinities)	2	181.3210	90.6605	9.483***
$A \times B$ (interaction)	2	23.9262	11.9631	1.251 *ns*
Within subgroups (error)	42	401.5213	9.5600	
Total	47	623.4066		

$F_{.05[1,42]} = 4.07 \qquad F_{.05[2,42]} = 3.22 \qquad F_{.001[2,42]} = 8.18$

Since this is a Model I anova, all mean squares are tested over the error *MS*. For a discussion of significance tests see Section 11.3.

Conclusions

Oxygen consumption does not differ significantly between the two species of limpets but differs with the salinity. At 50% seawater the O_2 consumption is increased. Salinity appears to affect the two species equally, for there is insufficient evidence of a species × salinity interaction.

that the differences in oxygen consumption are induced by varying salinities, and there does not appear to be sufficient evidence for species differences in oxygen consumption. The following tabulation of the relative magnitudes of the means (in the manner of the previous section)

	Seawater concentrations				
	100%		75%		50%
A. scabra		$>$		$<$	
A. digitalis		$>$		$<$	

shows that the pattern of signs in the two lines is identical. This representation may be misleading, however, since the mean of *A. scabra* is far higher at 100% seawater than at 75%, but that of *A. digitalis* is only very slightly higher. Although the oxygen consumption curves of the two species appear far from parallel (Figure 11.2), this suggestion of a species × salinity interaction cannot be shown to be significant when compared to the within-subgroups variance. Finding a significant difference among salinities does not conclude the analysis. The data suggest that at 75% seawater, oxygen consumption decreases. This can be tested by the methods of Section 9.7; however, a preferred method of analysis for response curves of this type will be presented in Section 16.6.

When we test the anova of fat consumption in rats (see Box 11.1), we notice again that only one mean square (between columns, representing fats) is

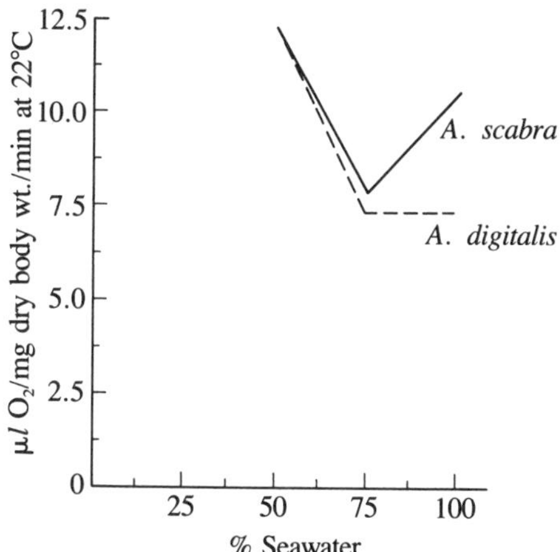

FIGURE 11.2 *Oxygen consumption by two species of limpets at three salinities. Data from Box 11.2.*

significant. We conclude, therefore, that the apparent difference in fat consumption between males and females in rats is not significant, but that the higher consumption of fresh as compared with rancid lard is highly significant. Absence of interaction means that this relation is the same in both sexes. When we analyze the results of the artificial example in Table 11.1, we find only the interaction *MS* significant. Thus we would conclude that the response to freshness of lard differs in the two sexes. This conclusion would be supported by inspection of the data, which would show that males preferred rancid fat and females preferred fresh fat.

In the artificial example the effects of the two factors (main effects) are not significant in any case. However, many statisticians would not even test them after finding the interaction mean square to be significant, since in such a case an overall statement for each factor would have little meaning, (e.g., Kempthorne, 1975). A simple statement of preference for either rancid or fresh lard would be unclear. The presence of interaction makes us qualify our statements: Rancid fat is preferred by males, and fresh fat is preferred by females. Similarly, interaction in the example of Box 11.2 would have meant that the pattern of response to changes in salinity differed in the two species. We would consequently have to describe separate, non-parallel response curves for the two species. In such a case planned and unplanned multiple-comparisons tests could be carried out to test differences among subgroup means separately for each species. When interaction is present, the relative magnitudes of the sums of squares for the main effects and the interaction depend on the particular levels of the main effects chosen. In the limpet experiment illustrated in Figure 11.2, if we had chosen seawater concentrations ranging only between 50% and 75%, there would have been an increased effect of salinity, little if any species effect, and only a small interaction effect. Had we selected the range from 75% to 100%, the salinity factor would have been diminished, while species and interaction effects would have been large.

It may occasionally become important to test for overall significance in a Model I anova in spite of the presence of interaction. Take, for example, the situation illustrated by Figure 11.3, which shows the effects of increasing proportions of two types of "conditioned" flour in fresh flour on the survival of wild-type *Tribolium castaneum* beetles. Conditioned flour is medium that has been heavily contaminated by waste products of the beetles. The two types of flour were conditioned by two different strains of beetles. The shapes of the response curves are quite different. Undoubtedly there is interaction between the strains conditioning the medium and the concentrations of the conditioned flour as these affect survival of the beetles. Yet the overall effects of increasing the proportion of conditioned flour are quite clear. As the concentration of conditioned flour increases, survival decreases in conditioned flour from both sources. To support this contention we might wish to test the mean square among concentrations (over the error *MS*), regardless of whether the interaction effect is significant.

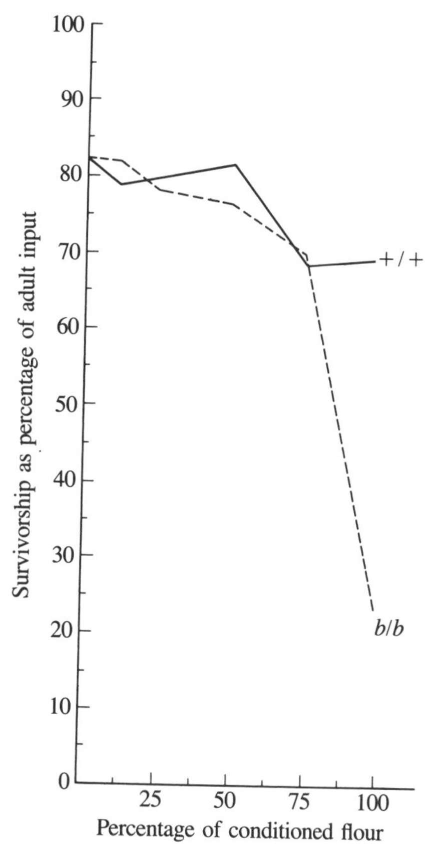

FIGURE 11.3 *Adult survivorship as percentage of egg input of wild-type* Tribolium castaneum *(a beetle) reared on medium conditioned by either wild-type (+/+) or black (b/b) individuals in varying concentrations. (Extracted from Karten, 1965.)*

Box 11.2 also lists expected mean squares for a Model II and a mixed model two-way anova. In the Model II anova note that the two main effects contain the variance component of the interaction as well as their own variance component. Such an arrangement is familiar from nested analysis of variance. When testing significance in a Model II anova, we first test **$(A \times B)$/*Error*.** If the interaction is significant we continue testing **$A/(A \times B)$** and **$B/(A \times B)$,** but when **$A \times B$** is not significant we must decide whether to pool the interaction *SS* with the error *SS* before testing the mean squares of the main effects. In such a case we proceed according to the rules of Box 10.3. Only one type of mixed model is shown in which factor *A* is assumed to be fixed and factor *B* to be random. If the situation is reversed, the expected mean squares change accordingly. Here it is the mean square representing the fixed treatment that carries with it the variance component of the interaction, while the mean square representing the random factor contains only the error variance and its own variance component and does not include the interaction component. We therefore test the *MS* of the random main effect over the error but test the fixed treatment *MS* over the interaction if the interaction is significant or if Box 10.3 contraindicates pooling; otherwise we test over the pooled error.

You may wonder how to write the formulas for expected mean squares for cases other than those given in the boxes. We will learn general formulas for these in Section 12.3.

The expected mean squares for mixed-model anovas presented earlier are not the only ones suggested for this case. Alternative models do exist that are beyond the level of this book. The formulation chosen here is the one most frequently encountered in the literature. However, readers are warned that they need to know which formulation is employed by the computer program they are using: The various formulations will yield different results that could lead to different conclusions.

An example of a Model II two-way anova is featured in Table 11.3. The data represent measurements of blood pH of mice. Five inbred strains were measured repeatedly. In experiment I, the mice were tested every other day for seven test periods, and in experiment II they were tested at six weekly intervals. The purpose of these experiments was to estimate the proportion of the variation that is genetic. This is an example of a Model II analysis because the experimenter was not concerned with the differences between specific strains but was considering the five strains to be a "random" sample of the inbred mouse strains available. The sampling periods are considered to be random as well, since the investigator did not wish to establish that day 1 produced blood pH values different from those of day 2, but merely tried to estimate the magnitude of fluctuations in blood pH over the period studied. There is no reason to believe in a directed trend of the blood pH over the period of the study.

Table 11.3 shows the expected mean squares of the two experiments, with the actual coefficients of the variance components. The symbolism differs somewhat from that given for Model II in Box 11.2. It is important that you become familiar with some variations in usage. Instead of using A and B as symbols for the two factors, we us M and D to represent mouse strains and days, respectively. Mnemonic notation like this is frequently used. In such cases investigators often also use a different symbolism for the coefficients; m might be used for a and d for b (to correspond to M and D, respectively). Notice also that the source-of-variation column (copied from the original publication) does not have prepositions such as "among" or "between" and also that the error term is described as "mice treated alike," which is what the error term represents in the present experiment: variation among mice of a given strain within a given day. You will have to become used to such differences in naming the sources of variation.

Significance testing in this example proceeds as follows. We first test interaction over the error; this is nonsignificant in both experiments. Following the rules in Box 10.3, assuming no knowledge about the relative magnitudes of the

Table 11.3 A MODEL II TWO-WAY ANOVA WITH REPLICATION.

Strain differences and daily differences in blood pH for 5 inbred strains of mice. In experiment I samples of 2 mice from each strain were tested every other day until 7 such groups had been tested; in experiment II samples of 5 mice from each strain were tested 6 times at one-week intervals. Only the anova tables are shown.

Experiment I ($a = 5, b = 7, n = 2$)

Source of variation	*df*	*MS*	F_s	Expected *MS*
Mouse strains	4	0.0427	11.24***	$\sigma^2 + 2\sigma^2_{MD} + 14\sigma^2_M$
Days of test	6	0.0148	3.89**	$\sigma^2 + 2\sigma^2_{MD} + 10\sigma^2_D$
Interaction	24	0.0038	1.03 *ns*	$\sigma^2 + 2\sigma^2_{MD}$
Mice treated alike (error)	35	0.0037		σ^2

Experiment II ($a = 5, b = 6, n = 5$)

Source of variation	*df*	*MS*	F_s	Expected *MS*
Mouse strains	4	0.0920	17.69***	$\sigma^2 + 5\sigma^2_{MD} + 30\sigma^2_M$
Days of test	5	0.0101	1.94 *ns*	$\sigma^2 + 5\sigma^2_{MD} + 25\sigma^2_D$
Interaction	20	0.0052	1.53 *ns*	$\sigma^2 + 5\sigma^2_{MD}$
Mice treated alike (error)	120	0.0034		σ^2

Variance components

	Experiment I		Experiment II	
Strains	0.00278	36.4%	0.00289	42.2%
Days	0.00110	14.4%	0.00020	2.9%
Interaction	0.00005	0.7%	0.00036	5.3%
Error	0.0037	48.5%	0.0034	49.6%

SOURCE: Data from Weir (1949).

interaction and error variances, we continue to test the main effects over the interaction. Actually, even had we pooled the two sums of squares, the results would not be substantially different. In experiment I, mouse strains and days of test have highly significant variance components. In experiment II, however, only mouse strains have such a component. The differences between the two experiments are reflected in the magnitudes of the estimated variance components given at the bottom of Table 11.3 as absolute values and as percentages.

The error variance is remarkably alike in the two experiments, showing that the techniques were well controlled and gave comparable results in both experiments. The major difference between the experiments is that a substantial percentage (14.4%) of variation is among days in experiment I, but this variance component comprises only 2.9% of the overall variation in experiment II. Clearly there are more fluctuations when readings are taken every two days. A far-fetched explanation might be that there is a rhythmic fluctuation in blood pH values with regular cycles every seven days. In such a case samples taken every other day would fluctuate widely, while those taken once a week would be very constant. There is, however, no evidence for such a phenomenon in these mice. Experiments I and II were run during different seasons, however, and it is quite possible that during the 14 days necessary for experiment I there were considerable day-to-day fluctuations in climate. An inquiry to the investigator revealed that the mice were not maintained in temperature-controlled rooms; hence daily fluctuations during experiment I might have resulted in this large value of the variance. During the 42 days of experiment II the weather might have been considerably more constant. Although these explanations may be incorrect, they illustrate how an anova furnishes insight into the structure of an experiment as well as clues to further questions to be asked and experiments to be undertaken.

A simple rearrangement of the data of Table 11.3 will provide some insight into the relations between two-way and nested anovas. In the design of experiment II in this table, 5 strains were observed on 6 separate days. In each strain 5 mice had their blood pH measured on one day. This design is shown at the top of Table 11.4. We arrange it as a two-way anova because the days are common factors throughout the analysis. Day 1 is the same day for strain 1 as it is for strain 3. The anova table for this experiment is reproduced once more in Table 11.4, showing also the numerical values for the sums of squares.

Suppose, however, that we had misinterpreted this experiment. We might have thought that the days were nested within the strains, as shown in the next design in Table 11.4. Such a design implies that we randomly chose six days on which to measure the blood pH of each strain and that the six days chosen for the tests on strain 1 were different from the test days for strain 2, and so on. If we had randomly chosen the days without any thought of making day 1 the same for all the strains, a nested analysis would have been appropriate. Taking the same data and analyzing them (incorrectly, in view of the actual nature of design) as a two-level nested anova, we obtain the anova table shown at the bottom of Table 11.4. Note that the subordinate level (subgroups within groups, which is days within strains in this example) is the summation of two levels from the two-way anova. Degrees of freedom and sums of squares of "days of test" and "interaction" add to the corresponding values for "days within strains." Thus in a hierarchic analysis of variance the subordinate level represents not only differences among the subgroups but also the interaction term. If differences among days exist but the nature of these differences varies from strain to strain, we would have a substantial contribution to interaction. This would not be separately detectable in the nested anova but would contribute to the magnitude of the subgroups-within-groups *SS*. When we incorrectly consider this example as a nested anova, the days within strains *MS* is significant at the 5% level, but in the

Table 11.4 EXPERIMENT II OF TABLE 11.3 REANALYZED TO SHOW RELATIONSHIPS BETWEEN A TWO-WAY ANOVA AND A NESTED ANOVA.

Correct design (two-way anova with replication = 5 mice [*] in each cell)

	Strains				
Days	1	2	3	4	5
1	*	*	*	*	*
2	*	*	*	*	*
3	*	*	*	*	*
4	*	*	*	*	*
5	*	*	*	*	*
6	*	*	*	*	*

Anova table

Source of variation	*df*	*SS*	*MS*	F_s
Mouse strains	4	0.3680	0.0920	17.69***
Days of test	5	0.0505	0.0101	1.94 *ns*
Interaction	20	0.1040	0.0052	1.53 *ns*
Error	120	0.4080	0.0034	
Total	149	0.9305		

Incorrect design (nested anova—days within strains: 5 mice [*] in each cell)

	Strains				
	1	2	3	4	5
Days	1 2 3 4 5 6	1 2 3 4 5 6	1 2 3 4 5 6	1 2 3 4 5 6	1 2 3 4 5 6
Mice	* * * * * *	* * * * * *	* * * * * *	* * * * * *	* * * * * *

Anova table

Source of variation	*df*	*SS*	*MS*	F_s
Mouse strains	4	0.3680	0.0920	14.84***
Days within strains	25	0.1545	0.0062	1.82*
Error	120	0.4080	0.0034	
Total	149	0.9305		

two-way anova neither days of test nor interaction are significant, although the appropriate F_s-ratios are suspiciously large.

When the treatment effects in a mixed-model two-way anova are significant, one can carry out unplanned multiple comparisons in the manner of Section 9.7. This method applies as well to the multiway analyses of variance in Chapter 12. Hochberg and Tamhane (1983) discuss such cases in detail. The error MS *to be used in the analyses is that used for the overall test of the main effects.*

11.4 Two-Way Anova Without Replication

In many experiments there will be no replication for each combination of factors represented by a cell in the data table. In such cases we cannot easily talk of "subgroups," since each cell contains a single reading only. At times it may be too difficult or too expensive to obtain more than one reading per cell, or the measurements may be known to be so repeatable that there is little point in estimating their error. As we will see in this section, a two-way anova without replication can be properly applied only with certain assumptions. For some models and tests in anova we must assume that no interaction is present.

Our illustration for this design is taken from limnology. In Box 11.3 we show temperatures of a lake taken at about the same time on four successive summer afternoons. These temperature measurements were made at ten different depths and were known to be very repeatable. Therefore only single readings were taken at each depth on any one day. What is the appropriate model for this anova? Clearly, the depths are Model I. The four days, however, are open to interpretation. If the days had been selected from among a wider span of time, say, April to October, we surely would have been interested in specific differences among the days, since these would reflect seasonal changes in temperature profiles. In such cases, days would represent fixed treatment effects as well. Since these four readings were taken during a period of stable midsummer weather, however, differences among the days are not likely to be of specific interest. It is improbable that an investigator would ask whether the water was colder on the 30th of July than on the 31st of July. A more meaningful way of looking at this problem would be to consider these four summer days as random samples that enable us to estimate the day-to-day variability of the temperature stratification in the lake during this climatically stable period.

The computations, shown in Box 11.3, are the same as those in Box 11.2 except that the expressions to be evaluated are simpler. The subgroup sum of squares in this example is the same as the total sum of squares. If this is not immediately apparent, consult Figure 11.4, which, when compared with Figure 11.1, illustrates that the error sum of squares based on variation within subgroups is missing in this example. Thus, after we subtract the sum of squares for columns (factor A) and for rows (factor B) from the total SS, we are left with only a

Box 11.3 TWO-WAY ANOVA WITHOUT REPLICATION.

Temperatures (°C) of Rot Lake on 4 early afternoons of the summer of 1952 at 10 depths. This is a mixed model anova.

Factor *B*: Depth in meters ($b = 10$)	Factor *A*: Days ($a = 4$)				
	29 July	30 July	31 July	1 August	$\bar{Y}_B$
0	23.8	24.0	24.6	24.8	24.300
1	22.6	22.4	22.9	23.2	22.775
2	22.2	22.1	22.1	22.2	22.150
3	21.2	21.8	21.0	21.2	21.300
4	18.4	19.3	19.0	18.8	18.875
5	13.5	14.4	14.2	13.8	13.975
6	9.8	9.9	10.4	9.6	9.925
9	6.0	6.0	6.3	6.3	6.150
12	5.8	5.9	6.0	5.8	5.876
15.5	5.6	5.6	5.5	5.6	5.575
$\bar{Y}_A$	14.890	15.140	15.200	15.130	603.6 = Grand sum

SOURCE: Data from Vollenweider and Frei (1953).

The four sets of readings are treated as replications (blocks) in this analysis. Depth is a fixed treatment effect, and days are considered random effects; hence this is a mixed model anova.

BOX 11.3 CONTINUED

Preliminary computation

1. Grand mean $\bar{\bar{Y}} = \frac{1}{ab}\sum^{a}\sum^{b} Y = \frac{1}{40}\,603.6 = 15.090$

2. SS_A (SS of columns) $= b\sum^{a}(\bar{Y}_A - \bar{\bar{Y}})^2 = 0.5620$

3. SS_B (SS of rows) $= a\sum^{b}(\bar{Y}_B - \bar{\bar{Y}})^2 = 2119.6510$

4. SS_{error} (remainder; discrepancy) $= \sum^{a}\sum^{b}(Y - \bar{Y}_A - \bar{Y}_B + \bar{\bar{Y}})^2 = 2.2430$

Anova table

	Source of variation	df	SS	MS	F_s	Expected MS
$\bar{Y}_A - \bar{\bar{Y}}$	A (columns; days)	3	0.5620	0.1873	2.255 *ns*	$\sigma^2 + b\sigma_A^2$
$\bar{Y}_B - \bar{\bar{Y}}$	B (rows; depths)	9	2119.6510	235.5168	2835.0***	$\sigma^2 + \sigma_{AB}^2 + \frac{a}{b-1}\sum\beta^2$
$Y - \bar{Y}_A - \bar{Y}_B + \bar{\bar{Y}}$	Error (remainder; discrepancy)	27	2.2430	0.0831		$\sigma^2 + \sigma_{AB}^2$
$Y - \bar{\bar{Y}}$	Total	39	2122.4560			

Conclusions

Highly significant decreases in temperature occur as depth increases. To test days we must assume interaction between days and depths to be zero. No added variance component among days can be demonstrated.

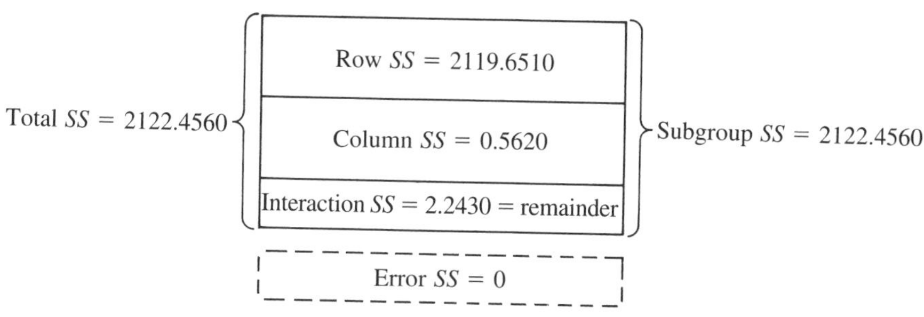

FIGURE 11.4 *Diagram of the partitioning of the total sums of squares (SS) in a two-way orthogonal anova without replication. The areas of the subdivisions are not shown proportional to the magnitudes of the sums of squares.*

single sum of squares, which is the equivalent of the previous interaction *SS* but which is now the only source for an error term in the anova. This *SS* is known as the **remainder *SS*** or the **discrepance.**

If you refer to the expected mean squares for the two-way anova in Box 11.2, you will discover why we made the statement earlier that for some models and tests in a two-way anova without replication we must assume that there is no interaction. If interaction is present, only a Model II anova can be entirely tested, while in a mixed model only the fixed level can be tested over the remainder mean square. In a pure Model I anova, or for the random factor in a mixed model, however, it would be improper to test the main effects over the remainder unless we could reliably assume that no added effect due to interaction is present. General inspection of the data in Box 11.3 convinces us that the temperature profile for any one day is faithfully reproduced on the other days. Thus, interaction is unlikely to be present. If, for example, a severe storm had churned up the lake on one day, changing the temperature relationships at various depths, interaction would have been apparent, and the test of the mean square among days carried out in Box 11.3 would not have been appropriate.

Since we assume no interaction, the row and column mean squares are tested over the error *MS*. The results are not surprising; casual inspection of the data would have predicted our findings. Added variance among days is no more than what could be expected by chance, but the differences in temperature due to depths are very large, yielding a value of $F_s = 2835.0$. It is extremely unlikely that chance alone could produce such differences.

A common application of two-way anova without replication is the *repeated testing of the same individuals*. By this we mean that the same group of individuals is tested repeatedly over a period of time. The individuals are one factor (usually considered as random and serving as replication), and the time dimension is the second factor, a fixed treatment effect. For example, we might measure growth of a structure in 10 individuals at regular intervals. When testing for the presence of an added variance component (due to the random factor) such a model again must assume that there is no interaction between time and the individuals; that is, the responses of the several individuals are parallel through time. This design is also used in physiological and psychological experiments in which we test the same group of individuals for the appearance of a response after treatment. Examples include increasing immunity after antigen inoculations, altered responses after conditioning, and measures of learning after a number of trials. For instance, we may study the speed with which 10 rats, repeatedly tested on the same maze, reach the end point. The fixed treatment effect would be the successive trials to which the rats have been subjected. The second factor, the 10 rats, is random, presumably representing a random sample of rats from the laboratory population. When the response variables are two-state attributes rather than being measurable, repeated testing of the same individuals is carried out by the method of Section 17.7.

The method of **randomized complete blocks** is a type of two-way anova (usually without replication) especially common in agricultural research, but also with applications in other biological field and laboratory experiments. When we apply a fertilizer to a small plot of land, we usually get a single reading expressed as yield of crop. Similarly, an application of an insecticide to a plot of land is measured as a single mortality among the insects in the plot or a single value for yield of crop. To test differences among five types of fertilizers (or insecticide treatments) we must have at least five plots available. The five resulting readings, however, would not yield an error variance useful for testing the significance of differences among treatments. Thus we need replication of the treatments to provide an error term. Let us assume for the sake of this discussion that we wish to replicate each treatment four times. We therefore need $5 \times 4 = 20$ plots to test five treatments. An area encompassing 20 plots of land is quite large relative to basic plot size and therefore is likely to be heterogeneous in its soil and in other microclimatic conditions. As a general rule adjacent plots are more like each other. We therefore do not wish to put all the replicates of treatment *A* into the same general area; otherwise what may appear to be a significant effect of treatment *A* may actually turn out to be due to the general area in which the plots of *A* are situated.

The method of randomized complete blocks is designed to overcome this difficulty. In our specific example we partition the land available for our experiment into four blocks, hoping that the environment within each of the blocks is as homogeneous as possible. We then subdivide each block into five plots and allocate the treatments at random to these plots. Such an arrangement is shown diagrammatically in Figure 11.5. The five treatments are called *A*, *B*, *C*, *D*, and *E*. The random arrangement of the treatments within each block can be obtained by permutating the five letters. There are $5! = 120$ permutations of the letters *A* through *E*. We can number these sequentially, and by consulting a table of random numbers, we can choose four permutations. The random arrangement of the treatments among the plots is necessary so that any given treatment does not always maintain the identical relative position in the field.

B	A	D	E	C	Block 1
B	C	E	D	A	Block 2
D	B	E	A	C	Block 3
C	B	A	D	E	Block 4

FIGURE 11.5 *A randomized-complete-blocks design. The field has been arranged into 4 blocks of 5 treatments each.*

For analysis the data are entered in a treatments × blocks table, all the treatments of one kind being entered in the appropriate block. If the $b = 4$ blocks are considered as rows (factor B) and the $a = 5$ treatments as columns (factor A), an anova table of this example would be constructed as follows:

Source of variation	df	Expected MS
Blocks	$b - 1$	$\sigma^2 \quad + a\sigma^2_B$
Treatments	$a - 1$	$\sigma^2 + [\sigma^2_{AB}] + \frac{b}{a-1}\sum^{a}\alpha^2$
Error	$(a - 1)(b - 1)$	$\sigma^2 + [\sigma^2_{AB}]$

The blocks are considered random effects. An individual observation in a randomized complete blocks design can therefore be decomposed as

$$Y_{ij} = \mu + \alpha_i + B_j + [(\alpha B)_{ij}] + \epsilon_{ij} \tag{11.2}$$

where α_i is the treatment effect and B_j is the random blocks effect. Note that in this expression and in the expected mean squares above, the interaction term has been placed within brackets. The brackets indicate that we assume the interaction to be absent. Testing of the treatment effect is not dependent on the validity of this assumption, since in the randomized-complete-blocks design we simply test the treatment MS over the error MS. The block effect is rarely tested, and we would have to assume that interaction is not present before we could test it. If the block MS is not significantly greater than the error MS, there was less heterogeneity among the blocks than anticipated or the blocks were laid out in such a way that most of the heterogeneity was within blocks instead of among blocks. Finding added variance among the blocks does not disturb us, since we expected to find such differences. In fact, if we found that the differences among blocks were small, then in future experiments we might want to use not a randomized-blocks design, but rather a single-classification anova.

The randomized-complete-blocks design may also be applied to experiments other than the agricultural ones for which it was first developed. If an experiment is too big to be undertaken at one time, it may have to be carried out over a period of days (each day representing a block), or it may have to be carried out in different laboratories (each serving as a block). When treatments are applied to different age classes of animals, these classes can become blocks. Whenever the environment in which the experiment is to be carried out or the research materials may be heterogeneous, these can be subdivided into more-homogeneous blocks and the experiment carried out and analyzed as randomized complete blocks.

One reason that the effects of blocks are rarely tested has already been explained: Unless we can be reasonably certain that there is no interaction variance component in our error term, we have no denominator mean square over which to test the block *MS*. But there is a second reason for our hesitancy. In our discussions so far, we have not emphasized an important consideration of two-way anova. In strict application, a two-way anova should be completely randomized. For example, to carry out the two-way anova on pea sections discussed earlier (see Section 11.1), we should assemble 100 dishes of pea sections and randomly allocate each to one of five sugar treatments and two pH levels so as to achieve 10 replicates for each treatment combination. Contrast this with the agricultural experiment. Before we apply our five fertilizer treatments to each of the four blocks of land, each block will have to be prepared by plowing, sowing seed, dressing soil, and so forth. This operation is unique for each block; given the vagaries of human performance it cannot be exactly repeated for the same block in succeeding years. Thus there is a random error deviation attached to each of the blocks due to the single and distinct instance of preparation each represents in the study. On assigning the various plots to the study, we are free to randomly assign a plot to any one treatment, but we are not free to randomly assign a plot to any one block since the five plots within any one block are limited to that block. This restriction on the variation causes a random error deviation peculiar to each block, which in turn generates an error variance component σ_R^2 that attaches itself to the block means square. This added error variance has been called **restriction error** by Anderson and McLean (1974b). The block means square is then represented by $\sigma^2 + a\sigma_R^2 + a\sigma_B^2$ and there is no suitable denominator mean square to test it over even if the interaction variance component is assumed to be absent. Thus caution should be exercised whenever a randomized-complete-blocks experiment is analyzed. Do not test the block *MS* when restriction error is present.

Restriction error is not limited to randomized complete blocks but also applies to two-way anovas with replication. In such cases, the variance component due to restriction error will attach itself to the mean square of the main effect within which the restriction error operates. Such a main effect can be either a fixed treatment or a random effect. Let us review the cases that we have encountered so far. The hypothetical example of the pea sections should not have any restriction error if carried out properly as discussed earlier. It is unlikely in most experimental setups, however, that the test conditions will be replicated 100 times. Hence there will be a restriction error attached to one of the mean squares in the study. The example of food consumption by male and female rats (see Box 11.1) has the possibility of restriction error. One cannot randomly assign a rat to one type of fat and one sex. Rats come prepackaged by nature into males and females; thus any randomization of the rats for the experiments must be carried out within each sex, not across the entire study. However, we would expect that restriction error in this example would be negligible. There is little error in choosing a male or a female and there should be little error in choosing males and females repeatedly, unlike error in the repeated tilling of the same block of agricultural land. The same sort of reasoning applies to the example from Box 11.2 in which two species of limpets were subjected to three salinities. Randomization can take place only within any one species. Nevertheless, again, there should be little if any restriction error due to sampling within one species. We would not hesitate, therefore, to test species as well as salinities, as we have done.

The analysis in Table 11.4 studying blood pH in mice is also a similar design. Assuming that all mice were living at the beginning of the experiment, any one mouse could be randomly assigned to the various time periods (days) when it is to be tested, but randomization could be only within any one strain. Assuming that the genetic factors determining blood pH are stable in these strains, we can ignore restriction error. But what if this were not the case? In a model such as this example, the presence of restriction error variance component σ_R^2 would simply be confounded with the strain variance component σ_M^2 and would inflate our estimate of the latter. There might be more hesitation in testing a fixed treatment main effect confounded with a restriction error variance component because in an extreme case an apparent treatment effect might be entirely due to restriction error, with no differences in means resulting from treatment differences.

In the example of water temperature for different depths on different days in Rot Lake (see Box 11.3), the error is restricted by depths having to be tested on any one day. Had a different test been run on any one date, different results would probably have been obtained. Since days are a random component, however, we need not be overly concerned about restriction error here other than to realize that our estimate of the variance component among days may be inflated by the presence of restriction error. In this particular example, an added variance component among days could not be demonstrated.

It is important to note that each treatment must be present in every block. If this cannot be done, or when one has too many treatments, different combinations of treatments are applied to a series of blocks. In such cases a more complex analysis must be carried out. Such designs, called ***incomplete blocks,*** *are described in Raktoe et al. (1981) and in Gill (1978, Chapter 6).*

Box 11.4 EXAMPLE OF A RANDOMIZED-COMPLETE-BLOCKS EXPERIMENT.

Mean dry weights (in milligrams) of 3 genotypes of beetles, *Tribolium castaneum,* reared at a density of 20 beetles per gram of flour. Four series of experiments represent blocks.

b series (b blocks; $b = 4$)	a genotypes ($a = 3$)			
	+ +	$+b$	bb	$\bar{Y}_B$
1	0.958	0.986	0.925	0.9563
2	0.971	1.051	0.952	0.9913
3	0.927	0.891	0.829	0.8823
4	0.971	1.010	0.955	0.9787
$\bar{Y}_A$	0.9568	0.9845	0.9153	11.426 = Grand sum

Anova table

	Source of variation	df	SS	MS	F_s
MS_B	Series	3	0.021391	0.007130	10.23**
MS_A	Genotypes	2	0.009717	0.004859	6.97*
$MS_{E(RB)}$	Error	6	0.004184	0.000697	

Estimate of error variance in a completely randomized design following Expression (11.3):

$$MS_{E(CR)} = \frac{b(a-1)MS_{E(RB)} + (b-1)MS_B}{ab-1}$$

$$= \frac{8 \times 0.000697 + 3 \times 0.007130}{11} = 0.002452$$

Relative efficiency of randomized-complete-blocks error variance $MS_{E(RB)}$ with respect to completely randomized error variance $MS_{E(CR)}$

$$RE = \frac{100 \times MS_{E(CR)}}{MS_{E(RB)}} = 351.6\%$$

When the degrees of freedom of the mean squares from which the relative efficiency is calculated are less than 20 (degrees of freedom of $MS_{E(RB)} = 6$ in our case), we use a correction factor for RE (Cochran and Cox, 1957). Calculate

$$\frac{(df_{RB}+1)(df_{CR}+3)}{(df_{RB}+3)(df_{CR}+1)}$$

BOX 11.4 CONTINUED

where df_{CR} and df_{RB} are the degrees of freedom of $MS_{E(CR)}$ and $MS_{E(RB)}$, respectively, and multiply it times the RE. In our case

$$\frac{(6 + 1)(9 + 3)}{(6 + 3)(9 + 1)} = \frac{(7 \times 12)}{(9 \times 10)} = 0.93333$$

$$RE = 351.6 \times 0.93333 = 328.2\%$$

SOURCE: Data from Sokal and Karten (1964).

An example of a randomized-complete-blocks experiment is shown in Box 11.4. This example is based on data extracted from a more extensive experiment by Sokal and Karten (1964) in which dry weights of three genotypes of *Tribolium castaneum* beetles were recorded. The experimental work involved measuring other variables and was very laborious. Therefore, the requisite amount of replication could not be carried out at one time, and four series of experiments, several months apart, were undertaken. Within one series of experiments the order of weighing different genotypes was random, thus yielding a randomized-complete-blocks design with four blocks (series) and three treatments (genotypes).

The anova table in Box 11.4 shows that there is a significant difference in weight among the genotypes: The heterozygotes are the heaviest, followed by the wild type, with the homozygous mutant weighing the least. In Box 11.4 we have not actually tested the differences between pairs of means by multiple-comparison tests. From the analysis shown we know only that the three means jointly are heterogeneous. Sokal and Karten, however, tested these differences and found them significant. In analogy to our earlier reasoning for the days mean square in the blood pH of mouse strains (see Table 11.3), we would consider series to be a random factor. The restriction error, if any, would be within genotypes, since samples of beetles cannot be randomly assigned to any one genotype. However, as in our earlier reasoning on sexes of rats, species of limpets, and strains of mice, we assume that the restriction error due to genotypes would be negligible. Beetles would be assigned to any one series on the assumption that the supply of beetles of any one genotype remains homogeneous throughout the study.

The significant effect of series implies that environmental factors among the series, minor changes in technique, or other factors were such that differences in weight resulted. By separating out this heterogeneity we were able to perform a much more sensitive test. Had we ignored differences among series and simply analyzed these data as four random replicates for each genotype, our error variance would have been that for a completely randomized design:

$$MS_{E(CR)} = \frac{b(a - 1)MS_{E(RB)} + (b - 1)MS_B}{ab - 1} \tag{11.3}$$

In this expression $MS_{E(CR)}$ is the expected error mean square in the completely randomized design, $MS_{E(RB)}$ is the observed error mean square in the randomized-blocks design, and MS_B is the observed mean square among blocks. This formula is derived in Section A.6 in the appendix. In Box 11.4 this value is $MS_{E(CR)} = 0.002452$, which is considerably greater than the observed error in the randomized-blocks design, $MS_{E(RB)} = 0.000697$. We can now calculate the relative efficiency of the randomized-blocks error variance with respect to the completely randomized error variance. This calculation is shown at the bottom of Box 11.4, where we note that randomized blocks in this case are 328.2% as efficient as the completely randomized design. Clearly it was well worthwhile to divide the experiment into series (blocks) and to segregate the variance due to series.

The computer program BIOM-pc can be used to perform the computations described in Sections 11.2 and 11.4. Section 12.5 gives a general description of the use of this program for anova.

11.5 Paired Comparisons

The randomized-complete-blocks designs studied in the previous section may differ in the number of rows (blocks) and treatment classes. One special case, however, is common enough to merit discussion in a separate section. This special case is that of randomized complete blocks in which there are only two treatments or groups ($a = 2$), known as **paired comparisons** because each observation for one treatment is paired with one for the other treatment. This pair is composed of the same individuals tested twice or of two individuals with common experiences so that we can legitimately arrange the data as a two-way anova.

Let us elaborate on this point. Suppose we test the muscle tone of a group of individuals, subject them to severe physical exercise, and measure their muscle tone once more. Since the same group of individuals will have been tested twice, we can arrange our muscle tone readings in pairs, each pair representing readings on one individual (before and after exercise). Such data are appropriately treated by a two-way anova without replication, which in this case would be a paired-comparison test because there are only two treatment classes. This "before-and-after treatment" comparison is a very frequent design leading to paired comparisons. Another design simply measures two stages in the development of a group of organisms, with time serving as the treatment intervening between the two stages. The example in Box 11.5 is of this nature. This study measures lower face width in a sample of girls five years old and in the same sample of girls when they are six years old. The paired comparison is for each individual girl, between her face width when she is five years old and her face width at six years.

Paired comparisons often result from dividing an organism or other individual unit so that half receives treatment 1 and the other half treatment 2, which may be

Box 11.5 PAIRED COMPARISONS (RANDOMIZED BLOCKS WITH $a = 2$).

Lower face width (skeletal bigonial diameter in centimeters) for 15 North American white girls measured at 5 and again at 6 years old.

Individuals	(1) 5-year-olds	(2) 6-year-olds	(3) $\bar{Y}_B$	(4) $D = Y_{i2} - Y_{i1}$ (difference)
1	7.33	7.53	7.430	0.20
2	7.49	7.70	7.595	.21
3	7.27	7.46	7.365	.19
4	7.93	8.21	8.070	.28
5	7.56	7.81	7.685	.25
6	7.81	8.01	7.910	.20
7	7.46	7.72	7.590	.26
8	6.94	7.13	7.035	.19
9	7.49	7.68	7.585	.19
10	7.44	7.66	7.550	.22
11	7.95	8.11	8.030	.16
12	7.47	7.66	7.565	.19
13	7.04	7.20	7.120	.16
14	7.10	7.25	7.175	.15
15	7.64	7.79	7.715	.15
$\bar{Y}_A$	7.4613	7.6613	$\sum^a\sum^b Y = 226.84$	$\bar{D} = 0.20$

SOURCE: Data from a larger study by Newman and Meredith (1956).

I. Analysis as a randomized-complete-blocks design

This example is a mixed model with the $a = 2$ ages as fixed treatment effects and the $b = 15$ individuals constituting the randomly chosen blocks.

Preliminary computation

1. Grand mean $\bar{\bar{Y}} = \frac{1}{ab}\sum^a\sum^b Y = \frac{1}{30}226.84 = 7.5613$

2. SS_A (SS of columns; treatments) $= b\sum^a(\bar{Y}_A - \bar{\bar{Y}})^2 = 0.3000$

3. SS_B (SS of rows; blocks) $= a\sum^b(\bar{Y}_B - \bar{\bar{Y}})^2 = 2.63675$

4. SS_{error} (remainder; discrepance; residual SS) $= \sum^a\sum^b(Y - \bar{Y}_A - \bar{Y}_B + \bar{\bar{Y}})^2 =$ 0.01080

Box 11.5 Continued

Anova table

Source of variation	df	SS	MS	F_s	Expected MS
Ages (columns; factor A)	1	0.30000	0.3000,000,0	388.89***	$\sigma^2 + \sigma^2_{AB} + \frac{b}{a-1}\sum \alpha^2$
Individuals (rows; factor B)	14	2.63675	0.188,339,0	244.14***	$\sigma^2 + a\sigma^2_B$
Remainder	14	0.01080	0.000,771,4		$\sigma^2 + \sigma^2_{AB}$
Total	29	2.94755			

$F_{.001[1,14]} = 17.1 \qquad F_{.001[14,14]} = 5.94$

Conclusions

The variance ratio for ages is highly significant. We conclude that faces of 6-year-old girls are wider than those of 5-year-olds. If we are willing to assume that the interaction σ^2_{AB} is zero, we may test for an added variance component among individual girls and would find it significant.

II. The same analysis carried out using the t-test for paired comparisons

$$t_s = \frac{\bar{D} - (\mu_1 - \mu_2)}{s_{\bar{D}}}$$

where $\bar{D}$ is the mean difference between the paired observations

$$\bar{D} = \sum \frac{D}{b} = \frac{3.00}{15} = 0.20$$

and $s_{\bar{D}} = s_D/\sqrt{b}$ is the standard error of $\bar{D}$ calculated from the observed differences in column (4).

$$s_D = \sqrt{\frac{\sum^b (D - \bar{D})^2}{b - 1}} = \sqrt{\frac{0.0216}{14}} = 0.039,279,2$$

$$s_{\bar{D}} = \frac{s_D}{\sqrt{b}} = \frac{0.039,279,2}{\sqrt{15}} = 0.010,141,9$$

We assume that the true difference between the means of the two groups, $\mu_1 - \mu_2$, equals zero.

$$t_s = \frac{\bar{D} - 0}{s_{\bar{D}}} = \frac{0.20 - 0}{0.010,141,9} = 19.7203$$

with $b - 1 = 14$ df, yields $P \ll 0.001$.

$$t_s^2 = 388.89$$

which equals the previous F_s.

the control. For example, if we wish to test the strength of two antigens or allergens in a random sample of volunteers, we might inject one into each arm of a single individual and measure the diameter of the red area produced. It would not be wise from the point of view of experimental design to test antigen 1 on half the individuals and antigen 2 on the other half. These individuals may be differentially susceptible to these antigens and we may learn little about the relative potency of the antigens, since this would be confounded by the differential responses of the subjects. This (undesirable) design would consequently result in a large variance, which in turn would make it difficult to reject the null hypothesis of no difference, unless that difference was very large.

A much better design is to inject antigen 1 into the left arm and antigen 2 into the right arm of a group of n individuals and to analyze the data as a randomized-complete-blocks design with n rows (blocks) and 2 columns (treatments). It is probably immaterial whether an antigen is injected into the right or left arm, but if we were designing such an experiment and knew little about the reaction of humans to antigens we might, as a precaution, randomly allocate antigen 1 to the left or right arm for different subjects and antigen 2 to the opposite arm. Note, however, that such a design should not be used if there is some interaction effect due to both antigens being injected in the same person. There could, for example, be a difference depending on which antigen is injected first. A similar example is the testing of some plant viruses by rubbing a certain concentration of the virus over the surface of a leaf and counting the resulting lesions. Since different leaves are differentially susceptible, a conventional way of measuring the strength of the virus is to wipe it over half of the leaf on one side of the midrib, rubbing the other half of the leaf with a control or standard solution.

Another design leading to paired comparisons occurs when the treatment is given to two individuals sharing a common experience, whether genetic or environmental. For example, a drug or a psychological test might be given to groups of twins or sibs, one of each pair receiving the treatment, the other one not.

Finally, the paired-comparison technique may be used when the two individuals to be compared share a single experimental unit and are thus subjected to common environmental experiences. If we have a set of rat cages, each of which holds two rats, and we are trying to compare the effect of a hormone injection with that of a control, we might inject one of each pair of rats with the hormone and use its cage mate as a control. This design would yield a $2 \times n$ anova in which the n cages would represent the blocks.

One reason for featuring the paired comparisons test in a separate section is that it alone among the two-way anovas without replication has an equivalent and traditional alternative method of analysis, the t-test for paired comparisons, which we will discuss after we have illustrated how to approach this problem by the method of randomized complete blocks.

The paired-comparison case in Box 11.5 analyzes face widths of five- and six-year-old girls, as already mentioned. The question posed is whether the faces of six-year-old girls are significantly wider than those of five-year-old girls. The data for 15 individual girls are shown in columns (1) and (2) of Box 11.5. The marginal means are furnished in column (3) and below columns (1) and (2). The computations for the two-way anova without replication in Box 11.5 are the same as those already shown for Box 11.3. Remember that since there is no replication, the total sum of squares is identical to the subgroup sum of squares. The anova table shows that there is a highly significant difference in face width between the two age groups. If interaction is assumed to be zero, there is a large added variance component among the individual girls, undoubtedly representing genetic as well as environmental differences.

The other method of analyzing paired comparisons designs is the well-known **t-test for paired comparisons.** It is simple to apply and is illustrated in part II of Box 11.5. This method tests whether the mean of sample differences between pairs of readings in the two columns is significantly different from a hypothetical mean, zero under the null hypothesis. The standard error over which this mean is tested is the standard error of the mean difference. The difference column (column (4) of the data table) must be calculated. The computations are straightforward, and the conclusions are the same as for the two-way anova. This is another instance in which we obtain the value of F_s when we square the value of t_s (see Section A.7 in the appendix for a proof).

Although the paired-comparisons t-test is the traditional method of solving this type of problem, we prefer the two-way anova. Its computation is no more time-consuming and has the advantage of providing a measure of the variance component among the blocks. This is useful knowledge because if there is no significant added variance component among blocks, one might simplify the analysis and design of future, similar studies by employing a completely randomized anova.

In the example in Box 11.5 we see again how important it is that we control the blocks (individuals) in this design. We could simply have analyzed these data as a completely randomized anova or as a t-test of the difference between two means. If we had done so we would have found that the two age groups are not significantly different. A completely randomized anova would yield an F_s-ratio for ages of 3.172, which is not greater than $F_{.05[1,28]}$. We could have predicted such an outcome. If we estimate the completely randomized error mean square by Expression (11.3) from the anova table in Box 11.5, we obtain a value of $MS_{E(CR)} = 0.091318$, which yields a relative efficiency of

$$RE = \frac{100 \times 0.091318}{0.000771} = 11{,}838.0\%$$

for the paired-comparisons (= randomized-blocks) design over the completely randomized design, a tremendous improvement. Consequently, it does not surprise us that failure to allow for differences among individual girls will lead us to erroneous results about the differences in face width.

11.6 Unequal Subclass Sizes

All of the examples of two-way anovas that we have considered thus far have had equal sample sizes for each subclass. Even with the best of intentions, however, an investigator often simply cannot produce such a balanced design. A given single replicate may be lost or ruined, or the data may vary naturally beyond the control of the investigator. For instance, in an experiment on the effects of treatment of parents on weight of their offspring, the number of offspring per parent is likely to differ among parents and will not be under the control of the investigator.

How can we analyze such data? As we will see, the analysis of two-way anovas with unequal subclass sizes is more complicated than any we have taken up so far. All other things being equal, we try to avoid such problems, sometimes even at the expense of losing some information. Thus if in an analysis with four rows and three columns we have at least seven replicates per subclass, but in some subclasses there are eight or nine replicates, we might be tempted to reduce the sample size of all subclasses to seven and proceed by the methods learned in the previous sections. Such a procedure is legitimate only if the subclass sample size itself does not affect the variable under study. For example, the number of offspring per mother might affect their weight. Another consideration is that in some experiments the number of replicates per subclass may itself be affected by the treatments and be a variable of interest to the investigator.

One way to deal with such problems is by the analysis of covariance, which we discuss in Section 14.9. Removal of any individuals from subclasses (to equalize sample sizes) must be done at random, and each individual within a subclass must have an equal likelihood of being removed in the process of reducing sample size. In some instances, reducing the subclasses to a common sample size is not advisable because error variance is very great and all available degrees of freedom are needed for the analysis of variance or because the original data were scarce and expensive to obtain and the time and expense of a complete analysis would be insignificant in comparison with the cost of obtaining the data. If the readings were measurements of fossils collected as the result of a long and expensive expedition to a distant land, clearly the investigator would wish to use every specimen, even though subclass sizes would be unequal.

One instance of unequal subclass sizes is relatively simple to analyze: a two-way anova with *unequal but proportional subclass sizes (numbers)* in which the ratios of the sample sizes of any one pair of rows are the same for all columns (or conversely, the ratios of sample sizes of any pair of columns are the same for all rows). Box 11.6 illustrates a simple example, an experiment studying the effect of thyroxin injections versus a control on the weight of seven-week-old chicks. These effects were studied for 12 male chicks and 8 female chicks each. Note that the ratio of number of males over females (the two rows), which is $12/8 = 1.5$, is the same for both columns—for the controls as well as for the experi-

Box 11.6 TWO-WAY ANOVA WITH UNEQUAL BUT PROPORTIONAL SUBCLASS NUMBERS.

Influence of thyroxin injections on seven-week weight of chicks (in grams). Sample sizes: $n_{11} = n_{12} = 12$; $n_{21} = n_{22} = 8$; total number of items $= \sum^a \sum^b n_{ij} = 40$. This is a Model I anova.

	Treatment ($a = 2$)		
Sexes ($b = 2$)	Control	Thyroxin injection	$\bar{Y}_B$
Males	560	410	
	500	540	
	350	340	
	520	580	
	540	470	
	620	550	
	600	480	
	560	400	
	450	600	
	340	450	
	440	420	
	300	550	
	$\bar{Y} = 481.6667$	$\bar{Y} = 485.8333$	483.7500
Females	530	550	
	580	420	
	520	370	
	460	600	
	340	440	
	640	560	
	520	540	
	560	520	
	$\bar{Y} = 518.7500$	$\bar{Y} = 500.0000$	509.3750
$\bar{Y}_A$	496.5000	491.5000	19,760 = Grand sum

SOURCE: Data from Radi and Warren (1938).

BOX 11.6 CONTINUED

Preliminary computation

1. $\bar{\bar{Y}} = \sum^{a}\sum^{b}\sum^{n_{ij}} Y \div \sum^{a}\sum^{b} n_{ij} = 19{,}760 \div 40 = 494.000$

2. SS_A (SS of columns) $= \sum^{a}\left(\sum^{b} n_{ij}\right)(\bar{Y}_A - \bar{\bar{Y}})^2 = (20 \times 6.25 + 20 \times 6.25) =$ 250.00

3. SS_B (SS of rows) $= \sum^{b}\left(\sum^{a} n_{ij}\right)(\bar{Y}_B - \bar{\bar{Y}})^2 = (24 \times 105.0625 + 16 \times$ 236.3906) = 6303.75

4. SS_{AB} (Interaction SS) $= \sum^{a}\sum^{b} n_{ij}(\bar{Y} - \bar{Y}_A - \bar{Y}_B + \bar{\bar{Y}})^2$
   $= (12 \times 21.00694) + (12 \times 21.00694) + (8 \times 47.265625) + (8 \times 47.265625)$
   $= 1260.42$

5. SS_{within} (within subgroups; error SS) $= \sum^{a}\sum^{b}\sum^{n_{ij}} (Y - \bar{Y})^2 = 292{,}545.83$

Anova table

Source of variation	df	SS	MS
A (columns; treatments)	1	250.00	250.00
B (rows; sex)	1	6,303.75	6,303.75
$A \times B$ (interaction)	1	1,260.42	1,260.42
Within subgroups (error)	36	292,545.83	8,126.27
Total	39	300,360.00	

Since the error MS is greater than all other mean squares, none of the latter can be greater than what one would expect by chance. We therefore conclude that thyroxin injections do not have any effect on the weight of seven-week-old chicks and also that the sexes do not differ in this respect at this stage.

The method for testing significance in such an anova is discussed in Section 11.6.

mentals (thyroxin-injected). We can also demonstrate proportionality of subclass sizes by the converse relationship. The ratio of the two columns (control and thyroxin) is the same ($12/12 = 8/8 = 1.0$) for both rows (sexes).

The computation of the sums of squares, outlined in Box 11.6, follows the familiar series of steps explained in Box 11.2. There is a minor difference in symbolism because of the inequality of the subclass sample sizes. The completed analysis of variance is shown at the end of the box. It needs no special testing because the error variance in this example is far greater than the treatment mean squares.

However, how should we have tested if the treatment mean squares had been greater than the error variance? We would follow the procedure of Section 11.3, illustrated in Box 11.2. There is only one other complication: The coefficients of the variance components differ for each level of variation. We first encountered this phenomenon in Section 10.3, when we worked out examples of nested analysis with unequal subclass sizes. Whenever a given mean square has to be tested over the error variance, this difference in coefficients is of no importance because it does not enter into the equation. However, as soon as we test over interaction (Model II and fixed treatment effects of a mixed model), we have to allow for the differences in coefficients. The evaluation of these coefficients is somewhat tedious and we will not present it here. The reader who runs into such a problem should consult Snedecor and Cochran (1967, section 16.4), where an example is worked out in great detail.

In two-way anovas with disproportional as well as unequal subclass sizes *the computational procedures are considerably more complicated and beyond the scope of this text. The reason for the complication is that such analyses are no longer orthogonal—that is, the separate effects are not independent and their sums of squares do not add up to yield the total sums of squares. The reader faced with an anova with disproportional and unequal sample sizes is advised to consult Steel and Torrie (1980, chapter 18) and Snedecor and Cochran (1989, chapter 20), where this subject is treated in some detail. These computations were once discouragingly tedious, but now they can be carried out routinely and efficiently by means of a computer.*

Several of the widely available statistical program libraries, such as SAS, BMDP, SYSTAT, or SPSS feature such programs. The solutions are based on the general linear hypothesis, a unified method for dealing with anova and regression, and on iterative approximate estimation techniques. The difficulty is that there are several (at least five) ways of testing null hypotheses concerning the main effects and the interactions. All these methods give identical results when the design is balanced, but as soon as it becomes unbalanced, the results differ. Commercially available programs usually allow one to choose among at least some of these alternative ways of testing the null hypothesis. For example, the widely used SAS programs provide four different types of sums of squares for the unbalanced case. Persons with unbalanced designs may wish to consult professional statisticians before embarking on their analyses. For a readable discussion of this case, see Herr and Gaebelein (1978). More recently, exact solutions have been furnished for the Model II case (Khuri and Littell, 1987) and the mixed model (Gallo and Khuri, 1990).

Box 11.7 SINGLE MISSING OBSERVATION IN A TWO-WAY ANOVA WITHOUT REPLICATION.

Homogenate glycolysis for different development stages of frog embryos. Measurement of the variate (lactic acid production) is made at 6 stages: 0, 360, 720, 1200, 1600, and 2000 minutes after first cleavage (temperature constant at 20°C). Total sample size minus missing value $= ab - 1 = 23$.

"Clutches" refers to egg clutches randomly chosen from material available for study. Since stages are fixed treatment effects, this is a mixed model.

	a clutches ($a = 4$)					
b stages ($b = 6$)	1	2	3	4	ΣY	$\bar{Y}$
I	21.4	. . .	7.0	9.5	(37.9)	(12.6)
II	14.3	13.5	5.4	6.6	39.8	10.0
III	23.4	14.1	5.9	7.1	50.5	12.6
IV	29.1	8.2	4.2	3.2	44.7	11.2
V	26.6	13.5	4.9	6.0	51.0	12.8
VI	21.7	5.2	6.6	5.9	39.4	9.8
ΣY	136.5	(54.5)	34.0	38.3	(263.3)	
$\bar{Y}$	22.8	(10.9)	5.7	6.4		

SOURCE: Data from Cohen (1954).

Before the analysis can be carried out, we must calculate an estimate for the missing observation for stage I, clutch 2:

$$\hat{Y}_{ij} = \frac{b\left(\sum^{a-1} Y\right)_j + a\left(\sum^{b-1} Y\right)_i - \sum^{ab-1} Y}{(a-1)(b-1)}$$

$$\hat{Y}_{1,2} = \frac{6(37.9) + 4(54.5) - 263.3}{(6-1)(4-1)} = \frac{182.1}{15} = 12.1$$

This value is then entered into the above table and the anova is carried out as usual, except that the degrees of freedom for error and total are each decreased by unity, and the row *SS* and column *SS* are both adjusted by a special correction before their *MS* is computed.

The anova table *before* the adjustment of *SS* (but using the estimated value of $\hat{Y}_{ij} = 12.1$):

Source of variation	*df*	*SS*
A(columns; blocks; clutches)	3	1121.57
B(rows; treatments)	5	36.22
Error	15	184.72
Total	23	1342.51

Box 11.7 Continued

$$\text{Adjusted } SS_A \text{ (blocks)} = SS_A - \frac{\left[\left(\sum^{a-1} Y\right)_j + a\left(\sum^{b-1} Y\right)_i - \sum^{ab-1} Y\right]^2}{a(a-1)(b-1)^2}$$

$$= 1121.57 - \frac{[37.9 + 4(54.5) - 263.3]^2}{4 \times 3 \times (5)^2}$$

$$= 1121.57 - \frac{(255.9 - 263.3)^2}{300} = 1121.57 - 0.18$$

$$= 1121.39$$

$$\text{Adjusted } SS_B \text{ (treatments)} = SS_B - \frac{\left[\left(\sum^{b-1} Y\right)_i + b\left(\sum^{a-1} Y\right)_j - \sum^{ab-1} Y\right]^2}{b(b-1)(a-1)^2}$$

$$= 36.22 - \frac{[54.5 + 6(37.9) - 263.3]^2}{6 \times 5 \times (3)^2}$$

$$= 36.22 - \frac{(281.9 - 263.3)^2}{270}$$

$$= 34.94$$

Completed anova

Source of variation	df	SS	MS	F_s
A (columns; blocks; clutches)	3	1121.39	373.797	28.34**
B (rows; treatments)	5	34.94	6.988	
Error	14	184.72	13.19	

Conclusions

Lactic acid production did not differ among developmental stages but was quite heterogeneous among clutches of eggs (the added variance component is significant if interaction is assumed to be absent).

11.7 Missing Values in a Randomized-Blocks Design

Another irregularity that may be encountered (for some of the same reasons as those explained in the previous section) is the problem of missing values in a two-way analysis without replication. Since there is no replication, a missing value means that one of the factor combinations is not represented in the analysis. Box 11.7 shows such an example, in which the amount of lactic acid production is measured in frog embryos of six developmental stages. The embryos had been obtained from four different egg clutches, which for the purposes of this experiment could be considered to have been randomly obtained; the developmental stages represent fixed treatments. The anova is therefore a mixed model.

In this example the missing value is that for stage I, clutch 2. Such a situation may have come about from accidental loss of this replicate or because in this clutch all the individuals had already developed into stage II. The first step in solving such a problem is to find an estimated value $\hat{Y}_{ij}$ for the missing variate, which we do by means of the estimation equation shown in Box 11.7. Evaluating this formula for the present example, we obtain an estimated lactic acid production of 12.1 units for stage I, clutch 2. The column and row sums used in this estimation equation are the sums of the row and column in which the missing value occurs. For this reason the summation signs carry the indices $a - 1$ and $b - 1$. After the missing value has been estimated, the computations proceed as previously indicated (see Box 11.3). The resulting anova table is in Box 11.7. However, a further adjustment to the sums of squares has to be made. This adjustment for both main effects is also shown in Box 11.7. The adjustment for SS_B is small and that for SS_A is quite trivial, but in other instances this would not necessarily be so. Note that in the completed anova one degree of freedom has been removed from the error SS and from the total SS because of the missing value. Since the missing value was estimated from the marginal totals, it obviously does not represent an independent degree of freedom. Thus, although this is a 6×4 table, the total degrees of freedom are only 22.

When two values are missing in a randomized-complete-blocks design, a similar but iterative procedure is used. This procedure is described in Steel and Torrie (1980, section 9.6) and Snedecor and Cochran (1989, section 15.3). Other possible approaches (such as the analysis of covariance) are suggested by these references as well.

EXERCISES 11

11.1 Swanson et al. (1921) determined soil pH for various soil samples from Kansas. An extract of their data (acid soils) is shown here.

County	Soil type	Surface pH	Subsoil pH
Finney	Richfield silt loam	6.57	8.34
Montgomery	Summit silty clay loam	6.77	6.13
Doniphan	Brown silt loam	6.53	6.32
Jewell	Jewell silt loam	6.71	8.30
Jewell	Colby silt loam	6.72	8.44
Shawnee	Crawford silty clay loam	6.01	6.80
Cherokee	Oswego silty clay loam	4.99	4.42
Greenwood	Summit silty clay loam	5.49	7.90
Montgomery	Cherokee silt loam	5.56	5.20
Montgomery	Oswego silt loam	5.32	5.32
Cherokee	Bates silt loam	5.92	5.21
Cherokee	Cherokee silt loam	6.55	5.66
Cherokee	Neosho silt loam	6.53	5.66

Do subsoils differ in pH from surface soils (assume that there is no interaction between localities and depth for pH reading)? *Answer:* $F_S = 0.894$ with 1 and 12 degrees of freedom.

11.2 The following data were extracted from a Canadian record book of purebred dairy cattle. Random samples of 10 mature (five-year-old and older) and 10 two-year-old cows were taken from each of 5 breeds (honor roll, 305-day class). The average butterfat percentages of these cows were recorded, yielding a total of 100 butterfat percentages, broken down into 5 breeds and into 2 age classes. The 100 butterfat percentages are given below. Analyze and discuss your results.

Breed									
Ayrshire		Canadian		Guernsey		Holstein-Friesian		Jersey	
Mature	2-yr	Mature	2-yr	Mature	2-yr	Mature	2-yr	Mature	2-yr
3.74	4.44	3.92	4.29	4.54	5.30	3.40	3.79	4.80	5.75
4.01	4.37	4.95	5.24	5.18	4.50	3.55	3.66	6.45	5.14
3.77	4.25	4.47	4.43	5.75	4.59	3.83	3.58	5.18	5.25
3.78	3.71	4.28	4.00	5.04	5.04	3.95	3.38	4.49	4.76
4.10	4.08	4.07	4.62	4.64	4.83	4.43	3.71	5.24	5.18
4.06	3.90	4.10	4.29	4.79	4.55	3.70	3.94	5.70	4.22
4.27	4.41	4.38	4.85	4.72	4.97	3.30	3.59	5.41	5.98
3.94	4.11	3.98	4.66	3.88	5.38	3.93	3.55	4.77	4.85
4.11	4.37	4.46	4.40	5.28	5.39	3.58	3.55	5.18	6.55
4.25	3.53	5.05	4.33	4.66	5.97	3.54	3.43	5.23	5.72

11.3 King et al. (1964) gave the following results for a study of the amount of cotton (in grams) used for nesting material in both sexes of two subspecies of the deer mouse *Peromyscus maniculatus.*

		P. m. gracilis		*P. m. bairdii*
♂	$\bar{Y}$	2.9	$\bar{Y}$	1.7
	s	1.4	s	0.9
	n	24	n	24
♀	$\bar{Y}$	2.6	$\bar{Y}$	2.1
	s	1.0	s	1.0
	n	26	n	26

Analyze and interpret. Note that you will have to compute the error mean square as a weighted average of the individual variances. *Answer:* $MS_{\text{sex}} = 0.0624$, $MS_{\text{within}} = 1.1845$.

11.4 Blakeslee (1921) studied length/width ratios of second seedling leaves of two types of Jimsonweed called globe (G) and nominal (N). Three seeds of each type were planted in 16 pots. Is there sufficient evidence to conclude that globe and nominal differ in length/width ratio?

Pot identification number	Types					
		G			*N*	
16533	1.67	1.53	1.61	2.18	2.23	2.32
16534	1.68	1.70	1.49	2.00	2.12	2.18
16550	1.38	1.76	1.52	2.41	2.11	2.60
16668	1.66	1.48	1.69	1.93	2.00	2.00
16767	1.38	1.61	1.64	2.32	2.23	1.90
16768	1.70	1.71	1.71	2.48	2.11	2.00
16770	1.58	1.59	1.38	2.00	2.18	2.16
16771	1.49	1.52	1.68	1.94	2.13	2.29
16773	1.48	1.44	1.58	1.93	1.95	2.10
16775	1.28	1.45	1.50	1.77	2.03	2.08
16776	1.55	1.45	1.44	2.06	1.85	1.92
16777	1.29	1.57	1.44	2.00	1.94	1.80
16780	1.36	1.22	1.41	1.87	1.87	2.26
16781	1.47	1.43	1.61	2.24	2.00	2.23
16787	1.52	1.56	1.56	1.79	2.08	1.89
16789	1.37	1.38	1.40	1.85	2.10	2.00

11.5 The mean length of developmental period (in days) for 3 strains of houseflies at 7 densities is given below (data from Sullivan and Sokal, 1963).

	Strains		
	OL	BELL	bwb
Density per container			
60	9.6	9.3	9.3
80	10.6	9.1	9.2
160	9.8	9.3	9.5
320	10.7	9.1	10.0
640	11.1	11.1	10.4
1280	10.9	11.8	10.8
2560	12.8	10.6	10.7

Do these flies differ in developmental period with density and among strains? You may assume lack of strain × density interaction. *Answer:* $MS_{\text{error}} = 0.3426$, $MS_{\text{strains}} = 1.3943$.

1.6 The following data were presented by French (1976) in a study on energy utilization of the pocket mouse *(Perognathus longimembris)* during hibernation at different temperatures.

Restricted food				Ad libitum food			
8°C		18°C		8°C		18°C	
Animal no.	Energy used (kcal/g)	Animal no.	Energy used (kcal/g)	Animal no.	Energy used (kcal/g)	Animal no.	Energy used (kcal/g)
1	62.69	5	72.60	13	95.73	17	101.19
2	54.07	6	70.97	14	63.95	18	76.88
3	65.73	7	74.32	15	144.30	19	74.08
4	62.98	8	53.02	16	144.30	20	81.40
		9	46.22			21	66.58
$\bar{Y}$	61.37	10	59.10	$\bar{Y}$	112.07	22	84.38
(SE)	(2.53)	11	61.79	*(SE)*	(19.71)	23	118.95
		12	61.89			24	118.95
		$\bar{Y}$	62.50			$\bar{Y}$	90.30
		(SE)	(3.49)			*(SE)*	(7.17)

Although the sample sizes are unequal, they happen to be proportional. Is there evidence that the amount of food available affects the amount of energy consumed at different temperatures during hibernation?

11.7 Suppose that the reading for density 640 and strain BELL are missing in the data of Exercise 11.5. Can you still analyze the data? Do your results correspond to those obtained in the earlier exercise? *Answer:* $MS_{\text{error}} = 0.3398$, $MS_{\text{strains}} = 1.5350$.

11.8 In a study of the infestation of potato tubers by the aphid *Rhopalosiphoninus latysiphon,* Haine (1955) investigated the following 14 varieties of potatoes, which differ in their time of maturation.

No.	Variety	Time of maturation
1	Biene	medium late
2	Heida	medium late
3	Böhms Mittelfrühe	medium early
4	Monika	medium late–late
5	Johanna	medium early–medium late
6	Condor	medium early–medium late
7	Ostbote	medium late
8	Falke	medium late–late
9	Bevelander	medium early–medium late
10	Erdgold	medium late
11	Ackersegen	late
12	Primula	very early
13	Sieglinde	early
14	Vera	very early

Mean counts of aphids [transformed as log $(Y + 1)$; see Section 13.7] at 8 times during one year are given in the table below.

Varieties	Oct.	Nov.	Dec.	Jan.	Feb.	Mar.	Apr.	May
1	0.23	1.06	1.63	1.76	1.60	2.04	2.04	1.93
2	0.00	0.31	0.63	0.66	0.93	1.51	2.67	2.31
3	0.34	0.55	0.40	0.55	0.36	0.64	2.09	2.14
4	1.13	1.43	1.40	1.45	1.51	2.13	1.92	2.02
5	0.21	1.22	1.28	1.10	0.86	1.36	2.26	2.44
6	0.00	0.00	0.08	0.24	0.24	1.16	1.79	2.01
7	0.12	0.65	0.37	0.60	0.94	1.41	2.30	1.98
8	0.11	0.41	0.87	0.88	1.28	1.68	1.87	2.17

(*Table continued on page 368.*)

Varieties	Oct.	Nov.	Dec.	Jan.	Feb.	Mar.	Apr.	May
9	0.06	0.38	0.04	0.27	0.26	0.70	2.20	2.34
10	0.06	0.60	0.28	0.63	0.61	0.97	2.17	2.28
11	0.22	0.66	1.02	1.08	0.91	1.31	1.93	1.77
12	0.75	1.48	1.42	1.39	1.18	1.79	1.93	2.09
13	0.23	0.79	0.37	0.43	0.65	1.03	2.34	2.25
14	1.19	1.62	1.90	1.65	1.90	2.02	2.27	2.22

Carry out an analysis of variance. Design an orthogonal set of comparisons to test whether potatoes differing in time of maturation differ in their aphid infestations.

12 MULTIWAY ANALYSIS OF VARIANCE

We now extend the two-way analysis of variance of the previous chapter to the simultaneous consideration of three or more factors. In Section 12.1 we briefly discuss the nature of the design and the new problems encountered during the analysis. A three-factor (three-way) anova is illustrated in Section 12.2. Section 12.3 treats multiway anovas with more than three main effects. Section 12.4 enumerates some of the more common experimental designs, other than those already learned, that might be employed by readers of this book. Appropriate instances for application are given and references to a detailed exposition of each design are cited. Finally, in Section 12.5, we discuss computer methods for the analysis of variance.

12.1 THE FACTORIAL DESIGN

In Chapter 11 we analyzed the effects of two factors on a sample. There is no reason, however, to restrict the design of such an analysis to a consideration of only two factors. Three or more factors may be analyzed simultaneously using a method often called **factorial analysis of variance.** In the example in the next section we will analyze the effects of temperature, oxygen concentration, and cyanide ion concentration on survival time in minnows. The number of main effects is theoretically unlimited. An analysis involving as many as five main effects, however, is a rarity because even without replication within a subgroup, the number of experimental units necessary becomes very large, and it is frequently impossible or prohibitive in cost to carry out such an experiment. By way of illustration, imagine an experiment with only three factors. If each of the three factors were divided into four classes or levels, we would have to carry out the experiment on $4 \times 4 \times 4 = 64$ experimental units to represent each combination of factors. This is a substantial number. You will realize that this would not permit any measurement of the basic experiment error and we would have to employ an interaction term as an estimate of experimental error (on the assumption that no added interaction effect is present). In the minimal case of two levels per factor, a five-factor factorial anova would require $2^5 = 32$ experimental units, even for a single replicate.

There are also logistical difficulties with such large experiments. It may not be possible to run all the tests in one day or to hold all of the material in a single controlled environmental chamber. Thus treatments may be confounded with undesired effects if different treatments are applied under not quite the same experimental conditions.

Another problem that accompanies a factorial anova with several main effects is the large number of possible interactions. We saw that a two-way anova (two-factor factorial) had only one interaction, $A \times B$. A three-factor factorial has three **first-order interactions,** $A \times B$, $A \times C$, and $B \times C$; it also has a **second-order interaction,** $A \times B \times C$. A four-factor factorial has six first-order interactions, $A \times B, A \times C, A \times D, B \times C, B \times D$, and $C \times D$; four second-order interactions, $A \times B \times C, A \times B \times D, A \times C \times D$, and $B \times C \times D$; and one **third-order interaction,** $A \times B \times C \times D$. These numbers go up rapidly. $k!/[(m + 1)!(k - m - 1)!]$ mth-order interactions exist in a k-factor factorial, representing the number of combinations of $m + 1$ items out of the k factors. Not only is the computation of these interactions tedious, but the testing of their significance and, more importantly, their interpretation become exceedingly complex.

The assumptions underlying multiway anova are the same as those for two-way anova. The expected value for a single observation in a (replicated) three-factor case in which all three main effects, A, B, and C, are fixed treatment effects (Model I) is

$$Y_{ijkl} = \mu + \alpha_i + \beta_j + \gamma_k + (\alpha\beta)_{ij} + (\alpha\gamma)_{ik} + (\beta\gamma)_{jk} + (\alpha\beta\gamma)_{ijk} + \epsilon_{ijkl}$$

where μ equals the parametric mean of the population; α_i, β_j, and γ_k are the fixed treatment effects for the ith, jth, and kth groups of treatments A, B, and C, respectively; $(\alpha\beta)_{ij}$, $(\alpha\gamma)_{ik}$, and $(\beta\gamma)_{jk}$ are first-order interaction effects in the subgroups represented by the indicated combinations of the ith group of factor A, the jth group of factor B, and the kth group of factor C; $(\alpha\beta\gamma)_{ijk}$ is the second-order interaction effect in the subgroup representing the ith, jth, and kth groups of factors A, B, and C, respectively; and ϵ_{ijkl} is the error term of the lth item in subgroup ijk. This expression is analogous to Expressions (8.2), (10.2), and (11.1). As before, all the main effects may be random (Model II), or only some of them may be, in which case the factorial anova is a mixed model.

In the next section we will analyze a three-way factorial to illustrate the procedures and the problems encountered.

12.2 A Three-Way Factorial Anova

The example we describe here comes from a study of pollution by factory effluents. The investigators measured the effects of different concentrations of cyanide ion on the survival time of minnows. Five concentrations of cyanide were combined with three concentrations of oxygen and three temperatures. The details of the experiment are given in Box 12.1. For each combination of factors, 10 replicate fishes were examined. The data shown here, however, are only the means of the 10 readings and thus do not provide an estimate of within-subgroup variance.

The basic data are tabulated in the second table of Box 12.1. Note that since this is a three-factor example, the third factor (oxygen concentration) is arranged within each class or level of the first factor (temperature) because we are limited by the two-dimensionality of the printed page. This arrangement, however, does not imply a nested anova; the factors define a three-dimensional table.

Next we arrange the data as three two-way tables of means. We will symbolize the means of these tables by $\bar{Y}_{AB}$, $\bar{Y}_{AC}$, and $\bar{Y}_{BC}$. These tables are produced by summing all levels of the third factor for each two-factor combination and dividing by the number of levels. Thus, the value for temperature level 2 against cyanide level 3 is the sum of all the oxygen concentrations for that combination of temperature and cyanide, divided by 3, the number of oxygen concentrations. In this case $\bar{Y}_{A_2B_3} = \bar{Y}_{T_2Cy_3} = (86 + 99 + 81) \div 3 = 88.6667$. The means along the margins of these tables symbolized by $\bar{Y}_A$, $\bar{Y}_B$, and $\bar{Y}_C$ are the means for each level of the indicated factor, averaged over all levels of the other two factors. They can be most easily obtained by averaging the means of type $\bar{Y}_{AB}$ over the appropriate row or column of the two-way tables. Necessarily, the marginal means for the same factor are identical in different tables. Finally we need a symbol for the observations in the body of the three-way table. Since they are means, we label them $\bar{Y}_{ABC}$.

The computations in Box 12.1 are longer than those of previous anovas but are not more complicated. The deviations for the sums of squares are obtained from values in both the original data tables and the two-way tables. Note that in this example there is no replication for each subgroup; hence there will be no error mean square and, as in the example of Box 11.3, the subgroup *SS* is the same as the total *SS*. Thus step 9 is not carried out in the present example but would be undertaken in a three-way anova *with* replication.

The layout of the anova table is shown next in Box 12.1. Note that the source of variation is subdivided into three main effects, *A*, *B*, and *C*; three first-order interactions, $A \times B$, $A \times C$, and $B \times C$; and one second-order interaction, $A \times B \times C$. Had there been replication in these data, we would also have had a within-subgroups (error) line below the second-order interaction. The experiment with the minnows is clearly Model I. Therefore, the expected mean squares in the table consist simply of the error variance σ^2 and the added effect due to treatment for the particular source of variation. The significance test is very simple. The mean square for any source of variation is tested over the error mean square. Since we do not have a pure measure of error in this example, we must use the second-order interaction mean square on the assumption that the added effect due to the $A \times B \times C$ interaction $\Sigma^{abc}(\alpha\beta\gamma)^2/(a-1)(b-1)(c-1)$ is

Box 12.1 THREE-WAY ANOVA WITHOUT REPLICATION (A 5 × 3 × 3 FACTORIAL ANOVA).

Time to intoxication by cyanide in *Phoxinus laevis,* a European minnow, using 5 concentrations of CN^- ion, 3 oxygen concentrations, and 3 temperatures. The variable is a transformation into logarithms of readings in minutes of survival time, coded and summed for 10 replicate fishes. Instead of presenting means, we furnish the sums, representing means coded × 10, as given in the original publication.

Factor *A* ($a = 3$) Temperatures in °C		Factor *B* ($b = 5$) CN^- concentrations in mg CN^-/ℓ		Factor *C* ($c = 3$) Oxygen concentrations in mg O_2/ℓ	
T_1	5	Cy_1	0.16	O_1	1.5
T_2	15	Cy_2	0.8	O_2	3.0
T_3	25	Cy_3	4.0	O_3	9.0
		Cy_4	20.0		
		Cy_5	100.0		

SOURCE: Data by Wuhrmann and Woker (1953).

In this example all three main effects are fixed treatments; it is therefore a Model I anova.

Factor A	Factor C	Factor B				
		Cy_1	Cy_2	Cy_3	Cy_4	Cy_5
T_1	O_1	201	150	131	130	97
	O_2	246	164	138	136	102
	O_3	271	170	149	127	99
		718	484	418	393	298
T_2	O_1	124	104	86	89	60
	O_2	158	111	99	91	74
	O_3	207	117	81	87	72
		489	332	266	267	206
T_3	O_1	79	63	50	51	32
	O_2	129	54	51	52	46
	O_3	142	93	62	51	52
		350	210	163	154	130

BOX 12.1 CONTINUED

For ease in computation, rearrange the data in three two-way tables of means as shown here. The value in each cell of a two-way table represents the mean of the items for the particular two-factor combination over all levels of the third factor. Thus, the value 239.3333 in the $A \times B$ table represents the combination $T_1 \times Cy_1$ averaged for the three oxygen concentrations O_1, O_2, and O_3 as follows: $(201 + 246 + 271) \div 3 = 239.3333$. The marginal means are computed from the bodies of the two-way tables and are used in computation.

Two-way tables of means

$A \times B$ (T × Cy in this example)

	Cy_1	Cy_2	Cy_3	Cy_4	Cy_5	$\bar{Y}_A$
T_1	239.3333	161.3333	139.3333	131.0000	99.3333	154.0667
T_2	163.0000	110.6667	88.6667	89.0000	68.6667	104.0000
T_3	116.6667	70.0000	54.3333	51.3333	43.3333	67.1333
$\bar{Y}_B$	173.0000	114.0000	94.1111	90.4444	70.4444	108.4000

$A \times C$ (T × O in this example)

	O_1	O_2	O_3	$\bar{Y}_A$
T_1	141.8000	157.2000	163.2000	154.0667
T_2	92.6000	106.6000	112.8000	104.0000
T_3	55.0000	66.4000	80.0000	67.1333
$\bar{Y}_C$	96.4667	110.0667	118.6667	108.4000

$B \times C$ (Cy × O in this example)

	O_1	O_2	O_3	$\bar{Y}_B$
Cy_1	134.6667	177.6667	206.6667	173.0000
Cy_2	105.6667	109.6667	126.6667	114.0000
Cy_3	89.0000	96.0000	97.3333	94.1111
Cy_4	90.0000	93.0000	88.3333	90.4444
Cy_5	63.0000	74.0000	74.3333	70.4444
$\bar{Y}_C$	96.4667	110.0667	118.6667	108.4000

The expressions for first-, second-, and third-order interaction SS in a four-factor factorial anova are

$$SS_{AB} = ncd \sum^{a}\sum^{b} (\bar{Y}_{AB} - \bar{Y}_A - \bar{Y}_B + \bar{\bar{Y}})^2$$

$$SS_{ABC} = nd \sum^{a}\sum^{b}\sum^{c} (\bar{Y}_{ABC} - \bar{Y}_{AB} - \bar{Y}_{AC} - \bar{Y}_{BC} + \bar{Y}_A + \bar{Y}_B + \bar{Y}_C - \bar{\bar{Y}})^2$$

$$SS_{ABCD} = n \sum^{a}\sum^{b}\sum^{c}\sum^{d} (\bar{Y}_{ABCD} - \bar{Y}_{ABC} - \bar{Y}_{ABD} - \bar{Y}_{ACD} - \bar{Y}_{BCD} + \bar{Y}_{AB} + \bar{Y}_{AC} + \bar{Y}_{AD} + \bar{Y}_{BC} + \bar{Y}_{BD} + \bar{Y}_{CD} - \bar{Y}_A - \bar{Y}_B - \bar{Y}_C - \bar{Y}_D + \bar{\bar{Y}})^2$$

Note the regular alternation of signs before the terms. Note also, that since the example in this box is a three-factor anova, only the first two expressions are used. In a three-factor anova, omit the coefficient d in SS_{AB} and SS_{ABC}.

In the formulas below we include n, sample size per subgroup, to make the presentation general. Because in the present example $n = 1$, it could have been omitted.

Preliminary computation

1. Grand mean $\bar{\bar{Y}} = \dfrac{1}{abcn} \sum^{a}\sum^{b}\sum^{c} = 4878 \div 45 = 108.4000$

2. SS_A (SS of temperatures) $= nbc \sum^{a} (\bar{Y}_A - \bar{\bar{Y}})^2 = 57{,}116.1333$

3. SS_B (SS of cyanide concentrations) $= nac \sum^{b} (\bar{Y}_B - \bar{\bar{Y}})^2 = 55{,}545.4667$

4. SS_C (SS of oxygen concentrations) $= nab \sum^{c} (\bar{Y}_C - \bar{\bar{Y}})^2 = 3758.8000$

5. SS_{AB} (temperature × cyanide interaction SS) $= nc \sum^{a}\sum^{b} (\bar{Y}_{AB} - \bar{Y}_A - \bar{Y}_B + \bar{\bar{Y}})^2 = 3685.8667$

6. SS_{AC} (temperature × oxygen interaction SS) $= nb \sum^{a}\sum^{c} (\bar{Y}_{AC} - \bar{Y}_A - \bar{Y}_C + \bar{\bar{Y}})^2 = 97.0667$

7. SS_{BC} (cyanide × oxygen interaction SS) $= na \sum^{b}\sum^{c} (\bar{Y}_{BC} - \bar{Y}_B - \bar{Y}_C + \bar{\bar{Y}})^2 = 5264.5333$

8. SS_{ABC} (temperature × cyanide × oxygen interaction SS)

$$= n \sum^{a}\sum^{b}\sum^{c} (\bar{Y}_{ABC} - \bar{Y}_{AB} - \bar{Y}_{AC} - \bar{Y}_{BC} + \bar{Y}_A + \bar{Y}_B + \bar{Y}_C - \bar{\bar{Y}})^2 = 1034.9333$$

Box 12.1 Continued

9. SS_{within} (within subgroups; error SS). Since this example is an anova without replication, there is no SS_{within}. However, if there had been replication, we could have calculated such an SS as follows:

$$SS_{\text{within}} = \sum^{a}\sum^{b}\sum^{c}\sum^{n}(Y - \bar{Y}_{ABC})^2$$

The computation of these quantities is self-evident.

Now fill in the anova table.

Source of variation		df	SS	MS	Expected MS (Model I)
Main effects					
$\bar{Y}_A - \bar{\bar{Y}}$	A	$a - 1$	**2**	$\frac{\mathbf{2}}{(a-1)}$	$\sigma^2 + \frac{bc}{a-1}\sum^{a}\alpha^2$
$\bar{Y}_B - \bar{\bar{Y}}$	B	$b - 1$	**3**	$\frac{\mathbf{3}}{(b-1)}$	$\sigma^2 + \frac{ac}{b-1}\sum^{b}\beta^2$
$\bar{Y}_C - \bar{\bar{Y}}$	C	$c - 1$	**4**	$\frac{\mathbf{4}}{(c-1)}$	$\sigma^2 + \frac{ab}{c-1}\sum^{c}\gamma^2$
First-order interactions					
$(\bar{Y}_{AB} - \bar{Y}_A - \bar{Y}_B + \bar{\bar{Y}})$	$A \times B$	$(a-1)(b-1)$	**5**	$\frac{\mathbf{5}}{(a-1)(b-1)}$	$\sigma^2 + \frac{c}{(a-1)(b-1)}\sum^{ab}(\alpha\beta)^2$
$(\bar{Y}_{AC} - \bar{Y}_A - \bar{Y}_C + \bar{\bar{Y}})$	$A \times C$	$(a-1)(c-1)$	**6**	$\frac{\mathbf{6}}{(a-1)(c-1)}$	$\sigma^2 + \frac{b}{(a-1)(c-1)}\sum^{ac}(\alpha\gamma)^2$
$(\bar{Y}_{BC} - \bar{Y}_B - \bar{Y}_C + \bar{\bar{Y}})$	$B \times C$	$(b-1)(c-1)$	**7**	$\frac{\mathbf{7}}{(b-1)(c-1)}$	$\sigma^2 + \frac{a}{(b-1)(c-1)}\sum^{bc}(\beta\gamma)^2$
Second-order interaction					
$(\bar{Y}_{ABC} - \bar{Y}_{AB} - \bar{Y}_{AC} - \bar{Y}_{BC} + \bar{Y}_A + \bar{Y}_B + \bar{Y}_C - \bar{\bar{Y}})$	$A \times B \times C$	$(a-1)(b-1)(c-1)$	**8**	$\frac{\mathbf{8}}{(a-1)(b-1)(c-1)}$	$\sigma^2 + \frac{1}{(a-1)(b-1)(c-1)}\sum^{abc}(\alpha\beta\gamma)^2$

If the anova had been replicated, all expected mean squares would have their added components multiplied by n. We would also have had an error term as follows:

$Y - \bar{Y}$	Within subgroups	$abc(n-1)$	**9**	$\frac{\mathbf{9}}{abc(n-1)}$	σ^2
$Y - \bar{\bar{Y}}$	Total	$abcn - 1$	$\sum$ **(2 . . . 9)**		

The expected mean squares shown above are for a Model I anova. We now show expected mean squares for Model II and mixed models. Since it is the most general case, the replicated design is shown. If you have a case without replication, simply eliminate the within-subgroups (error) term and set $n = 1$.

Expected mean squares for a three-way factorial anova (other models)

Source of variation	Model II	Mixed Model (A and B fixed, C random)
A	$\sigma^2 + n\sigma^2_{ABC} + nc\sigma^2_{AB} + nb\sigma^2_{AC} + nbc\sigma^2_{A}$	$\sigma^2 + nb\sigma^2_{AC} + \frac{nbc}{a-1}\sum^{a}\alpha^2$
B	$\sigma^2 + n\sigma^2_{ABC} + nc\sigma^2_{AB} + na\sigma^2_{BC} + nac\sigma^2_{B}$	$\sigma^2 + na\sigma^2_{BC} + \frac{nac}{b-1}\sum^{b}\beta^2$
C	$\sigma^2 + n\sigma^2_{ABC} + nb\sigma^2_{AC} + na\sigma^2_{BC} + nab\sigma^2_{C}$	$\sigma^2 + nab\sigma^2_{C}$
$A \times B$	$\sigma^2 + n\sigma^2_{ABC} + nc\sigma^2_{AB}$	$\sigma^2 + n\sigma^2_{ABC} + \frac{nc}{(a-1)(b-1)}\sum^{ab}(\alpha\beta)^2$
$A \times C$	$\sigma^2 + n\sigma^2_{ABC} + nb\sigma^2_{AC}$	$\sigma^2 + nb\sigma^2_{AC}$
$B \times C$	$\sigma^2 + n\sigma^2_{ABC} + na\sigma^2_{BC}$	$\sigma^2 + na\sigma^2_{BC}$
$A \times B \times C$	$\sigma^2 + n\sigma^2_{ABC}$	$\sigma^2 + n\sigma^2_{ABC}$
Within subgroups	σ^2	σ^2

BOX 12.1 CONTINUED

Anova table

Source of variation		df	SS	MS	F_s
A	Temperature	2	57,116.1333	28,558.0667	
B	Cyanide	4	55,545.4667	13,886.3667	
C	Oxygen	2	3,758.8000	1,879.4000	
$A \times B$	T × Cy	8	3,685.8667	460.7333	7.123***
$A \times C$	T × O	4	97.0667	24.2667	< 1 *ns*
$B \times C$	Cy × O	8	5,264.5333	658.0667	10.174***
$A \times B \times C$	T × Cy × O	16	1,034.9333	64.6833	
Total		44	126,502.8000		

$F_{.001[8,16]} = 6.19$

Since this is a Model I anova, the presence of each effect is tested by the ratio of its mean square over MS_{ABC}, which represents the error variance σ^2, on the assumption that the temperature × cyanide × oxygen interaction is zero.

Conclusions

Temperature × cyanide interaction and cyanide × oxygen interaction are significant; that is, the effect of cyanide ion on survival time depends on the oxygen concentration and the temperature of the water. When, as in this instance, interactions involving all main effects are significant, it is usually of no interest to test the main effect mean squares. A statement about the differences due to cyanide will not mean much unless we qualify it by specifying temperature and oxygen concentration. If we wish to assess the gross effects of these three variables, however, we can test them over the error MS; in this example we would find them highly significant.

zero. The meaning of a second-order interaction will be explained later in this section. As mentioned in Chapter 11, main effects in a Model I anova are rarely tested if the interactions are significant, but if desired, such a test is simply the ratio of the main effect mean square over the error mean square.

Box 12.1 also shows expected mean squares of a three-factor factorial assuming Model II and a mixed model in which factors A and B are fixed and factor C is random. In these expected mean squares we have taken the general case in which there is replication within subgroups. For anovas without replication one simply eliminates the last line of the table (the within-subgroups mean square) and sets the coefficient $n = 1$ throughout the table.

Testing significance in Model II and mixed model anovas is more complicated than for Model I. Let us look at a Model II anova. Inspection of the expected mean squares in Box 12.1 shows that $(\boldsymbol{A} \times \boldsymbol{B} \times \boldsymbol{C})/\boldsymbol{Error}$ is the appropriate test for the second-order interaction. If second-order interaction is present, we test first-order interactions over the second-order interactions. However, there is no exact test for the main effects. There is no mean square in the table that, when subtracted from the mean square of a main effect, will leave only the added variance component due to the main effect. Thus, to test main effects we have to create a synthetic denominator mean square, which can be done in several ways. For example, to test the mean square of A we could construct a denominator MS as the sum of the mean square of $A \times B$ and $A \times C$ minus the mean square of $A \times B \times C$ (in order to subtract the extra values of $\sigma^2 + n\sigma^2_{ABC}$ from the denominator). Significance tests involving such synthetic mean squares are carried out by the approximate method learned in Chapter 10 (see Boxes 10.6 and 10.7) and are subject to the same limitations. We will illustrate such a test in the next section.

In a mixed model some mean squares can be tested directly over an appropriate MS, but others may need construction of a synthetic denominator mean square (in the example in Box 12.1 all mean squares can be tested over an appropriate error MS).

The analysis of the minnow data is shown in an anova table at the end of Box 12.1. Results of the significance tests indicate that there is a highly significant temperature × cyanide interaction, as well as a cyanide × oxygen interaction, but no temperature × oxygen interaction. These relationships are shown in Figure 12.1, in which the data from the three two-factor tables of Box 12.1 have been graphed. Note that lines depicting the relation between cyanide and temperature and cyanide and oxygen are not parallel, but those depicting the temperature × oxygen relationship are more or less parallel. At the low concentrations of cyanide the effects of oxygen and temperature are much more marked than they are at the higher concentrations. This illustrates the meaning of interaction in these data. We could test the significance of the three main effects, all of which would be highly significant, but in view of the interactions this result would not be very meaningful. From Figure 12.1 it is clear that there is a

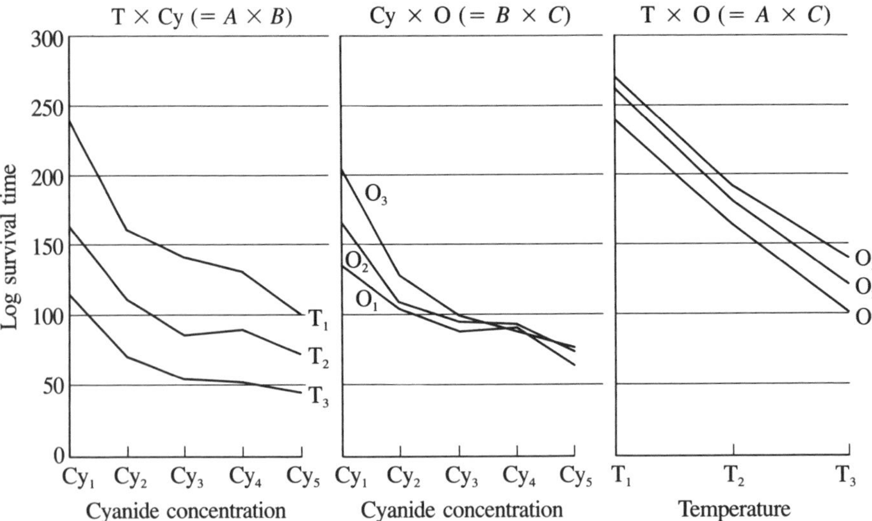

FIGURE 12.1 *Graphic demonstration of first-order interaction. Data from Box 12.1.*

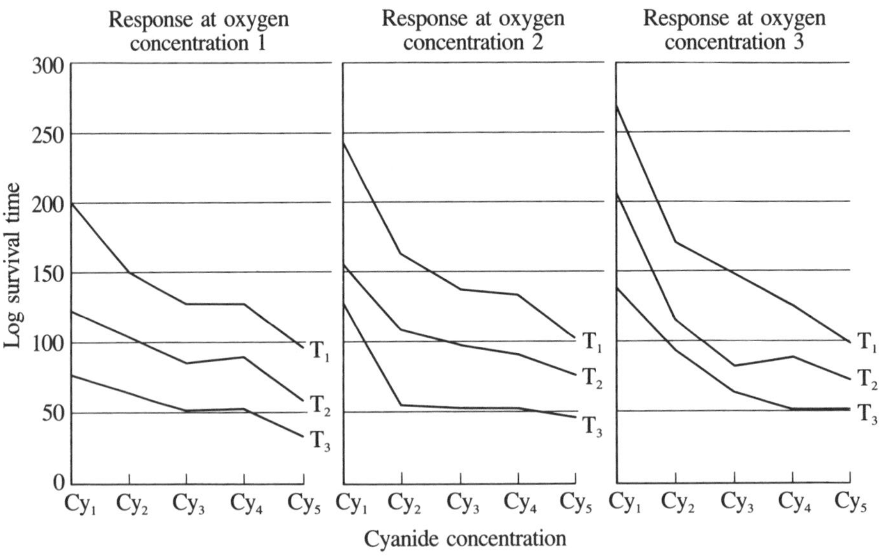

FIGURE 12.2 *Graphic demonstration of second-order interaction. Data from Box 12.1.*

substantial overall effect, at least of cyanide and temperature. As cyanide concentration increases, survival time of the fish decreases; also as temperature increases, survival time decreases.

The meaning of a second-order interaction requires some explanation. Interaction $A \times B \times C$ can be interpreted as indicating that interaction $A \times B$ differs depending on the level of factor C (or $A \times C$ depending on level of factor B, and so on). We can see this in Figure 12.2. The three panels of this figure depict the relationship between cyanide concentration and temperature separately for each oxygen concentration. Contrast this figure with the first panel of Figure 12.1, in which these relations were summarized for the whole experiment. One way of interpreting second-order interactions is to compare the three panels of Figure 12.2. Strong interaction would be reflected in marked differences in the patterns of the three temperature lines among the panels. Although some slight differences are evident, we assume that they were not of sufficient magnitude to have led to significant second-order interaction effects. Of course, we have no way of testing this assumption without an estimate of the error variance.

12.3 HIGHER-ORDER FACTORIAL ANOVAS

It follows from the previous section that raising the number of factors above three will increase the complexity of the analysis considerably. Not only must there be more experimental units to provide at least one replicate per experimental condition, but there will be many interaction terms to be evaluated and interpreted. Writing down the expected mean square and testing the significance of each source of variation in the anova is more complex than in any example that we have tackled so far. If you carefully follow the rules we will describe, however, you should not have too much difficulty.

Assume that you have a four-factor factorial without replication. We label the factors A, B, C, and D. Assume further that factors A, B, and C are fixed treatment effects (Model I) and factor D is random. The number of levels for each corresponding main effect will be a, b, c, and d.

One problem is to write down the expected mean squares for any model anova involving any number of factors. You were shown these mean squares for three cases of a three-way anova in Box 12.1. We have found the following scheme helpful for writing down the expected mean squares, as well as for subsequent significance testing. The arrangement is illustrated in Table 12.1, and the steps are as follows. First write the sources of variation along the left margin of the intended table, leaving three lines free at the top. Next copy the same sources across the top line labeled Identification, starting with the error at the left and going to the topmost main effect at the right. Each symbolizes a component of variation for either an error, an interaction, or a main effect. Note that we left out the multiplication sign between letters in interaction terms for simplicity of notation. Block out by means of heavy lines the various levels of the sources of variation (main effects, first-order interactions, second-order interactions, third-order interaction, and error) in columns as well as in rows of the table.

The second line across the top of the table indicates whether the component in question is an added component due to treatment (symbolized by Σ^2) or a variance component (symbolized by σ^2). In a pure Model I anova all components,

Table 12.1 EXPECTED MEAN SQUARES IN A FOUR-FACTORIAL ANOVA.

Mixed model: A, B, C are fixed treatments; D is random.

Identification[a]	Error	$ABCD$	BCD	ACD	ABD	ABC	CD	BD	BC	AD	AC	AB	D	C	B	A
Component	σ^2	σ^2	σ^2	σ^2	σ^2	$\sum^2$	σ^2	σ^2	$\sum^2$	σ^2	$\sum^2$	$\sum^2$	σ^2	$\sum^2$	$\sum^2$	$\sum^2$
Coefficient	1	n	na	nb	nc	nd	nab	nac	nad	nbc	nbd	ncd	$nabc$	$nabd$	$nacd$	$nbcd$
Source of variation																
A	*									*						*
B	*							*							*	
C	*						*							*		
D	*												*			
AB	*				*							*				
AC	*			*							*					
AD	*									*						
BC	*		*						*							
BD	*							*								
CD	*						*									
ABC	*	*				*										
ABD	*				*											
ACD	*			*												
BCD	*		*													
$ABCD$	*	*														
Error	*															

[a] Identifying letters are used as subscripts.

except the one representing error, are Σ^2. In a pure Model II, all components are σ^2. In a mixed model, any component whose identification does not contain a letter representing a random effect is an added component due to treatment; all others are variance components. Thus, in the example of Table 12.1, since the interaction component AC does not contain the random effect D, it is an added component due to treatment and is identified by Σ^2. By contrast, interaction component ACD, which contains D, is a variance component identified by σ^2.

Next we work out the coefficients for each component other than the error variance. These coefficients can be written down by a very simple rule. The maximum possible coefficient is *nabcd.* Omit from this formula any letters whose corresponding capitals occur in the first line. Thus, for the component *ABD,* the coefficient is *nc*. Similarly, for the component of *B*, the coefficient is *nacd.* In anovas without replication, we can set $n = 1$ and ignore it when writing down the coefficients. We are now able to write down any component in its proper form with the correct coefficients. Thus the component of the interaction ACD from Table 12.1 is $nb\sigma^2_{ACD}$, and that component for the main effect C is $nabd \sum \gamma^2/(c - 1)$. Remember that added components due to treatment are labeled by Greek letters and are always divided by the corresponding degrees of freedom.

We now determine the complete expected mean square for any source of variation by considering the following five rules. (1) *All sources of variation contain the error variance* as shown by an asterisk in that column. (2) *They also contain the component corresponding to their name.* For example, the main effect B contains the component for B, and interaction $A \times B \times C$ contains the component for *ABC.* The appropriate asterisks form a diagonal across the table starting at the lower left with the error mean square, and no mean square will have any component to the right of this diagonal.

Next, examine all components located between the error component and the final component in each line. (3) *Consider only those in blocks to the left of the block of the final component.* Thus, if you are working out a first-order interaction, consider only second- and higher-order interactions, and so on. (4) *In the blocks of interest, consider only the components that contain within them all the letters identifying the final component.* Thus, for main effect B consider only *AB*, *BC*, and *BD*, as well as all higher-order interactions containing B; do not consider *AC*, AD, or *CD.* For interaction *BC* consider only *ABC, BCD,* and *ABCD.* The final rule: (5) *If any of the letters in the interaction terms being considered (other than those identifying the final component in each line) represent a fixed treatment effect, this component is not included in the expected MS; if all letters represent random effects, include the component.*

Now let us apply these rules. In a pure Model I anova all letters represent fixed effects. It is obvious therefore by rule (5) that all components except the final component in the line [rule (2)] and the error variance [rule (1)] will be omitted from each expected mean square. The result is the characteristic expected mean squares of a Model I anova containing only the error variance and the component representing a given source of variation (see Box 12.1). By contrast, a pure Model II anova contains all possible interaction terms. In a four-factor Model II anova with replication, the expected mean square for factor B would therefore read

$$\sigma^2 + n\sigma^2_{ABCD} + na\sigma^2_{BCD} + nc\sigma^2_{ABD} + nd\sigma^2_{ABC} + nac\sigma^2_{BD} + nad\sigma^2_{BC} + ncd\sigma^2_{AB} + nacd\sigma^2_{B}$$

Expected mean squares of mixed model anovas depend on the particular combination of fixed and random factors. We will compute several mean squares in the mixed model example of Table 12.1 to make certain that the rules are understood. The mean square of A, a fixed treatment effect, will lead to consideration of three first-order interactions containing an A [rules (3) and (4)]. AB and AC include letters representing fixed factors B and C and hence are omitted [rule (5)], but since D is a random factor, the AD term is retained. Any second- or third-order interaction including A also contains at least one letter representing a fixed effect. Thus, ABD contains BD in addition to A; although D is random, B is fixed, so the component is excluded [rule (5)]. Therefore, the expected mean square of A (EMS_A) consists of three components, as indicated by the asterisks in Table 12.1.

$$EMS_A = \sigma^2 + nbc\sigma^2_{AD} + \frac{nbcd}{a-1}\sum^{a}\alpha^2$$

The expected mean square for main effect D contains only the error variance and the final component, since all other letters represent fixed effects and their components would be omitted [rule (5)]. Finally, let us work out the expected mean square for the first order interaction $A \times B$. We find AB contained in the second order terms ABC and ABD [rule (4)]. In ABC the letter C represents a fixed effect; hence the term is omitted [rule (5)]. In ABD the D is a random factor, so the term is included.

All included terms are represented by asterisks in Table 12.1. When filled in completely, the table is useful for pointing out how significance tests for given sources of variation should be carried out. From Table 12.1 it is clear that we would test the main effect of C over the $C \times D$ mean square or the interaction $B \times C$ over the $B \times C \times D$ interaction MS. The mixed model in Table 12.1 is a relatively simple case. If two of the main effects had been random, the situation would have been more complicated. One of the problems frequently encountered in Model II or mixed factorials is that there is no appropriate error term against which to test a given mean square. For instance, in the expected mean squares of the Model II anova shown near the end of Box 12.1, there is no exact test for the expected mean square of main effect B. The appropriate significance test must be carried out by an approximate procedure balancing the variance components as follows:

$$F_s = \frac{MS_B}{MS_{AB} + MS_{BC} - MS_{ABC}}$$

where the appropriate degrees of freedom for the denominator mean square have to be calculated, as in Box 10.6.

Another complication in calculating multifactor factorials is the computation of the higher-order interactions. We saw in Box 12.1 how this is done for a three-way anova. For higher-order interactions similar arrangements into subtables of the original data table may be useful for interpreting direct computational formulas for the interaction sums of squares. Box 12.1 gives formulas for calculating first-through third-order interactions (in a four-factor factorial), from which the reader should have no difficulty generating yet higher-order interactions in factorial anovas of any number of factors. Note the regular alternation of plus and minus signs.

The following shorthand notation for indicating the nature of the factorial design is frequently used. A three-factor factorial with each factor expressed at two levels is indicated as a $2 \times 2 \times 2$ factorial or 2^3 factorial. A two-way anova with each factor at three levels is a 3^2 factorial. A $2 \times 3 \times 4$ factorial has three factors—one at two levels, another at three levels, and the third at four levels. For many standard factorials, especially efficient computational methods have been developed, although their importance is lessening with the increasing use of computers. Factorial anovas involving treatments at only two levels can be computed with special elegance. Complete computations for a 2×2 factorial two-way anova, each factor at two levels with replication, are illustrated in Steel and Torrie (1980, section 15.3). Raktoe et al. (1981) discuss the general 2^k case, as do Dunn and Clark (1974, chapter 8).

12.4 OTHER DESIGNS

The analyses of variance that we have studied up to now are only an introduction to the field of **experimental design,** *which covers a great variety of designs developed in response to the needs of experimental scientists. A detailed consideration of these is beyond the scope of this book. Therefore, if your experiments do not seem to fit the simple designs already discussed, we strongly urge you to consult a professional statistician. However, we will briefly outline some of the more common designs, the reasons for using them, and sources for in-depth discussion and computational outline.*

In Chapter 11 we first became acquainted with the randomized-complete-blocks design, in which the rows of a two-way anova represent blocks of an experimental layout. In many instances it becomes necessary to lay out two-dimensional blocks. The most obvious cases are agricultural experiments, where fields may be heterogeneous not only in a north–south direction, but also in an east–west direction. In such a case, to test a single treatment (factor) at k *levels, the* **Latin square design** *is especially appropriate. We divide the field into* k *row blocks and* k *column blocks as shown in Figure 12.3. Each level of the treatment*

	Drier →			Wetter
Low N content	T_1	T_4	T_3	T_2
↓	T_3	T_2	T_1	T_4
↓	T_2	T_3	T_4	T_1
High N content	T_4	T_1	T_2	T_3

FIGURE 12.3 *An example of four chemical treatments (T_1 through T_4) laid out in a field according to a Latin square design to control known nitrogen content and moisture gradients.*

T_i *is replicated only once in a given row or column. For analysis of a Latin square experiment, see Steel and Torrie (1980, sections 9.10 to 9.14) and Dunn and Clark (1974, chapter 6).*

This design need not be restricted to agricultural research. Suppose we wish to make quantitative chemical analyses of four related substances in a search for chemical differences among them. The analysis of a single sample is a tedious procedure that takes all day and ties up one analytical apparatus. Assuming that we have four such apparatuses in our laboratory, we can now run our analyses over four days, arranging the four treatments in the manner of a Latin square; that is, any one substance would be tested only once on any one day and only once in any one apparatus. The resulting anova table would yield effects representing day-to-day differences as well as differences among apparatuses, but the mean square of greatest interest would be that relating to differences among the four substances.

Another common design is the **split plot design,** *which is for two (or more) factors. Each level of one factor (factor* A*) is assigned to a whole plot within a block, and each plot is subdivided into as many subplots as there are levels in* B*, which are then randomly allocated to these subplots. Two different error mean squares are computed: one, expected to be larger, for the plots, and a smaller one for the subplots. This design yields more precise information on the factor allocated to the split plots or subplots at the expense of losing information on the factor assigned to the whole plots. For instance, we may be studying the differences between five genetic strains at three temperatures. The three temperatures may be produced in different temperature chambers in one laboratory. The five genetic strains may be in separate culture bottles in the chambers. The reader may ask why such an analysis could not be done as a simple factorial of strains against temperatures. It could indeed, but it would require 15 different temperature chambers, one for each combination of temperature and strains. The discussion of restriction errors from the previous chapter applies here (see Section 11.4). Split plot designs are quite often incorrectly analyzed as factorial anovas. They are discussed in many texts. We find the discussion in Snedecor and*

Cochran (1989, section 16.15) quite clearly presented, and Steel and Torrie (1980) devote a whole chapter (16) to this subject.

Many other designs are found in textbooks of experimental design. For example, when in a design for which a randomized block is appropriate, we find it impossible to include all treatments in every block because of the amount of work or space, we must use an **incomplete block design.** *If only a fraction of combinations of factors (not all of them) are tested in a factorial design, it is called a* **fractional factorial.** *The proper design of such experiments is complex, requiring certain assumptions for the analyses to be meaningful. Restriction errors (see Section 11.4) are especially important in more complicated designs. A simple book on experimental design is Pearce (1965). Of the more comprehensive texts on experimental design, we expect that nonstatistical readers will find Raktoe et al. (1981), Montgomery (1966), and Gill (1978) the most readable. Other texts at an intermediate level are Ogawa (1974), Anderson and McLean (1974a), and Lindeman (1974).*

12.5 Anova by Computer

By now you should be impressed by the number of possible designs in analysis of variance. For most of the common designs the computations are relatively simple and straightforward when there are equal sample sizes. The main difficulty in writing programs for analysis of variance is to make them general enough. One would not wish to write separate programs for each possible type of analysis of variance.

Hartley (1962) has suggested the simple method of using a factorial anova without replication and with an indefinite number of factors (as many as necessary) to represent many other designs on the computer—that is, although our actual analysis may follow a different design, we present our data laid out as though they were a factorial anova and proceed to analyze them in that manner. This method results in replication being represented as a (dummy) factor. One can analyze the data as a factorial anova without replication and use the results of that anova to reconstitute the proper anova table for the design we had originally intended.

This strategy is made clear by Table 12.2, which lists the steps to be taken when using the BIOM-pc computer programs for some of the common designs we have studied. Look at the single-classification anova with equal n in the first row of this table. As you will recall from Section 9.3, such an anova has a groups and n items per group. We set up the data as though the groups were factor A of the factorial and the replicated items factor B, giving them the appearance of a two-way anova without replication, as in Box 11.2. The simulated factors are listed in the second column of Table 12.2. In the last column we are told how to reconstitute the degrees of freedom and sums of squares of the intended anova table. For our example, df and SS among groups are simply the corresponding

Table 12.2 COMPUTATION OF VARIOUS ANOVA DESIGNS IN THE FORM OF A FACTORIAL ANOVA WITHOUT REPLICATION.

For use with computer programs such as BIOM-pc.

Design	Simulated factors	Intended anova table in terms of *SS* and *df* of factorial anova
Single-classification anova (equal *n*)		
a groups	Groups $= A$	Groups $= A$
n items per group	Items $= B$	Within (error) $= B + (AB)$
Two-way anova with replication		
a columns	Columns $= A$	Columns $= A$
b rows	Rows $= B$	Rows $= B$
n replicates	Replicates $= C$	Interaction $= (AB)$
		Within subgroups $= C + (AC) + (BC) + (ABC)$
Nested anova		
a groups	Groups $= A$	Groups $= A$
b subgroups per group	Subgroups $= B$	Subgroups within groups $= B + (AB)$
n items per subgroup	Items $= C$	Within subgroups $= C + (AC) + (BC) + (ABC)$

SOURCE: Modified from Hartley (1962).

degrees of freedom and sums of squares for A; the within (error) term for degrees of freedom or sums of squares is the sum of the B term and the AB interaction. The other designs can be analyzed similarly. A little study of the relationships will show you how they were derived and will also give you interesting information about the relationships among the various designs. This scheme permits the computation of a great variety of balanced designs (with no missing replicates), using a single simple anova program. For further details, consult Hartley (1962).

More-comprehensive programs also exist that automatically evaluate the various *SS* and *df* for many different designs. They usually allow the design to be expressed as a linear model using a simple notation similar to that of Table 12.2. They also allow for unequal numbers of observations in each cell (a topic beyond the scope of this book). Widely available programs include those that are part of the Bio-medical Computer Programs (BMDP) (Dixon, 1985), those in SPSS (1990), those in SAS (1988), and those in SYSTAT (Wilkinson, 1989).

EXERCISES 12

12.1 Price (1954) investigated the survival of *Bacterium tularense* in lice under three relative humidities (RH 100%, 50%, and 0%), four temperatures (37, 29, 20, and 4°C), and on three substrates (killed lice, starved lice, and louse feces). Two replicate experiments were run. Write out the expected mean squares and make the appropriate tests of significance. Interpret. *Answer: EMS* for MS_T is $\sigma^2 + hs\sigma^2_{ET} + ehs\Sigma\tau^2/(t - 1)$.

Analysis of variance		
Source of variation	*df*	*MS*
Main effects		
Experiments *(E)*	1	0.1108
Humidities *(H)*	2	0.7613
Substrates *(S)*	2	0.0376
Temperatures *(T)*	3	3.1334
First-order interactions		
$E \times H$	2	0.1185
$E \times S$	2	0.1266
$E \times T$	3	0.0060
$H \times S$	4	0.1985
$H \times T$	6	0.0666
$S \times T$	6	0.0654
Second-order interactions		
$E \times H \times S$	4	0.0130
$E \times H \times T$	6	0.0601
$E \times S \times T$	6	0.0206
$H \times S \times T$	12	0.0475
Third-order interaction		
$E \times H \times S \times T$	12	0.0517

12.2 Analyze the data of Box 11.1 as if the experiment had been in the form of a three-factor anova without replication; that is, let three rats in each subgroup represent a dummy factor. In the resulting anova table show which *SS* and *df* must be pooled in order to obtain the correct results shown at the end of Box 11.1

12.3 Analyze the data given in Box 10.1 as if the experiment followed the form of a three-factor anova without replication. To do this you must introduce the dummy factors "females" (with four levels) and "measurements" (with two levels). In the resulting anova table indicate which *SS* and *df* must be pooled in order to obtain the correct anova. *Answer: SS* among females within cages = *SS* females + $SS_{F \times C}$.

12.4 Write out the expected mean squares for a five-factor experiment in which factors A, B, and C are random and factors D and E are fixed treatment effects. Let $n = 1$.

12.5 Skinner and Allison (1923) studied the effect on cotton of date of planting and of the addition of borax to fertilizer. For each treatment combination a single measurement was made of the weight of the green plants in pounds. Analyze and interpret. For the analysis, assume that the second-order interaction has negligible effects. Did the addition of borax affect the weight of harvested plants? *Answer:* $MS_{\text{error}} = 16.8741$, $MS_{\text{borax}} = 167.6296$.

		Method of application		
Date of planting	Amount of borax (pounds)	I In drill; seed planted 1 week later	II In drill; seed planted immediately	III Broadcast; seed planted immediately
	0	61	60	73
June 2	5	56	59	80
	10	58	61	55
	0	67	72	72
June 9	5	69	62	79
	10	67	58	68
	0	62	69	68
June 18	5	62	70	71
	10	57	72	64
	0	40	45	54
July 7	5	37	49	51
	10	29	33	41
	0	26	28	35
July 15	5	26	24	36
	10	21	23	33
	0	10	13	17
Aug. 3	5	10	11	14
	10	8	12	14

12.6 The oven-dry weights (in grams) of new growth in hybrid poplars grown in concrete soil frames and treated with lime (L), nitrogen (N), phosphorus (P), and potassium (K) are given below. The frames were laid out in three blocks. These data can be interpreted as a five-way orthogonal anova without replication. Carry out the analysis and test significance, pooling mean squares where desirable. Data from Lunt (1947). O indicates absence of treatment (controls).

	Treatments							
Blocks	O	P	PK	K	NK	N	NP	NPK
1	13.9	14.2	14.7	13.6	31.7	57.9	49.5	49.7
2	14.3	22.8	12.8	12.7	25.6	21.7	35.5	38.1
3	15.8	22.1	13.3	15.5	25.7	31.0	30.7	36.3

	Treatments							
Blocks	L	LP	LPK	LK	LNK	LN	LNP	LNPK
1	15.3	11.8	17.8	16.6	41.2	43.0	63.8	53.4
2	19.4	23.2	21.4	20.1	59.3	62.5	59.7	53.5
3	15.9	22.7	20.6	15.1	32.0	37.1	41.3	58.5

13 Assumptions of Analysis of Variance

We have blithely proceeded to study design after design in analysis of variance without bothering to discuss whether the analysis is appropriate and its test of significance reliable for a given set of data. In this chapter we will examine the underlying assumptions of the analysis of variance, methods for testing whether these assumptions are valid, the consequences if the assumptions are violated, and steps to be taken if the assumptions cannot be met. By taking up the fundamental assumptions of anova last in our consideration of the subject, we may seem to be building the superstructure before the foundation. Rigorous mathematical treatment clearly demands statement of the model prior to a treatment of the subject. Yet we have found that the generally nonmathematical audiences we have been addressing learn the concepts more easily if they first understand the structure and purpose of analysis of variance and are able to carry out the computations. We should stress, however, that before carrying out an anova on an actual research problem you should reassure yourself that the assumptions listed in this chapter seem reasonable, and if they are not, you should carry out one of several possible alternative steps to remedy the situation.

Section 13.1 refreshes your memory on a fundamental assumption, random sampling. In each of the next four sections we explain a special assumption of analysis of variance, describe a procedure for testing it, briefly discuss the consequences if the assumption does not hold, and (if it doesn't) give instructions on how to proceed. Section 13.2 treats the assumption of independence, 13.3 homogeneity of variance, 13.4 normality, and 13.5 additivity.

In many cases departure from the assumptions of analysis of variance can be rectified by transformation of the original data into a new scale. Section 13.6 gives a brief general introduction to transformations. The next four sections discuss some useful transformations. Section 13.7 decribes the logarithmic transformation, 13.8 the square root transformation, 13.9 the Box–Cox transformation, and 13.10 the arcsine or angular transformation.

When transformations are unable to make the data conform to the assumptions of analysis of variance, other techniques of analysis, analogous to the intended anova, must be employed. These are the nonparametric or distribution-free techniques, which are sometimes used by preference even when the parametric method (anova in this case) can be legitimately employed. Ease of computation and a preference for the generally simple assumptions of the nonparametric analyses cause many research workers to turn to them. When the assumptions of the anova are met, however, these methods are less powerful than analysis of variance. Section 13.11 examines several nonparametric methods in lieu of single-classification anova and Section 13.12 features nonparametric methods in lieu of two-way anova.

13.1 A Fundamental Assumption

All anovas require that sampling of individuals be random. Thus, in a study of the effects of three doses of a drug (plus a control) on five rats each, the five rats allocated to each treatment must be selected at random. If the five rats employed as controls are either the youngest or the smallest or the heaviest rats, while those allocated to some other treatment are selected in some other way, the results are not apt to yield an unbiased estimate of the true treatment effects. Nonrandomness of sample selection may well be reflected in lack of independence of the items (see Section 13.2), in heterogeneity of variances (Section 13.3), or in nonnormal distribution (Section 13.4). Adequate safeguards to ensure random sampling during the design of an experiment or when sampling from natural populations are essential.

13.2 Independence

An assumption stated in each explicit expression for the expected value of a variate [for example, Expression (8.2) was $Y_{ij} = \mu + \alpha_i + \epsilon_{ij}$] is that the error term ϵ_{ij} is a random normal variable. In addition, for completeness we should add that it is assumed that the ϵ's are independently and identically (see Section 13.3) distributed.

Thus, if the variates within any one group are arranged in a logical order independent of their magnitude (such as the order in which the measurements were obtained), we expect the ϵ_{ij}'s to succeed each other in a random sequence. Consequently, we assume a long sequence of large positive values followed by an equally long sequence of negative values to be quite unlikely. We would also not expect positive and negative values to alternate with regularity.

How could departures from independence arise? An obvious example is an experiment in which the experimental units are plots of ground laid out in a field. In such a case adjacent plots of ground often give similar yields. It would thus be important not to group all the plots containing the same treatment into an adjacent series of plots, but rather to randomize the allocation of treatments among the experimental plots. The physical process of randomly allocating the treatments to the experimental plots ensures that the ϵ's will be independent.

Lack of independence of the ϵ's can result from correlation in time rather than in space. In an experiment we might measure the effect of a treatment by recording weights of ten individuals. The balance we use may suffer from a maladjustment that results in giving successive underestimates, compensated for by several overestimates. Conversely, compensation by the operator of the balance may result in regularly alternating over- and underestimates of the true weight. Here again randomization may overcome the problem of nonindependence of errors. For example, we may determine the sequence in which individuals of the various groups are weighed according to some random procedure.

Both of these examples—the spatial and the temporal—are instances of positive **autocorrelation,** the self-similarity of variates adjacent in space or time. Regular alternation of positive and negative errors is a manifestation of negative autocorrelation.

Independence of errors in a sequence of continuous variates may be tested as first proposed by the well-known mathematician John von Neumann (von Neumann et al., 1941), with critical values tabulated by Young (1941). The test is based on successive differences between normal variates, $d_i = Y_{i+1} - Y_i$, which are squared. In Section 15.3 you will learn why the expected sum of such squared differences is twice the sum of squares of variable Y if the variates are independent. Thus in the case of independent errors the ratio $\eta = \Sigma d^2/\Sigma y^2$ should approximate 2. If there are sequences of similar variates, their differences will be less than what they would have been if the variates were randomly ordered, and the ratio η will be less than 2. Conversely, if there is a nonrandom alternation of the magnitudes of the variates, the variance of the differences will be greater than expected and η will be greater than 2. Young (1941) has tabulated critical values for $|1 - \eta/2|$, which we reproduce in Statistical Table **HH** up to a sample size of $n = 25$. When $n > 25$ we can use the normal approximation

$$t_s = \frac{|1 - \eta/2|}{\sqrt{(n - 2)/(n^2 - 1)}}$$

The computations are summarized in Box 13.1, where we examine the sequence of 25 numbers representing the aphid stem mother femur lengths from Box 2.1. We compute first differences to match all but the last observation; then we square and sum these d's. The result is an estimate of $\Sigma d^2 = 9.3700$. When we divide this value by the sum of squares of the femur lengths ($\Sigma y^2 = 0.1337$), we obtain $\eta = 2.9194$. Since, $\eta > 2$, a nonrandom alternation of variates is indicated. Computing $|1 - \eta/2|$, we obtain 0.459683, which in Table **HH** yields a two-tailed $P < 0.02$ for $n = 25$. Had this example been based on more than 25 observations, we could have tested it by using the normal approximation. If we do so in any case, we obtain $t_s = 2.394$, which yields $0.01 < P < 0.02$. We conclude that the observations occur in a sequence that appears to be nonrandom,

Box 13.1 TEST FOR SERIAL INDEPENDENCE OF A CONTINUOUS VARIABLE.

Twenty-five aphid stem mother femur lengths. Data from Box 2.1.

Y_i	$d_i^2 = (Y_{i+1} - Y_i)^2$
3.8	0.04
3.6	0.49
4.3	0.64
3.5	0.64
4.3	1.00
3.3	1.00
4.3	0.16
3.9	0.16
4.3	0.25
3.8	0.01
3.9	0.25
4.4	0.36
3.8	0.81
4.7	1.21
3.6	0.25
4.1	0.09
4.4	0.01
4.5	0.81
3.6	0.04
3.8	0.36
4.4	0.09
4.1	0.25
3.6	0.36
4.2	0.09
3.9	
$\Sigma y^2 = 3.2096$	$\Sigma d^2 = 9.3700$

Computation

1. Make a column of the observations. Construct a second column of first differences between the observations and square them as shown.
2. Compute the sum of squares of the observations and the sum of the squared differences (shown at the bottom of the columns).
3. Compute $\eta = \Sigma d^2/\Sigma y^2 = 9.3700/3.2096 = 2.9194$.

BOX 13.1 CONTINUED

4. Evaluate $|1 - \eta/2| = 0.459683$. If $n \leq 25$, consult Statistical Table **HH** for significance. In our case the two-tailed probability is $0.01 < P < 0.02$. For illustrative purposes we also evaluate

$$\frac{|1 - \eta/2|}{\sqrt{(n - 2)/(n^2 - 1)}}$$

and compare with $t_{\alpha[\infty]}$. This is the approximation we would use with $n > 60$. Since n for this example is 25, the approximation should be close. We find that $t_s = 0.459683/\sqrt{23/(25^2 - 1)} = 2.394348$, which in Statistical Table **B** is significant at $0.01 < P < 0.02$. The observations are not serially independent. The fact that η is greater than 2 suggests a nonrandom alternation of the observations. Values of $\eta < 2$ indicate serial correlation (= autocorrelation) between adjacent variates.

We used a two-tailed significance test here, since we had no a-priori notion of the nature of the departure from serial independence. In some instances our alternative hypothesis would be one-tailed, in which case the probabilities at the head of the columns of Table **HH,** or of Table **B** when using the normal approximation, should be halved.

so we question the assumption of independence in these data. Recent work has indicated that the femur lengths may have come from a dimorphic sample. Possibly the technician mounting the aphids on slides for measurement alternated between the two types of galls in a conscious (but misguided) attempt to strike a balance. A ratio of η significantly less than 2 would have indicated some serial correlation (= autocorrelation)—succeeding variates would be more similar to each other because of technician or instrument bias.

For a nonparametric serial correlation test of continuous variates, or when the variates are nominal, employ a runs test (see Section 18.2).

There is no simple adjustment or transformation to overcome the lack of independence or errors. The basic design of the experiment or the way in which it was performed must be changed. We have seen how a randomized-blocks design often overcomes lack of independence of error by randomizing the effects of differences in soils or cages. Similarly, in the experiment with the biased balance we could obtain independence of errors by redesigning the experiment, using different times of weighing as blocks. Of course, if a source of error is suspected or known, attempts can be made to remove it; if we know, for example, that the balance is biased, we may have it fixed. If the ϵ's are not independent, the validity of the usual F-test of significance can be seriously impaired.

13.3 HOMOGENEITY OF VARIANCES

In Section 9.4 and Box 9.6, in which we described the t-test for the difference between two means, we said that the statistical test was valid only if we could assume that the variances of the two samples were equal. Although we have not stressed it so far, this assumption that the ϵ_{ij}'s have identical variances also underlies the equivalent anova test for two samples—and in fact any type of anova. *Equality of variances* in a set of samples is an important precondition for several statistical tests. Synonyms for this condition are *homogeneity of variances* or **homoscedasticity,** a jawbreaker that makes students in any biometry class sit up and take notice. The term is coined from Greek roots meaning equal scatter; the converse condition (inequality of variances among samples) is called **heteroscedasticity.** Because we assume that each sample variance is an estimate of the same parametric error variance, the assumption of homogeneity of variances makes intuitive sense.

We have already seen how to test whether two samples are homoscedastic prior to a t-test of the differences between two means or a two-sample analysis of variance: We use an F-test for the hypotheses $H_0: \sigma_1^2 = \sigma_2^2$ and $H_1: \sigma_1^2 \neq \sigma_2^2$ (see Section 8.3 and Box 8.1). Markowski and Markowski (1990) point out that the reliability of these tests depends on the normality of the underlying distributions, but they make no suggestions for cases in which the normality assumption is violated. When the two sample sizes are equal, these authors also state that the t-test is insensitive to the heterogeneity of the variances, making preliminary tests of equality of the two variances unnecessary.

When there are more than two groups, **Bartlett's test for homogeneity of variances** is frequently suggested. This test is illustrated in Box 13.2 on lengths of molars for samples of eight species of fossil mammals. The computation is explained in the box. The final statistic of significance is X^2, which is compared with a critical value of chi squared. Note that the samples are highly heteroscedastic and that an ordinary analysis of variance is therefore not appropriate in this case. Regrettably, Bartlett's test is unduly sensitive to departures from normality in the data (see Section 13.4), and a significant X^2 may therefore indicate nonnormality rather than heteroscedasticity. For this reason many statisticians no longer recommend Bartlett's test when the normality of the distribution is in doubt.

There is also a "quick and dirty" method which, while not quite as efficient, is preferred by many because of its simplicity. This method is Hartley's (1950) $\mathbf{F_{max}}$**-test.** This test uses a statistic that is the ratio of the largest to the smallest of several sample variances. Critical values are given in Statistical Table **G.** The $F_{\max}$-test shown in Box 13.2 corroborates the findings of the more exact test by Bartlett. The variances are significantly heterogeneous.

An alternative test for heteroscedasticity, which is less sensitive to departures from normality (Martin and Games, 1977), is the *log-anova* or **Scheffé–Box test.** To carry out this test one forms subsamples of the variates in each group of an anova and separately calculates the variance of each subsample. These resulting variances are transformed to their natural logarithms, upon which a single-classification anova is carried out. If the resulting anova is significant, the variances among groups are significantly greater than would be expected on the

Box 13.2 TESTS OF HOMOGENEITY OF VARIANCES.

Lengths of the third molar of 8 species of the condylarth *Hyopsodus*.

Samples ($a = 8$)	(1) $df = n_i - 1$	(2) s_i^2	(3) $\ln s_i^2$
1	17	0.0707	−2.649310
2	12	0.1447	−1.933093
3	16	0.0237	−3.742280
4	15	0.0836	−2.481712
5	7	0.2189	−1.519140
6	10	0.1770	−1.731606
7	9	0.0791	−2.537042
8	9	0.2331	−1.456288
	95		

SOURCE: Data from Olson and Miller (1958).

Bartlett's test for homogeneity of variances

1. Convert the $a = 8$ variances to their natural logarithms.

2. Sum the degrees of freedom:

$$\sum^{a} (n_i - 1) = 17 + 12 + \cdots + 9 = 95$$

3. Compute a weighted average variance:

$$s^2 = \frac{\sum^{a} (n_i - 1)s_i^2}{\sum^{a} (n_i - 1)} = \frac{17(0.0707) + 12(0.1447) + \cdots + 9(0.2331)}{95}$$
$$= 0.112459$$

and finds it logarithm, $\ln 0.112459 = -2.185167$

4. Compute a weighted sum of the logarithms of the variances:

$$\sum^{a} (n_i - 1) \ln s_i^2 = 17(-2.649310) + 12(-1.933093) + \cdots + 9(-1.456288)$$
$$= -229.227545$$

5.

$$X^2 = \left[\sum^{a} (n_i - 1)\right] \ln s^2 - \sum^{a} (n_i - 1) \ln s_i^2$$
$$= (\text{quantity } \mathbf{2} \times \text{quantity } \mathbf{3}) - \text{quantity } \mathbf{4}$$
$$= 95\,(-2.185167) - (-229.227545) = 21.636721$$

Box 13.2 Continued

After X^2 is divided by a correction factor, quantity **6** below, it is distributed as $\chi^2_{[a-1]}$ when H_0 is true. Since this factor is usually only slightly greater than unity, there is no need to apply it unless the X^2 lies slightly above the borderline of significance.

Since $\chi^2_{.01[7]} = 18.475$, our value of X^2 is clearly significant, and the correction factor is unlikely to change this decision. We compute it simply to illustrate the formula.

6. Correction factor $$C = 1 + \frac{1}{3(a-1)}\left[\sum^{a} \frac{1}{n_i - 1} - \frac{1}{\sum^{a}(n_i - 1)}\right]$$

$$= 1 + \frac{1}{3(8-1)}\left[\frac{1}{17} + \frac{1}{12} + \cdots + \frac{1}{9} - \frac{1}{95}\right]$$

$$= 1 + \frac{1}{21}[0.725877] = 1.034566$$

$$\text{Adjusted } X^2 = \frac{X^2}{C} = \frac{21.63672}{1.034566} = 20.914 \qquad \text{with } a - 1 = 7\ df.$$

Since $\chi^2_{.01[7]} = 18.475$, we conclude that the variances of the eight samples are heterogeneous.

The F_{max}-test

Find the greatest variance, $s^2_{max} = 0.2331$, and the smallest, $s^2_{min} = 0.0237$. Then compute the maximum variance ratio: $s^2_{max}/s^2_{min} = 0.2331/0.0237 = 9.84$. Table **G** (cumulative probability distribution of $F_{max\ \alpha[a,\ n-1]}$) assumes that the *df* of all variances are equal, but we can make an approximate test using the lesser of the degrees of freedom of the two variances needed in computing the variance ratio ($n_8 - 1 = 9$). $F_{max.05[8,9]} = 8.95$, so the observed variance ratio would be considered significant at $P < 0.05$. The variances of the eight samples are heterogeneous.

The log-anova test for homogeneity of variance (Scheffé-Box test)

Because this test requires original variates, it cannot be applied to data such as the molar lengths featured earlier in this box, in which only means and variances of each sample are given. The test is therefore illustrated with another example, the scutum widths of tick larvae from Box 9.1.

1. The n_i observations in each groups are randomly divided into m_i subsamples. The number of subsamples, m_i, should approximate $\sqrt{n_i}$, and the n_{ij} for each subsample (sample size of the *j*th subsample of group *i*) should be taken as equal as possible. Applying these rules to the data in Box 9.1, the following arrangement of the variates is obtained.

Box 13.2 Continued

	Hosts ($a = 4$)			
	1	2	3	4
	372	364	348	342
	380	358	351	372
	382	342	362	374
	368	356	372	376
	374	338	344	344
	366	344	352	360
	360	350	360	
	376	376	362	
		366	366	
		350		
			354	
			342	
			358	
			348	
n_i	8	10	13	6
m_i	3	3	4	2
$Z_{ij} = \ln s^2_{ij}(\nu_{ij})$	3.332(2)	4.862(2)	3.995(2)	5.772(2)
	2.853(2)	4.431(2)	5.338(2)	5.545(2)
	4.852(1)	5.098(3)	2.234(2)	
			3.892(3)	
$\bar{Z}_i$	3.444	4.840	3.868	5.658
$\bar{\bar{Z}}$	4.342			

2. Compute the sample variance s^2_{ij} separately for each of the subsamples and let $Z_{ij} = \ln s^2_{ij}$. These values of Z_{ij} are shown below for each group followed by their degrees of freedom $\nu_{ij} = n_{ij} - 1$ in parentheses.
3. Carry out an analysis of variance of the Z_{ij} weighted by their respective degrees of freedom

$$MS_{\text{among}} = \frac{\sum^{a}(n_i - m_i)(\bar{Z}_i - \bar{\bar{Z}})^2}{a - 1} = 4.9060$$

$$MS_{\text{within}} = \frac{\sum^{a}\sum^{ni} \nu_{ij}(Z_{ij} - \bar{Z}_i)^2}{\sum^{a}(m_i - 1)} = 1.6240$$

BOX 13.2 CONTINUED

where

$$\overline{Z}_i = \frac{\sum^{m_i} v_{ij} Z_{ij}}{\sum^{m_i} v_{ij}} \text{ and } \overline{\overline{Z}} = \frac{\sum^{a} \sum^{m_i} v_{ij} Z_{ij}}{\sum^{a} \sum^{m_i} v_{ij}}$$

4.

$$F_s = \frac{MS_{\text{among}}}{MS_{\text{within}}} = \frac{4.9060}{1.6240} = 3.021$$

Since $F_{.05[3,8]} = 4.07$, the observed ratio is not significant, and we conclude that the variances are homogeneous.

basis of the average variances within groups—in other words, the groups are heteroscedastic. There is one minor complication in carrying out the anova of the logarithmic transforms: they are weighted by their degrees of freedom to allow for unequal subsample sizes. The details of the computation are shown at the end of Box 13.2, where the Scheffé–Box test is applied to the scutum widths of tick larvae from Box 9.1. The significance test at the end of the procedure indicates acceptance of the null hypothesis—the variances of the measurements do not appear to differ among the four hosts. The analysis of these data by anova is thus justified. The BIOM-pc package of computer programs carries out the three tests of homogeneity of variances discussed here.

What could be the reasons for the heteroscedasticity in the fossil mammals? Some of the populations may inherently vary more than others. Some species may be relatively uniform for one character, while others are quite variable for the same character. In an anova representing the results of an experiment, one sample may well have been obtained under less standardized conditions than the others and hence has a greater variance. There are also many cases in which the heterogeneity of variances is a function of an improper choice of measurement scale. With some measurement scales, variances vary as functions of means. Thus differences among means bring about heterogeneous variances. For example, in variables following the Poisson distribution, the variance is equal to the mean, so populations with greater means therefore have greater variances. Such departures from the assumption of homoscedasticity can often be easily corrected by a suitable transformation, as discussed later in this chapter.

A very rapid first inspection for heteroscedasticity checks for correlation between the means and variances or between the means and the ranges of the ratio $s^2/\overline{Y}$ or $s/\overline{Y} = V$ will be approximately constant for the samples. If means and variances are independent, these ratios will vary widely.

The consequences of moderate heterogeneity of variances are not too serious for the overall test of significance, but single degree-of-freedom comparisons may be far from accurate.

Box 13.3 APPROXIMATE TEST OF EQUALITY OF MEANS WHEN THE VARIANCES ARE HETEROGENEOUS. UNPLANNED COMPARISONS AMONG PAIRS OF MEANS USING THE GAMES–HOWELL METHOD.

Length of the third molar of 8 species of the condylarth *Hyopsodus*. Samples are listed by magnitude of $\bar{Y}$. The data were given in part in Box 13.2.

(1) Samples ($a = 8$)	(2) n_i	(3) $\bar{Y}_i$	(4) s_i^2	(5) $s_{\bar{Y}_i}^2 = \dfrac{s_i^2}{n_i}$	(6) $\dfrac{(s_{\bar{Y}_i}^2)^2}{(n_i - 1)}$
1	18	3.88	.0707	.00393	.000,000,907
2	13	4.61	.1447	.01113	.000,010,325
4	16	4.73	.0836	.00522	.000,001,820
3	17	4.79	.0237	.00139	.000,000,121
5	8	4.92	.2189	.02736	.000,106,958
6	11	4.96	.1770	.01609	.000,025,892
7	10	5.20	.0791	.00791	.000,006,952
8	10	6.58	.2331	.02331	.000,060,373

Computation

1. Make a table giving n, $\bar{Y}$, and s^2 for each sample as shown above in columns (2) through (4). It is most convenient to arrange the samples according to the magnitude of their means.
2. Compute the variance of each mean, $s_{\bar{Y}}^2$, and the quantity $(s_{\bar{Y}}^2)^2/(n - 1)$ for each sample as shown in columns (5) and (6).
3. The minimum significant difference, MSD_{ij}, between any pair (i, j) of means is given by the following expression (from Games and Howell, 1976).

$$MSD_{ij} = Q_{\alpha[k,\nu^*]}\,(s_{\bar{Y}_i}^2 + s_{\bar{Y}_j}^2)^{1/2}$$

where

$$\nu^* = \frac{(s_{\bar{Y}_i}^2 + s_{\bar{Y}_j}^2)^2}{\dfrac{(s_{\bar{Y}_i}^2)^2}{(n_i - 1)} + \dfrac{(s_{\bar{Y}_j}^2)^2}{(n_j - 1)}}$$

is the special weighted average degrees of freedom based on the quantities in columns (5) and (6) above. The critical value $Q_{\alpha[k,\nu^*]}$ is obtained from the table of the studentized range, Statistical Table **J.**

For the comparison $\bar{Y}_8 - \bar{Y}_1 = 6.58 - 3.88 = 2.70$ (the most deviant pair of means) the $MSD_{1,8}$ is computed as follows:

$$\nu^* = \frac{(0.00393 + 0.02331)^2}{(0.000{,}000{,}907 + 0.000{,}060{,}373)}$$

$$= \frac{(0.027424)^2}{0.000{,}061{,}280}$$

$$= 12.11$$

Box 13.3 Continued

In Table **J** we interpolate to obtain $Q_{.05[8,12.11]} = 5.111$. The *MSD* is then

$$MSD_{1,8} = 5.119(0.00393 + 0.02331)^{1/2}$$
$$= 5.119(0.16505)$$
$$= 0.8435$$

The means of samples 1 and 8 are clearly significantly different at a 5% experimentwise level of significance.

4. One may systematically test the differences between all pairs of means, most conveniently by making a table of the difference between each pair of means and then comparing it with a table of MSD_{ij} values, as was done in Box 9.11.

		j							
		1	2	4	3	5	6	7	8
	1	0	.5851	.4403	.3417	.9609	.7060	.5548	.8435
	2	.73*	0	.6098	.5528	1.0043	.7867	.6750	.9088
	4	.85*	.12	0	.3879	.9806	.7150	.5735	.8529
i	3	.91*	.18	.06	0	.9861	.6879	.5281	.8338
	5	1.04*	.31	.19	.13	0	1.0402	.9907	1.1120
	6	1.08*	.35	.23	.17	.04	0	.7628	.9576
	7	1.32*	.59	.47	.41	.28	.24	0	.8867
	8	2.70*	1.97*	1.85*	1.79*	1.66*	1.62*	1.38*	0

The differences $\bar{Y}_i - \bar{Y}_j$ are given below the diagonal, and the corresponding MSD_{ij} values for an experimentwise 0.05 level of significance are given above the diagonal. Differences greater than *MSD* are indicated with an asterisk.

From this table we conclude that the means for samples 1 and 8 differ from each other and from all other means and that the differences between the means for samples 2 through 7 do not differ significantly at the $\alpha = 0.05$ experimentwise level of significance.

What can we do if our data are inherently heteroscedastic? We may carry out an approximate test of equality of means on the assumption of heterogeneity of variances as shown in Box 13.3 (based on Games and Howell, 1976). The method performs unplanned comparisons between pairs of means using a studentized range with specially weighted average degrees of freedom and a standard error based on the averages of the variances of the means. We find that means for samples 1 and 8 differ from each other and from the other means (samples 2 to 7), which do not differ among themselves. The BIOM-pc package of computer programs carries out this test.

Box 13.4 WELCH'S APPROXIMATE *t*-TEST OF EQUALITY OF THE MEANS OF TWO SAMPLES WHOSE VARIANCES ARE ASSUMED TO BE UNEQUAL.

Comparison of chemical composition of the urine of apes. Milligrams of glutamic acid per milligram of creatinine.

	n	$\bar{Y}$	$s_{\bar{Y}}$
Chimpanzees	37	0.115	0.017
Gorillas	6	0.511	0.144

SOURCE: Data from Gartler et al. (1956).

First we reconstruct the variances of the two samples: $s^2 = s_{\bar{Y}}^2 \times n$

$$\text{Chimpanzees} \qquad s^2 = (0.017)^2 \times 37 = 0.010693$$

$$\text{Gorillas} \qquad s^2 = (0.144)^2 \times 6 = 0.124416$$

By the method of Box 8.1, $F_s = 0.124416/0.010693 = 11.64$. Since $F_{.001[5,30]} = 5.53$, it is not necessary to interpolate for the exact degrees of freedom. It is most improbable ($P \ll 0.001$) that the two variances were sampled from the same population. We therefore cannot use the t-tests of Section 9.4 and Box 9.6 and use the following approximate test instead:

$$t'_s = \frac{(\bar{Y}_1 - \bar{Y}_2) - (\mu_1 - \mu_2)}{\sqrt{\dfrac{s_1^2}{n_1} + \dfrac{s_2^2}{n_2}}}$$

where t'_s is expected to be distributed approximately as t' when the null hypothesis is true and the other quantities have their conventional meaning.

The critical value of t'_α for type I error α is computed as

$$t'_\alpha = \frac{t_{\alpha[\nu_1]}\dfrac{s_1^2}{n_1} + t_{\alpha[\nu_2]}\dfrac{s_2^2}{n_2}}{\dfrac{s_1^2}{n_1} + \dfrac{s_2^2}{n_2}} = \frac{t_{\alpha[\nu_1]}\, s_{\bar{Y}_1}^2 + t_{\alpha[\nu_2]}\, s_{\bar{Y}_2}^2}{s_{\bar{Y}_1}^2 + s_{\bar{Y}_2}^2}$$

We compute

$$t'_s = \frac{(0.115 - 0.511)}{\sqrt{(0.017)^2 + (0.144)^2}} = \frac{-0.396}{\sqrt{0.000289 + 0.020736}}$$

$$= \frac{-0.396}{\sqrt{0.021025}} = \frac{-0.396}{0.1450} = -2.73$$

Since $t_{.05[36]} = 2.028$ and $t_{.05[5]} = 2.571$, $s_{\bar{Y}_1}^2 = 0.000289$, and $s_{\bar{Y}_2}^2 = 0.020736$,

$$t'_{.05} = \frac{2.028(0.000289) + 2.571(0.020736)}{0.021025} = 2.564$$

BOX 13.4 CONTINUED

Since $|t'_s| > t'_{.05}$, we conclude that the two means are significantly different. The gorillas excrete more glutamic acid in their urine than do the chimpanzees.

Note that the formula for the critical value t'_α is a weighted average whose value must lie between $t_{\alpha[\nu_1]}$ and $t_{\alpha[\nu_2]}$. Therefore, if, as in our case, t'_s is greater than the t_α for the smaller number of degrees of freedom, we need not calculate t'_α but can accept t'_s to be significant. Conversely, if t'_s is smaller than t_α for the larger number of degrees of freedom, we can declare it not significant without further test. Only when t'_s lies between $t_{\alpha[\nu_1]}$ and $t_{\alpha[\nu_2]}$ do we need calculate t'_α.

When the sizes of the two samples are equal, $n_1 = n_2 = n$, we can simplify the formula for t'_s to

$$t'_s = \frac{(\bar{Y}_1 - \bar{Y}_2) - (\mu_1 - \mu_2)}{\sqrt{\frac{1}{n}(s_1^2 + s_2^2)}}$$

which equals Expression (9.3) (see Section 9.4), where the variances were assumed to be equal. For the homoscedastic case, however, $df = 2(n - 1)$. In this case with unequal variances the appropriate $df = n - 1$.

We apply the formula to the survival time of the female and male roaches of Box 8.1. Although we found these variances not to be significantly different, we might have reason to suspect that the variances of males and females were in fact different. Since the test of the present box is always more conservative than that of Box 9.6, in which we assume homoscedasticity, we might feel safer to assume $\sigma_1^2 \neq \sigma_2^2$. We repeat the original data:

Females ($n_1 = 10$)	$\bar{Y}_1 = 8.5$ days	$s_1^2 = 3.6$
Males ($n_2 = 10$)	$\bar{Y}_2 = 4.8$ days	$s_2^2 = 0.9$

$$t'_s = \frac{(8.5 - 4.8)}{\sqrt{\frac{1}{10}(3.6 + 0.9)}} = \frac{3.7}{\sqrt{0.45}} = \frac{3.7}{0.6708} = 5.52$$

In view of the critial value $t_{.001[9]} = 4.781$, the longer survival of females is clearly significant.

When there are only two means to be tested (see Section 9.4), we frequently test the hypothesis of equality of means by a t-test. Such a test assumes the equality of the two sample variances. When this assumption is not valid, we can perform **Welch's approximate t-test,** which is illustrated in Box 13.4. This test calculates an approximate t-value, t'_s, for which the critical value is calculated as a weighted average of the critical values of t based on the corresponding degrees of freedom of the two samples. The computation is quite simple. Gans (1991) recommends using Welch's test in cases of two-sample tests, whether heteroscedastic or not, and Markowski and Markowski (in a comment published below Gans's letter) concur that this test might be advisable for normally distributed data, but that other tests are better when the parent distributions differ from the normal.

A related test based on fiducial probability is the **Behrens–Fisher test,** which can be found in the explanatory comments for table VI in Fisher and Yates (1963). Unplanned multiple-comparison tests based on unequal variances are presented by Hochberg (1976) and Games and Howell (1976).

13.4 Normality

We have assumed that the error terms, ϵ_{ij}, of the variates in each sample will be independent, that the variances of the error terms of the several samples will be equal, and, finally, that the error terms are distributed normally. To test this last assumption we compute normal expected frequencies from the observed data and test the departure of the observed from the expected distributions. In Section 17.2 you will be introduced to two tests, the G-test and the Kolmogorov–Smirnov test for goodness of fit, either of which could routinely be applied to each sample in an anova. The BIOM-pc package of computer programs does just that. Alternatively, a graphic test as illustrated in Section 6.7 might be applied to each sample separately. Testing for $\gamma_1 = 0$ and $\gamma_2 = 0$ (see Section 7.9) is a partial test for normality. When the size of each sample is quite small, one can test normality for the entire analysis by computing the standardized deviates $(Y_{ij} - \bar{Y}_i)/s_i$ separately for each sample i, pooling the resulting deviates, and testing them for normality by any of the techniques mentioned here.

When examining a sample for normality, an investigator may notice single extreme variates, or **outliers.** These may be due to errors of measurement, to recording or transcription errors, to admixture of one or more individuals from a population different from the one under study, and so forth. If the investigator can assume that the population is normally distributed, then tests are available for detecting outliers. For sample sizes up to $n = 25$ we can use a test devised by Dixon (1950). We show how to apply this test using the data from Exercise 9.5. On examining the variates for the five samples, we discover that the lowest variate for sample 1 is 3.8, considerably below most other variates in the sample. Might this be an outlier?

To carry out Dixon's test, order the n variates from low to high or from high to low, so that the first variate, Y_1, is the suspected outlier. Then consult Statistical Table **CC.** Here we find that for samples of size 20, the critical ratio should be computed as

$$r_{22} = \frac{Y_3 - Y_1}{Y_{n-2} - Y_1}$$

where Y_1 is the suspected outlier and the other subscripted values of Y are the respective variates in an ordered array of the variates with the first being the suspected outlier. For sample 1 we obtain

$$r_{22} = \frac{4.5 - 3.8}{5.8 - 3.8} = 0.350$$

For $\alpha = 0.10$ Table **CC** shows that the critical value of this ratio is 0.401. Hence, we conclude that the suspected outlier is not sufficiently deviant to justify its deletion from the sample. Table **CC** is one-tailed. For a two-tailed test, double the values of α.

For sample sizes beyond the scope of Table **CC** ($n > 25$), we can use the test $(Y_1 - \bar{Y})/s$, where Y_1 is the suspected outlier, $\bar{Y}$ is the sample mean, and s is the sample standard deviation. Critical values for this ratio have been computed by Grubbs (1969) and can be looked up in Statistical Table **DD.** As an example we may examine the 37 urine samples of chimpanzees from Exercise 4.2. Is either the smallest or the largest extreme variate an outlier? We test $(0.008 - 0.115)/0.10404 = -1.028$ and $(0.440 - 0.115)/0.10404 = 3.124$. The second of these deviations is significant at a two-tailed probability of 0.05.

When outliers have been detected, they can be removed if external consideration of the sampling procedure and of the experimental design warrant it; an alternative procedure would be to **winsorize** the data. In winsorization the outliers in an ordered array are replaced by their neighboring values. Thus in the chimpanzee example, the abnormally high reading of 0.440 from Exercise 4.2 would be replaced by the next highest adjacent value, 0.370. The mean is then computed from the windsorized sample. It has been shown that for samples from normal distributions, the relative efficiency of estimating means by this technique is above 90% whenever sample sizes are greater than 5.

The consequences of nonnormality of error are not too serious, since means will follow the normal distribution more closely than the distribution of the variates themselves (a consequence of the central limit theorem; see Section 7.1). Only very skewed distributions would have a marked effect on the significance level of the F-test or on the efficiency of the design. The best way to correct for lack of normality is to carry out a transformation that will make the data normally distributed, as explained in later sections of this chapter. If no simple transformation is satisfactory, a nonparametric test, as carried out in Sections 13.11 and 13.12, should be substituted for the analysis of variance.

13.5 ADDITIVITY

In two-way or higher-order anova without replication it is necessary to assume that interaction is not present if one is to make tests of the main effects in a Model I anova. This assumption of no interaction in a two-way anova is sometimes also referred to as the assumption of **additivity** of the main effects. By this we mean that any single observed variate can be decomposed into additive components representing the treatment effects of a particular row and column, as well as a random term special to it. If interaction is present, the F-test will be very inefficient and possibly misleading if the effect of the interaction is very large. A check of this assumption requires either more than a single observation per cell (so that an error mean square can be computed) or an independent estimate of the error mean square from previous *comparable* experiments.

Interactions can be due to a variety of causes. Most frequently interaction results from a given treatment combination (such as level 2 of factor A when combined with level 3 of factor B) that makes a variate deviate from the expected sum of the treatment effects. Such a deviation is regarded as an inherent property of the natural system under study, as in examples of synergism or interference (see Section 11.2). Similar effects occur when a given replicate is quite aberrant, as may happen if an exceptional plot is included in an agricultural experiment, if a diseased individual is included in a physiological experiment, or if by mistake an individual from a different species is included in a biometric study. Finally, an interaction term will result if the effect of the two factors A and B on the response variable Y are multiplicative rather than additive. An example will make this clear.

In Table 13.1 we show the additive and multiplicative treatment effects in a hypothetical two-way anova. Let us assume that the expected population mean μ is zero. Then the mean of the sample subjected to treatment 1 of factor A and treatment 1 of factor B should be 2 by the conventional additive model. This is so because each factor at level 1 contributes unity to the mean. Similarly, the expected subgroup mean subjected to level 3 for factor A and level 2 for factor B is 8, since the respective contributions to the mean are 3 and 5. If the process is multiplicative rather than additive, however, as occurs in a variety of physicochemical and biological phenomena, the expected values are quite different. For treatment A_1B_1, the expected value equals 1, which is the product of 1 and 1. For treatment A_3B_2, the expected value is 15, the product of 3 and 5. If we were to

Table 13.1 ADDITIVE AND MULTIPLICATIVE EFFECTS.

	Factor A			
Factor B	$\alpha_1 = 1$	$\alpha_2 = 2$	$\alpha_3 = 3$	
	2	3	4	Additive effects
$\beta_1 = 1$	1	2	3	Multiplicative effects
	0	0.30	0.48	Log of multiplicative effects
	6	7	8	Additive effects
$\beta_2 = 5$	5	10	15	Multiplicative effects
	0.70	1.00	1.18	Log of multiplicative effects

analyze multiplicative data of this sort by a conventional anova, we would find that the interaction sum of squares would be greatly augmented because of the nonadditivity of the treatment effects. In this case, there is a simple remedy. By transforming the variable into logarithms (see Table 13.1), we are able to restore the additivity of the data. The third item in each cell gives the logarithm of the expected value, assuming multiplicative relations. Notice that the increments are strictly additive again ($SS_{A\times B} = 0$). As a matter of fact, on a logarithmic scale we could simply write $\alpha_1 = 0$, $\alpha_2 = 0.30$, $\alpha_3 = 0.48$, $\beta_1 = 0$, $\beta_2 = 0.70$. This is a good illustration of how transformation of scale, discussed in detail in later sections of this chapter, helps us meet the assumptions of analysis of variance.

If we wish to ascertain whether the interaction found in a given set of data can be explained in terms of multiplicative main effects, we can perform a test devised by Tukey (1949), which is illustrated in Box 13.5. This test is also useful when testing for nonadditivity in a two-way Model I anova without replication in experiments where it is reasonable to assume that intraction, if present at all, could be due only to multiplicative main effects. Tukey's test partitions the interaction sum of squares into one degree of freedom due to multiplicative effects of the main effects and a residual sum of squares to represent the other possible interactions or to serve as error in case the anova has no replication. The computations are straightforward. We find that in this example (oxygen consumption of two species of limpets), which did not have any significant interaction, the single-degree-of-freedom *SS* for nonadditivity due to multiplicative effects is also not significant. If the test for nonadditivity had been significant and the residual *SS* nonsignificant, then an analysis of log oxygen consumption would be preferred, since it would yield a simpler analysis. The BIOM-pc program contains Tukey's test for nonadditivity.

13.6 TRANSFORMATIONS

If the evidence indicates that the assumptions for an analysis of variance or a *t*-test cannot be maintained, two courses of action are open to us. We may carry out a different test not requiring the rejected assumptions, such as the distribution-free tests in lieu of anova discussed at the end of this chapter; or we may transform the variable to be analyzed in such a manner that the resulting transformed variates meet the assumptions of the analysis. Let us look at a simple example of what transformation will do. A single variate of the simplest type of anova (completely randomized, single classification, Model I) decomposes as follows: $Y_{ij} = \mu + \alpha_i + \epsilon_{ij}$. In this model the components are additive with the error term ϵ_{ij} distributed normally. However, we might encounter a situation in which the components were multiplicative in effect, where $Y_{ij} = \mu\alpha_i\epsilon_{ij}$, the product of these three terms. If we fitted the standard anova model as given here to these data, the observed deviations from the group means would lack normality and homoscedasticity.

Box 13.5 TUKEY'S TEST FOR NONADDITIVITY.

Oxygen consumption of 2 species of limpets *(Acmaea)* at 3 salinities. Data from Box 11.2. Since only the subgroup cell means are needed for computation, the individual observations are not given here. Each subgroup cell mean is based on $n = 8$ observations.

Factor B: Seawater concentration ($b = 3$)	(1) Factor A: Species ($a = 2$) *A. scabra*	(2) *A. digitalis*	(3) $\bar{Y}_B$	(4) $n\bar{Y}_B - n\bar{\bar{Y}}$
100%	10.5613	7.4288	8.9950	−4.9967
75%	7.8900	7.3375	7.6138	−16.0467
50%	12.1738	12.3263	12.2500	21.0433
$\bar{Y}_A$	10.2083	9.0308	$9.6196 = \bar{\bar{Y}}$	
$n\bar{Y}_A - n\bar{\bar{Y}}$	4.7100	−4.7100		
Q_i	614.3681	836.1846		

Computation

1. Form a two-way table of subgroup (cell) means. Compute row means divided by a, column means divided by b, and the grand mean divided by ab. These calculations yield values of $\bar{Y}_B$, $\bar{Y}_A$, and $\bar{\bar{Y}}$, respectively.
2. Evaluate the following expressions:

$$Q = n^3\sum^{a}\sum^{b} \bar{Y}_{ij}(\bar{Y}_{Ai} - \bar{\bar{Y}})(\bar{Y}_{Bj} - \bar{\bar{Y}}) = -1044.7515$$

$$K = n^4\sum^{a}(\bar{Y}_{Ai} - \bar{\bar{Y}})^2\sum^{b}(\bar{Y}_{Bj} - \bar{\bar{Y}})^2 = 32{,}179.5485$$

In these expressions $\bar{Y}_{ij}$ stands for the subgroup mean for the ith column and the jth row, and $\bar{Y}_{Ai}$ and $\bar{Y}_{Bj}$ represent the $\bar{Y}_A$ mean for column i and the $\bar{Y}_B$ mean for column j, respectively.

3. *SS* for nonadditivity $SS_{\text{nonadd}} = \dfrac{Q^2}{Kn} = \dfrac{(-1044.7515)^2}{(32{,}179.5485 \times 8)} = 4.2399$

This *SS* with one degree of freedom is a part of the interaction *SS* (calculated in Box 11.2).

BOX 13.5 CONTINUED

Source of variation	df	SS	MS
$A \times B$	2	23.9262	
Nonadditivity	1	4.2399	4.2399
Residual	1	19.6863	19.6863

$F_s = MS_{nonadd}/MS_{resid}$ is less than one and hence clearly not significant. We have no evidence of nonadditivity (multiplicative effects of the factors A and B) in these data.

These quantities may appear to have been pulled out of a hat. Actually, the SS for nonadditivity is the explained sum of squares for the regression of the subgroup means on the deviations of their marginal means. The meaning of this sentence may not be clear until you have mastered the regression chapter through Section 14.4.

The general parametric mean μ is constant in any one anova, but the treatment effect α_i differs from group to group. Clearly, the scatter among the variates Y_{ij} would double in a group in which α_i was twice as great as in another. Assume that $\mu = 1$, the smallest $\epsilon_{ij} = 1$, and the greatest 3; then if $\alpha_i = 1$, the range of the Y's will be $3 - 1 = 2$. When $\alpha_i = 4$, however, the corresponding range will be four times as wide, from $4 \times 1 = 4$ to $4 \times 3 = 12$, a range of 8. Such data will be heteroscedastic. We can correct this situation simply by transforming our model into logarithms. We would therefore obtain $\log Y_{ij} = \log \mu + \log \alpha_i + \log \epsilon_{ij}$, which is additive and homoscedastic. The entire analysis of variance would then be carried out on the transformed variates.

At this point many of you will feel more or less uncomfortable about what we have done. Transformation seems to much like "data grinding." When you learn that often a statistical test may be made significant after transformation of a set of data, though it would not have been so without such a transformation, you may feel even more suspicious. What is the justification for transforming the data? It takes some getting used to the idea, but there is really no scientific necessity to employ the common linear or arithmetic scale to which we are accustomed. Teaching the "new math" in elementary schools has done much to dispel the naive notion that the decimal system of numbers is the only "natural" one. It takes extensive experience in science and in the handling of statistical data to appreciate the fact that the linear scale, so familiar to all of us from our earliest experience, occupies a similar position with relation to other scales of measurement as does the decimal system of numbers with respect to the binary or octal numbering systems and others. If a relation is multiplicative on a linear scale, it may make much more sense to think of it as an additive system on a logarithmic scale. The square root of a variable is another frequent transformation. The square root of the surface area of an organism may be a more appropriate measure of the fundamental biological variable subjected to physiological and evolutionary forces than is the area. This concept is reflected in the normal distribution of the square root of the variable as compared to the skewed distribution of areas. In many cases experience has taught us to express experimental variables not in linear scale but as logarithms, square roots, reciprocals, or angles. Thus, pH values are logarithms, and dilution series in microbiological titrations are expressed as reciprocals. As soon as you are ready to accept the idea that the scale of measurement is arbitrary, you simply have to look at the distributions of transformed variates to decide which transformation most closely satisfies the assumptions of the analysis of variance before carrying out an anova.

A fortunate fact about transformations is that very often several departures from the assumptions of anova are simultaneously cured by the same transformation to a new scale. Simply by making the data homoscedastic, we often also make them approach normality and insure additivity of the treatment effects.

When a transformation is applied, tests of significance are performed on the transformed data, but estimates of means are usually given in the familiar untransformed scale. Since the transformations discussed in this chapter are nonlinear, confidence limits computed in the transformed scale and changed back to the original scale would be asymmetrical. Stating the standard error in the original scale would therefore be misleading. In reporting results of research with variables that require transformation, you must furnish means in the untransformed scale followed by their (asymmetrical) confidence limits rather than by their standard errors.

An easy way to find out whether a given transformation will yield a distribution satisfying the assumptions of the anova is to plot the cumulative distributions of the several samples against a probability scale. By changing the scale of the second coordinate axis from linear to logarithmic, square root, or any other scale, we can see whether a previously curved line, indicating skewness, straightens out to indicate normality (you may wish to refresh your memory on these graphic techniques, which we studied in Section 6.7). We can look up class limits on transformed scales or, easier still, use a variety of available probability graph papers whose second axis is in logarithmic, angular, or other scale. Thus we not only test whether the data become more normal through transformation, but we can also get an estimate of the standard deviation under transformation as measured by the slope of the estimated line. The assumption of homoscedasticity implies that the slopes for the several samples should be the same. If the slopes are very heterogeneous, homoscedasticity has not been achieved. Alternatively, we can examine goodness-of-fit tests for normality (see Chapter 17) for the samples under various transformations. The transformation that yields the best fit over all samples will be chosen for the anova. It is important that the transformation not be selected on the basis of giving the best anova results, since such a procedure would distort the significance level.

Four transformations will be discussed here: the logarithmic transformation (Section 13.7), the square root transformation (Section 13.8), the Box–Cox transformation (Section 13.9), and the angular or arcsine transformation (Section 13.10). Two other transformations will be taken up in connection with regression analysis in Section 14.11.

13.7 THE LOGARITHMIC TRANSFORMATION

The most common transformation is the conversion of all variates into logarithms, usually common logarithms. Whenever the mean is positively correlated with the variance (greater means are accompanied by greater variances), the logarithmic transformation is likely to remedy the situation and make the variance independent of the mean. Frequency distributions skewed to the right are often made more symmetrical by transformation to logarithmic scale. We saw in the previous section and in Table 13.1 that logarithmic transformation is also called for when effects are multiplicative.

Table 13.2 illustrates the effect of a logarithmic transformation. These data are lengths of juvenile silver salmon, checked in a downstream trap during four different two-week periods. Note that the variances increase as the means do. This relation could have been foreseen from the increase in range of each sample with increase in mean if one compared samples based on approximately the same number of fish, such as the first and third samples. As soon as the variates are transformed to logarithms, the variances of the samples become independent of their means. In most cases this transformation also removes heteroscedasticity; in this particular example the heterogeneity among the variances is reduced but not eliminated—differences among the periods could be tested by the nonparametric techniques discussed in Section 13.11. To report these data, you should transform the means back into linear scale by looking up their antilogarithms. Remember that a statement of the standard error is of no value in these data. Rather, you first compute confidence limits in logarithmic scale, transform the confidence limits back to linear scale, and report the reliability of your sample estimate as such. Thus for the first sample ($n = 421$) we would report the mean and 95% confidence limits as follows:

$$\bar{Y} = \text{antilog}\,(\overline{\log Y}) = \text{antilog}\,(1.9068) = 80.681$$

$$\begin{aligned} L_1 &= \text{antilog}\,[\overline{\log Y} - t_{.05[420]}\sqrt{S^2_{\overline{\log Y}}/n}] = \text{antilog}\,[1.9068 - 1.966(0.001768)] \\ &= \text{antilog}\,(1.9033) = 80.038 \end{aligned}$$

$$L_2 = \text{antilog}\,[1.9068 + 1.966(0.001768)] = \text{antilog}\,(1.9102) = 81.329$$

Table 13.2 AN APPLICATION OF THE LOGARITHMIC TRANSFORMATION.

Length of the juvenile silver salmon Oncorhynchus kisutch *checked in a downstream trap during four different two-week periods (pooled over period 1933 to 1942).*

(1) Length in mm Y	(2) log of length $\log Y$	(3) Nov. 12–25 f	(4) Jan. 7–20 f	(5) Apr. 1–14 f	(6) Apr. 29–May 12 f
60	1.778	1	2	—	—
65	1.813	4	—	—	—
70	1.845	33	1	—	—
75	1.875	94	4	—	1
80	1.903	125	2	1	—
85	1.929	98	12	5	5
90	1.954	44	5	6	20
95	1.978	18	5	5	105
100	2.000	3	—	16	328
105	2.021	1	—	25	757
110	2.041	—	—	45	1371
115	2.061	—	—	57	1562
120	2.079	—	—	87	1372
125	2.097	—	—	65	781
130	2.114	—	—	20	388
135	2.130	—	—	5	159
140	2.146	—	—	5	51
145	2.161	—	—	5	18
150	2.176	—	—	1	6
155	2.190	—	—	—	4
160	2.204	—	—	1	1
165	2.217	—	—	—	1
n		421	31	349	6930
Untransformed variable					
$\bar{Y}$		80.962	83.710	116.977	115.551
s^2		45.176	83.2796	124.6719	82.5407
log transformation					
$\overline{\log Y}$		1.9068	1.9200	2.0661	2.0614
$s^2_{\log Y}$		0.001316	0.002603	0.001825	0.001166
Back-transformed means					
antilog $(\overline{\log Y})$		80.681	83.181	116.427	115.194

SOURCE: Data from Shapovalov and Taft (1954).

Note that these limits are asymmetrical around the mean: $L_2 - \bar{Y} = 81.329 - 80.681 = 0.648$, $\bar{Y} - L_1 = 80.681 - 80.038 = 0.643$. These limits differ from confidence limits computed from the original measurements, which would be as follows:

$$\bar{Y} = 80.962$$

$$L_1 = \bar{Y} - t_{.05[420]}\sqrt{s^2/n} = 80.962 - 1.966(0.3303) = 80.313$$

$$L_2 = 80.962 + 1.966(0.3303) = 81.611$$

When the variates to be transformed include zeros, a technical problem arises, since the logarithm of zero is negative infinity. In such cases the transformation $\log(Y + 1)$ avoids the problem. When values include numbers between zero and one, it may be desirable to code the variates by multiplying by 10,000, or some higher power of 10 to avoid negative characteristics in the logarithms. Although logarithms to any base are acceptable, common logarithms may be simplest.

Logarithmic transformations are frequently needed in the analysis of variables related to the growth of organisms, such as the example in Table 13.2.

13.8 The Square Root Transformation

When the data are counts—insects on a leaf or blood cells in a hemacytometer, for example—we frequently find the square root transformation of value. Remember that such distributions are likely to be Poisson rather than normally distributed and that in a Poisson distribution the variance is the same as the mean. Therefore, the mean and variance cannot be independent but will vary identically. Transforming the variates to square roots generally makes the variances independent of the means. When the counts include zero values, it is desirable to code all variates by adding 0.5. The transformation then is $\sqrt{Y + \frac{1}{2}}$. Improved transformations have been suggested by Anscombe (1948; $Y' = \sqrt{Y + \frac{3}{8}}$) and by Freeman and Tukey (1950; $Y' = \sqrt{Y} + \sqrt{Y + 1}$), but they are less widely used.

Table 13.3 shows an application of the square root transformation. Again, the sample with the greater mean has a significantly greater variance prior to transformation. After transformation the variances are not significantly different. To report means, the transformed means are squared again and confidence limits are reported in lieu of standard errors.

If there are only two counts that need to be compared, one cannot test means of transformed variables. In such a case we take advantage of the fact that the difference between two independent variates sampled from the same distribution has a variance twice that of the variates (see Section 15.3). We set

$$t_{\alpha[\infty]} = \frac{Y_1 - Y_2}{\sqrt{Y_1 + Y_2}} \tag{13.1}$$

Table 13.3 AN APPLICATION OF THE SQUARE ROOT TRANSFORMATION.

The data represent the number of adult Drosophila *emerging from 15 single-pair cultures for two different medium formulations (medium A contained DDT).*

(1) Number of flies emerging Y	(2) Square root of number of flies $\sqrt{Y}$	(3) Medium A f	(4) Medium B f
0	0.000	1	—
1	1.000	5	—
2	1.414	6	—
3	1.732	—	—
4	2.000	3	—
5	2.236	—	—
6	2.449	—	—
7	2.646	—	2
8	2.828	—	1
9	3.000	—	2
10	3.162	—	3
11	3.317	—	1
12	3.464	—	1
13	3.606	—	1
14	3.742	—	1
15	3.873	—	1
16	4.000	—	2
		15	15

Untransformed variable

	Medium A	Medium B
$\bar{Y}$	1.933	11.133
s^2	1.495	9.410

Square root transformation

	Medium A	Medium B
$\overline{\sqrt{Y}}$	1.299	3.307
$s^2_{\sqrt{Y}}$	0.2634	0.2099

Tests of equality of variances

$$F_s = \frac{s_2^2}{s_1^2} = \frac{9.410}{1.495} = 6.294^{**} \qquad F_{.025[14,14]} = 2.98 \quad F_{.005[14,14]} = 4.30 \qquad F_s = \frac{s^2_{\sqrt{Y_1}}}{s^2_{\sqrt{Y_2}}} = \frac{0.2634}{0.2099} = 1.255\ ns$$

TABLE 13.3 CONTINUED

Back-transformed (squared) means		
$(\overline{\sqrt{Y}})^2$	1.687	10.937
95% confidence limits		
$L_1 = \overline{\sqrt{Y}} - t_{.05} s_{\overline{\sqrt{Y}}}$	$1.299 - 2.145\sqrt{\dfrac{0.2634}{15}}$	$3.307 - 2.145\sqrt{\dfrac{0.2099}{15}}$
	$= 1.015$	$= 3.053$
$L_2 = \overline{\sqrt{Y}} + t_{.05} s_{\overline{\sqrt{Y}}}$	1.583	3.561
Back-transformed (squared) confidence limits		
L_1^2	1.030	9.324
L_2^2	2.507	12.681

SOURCE: Data from R. R. Sokal (unpublished results).

In this equation Y_1 and Y_2 are the two counts; they are also the best estimates of the parametric means μ_1 and μ_2 of counts from which the individual counts were sampled. Recall that for Poisson distributions $\mu = \sigma^2$; thus Expression (13.1) becomes

$$t = \frac{Y_1 - Y_2}{\sqrt{2s^2}}$$

Let us illustrate this with an example. During a given week two counties with approximately equal populations reported 17 and 32 new influenza cases. Do these figures indicate different morbidities in these two counties? Applying Expression (13.1), we obtain

$$t_s = \frac{17 - 32}{\sqrt{17 + 32}} = -\frac{15}{7} = -2.143$$

which is significant at $0.02 < P < 0.05$ (two-tailed test).

13.9 THE BOX–COX TRANSFORMATION

Often one has no a-priori reason for selecting a specific transformation. Rather than simply trying various transformations to find out which one works best, Box and Cox (1964) developed a procedure for estimating the best transformation to normality within the family of power transformations:

$$Y' = (Y^\lambda - 1)/\lambda \quad (\text{for } \lambda \neq 0) \tag{13.2}$$

$$Y' = \ln Y \qquad (\text{for } \lambda = 0) \tag{13.2a}$$

The generality of the Box–Cox transformation is evident from the fact that setting $\lambda = 1$ gives a simple linear transformation (which does not change the shape of the distribution), and that $\lambda = \frac{1}{2}$ is equivalent to the square root transformation, $\lambda = 0$ corresponds to the logarithmic transformation, and $\lambda = -1$ is equivalent to the reciprocal transformation. The value, $\hat{\lambda}$, which maximizes

$$L = -\frac{\nu}{2}\ln s_T^2 + (\lambda - 1)\frac{\nu}{n}\sum \ln Y \tag{13.3}$$

known as the **log-likelihood function,** yields the best transformation to normality within this family of transformations. In this expression s_T^2 is the variance of the *transformed Y* values (based on ν degrees of freedom). For a single sample s_T^2 would be the usual sample variance, but for a series of samples it could be the error mean square. The second term in the equation involves the sum of the natural logarithms of the original *untransformed* variates. The other terms are ν and n, which are degrees of freedom and sample sizes, as usual. Since $\hat{\lambda}$ must be found by an iterative procedure, the transformation is feasible only if performed by computer (it is included as an option in the BIOM-pc package of computer programs).

For the milk yield data of Table 6.1, the Box–Cox transformation after several iterations yields an estimate of $\hat{\lambda} = -2.1089$. Figure 13.1 shows L as a function of λ; L is maximized at $\hat{\lambda} = -2.1089$. The Box–Cox algorithm also estimates confidence limits to $\hat{\lambda}$. Since $-2L_{\max}$ approximately follows the $\chi^2_{[1]}$ distribution, the lower and upper confidence limits (λ_1 and λ_2) correspond to the pair of values of λ that have L values equal to $L_{\max} - \frac{1}{2}\chi^2_{\alpha[1]}$. Solutions for λ_1 and λ_2 can be obtained by solving for the two roots of $L - (L_{\max} - \frac{1}{2}\chi^2_{\alpha[1]})$ iteratively. For the milk yield data one obtains the limits $\lambda_1 = -3.412$ and $\lambda_2 = -0.828$ (see Figure 13.1). Since these limits exclude $\lambda = 1.0$ (linear transformation), we conclude that a transformation should be applied. Note that the

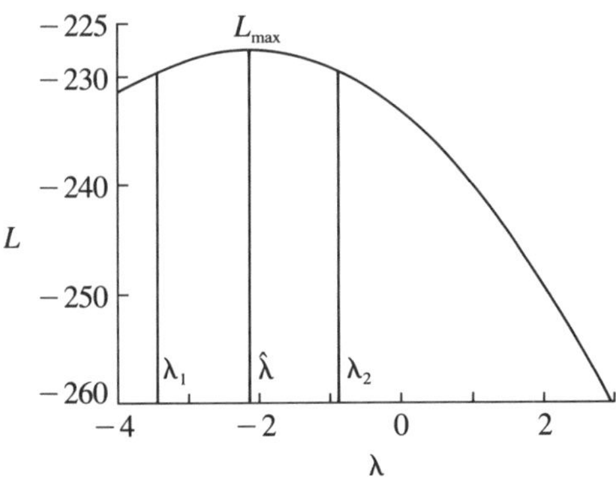

FIGURE 13.1 *Plot of the log-likelihood function L versus the Box–Cox parameter λ for the milk yield data of Table 6.1. The 95% confidence limits (λ_1, λ_2) are also shown.*

optimal transformation $\hat{\lambda} = -2.1089$ is approximately an inverse square transformation. There is no theoretical reason to suggest an inverse square law for milk yields—it is simply an empirical result. Its effectiveness is demonstrated by the fact that it reduces the significant $g_1 = 0.941$ for the untransformed milk yields to a nonsignificant $g_1 = 0.105$ after the Box–Cox transformation. Figure 13.2 shows the data before and after transformation plotted on arithmetic probability paper in the manner of Section 6.7. It is quite clear that the transformation has made the sample approximate the normal distribution better than it did before. The Kolmogorov–Smirnov test for goodness of fit to a normal distribution (see Section 17.2)—which previously yielded a value of $D = 0.150$ (significant at $P < 0.01$), implying that the milk yields are not normally distributed—after transformation yielded $D = 0.092$ (just significant at $P = 0.05$), a clear improvement.

One may wish to transform data not only to achieve normality but also to induce homogeneity of variances. The Box–Cox transformation can be extended simply by defining a new log-likelihood function:

$$L' = L - \tfrac{1}{2}X^2 \tag{13.4}$$

where L is as defined earlier and X^2 is the test statistic defined in Box 13.2 for Bartlett's test of homogeneity of variances. The value of λ that maximizes L' corresponds to the transformation that attempts to normalize the distribution as well as to equalize the variances. An iterative procedure for finding the value of λ that maximizes L' is included in the BIOM-pc package of computer programs. When we applied this procedure to the *Hyopsodus* data from Box 13.2, we found that L' is maximized at $\lambda = 0.1098$. A value of λ near 0 maximizes both L ($\lambda = 0.1581$) and $-X^2$ ($\lambda = 0.0543$). The 95% confidence limits for λ and $\lambda_1 = -0.543$ and $\lambda_2 = 0.728$. Since λ is so close to 0 (and the confidence limits include the value of 0), one is tempted to use the logarithmic transformation, as is often done on such data. The transformed data yield an adjusted X^2 value of 16.730, which is just significant at the 5% level. The untransformed data yield an adjusted X^2 value that is significant at the 1% level (see Box 13.2), so the amount of heterogeneity has been reduced, although perhaps not sufficiently to warrant carrying out an anova.

If a program for the Box–Cox transformation is not available, the following rule of thumb may be helpful. Try the series of power transformations $1/\sqrt{Y}$, $\sqrt{Y}$, $\ln Y$, $1/Y$ for samples skewed to the right (g_1 positive) and the transformation Y^2, Y^3, . . . for samples skewed to the left (g_1 negative).

13.10 The Arcsine Transformation

This transformation, also known as the *angular transformation,* is especially appropriate to percentages and proportions. You may remember from Section 5.2 that the standard deviation of a binomial distribution is $\sigma = \sqrt{pq/k}$. Since

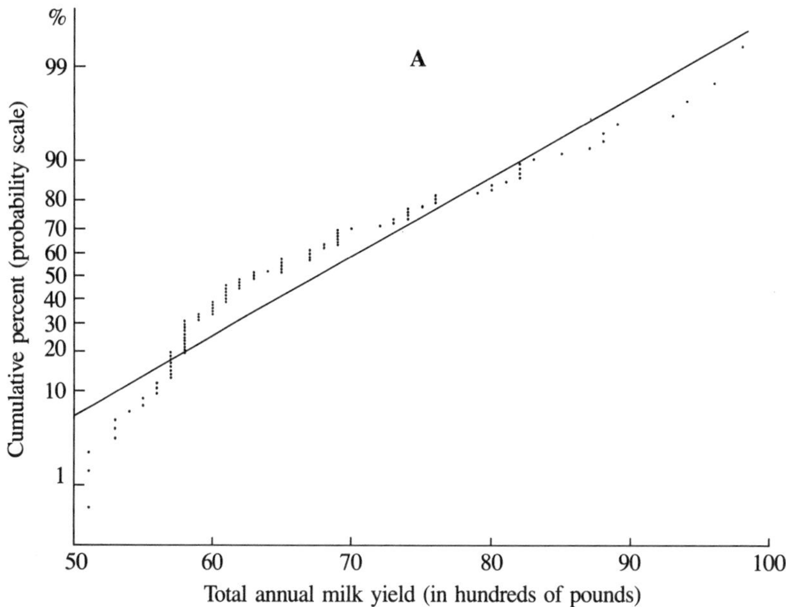

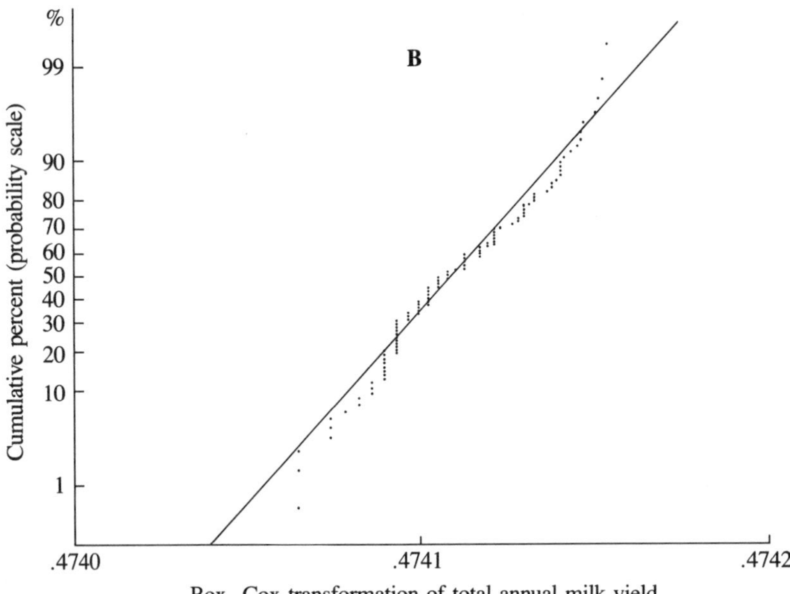

FIGURE 13.2 *Normal probability plots for the milk yield data of Table 6.1.* **A.** *Untransformed data.* **B.** *Data transformed using the Box–Cox transformation. Straight lines fitted by eye.*

$\mu = p$, $q = 1 - p$, and k is constant for any one problem, in a binomial distribution the variance is a function of the mean. The arcsine transformation prevents this dependence.

The arcsine transformation finds $\theta = \arcsin \sqrt{p}$, where p is a proportion. The term arcsin is synonymous with inverse sine or $\sin^{-1}$, which stands for the angle whose sine is the given quantity. The function is available in most pocket calculators and computer languages. Thus, for the proportion 0.431 we find 41.03°, the angle whose sine is $\sqrt{0.431}$. The arcsine transformation stretches out both tails of a distribution of percentages or proportions and compresses the middle. An example of its application is shown in Table 13.4, where percent fertilities in a sample of 100 vials of 10 eggs each are shown for a strain of *Drosophila*. Figure 13.3 shows that the distribution is not normal (it is platykurtic; see Section 6.6). When the percentages are transformed to angles in column (2) of Table 13.4, the distribution closely approximates the normal. The expected variance of such a distribution of arcsines is $\sigma_\theta^2 = 180^2/4\pi^2 n = 820.7/n$, where θ represents the angles in which the arcsines are expressed. When θ is measured in radians rather than degrees, $\sigma_\theta^2 = 1/4n$. To report means, we need to convert them back

Table 13.4 AN APPLICATION OF THE ARCSINE TRANSFORMATION.

Percent fertility of eggs of the CP strain of Drosophila melanogaster *raised in 100 vials of 10 eggs each.*

(1) % fertility $100p$	(2) $\arcsin \sqrt{p}$	(3) f	(4)[a] Cumulative frequencies F
0	0	1	1
10	18.43	3	4
20	26.57	8	12
30	33.21	10	22
40	39.23	6	28
50	45.00	15	43
60	50.77	14	57
70	56.79	12	69
80	63.43	13	82
90	71.57	9	91
100	90.00	9	100
		100	

SOURCE: Data from Sokal (1966); see Figure 13.3.
[a] Needed for preparation of Figure 13.3.

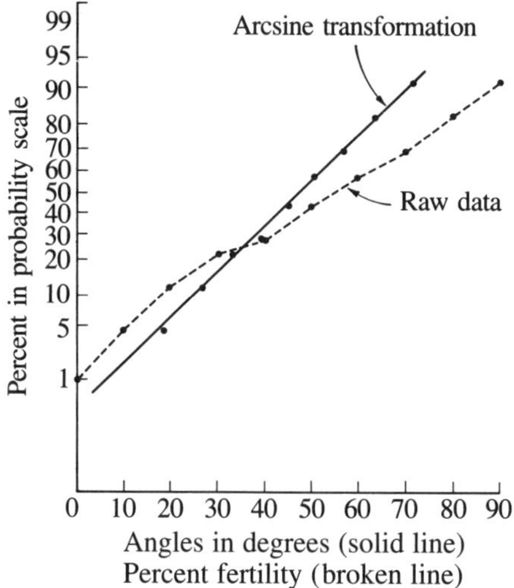

FIGURE 13.3 *Data from Table 13.4 plotted on arithmetic probability paper to show the effect of the arcsine transformation.*

to proportions or percentages. This conversion is accomplished by converting the angle θ to the proportion p by $p = [\sin(\theta)]^2$. When the percentages in the original data fall between 30 and 70%, it is generally not necessary to apply the arcsine transformation.

Improved angular transformations for proportions have been suggested by Johnson and Kotz (1969) and Freeman and Tukey (1950). These transformations are, respectively,

$$\theta = \arcsin \sqrt{\frac{Y + \frac{3}{8}}{n + \frac{3}{4}}}$$

and

$$\theta' = 2\sqrt{n}\left[\arcsin \sqrt{\frac{Y + \frac{3}{8}}{n + \frac{3}{4}}} - \arcsin \sqrt{p}\right]$$

where Y is the observed number possessing the property and n is the sample size.

13.11 Nonparametric Methods in Lieu of Single-Classification Anova

If none of the transformations already described manage to make our data meet the assumptions of analysis of variance, we may resort to an analogous **non-parametric method.** These techniques are also called *distribution-free methods,* since they are not dependent on a given distribution (such as the normal in the case of anova) but usually will work for a wide range of different distributions. They are called nonparametric methods because their null hypothesis is not concerned with specific parameters (such as the mean in analysis of variance) but only with the distribution of the variates. In recent years, nonparametric analysis of variance has become popular because it is simple to compute and permits freedom from worry about the distributional assumptions of an anova. Yet in cases where these assumptions hold entirely or even approximately, the analysis of variance is generally the more powerful statistical procedure for detecting departures from the null hypothesis.

The easiest way to get a feel for these nonparametric tests is to become acquainted with one. The first tests to be taken up are all based on the idea of *ranking* the variates in an example after pooling all groups and considering them as a single sample for purposes of ranking. A convenient method for arraying the samples in rank order is by the stem-and-leaf displays introduced in Section 2.5. The present section considers analogues of single-classification analysis of variance. For the general case with a samples and n_i variates per sample, the **Kruskal–Wallis test** is generally recommended.

Box 13.6 illustrates the Kruskal–Wallis test applied to the rate of growth of pea sections, introduced in Box 9.4. This is an example of single classification with equal sample sizes, but the computational formulas shown in Box 13.6 are general and apply to varying sample sizes as well. First we rank all variates from the smallest to the largest, ignoring the division into groups. During this procedure ties are a frequent problem. You will note several ties for almost every value on the measurement scale of the variable. For these ties we calculate the average of the ranks occupied by the tied values, as shown in Box 13.6. Next we reconstitute the original data table but replace each original variate by its rank or average rank as appropriate; this process is shown in Box 13.6. Then we evaluate the following expression, which is the expected value of the variance of the sums of the ranks:

$$H = \left[\frac{12}{\sum^{a} n_i\left(\sum^{a} n_i + 1\right)} \sum^{a} \frac{\left(\sum^{n_i} R\right)_i^2}{n_i}\right] - 3\left(\sum^{a} n_i + 1\right) \qquad (13.5)$$

This formula seems formidable, but most of the quantities are familiar. $\Sigma^a n_i$ is simply the sum of the sample sizes of the entire analysis. The numbers 12 and 3 are constants in every problem, and $(\Sigma^{n_i} R)_i$ is the sum of the ranks for the ith group.

The statistic H as shown by Expression (13.5) is appropriate for data without ties but must be divided by a correction factor D when ties are present. This correction factor is computed from a formula shown in Box 13.6. Because H is a variance, it is distributed approximately as $\chi^2_{[a-1]}$ for large samples when the null

Box 13.6 KRUSKAL–WALLIS TEST (A TEST FOR DIFFERENCES OF LOCATION IN RANKED DATA GROUPED BY SINGLE CLASSIFICATION).

Effect of different sugars on growth of pea sections. Data from Box 9.4: $a =$ 5 groups; n_i = number of items in group i (in this example sample sizes are equal, $n_i = n = 10$).

Computation

1. Rank all observations from smallest to largest when pooled together into a single sample. In case of ties, compute the average ranks. For example, the 4 variates $Y = 59$ represent ranks 18, 19, 20, and 21. Their average rank therefore equals $(18 + 19 + 20 + 21)/4 = 19.5$.

Average Rank	Rank	Y	Average Rank	Rank	Y	Average Rank	Rank	Y
2	1	56	19.5	18	59		35	64
2	2	56	19.5	19	59	37.5	36	65
2	3	56	19.5	20	59	37.5	37	65
7	4	57	19.5	21	59	37.5	38	65
7	5	57	23.5	22	60	37.5	39	65
7	6	57	23.5	23	60		40	66
7	7	57	23.5	24	60	42.5	41	67
7	8	57	23.5	25	60	42.5	42	67
7	9	57	27.5	26	61	42.5	43	67
7	10	57	27.5	27	61	42.5	44	67
14	11	58	27.5	28	61		45	68
14	12	58	27.5	29	61		46	70
14	13	58	31.5	30	62		47	71
14	14	58	31.5	31	62	48.5	48	75
14	15	58	31.5	32	62	48.5	49	75
14	16	58	31.5	33	62		50	76
14	17	58		34	63			

BOX 13.6 CONTINUED

2. Replace each observation in the original data table by its rank or average rank.

		Treatments							
Control		2% glucose added		2% fructose added		1% glucose + 1% fructose added		2% sucrose added	
Y	Rank	Y	Rank	Y	Rank	Y	Rank	Y	Rank
75	48.5	57	7	58	14	58	14	62	31.5
67	42.5	58	14	61	27.5	59	19.5	66	40
70	46	60	23.5	56	2	58	14	65	37.5
75	48.5	59	19.5	58	14	61	27.5	63	34
65	37.5	62	31.5	57	7	57	7	64	35
71	47	60	23.5	56	2	56	2	62	31.5
67	42.5	60	23.5	61	27.5	58	14	65	37.5
67	42.5	57	7	60	23.5	57	7	65	37.5
76	50	59	19.5	57	7	57	7	62	31.5
68	45	61	27.5	58	14	59	19.5	67	42.5
$\left(\sum^{n_i} R\right)_i$	450.0		196.5		138.5		131.5		358.5

3. Sum the ranks separately for each group. Enter in row $(\Sigma^{n_i} R)_i$. For example, $\left(\sum^{n_i} R\right)_1 = 48.5 + 42.5 + \cdots + 50 + 45 = 450.0$

4. Compute Expression (13.5). The numbers 12 and 3 are constants.

$$H = \left[\frac{12}{\left(\sum^a n_i\right)\left(\sum^a n_i + 1\right)} \sum^a \frac{\left(\sum^{n_i} R\right)_i^2}{n_i}\right] - 3\left(\sum^a n_i + 1\right)$$

$$= \frac{12}{50(50+1)}\left[\frac{450.0^2}{10} + \frac{196.5^2}{10} + \cdots + \frac{358.5^2}{10}\right] - 3(50+1)$$

$$= \frac{12}{50(51)}(40{,}610.9) - 3(51) = 191.110 - 153 = 38.110$$

5. Since there were ties, this H value must be corrected by dividing it by

$$D = 1 - \frac{\sum^m T_j}{\left(\sum^a n_i - 1\right)\sum^a n_i\left(\sum^a n_i + 1\right)}$$

Box 13.6 Continued

where T_j is a function of the t_j, the number of variates tied in the jth group of ties. (This t has no relation to Student's t.) The function is $T_j = t_j^3 - t_j$, computed most easily as $(t_j - 1)t_j(t_j + 1)$. Since in most cases the tied group will range from $t = 2$ to $t = 10$ ties, we give a small table of T over this range; the summation of T_j is over the m different ties.

t_j	2	3	4	5	6	7	8	9	10
T_j	6	24	60	120	210	336	504	720	990

For example, for the first tied group in the table of ranks, $t_j = 3$, since there are 3 variates of equal magnitude.

The t_j's for the present example are shown below, together with the corresponding T_j's:

t_j	3	7	7	4	4	4	4	4	4	2
T_j	24	336	336	60	60	60	60	60	60	6

$$\sum^{m} T_j = 24 + 336 + \cdots + 6 = 1062$$

$$D = 1 - \frac{1062}{(50 - 1)50(50 + 1)} = 1 - \frac{1062}{49(50)51} = 1 - \frac{1062}{124{,}950}$$

$$= 1 - 0.00850 = 0.99150$$

$$\text{Adjusted } H = \frac{H}{D} = \frac{38.110}{0.99150} = 38.437$$

If the null hypothesis (that the a groups do not differ in "location") is true, H is distributed approximately as $\chi^2_{[a-1]}$. Since H is much greater than $\chi^2_{.005[4]} = 14.860$, we may confidently reject the null hypothesis and conclude that different sugars affect the rate of growth of pea sections differentially. Box 13.8 gives a method for unplanned testing of differences among treatments.

hypothesis is true. For tests at $\alpha = 0.10$ or $\alpha = 0.05$, the χ^2 approximation is very good even with n as small as 5. For $\alpha = 0.01$, however, the test is conservative for small values of n; it rejects less than 1% of the tests if the hypothesis is true (personal communication from K. R. Gabriel, based on an unpublished Monte Carlo study). When $a = 3$ and the sample sizes n_i are each less than five, the statistic H is not distributed as chi square. An exact distribution has been tabled and can be found in Kruskal and Wallis (1952) or in Siegel and Castellan (1988; table O).

If the populations are not different from each other, we expect their rank sums to be approximately the same (allowing for differences in sample size where such exist). Note in the example of Box 13.6 that we decisively reject the null hypothesis, which is that the true "location" of the several populations is the same. By location in this context we refer to the rank-ordered positions of the individual variates along the measurement- or Y-axis. The null hypothesis is that each of the a groups contains a random allocation of all available ranks. The alternative hypothesis is that the ranks differ among the groups. The conclusion reached by the Kruskal–Wallis test is the same as that reached by the regular anova in Box 9.4. The BIOM-pc program package performs the Kruskal–Wallis test.

When the test is between only two samples (such a design would give rise to a t-test or anova with two classes), we employ either of two nonparametric tests, which yield the same statistic and give the same results the **Mann–Whitney U-test** or the **Wilcoxon two-sample test.** The null hypothesis is that the two samples come from populations having the same location. Each of these is applied in Box 13.7 to data that are morphological measurements on two samples of chigger nymphs. The Mann–Whitney U-test as illustrated in Box 13.7 is a semigraphical test and is quite simple to apply. It is especially convenient when the data are already graphed and there are few items in each sample. The display of data for anovas as a series of dots has other advantages as well. It permits the investigator to determine at a glance whether the assumptions of normality and homoscedasticity are reasonable, and also whether the null hypothesis of identical group means appears to hold. Systematic testing of data by such an approach (especially for randomized-blocks designs) has been suggested by Fawcett (1990), which should be consulted for further details.

Note that the methods of Box 13.7 do not require that each individual observation represent a precise measurement. As long as you can order the observations you are able to perform these tests. Thus, for example, suppose you placed some meat out in the open and studied the arrival times of individuals of two species of blowflies. You could record exactly the time of arrival of each fly, starting from a point zero in time when the meat was set out. On the other hand, you might simply rank arrival times of the two species, noting that individual 1 of species B came first, 2 individuals from species A next, then 3 individuals of B, followed by the simultaneous arrival of one of each of the two species (a tie), and so forth. Although such ranked or ordered data cannot be analyzed by the parametric methods studied earlier, the techniques of Box 13.7 are entirely applicable.

The method of calculating the sample statistic U_s for the Mann–Whitney and the Wilcoxon tests is straightforward, as shown in Box 13.7. When there are no tied observations across the two groups, the critical values for $U_{\alpha[n_1, n_2]}$ are given by Statistical Table **U,** which is adequate for cases in which the larger sample size $n_1 \leq 20$. The probabilities in Table **U** assume a one-tailed test. For a two-tailed test you should double the value of the probability shown in that table. When $n_1 > 20$, compute the expression shown near the bottom of Box 13.7. Since this expression is distributed as a normal deviate, consult the table of t (Table **B**), for $t_{\alpha[\infty]}$ using one- or two-tailed probabilities, depending on your hypothesis. A further complication arises from tied observations across the two

Box 13.7 MANN-WHITNEY *U*-TEST AND WILCOXON TWO-SAMPLE TEST FOR TWO SAMPLES (RANKED OBSERVATIONS, NOT PAIRED).

Two samples of nymphs of the chigger *Trombicula lipovskyi.* Variate measured is length of cheliceral base stated as micrometer units.

(1)	(2)	(3)	(4)
Sample A		Sample B	
Y	Rank (R)	Y	Rank (R)
104	2	100	1
109	7	105	3
112	9	107	4.5
114	10	107	4.5
116	11.5	108	6
118	13.5	111	8
118	13.5	116	11.5
119	15	120	16
121	17.5	121	17.5
123	19.5	123	19.5
125	21		$91.5 = \sum^{n_2} R$
126	22.5		
126	22.5		
128	25		
128	25		
128	25		
	$259.5 = \sum^{n_1} R$		

SOURCE: Data from D. A. Crossley (unpublished results).

Designate sample size of the larger sample as n_1 and that of the smaller sample as n_2. In this case, $n_1 = 16$, $n_2 = 10$. If the two samples are of equal size, it does not matter which is designated as 1.

There are two equivalent procedures for carrying out a test of equality of "location" of two samples.

The Mann–Whitney U-test

1. List the observations from the smallest to the largest in such a way that the two samples may be compared easily. A convenient method is to make a graph as shown here.

Box 13.7 Continued

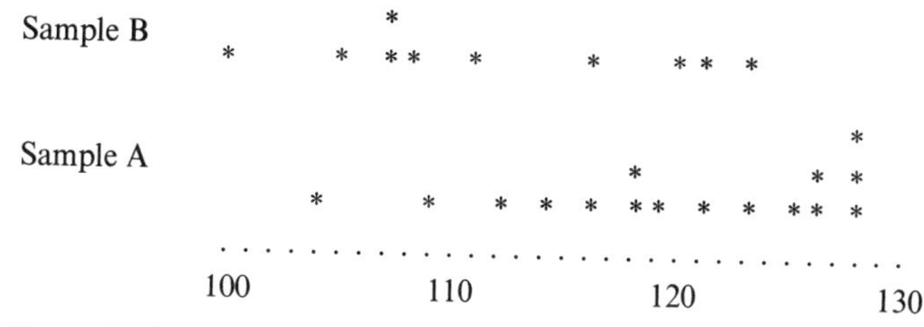

2. For each observation in one sample (it is convenient to use the smaller sample), count the number of observations in the other sample that are lower in value (to the left). Count $\frac{1}{2}$ for each tied observation. For example, there are zero observations in sample A less than the first observation in sample B; 1 observation less than the second, third, fourth, and fifth observations in sample B; 2 observations in A less than the sixth in B; 4 observations in A less than the seventh in B, but one is equal (tied) with it, so we count $4\frac{1}{2}$. Continuing in a similar manner, we obtain counts of 8, $8\frac{1}{2}$, and $9\frac{1}{2}$. The sum of these counts $C = 36\frac{1}{2}$. The Mann–Whitney statistic U_s is the greater of the two quantities C and $n_1 n_2 - C$, in this case $36\frac{1}{2}$ and $(16 \times 10) - 36\frac{1}{2} = 123\frac{1}{2}$.

The Wilcoxon two-sample test

1. Rank all of the observations together from low to high, as shown in columns (2) and (4) above. Give average ranks in the case of ties.
2. Sum the ranks of the smaller sample (sample size n_2).
3. Compute the Wilcoxon statistic as

$$C = n_1 n_2 + \frac{n_2(n_2 + 1)}{2} - \sum^{n_2} R = 16(10) + \frac{10(10 + 1)}{2} - 91.5$$

$$= 160 + 55 - 91.5 = 123.5$$

where n_1 is the size of the larger sample and n_2 is the size of the smaller. This statistic must be compared with $n_1 n_2 - C$ and the greater of the two quantities chosen as a test statistic U_s. In this case, $160 - 123.5 = 36.5$, so we use 123.5, which is identical to the value obtained in the Mann–Whitney test.

Testing the significance of U_s

No tied variates in samples (or variates tied within one or both groups only): When $n_1 \le 20$, compare U_s with the critical value for $U_{\alpha[n_1, n_2]}$ in Statistical Table **U**. The null hypothesis is rejected if the observed value is too large.

In cases where $n_1 > 20$, calculate the following quantity:

$$t_s = \frac{\left(U_s - \dfrac{n_1 n_2}{2}\right)}{\sqrt{\dfrac{n_1 n_2(n_1 + n_2 + 1)}{12}}}$$

Box 13.7 Continued

which is approximately normally distributed. The denominator 12 is a constant. Look up the significance of t_s in Table **B** against critical values of $t_{\alpha[\infty]}$ for a one-tailed test or two-tailed test as required by the hypothesis.

Tied variates (ties occur across both samples): There is no exact test. When $n_1 \leq 20$, the critical values in Table **U** are conservative. That is, the actual probability value will be less than the tabled one. Our example is a case in point. Ranks for variates 116, 121, and 123 are tied *across* samples (ranks for variates 107, 118, 126, and 128 are tied *within* samples). We consult Table **U** and find $U_{.025[16,10]} = 118$ and $U_{.01[16,10]} = 124$. Hence the two samples are significantly different at least at $0.05 > P > 0.02$ (we double the probabilities because this is a two-tailed test).

For sample sizes $n_1 > 20$, and when an approximation to the correct probability value is desired, evaluate

$$t_s = \frac{\left(U_s - \frac{n_1 n_2}{2}\right)}{\sqrt{\left(\frac{n_1 n_2}{(n_1 + n_2)(n_1 + n_2 - 1)}\right)\left(\frac{(n_1 + n_2)^3 - (n_1 + n_2) - \sum^m T_j}{12}\right)}}$$

where $\Sigma^m T_j$ has the same meaning as in Box 13.6. Compare t_s with $t_{\alpha[\infty]}$. For our example we compute

$$t_s = \frac{\left(123.5 - \frac{16 \times 10}{2}\right)}{\sqrt{\left(\frac{16 \times 10}{(16+10)(16+10-1)}\right)\left(\frac{(16+10)^3 - (16+10) - (6+6+6+6+6+6+24)}{12}\right)}}$$

$$= \frac{43.5}{\sqrt{358.7692}} = 2.297$$

The two samples are significantly different at $0.05 > P > 0.02$. The corresponding value of t_s without the correction for ties is 2.293, so in this case the conclusions are the same.

groups. There is no exact test. For sample sizes $n_1 \leq 20$ use Table **U,** which will then be conservative. Larger sample sizes require the more elaborate formula shown at the bottom of Box 13.7. It takes a substantial number of ties to affect the outcome of the test appreciably. Corrections for ties increase the t_s-value slightly; hence the uncorrected formula is more conservative.

It is desirable to obtain an intuitive understanding of the rationale behind these tests. In the Mann–Whitney test we can conceive of two extreme situations: In one case the two samples overlap and coincide entirely; in the other they are quite separate. In the latter case, if we take the sample with the lower-valued variates, there will be no points of the contrasting sample to the left of it; that is, we can go through every observation in the lower-valued sample without having any items of the higher-valued one to the left of it. Conversely, all the points of the lower-valued sample would be to the left of every point of the higher-valued one if we had started out with the latter. Our total count would therefore be the total count of one sample multiplied by every observation in the second sample, which yields n_1n_2. Thus, since we are told to take the greater of the two values, the sum of the counts C or $n_1n_2 - C$, our result in this case would be n_1n_2. But if two equal-sized samples coincide completely, then for each point in one sample we would have those points below it plus a half point for the tied value representing the observation in the second sample that is at exactly the same level as the observation under consideration. A little experimentation will show C to be $[n(n-1)/2] + (n/2) = n^2/2$. Clearly the range of possible U-values must be between this and n_1n_2, and the critical value must be somewhere within this range.

In computing the Wilcoxon statistic, the magnitude of the sum of the ranks determines the outcome of the test. If this sum is unusually large or unusually small, the sample under consideration is near the upper or lower extreme of the overall distribution of ranks of the two samples considered together.

Our conclusion as a result of the tests in Box 13.7 is that the two samples differ in the distribution of cheliceral base length. The chiggers of sample four have longer cheliceral bases than those of sample B.

Alternatively, the nonparametric test may be targeted at a comparison of the medians of the two samples only. Fligner and Policello (1981) describe such a test as a modification of the Mann–Whitney–Wilcoxon test. Tests on differences among medians can also be carried out by means of two-way tests of independence (see Section 17.4).

Now that you have learned a technique for testing two groups, we can return to the case of several groups and show how we could carry out multiple comparisons between pairs of treatments based on a nonparametric test of significance. This is the nonparametric analogue to the unplanned tests we studied in Section 9.7. The approach we take here is the simultaneous test procedure of Dwass (1960) as further developed by K. R. Gabriel (unpublished). The technique, described in Box 13.8, finds the Mann–Whitney–Wilcoxon U-statistic for pairwise comparisons for each of the treatments and compares these with a critical value found by the formula shown in Box 13.8. The computation of the statistic, U_s, is quite simple. We again take the greater of two quantities C or $(n^2 - C)$. In Box 13.7 in place of n^2 we used n_1n_2, but the *STP* (simultaneous test procedure) method of Box 13.8 applies only to equal sample sizes; hence n_1n_2 becomes n^2. The method is recommended only for n equal to or greater than 8.

In our discussion of unplanned comparisons in Section 9.7 we pointed out that although the experimentwise type I error is α, tests of subsets have actual error rates of $\alpha' < \alpha$. Thus in this test the type I error probabilities for tests involving any set of four samples is 0.0323, that for any set of three samples is 0.0175, and that for a pair of samples is 0.0064. Our tests of pairs are therefore quite

Box 13.8 NONPARAMETRIC MULTIPLE COMPARISONS BY STP (AN UNPLANNED TEST FOR EQUAL SAMPLE SIZES).

This method is based on U, the Mann–Whitney statistic, and requires equal sample sizes, with $n > 8$.

The technique is applied to the data on pea sections in Box 13.6. As in the ordinary Mann–Whitney test, it is simplest to prepare a graph of the data as shown below (see also Box 13.7).

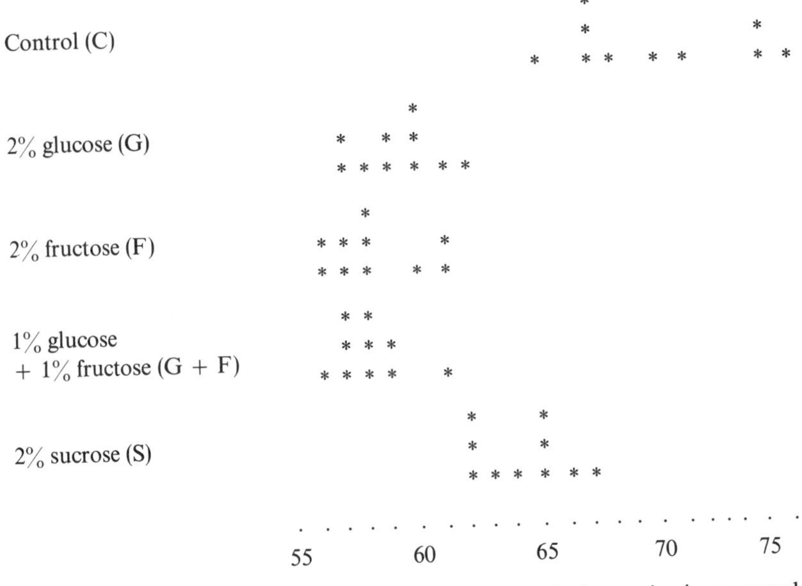

For each pair of samples, compute U_s as follows. For each observation in one sample, count the number of observations in the other samples that are lower in value (to the left). Count a $\frac{1}{2}$ for each tied observation. For example, control (C) versus glucose (G): 10 observations in G are below the first observation in C, 10 observations in G are less than the second observation in C, and so on. The sum of the counts is $C(\text{C, G}) = 10 + 10 + \cdots + 10 = 100$. Obviously, the same sum will be found for $C(\text{C, F})$ and $C(\text{C, G + F})$.

$$C(\text{C, S}) = 6\tfrac{1}{2} + 9\tfrac{1}{2} + 9\tfrac{1}{2} + 9\tfrac{1}{2} + 10 + 10 + 10 + 10 + 10 + 10 = 95$$

$$C(\text{G, F}) = 3 + 3 + 5\tfrac{1}{2} + 7 + 7 + 7\tfrac{1}{2} + 7\tfrac{1}{2} + 7\tfrac{1}{2} + 9 + 10 = 67$$

$$C(\text{G, G + F}) = 2\tfrac{1}{2} + 2\tfrac{1}{2} + 5\tfrac{1}{2} + 8 + 8 + 9 + 9 + 9 + 9\tfrac{1}{2} + 10 = 73$$

$$C(\text{G, S}) = 0 + 0 + 0 + 0 + 0 + 0 + 0 + 0 + 0 + 1\tfrac{1}{2} = 1\tfrac{1}{2}$$

$$C(\text{F, G + F}) = \tfrac{1}{2} + \tfrac{1}{2} + 2\tfrac{1}{2} + 2\tfrac{1}{2} + 5\tfrac{1}{2} + 5\tfrac{1}{2} + 5\tfrac{1}{2} + 9 + 9\tfrac{1}{2} + 9\tfrac{1}{2} = 50\tfrac{1}{2}$$

$C(\text{F, S})$: All points of F are below those of S, so the count must be the minimum possible; namely, zero. This is also true of $C(\text{G + F, S})$.

The greater of the two quantities C or $(n^2 - C)$ is now entered as U_s into the table below (n is the sample size for each group).

BOX 13.8 CONTINUED

Summary of U_s values

	C	G	F	G+F	S
C	╳				
G	100*	╳			
F	100*	67	╳		
G + F	100*	73	50½	╳	
S	95*	98½*	100*	100*	╳

*Significant at an experimentwise error rate of 0.05 by test described below

The critical value for U_s is:

$$U_{\alpha[a,n]} = \frac{n^2}{2} + Q_{\alpha[a,\infty]} n \sqrt{\frac{(2n + 1)}{24}}$$

where a is the number of samples, n is the number of observations within each sample, and $Q_{\alpha[a,\infty]}$ is the α significance level for the studentized range for a groups and ∞ df (see Statistical Table **J**).

In our case

$$U_{.05[5,10]} = \frac{10^2}{2} + (3.858)10 \sqrt{\frac{[2(10) + 1]}{24}} = 50 + 38.58 \sqrt{\frac{21}{24}}$$

$$= 50 + 38.58 \sqrt{0.875} = 50 + 36.09 = 86.09$$

All U_s-values equal to or greater than this are significant and have been marked with an asterisk in the table above. All sets of groups containing a significant U_s between at least one pair are significantly heterogeneous. Thus, we find that the set (G, F, G + F) is not significantly heterogeneous. Since C and S were heterogeneous with any other treatment, any set containing either C or S (or both) would be significantly heterogeneous. The data therefore break up into the following partition:

(G, F, G + F) (S) (C)

This is the same conclusion that was reached in Box 9.10 and Section 9.7, although such complete agreement cannot always be expected.

When there are tied observations across groups as in the example here, the formula for the critical value is not quite correct. The square rooted term should be changed to

$$\sqrt{\frac{(2n)^3 - 2n - \sum^{m} T_j}{2n(2n - 1)24}}$$

requiring a recomputation of each value of $U_{\alpha[a,n]}$ for each pair of samples. However, since the correction due to ties is generally quite small, and the critical values computed by the simpler formula are conservative, we did not trouble to adjust for ties.

conservative. The computations are simple and lead to the conclusion that glucose, fructose, and the mixed sugar are not different from each other, but are different from sucrose, which in turn is different from the control. This is the same conclusion that was reached in Box 9.10, but of course this nonparametric method need not always give the same conclusion as a corresponding anova.

Two additional points should be noted. We tested all pairwise comparisons in this example because of certain statistical properties of the *STP* in the nonparametric case. When significant pairs are members of larger sets, these larger sets must themselves be significant and, conversely, significant larger sets cannot be broken up into pairs all of which are nonsignificant. That is, there must be at least one significant pair causing the significance of the larger set. Therefore, by knowing which pairs are nonsignificant we can construct the maximally nonsignificant sets, as we have done in this example. The second point is that while the Kruskal–Wallis test of Box 13.6 is a generalization of the Wilcoxon two-sample test employed in the *STP*, it is not based on the same test criterion. Therefore, although this is unlikely, a nonsignificant Kruskal–Wallis analysis among several groups might yield at least some significant pairs when investigated by the nonparametric *STP* of Box 13.8.

The Kruskal–Wallis and Mann–Whitney tests are based on ranks and measure differences in location. A nonparametric test that tests differences between two distributions is the **Kolmogorov–Smirnov two-sample test.** Its null hypothesis is that the two samples are distributed identically; thus the test is sensitive to differences in location, dispersion, skewness, and so forth. This test is quite simple to carry out. It is based on the unsigned differences between the relative cumulative frequency distributions of the two samples. Expected critical values can be looked up in a table or approximated. Comparison between observed and expected values leads to decisions about whether the maximum difference between the two cumulative frequency distributions is significant. Box 13.9 gives some examples.

First we show the application of the method to small samples (both n_1 and $n_2 \leq 25$). We apply the test to the example from Box 13.7, where two samples of chigger nymphs were measured for length of cheliceral base. We use the symbol F for cumulative frequencies, which are summed with respect to the class marks shown in column (1) and give the cumulative frequencies of the two samples in columns (2) and (3). Relative expected frequencies are obtained by division by the respective sample sizes in columns (4) and (5), and column (6) features the absolute difference between relative cumulative frequencies. The maximum unsigned difference is 0.475 which is multiplied by $n_1 n_2$, yielding 76. The critical value for this statistic can be found in Statistical Table **W,** which furnishes critical values for the two-tailed two-sample Kolmogorov–Smirnov test. We obtain $n_1 n_2 D_{.10} = 76$ and $n_1 n_2 D_{.05} = 84$. There is a 10% probability of obtaining the observed difference by chance alone, and we conclude that the two samples do not differ significantly in their distributions. This contradicts the findings of a significant difference by the Mann–Whitney U-test in Box 13.7.

Box 13.9 KOLMOGOROV–SMIRNOV TWO-SAMPLE TEST, TESTING DIFFERENCES IN DISTRIBUTIONS OF TWO SAMPLES OF CONTINUOUS OBSERVATIONS.

I. *Test for two small samples (both n_1 and $n_2 \leq 25$).*

Data on length of cheliceral base in chigger nymphs from Box 13.7. The sample sizes are $n_1 = 16$, $n_2 = 10$.

(1) Y	(2) Sample A F_1	(3) Sample B F_2	(4) $\frac{F_1}{n_1}$	(5) $\frac{F_2}{n_2}$	(6) $d = \left\lvert \frac{F_1}{n_1} - \frac{F_2}{n_2} \right\rvert$
100	0	1	0	0.100	0.100
101	0	1	0	0.100	0.100
102	0	1	0	0.100	0.100
103	0	1	0	0.100	0.100
104	1	1	0.062	0.100	0.038
105	1	2	0.062	0.200	0.138
106	1	2	0.062	0.200	0.138
107	1	4	0.062	0.400	0.338
108	1	5	0.062	0.500	0.438
109	2	5	0.125	0.500	0.375
110	2	5	0.125	0.500	0.375
111	2	6	0.125	0.600	0.475 ← D
112	3	6	0.188	0.600	0.412
113	3	6	0.188	0.600	0.412
114	4	6	0.250	0.600	0.350
115	4	6	0.250	0.600	0.350
116	5	7	0.312	0.700	0.388
117	5	7	0.312	0.700	0.388
118	7	7	0.438	0.700	0.262
119	8	7	0.500	0.700	0.200
120	8	8	0.500	0.800	0.300
121	9	9	0.562	0.900	0.338
122	9	9	0.562	0.900	0.338
123	10	10	0.625	1.000	0.375
124	10	10	0.625	1.000	0.375
125	11	10	0.688	1.000	0.312
126	13	10	0.812	1.000	0.188
127	13	10	0.812	1.000	0.188
128	16	10	1.000	1.000	0

BOX 13.9 CONTINUED

Procedure

1. Form cumulative frequencies F of the items in samples 1 and 2. Thus in column (2) note that there are 3 measurements in sample A at or below 112.5 micrometer units. By contrast there are 6 such measurements in sample B [column (3)].
2. Compute relative cumulative frequencies by dividing frequencies in columns (2) and (3) by n_1 and n_2, respectively, and enter in columns (4) and (5).
3. Compute d, the absolute value of the difference between the relative cumulative frequencies in columns (4) and (5), and enter in column (6).
4. Locate the largest unsigned difference D. It is 0.475.
5. Multiply D by $n_1 n_2$. We obtain $(16)(10)(0.475) = 76$.
6. Compare $n_1 n_2 D$ with its critical value in Table **W**, where we obtain a value of 84 for $P = 0.05$. We accept the null hypothesis that the two samples have been taken from populations with the same distribution. The results of the Kolmogorov–Smirnov two-sample test contrast with the moderately significant difference by the Mann–Whitney U-test in Box 13.7. The Kolmogorov–Smirnov test is less powerful than the Mann–Whitney U-test with respect to the alternative hypothesis of the latter, i.e., differences in location. However, the Kolmogorov–Smirnov test examines differences in both shape and location of the distributions and is thus a more comprehensive test.

An alternative computational scheme, which is more efficient but perhaps less instructive for the novice, was proposed by Gideon and Mueller (1978). Pool both samples into a single array. For the chigger data just considered this would yield

Y	Running sum	Y	Running sum
100	−16	118	−42
104	−6	119	−32
105	−22	120	−48
107	−38	121	−38
107	−54	121	−54
108	−70	123	−44
109	−60	123	−60
111	−76	125	−50
112	−66	126	−40
114	−56	126	−30
116	−46	128	−20
116	−62	128	−10
118	−52	128	0

Note that all members of sample A have been underlined for convenience.

BOX 13.9 CONTINUED

Next compute a running sum for which you add n_2 whenever an item from sample A is encountered and subtract n_1 whenever the item is from sample B. The running sum is shown alongside the arrayed variates. Cross out running sums for all but the last of each set of tied values. The largest unsigned running sum is $n_1 n_2 D = 76$, as before.

II. *Approximate test for two larger samples (both sample sizes between 26 and 40, or n_1 or $n_2 > 40$).*

Compute D as shown in part **I** and test as follows. Using Expression (13.6) compute

$$D_\alpha = K_\alpha \sqrt{\frac{n_1 + n_2}{n_1 n_2}}$$

where

$$K_\alpha = \sqrt{\frac{1}{2}\left[-\ln\left(\frac{\alpha}{2}\right)\right]}$$

For equal sample sizes the formula for D_α simplifies to $K_\alpha\sqrt{2/n}$. These critical values are for two-tailed tests.

Whenever $n_1 = n_2 = n$ and $26 < n \leq 40$, the exact distribution of D can be looked up in Table 15.4 of Owen (1962). The quantity tabled there is actually the cumulative probability distribution corresponding to values of nD (called nr in Owen's table). This table is intended for a two-tailed test.

III. *Approximate test for two frequency distributions (large sample sizes). Length of silver salmon at two time periods. Data from Table 13.2.*

$n_1 = 421$, $n_2 = 349$

(1) Length in mm	(2) Upper class limit	(3) Nov. 12–25 F_1	(4) Apr. 1–14 F_2	(5) $\frac{F_1}{n_1}$	(6) $\frac{F_2}{n_2}$	(7) $d = \left\lvert\frac{F_1}{n_1} - \frac{F_2}{n_2}\right\rvert$
60	62.5	1	0	0.002	0	0.002
65	67.5	5	0	0.012	0	0.012
70	72.5	38	0	0.090	0	0.090
75	77.5	132	0	0.314	0	0.314
80	82.5	257	1	0.610	0.003	0.607
85	87.5	355	6	0.843	0.017	0.826
90	92.5	399	12	0.948	0.034	0.914
95	97.5	417	17	0.990	0.049	0.941 ← D

Box 13.9 Continued

(1) Length in mm	(2) Upper class limit	(3) Nov. 12–25 F_1	(4) Apr. 1–14 F_2	(5) $\frac{F_1}{n_1}$	(6) $\frac{F_2}{n_2}$	(7) $d = \left\lvert \frac{F_1}{n_1} - \frac{F_2}{n_2} \right\rvert$
100	102.5	420	33	0.998	0.095	0.903
105	107.5	421	58	1.000	0.166	0.834
110	112.5	421	103	1.000	0.295	0.705
115	117.5	421	160	1.000	0.458	0.542
120	122.5	421	247	1.000	0.708	0.292
125	127.5	421	312	1.000	0.894	0.106
130	132.5	421	332	1.000	0.951	0.049
135	137.5	421	337	1.000	0.966	0.034
140	142.5	421	342	1.000	0.980	0.020
145	147.5	421	347	1.000	0.997	0.003
150	152.5	421	348	1.000	0.997	0.003
155	157.5	421	348	1.000	0.997	0.003
160	162.5	421	349	1.000	1.000	0

Procedure

1. Form cumulative frequency distributions F from the frequencies in the two samples (shown in Table 13.2). Column (2) gives the upper class limits. In column (3) we note 5 fish at or below 67.5 mm. These were obtained by adding 1 (for 60 mm) and 4 (for 65 mm).
2. Compute relative cumulative frequencies by dividing frequencies in columns (3) and (4) by n_1 and n_2, respectively, and enter in columns (5) and (6).
3. Compute d, the absolute difference between the relative cumulative frequencies in columns (5) and (6) and enter in column (7).
4. Locate the largest unsigned difference D. In this case $D = 0.941$.
5. Critical values D_α for D are found from Expression (13.6) as given in part **II**.

$$D_\alpha = K_\alpha \sqrt{\frac{n_1 + n_2}{n_1 n_2}}$$

where

$$K_\alpha = \sqrt{\frac{1}{2}\left[-\ln\left(\frac{\alpha}{2}\right)\right]}$$

Thus for $\alpha = 0.05$ $K_{.05} = 1.35810$, and for $\alpha = 0.01$ $K_{.01} = 1.62762$.

BOX 13.9 CONTINUED

In this case

$$D_{.05} = 1.35810\sqrt{\frac{421 + 349}{421(349)}} = 1.35810(0.072392)$$
$$= 0.098316$$
$$D_{.01} = 1.62762(0.072392) = 0.11783$$

The unsigned difference $D = 0.941 \gg D_{.01} = 0.118$, so we conclude that the two samples come from populations with (very) different distributions. The critical value D_α is intended for a two-tailed test. When $n_1 = n_2 = n$ the formula for D_α simplifies to

$$D_\alpha = K_\alpha\sqrt{\frac{2}{n}}$$

However, the two tests differ in their sensitivities to different alternative hypotheses—the Mann–Whitney U-test is sensitive to the number of interchanges in rank necessary to separate the two samples (shift in location), whereas the Kolmogorov–Smirnov test measures differences in the entire distributions of the two samples and is thus less sensitive to differences in location only.

A more efficient algorithm for the same test is featured next (see Box 13.9). The instructions are given for the exact test and for an approximate test that can be used when sample sizes exceed 25. An approximate two-tailed critical value for the test statistic D can be computed as

$$D_\alpha = K_\alpha\sqrt{\frac{n_1 + n_2}{n_1 n_2}} \tag{13.6}$$

where

$$K_\alpha = \sqrt{\frac{1}{2} - \ln\left(\frac{\alpha}{2}\right)}$$

The final example in Box 13.9 is for large samples arrayed in frequency distributions. The data are the lengths of silver salmon from Table 13.2. We have chosen the samples for two time periods, November and April, and are testing whether these differ in distribution of body length. The columns are computed as in the earlier example, and the critical value D_α is computed by Expression (13.6). We see from Box 13.9 that the maximum unsigned difference $D = 0.941 \gg D_{.01} = 0.118$, and thus we reject the null hypothesis that the two samples come from populations with the same distribution. An underlying assumption of all Kolmogorov–Smirnov tests (goodness-of-fit tests by means of this statistic are treated in Chapter 17) is that the variables studied are continuous.

13.12 Nonparametric Methods in Lieu of Two-Way Anova

Nonparametric tests exist for randomized blocks, the most common design for two-way analysis of a variance. The tests are for designs in which, if they were to be done by anova, one factor would be Model I, and the other factor would represent the blocks in the experiment. A Model II design, estimating the variance components, would clearly not be appropriate in a test based on ranks, since we are concerned only with location of the various samples in the overall ranking.

The test illustrated in Box 13.10 is **Friedman's method for randomized blocks.** This example involves the lake temperature data from Box 11.3, where they were used to illustrate the computation of a randomized-blocks anova without replication. Recall that the four dates are considered blocks and the depths are the treatment effects. In Friedman's method the variates are ranked *within each block,* as shown in Box 13.10. Note that because of the regularity of the thermal stratification of the lake, the ranking for the four days is entirely uniform. We sum the ranks for each of the treatment groups and compute a statistic X^2, identical to H in Box 13.6. Again X^2 is distributed as $\chi^2_{[a-1]}$, where a is the number of treatments (depths in this example). This value will be maximal in our example because the ranks are ordered uniformly in each of the blocks. In an example in which the variates were not as uniformly ranked in each block, the marginal sums (row sums in this case) would not be as high or as low as in this instance but would tend more toward intermediate values. Since these sums are being squared, the large values exert a disproportionate influence upon the sample statistic X^2.

We conclude that the effect of depth on water temperature is highly significant, which agrees with our earlier finding by analysis of variance. At the end Box 13.6 we also show how to carry out multiple comparisons among the groups, and we find that, despite the high overall significance of the data, none of the comparisons can be declared significant. This result illustrates the sharp decrease in power due to both the recoding of the variates as ranks and the decrease in pairwise α in the multiple-comparisons test. Had we carried out the multiple-comparisons test parametrically using the T-method, all but 4 of the 45 pairwise comparisons would have been significant at $P < 0.01$.

A special case of randomized blocks is the paired-comparisons design, discussed in Section 11.5 and illustrated in Box 11.5. This design can be carried out by Friedman's test, but a more widely used method is **Wilcoxon's signed-ranks test,** which is applied in Box 13.11 to a new example. The data are mean litter sizes in two strains of guinea pigs kept in large colonies during the years 1916 through 1924. Each of these values is the average of many litters. Note the parallelism in the changes in the variable in the two strains. During 1917 and 1918 (war years for the United States), a shortage of caretakers and food resulted in a decrease in the number of offspring per litter. As soon as better conditions returned, the mean litter size increased again. Notice that a subsequent drop in

Box 13.10 FRIEDMAN'S METHOD FOR RANDOMIZED BLOCKS.

Temperatures (°C) of Rot Lake on 4 early afternoons of the summer of 1952 at 10 depths. Data from Box 11.3.

Depth in meters ($a = 10$)	Days ($b = 4$)			
	29 July	30 July	31 July	1 August
0	23.8	24.0	24.6	24.8
1	22.6	22.4	22.9	23.2
2	22.2	22.1	22.1	22.2
3	21.2	21.8	21.0	21.2
4	18.4	19.3	19.0	18.8
5	13.5	14.4	14.2	13.8
6	9.8	9.9	10.4	9.6
9	6.0	6.0	6.3	6.3
12	5.8	5.9	6.0	5.8
15.5	5.6	5.6	5.5	5.6

Procedure

The four sets of readings are treated as blocks in this analysis.

1. Assign ranks to items within each of the $b = 4$ blocks (columns in this case) separately, resulting in the table that follows. If there are ties, treat them in the conventional manner (by average ranks). There are $a = 10$ treatments (depths; rows in this case).

Depth in meters	Days				
	29 July	30 July	31 July	1 August	$\sum^{b} R_{ij}$
0	10	10	10	10	40
1	9	9	9	9	36
2	8	8	8	8	32
3	7	7	7	7	28
4	6	6	6	6	24
5	5	5	5	5	20
6	4	4	4	4	16
9	3	3	3	3	12
12	2	2	2	2	8
15.5	1	1	1	1	4

Box 13.10 Continued

2. Sum the ranks for each of the $a = 10$ groups (rows in the present case) $= \Sigma^b R_{ij}$. This yields values such as 40, 36, 32, ...

3. Compute

$$X^2 = \left[\frac{12}{ab(a+1)} \sum^a \left(\sum^b R_{ij}\right)^2\right] - 3b(a+1)$$

$$= \frac{12}{10(4)(10+1)}(40^2 + 36^2 + \cdots + 8^2 + 4^2) - 3(4)(10+1)$$

$$= \frac{12}{440}(6160) - 132 = 36.000$$

This value is to be compared with $\chi^2_{[a-1]}$. Since $\chi^2_{0.01[9]} = 27.877$, our observed value is significant at $P < 0.001$, as was to be expected in view of the consistency of the data.

Multiple comparisons

We can again carry out multiple comparisons among the treatment means. The procedure is that of Box 13.8. The only difference is that in the formula for the critical value of U_s we replace the quantity n by bn, where b is the number of blocks and n is the number of replicates per subgroup (block per treatment). In the frequent cases where each treatment–block combination is not replicated (as in the example of this box), $n = 1$ and b replaces the n of the formula in Box 13.8. In this example the formula for the critical value of U_s is

$$U_{\alpha[a,b]} = \frac{b^2}{2} + Q_{\alpha[a,\infty]}b\sqrt{\frac{2b+1}{24}}$$

$$= \frac{4^2}{2} + 4.474(4)\sqrt{\frac{2(4)+1}{24}}$$

$$= 18.9590$$

Inspection of the values in the table reveals that the observations are virtually nonoverlapping between adjacent depths. Therefore the value of C is $4 + 4 + 4 + 4 = 16$ between each pair [except for $C(9,12)$ which is $3\frac{1}{2} + 3\frac{1}{2} + 4 + 4 = 15$]. It therefore appears that none of the comparisons can be declared significant, despite the high overall significance of the analysis.

1922 is again mirrored in both lines, suggesting that these fluctuations are environmentally caused. It is therefore appropriate that the data be treated as randomized blocks, with years as blocks and the strain differences as the fixed treatments.

Column (3) in Box 13.11 lists the differences on which a conventional paired-comparisons t-test could be performed. For Wilcoxon's test these

Box 13.11 WILCOXON'S SIGNED-RANKS TEST FOR TWO GROUPS, ARRANGED AS PAIRED OBSERVATIONS.

Mean litter size of two strains of guinea pigs, compared over $n = 9$ years.

Year	(1) Strain B	(2) Strain 13	(3) D	(4) Rank (R)
1916	2.68	2.36	+0.32	+9
1917	2.60	2.41	+0.19	+8
1918	2.43	2.39	+0.04	+2
1919	2.90	2.85	+0.05	+3
1920	2.94	2.82	+0.12	+7
1921	2.70	2.73	−0.03	−1
1922	2.68	2.58	+0.10	+6
1923	2.98	2.89	+0.09	+5
1924	2.85	2.78	+0.07	+4
	Absolute sum of negative ranks			1
	Sum of positive ranks			44

SOURCE: Data from S. Wright (unpublished results).

Procedure

1. Compute the differences between the n pairs of observations. These are entered in column (3), labeled D.
2. Rank these differences from the smallest to the largest *without regard to sign.*
3. Assign to the ranks the original signs of the differences.
4. Sum the positive and negative ranks separately. The sum that is smaller in absolute value, T_s, is compared with the values in Statistical Table **V** for $n = 9$.

Since $T_s = 1$, which is equal to or less than the entry for one-tailed $\alpha = 0.005$ in the table, our observed difference is significant at the 1% level. Litter size in strain B is significantly different from that of strain 13.

For large samples ($n > 50$) compute

$$t_s = \frac{T_s - \dfrac{n(n+1)}{4}}{\sqrt{\dfrac{n(n+\frac{1}{2})(n+1)}{12}}}$$

where T_s is as defined in step **4** above. Compare the computed value with $t_{\alpha[\infty]}$ in Statistical Table **B.**

differences are ranked *without regard to sign,* so the smallest absolute difference is ranked 1, and the largest absolute difference (of the nine differences) is ranked 9. Tied ranks are computed as averages as usual. After the ranks have been computed the original sign of each difference is assigned to the corresponding rank. The sum of the positive or of the negative ranks, whichever one is smaller in absolute value, is then computed (it is labeled T_s) and is compared with the critical value T in Statistical Table **V** for the corresponding sample size. In view of the significance of the rank sum, it is clear that strain B has a litter size different from that of strain 13. This test is very simple to carry out, but it is, of course, not as efficient as the corresponding parametric t-test, which is preferred if the necessary assumptions hold. Note that one needs at least six differences in order to carry out Wilcoxon's signed-ranks test. In six paired comparisons, all differences must be of like sign for the test to be significant at the 5% level.

For a large sample an approximation to the normal curve is available which is given in Box 13.11. Note that the absolute magnitudes of these differences play a role only as they affect the ranks of the differences.

A still simpler test is the **sign test,** in which we count the number of positive and negative signs among the differences (omitting all differences of zero). We then test the hypothesis that the n plus and minus signs are sampled from a population in which the two kinds of signs are present in equal proportions, as might be expected if there were no true difference between the two paired samples. Such sampling should follow the binomial distribution, and the test of the hypothesis that the parametric frequency of the plus signs is $\hat{p} = 0.5$ can be made in a number of ways. Let us learn these methods by applying the sign test to the guinea pig data of Box 13.11. There are nine differences, of which eight are positive and one is negative. We could follow the methods of Section 5.2 (illustrated in Table 5.3) in which we calculate the expected probability of sampling one minus sign in a sample of nine on the assumption of $\hat{p} = \hat{q} = 0.5$. The probability of such an occurrence and all "worse" outcomes equals 0.0195. Since we have no a-priori notions that one strain should have a greater litter size than the other, this is a two-tailed test and we double the probability to 0.0390. Clearly, this is an improbable outcome, and we reject the null hypothesis that $\hat{p} = \hat{q} = 0.5$.

Since the computation of the exact probabilities is tedious, a second approach is to make use of Statistical Table **Q,** which furnishes critical values for selected parametric proportions. The first part of Table **Q** gives values for $p = 0.5$, appropriate for the sign test. We reduce sample size n by one for each omitted zero difference sign. There are none in our example. We find for $n = 9$ that 1 minus and 8 plus signs are critical values at 5%. To yield a 2% significance level we would have required 0 minus and 9 plus signs. We conclude that the guinea pig strains differ in mean litter size at the 5% level. The percentages at the head of Table **Q** are for a two-tailed test. Had our results been 1 plus sign and 8 minus signs, we could have employed the table in an identical manner. If our alternative hypothesis had been that strain B has larger litters than strain 13, we could have halved the percentages at the head of the table, used the critical value for the appropriate tail of the distribution, and recorded a one-tailed significance level of 2.5%. Obviously, such a one-tailed test would be carried out only if the results were in the direction of the alternative hypothesis. Thus, if the alternative hypothesis were that strain 13 in Box 13.11 had greater litter size than strain B, we would not have bothered testing this example at all, since the observed proportion of years showing this relation was less than half. For samples greater than 100 one may need to interpolate in Table **Q.**

For larger samples, we can also use the normal approximation to the binomial distribution as follows:

$$t_s = \frac{Y - \mu}{\sigma_Y} = \frac{Y - kp}{\sqrt{kpq}}$$

substituting the mean and standard deviation of the binomial distribution (see Section 5.2). In our case, we let n stand for K and assume that $\hat{p} = \hat{p} = 0.5$. Therefore,

$$t_s = \frac{Y - \frac{1}{2}n}{\sqrt{\frac{1}{4}n}} = \frac{Y - \frac{1}{2}n}{\frac{1}{2}\sqrt{n}}$$

The value of t_s is then compared with $t_{\alpha[\infty]}$ in Statistical Table **B,** using one tail or two tails of the distribution as warranted. When the sample size $n \geq 12$, this approximation is satisfactory. For a more accurate approximation of the binomial confidence limits consult Burstein (1973).

A third approach would be to test the departure from expectation $\hat{p} = \hat{q} = 0.5$ by one of the methods of Chapter 17.

We can employ the sign test also for slightly more complicated hypotheses. If the two columns of data in a paired-comparisons case are labeled Y_1 and Y_2, respectively, we can test whether the response Y_1 is greater than Y_2 by K units. Simply evaluate the difference $D = Y_1 - (Y_2 + K)$ and carry out a sign test on the signs of the D-values. A second hypothesis might be that the response Y_1 is greater than Y_2 by $100p\%$. In such a case the sign test should be performed on the differences $D = Y_1 - (1 + p)Y_2$.

When the two-way anova is replicated and both factors are fixed treatment effects, we can employ an extension of the Kruskal–Wallis test devised by Scheirer et al. (1976). The procedure is simple. Treat all the observations in the basic data table as a single array and rank them. Then replace the ranks in the data table and carry out a two-way anova on the ranks as though they were the variates. The expected variance of any array of n ranks is $n(n + 1)/12$. Convince yourself that this statement is true by calculating the sum of squares of a small array and dividing it by $n - 1$ to obtain the mean square. Since the MS computed by this formula is constant for any given number n of ranks, we can consider it a parameter. Recall from Section 7.6 that the chi-square distribution is the ratio of a sum of squares divided by a parametric variance. Therefore we can test the

Box 13.12 A TWO-WAY ANOVA DESIGN FOR RANKED DATA. THE SCHEIRER–RAY–HARE EXTENSION OF THE KRUSKAL–WALLIS TEST.

Food consumption related to fat quality and sex in rats. Data from Box 11.1. $a = 2$ (fats), $b = 2$ (sexes), $n = 3$ (number of rats per subgroup).

	Fats ($a = 2$)			
	Fresh		Rancid	
Sex ($b = 2$)	Y	Rank	Y	Rank
♂	709	12	592	6
	679	10	538	4
	699	11	476	1
♀	657	8	508	3
	594	7	505	2
	677	9	539	5

Computation

1. Rank all observations from smallest to largest when pooled together into a single sample. Replace each variate in the original data table by its rank.
2. Carry out an analysis of variance of the ranks following the outline in Box 11.1. The following table results.

Source of variation	df	SS	MS
A (columns; fats)	1	108.0000	108.0000
B (rows; sexes)	1	8.3333	8.3333
$A \times B$ (interaction)	1	5.3333	5.3333
Error	8	21.3334	2.6667

3. Compute the total MS as $abn(abn + 1)/12$. If there are ties in the ranks it is easiest to compute the total MS from the total SS, by adding all sums of squares in the table and

BOX 13.12 CONTINUED

dividing by the total degrees of freedom. Alternatively, we could compute the correction factor D from Box 13.6 and use it to adjust H (computed in step **4**), as shown in that box.

4. Compute H as SS/MS_{total}

For factor A

$$H = \frac{108.0000}{13.0000} = 8.3077$$
$$0.001 < P < 0.005$$

For factor B

$$H = \frac{8.3333}{13.0000} = 0.6410$$
$$0.50 < P < 0.10$$

For interaction AB

$$H = \frac{5.3333}{13.0000} = 0.4103$$
$$0.90 < P < 0.50$$

The significance for H is tested as a χ^2-variable, with the degrees of freedom pertaining to the SS being tested.

We conclude that only factor A, the quality of the fat consumed, exerts a significant effect.

sums of squares that emerged from the anova of the ranks by dividing them by the total MS of the ranks and considering the ratio H as a χ^2-distributed variable with the degrees of freedom pertaining to the sum of squares in the numerator. This is entirely analogous to what the Kruskal–Wallis test does for the single-classification design, and we could have computed the significance of a Kruskal–Wallis test by using the procedures outlined here.

In Box 13.12 we employ the "rats and fats" example of Box 11.1 to illustrate the method just described. When analyzed nonparametrically, the results do not differ from those of the earlier anova. There is a significant treatment effect for fats, but neither sex nor the interaction is significant. This approach can be extended to multiway anovas in a straightforward manner. A general but computationally more complex formula for dealing with such cases is given by Groggel and Skillings (1986).

EXERCISES 13

13.1 In a study of flower color in butterfly weed *(Asclepias tuberosa)*, Woodson (1964) obtained the following results:

Geographic region	$\bar{Y}$	n	s
C1	29.3	226	4.59
SW2	15.8	94	10.15
SW3	6.3	23	1.22

The variable recorded was a color score (ranging from 1 for pure yellow to 40 for deep orange red) obtained by matching flower petals to sample colors in Maerz and Paul's *Dictionary of Color*. Analyze and interpret (note that the standard deviations appear unequal). *Answer:* $v^*_{1,2} = 108.8$; all are significantly different at $P = 0.05$ using the method of Games and Howell (see Box 13.3).

13.2 The number of bacteria in 1 cm³ of milk from 3 cows was counted at 3 intervals (data from Park et al., 1924), yielding the following results:

	At time of milking	After 24 hours	After 48 hours
Cow No. 1	12,000	14,000	57,000
2	13,000	20,000	65,000
3	21,500	31,000	106,000

(a) Calculate means and variances for the three periods and examine the relation between these two statistics. Transform the variates to logarithms and compare means and variances based on the transformed data. Discuss.
(b) Carry out an anova on transformed and untransformed data. Discuss your results.
(c) Analyze the data by Friedman's method for randomized blocks.

13.3 Allee and Bowen (1932) studied survival time of goldfish (in minutes) when placed in colloidal silver suspensions. Experiment no. 9 involved 5 replications, and experiment no. 10 involved 10 replicates. Do the results of the two experiments differ? Addition of urea, NaCl, and Na_2S to a third series of suspensions apparently prolonged the life of the fish.

Colloidal silver		
Experiment No. 9	Experiment No. 10	Urea and salts added
210	150	330
180	180	300
240	210	300
210	240	420

Colloidal silver		
Experiment No. 9	Experiment No. 10	Urea and salts added
210	240	360
	120	270
	180	360
	240	360
	120	300
	150	120

Analyze and interpret. Check assumptions of normality and equality of variances. Compare anova results with those obtained using the Kruskal–Wallis test. *Answer:* $H = 12.0047$.

13.4 In a study of a sea cucumber (*Stichopus* sp.) measurements were made of the lengths of "buttons" (in micrometer units) in samples of tissue from 3 body regions for 2 specimens (data from David Olsen, unpublished results).

Specimen	Body regions									
	Dorsal				Lateral		Ventral			
No. 2	115	91	102	89	92	90	96	90	72	93
	88	84	80	73	90	91	98	114	89	81
	100	98	77	100	90	91	80	90	91	124
	96	87	91	111	93	82	93	87	111	92
	88	91	85	94	85	91	96	128	91	108
No. 9	81	108	92	88	113	112	91	104	96	107
	75	90	90	110	92	90	117	105	85	98
	90	96	100	91	96	105	104	91	90	80
	95	119	100	97	85	91	91	85	82	83
	99	88	91	95	88	92	81	80	89	90

Are the variances homogeneous? Are the distributions normal? Make histograms for the data in each cell of the two-way table and perform graphic analyses. Analyze by a two-way anova. Why can you show the effect of body regions to be significant? Is there anything disturbing about these data?

13.5 Test for a difference in surface and subsoil pH in the data of Exercise 11.1, using Wilcoxon's signed-ranks test. *Answer:* $T_s = 38$; $P > 0.10$.

13.6 Reanalyze the birth weight data of Exercise 9.3 by using the Kolmogorov–Smirnov two-sample test.

13.7 Reanalyze the reproduction data of Box 9.5 by nonparametric methods (Mann–Whitney U-test and Kolmogorov–Smirnov two-sample test). Do your conclusions differ from those arrived at in Box 9.5? *Answer:* No; $U_s = 31$ and $n_1 n_2 D = 21$.

13.8 Check whether the butterfat data of Exercise 4.3 should be transformed by using the Box–Cox procedure. Perform a graphic analysis on both the original and the transformed data.

13.9 Allee et al. (1934) studied the rate of growth of *Ameiurus melas* in conditioned and unconditioned well water and obtained the following results for the gain in average length of sample fish. Although the original variates are not available, we may still test for differences between the two treatment classes. *Answer:* $T_s = 3$, $P = 0.0049$ using Wilcoxon's signed-ranks test.

	Average gain in length (in millimeters)	
Replicate	Conditioned water	Unconditioned water
1	2.20	1.06
2	1.05	0.06
3	3.25	3.55
4	2.60	1.00
5	1.90	1.10
6	1.50	0.60
7	2.25	1.30
8	1.00	0.90
9	−0.09	−0.59
10	0.83	0.58

14 LINEAR REGRESSION

We now turn to the simultaneous analysis of two variables. Even though we may have considered more than one variable at a time in the studies we have described thus far (e.g., seawater concentration and oxygen consumption in Box 11.2, or depth and water temperature in Box 11.3), our actual analyses were of only one variable. Frequently, however, we measure two or more variables on each individual; we would therefore like to be able to express more precisely the nature of the relationships between these variables. Thus we come to the subjects of **regression** and **correlation.** In regression we estimate the relationship of one variable with another by expressing the one in terms of a linear (or a more complex) function of the other. We also use regression to predict values of one variable in terms of the other. In correlation analysis, which is sometimes confused with regression, we estimate the degree to which two variables vary *together.* Chapter 15 deals with correlation, and we will postpone our effort to clarify the relation and distinction between regression and correlation until then. The variables involved in regression and correlation are continuous or, if meristic, are treated as though they were continuous (see Chapter 2). If the variables are qualitative (that is, if they are attributes), the methods of regression and correlation cannot be used.

In Section 14.1 we review mathematical functions and introduce the new terminology required for regression analysis. Section 14.2 follows with a discussion of the appropriate statistical models for regression analysis. The basic computations in simple linear regression are shown in Section 14.3, which outlines the computation for the case of one dependent variate for each independent variate. Tests of significance and computation of confidence intervals for regression problems are discussed in Section 14.4, where we also discuss how to fit a regression line through the origin. Section 14.5 presents the case with several dependent variates for each independent variate.

Section 14.6 summarizes regression and discusses the various uses of regression analysis in biology. In this section we also show how to estimate the relative efficiency of experimental designs in which more than one variable is measured. Section 14.7 examines a procedure known as inverse prediction, in which the most probable value of the independent variate is estimated from an observed dependent variate. In Section 14.8 we test differences among two or more regression coefficients. These tests lead to the analysis of covariance featured in Section 14.9. In this technique, group means in an anova of a dependent variable are first adjusted for differences in a second, independent variable before being tested for homogeneity.

In Section 14.10 we relate regression to the analysis of variance, showing how the sums of squares and degrees of freedom among groups can be partitioned into single-degree-of-freedom contrasts, and we demonstrate that these contrasts are really regression computations. Linear regression can be treated as such a contrast, and the residual degrees of freedom can be used to test for departures from linearity in the regression relationship. Section 14.11 describes how to examine residuals from regression to detect outliers or systematic departures from the regression line. We also show how transformation of scale can straighten out curvilinear relationships for ease of analysis. When transformation cannot linearize the relation between variables, an alternative approach is a nonparametric test for regression. Such tests are illustrated in Section 14.12.

Most methods we have described here assume that the independent variable is measured without error. When both variables are measured with error, the so-called Model II case, techniques for estimating regression equations differ, as we will see in Section 14.13.

14.1 Introduction to Regression

Much scientific thought concerns the relations between pairs of variables hypothesized to be in a cause-and-effect relationship. We will not discuss here the philosophical requirements for establishing whether the relationship between two variables is really one of cause and effect. We will be content with establishing the form and significance of *functional relationships* between two variables, leaving the demonstration of cause-and-effect relationships to the established procedures of the scientific method. A **function** is a mathematical relationship enabling us to predict what values of variable Y correspond to given values of a variable X. Such a relationship, generally written as $Y = f(X)$, is familiar to all of us from our general scientific and cultural experience. Nevertheless, as an introduction to the subject of regression, we will briefly review functions.

The simplest type of regression follows the equation $Y = X$, as illustrated in Figure 14.1. This graph shows the number of growth rings on a tree as a function of age in years (by drawing this as a continuous function we are assuming we can estimate fractions of growth rings corresponding to fractions of years). Although the graph is self-explanatory, let us nevertheless examine the properties of the function $Y = X$. Whatever the value of X (age in years), the value of Y (number of growth rings) will be of corresponding magnitude. Thus, if the tree is 10 years

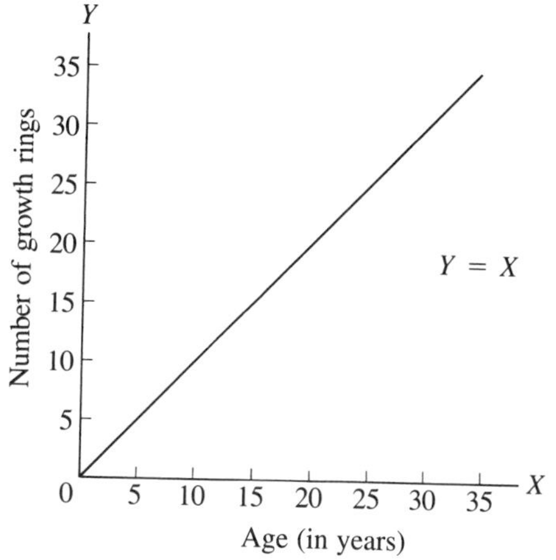

FIGURE 14.1 *Number of growth rings on a tree as a function of age. An example of the simplest type of regression.*

old, it will have 10 growth rings. When $X = 0$, Y also equals zero, and the line describing the function goes through the origin of the coordinate system illustrated in Figure 14.1, indicating that a tree zero years old has no growth rings, a reasonable fact in terms of the biology of trees. Clearly, with the relationship $Y = X$ we can predict accurately the number of growth rings, given the age of a tree. We call the variable Y the **dependent variable** and X the **independent variable.** The magnitude of Y depends on the magnitude of X and can therefore be predicted from the independent variable, which presumably is free to vary. Remember that although a cause is always considered an independent variable and an effect a dependent variable, a functional relationship observed in nature may not be a cause-and-effect relationship. It is conventional in statistics to label the dependent variable Y and the independent variable X.

Figure 14.2 shows another functional relationship, indicated by the equation $Y = bX$. Here the independent variable X is multiplied by a coefficient b, a slope factor. In this example, which relates the height of a plant in centimeters to its age in days, the slope factor $b = \frac{1}{7} = 0.143$, which means that for an increase of

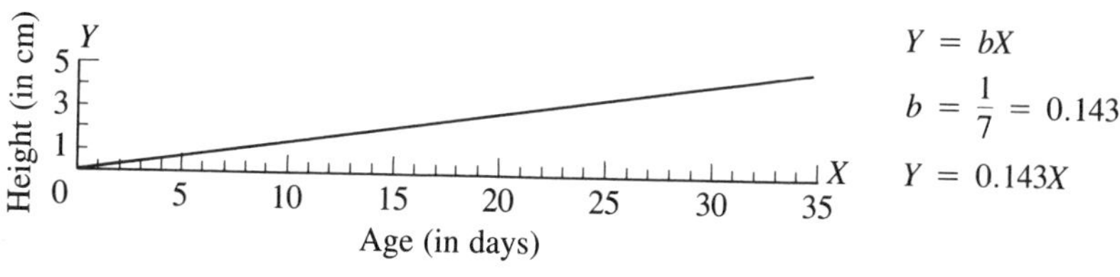

FIGURE 14.2 *Height of a plant as a function of age.*

7 units of X, there will be an increase of 1 unit of Y. Thus, a 14-day-old plant will be 2 cm tall. For any fixed scale, the slope of the function line depends on the magnitude of b, the slope factor. As b increases, the slope of the line becomes steeper. Note again that when $X = 0$, the dependent variable also equals zero. This makes sense; a plant that is zero days old has no height.

In biology, such a relationship is appropriate over only a limited range of values of X. Negative values of X are meaningless, and it is unlikely that the plant will continue to grow at a uniform rate. Quite probably the slope of the functional relationship will flatten out as the plant gets older and approaches its mature size. For a limited portion of the range of variable X (age in days), however, the linear relationship $Y = bX$ may be an adequate description of the functional dependence of Y on X.

A third type of function is depicted in Figure 14.3, which illustrates the effect of two drugs on blood pressure in two species of animals. The relationships shown in this graph can be expressed by the formula $Y = a + bX$. The highest line is of the relationship $Y = 20 + 15X$, which represents the effect of drug A on animal P. The quantity of drug is measured in micrograms, the blood pressure in millimeters of mercury. From the equation, it is easy to calculate the blood pressure to be expected in the animal after a given amount of drug has been administered. Thus, after 4 μg of the drug have been given, the blood pressure will be $Y = 20 + (15)(4) = 80$ mm Hg. Note that by this formula, when the independent variable equals zero, the dependent variable does not also equal zero, but equals a. When $X = 0$, the blood pressure is 20 mm, which is the normal blood pressure of animal P in the absence of the drug. It would not be biologically reasonable to assume that the animal had no blood pressure in the absence of the drug. The magnitude 20 mm can be obtained by solving the equation after substituting zero for X or by examining the graph for the

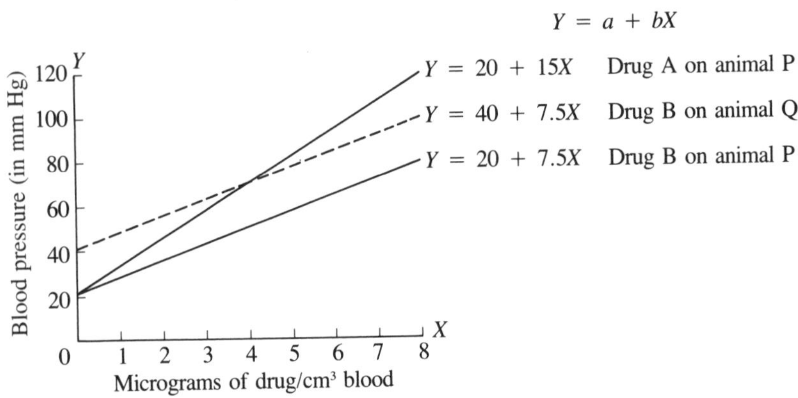

FIGURE 14.3 *Blood pressure of an animal as a function of drug concentration in the blood.*

intersection of the function line with the Y-axis. This point yields the magnitude of a, which is therefore called the **Y-intercept.**

The two other functions in Figure 14.3 show the effects of varying both a, the Y-intercept, and b, the slope. In the lowest line, $Y = 20 + 7.5X$, the Y-intercept remains the same, but the slope has been halved. This line represents the effect of a different drug, B, on the same organism, P. Obviously, when no drug is administered, the blood pressure should be at the same Y-intercept, since the identical organism is being studied. However, different drugs, are likely to exert different hypertensive effects, as reflected by the different slopes. The third relationship (the broken line) also describes the effect of drug B, which is assumed to remain the same, but the experiment is carried out on a different species, Q, whose normal blood pressure is assumed to be 40 mm Hg. Thus, the equation for the effect of drug B on species Q is written as $Y = 40 + 7.5X$. Note that this line is exactly parallel to the one studied previously, since it has the same slope.

You have probably noticed that the equation represented in Figure 14.3, $Y = a + bX$, is the most general linear equation and contains within it the two simpler versions from Figures 14.1 and 14.2. In $Y = bX$ we simply assume that $a = 0$; in $Y = X$, $a = 0$ and $b = 1$.

Those of you familiar with analytical geometry will have recognized the slope factor b as the *slope* of the function $Y = a + bX$, generally symbolized by m. If you know calculus, you will, of course, recognize b as the *derivative* of that same function ($dY/dX = b$). In biometry, b is called the **regression coefficient** and the function is called a **regression equation.** The origin of the word "regression" lies in biometric studies made by Francis Galton in which the literal meaning of regression as "returning to" or "going back to" was appropriate; however, the accepted meaning in statistics now relates to the dependence of the means of variable Y on the independent variable X by some mathematical equation. When we wish to stress that the regression coefficient is of variable Y on variable X, we write $b_{Y \cdot X}$. If ever we wish to regress X on Y (when this is legitimate), the proper symbol for the regression coefficient is $b_{X \cdot Y}$.

14.2 Models in Regression

In any real example, observations do not lie perfectly along a regression line but scatter along both sides of the line. This scatter is usually due to inherent, natural variation of the items (genetically and environmentally caused) and to measurement error. Thus in regression a functional relationship does not mean that given X the value of Y must be $a + bX$, but rather that the mean (or expected value) of Y is $a + bX$.

The appropriate computations and significance tests in regression involve the following two models. The more common of these, **Model I regression,** is especially suitable in experimental situations. It is based on four assumptions:

1. The independent variable X is measured without error. We therefore say that the X's are "fixed," which means that whereas Y, the dependent variable, is a random variable, X does not vary at random, but is under the control of the investigator. In the example of Figure 14.3, for instance, the dose of drug which we varied at will, was X, whereas the variable whose response we studied, blood pressure, was the random variable, Y. We can manipulate X in the same way that we were able to manipulate the treatment effect in a Model I anova. In fact, as you will see later, there is a very intimate relationship between Model I anovas and Model I regression.

2. The expected value for the variable Y for any given value X is described by the linear function $\mu_Y = \alpha + \beta X$. This relation is the same function we encountered in the previous section, but here we use Greek letters for a and b, since we are describing a parametric relationship. Another way of stating this assumption is that the parametric means μ_Y of the values of Y are a function of X and lie on a straight line described by this equation.

3. For any given value X_i of X, the Y's are independently and normally distributed. This relationship can be represented by the equation $Y_{ij} = \alpha + \beta X_i + \epsilon_{ij}$, where the ϵ_{ij}'s are assumed to be normally distributed error terms with a mean of zero. Figure 14.4 illustrates this concept with a regression line similar to the ones in Figure 14.3. A given experiment can be repeated several times. Thus, for instance, we could administer 2, 4, 6, 8, and 10 μg of the drug to each of 20 individuals of an animal species and obtain a frequency distribution of blood pressure responses Y to the independent variates $X = 2, 4, 6, 8$, and 10 μg. The inherent variability of biological material means that the responses to each dosage would not be the same in every individual; thus we would obtain a distribution of values of Y (blood pressure) around the expected value. Assumption 3 states that these sample values will be independently and normally distributed, as is indicated by the normal curves superimposed about several points in the regression line in Figure 14.4. A few curves are shown to give you an idea of the scatter about the regression line. In actuality there is, of course, a continuous scatter, as though these separate normal distributions were stacked right next to

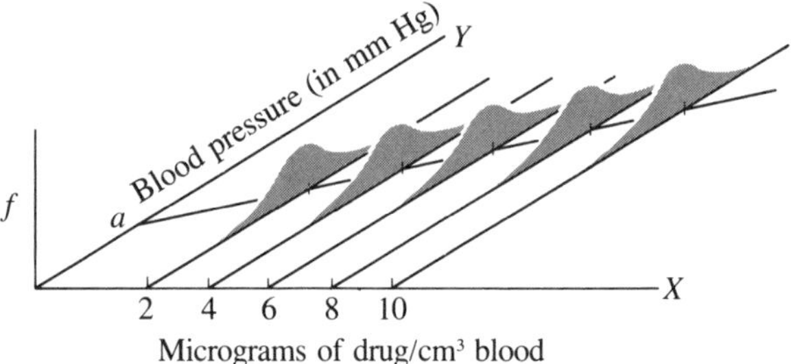

FIGURE 14.4 *Blood pressure of an animal as a function of drug concentration in the blood. Repeated sampling for a given drug concentration.*

each other, since there are an infinite number of possible intermediate values of X between any two dosages. In the rare cases in which the independent variable is discontinuous, the distributions of Y are physically separate from each other and occur only along the points of the abscissa that correspond to independent variates. An example of such a case is the weight of offspring (Y) in litters of mice. There may be three or four offspring per litter, but there would be no intermediate value of X representing 3.25 mice per litter.

Not every experiment has more than one value of Y for each value of X. In fact, the basic computations we will learn in the next section apply to only one value of Y per X, since this is the more common case. Even in such instances, however, the basic assumption of Model I regression is that the single variate of Y corresponding to the given value of X is a sample of size 1 from a population of independently and normally distributed variates.

4. The final assumption is familiar: The samples along the regression line are homoscedastic; that is, they have a common variance, σ^2, which is the variance of the ϵ's in the expression for assumption 3. Thus we assume that the variance around the regression line is constant and hence is independent of the magnitude of X or Y.

Many regression analyses in biology do not meet the assumptions of Model I regression. Frequently both X and Y are subject to natural variation and/or measurement error. Also the variable X may not be fixed, that is, under control of the investigator. Suppose we sample a population of female flies and measure wing length and total weight of each individual. We might be interested in studying wing length as a function of weight or we might wish to predict wing length for a given weight. In this case the weight, which we treat as an independent variable, is not fixed and certainly not the "cause" of differences in wing length. The weights of the flies vary for genetic and environmental reasons and are also subject to measurement error. The general case in which both variables show random variation is called **Model II regression.** Some of the computations for estimating functional relationships and prediction are modified in this model, which is discussed further in Section 14.13.

In Model II regression, the appropriate regression line may vary depending on whether functional relationship or prediction is the aim of the investigator. It is therefore important to keep this aim clearly in mind. In Model I regression (the familiar "regression" of statistics texts and research articles), however, the same equation serves both purposes. In the discussion that follows, which is devoted to Model I regression, we therefore do not stress the distinction between these aims. We now turn to the rationale behind the computational steps in Model I regression.

14.3 THE LINEAR REGRESSION EQUATION

To learn the basic computations of a Model I linear regression, we choose an example with only one Y-value per independent variate X, since this example is computationally simpler than dealing with samples and is by far the more common case encountered in research work. The extension to a sample of values of Y for each X is explained in Section 14.5. Just as in the case of the previous analyses, there are also simple computational formulas, which will be presented at the end of this section.

The data for our discussion of regression come from a study of water loss in *Tribolium confusum,* the confused flour beetle. Nine batches of 25 beetles were weighed (individual beetles could not be weighed with available equipment), kept at different relative humidities, and weighed again after six days of starvation. Weight loss in milligrams was computed for each batch. This example is a Model I regression in which the weight loss is the dependent variable Y and the relative humidity is the independent variable X, a fixed treatment effect under the control of the experimenter. The purpose of the analysis is to establish whether the relationship between relative humidity and weight loss can be described adequately by a linear regression of the general form $Y = a + bX$. The original data are shown in columns (1) and (2) of Table 14.1. They are plotted in Figure 14.5, from which it appears that a negative relationship exists between weight loss and humidity: as humidity increases, weight loss decreases. The means of weight loss and relative humidity, $\bar{Y}$ and $\bar{X}$, respectively, are marked along the coordinate axes. The average humidity is 50.39%, and the average weight loss is 6.022 mg. How can we fit a regression line to these data, permitting us to estimate a value of Y for a given value of X?

Unless the observations lie exactly on a straight line, we need a criterion for determining the best possible placing of the regression line. Statisticians generally follow the principle of least squares, which we first encountered in Chapter

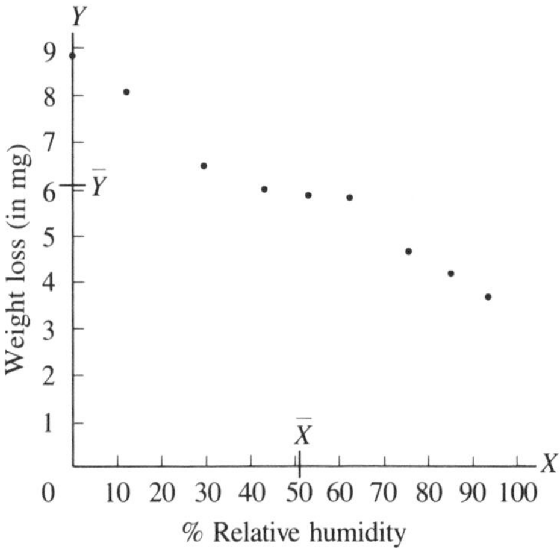

FIGURE 14.5 *Weight loss of nine batches of 25* Tribolium *beetles after six days of starvation at nine different relative humidities. Data from Table 14.1.*

Table 14.1 BASIC COMPUTATIONS IN REGRESSION.

Weight loss (in milligrams) of nine batches of 25 Tribolium *beetles after six days of starvation at nine different humidities.*

	(1) Percent relative humidity X	(2) Weight loss in mg Y	(3) $x = (X - \bar{X})$	(4) $y = (Y - \bar{Y})$	(5) x^2	(6) xy	(7) y^2	(8) $\hat{Y}$	(9) $d_{Y \cdot X} = Y - \hat{Y}$	(10) $d^2_{Y \cdot X}$	(11) $\hat{y} = \hat{Y} - \bar{Y}$	(12) $\hat{y}^2$
	0	8.98	−50.39	2.958	2539.1521	−149.0536	8.7498	8.7038	0.2762	0.0763	2.6818	7.1921
	12	8.14	−38.39	2.118	1473.7921	−81.3100	4.4859	8.0652	0.0748	0.0056	2.0432	4.1747
	29.5	6.67	−20.89	0.648	436.3921	−13.5367	0.4199	7.1338	−0.4638	0.2151	1.1118	1.2361
	43	6.08	−7.39	0.058	54.6121	−0.4286	0.0034	6.4153	−0.3353	0.1124	0.3933	0.1547
	53	5.90	2.61	−0.122	6.8121	−0.3184	0.0149	5.8831	0.0169	0.0003	−0.1389	0.0193
	62.5	5.83	12.11	−0.192	146.6521	−2.3251	0.0369	5.3776	0.4524	0.2047	−0.6444	0.4153
	75.5	4.68	25.11	−1.342	630.5121	−33.6976	1.8010	4.6857	−0.0057	0.0000	−1.3363	1.7857
	85	4.20	34.61	−1.822	1197.8521	−63.0594	3.3197	4.1801	0.0199	0.0004	−1.8419	3.3926
	93	3.72	42.61	−2.302	1815.6121	−98.0882	5.2992	3.7543	−0.0343	0.0012	−2.2677	5.1425
Sum	453.5	54.20	−0.01	0.002	8301.3889	−441.8176	24.1307	54.1989	0.0011	0.6160	0.0009	23.5130
Mean	50.39	6.022						6.022				
Sum/$(n-1)$					1037.6736	−55.2272	3.0163			0.0880[a]		

SOURCE: Nelson (1964).
[a] Sum divided by $n - 2$.

4. If we were to draw a horizontal line through $\bar{X}$, $\bar{Y}$ (that is, a line parallel to the X-axis at the level of $\bar{Y}$), then deviations to that line drawn parallel to the Y-axis would represent the deviations from the mean for these observations with respect to variable Y, (as in Figure 14.6). We learned in Chapter 4 that the sum of these observations, $\Sigma(Y - \bar{Y}) = \Sigma y = 0$. The sum of squares of these deviations, $\Sigma(Y - \bar{Y})^2 = \Sigma y^2$, are less than those from any other horizontal line. Any horizontal line drawn through the data at a point other than $\bar{Y}$ would yield a sum of deviations other than zero and a sum of deviations squared greater than Σy^2. Therefore a mathematically correct but impractical method for finding the mean of Y would be to draw a series of horizontal lines across a graph, calculate the sum of squares of deviations from it, and choose the line yielding the smallest sum of squares.

In linear regression, we draw a straight line through our observations, but the line is not necessarily horizontal. A sloped regression line indicates for each value of the independent variable X_i an estimated value of the dependent variable. To distinguish between the estimated values of Y_i and the observed values, we will hereafter designate the former as $\hat{Y}_i$ (read: y-hat or Y-caret); the latter will remain Y_i. The regression equation therefore should read

$$\hat{Y} = a + bX \tag{14.1}$$

which indicates that for given values of X, this equation provides estimated values $\hat{Y}$ (as distinct from the observed values Y in any actual case). The deviation of an observation Y_i from the regression line is $(Y_i - \hat{Y}_i)$ and is generally symbolized as $d_{Y \cdot X}$. These deviations are drawn parallel to the Y-axis, and they meet the sloped regression line at an angle (Figure 14.7). The sum of these deviations is again zero ($\Sigma d_{Y \cdot X} = 0$), and the sum of their squares yields a

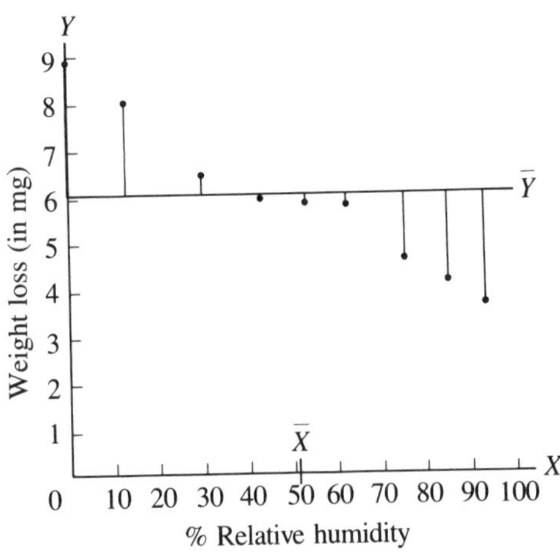

FIGURE 14.6 *Deviations from the mean of* Y *for the data of Figure 14.5.*

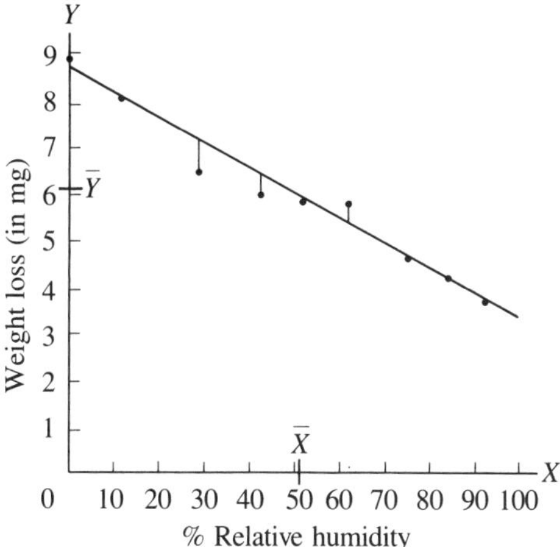

FIGURE 14.7 *Deviations from the regression line for the data of Figure 14.5.*

quantity $\Sigma(Y - \hat{Y})^2 = \Sigma d^2_{Y \cdot X}$ analogous to the sum of squares Σy^2. For reasons that will become clear later, $\Sigma d^2_{Y \cdot X}$ is called the unexplained or residual sum of squares. The **least squares linear regression line** through a set of points is defined as the straight line that results in $\Sigma d^2_{Y \cdot X}$ being at a minimum. Geometrically, the basic idea is that we prefer using a line that is in some sense close to as many points as possible. For ordinary Model I regression analysis, it is most useful to define closeness in terms of the vertical distances from the points to a line, using the line that makes the sum of these squared deviations as small as possible.

A convenient consequence of this criterion is that the line must pass through the point $\bar{X}$, $\bar{Y}$. It would be possible but impractical to calculate the correct regression slope by pivoting a ruler around the point $\bar{X}$, $\bar{Y}$, and calculating the unexplained sum of squares, $\Sigma d^2_{Y \cdot X}$, for each of the innumerable possible positions. Whichever position gave the minimal value of $\Sigma d^2_{Y \cdot X}$ would be the least squares regression line. The usual derivation of the formula for the slope of a line yielding a minimal value of $\Sigma d^2_{Y \cdot X}$ uses calculus (see Section A.8 in the appendix). We can obtain equations for b and a by plain algebra as well (see Section A.9 in the appendix). Either results in the formula for the regression coefficient of Y on X, $b_{Y \cdot X} = \Sigma xy / \Sigma x^2$. We will now calculate this quantity for our weight loss data.

We first compute the deviations from the respective means of X and Y as shown in columns (3) and (4) of Table 14.1. The sums of these deviations, Σx and Σy, differ slightly from their expected value of zero because of rounding errors. The squares of these deviations yield sums of squares and variances in columns (5) and (7). In column (6) are the products xy, which in this example are all negative because the x- and y-deviations are of unlike sign. The sum of these products, $\Sigma^n xy$, is a new quantity, called the **sum of products.** This poor but well-established term refers to Σxy, the sum of the products of the deviations rather than ΣXY, the sum of the products of the variates. Recall that Σy^2 is called the sum of squares, whereas ΣY^2 is the sum of the squared variates. The sum of products is analogous to the sum of squares. When divided by the degrees of freedom, the sum of products yields the **covariance,** just as the variance results from a similar division of the sum of squares. (We first encountered covariances in Section 8.4.) Note that the sum of products can be negative as well as positive. A negative sum of products indicates a negative slope of the regression line: as X increases, Y decreases. In this respect the sum of products differs from a sum of squares, which can be only positive. From Table 14.1 we find that $\Sigma xy = -441.8176$, $\Sigma x^2 = 8301.3889$, and $b = \Sigma xy/\Sigma x^2 = -0.05322$. Thus, for a one-unit increase in X, there is a decrease of 0.05322 units in Y. In terms of the present example, a 1% increase in relative humidity indicates a weight loss of 0.05322 mg.

You may wish to convince yourself that the formula for the regression coefficient is intuitively reasonable. This formula is the ratio of the sum of *products* of deviations for X and Y over the sum of *squares* of deviations for X. The product for X_i, a single value of X, is $x_i y_i$. Similarly, the squared deviation for X_i is x_i^2 or $x_i x_i$. The ratio $x_i y_i / x_i x_i$ thus reduces to y_i/x_i. Although $\Sigma xy/\Sigma x^2$ only approximates the average of y_i/x_i for the n values of X_i, the latter ratio indicates the direction and magnitude of the change in Y for a unit change in X. Thus if y_i on the average equals x_i, b will equal 1. When $y_i = -x_i$, $b = -1$. Also when $|y_i| > |x_i|$, $b > |1|$, and conversely when $|y_i| < |x_i|$, $b < |1|$.

How can we complete the equation $\hat{Y} = a + bX$? We have stated that the regression line passes through the point $\bar{X}, \bar{Y}$. At $\bar{X} = 50.39$, $\hat{Y} = 6.022$; that is, we use $\bar{Y}$, the observed mean of Y, as an estimate $\hat{Y}$ of the mean. We can substitute these means into Expression (14.1):

$$\hat{Y} = a + bX$$

$$\bar{Y} = a + b\bar{X}$$

$$a = \bar{Y} - b\bar{X}$$

$$\begin{aligned} a &= 6.022 - (-0.05322)(50.39) \\ &= 8.7038 \end{aligned}$$

$$\text{Therefore } \hat{Y} = 8.7038 - 0.05322X.$$

This equation predicts weight loss from relative humidity. Note that when X is zero (0% relative humidity), the estimated weight loss is greatest, equal to $a = 8.7038$ mg. As X increases to a maximum of 100, however, the weight loss decreases to 3.3818 mg.

We can use the regression formula to draw the regression line. Simply estimate $\hat{Y}$ at two convenient points of X such as $X = 0$ and $X = 100$ and draw a straight line between them. This line, added to the observed data, is shown in

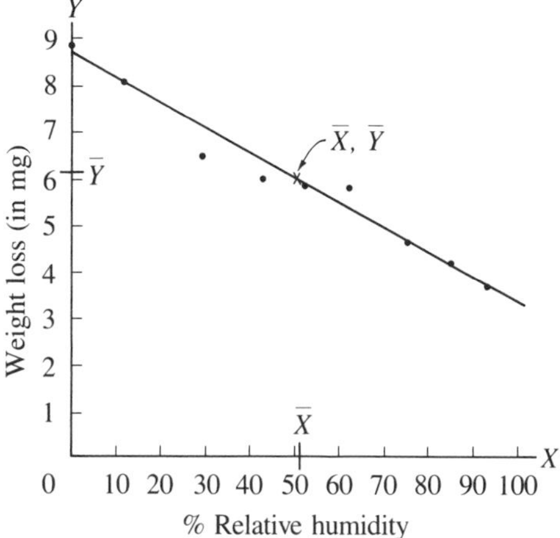

FIGURE 14.8 *Linear regression fitted to data of Figure 14.5.*

Figure 14.8. Note that it goes through the point $\overline{X}, \overline{Y}$. In fact, to draw the regression line, we frequently use the intersection of the two means and one other point.

Now bear with us through some very elementary algebra. Since

$$a = \overline{Y} - b\overline{X}$$

we can write Expression (14.1), $\hat{Y} = a + bX$, as

$$\begin{aligned}\hat{Y} &= (\overline{Y} - b\overline{X}) + bX \\ &= \overline{Y} + b(X - \overline{X})\end{aligned}$$

Therefore

$$\hat{Y} = \overline{Y} + bx \tag{14.2}$$

Also,

$$\hat{Y} - \overline{Y} = bx$$

$$\hat{y} = bx \tag{14.3}$$

where $\hat{y}$ is defined as the deviation $\hat{Y} - \overline{Y}$. Next, using Expressions (14.1) or (14.2), we estimate $\hat{Y}$ for every one of our given values of X. The estimated values $\hat{Y}$ are shown in column (8) of Table 14.1. Compare them with the observed values of Y in column (2). Overall agreement between the two columns of values is good. Note that except for rounding errors, $\Sigma\hat{Y} = \Sigma Y$ and hence $\hat{\overline{Y}} = \overline{Y}$ Observed Y-values, however, usually are different from the estimated values $\hat{Y}$ because of individual variation around the regression line.

The deviations of each observed Y-value from its estimated value $(Y - \hat{Y}) = d_{Y \cdot X}$ are listed in column (9) of Table 14.1. Note that these deviations exhibit one of the properties of deviations from a mean: They sum to zero except for rounding errors. Thus $\Sigma d_{Y \cdot X} = 0$, just as $\Sigma y = 0$. Next, we compute in column (10) the squares of these deviations and sum them to give a new sum of squares, $\Sigma d^2_{Y \cdot X} = 0.6160$. A comparison of $\Sigma(Y - \bar{Y})^2 = \Sigma y^2 = 24.1307$ with $\Sigma(Y - \hat{Y})^2 = \Sigma d^2_{Y \cdot X} = 0.6160$ shows that the new sum of squares is much less than the previous one. What has caused this reduction? Allowing for its relationship with X has eliminated most of the variance of Y from the sample. Remaining is the *residual* or **unexplained sum of squares** $\Sigma d^2_{Y \cdot X}$, which expresses that portion of the total SS of Y which is not accounted for by differences in X. This value is unexplained with respect to X. The difference between the total, SS, Σy^2, and the unexplained SS, $\Sigma d^2_{Y \cdot X}$, is not surprisingly called the **explained sum of squares** or the *sum of squares due to regression,* $\Sigma \hat{y}^2$, and is based on the deviations $\hat{y} = \hat{Y} - \bar{Y}$. The computation of these deviations and their squares is shown in columns (11) and (12) of Table 14.1. Note that $\Sigma \hat{y}$ approximates zero and that $\Sigma \hat{y}^2 = 23.5130$. Adding the unexplained SS (0.6160) to this value results in $\Sigma y^2 = \Sigma \hat{y}^2 + \Sigma d^2_{Y \cdot X} = 24.1290$, which is equal (except for rounding errors) to the independently calculated value of 24.1307 in column (7).

We will return to the meaning of the unexplained and explained sums of squares later. We retain the terms ''explained'' and ''unexplained'' for regression sums of squares in this book because they are well established, but we must strongly caution the reader not to read too much into them. The explanation refers to how well the function $\hat{Y} = a + bX$ fits the observed data, not necessarily to any causal explanation of variable Y by variable X.

For the regression equation in case where there is a single value of Y for each value of X, the regression coefficient $\Sigma xy / \Sigma x^2$ can be computed as follows:

$$b_{Y \cdot X} = \frac{\sum^{n} (X - \bar{X})(Y - \bar{Y})}{\sum^{n} (X - \bar{X})^2} \tag{14.4}$$

We have already learned that the numerator of this expression is called the sum of products of X and Y, and we recognize the denominator as the sum of squares of X.

To compute regression statistics, we need eight quantities initially: n, ΣX, ΣY, $\bar{X}$, $\bar{Y}$, Σx^2, Σy^2, and Σxy. From these quantities the regression equation is calculated as shown in Box 14.1, which also illustrates computation of the explained sum of squares, $\Sigma \hat{y}^2 = \Sigma(\hat{Y} - \bar{Y})^2$, and the unexplained sum of squares, $\Sigma d^2_{Y \cdot X} = \Sigma(Y - \hat{Y})^2$. There are also convenient formulas for these two quantities, but they have to be used with caution because they may be sensitive to rounding errors. That

$$\sum d^2_{Y \cdot X} = \sum y^2 - \frac{\left(\sum xy\right)^2}{\sum x^2} \tag{14.5}$$

Box 14.1 COMPUTATION OF REGRESSION STATISTICS. SINGLE VALUE OF Y FOR EACH VALUE OF X.

Data from Table 14.1.

Weight loss in mg (Y)	8.98	8.14	6.67	6.08	5.90	5.83	4.68	4.20	3.72
Percent relative humidity (X)	0	12.0	29.5	43.0	53.0	62.5	75.5	85.0	93.0

Computation

1. Compute sample size, sums, means, sums of squares, and the sum of products.

$$n = 9$$

$$\sum X = 453.5 \qquad \sum Y = 54.20$$

$$\bar{X} = 50.389 \qquad \bar{Y} = 6.022$$

$$\sum x^2 = 8301.3889 \qquad \sum y^2 = 24.1306$$

$$\sum xy = \sum (X - \bar{X})(Y - \bar{Y}) = -441.8178$$

2. The regression coefficient is

$$b_{Y \cdot X} = \frac{\sum xy}{\sum x^2} = \frac{-441.8178}{8301.3889} = -0.05322$$

3. The Y-intercept is

$$a = \bar{Y} - b_{Y \cdot X}\bar{X} = 6.022 - (-0.05322)(50.389) = 8.7040$$

4. The explained sum of squares is

$$\sum \hat{y}^2 = \sum (\hat{Y} - \bar{Y})^2 \text{ or}$$

$$\frac{\left(\sum xy\right)^2}{\sum x^2} = \frac{(-441.8178)^2}{8301.3889} = 23.5145$$

5. The unexplained sum of squares is

$$\sum d^2_{Y \cdot X} = \sum (Y - \hat{Y})^2 = 0.6161$$

is demonstrated at the end of Section A.9 in the appendix. The term subtracted from Σy^2 is the explained sum of squares, as we show here:

$$\sum \hat{y}^2 = \sum b^2 x^2 = b^2 \sum x^2 = \frac{\left(\sum xy\right)^2}{\left(\sum x^2\right)^2} \sum x^2$$

$$\sum \hat{y}^2 = \frac{\left(\sum xy\right)^2}{\sum x^2} \tag{14.6}$$

The BIOM-pc package of computer programs calculates linear regression with one or more values of Y per value of X.

14.4 Tests of Significance in Regression

Up to now we have interpreted regression as a method for providing an estimate, $\hat{Y}_i$, given a value of X_i. Regression can also be interpreted as a method for accounting for some of the variation of the dependent variable Y in terms of variation of the independent variable X. The SS of a sample of Y-values, Σy^2, is computed by summing and squaring deviations, $y = Y - \bar{Y}$. Figure 14.9 shows that the deviation y can be decomposed into two parts, $\hat{y}$ and $d_{Y \cdot X}$. Figure 14.9 also shows that the deviation $\hat{y} = \hat{Y} - \bar{Y}$ represents the deviation of the estimated value $\hat{Y}$ from the mean of Y. The height of $\hat{y}$ is clearly a function of x. We have already seen that $\hat{y} = bx$ [see Expression (14.3)]. Those of you familiar with analytical geometry will recognize this expression as the point-slope form of the equation. If b, the slope of the regression line, were steeper, $\hat{y}$ would be

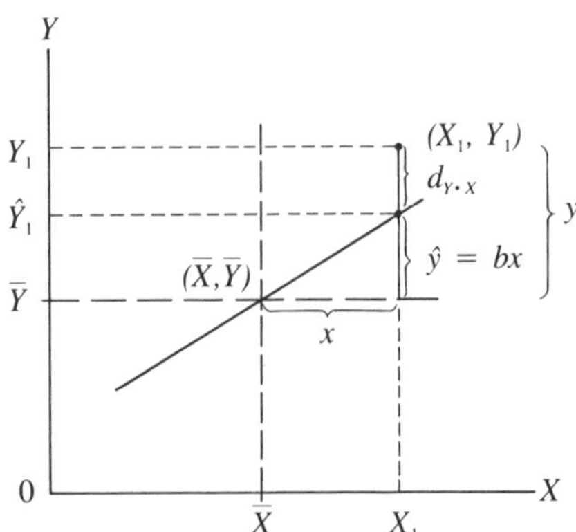

FIGURE 14.9 *Schematic diagram showing the relations involved in partitioning the sum of squares of the dependent variable.*

relatively larger for a given value of x. The remaining portion of the deviation y, $d_{Y \cdot X}$, represents the residual variation of the variable Y after the explained variation has been subtracted. We can see that $y = \hat{y} + d_{Y \cdot X}$ by writing out these deviations explicitly: $Y - \bar{Y} = (\hat{Y} - \bar{Y}) + (Y - \hat{Y})$.

For each of these deviations we can compute a corresponding sum of squares. From Expression (14.5) we obtain

$$\sum y^2 = \frac{\left(\sum xy\right)^2}{\sum x^2} + \sum d_{Y \cdot X}^2$$

Of course, Σy^2 corresponds to y, $\Sigma d_{Y \cdot X}^2$ to $d_{Y \cdot X}$, and

$$\sum \hat{y}^2 = \frac{\left(\sum xy\right)^2}{\sum x^2}$$

corresponds to $\hat{y}$ (as shown in the previous section). Thus we are able to partition the sum of squares of the dependent variable in regression in a way analogous to the partition of the total *SS* in analysis of variance (see Section 8.5). You may wonder how the additive relation of the deviations can be matched by an additive relation of their squares without the presence of any cross products. Simple algebra (see Section A.10 in the appendix) shows that the cross products cancel out. The magnitude of the unexplained deviation ($d_{Y \cdot X}$) is independent of the magnitude of the explained deviation ($\hat{y}$) just as in anovas the magnitude of the deviation of an item from the sample mean is independent of the magnitude of the deviation of the sample mean from the grand mean. This relationship between regression and analysis of variance can be carried further. We can undertake an analysis of variance of the partitioned sums of squares as follows:

	Source of variation	*df*	*SS*	*MS*	Expected *MS*
$\hat{Y} - \bar{Y}$	Explained (estimated Y from mean of Y)	1	$\sum \hat{y}^2 = \sum (\hat{Y} - \bar{Y})^2$	$s_{\hat{Y}}^2$	$\sigma_{Y \cdot X}^2 + \beta^2 \sum x^2$
$Y - \hat{Y}$	Unexplained, error (observed Y from estimated Y)	$n - 2$	$\sum d_{Y \cdot X}^2 = \sum (Y - \hat{Y})^2$ $= \sum y^2 - \sum \hat{y}^2$	$s_{Y \cdot X}^2$	$\sigma_{Y \cdot X}^2$
$Y - \bar{Y}$	Total (observed Y from mean of Y)	$n - 1$	$\sum y^2 = \sum (Y - \bar{Y})^2$	s_Y^2	

The **explained mean square** or *mean square due to linear regression,* measures the amount of linear variation in Y accounted for by variation of X. The explained mean square is tested over the **unexplained mean square,** which measures the residual variation and is used as an error *MS*. The mean square due

Box 14.2 STANDARD ERRORS OF REGRESSION STATISTICS AND THEIR DEGREES OF FREEDOM.

For explanation of this box, see Sections 14.4 and 14.5: ν = degrees of freedom; a = number of values of X when there are n_i Y-values for each X; n = sample size when there is a single Y-value for each value of X.

Statistic	s	More than one Y-value for each value of X		Single Y-value for each value of X	
Regression coefficient $b_{Y \cdot X}$ (derivation in appendix Section A.11)	s_b	$\sqrt{\frac{s^2_{Y \cdot X}}{\sum x^2}}$	$\nu = a - 2$	$\sqrt{\frac{s^2_{Y \cdot X}}{\sum x^2}}$	$\nu = n - 2$
Sample mean $\bar{Y}_i$ or $\bar{Y}$	$s_{\bar{Y}}$	$\bar{Y}_i$ for group i $\sqrt{\frac{MS_{\text{within}}}{n_i}}$	$\nu = \sum n_i - a$	$\bar{Y}$ at $\bar{X}$ $\sqrt{\frac{s^2_{Y \cdot X}}{n}}$	$\nu = n - 2$
Estimated Y for a given value $X_i\hat{Y}_i$ (derivation in appendix Section A.12)	$s_{\hat{Y}}$	$\sqrt{s^2_{Y \cdot X}\left[\frac{1}{\sum n_i} + \frac{(X_i - \bar{X})^2}{\sum x^2}\right]}$ ($\nu = \sum n_i - 2$ if pooled $s^2_{Y \cdot X}$ is employed)	$\nu = a - 2$	$\sqrt{s^2_{Y \cdot X}\left[\frac{1}{n} + \frac{(X_i - \bar{X})^2}{\sum x^2}\right]}$	$\nu = n - 2$

Predicted Y_i for a given value X_i	$\hat{s}_Y$	$\sqrt{s^2_{Y \cdot X}\left[\frac{1}{\sum n_i} + \frac{(X_i - \overline{X})^2}{\sum x^2}\right] + MS_{\text{within}}}$ $\nu = \sum n_i - 2$	$\sqrt{s^2_{Y \cdot X}\left[1 + \frac{1}{n} + \frac{(X_i - \overline{X})^2}{\sum x^2}\right]}$ $\nu = n - 2$
Predicted mean $\overline{Y}_i$ of k items for a given value X_i	$\hat{s}_{\overline{Y}}$	$\sqrt{s^2_{Y \cdot X}\left[\frac{1}{\sum n_i} + \frac{(X_i - \overline{X})^2}{\sum x^2}\right] + \frac{MS_{\text{within}}}{k}}$ $\nu = \sum n_i - 2$	$\sqrt{s^2_{Y \cdot X}\left[\frac{1}{k} + \frac{1}{n} + \frac{(X_i - \overline{X})^2}{\sum x^2}\right]}$ $\nu = n - 2$

NOTE: If $F_s = s^2_{Y \cdot X}/MS_{\text{within}}$ is not significant, you may wish to pool the two mean squares as follows:

$$\text{pooled } s^2_{Y \cdot X} = \frac{SS_{\text{within}} + \sum d^2_{Y \cdot X}}{\sum n_i - 2}$$

to linear regression, $s_{\hat{Y}}^2$, is based on one degree of freedom, and consequently $n - 2$ *df* remain for the error *MS,* since the total sum of squares possesses $n - 1$ degrees of freedom. The test is of the null hypothesis $H_0{:}\beta = 0$. Carrying out such an anova on the weight loss data of Box 14.1 produces the following results:

Source of variation	*df*	*SS*	*MS*	F_s
Explained—due to linear regression	1	23.5145	23.5145	267.18***
Unexplained—error around regression line	7	0.6161	0.08801	
Total	8	24.1306		

The significance test is $F_s = s_{\hat{Y}}^2/s_{Y \cdot X}^2$. The observed value of F_s shows that a large and statistically significant portion of the variation of Y has been explained by regression on X. We can estimate this portion by the fraction $\Sigma\hat{y}^2/\Sigma y^2 = 23.5145/24.1306 = 0.9745$. Thus 97.45% of the variation of Y can be explained by the variation of X. This ratio is known as the *coefficient of determination* (see Section 15.2), and symbolized by r^2 or R^2.

We now proceed to the computation of the standard errors for various regression statistics, their employment in tests of hypotheses, and the computation of confidence limits. Box 14.2 lists these standard errors in two columns. The left column is for the case of more than one Y-value for each value of X, which will be discussed in Section 14.5. The right column is for the case with a single Y-value for each value of X. The first row of the table gives the *standard error of the regression coefficient,* which is simply the square root of the quotient of the unexplained variance divided by the sum of squares of X. Note that the unexplained variance $s_{Y \cdot X}^2$ is a fundamental quantity that is a part of all standard errors in regression. The standard error of the regression coefficient permits us to test various hypotheses and to set confidence limits to our sample estimate of β, since it is normally distributed if the $d_{Y \cdot X}$ are normally distributed. The computation of s_b is illustrated in step **1** of Box 14.3, using the weight loss example of Box 14.1.

The significance test illustrated in step **2** tests the *"significance" of the regression coefficient;* that is, it tests the null hypothesis that the sample value of b comes from a population with a parametric value $\beta = 0$ for the regression coefficient. This is a t-test, the appropriate degrees of freedom being $n - 2 = 7$. If we cannot reject the null hypothesis, there is no evidence that the regression is significantly deviant from zero in either the positive or negative direction. Our conclusions for the weight loss data are that a highly significant negative regression is present. In Section 9.4 we saw that $t^2 = F$. The square of $t_s = -16.345$ from Box 14.3 is 267.16, which (within rounding error) equals the value of F_s found in the anova earlier in this section. The significance test in step **2** of Box

Box 14.3 SIGNIFICANCE TESTS AND COMPUTATION OF CONFIDENCE LIMITS OF REGRESSION STATISTICS. SINGLE VALUE OF Y FOR EACH VALUE OF X.

Based on standard errors and degrees of freedom of Box 14.2; using example of Box 14.1.

$$n = 9 \qquad \bar{X} = 50.389 \qquad \bar{Y} = 6.022$$

$$b_{Y \cdot X} = -0.05322 \qquad \sum x^2 = 8301.3889$$

$$s^2_{Y \cdot X} = \frac{\sum d^2_{Y \cdot X}}{(n-2)} = \frac{0.6161}{7} = 0.08801$$

Computation

1. Standard error of the regression coefficient:

$$s_b = \sqrt{\frac{s^2_{Y \cdot X}}{\sum x^2}} = \sqrt{\frac{0.08801}{8301.3889}} = \sqrt{0.000{,}010{,}602} = 0.003{,}256{,}1$$

2. Testing whether regression coefficient $\beta = 0$:

$$t_s = \frac{(b-0)}{s_b} = \frac{-0.053{,}22}{0.003{,}2561} = -16.345$$

$$t_{.001[7]} = 5.408 \qquad P < 0.001$$

3. 95% confidence limits for regression coefficient:

$$t_{.05[7]}s_b = 2.365(0.003{,}256{,}1) = 0.00770$$

$$L_1 = b - t_{.05[7]}s_b = -0.05322 - 0.00770 = -0.06092$$

$$L_2 = b + t_{.05[7]}s_b = -0.05322 + 0.00770 = -0.04552$$

4. Standard error of the sampled mean $\bar{Y}$ (at $\bar{X}$):

$$s_{\bar{Y}} = \sqrt{\frac{s^2_{Y \cdot X}}{n}} = \sqrt{\frac{0.08801}{9}} = 0.098{,}888{,}3$$

5. 95% confidence limits for the mean μ_Y corresponding to $\bar{X}$ ($\bar{Y} = 6.022$):

$$t_{.05[7]}s_{\bar{Y}} = 2.365(0.098{,}888{,}3) = 0.233871$$

$$L_1 = \bar{Y} - t_{.05[7]}s_{\bar{Y}} = 6.022 - 0.2339 = 5.7881$$

$$L_2 = \bar{Y} + t_{.05[7]}s_{\bar{Y}} = 6.022 + 0.2339 = 6.2559$$

6. Standard error of $\hat{Y}$, an estimated Y for a given value of X_i:

$$s_{\hat{Y}} = \sqrt{s^2_{Y \cdot X}\left[\frac{1}{n} + \frac{(X_i - \bar{X})^2}{\sum x^2}\right]}$$

Box 14.3 Continued

For example, for $X_i = 100\%$ relative humidity,

$$s_{\hat{Y}} = \sqrt{0.08801\left[\frac{1}{9} + \frac{(100 - 50.389)^2}{8301.3889}\right]}$$
$$= \sqrt{0.08801[0.40760]} = \sqrt{0.035873} = 0.18940$$

7. 95% confidence limits for μ_{Y_i} corresponding to the estimate $\hat{Y}_i = 3.3817$ at $X_i =$ 100% relative humidity:

$$t_{.05[7]}s_{\hat{Y}} = 2.365(0.18940) = 0.44793$$
$$L_1 = \hat{Y}_i - t_{.05[7]}s_{\hat{Y}} = 3.3817 - 0.4479 = 2.9338$$
$$L_2 = \hat{Y}_i + t_{.05[7]}s_{\hat{Y}} = 3.3817 + 0.4479 = 3.8296$$

8. Standard error of a predicted mean $\overline{Y}_i$ to be obtained in a new experiment run at $X_i = 100\%$ relative humidity. Our best prediction for this mean would be $\overline{Y}_i = \hat{Y}_i = 3.3817$. If the new experiment were based on a sample size of $k = 5$, the standard error of the predicted mean would be

$$\hat{s}_{\overline{Y}} = \sqrt{s^2_{Y \cdot X}\left[\frac{1}{k} + \frac{1}{n} + \frac{(X_i - \overline{X})^2}{\sum x^2}\right]}$$
$$= \sqrt{0.08801\left[\frac{1}{5} + \frac{1}{9} + \frac{(100 - 50.389)^2}{8301.3889}\right]}$$
$$= \sqrt{0.08801[0.60760]} = \sqrt{0.05347} = 0.23124$$

9. 95% prediction limits for a sample mean of 5 weight losses at 100% relative humidity (using the standard error computed above):

$$t_{.05[7]}\hat{s}_{\overline{Y}} = 2.365(0.23124) = 0.5469$$
$$L_1 = \hat{Y}_i - t_{.05[7]}\hat{s}_{\overline{Y}} = 3.3817 - 0.5469 = 2.8348$$
$$L_2 = \hat{Y}_i + t_{.05[7]}\hat{s}_{\overline{Y}} = 3.3817 + 0.5469 = 3.9286$$

14.3 could, of course, also have been used to test whether b is significantly different from a parametric value β other than zero.

Setting confidence limits to the regression coefficient presents no new features. The computation is shown in step **3** of Box 14.3. Since s_b is small, the confidence interval is narrow. The confidence limits are shown in Figure 14.10 as dashed lines representing the 95% bounds of the slope. Note that the regression line, as well as its confidence limits, pass through the means for X and Y. Variation in b therefore rotates the regression line about the point $\overline{X}$, $\overline{Y}$.

Next we calculate a *standard error for the observed sample mean* $\overline{Y}$. Recall from Section 7.1 that $s^2_{\overline{Y}} = s^2_Y/n$. Now that we have regressed Y on X, however, we are able to account for (that is, hold constant) some of the variation of Y in terms of the variation of X. The variance of Y around the point $\overline{X}$, $\overline{Y}$ on the

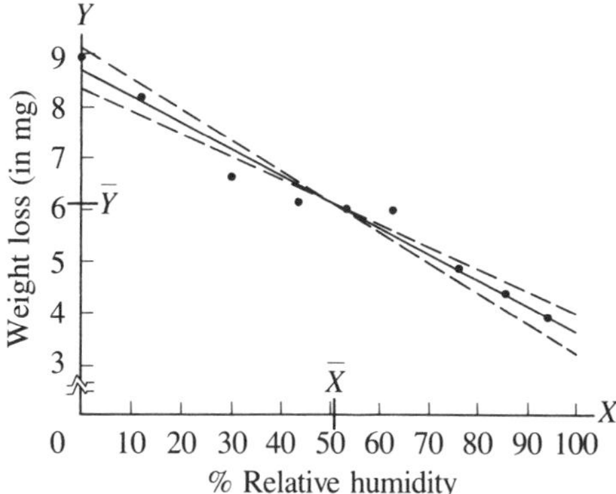

FIGURE 14.10 *95% confidence limits to the slope of the regression line of Figure 14.8.*

regression line is less than $s_{\bar{Y}}^2$; it is $s_{\bar{Y}\cdot X}^2$. At $\bar{X}$ we may therefore compute confidence limits of $\bar{Y}$, using as a standard error of the mean $s_{\bar{Y}} = \sqrt{s_{Y\cdot X}^2/n}$ with $n - 2$ degrees of freedom. This standard error is computed in step **4** of Box 14.3, and 95% confidence limits for the sampled mean $\bar{Y}$ at $\bar{X}$ are calculated in step **5**. These limits (5.7881–6.2559) are considerably narrower than the confidence limits for the mean based on the conventional standard error $s_{\bar{Y}}$, which would be from 4.687 to 7.357. Thus, knowing the relative humidity greatly reduces the uncertainty in weight loss.

The standard error for $\bar{Y}$ is only a special case of the *standard error for any estimated value $\hat{Y}$ along the regression line.* A new factor now enters the error variance. The magnitude of this factor is in part a function of the distance of a given value X_i from its mean $\bar{X}$. The farther away X_i is from its mean, the greater will be the error of estimate. This factor is shown in the third row of Box 14.2 as the deviation $X_i - \bar{X}$, squared and divided by the sum of squares of X. The standard error for an estimate $\hat{Y}_i$ for a relative humidity $X_i = 100\%$ is given in step **6** of Box 14.3. The 95% confidence limits for $\mu_{\hat{Y}_i}$, the parametric value corresponding to the estimate $\hat{Y}_i$, are shown in step **7**. Note that the width of the confidence interval is $3.8296 - 2.9338 = 0.8958$, considerably wider than the confidence interval at $\bar{X}$ calculated in step **5**, which was $6.2559 - 5.7881 = 0.4678$. If we calculate a series of confidence limits for different values of X_i, we obtain a biconcave confidence belt (Figure 14.11). The farther away from the mean we get, the less reliable are our estimates of Y because of the uncertainty about the true slope, β, of the regression line.

Furthermore, the linear regressions that we fit are often only rough approximations to more complicated functional relationships between biological variables. Very often there is an approximately linear relation along a certain range of the independent variable, beyond which range the slope changes rapidly. For example, the pulse of a poikilothermic animal is directly proportional to temperature over a range of tolerable temperatures, but beneath and above this range the

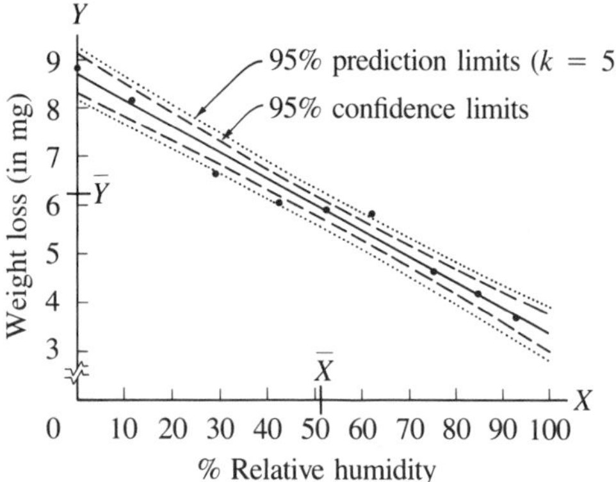

FIGURE 14.11 *95% confidence limits to regression estimates for data of Figure 14.8.*

pulse eventually decreases as the animal freezes or suffers heat prostration. Hence common sense indicates that one should be cautious about extrapolating from a regression equation, especially if one has any doubts about the linearity of the relationship.

The confidence limits for α, the parametric value of a, are a special case ($X_i = 0$) of the confidence limits for $\mu_{\hat{Y}_i}$, and the standard error of a is therefore

$$s_a = \sqrt{s^2_{Y\cdot X}\left[\frac{1}{n} + \frac{\overline{X}^2}{\sum x^2}\right]}$$

In certain situations, we may require that a regression line pass through the origin. This would be proper for cases in which the parametric Y-intercept α is really zero, as in growth curves where the size of a structure is effectively zero at age zero, or in biochemical work where none of a substance is produced by time zero. Fitting a regression through the origin would be supported if the confidence limits for α include zero. In such a case the computations are modified so that the regression line, instead of passing through the bivariate mean ($\overline{X}$, $\overline{Y}$), is forced to pass through the origin of the coordinate system (0, 0). The deviations in the regression formulas, therefore, are taken from the origin; thus a deviation such as $Y - \overline{Y}$ becomes $Y - 0 = Y$. The formula for the regression coefficient consequently modifies Expression (14.4) as follows:

$$b_{Y\cdot X} = \frac{\sum XY}{\sum X^2} \tag{14.7}$$

and the regression equation becomes $\hat{Y} = b_{Y\cdot X}X$. For the unexplained sum of squares in regression through the origin, we modify Expression (14.5):

$$\sum d^2_{Y\cdot X} = \sum Y^2 - \frac{\left(\sum XY\right)^2}{\sum X^2} \tag{14.8}$$

and for the explained *SS* we do the same with Expression (14.6):

$$\sum \hat{y}^2 = \frac{\left(\sum XY\right)^2}{\sum X^2} \tag{14.9}$$

The total *SS* for *Y* necessarily becomes ΣY^2. Since the sums of squares do not center on the mean, we do not lose a degree of freedom for that parameter. Thus, for regression through the origin, the three mean squares s_Y^2, $s_{\hat{Y}}^2$, and $s_{Y\cdot X}^2$ possess n, 1, and $n - 1$ degrees of freedom, respectively.

To evaluate standard errors for regression statistics, modify the formulas in Box 14.2 by replacing Σx^2 and $\bar{X}$ with ΣX^2 and 0, respectively, and employ them with $n - 1$ degrees of freedom. Keep in mind the following caution about using the explained proportion of the total sum of squares in cases where the intercept is zero: The statistic has to be computed in a different way, with the formula $\Sigma\hat{Y}^2/\Sigma Y^2$, and cannot be compared with its counterpart for the model with a nonzero intercept. For more on this topic see Kvålseth (1985).

The last two standard errors in Box 14.2 have somewhat different applications. They relate to *predications made on the basis of the regression equation.* Once you have established a functional relationship between two variables, as between weight loss and relative humidity in this example, you can use this equation to predict the outcome of future experiments. If, for example, you wish to run another batch of *Tribolium* beetles at 100% relative humidity, your best estimate for the weight loss would be $\hat{Y} = 3.3817$ mg, based on the regression equation of Box 14.1. You should expect a standard error greater than that of the estimate $\hat{Y}$ because the error variance of weight losses within the new sample at $X_i = 100\%$ relative humidity has been added. The formula in the fourth row of Box 14.2 is what is given most frequently in statistics texts, but it is of limited usefulness because it describes the expected variance of the sample items—that is, your prediction of the variance of future individual weight loss readings.

More useful is the formula in the last row of the box, which describes the *standard error of the predicted sample mean based on k items.* Thus, if you were to repeat the experiment on weight loss and run five batches of beetles at 100% relative humidity, you could expect the standard error of your sample mean of the five batches to be 0.23124 as calculated in step **8** of Box 14.3. The prediction limits for the sample mean are shown in step **9**. Note that they are wider than the confidence limits of the corresponding estimated value $\hat{Y}$ for the relative humidity $X_i = 100\%$. The biconcave prediction belts, graphed in Figure 14.11, are wider than the confidence belts of the estimated *Y*-values. As the size of the intended sample increases, the standard error diminishes and the prediction limits approach the confidence limits shown in step 7. When the sample size decreases to unity, one computes the prediction limits from the standard error in the fourth row in Box 14.2. These prediction limits yield bounds for the sample items.

In the next section we show how to compute regression when there is more than one Y-value per value of X. We also show the computation of standard errors for such cases based on the formulas in the third column of Box 14.2.

14.5 More Than One Value of Y For Each Value of X

We now take up the case of Model I regression as originally defined in Section 14.2 and illustrated by Figure 14.4. For each value of the treatment X we sample Y repeatedly, obtaining a sample distribution of Y-values at each of the chosen points of X. The example we use here is an experiment from Sokal's laboratory in which *Tribolium* beetles were reared from eggs to adulthood at four different densities. The percentage survival to adulthood was calculated for varying numbers of replicates at these densities. Following Section 13.10, these percentages were given arcsine transformations which are listed in Box 14.4. These values are more likely to be normal and homoscedastic than are percentages. The arrangement of these data is very much like that of a single-classification Model I anova. There are four different densities and several replicated survival values at each density. We would like to determine whether there are differences in survival among the four groups and whether we can establish a regression of survival on density.

Our first approach is to carry out an analysis of variance, using the methods of Section 9.2 and Box 9.1. The possible outcomes are illustrated in Figure 14.12. An analysis of variance that is not significant indicates that the means are not significantly different from each other (see Figure 14.12A). It would thus seem unlikely that a regression line fitted to these data would have a slope significantly different from zero. However, although both analysis of variance and linear regression test the same null hypothesis—equality of means—the regression test is more powerful (less type II error; see Section 7.8) against the alternative hypothesis that there is a linear relationship between the group means and the independent variable X. Thus, when the means increase or decrease slightly as X increases, they may not be different enough for the mean square among groups to be significant by an anova, yet a significant regression may be found. When we find a marked regression of the means on X (Figure 14.12B), we usually find a significant difference among the means by an anova. We cannot turn this argument around, however, and say that a significant difference among means as shown by an anova necessarily indicates that a significant linear regression can be fitted to these data. In Figure 14.12C the means follow a U-shaped function (a parabola). Although the means would likely be significantly different from each other, a straight line fitted to these data would be a horizontal line halfway between the upper and the lower points. For such data, *linear* regression can explain only very little of the variation of the dependent variable. A curvilinear regression (see Section 16.6), however, would fit these data and remove most of the variance of Y. A similar case is shown in Figure 14.12D, in which the means describe a periodically changing phenomenon, rising and falling alternatively.

Box 14.4 COMPUTATION OF REGRESSION WITH MORE THAN ONE VALUE OF *Y* PER VALUE OF *X*. GENERAL CASE WITH UNEQUAL SAMPLE SIZES.

The variates Y are arcsine transformations of the percentage survival of the beetle *Tribolium castaneum* at 4 densities (X = number of eggs per gram of flour medium).

	Density = X ($a = 4$)			
	5/g	20/g	50/g	100/g
	61.68	68.21	58.69	53.13
	58.37	66.72	58.37	49.89
Survival (in degrees)	69.30	63.44	58.37	49.82
	61.68	60.84		
	69.30			
n_i	5	4	3	3
$\bar{Y}_i$	64.0660	64.8025	58.4767	50.9467
$\sum^{a} n_i = 15$				

SOURCE: Data from Sokal (1967).

Anova computation

The steps are the same as those described in Box 9.1.

1. Grand mean $\bar{\bar{Y}} = \dfrac{1}{\sum^{a} n_i} \sum^{a} \sum^{n_i} Y = 60.5207$

2. $SS_{\text{among}} = \sum^{a} n_i(\bar{Y}_i - \bar{\bar{Y}})^2 = 423.7016$

3. $SS_{\text{within}} = \sum^{a} \sum^{n_i} (Y_{ij} - \bar{Y}_i)^2 = 138.6867$

4. SS_{total} = quantity 2 + quantity 3 = 562.3883

Anova table

	Source of variation	df	SS	MS	F_s
$\bar{Y} - \bar{\bar{Y}}$	Among groups	3	423.7016	141.2339	11.20**
$Y - \bar{Y}$	Within groups	11	138.6867	12.6079	
$Y - \bar{\bar{Y}}$	Total	14	562.3883		

The groups differ significantly.

Box 14.4 Continued

We proceed to test whether the differences among the survival values can be accounted for by linear regression on density. If

$$F_s < \frac{1}{a-1} F_{\alpha[1, \sum^a n_i - a]}$$

regression cannot be significant.

Computation for regression analysis

5. Sum of squares of $X = \sum x^2 = \sum^a n_i(X_i - \bar{X})^2 = 18{,}690$

6. Sum of products $= \sum xy = \sum^a n_i(X_i - \bar{X})(\bar{Y}_i - \bar{\bar{Y}}) = -2747.62$

7. Explained sum of squares $= \sum \hat{y}^2 = \sum^a n_i(\hat{Y}_i - \bar{\bar{Y}})^2 = \dfrac{\left(\sum xy\right)^2}{\sum x^2}$

$$= \frac{(\text{quantity } \mathbf{6})^2}{\text{quantity } \mathbf{5}} = \frac{(-2747.62)^2}{18{,}690} = 403.9281$$

8. Unexplained sum of squares $= \sum d^2_{Y \cdot X}$

$$= \sum^a \sum^{n_i} (Y_{ij} - \hat{Y}_i)^2 = SS_{\text{among}} - \sum \hat{y}^2$$

$= \text{quantity } \mathbf{2} - \text{quantity } \mathbf{7}$

$= 423.7016 - 403.9281 = 19.7735$

Completed anova table with regression

Source of variation		df	SS	MS	F_s	Expected MS
$\bar{Y} - \bar{\bar{Y}}$	Among densities (groups)	3	423.7016	141.2339	11.20**	$\sigma^2 + \frac{n_0}{a-1}\sum \alpha^2$
$\hat{Y} - \bar{\bar{Y}}$	Linear regression	1	403.9281	403.9281	40.86*	$\sigma^2 + n_0\sigma^2_D + \beta^2 \sum x^2$
$\bar{Y} - \hat{Y}$	Deviations from regression	2	19.7735	9.8868	<1 *ns*	$\sigma^2 + n_0\sigma^2_D$
$Y - \bar{Y}$	Within groups	11	138.6867	12.6079		σ^2
$Y - \bar{\bar{Y}}$	Total	14	562.3883			

In addition to the familiar mean squares, MS_{among} and MS_{within}, we now have the mean square due to linear regression, $MS_{\hat{Y}}$, and the mean square for deviations

BOX 14.4 CONTINUED

from regression, $MS_{Y \cdot X}$ $(= s^2_{Y \cdot X})$. To test if the deviations from linear regression are significant, compare the ratio $F_s = MS_{Y \cdot X}/MS_{\text{within}}$ with $F_{\alpha[a-2, \sum^a n_i - a]}$. Since we find that $F_s < 1$, we accept the null hypothesis that the deviations from linear regression are zero. Using the rules of Box 10.3, we do not pool MS_{within} with $s^2_{Y \cdot X}$.

To test for linear regression, we therefore test $MS_{\hat{Y}}$ over the mean square of deviations from regression $s^2_{Y \cdot X}$, and since $F_s = 403.9281/9.8868 = 40.86$ is greater than $F_{.05[1,2]} = 18.5$, we reject the null hypothesis that there is no regression, or that $\beta = 0$. If we had pooled, we would have compared the ratio $F_s = MS_{\hat{Y}}/s^2_{Y \cdot X} = 403.9281/12.1892 = 33.14$ with $F_{.001[1,13]} = 17.8$ and would have reached the same conclusion.

9. Regression coefficient (slope of regression line) $= b_{Y \cdot X} = \dfrac{\sum xy}{\sum x^2}$

$$= \frac{\text{quantity } \mathbf{6}}{\text{quantity } \mathbf{5}} = \frac{-2747.62}{18{,}690}$$

$$= -0.14701$$

10. Y-intercept $= a = \bar{\bar{Y}} - b_{Y \cdot X}\bar{X}$

$$= \text{quantity } \mathbf{1} - \left(\text{quantity } \mathbf{9} \times \frac{\sum^a n_i X}{\sum^a n_i}\right)$$

$$= 60.5207 - (-0.14701)(37.0000) = 65.96004$$

Thus, the regression equation is $\hat{Y} = 65.96004 - 0.14701X$.

Again the regression line for these data has slope zero. A curvilinear (cyclical) regression could also be fitted to such data, but our main purpose in showing this example is to indicate that there could be heterogeneity among the means of Y apparently unrelated to the magnitude of X. Remember that in real examples you rarely get a regression as clear-cut as the linear case in 14.12B or the curvilinear one in 14.12C, nor will you necessarily get heterogeneity of the type shown in 14.12D, in which any straight line fitted to the data would be horizontal. You are more likely to get data in which linear regression can be demonstrated, but which will not fit a straight line well. The residual deviations of the means around linear regression might be removed by changing from linear to curvilinear regression (as is suggested by the pattern of points in Figure 14.12E) or they remain as inexplicable residual heterogeneity around the regression line, as suggested in Figure 14.12F.

We carry out the computations by first following the familiar outline for analysis of variance and obtain the anova table in Box 14.4. The three degrees of freedom among the four groups yield a mean square that would be highly significant if tested over the within-groups mean square. Next we carry out the additional steps for the regression analysis. We compute the sum of squares of X, the sum of products of X and Y, the explained sum of squares of Y, and the

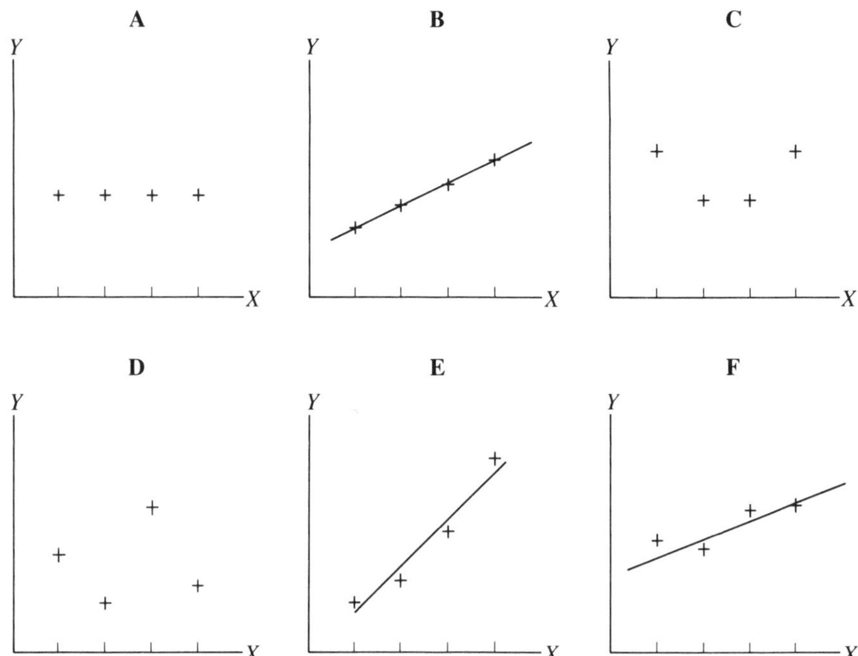

FIGURE 14.12 *Differences among means and linear regression. General trends only are indicated by these figures. Significance of any of these would depend on the outcomes of appropriate tests. (For further explanation see text.)*

unexplained sum of squares of Y. The formulas may look unfamiliar because of the complication of the several Y's per value of X. The computations for the sum of squares of X involve the multiplication of X by the number of items in the study. Thus, although there may appear to be only four densities, there are as many densities (although of only four magnitudes) as there are values of Y in the study. Having completed these computations, we present the results in another anova table (see Box 14.4). Note that the major quantities in this table are the same as in a single-classification anova, but in addition we now have a sum of squares representing linear regression, which is always based on one degree of freedom. This sum of squares is subtracted from the SS among groups, leaving a residual sum of squares (of two degrees of freedom in this case) representing the deviations from linear regression.

We should understand what these sources of variation represent. The linear model for regression with replicated Y per X is derived directly from Expression (8.2), which is

$$Y_{ij} = \mu + \alpha_i + \epsilon_{ij}$$

The treatment effect $\alpha_i = \beta x_i + D_i$, where βx is the component due to linear regression and D_i is the deviation of the mean $\bar{Y}_i$ from regression, which is assumed to have a mean of zero and a variance of σ_D^2. Thus we can write

$$Y_{ij} = \mu + \beta x_i + D_i + \epsilon_{ij} \tag{14.10}$$

The SS due to linear regression represents the portion of the SS among groups that can be explained by linear regression on X. The SS due to deviations from regression represents the residual variation or scatter around the regression line (see Figure 14.12). The SS within groups is a measure of the variation of the items around each group mean. The expected mean squares for each source of variation are given in the completed anova table in Box 14.4.

We first test whether the mean square for deviations from regression ($MS_{Y \cdot X} = s^2_{Y \cdot X}$) is larger than the within-groups MS by computing the variance ratio of $MS_{Y \cdot X}$ over the within-groups MS. In our case, the deviations from regression are clearly not significant, since the mean square for deviations is less than that within groups. This situation is analogous to testing mean squares for subgroups in a hierarchic anova, and in view of the nonsignificance of the mean square for deviations, we might consider pooling it with the error MS. We follow the rules in Box 10.3. Since we have no prior knowledge of the variance ratio, and $F_s > F_{.75}$, we do not pool. We now test the mean square for regression, $MS_{\hat{Y}}$, over the mean square for deviations from regression and find $MS_{\hat{Y}}$ to be significantly greater. Thus linear regression on density has clearly removed some of the variation of survival values. When pooling, we test the regression mean square over the pooled mean square representing the SS of deviations from regression and within-groups SS. Had we done so in this case, we would have obtained the same result. We should caution you again about confusing significance and importance when reporting a test result. If the result is not (statistically) significant, then you have not substantiated that any difference exists. If it is significant, it is important only if its effect is large. In this case the effect of density is both significant and important, since most of the variance of percentage survival is accounted for by the variation in density.

Significance of the mean square for deviations from regression could mean either that Y is a curvilinear function of X or that there is a large amount of random heterogeneity around the regression line (see Figure 14.12); actually a mixture of both conditions may prevail. If the $MS_{Y \cdot X}$, is significant and curvilinearity is suspected, a curvilinear regression (see Section 16.6) may be fitted. If heterogeneity is responsible for a significant $MS_{Y \cdot X}$, we use it as the denominator mean square against which to test the significance of linear regression.

You may wonder how the terms in this example are related to those in the earlier example with a single Y for each value of X (see Box 14.1). Obviously the single degrees of freedom for sum of squares due to linear regression correspond in the two examples because there is no replication in the earlier example, however, we have no independent estimate of error of the Y's for a given value of X. Thus the unexplained sum of squares with $n - 2$ degrees of freedom in the nonreplicated example subsumes the deviation sum of squares with $a - 2$ degrees of freedom, as well as the error sum of squares with $\sum^a n_i - a$ degrees of freedom in the replicated case. The total sum of squares with $n - 1$ degrees of

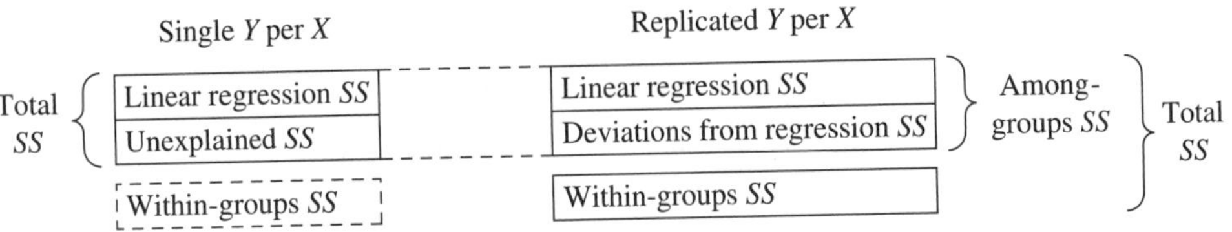

FIGURE 14.13 *Diagrammatic representation of the partitioning of the total sums of squares* (SS) *in two regression designs—single* Y *per* X *and more than one* Y *per* X*—corresponding to Boxes 14.1 and 14.4. The areas of the subdivisions are not shown proportional to the magnitudes of the sums of squares.*

freedom in the nonreplicated case is equivalent to the sum of squares among groups with $a - 1$ degrees of freedom in the replicated example. Thus in the nonreplicated case we are unable to separate the two sources of variation—deviations from regression and error among Y's for a given value of X—just as in an analogous case in a two-way anova we were unable to separate interaction and error variance in a completely randomized blocks design, whereas we were able to achieve this separation in a two-way anova with replication (see Chapter 11). These relations are shown in Figure 14.13.

We complete computation of the regression coefficient and regression equation as shown at the end of Box 14.4. Our conclusions are that as density increases, survival decreases and that this relationship can be expressed by a linear regression of the form $\hat{Y} = 65.9601 - 0.14701X$, where X is density per gram and $\hat{Y}$ is the arcsine transformation of percentage survival (Figure 14.14).

We will now compute standard errors and confidence limits for the example we have just discussed. The pertinent standard errors are found in the third column of Box 14.2. Since we decided not to pool the mean square for deviations from regression with the within-groups mean square, we use the former with its 2 degrees of freedom and the MS_{within} with its 11 *df*. The appropriate degrees of

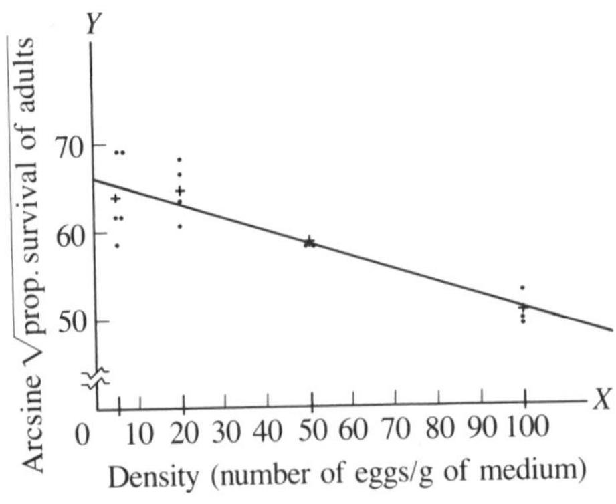

FIGURE 14.14 *Linear regression fitted to data of Box 14.4. Sample means are identified by + signs.*

freedom for the standard errors are given in Box 14.2. In other cases, if we did pool, we would use the pooled mean square of deviations around regression throughout when $s^2_{Y \cdot X}$ is called for in Box 14.2. The computation of the *standard error of the regression coefficient* and the 95% *confidence limits of* β is shown in steps **1** and **2** of Box 14.5. We calculate the *standard error of the sample mean* $\bar{Y}_i$ for any one group i based on the MS_{within}. Thus in step **3** we compute the standard error for the mean survival of 64.0660 at density 5/g as 1.58795 (expressed in degrees; remember this is an arcsine transformation), and the 95% confidence limits computed in step **4** are 60.5709 to 67.5611.

Next we estimate the mean survival $\hat{Y}_i$ for another value of the density, say $X_i = 59/\text{g}$. Note that this value of X_i was not one of the four employed in the original experiment. As long as X (density) is a continuous variable, we can estimate Y for any biologically reasonable value of X that does not involve undue extrapolation beyond the range of observed X-values. The estimate of $\hat{Y}_i$ for $X_i = 59/\text{g}$ can be obtained from the regression equation and yields a survival value of 57.2865°. In step **5** we calculate the *standard error of the estimated mean,* followed by the *confidence limits for the estimated mean* (53.1702 to 61.4028) in step **6**. Next we predict the mean that we would obtain in another experiment. At density 59/g our best estimate for such a survival value is, of course, again the regression estimate of 57.2865°. However, the standard error is as shown in the last row of the table in Box 14.2. Let us assume that the mean to be obtained in our next experiment will be based on 10 replicates. In steps **7** and **8** we obtain the *standard error for the predicted mean* ($\hat{s}_{\bar{Y}} = 1.4751$) and *prediction limits* of 54.1003 and 60.4727, which in this unusual case are narrower than the confidence limits for the estimated mean obtained previously because of the greater number of degrees of freedom of the standard error and consequent decrease of t. Confidence and prediction bands around the regression line are shown in Figure 14.15.

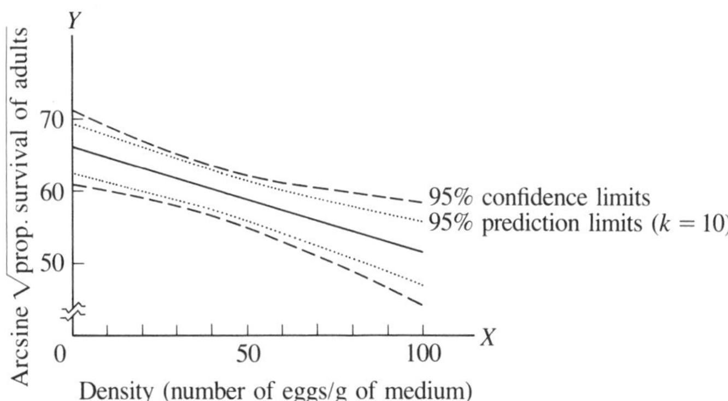

FIGURE 14.15 *Confidence limits to regression estimates for data of Figure 14.14. Note the unusual circumstance in which the confidence limits are outside the prediction limits, a phenomenon that occurs here because in this problem the latter are based on substantially more degrees of freedom.*

Box 14.5 COMPUTATION OF STANDARD ERRORS AND CONFIDENCE LIMITS OF REGRESSION STATISTICS. MORE THAN ONE VALUE OF *Y* FOR EACH VALUE OF *X*.

Data from Box 14.4.

$b_{Y \cdot X} = -0.14701$	$s^2_{Y \cdot X} = 9.8868$
$\sum x^2 = 18{,}690$	$MS_{\text{within}} = 12.6079$

Computation

In all of these computations the *MS* for deviations from regression ($s^2_{Y \cdot X} = 9.8868$ with 2 *df*) is used. The formulas for the standard errors are from the third column of Box 14.2.

1. Standard error of the regression coefficient:

$$s_b = \sqrt{\frac{s^2_{Y \cdot X}}{\sum x^2}} = \sqrt{\frac{9.8868}{18{,}690}} = \sqrt{0.000{,}529{,}0} = 0.02300$$

2. 95% confidence limits for regression coefficient β:

$$t_{.05[2]}s_b = 4.303(0.02300) = 0.09897$$

$$L_1 = b - t_{.05[2]}s_b = -0.14701 - 0.09897 = -0.24598$$

$$L_2 = b + t_{.05[2]}s_b = -0.14701 + 0.09897 = -0.04804$$

3. Standard error of a sample mean $\bar{Y}_i$ for any group i:

$$s_{\bar{Y}} = \sqrt{\frac{MS_{\text{within}}}{n_i}}$$

For the first sample at density $X_1 = 5/\text{g}$, we find $n_1 = 5$. The standard error would be

$$s_{\bar{Y}} = \sqrt{\frac{12.6079}{5}} = \sqrt{2.52158} = 1.58795$$

4. The 95% confidence interval for the mean of the first sample ($X_1 = 5/\text{g}$; $\bar{Y}_1 = 64.0660$):

$$t_{.05[11]}s_{\bar{Y}} = 2.201(1.58795) = 3.4951$$

$$L_1 = \bar{Y}_1 - t_{.05[11]}s_{\bar{Y}} = 64.0660 - 3.4951 = 60.5709$$

$$L_2 = \bar{Y}_1 + t_{.05[11]}s_{\bar{Y}} = 64.0660 + 3.4951 = 67.5611$$

5. Standard error of $\hat{Y}_i$, an estimated mean for a given value of X_i. As an example we will compute $\hat{Y}_i$ for $X_i = 59/\text{g}$. Since $\hat{Y} = a + bx$, for $X_i = 59$, $\hat{Y}_i = 65.9601 -$

BOX 14.5 CONTINUED

0.14701(59) = 57.2865. We use the following formula for the standard error:

$$s_{\hat{Y}} = \sqrt{s^2_{Y \cdot X}\left[\frac{1}{\sum n_i} + \frac{(X_i - \bar{X})^2}{\sum x^2}\right]}$$

$\bar{X}$ in this formula refers to

$$\frac{\sum^a n_i X}{\sum^a n_i} = \frac{555}{15} = 37.0 \qquad \text{(values from Box 14.4).}$$

For a density of $X_i = 59/\text{g}$

$$s_{\hat{Y}} = \sqrt{9.8868\left[\frac{1}{15} + \frac{(59 - 37)^2}{18{,}690}\right]} = \sqrt{9.8868[0.09256]}$$
$$= \sqrt{0.91512} = 0.95662$$

6. 95% confidence interval for μ_{Y_i} corresponding to the estimate $\hat{Y}_i = 57.2865$ at $X_i =$ 59/g:

$$t_{.05[2]}s_{\hat{Y}} = 4.303(0.95662) = 4.1163$$

$$L_1 = \hat{Y}_i - t_{.05[2]}s_{\hat{Y}} = 57.2865 - 4.1163 = 53.1702$$

$$L_2 = \hat{Y}_i + t_{.05[2]}s_{\hat{Y}} = 57.2865 + 4.1163 = 61.4028$$

95% of such intervals should contain the true mean μ_{Y_i} for $X_i = 59$. These limits are plotted in Figure 14.15 for values of X_i from 0 to 100.

7. Standard error of a predicted mean $\bar{Y}_i$ to be obtained in a new experiment run at a density of $X_i = 59/\text{g}$. Our best prediction for this mean would be $\bar{Y}_i = \hat{Y} = 57.2865$. If the new experiment were based on a sample of size $k = 10$, the standard error of the predicted mean would be

$$\hat{s}_{\bar{Y}} = \sqrt{s^2_{Y \cdot X}\left[\frac{1}{\sum n_i} + \frac{(X_i - \bar{X})^2}{\sum x^2}\right] + \frac{MS_{\text{within}}}{k}}$$
$$= \sqrt{9.8868\left[\frac{1}{15} + \frac{(59 - 37)^2}{18{,}690}\right] + \frac{12{,}6079}{10}}$$
$$= \sqrt{0.91514 + 1.26079} = \sqrt{2.17593} = 1.47510$$

8. 95% prediction limits for the mean of a sample of 10 survival readings at density 59/g (using the standard error computed above):

$$t_{.05[13]}\hat{s}_{\bar{Y}} = 2.160(1.4751) = 3.1862$$

$$L_1 = \hat{Y}_i - t_{.05[13]}\hat{s}_{\bar{Y}} = 57.2865 - 3.1862 = 54.1003$$

$$L_2 = \hat{Y}_i + t_{.05[13]}\hat{s}_{\bar{Y}} = 57.2865 + 3.1862 = 60.4727$$

Although the standard error of these prediction limits is greater than that of the estimated $\hat{Y}_i$ in step **5**, these limits (also shown in Figure 14.15) are narrower than those found in step **6** because of the greater number of degrees of freedom and consequent decrease in magnitude of t.

If the sample sizes for the separate values of X are equal, we can simplify the formulas for the various sums of squares. For example, the formula for the sum of products (step **6** in Box 14.4) becomes $n \sum^a (X_i - \bar{X})(\bar{Y}_i - \bar{\bar{Y}})$. Almost all such computations are carried out by computers nowadays, making these simplifications largely irrelevant but significance tests for the various regression statistics could be carried out by using the formulas in Box 14.2, simplifying whenever possible to take advantage of the equality of sample sizes.

14.6 The Uses of Regression

We have been so busy learning the mechanics of regression analysis that we have not given much thought to the applications of regression. In this section we will discuss six more or less distinct types of applications. All are discussed in terms of Model I regression. When applying these techniques, be careful that the four assumptions listed in Section 14.2 are correct. From a logical and philosophical point of view, some of these applications also fit a Model II case. The computational procedures would then differ as discussed in Section 14.13.

The first application of regression is the *study of causation*. If we wish to know whether variation in a variable Y is caused by changes in another variable X, we manipulate X in an experiment to see whether we can find statistical evidence for a linear relation of Y on X. Causation is a complex, philosophical idea that we will not go into here. You have undoubtedly been cautioned from your earliest scientific experience not to confuse concomitant variation with causation. When variables vary together, this covariation may be by chance or both variables may be functions of another variable. These cases are usually examples of Model II regression. When we manipulate one variable and find that such manipulations affect a second variable, we generally are satisfied that the variation of the independent variable X is the cause of the *variation* of the dependent variable Y (not the cause of the variable!). Even here however, it is best to be cautious. When we find that heart rate in a cold-blooded animal is a function of ambient temperature, we may conclude that temperature is one of the causes of differences in heart rate. Other factors may also be affecting heart rate.

Sometimes we make the mistake of inverting the cause-and-effect relationship. Such a mistake in this case is unlikely, since suggesting that heart rate affects the temperature of the general environment seems immediately unreasonable. If we were testing two chemical substances in the blood, however, we might more easily be mistaken about the cause-and-effect relationship. If these two chemicals are related by a feedback relationship, then administration of chemical A, the precursor, might increase production of chemical B, the product. Injection of chemical B, however, might inhibit production of the precursor. If we did not know which was the precursor and which the product, we might be confused about cause-and-effect relationships here. Despite these cautions, regression analysis is a common device for screening out causal relationships.

Although a significant regression of Y on X does not prove that changes in X are the cause of variations in Y, the converse statement is true. When we find no significant regression of Y on X, we may infer that deviations of X do not affect Y. Note, however, the possibility of type II error. There may indeed be an effect of X on Y that cannot be substantiated because the sample size of the study is inadequate.

The *description of scientific laws* is a second general application of regression analysis. The aim of science is mathematical description of relations between variables in nature, and regression analysis permits us to estimate functional relationships between variables. These functional relationships do not always have clearly interpretable biological meaning. Thus, in many cases it may be difficult to assign a biological interpretation to the statistics a and b, or their corresponding parameters α and β. When we are able to do so, we have a **structural mathematical model,** one whose components have clear scientific meaning. Mathematical curves that are not structural models are also of value in science. Most regression lines are **empirically fitted curves,** in which the functions simply represent the best mathematical fit (by a criterion such as least squares) to an observed set of data. The constants necessary to fit these curves may not possess any clear inherent meaning. Biologists generally do not like empirically fitted curves. When they adequately describe the relations between natural phenomena, however, empirically fitted curves are of value as temporary devices until enough insight into the phenomena is obtained to postulate a hypothesis from which a structural model can be constructed. Structural models are preferred for many reasons, although they may actually not be markedly better predictors than are empirically fitted curves over the observed range in which the line is being fitted.

Successful *prediction* of an expected value of Y for a given value of X is a third aspect of regression analysis and has been suggested as one justification for employing empirically fitted curves. Prediction is frequently an objective of applied research. Associated with prediction is estimating the difference between observed and expected values in the dependent variate; these differences can lead to the discovery of atypical values, that is, outliers, which are due to inhomogeneity of samples or measurement errors. Although correct structural mathematical models should be better at prediction than empirically fitted curves are—especially for extrapolation beyond the range of X's observed in the sample—empirical curves are frequently of great utility. Examples from the history of science are not hard to find. Good empirical formulas for the motions of the heavenly bodies were available in the time of Copernicus and Kepler; the Newtonian ideas on gravitation altered the nature of the formulas but improved the prediction only relatively mildly. Subsequent improvements in the structural formulas have continued to lessen the error of prediction. A biological analogue relates to the use for many years of heart-muscle extract as a simulated antigen in the Wassermann test for syphilis. Although there seems to be no logical or biological reason why such an extract should provide an antigen for a spirochete-induced disease, the method proved to be reasonably reliable. It is with the same sort of philosophy that we often use regression equations to describe scientific laws, knowing full well that the structure of the equations may not represent the workings of nature.

Comparison of dependent variates is another application of regression. As soon as it is established that a given variable is a function of another one, as in Box 14.4, where we found the survival of beetles to be a function of density, we can ask how strong this relation is. In the beetle example, for instance, we ask to what degree the observed difference in survival between two samples of beetles is a function of the density at which they have been raised. Comparing beetles raised at very high density (and expected to have low survival) with those raised under optimal conditions of low density would be unfair. This same point of view makes us hesitate to compare the mathematical knowledge of a fifth-grader with that of a college student. Since we could undoubtedly obtain a regression of mathematical knowledge on years of schooling in mathematics, we should be comparing how far a given individual deviates from his expected value based on such a regression. Thus, relative to other classmates and to age group, the fifth-grader may be far better than is the college student relative to his/her peer group. This consideration suggests that we calculate *adjusted Y-values* that allow for the magnitude of the independent variable X. A conventional way of calculating such adjusted Y-values is to estimate the Y-value one would expect if the independent variable were equal to its mean $\bar{X}$ and the observation retained its observed deviation ($d_{Y \cdot X}$) from the regression line. Since $\hat{Y} = \bar{Y}$ when $X = \bar{X}$, the adjusted Y-value can be computed as

$$Y_{\text{adj}} = \bar{Y} + d_{Y \cdot X} = Y - bx \tag{14.11}$$

Let us apply this formula to the data of Table 14.1. Comparing the weight loss in the *Tribolium* sample raised at 29.5% relative humidity ($Y = 6.67$ mg) and that at 62.5% relative humidity ($Y = 5.83$ mg), we are led to conclude that weight loss in the more humid environment was less by 0.84 mg. According to our regression line, we expect 7.1338 mg weight loss at 29.5% RH, while at 62.5% RH we expect a loss of 5.3776 mg. Thus, the expected difference should be greater (and it is, 1.76 mg). To calculate adjusted Y values for these two relative humidities, we need the following quantities:

X_i	29.5	62.5	%RH
Y_i	6.67	5.83	mg
$\hat{Y}_i$	7.1338	5.3776	mg
$d_{Y \cdot X}$	−0.4638	0.4524	mg
$Y_{\text{adj}} = \bar{Y} + d_{Y \cdot X}$	5.558	6.474	mg

Remembering that $\bar{Y} = 6.022$, we obtain the adjusted Y-values using Expression (14.11). These adjusted values lead to interpretations different from those based on the observed Y-values. We now find a higher adjusted weight loss at the higher humidity. What does this mean? It indicates that the sample used in the experiment at the higher humidity happened to deviate strongly above the regression line. It lost more weight than was expected. On the other hand, the sample at 29.5% relative humidity had a lower weight loss than expected. Thus, although overall weight loss was greater at the lower humidity, adjusted weight loss allowing for the regression relationship is quite the reverse for these two samples: The sample of beetles tested at 62.5% RH lost relatively more weight than the sample at the lower humidity.

In this manner, adjusted means are useful for making comparisons among dependent variates after allowing for differences among independent variates that affect them. For the standard error of an adjusted mean, we use the standard error of estimated Y for a given value of X_i, shown in the third row of Box 14.2. Sometimes we do not even bother to calculate adjusted Y. We can simply state $d_{Y \cdot X}$ and compare the unexplained deviations among the variates. A large and positive unexplained deviation means that this individual or sample lies considerably above expectation by regression; a negative value of $d_{Y \cdot X}$ indicates the converse. Since the deviations $d_{Y \cdot X}$ are in absolute measurement units, it may be desirable to divide them by $s_{Y \cdot X}$ to obtain standardized scores. These are quite frequently used in educational statistical research to provide an estimate in standard deviation units of an individual's relative standing with respect to his/her group, allowing for an independent variable such as age or amount of education.

Differences among adjusted Y-values are not necessarily scientifically meaningful. They may merely represent random error around the regression line—the result of unknown and probably unknowable factors differentiating the responses of the individuals concerned. Such differences may, however, reflect biologically meaningful distinctions among the individuals leading to recognition of new and important causal factors. Statistical analysis cannot by itself distinguish between these alternatives. Further experimentation and analysis are required to reveal these new insights.

Statistical control is an application of regression that is not widely known among biologists and represents a scientific philosophy that is not well established in biology outside agricultural circles. Biologists frequently categorize work as descriptive or experimental, implying that only the latter can be analytical. Statistical approaches applied to descriptive work can in some instances, however, take the place of experimental techniques quite adequately—occasionally they are even to be preferred. These approaches are attempts to substitute statistical manipulation of a concomitant variable for control of the variable by experimental means. An example will clarify this technique.

Assume that we are studying the effects of diet on blood pressure in rats. We find that the variability of blood pressure in our rat population is considerable, even before we introduce differences in diet. Further study reveals that the variability is due largely to differences in age among the rats of the experimental population. This relationship can be demonstrated by a significant linear regression of blood pressure on age. Thus, to reduce the variability of blood pressure in the population, we should keep the age of the rats constant. The reaction of most biologists at this point would be to repeat the experiment using rats of only one age group; this is a valid, commonsense approach, which is part of the experimental method. An alternative approach using the analysis of covariance (see Section 14.9) is superior in cases in which holding the variable constant is impractical or too costly. For example, we could continue to use rats of variable ages and simply record the age of each rat as well as its blood pressure. Then we could regress blood pressure on age and use an adjusted mean as the basic blood pressure reading for each individual. We could then evaluate the effect of differences in diet on these adjusted means, or we could analyze the effects of diet on unexplained deviations, $d_{Y \cdot X}$, after the experimental blood pressures were regressed on age (which amounts to the same thing).

What are the advantages of such an approach? Sometimes it is impossible to secure adequate numbers of individuals all of the same age. By using regression we can use all the individuals in the population. Employing statistical control assumes that recording the independent variable X is relatively easy and that this variable can be measured without error, which would be generally true of such a variable as age of a laboratory animal. Statistical control may also be preferable because we obtain information over a wider range of both Y and X and because we add to our knowledge about the relations between these two variables, which would not happen if we restricted ourselves to a single age group. Furthermore, when the assumed additivity of treatment effects does not hold, this fact can be learned only over a range of age groups. Suppose, for example, that the effect of diet on blood pressure is noticeable only in older rats and not in younger ones. Clearly we would not discover this unless we used rats of different age groups.

Regression permits us to predict what the variance of an organism would be under statistical or experimental control. Instead of s_Y^2, the previous variance of Y, we can now use the unexplained residual variance $s_{Y \cdot X}^2$. By regressing Y on X we increase the information per individual by $100(s_Y^2/s_{Y \cdot X}^2 - 1)\%$. This quantity is very much like the relative efficiency in design discussed in Section 10.4. In calculating relative efficiency in regression, we customarily take the ratio of total variance over unexplained variance.

Let us work out an example with the survival values of Box 14.4. Since these data are structured into variance among densities and within densities, we have to be careful which variance we use here. The variance among densities was reduced by regression on density, and we should compare it with the unexplained mean square around regression, referred to in Box 14.4 as "the mean square of deviations from regression." The relative efficiency is computed as $RE = (MS_{\text{among}} \times 100)/MS_{Y \cdot X} = (141.2339 \times 100)/9.8868 = 1428.5\%$. We could improve the efficiency of our design by $1428.5 - 100 = 1328.5\%$ by recording the density at which the beetles were raised and by regressing survival on density. If information on density were easily available, this procedure would be well worthwhile. A similar increase in efficiency would result from using only beetles reared at the previous mean density.

Finally, *substitution of variables* is a special application of regression that may occasionally be useful. Suppose we are interested in a response variable that is very difficult or expensive to measure. For example, assume that we wish to measure blood pressure in mice, which might be quite complex with the equipment at our disposal. Assume further that after much effort we have accurately measured the blood pressures of 25 mice of known, but differing ages. After regressing blood pressure on age, we find that a substantial portion of the variance of blood pressure was a function of age. Instead of continuing to measure blood pressure, we could simply record the ages of the mice and predict blood pressure from them. Obviously this approach would not be as efficient as measuring the blood pressure directly. If our initial experiment has been carried out properly, however, we might be able to predict to a satisfactory degree of accuracy what the blood pressures of mice should be, given their age distribution. We might, therefore, have to use a greater number of mice to predict blood pressure from age than if we measured the blood pressure, but it may be much more economical to record the ages of large samples of mice than to measure the blood pressures of smaller samples. Whenever the dependent variable is difficult and costly to measure, consider substituting variables.

14.7 ESTIMATING X FROM Y

Occasionally we have a problem in which we know the value of Y for an individual and wish to estimate the corresponding value of X. This problem may appear simple, requiring only that we reverse the regression equation to write an equation of the type $\hat{X} = a' + b_{X \cdot Y}Y$, where $a \neq a'$ and $b_{X \cdot Y} = \Sigma xy/\Sigma y^2$. Such an approach would be improper, however, because our initial assumptions were that X is measured without error and Y is the dependent, random, and normally distributed variable. Unless the situation is reversed, it is not legitimate to regress X on Y. The appropriate procedure is simple, although the computations (outlined in Box 14.6) are tedious.

Since $\hat{Y}_i = a + b_{Y \cdot X}X_i$, we can estimate X_i by rearranging this equation to yield $\hat{X}_i = (Y_i - a)/b_{Y \cdot X}$. In Box 14.6 we assume that the weight loss in a sample of 25 *Tribolium* beetles is 7 mg. We wish to estimate the relative humidity $\hat{X}_i$ at which these beetles were kept, assuming that the experimental setup was the same as before. Following the formula just given, we estimate the relative humidity at 32.0124%. It is not appropriate to assign standard errors to such an estimate, but there is a method for providing it with confidence limits. Since computational formulas for these limits are unwieldy, we divide the computation into three steps. First we compute quantity D as defined in Box 14.6; then we proceed to a second quantity, H. The limits are relatively simple functions of D

Box 14.6 ESTIMATING *X* FROM *Y*.

Data from Table 14.1 and Box 14.1.

Given a weight loss reading of $Y_i = 7$ mg for a sample of 25 *Tribolium* beetles, what can we infer about the relative humidity under which they were kept (assuming that the experimental setup was identical to that previously used)?

Since $\hat{Y}_i = a + b_{Y \cdot X}X_i$,

$$\hat{X}_i = \frac{(Y_i - a)}{b_{Y \cdot X}} = \frac{7.0 - 8.7037}{-0.05322} = 32.0124$$

The 95% confidence limits of this estimate are computed as follows:

We define a quantity D as

$$D = b_{Y \cdot X}^2 - t_{.05[n-2]}^2 s_b^2 = (-0.05322)^2 - (2.365)^2(0.0032561)^2 = 0.002773$$

where $n = 9$, the number of samples (of beetles) in the previous study (see Table 14.1), and another quantity H as

$$H = \frac{t_{.05[n-2]}}{D}\sqrt{s_{Y \cdot X}^2\left[D\left(1 + \frac{1}{n}\right) + \frac{(Y_i - \bar{Y})^2}{\sum x^2}\right]}$$

$$= \frac{2.365}{0.002773}\sqrt{0.08801\left[(0.002773)\left(1 + \frac{1}{9}\right) + \frac{(7.0 - 6.022)^2}{8301.3889}\right]}$$

$$= 852.8669\sqrt{0.08801[0.0030811 + 0.0001152]}$$

$$= 852.8669\sqrt{0.00028131} = 852.8669(0.016772) = 14.30428$$

The 95% confidence limits are

$$L_1 = \bar{X} + \frac{b_{Y \cdot X}(Y_i - \bar{Y})}{D} - H$$

$$= 50.389 + \frac{(-0.05322)(7 - 6.022)}{0.002773} - 14.30428$$

$$= 50.389 - 18.76998 - 14.30428 = 31.619 - 14.30428 = 17.315$$

$$L_2 = \bar{X} + \frac{b_{Y \cdot X}(Y_i - \bar{Y})}{D} + H = 31.619 + 14.30428 = 45.923$$

Note that the limits are symmetrical about $\bar{X} + [b_{Y \cdot X}(Y_i - \bar{Y})/D]$, not about $\hat{X}_i = (Y_i - a)/b_{Y \cdot X}$.

If we wish to estimate X_i for a $\bar{Y}_i$ based on a sample of size n in a regression analysis with more than one Y-value per value of X (as in Box 14.4, for example), H becomes

$$H = \frac{t_{\alpha[\nu]}}{D}\sqrt{s_{Y \cdot X}^2\left[D\left(\frac{1}{n} + \frac{1}{a}\right) + \frac{(\bar{Y}_i - \bar{\bar{Y}})^2}{\sum x^2}\right]}$$

where α is the significance level chosen, $s_{Y \cdot X}^2$ is the *MS* of deviations around regression (or a pooled *MS*), a is the number of groups in the anova, and ν is the number of degrees of freedom for $s_{Y \cdot X}^2$.

and H, as shown in the box. However, note that these confidence limits are unusual in one respect—they are *not* symmetrical around the estimate $\hat{X}_i$, but are symmetrical around another value, $\bar{X} + [b_{Y \cdot X}(Y_i - \bar{Y})/D]$, which is close to $\hat{X}_i$ but not identical to it. By the way, all values of t in Box 14.6 are shown as $t_{.05}$ to provide 95% confidence limits. For other $100(1 - \alpha)\%$ confidence limits, you should use t_α. Note the wide confidence interval: 17.3 to 45.9% relative humidity after rounding (the estimate $\hat{X}_i$ should also be rounded to 32.0% RH).

Box 14.6 also provides a formula that is appropriate when the estimate of X is based on an initial study of replicated Y-values per single value of X and there is more than one value of Y from which to estimate $\hat{X}_i$. This method of estimating X from Y, also called ***inverse prediction*** is applied frequently in the statistical analysis of dosage–mortality problems in bioassay. Such a study involves a regression of cumulative mortalities of organisms on dosage of a substance. Thus dosage X_1 will cause a mortality of Y_1 percent; dosage $X_2 > X_1$, a mortality of $Y_2 > Y_1$ percent; $X_3 > X_2$, a mortality of $Y_3 > Y_2$ percent; and so forth. At a certain dosage, mortality is 100%; the entire sample of organisms is killed. Frequently, such data are transformed—the mortalities to a so-called probit scale (see Section 14.11) and the dosages to logarithms to make the regression of mortality on dosage linear. A common measure of the potency of the substance (or of the tolerance of the organisms) is the dosage required to kill 50% or 95% of the organisms. Such a point is called an LD_{50} or LD_{95} of the organisms, the 50% or 95% lethal dose. This problem is clearly one of inverse prediction: Given a value of Y, namely 50% or 95% mortality, estimate a corresponding value of X, the dosage. Confidence limits to the estimate are set in the manner indicated above.

If the inverse prediction is applied to a regression equation through the origin, the formula changes. Consult Seber (1977, Section 7.3) in this instance.

14.8 COMPARING REGRESSION LINES

Often an investigator obtains two or more regression lines from similar data and wishes to know whether the functional relationships described by the regression equations are the same. For example, the scientist may have established a regression of blood pressure on age in a sample of animals and may now wish to compare this regression equation with those in three other samples, each of which has been subjected to a different diet or drug. The basic design of such a test is the single-classification analysis of variance. There will be a samples, representing the treatment groups and the control. There is one major new aspect, however. In previous analyses we encountered only one variable, Y. In this example Y would be the blood pressure. In addition, however, for each reading of Y we also have a reading of X, the age of the animal. Thus two separate analyses of variance are possible, one for each variable, as well as a joint analysis—the analysis of covariance between X and Y. Such an analysis in its complete form is called the analysis of covariance and is discussed in Section 14.9.

In this section we examine a preliminary step in an analysis of covariance: testing two or more regression lines for homogeneity of slope. Why would we be interested in testing differences between regression slopes? We might find that different toxicants yield different dosage—mortality curves or that different drugs yield different relationships between dosage and response (see, for example, Figure 14.3). In the example we discuss here, genetically differing cultures yielded different responses to increasing density, an important fact in understanding the effect of natural selection in these cultures. The regression slope of one variable on another is as fundamental a statistic of a sample as is the mean or the standard deviation; when comparing samples it may be as important to compare regression coefficients as it is to compare these other statistics. Another reason for testing differences among slopes is that the homogeneity of regression slopes is an important assumption in the analysis of covariance.

We have deliberately chosen a complicated example so that you can review some of the things you have learned about regression in previous sections. The steps in testing the equality of slopes are explained extensively in Box 14.7; only a few comments are necessary here. The example is taken from a study of the effects of increasing densities of eggs (independent variable X) on survival to adulthood (dependent variable Y) in *Tribolium* beetles. This experiment was performed in cultures at five gene frequencies of b^+, the wild-type allele of the *black* locus. Varying numbers of replicates, somewhat below 100, were used for each gene frequency. Some of the replicates were run at the same densities; therefore the methods of Box 14.4 for repeated Y-values for the same value of X had to be employed for the separate regression analyses at each gene frequency. Some preliminary computations and the basic regression analyses are shown in Box 14.7. From these we learn that only gene frequency 0.00 has a significant added component in survival values due to differences among densities (groups), but a significant regression of survival on density could be demonstrated for the 0.00, 0.25, and 0.50 gene frequencies. Heterogeneity among the means of residuals around the regression line was observed at gene frequency 0.00, which showed marked departure from linearity in the response of survival to density. The remaining statistics necessary for the test of equality among the regression coefficients are also shown in the table in Box 14.7.

To test for equality among k regression coefficients, we need a sum of squares among regression coefficients, which we obtain by the following formula:

$$SS_{\text{among } b\text{'s}} = \sum^{k} \left[\sum x^2 (b - \bar{b})^2 \right] \tag{14.12}$$

where $\bar{b}$ is the pooled or common slope and is equal to

$$\bar{b} = \frac{\sum^{k} \left(b \sum x^2 \right)}{\sum^{k} \sum x^2} = \frac{\sum^{k} \sum xy}{\sum^{k} \sum x^2}$$

Box 14.7 TESTS FOR EQUALITY OF SLOPES OF SEVERAL REGRESSION LINES.

The effect of varying density on survival of beetles to adulthood. Angular transformations of proportion surviving to adulthood (Y) in *Tribolium castaneum* regressed on density of egg input (X). Beetles reared at 5 gene frequencies in 40 g of flour. These data had more than one replicate (Y-value) per density X and were initially analyzed in the manner of Box 14.4, separately for each gene frequency.

Preliminary computation

The voluminous original data are not shown here, but the basic statistics, as well as the results of separate regression analyses for each of the five gene frequencies of b^+, are supplied in the following table. The computations are as explained in Box 14.4. Asterisks, representing significance values as employed conventionally, are appended to sums of squares. To conserve space, neither mean squares nor degrees of freedom are shown. Significance of the SS_{among} (of Y) indicates that survival proportions differ among the various densities. If the explained SS is significant, then the regression of survival on density is significant for that gene frequency. Significance of the unexplained SS (tested over SS_{error}) indicates residual heterogeneity of means around the regression line, which could be due to added effects that influence various densities differentially. In this case, the heterogeneity of $\sum d^2_{Y \cdot X}$ at gene frequency 0.00 is due to nonlinearity of survival response to density.

Regression statistics (computed by the methods of Box 14.4)

	Gene frequencies of b^+				
	0.00	0.25	0.50	0.75	1.00
$\sum^a n_i$	68	94	91	90	64
a (no. of groups)	25	30	27	26	20
$\sum x^2$	6606.235	8346.426	6452.440	5957.822	4071.359
$\sum xy$	−4559.436	−2052.158	−1561.988	−441.253	−70.830
SS_{among}	4976.891***	1713.770	1542.882	789.792	508.624

BOX 14.7 CONTINUED

Regression statistics (computed by the methods of Box 14.4)

	Gene frequencies of b^+				
	0.00	0.25	0.50	0.75	1.00
$\sum \hat{y}^2$	3146.793***	504.570***	378.122***	32.680	1.232
$\sum d^2_{Y \cdot X}$	1830.098*	1209.200	1164.760	757.112	507.392
SS_{within}	1589.038	2456.035	3251.303	1498.026	1190.329
$b_{Y \cdot X}$	−0.6902	−0.2459	−0.2421	−0.0741	−0.0174
$\overline{\overline{Y}}$	40.116	51.593	54.162	55.894	56.800
$\overline{X}$	24.765	26.681	24.341	22.956	18.297
a (Y-intercept)	57.209	58.154	60.055	57.595	57.118

SOURCE: Data from Sokal and Sonleitner (1968).

Test of equality among k *regression coefficients*

The following required quantities are computed [see Expression (14.12)]:

1. $\sum^{k} \sum xy = (-4559.436) + \cdots + (-70.830) = -8685.665$

2. $\sum^{k} \sum x^2 = 6606.235 + \cdots + 4071.359 = 31{,}434.282$

3. $\bar{b} = \dfrac{\sum^{k} \sum xy}{\sum^{k} \sum x^2} = \dfrac{\text{quantity } \mathbf{1}}{\text{quantity } \mathbf{2}} = \dfrac{-8685.665}{31{,}434.282} = -0.27631$

4. $SS_{\text{among } b\text{'s}} = \sum^{k}\left[\sum x_i^2(b - \bar{b})^2\right] = 1663.4803$

This sum of squares has $k - 1 = 5 - 1 = 4$ *df.*

5. $MS_{\text{among } b\text{'s}} = \dfrac{\text{quantity } \mathbf{4}}{(k-1)} = \dfrac{1663.4447}{4} = 415.8701$

6. $\bar{s}^2_{Y\cdot X}$, a weighted average $s^2_{Y\cdot X}$ for all groups $= \dfrac{\sum^{k}\sum d^2_{Y\cdot X}}{\left(\sum^{k} a - 2k\right)}$ (where a refers to the number of groups)

$$= \frac{1830.098 + \cdots + 507.392}{(128 - 10)} = \frac{5468.562}{118} = 46.3437$$

The degrees of freedom pertaining to this mean square are $\Sigma^k\, a - 2k = 118$. We can now display our results in the form of an anova table.

Source of variation	*df*	*SS*	*MS*	F_s
Among b's (variation among regressions)	4	1663.4803	415.8701	8.974***
Weighted average of deviations from regression (average variation within regressions)	118	5468.562	46.3437	

7. $F_s = \dfrac{MS_{\text{among } b\text{'s}}}{\bar{s}^2_{Y\cdot X}} = \dfrac{\text{quantity } \mathbf{7}}{\text{quantity } \mathbf{8}} = \dfrac{415.8701}{46.3437} = 8.974 \gg F_{.001[4,118]}$, since $F_{.001[4,120]} = 4.95$

We conclude that the five groups were not sampled from populations of equal slopes. The beetles differ in their survival response to density depending on their gene frequencies.

BOX 14.7 CONTINUED

If the sample regression coefficients had been homogeneous we could have used $\bar{b}$, the average regression coefficient computed in step **3,** as an estimate of the common slope. Since we have evidence that the regression coefficients are not all equal, we are interested in a detailed analysis of the differences among the b's. As discussed in Section 9.6, we can take one of two approaches: planned or unplanned comparisons. The experimenter might wish to test certain hypotheses specified before the data were examined (planned comparisons). For example, in the present case we are interested in testing whether the magnitude of the slopes is a function of the initial gene frequency in the population (unplanned comparisons). One may wish to test for the difference in slopes of a particular pair of regression lines that the investigator has reason to believe may differ. The other approach is to test all possible pairs of regression lines in order to ascertain which differ from which. Such a procedure (see Section 9.7) is illustrated in Box 14.8.

F-*test for difference between two regression coefficients*

For $k = 2$ regression coefficients, we can simplify the test for differences among b's to

$$F_s = \frac{(b_1 - b_2)^2}{\dfrac{\sum x_1^2 + \sum x_2^2}{\left(\sum x_1^2\right)\left(\sum x_2^2\right)} \bar{s}^2_{Y \cdot X}}$$

where $\bar{s}^2_{Y \cdot X}$ is the weighted average $s^2_{Y \cdot X}$ from step **6** in the previous computation. Since there are only two groups, we can write its formula as

$$\frac{\sum d^2_{Y \cdot X(1)} + \sum d^2_{Y \cdot X(2)}}{a_1 + a_2 - 4}$$

Compare F_s with $F_{\alpha[1, a_1 + a_2 - 4]}$. For cases with single Y per X, employ n_1 and n_2 in place of a_1 and a_2.

Following the outline in Box 14.7 we are led to the conclusion that the five groups were not sampled from populations with equal slopes. The layout in Box 14.7 can, of course, be adapted to simpler types of problems in which each sample for which a regression coefficient has been computed has only one Y per value of X. Such adaptations are not difficult.

Since we know that the regressions are different, we might wish to test whether the regression slopes are themselves a function of the differences in the criterion variable defining each class. In this case the criterion variable is gene frequency. We perform the test by calculating a linear regression of each value of b on the gene frequency of its sample. (If you are interested in carrying out this specialized operation, refer to Box 14.8 in previous editions of this book.) A similar example is regression of growth (Y) on time (X) at different temperatures. Growth rates ($b_{Y \cdot X}$) might differ and be linear functions of the temperatures.

At the end of Box 14.7 is a simple test for the significance of the *difference between two regression coefficients*. This significance can be tested either by a t-test or by an F-test. The test shown there is an F-test.

In Box 14.8 we show an *unplanned test of all k regression coefficients* to examine their differences for significance. In this special example, there is very little purpose in performing the unplanned test because the planned tests were adequate to interpret the data. However, we show how this test can be done when no planned tests suggest themselves, such as regressions induced in different samples by qualitatively different treatments (no criterion variable), and when it is still important to screen the response curves for parallelisms and differences. The method involves minimum significant differences obtained by the T′-, GT2-, and Tukey–Kramer methods by analogy to the procedures illustrated in Box 9.11; the simple computations are laid out completely in Box 14.8.

14.9 Analysis of Covariance

An ***analysis of covariance*** (abbreviated *ancova,* for consistency with "anova") tests a dependent variable Y for homogeneity among group means in a design similar to that of the closely related analysis of variance. Before being tested for homogeneity, however, the means are adjusted for the groups' differences in the independent variable X, now also known as the **covariate**. This adjustment is carried out using linear regression procedures. One way of visualizing the structure of analysis of covariance is to think of the variates for each group as being separately regressed on X, with all the regression lines of the groups having a common (pooled) slope. The main significance test in an ancova is a test of homogeneity of the Y-intercepts, a_i, for all the groups. Since an ancova requires the slopes of the regression lines being fitted in each group to be parallel, we first have to test the assumption of parallelism of slopes of Y on X for all the groups, as we did in Section 14.8.

Box 14.8 UNPLANNED COMPARISONS AMONG A SET OF REGRESSION COEFFICIENTS.

Data from Box 14.7.

Computation

1. Calculate the standard error of a regression coefficient b_i using $\bar{s}^2_{Y \cdot X}$, the weighted average unexplained variance for all groups (quantity 6 from Box 14.7).

$$s_{b_i} = \sqrt{\frac{\bar{s}^2_{Y \cdot X}}{\sum x_i^2}}$$

where Σx_i^2 is the sum of squares of X for the ith group. Thus

$$s_{b_i} = \sqrt{\frac{46.3437}{6606.235}} = 0.08376$$

The other standard errors are computed in a similar manner and are shown in the table below.

	Gene frequencies of b^+				
	0.00	0.25	0.50	0.75	1.00
b_i	−0.6902	−0.2459	−0.2421	−0.0741	−0.0174
$\sum x_i^2$	6606.235	8346.426	6452.440	5957.822	4071.359
s_{b_i}	0.08376	0.07452	0.08475	0.08820	0.10669

2. By analogy with Box 9.11, we have to decide among the T′-, the GT2-, and the Tukey–Kramer methods. When the Σx_i^2's are very similar, T′ will be more powerful than the GT2-method. An alternative method, for unequal Σx_i^2's, is the Tukey–Kramer method. In borderline situations all three methods should be tried, and the one yielding the shortest minimum significant difference, *MSD*, should be adopted. In the rare case in which all the Σx_i^2's are equal we can employ the T-method (see Box 9.10).

3. Substitute into Expression (9.8), MSD = (critical value) × SE, as follows. For the T′-method

$$MSD_{ij} = Q'_{\alpha[k,\nu]}\max[s_{b_i}, s_{b_j}]$$

where Q' is a value from the studentized augmented range distribution in statistical Table **L**.

For the GT2-method

$$MSD_{ij} = m_{\alpha[k^*,\nu]}(s^2_{b_i} + s^2_{b_j})^{1/2}$$

where $m_{\alpha[k^*,\nu]}$ is a value from the studentized maximum modulus distribution in Table **M.** As in our earlier consideration of these distributions (see Box 9.11), $k = a$,

BOX 14.8 CONTINUED

$k^* = k(k-1)/2 = a(a-1)/2$, and ν is the degrees of freedom of $\bar{s}^2_{Y \cdot X}$ (in this case 118).

For the Tukey–Kramer method

$$MSD_{ij} = Q_{\alpha[k,\nu]}\left[\frac{(s^2_{b_i} + s^2_{b_j})}{2}\right]^{1/2}$$

where $Q_{\alpha[k,\nu]}$ is a value from the studentized range distribution in Table **J.** The arguments k and ν assume the same values as in step **3.**

4. A pair of regression coefficients, b_i and b_j, is declared significantly different at the experimentwise error rate α if and only if their difference equals or exceeds the critical difference MSD_{ij}, that is, if $|b_i - b_j| \geq MSD_{ij}$. For example, comparing regression coefficients 1 and 2 by the T′-method yields

$$\begin{aligned} MSD_{12} &= Q'_{.05[5,118]} \max[0.08376, 0.07452] \\ &= 3.917(0.08376) \\ &= 0.32809 \end{aligned}$$

The same comparison by the GT2-method yields

$$\begin{aligned} MSD_{12} &= m_{.05[10,118]}(s^2_{b_1} + s^2_{b_2})^{1/2} \\ &= 2.849(0.08376^2 + 0.07452^2)^{1/2} \\ &= 2.849(0.11211) \\ &= 0.31941 \end{aligned}$$

By the Tukey–Kramer method we get

$$\begin{aligned} MSD_{12} &= Q_{.05[5,118]}\left[\frac{(s^2_{b_1} + s^2_{b_2})}{2}\right]^{1/2} \\ &= 3.918[0.079275] \\ &= 0.31060 \end{aligned}$$

The critical value for the GT2-method is smaller than that given by the T'-method, since the Σx^2's are sufficiently unequal. The value for the Tukey–Kramer method is smallest. All other pairs of regression coefficients can be tested in a similar manner.

5. To avoid the recomputation of MSD_{ij} for each comparison, we can conveniently use Gabriel's approximate method (see Box 9.11), in which comparison intervals are constructed such that two regression coefficients are significantly different if and only if their intervals do not overlap.

For the T′-method

$$l_i = b_i - \tfrac{1}{2}Q'_{\alpha[k,\nu]}s_{b_i}$$

$$u_i = b_i + \tfrac{1}{2}Q'_{\alpha[k,\nu]}s_{b_i}$$

For the GT2-method

$$l_i = b_i - \sqrt{\tfrac{1}{2}}m_{\alpha[k^*,\nu]}s_{b_i}$$

$$u_i = b_i + \sqrt{\tfrac{1}{2}}m_{\alpha[k^*,\nu]}s_{b_i}$$

Box 14.8 Continued

In the present example the GT2-method yields narrower intervals:

l_i	−0.8589	−0.3960	−0.4128	−0.2518	−0.2323
u_i	−0.5215	−0.0958	−0.0714	0.1036	0.1975

These limits are plotted in the figure that follows. It is clear that b_1 differs from the others, which are not significantly different from one another at the 5% experimentwise error rate.

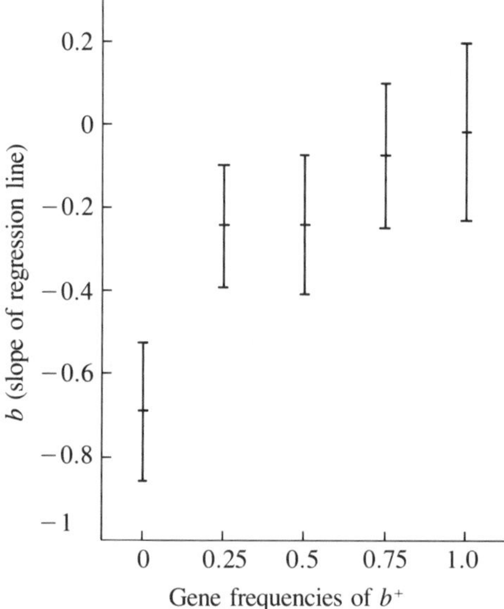

95% comparison intervals by the GT2-method for the regression coefficients of survival on density (see Box 14.7). The computation of these intervals is shown in step **5** *of this box. Regression coefficients whose intervals do not overlap are significantly different.*

The example we have chosen is based on an experiment by Yamauchi and Kimizuka (1971) in which the membrane potential was measured for four different cation systems (groups) as a function of the logarithm of activity ratio of various electrolytes. The activity ratio was varied by using different concentrations of the cations. Although the purpose of the original study was simply to determine the slope of the regression of membrane potential (in millivolts) on the logarithm of the activity ratio, the data can also serve as an example for the analysis of covariance. We wish to test whether the mean membrane potential is different for the four different groups. As Figure 14.16 shows, the four groups appear to be quite distinct. As we will see, however, when we analyze these data

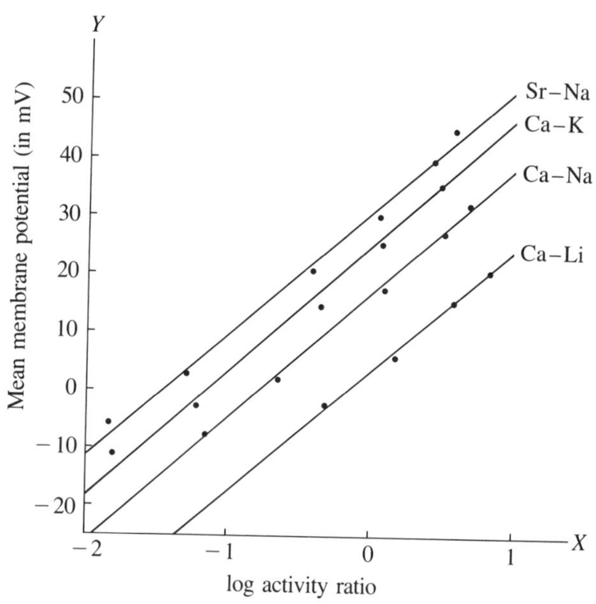

FIGURE 14.16 *Membrane potential for four different cation systems as a function of the logarithm of the activity ratio of the various electrolytes at various concentrations. Separate linear regression lines have been fitted to each cation system. (Data from Yamauchi and Kimizuka, 1971).*

using a single-classification anova, we are unable to demonstrate differences among the means because of a broad overlapping of the millivolt values of the membrane potential caused by the large range of activity ratios used for each group (cation system). It is clear, however, that if the activity ratios had been held constant in the four groups, demonstrating the difference among the means of membrane potential would not be difficult. One way of looking at an analysis of covariance is that it attempts to simulate the results that would have been obtained had a constant value been used for the independent variable X. If the regression lines for each group are parallel, then a test for equality of the Y-intercepts is equivalent to testing the differences of the means in the simulated experiment with constant X.

We limit our treatment of analysis of covariance in this text to the single-classification, completely randomized design. Each group is a sample of n_i pairs of values of Y and X, as shown at the beginning of Box 14.9. The arrangement in the box allows for unequal sample sizes n_i, which exist in this example; the computations can be simplified when sample sizes are equal. The expression symbolizing the decomposition of a single variate in a single-classification ancova in the manner of Expressions (8.2), (10.2) and (11.1) is as follows:

$$Y_{ij} = \mu + \alpha_i + \beta_{\text{within}}(X_{ij} - \overline{X}_i) + \epsilon_{ij} \qquad (14.13)$$

Box 14.9 ANALYSIS OF COVARIANCE. SINGLE CLASSIFICATION.

The data in this box are based on an experiment by Yamauchi and Kimizuka (1971) in which they measured membrane potential (in millivolts) for 4 different cation systems (groups) as a function of the logarithm of the activity ratio of various electrolytes at various concentrations. We wish to test whether the mean membrane potential Y is different for the different systems, taking the covariate X (log of activity ratio) into account. The data will be arranged in the manner of single-classification anova.

	a groups						
	1		2			a	
	Y	X	Y	X		Y	X
	Y_{11}	X_{11}	Y_{21}	X_{21}	$\cdots$	Y_{a1}	X_{a1}
	Y_{12}	X_{12}	Y_{22}	X_{22}	$\cdots$	Y_{a2}	X_{a2}
	Y_{13}	X_{13}	Y_{23}	X_{23}	$\cdots$	Y_{a3}	X_{a3}
	$\vdots$	$\vdots$	$\vdots$	$\vdots$		$\vdots$	$\vdots$
	Y_{1n_i}	X_{1n_i}	Y_{2n_i}	X_{2n_i}	$\cdots$	Y_{an_i}	X_{an_i}
Sample sizes	n_1		n_2			n_a	
Means	$\bar{Y}_1$	$\bar{X}_1$	$\bar{Y}_2$	$\bar{X}_2$		$\bar{Y}_a$	$\bar{X}_a$

	a = 4 groups (cation systems)							
	Ca–Li		Ca–Na		Ca–K		Sr–Na	
	Y	X	Y	X	Y	X	Y	X
	−2.4	−0.31	−7.0	−1.18	−10.8	−1.79	−5.4	−1.83
	6.3	0.17	2.1	−0.65	−2.8	−1.21	3.0	−1.25
	15.8	0.58	17.8	0.10	14.2	−0.35	20.7	−0.41
	20.5	0.81	27.3	0.50	25.5	0.08	30.5	0.05
			32.0	0.67	35.7	0.49	39.9	0.43
					41.2	0.65	45.0	0.59
Sample sizes	4		5		6		6	
Means	10.05	0.3125	14.44	−0.1120	17.167	−0.3550	22.283	−0.4033

Preliminary computation

The following quantities are the familiar ones from a single-classification anova. They follow the outline of Box 9.1.

	Y	X
1. Grand means $\bar{\bar{Y}}$ or $\bar{\bar{X}}$	16.6238	−0.1838
2. SS_{among}	390.65443	1.47606
3. SS_{within}	5623.06362	12.53044
4. SS_{total} = quantity **2** + quantity **3**	6013.71805	14.00650

BOX 14.9 CONTINUED

Preliminary anova of the dependent variable

Source of variation	*df*	*SS*	*MS*	F_s
Among	3	390.65443	130.21814	0.394 *ns*
Within	17	5623.06362	330.76845	
Total	20	6013.71805		

The preliminary anova (without regression on the covariate) is unable to detect differences among the means.

Analysis of covariance

1. For each group separately compute SS_Y, SP_{XY}, SS_X, $b_{Y \cdot X}$, $SS_{\hat{Y}}$, $SS_{Y \cdot X}$, and $MS_{Y \cdot X}$. The familiar formulas from Box 14.1 apply. We illustrate here the computations for group 1 (Ca–Li):

$$\sum^{n_1} y^2 = \sum^{n_1} (Y_1 - \bar{Y}_1)^2 = 311.33000$$

$$\sum^{n_1} xy = \sum^{n_1} (X_1 - \bar{X}_1)(Y_1 - \bar{Y}_1) = 15.02150$$

$$\sum^{n_1} x^2 = \sum^{n_1} (X_1 - \bar{X}_1)^2 = 0.726875$$

$$b_1 = \frac{\sum^{n_1} xy}{\sum^{n_1} x^2} = \frac{15.0215}{0.726875} = 20.66586$$

$$\sum^{n_1} \hat{y}^2 = \frac{\left(\sum^{n_1} xy\right)^2}{\sum^{n_1} x^2} = \frac{(15.0215)^2}{0.726875} = 310.43228$$

$$\sum^{n_1} d^2_{Y\cdot X} = \sum^{n_1} y^2 - \sum^{n_1} \hat{y}^2 = 311.33000 - 310.43228 = 0.89772$$

$$s^2_{Y\cdot X} = \frac{\sum^{n_1} d^2_{Y\cdot X}}{(n_1 - 2)} = \frac{0.8977}{2} = 0.44886$$

$$F_s = \frac{\sum^{n_1} \hat{y}^2}{s^2_{Y\cdot X}} = \frac{310.43228}{0.44886} = 691.602$$

The regression of membrane potential on log activity ratio is highly significant for group 1. It is similarly significant for the other three groups. These values for the a groups are entered in the first a lines of the layout for an ancova (see below).

2. Pool SS_Y, SP_{XY}, and SS_X as in an anova of a groups. The value for SS_Y and SS_X can be obtained from the preliminary computations, or all three quantities result from summing the corresponding statistics for the groups.

$$\sum y^2_{\text{within}} = \sum^{a}\sum^{n} y^2 = 311.33000 + 1096.97201 + 2180.13330 + 2034.62832$$
$$= 5623.06362$$

$$\sum xy_{\text{within}} = \sum^{a}\sum^{n} xy = 15.02150 + 51.85140 + 100.62800 + 96.80267$$
$$= 264.30357$$

$$\sum x^2_{\text{within}} = \sum^{a}\sum^{n} x^2 = 0.72687 + 2.46108 + 4.70355 + 4.63893$$
$$= 12.53044$$

Enter these values in the row entitled "Pooled within."

3. Compute the pooled regression coefficient within, b_{within}, and the explained and unexplained sums of squares for this level of variation, as well as the mean square for the unexplained SS, and enter these values in the "Pooled within" line of the ancova table as well.

$$b_{\text{within}} = \frac{\sum^{a}\sum^{n} xy}{\sum^{a}\sum^{n} x^2} = \frac{264.30357}{12.53044} = 21.09292$$

BOX 14.9 CONTINUED

$$\sum \hat{y}^2_{\text{within}} = \frac{\left(\sum^a \sum^n xy\right)^2}{\sum^a \sum^n x^2} = \frac{(264.30357)^2}{12.53044} = 5574.93465$$

$$\sum d^2_{Y \cdot X(\text{within})} = \sum^a \sum^n y^2 - \sum \hat{y}^2_{\text{within}} = 5623.06362 - 5574.93465 = 48.12896$$

$$s^2_{Y \cdot X(\text{within})} = \frac{\sum d^2_{Y \cdot X(\text{within})}}{\sum^a n_i - a - 1} = \frac{48.12896}{16} = 3.00806$$

$$F_s = \frac{\sum \hat{y}^2_{\text{within}}}{s^2_{Y \cdot X(\text{within})}} = \frac{5574.93465}{3.00806} = 1853.332^{***} \qquad F_{.001[1,16]} = 16.1$$

The pooled regression of membrane potential on log activity ratio is highly significant.

4. Sum the unexplained sums of squares of the separate groups and their degrees of freedom. Compute the appropriate mean square. Enter values in the "Sum of groups" line of the anova table.

$$\sum^a \sum^n d^2_{Y \cdot X} = 0.89772 + 4.53791 + 27.29251 + 14.60442 = 47.33257$$

$$\sum^a n_i - 2a = 21 - 8 = 13$$

$$\bar{s}^2_{Y \cdot X} = \frac{\sum^a \sum^n d^2_{Y \cdot X}}{\sum^a n_i - 2a} = \frac{47.33257}{13} = 3.64097$$

5. Compute $SS_{\text{among } b\text{'s}}$ and $MS_{\text{among } b\text{'s}}$ and test significance.

$$SS_{\text{among } b\text{'s}} = \sum d^2_{Y \cdot X(\text{within})} - \sum^a \sum^n d^2_{Y \cdot X} = 48.12896 - 47.33257 = 0.79640$$

$$MS_{\text{among } b\text{'s}} = \frac{SS_{\text{among } b\text{'s}}}{a - 1} = \frac{0.794640}{3} = 0.26547$$

$$F_s = \frac{MS_{\text{among } b\text{'s}}}{\bar{s}^2_{Y \cdot X}} = \frac{0.26547}{3.64097} = 0.0729\ ns \qquad F_{.001[3,13]} = 10.2$$

Computational layout for ancova

Source of variation	df	SS_Y	SP_{XY}	SS_X	$b_{Y \cdot X}$	$SS_{\hat{Y}}$	df	$SS_{Y \cdot X}$	$MS_{Y \cdot X}$
Group 1	$n_1 - 1$	$\sum^{n_1} y^2$	$\sum^{n_1} xy$	$\sum^{n_1} x^2$	b_1	$\sum^{n_1} \hat{y}^2$	$n_1 - 2$	$\sum^{n_1} d^2_{Y \cdot X}$	$s^2_{Y \cdot X(1)}$
Group 2	$n_2 - 1$	$\sum^{n_2} y^2$	$\sum^{n_2} xy$	$\sum^{n_2} x^2$	b_2	$\sum^{n_2} \hat{y}^2$	$n_2 - 2$	$\sum^{n_2} d^2_{Y \cdot X}$	$s^2_{Y \cdot X(2)}$
⋮									
Group a	$n_a - 1$	$\sum^{n_a} y^2$	$\sum^{n_a} xy$	$\sum^{n_a} x^2$	b_a	$\sum^{n_a} \hat{y}^2$	$n_a - 2$	$\sum^{n_a} d^2_{Y \cdot X}$	$s^2_{Y \cdot X(a)}$
Sum of groups							$\sum n_i - 2a$	$\sum^{a} \sum^{n} d^2_{Y \cdot X}$	$\bar{s}^2_{Y \cdot X}$
Among b_i's							$a - 1$	$SS_{\text{among } b\text{'s}}$	$MS_{\text{among } b\text{'s}}$
Pooled within	$\sum^{a} n_i - a$	$\sum^{a} \sum^{n} y^2 = \sum y^2_{\text{within}}$	$\sum^{a} \sum^{n} xy = \sum xy_{\text{within}}$	$\sum^{a} \sum^{n} x^2 = \sum x^2_{\text{within}}$	b_{within}	$\sum \hat{y}^2_{\text{within}}$	$\sum^{a} n_i - a - 1$	$\sum d^2_{Y \cdot X(\text{within})}$	$s^2_{Y \cdot X(\text{within})}$
Adjusted means							$a - 1$	$\sum d^2_{Y \cdot X(\text{adj})}$	$s^2_{Y \cdot X(\text{adj})}$
Total	$\sum^{a} n_i - 1$	$\sum y^2_{\text{total}}$	$\sum xy_{\text{total}}$	$\sum x^2_{\text{total}}$			$\sum^{a} n_i - 2$	$\sum d^2_{Y \cdot X(\text{total})}$	
Among groups	$a - 1$	$\sum y^2_{\text{among}}$	$\sum xy_{\text{among}}$	$\sum x^2_{\text{among}}$	b_m	$\sum \hat{y}^2_{\text{among}}$	$a - 2$	$\sum d^2_{Y \cdot X(\text{among})}$	$s^2_{Y \cdot X(\text{among})}$

Box 14.9 Continued

Computed quantities corresponding to the layout

Source of variation	df	SS_Y	SP_{XY}	SS_X	$b_{Y \cdot X}$	$SS_{\hat{Y}}$	df	$SS_{Y \cdot X}$	$MS_{Y \cdot X}$
Group 1 (Ca–Li)	3	311.33000	15.02150	0.726875	20.66586	310.43228	2	0.89772	0.44886
Group 2 (Ca–Na)	4	1096.97201	51.85140	2.46108	21.06856	1092.43410	3	4.53791	1.51264
Group 3 (Ca–K)	5	2180.13330	100.62800	4.70355	21.39405	2152.84078	4	27.29251	6.82313
Group 4 (Sr–Na)	5	2034.62832	96.80267	4.63893	20.86744	2020.02390	4	14.60442	3.65111
Sum of groups							13	47.33257	3.64097
Among b's							3	0.79640	0.26547
Pooled within	17	5623.06362	264.30357	12.53044	21.09292	5574.93465	16	48.12896	3.00806
Adjusted means:							3	1768.58266	589.52755
Total	20	6013.71805	242.45690	14.00650			19	1816.71162	
Among groups	3	390.65443	−21.84666	1.47606	−14.80069	323.34568	2	67.30874	33.65437

6. Fill in SS_X and SS_Y for Total and Among Groups using values already obtained in the preliminary computations (steps **4** and **2** thereof, respectively). The product terms SP_{XY} for these two levels can be obtained in a similar manner. We need evaluate only Σxy_{total}.

$$\sum xy_{\text{total}} = \sum^{a} \sum^{n} (X - \bar{\bar{X}})(Y - \bar{\bar{Y}}) = 242.45690$$

We obtain $\Sigma\, xy_{\text{among}}$ by subtraction

$$\sum xy_{\text{among}} = \sum xy_{\text{total}} - \sum xy_{\text{within}} = 242.45690 - 264.30357 = -21.84666$$

which could have been computed independently as

$$\sum xy_{\text{among}} = \sum^{a} n_i(\bar{X}_i - \bar{\bar{X}})(\bar{Y}_i - \bar{\bar{Y}})$$

7. Calculate unexplained sums of squares for these two levels of variation, and for the among-groups level compute the unexplained mean square, the regression coefficient, b_{among}, and the explained sum of squares for the groups.

$$\sum d^2_{Y \cdot X(\text{total})} = \sum y^2_{\text{total}} - \frac{\left(\sum xy_{\text{total}}\right)^2}{\sum x^2_{\text{total}}} = 6013.71805 - \frac{(242.45690)^2}{14.00650}$$
$$= 1816.71162$$

$$\sum d^2_{Y \cdot X(\text{among})} = \sum y^2_{\text{among}} - \frac{\left(\sum xy_{\text{among}}\right)^2}{\sum x^2_{\text{among}}} = 390.65443 - \frac{(-21.84666)^2}{1.47606}$$
$$= 67.30874$$

$$s^2_{Y \cdot X(\text{among})} = \frac{\sum d^2_{Y \cdot X(\text{among})}}{a - 2} = \frac{67.30874}{2} = 33.65437$$

$$b_{Y \cdot X(\text{among})} = \frac{\left(\sum xy_{\text{among}}\right)}{\left(\sum x^2_{\text{among}}\right)} = \frac{(-21.84666)}{(1.47606)} = -14.80069$$

8. Test the null hypothesis that there are no differences among sample means ($\bar{Y}_i$) when these are adjusted for a common $\bar{X}$ and a common regression line. This is equivalent to testing for homogeneity of the Y-intercepts, a_i. Compute

$$\sum d^2_{Y \cdot X(\text{adj})} = \sum d^2_{Y \cdot X(\text{total})} - \sum d^2_{Y \cdot X(\text{within})} = 1816.71162 - 48.12896 = 1768.58266$$

$$s^2_{Y \cdot X(\text{adj})} = \frac{\sum d^2_{Y \cdot X(\text{adj})}}{a - 1} = \frac{1768.58226}{3} = 589.52755$$

$$F_s = \frac{s^2_{Y \cdot X(\text{adj})}}{s^2_{Y \cdot X(\text{within})}} = \frac{589.52755}{3.0080} = 195.980^{***} \qquad F_{.001[3,16]} = 9.00$$

Box 14.9 Continued

The results can be summarized in an analysis of covariance table as shown here.

Final ancova table

Source of variation	*df*	*SS*	*MS*	F_s
Adjusted means (among a_i's)	3	1768.5827	589.5276	195.980***
Error (deviations from a common slope)	16	48.1290	3.0081	

Conclusions

The membrane potentials differ very significantly among the four cation systems when log activity ratios are kept constant. There is a significant heterogeneity of the means around their regression slope.

Variate Y_{ij} is the jth item in the ith group; μ is the grand mean of the population, α_i is the fixed treatment effect for group i; $\beta_{\text{within}}(X_{ij} - \overline{X}_i)$ is the effect explained by the difference of the variate X_{ij} from $\overline{X}_i$, its mean for X; and ϵ_{ij} is, as usual, the random deviation. Note that the slope is the parametric value for the pooled regression within groups.

The symbolic table of variates is followed by the symbolic sample sizes and means of each group. The raw data (obtained by digitizing a figure in the original paper) are given next in Box 14.9, together with their sample sizes and means. We then carry out a preliminary anova of the dependent variable, membrane potential. The computations are not shown in Box 14.9, but the numbered steps indicated there are the same as those in Box 9.1 (single-classification anova with unequal sample sizes); they should present no difficulty. The result of the preliminary analysis of variance is shown in the next table. The groups (cation systems) do not differ significantly, backing up our earlier statement that mean membrane potentials for the four cation systems were not differentiated, presumably because of overlap of the variates.

The computational layout of an analysis of covariance (shown in tabular form later in Box 14.9) is more complicated than that for an anova. We will learn its construction by filling in the required values systematically. In step **1**, we enter the necessary statistics for computing regression separately for each group. We show formulas for group 1 only. The computations are straightforward, as is the significance test for the amount of variance explained by regression. The resulting values for the four groups are entered in the table of computed quantities corresponding to the layout for an ancova.

What have we learned so far? We have studied the relation between membrane potential and log activity ratio separately for each of the four cation systems. We find a highly significant regression in all four cases. This level of significance, is a prerequisite to an analysis of covariance, and we detail it here for the sake of completeness. Had there been no significant regression in any of the groups, it would be less likely (though not impossible) for this particular independent variable to account for differences in the dependent variable.

Let us return to the central question of an ancova: Are there differences among groups for the dependent variable when differences among groups in the independent variable are taken into account? Suppose the regressions for the four groups had yielded results such as those in Figure 14.17A. We cannot predict the magnitude of the variable Y for a given value of X without specifying the group for which the prediction is to be made. This situation is the classic picture of interaction, first encountered in Chapter 11. Any meaningful comparison of the sample means $\overline{Y}_i$ must assume that the regression slopes are parallel as shown in Figure 14.17B. Here the expected differences among the four groups are constant for any value of X. Thus we might usefully carry out an analysis of variance on adjusted means, which can be most conveniently represented by the Y-intercepts, a_i.

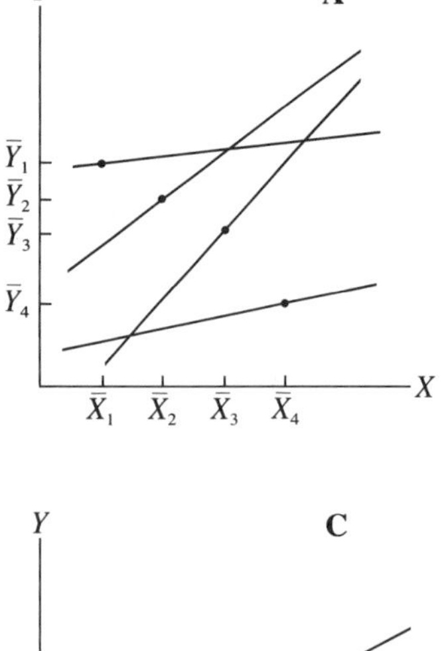

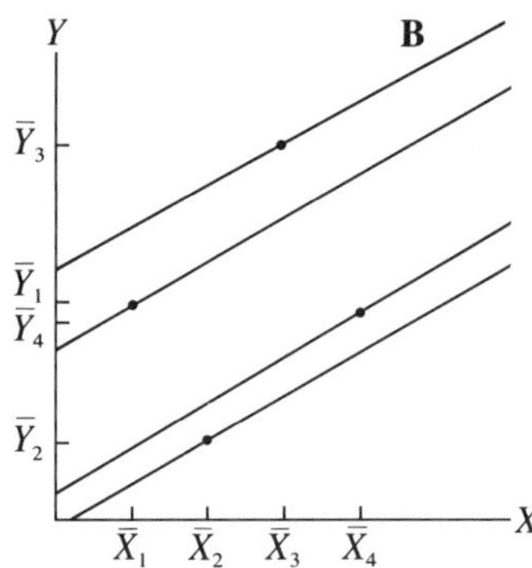

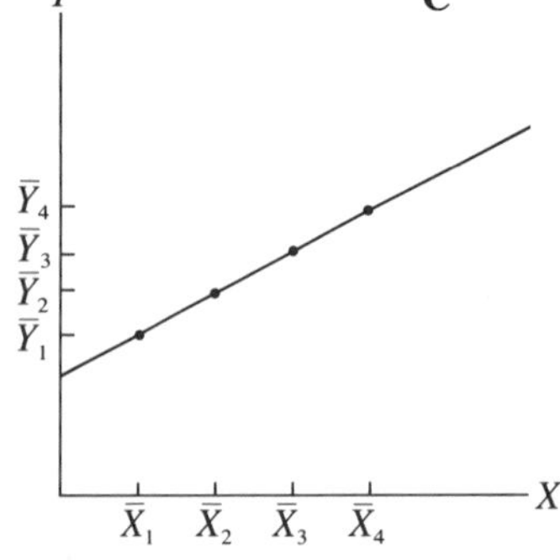

FIGURE 14.17 *Schematic diagrams to illustrate null hypotheses tested in analysis of covariance. Four samples representing the same pair of variables are shown. The means of each variable for each sample are indicated along the axes of each graph.* ***A.*** *Separate regression lines fitted to each sample are not parallel.* $H_0{:}\beta_1 = \beta_2 = \beta_3 = \beta_4$ *is rejected.* ***B.*** *Separate regression lines fitted to each sample are parallel but their* Y*-intercepts differ.* $H_0{:}\alpha_1 = \alpha_2 = \alpha_3 = \alpha_4$ *is rejected.* ***C.*** *All four regression lines coincide, although bivariate means do not. The samples have a common regression slope* β *and a common* Y*-intercept* α.

The assumption that the separate regression slopes of the groups do not differ but represent a single slope is fundamental to the analysis of covariance, and its test precedes any further tests. We test this hypothesis by means of steps **2** to **5** in Box 14.9. In an analysis of covariance by computer, such as by the BIOM-pc package of computer programs, all possible tests are carried out because such a procedure is simple and practical. In our discussion here, however, we will detail each step for instructive purposes.

First we pool the data for the four groups, obtaining pooled sums of squares and sums of products within groups, as shown in step **2**. These values are entered in the ancova table in the labeled ''Pooled within.'' From these values, we compute the pooled regression coefficient in the customary manner (step **3**), obtaining 21.09292, a weighted average of the separate values for the four groups. This regression is highly significant when tested by computing the explained and unexplained sums of squares and mean squares and performing the F-test. The sums of squares and mean squares are also entered in the "Pooled within" line of the ancova table. Next we sum the unexplained sums of squares of the four separate groups as well as their degrees of freedom (step **4**). This quantity and its mean square are entered in the "Sum of groups" line of the ancova table.

When we compare this sum of squares, labeled $\Sigma^a\Sigma^n d^2_{Y \cdot X}$, with the unexplained sum of squares of the pooled data within groups, $\Sigma d^2_{Y \cdot X(\text{within})}$, we find the latter to be greater than the former. Why? The sum of the separate unexplained sums of squares represents least squares estimates around separate regression lines, one for each group. The least squares criterion ensures that fitting the separate regression lines to each group produces a better fit than does fitting a line of average slope to each of the four groups. The difference between these sums of squares is due to differences among the regression slopes for the separate groups and is computed in step **5**. If the slopes were identical, the two unexplained sums of squares would also be the same and the sums of squares due to differences among b's would be zero. We test the mean square among b's against the weighted average unexplained MS for groups, $\bar{s}^2_{Y \cdot X}$, and obtain a nonsignificant value of $F_s = 0.0729$. (In fact, the fit is almost too good to be true: The probability of obtaining regression slopes in such close agreement or better is less than 0.001.) We accept the null hypothesis that the separate regressions of membrane potential on log activity ratio for each group do not differ in slope. The regression equation for the average regression based on the pooled-within covariance is found by substituting the grand means for X and Y obtained from the grand totals (step **1** of the preliminary computation) in the customary formula for a:

$$a = \bar{\bar{Y}} - b_{\text{within}}\bar{\bar{X}} = 16.6238 - 21.09292(-0.1838) = 20.50075$$

Fitting a single regression line through the entire data set is not usually of interest in itself, but the quantities obtained are needed for the computations that follow. Using the same equation, one can compute Y-intercepts for each of the groups separately by passing a regression line with the common slope through the group means. Thus $a_i = \bar{Y}_i - b_{\text{within}}\bar{X}_i$. The results in this example are

$$a_1 = 3.4585$$

$$a_2 = 16.8024$$

$$a_3 = 24.6550$$

$$a_4 = 30.7898$$

The analysis of covariance will test whether these intercepts are homogeneous, that is, whether they could have been sampled from a common population.

Before we carry out this test we need to calculate some other quantities. In step **6** we find the total sum of squares and the sum of squares among groups for Y and X according to the analysis of variance. These quantities have already been obtained in the preliminary computations and need only be entered in the appropriate lines of the ancova table. Similarly, we need to compute the total sum of products and the sum of products among groups. To do this we use a formula analogous to that used in analysis of variance, which is shown in step **6**. Next we calculate unexplained sums of squares for both of these levels (Total and Among Groups). For the among-groups level we can also compute an unexplained mean square, a regression coefficient, and an explained sum of squares. These straightforward computations are shown in step **7**.

The main null hypothesis in analysis of covariance is that there are no differences among sampled means $\bar{Y}_i$ when these are adjusted for a common mean of X and a common regression line. This hypothesis is illustrated in Figure 14.17C, which shows a single regression line with, necessarily, a single Y-intercept. In such a case, the means of the dependent variable Y differ only as the means of the independent variable X differ. For the test we need to know the sum of squares for adjusted means $\Sigma d^2_{Y \cdot X(\text{adj})}$, which is the difference between the unexplained sum of squares of the totals $\Sigma d^2_{Y \cdot X(\text{total})}$ and the unexplained sum of squares pooled within groups $\Sigma d^2_{Y \cdot X(\text{within})}$. Remember that $\Sigma d^2_{Y \cdot X(\text{total})}$ represents deviations from a single regression line fitted to the entire data set, whereas $\Sigma d^2_{Y \cdot X(\text{within})}$ is the sum of deviations around four separate regression lines with a common slope. Thus the difference between these sums of squares represents the only contrast between these two models—the potential differences in Y-intercepts of the four separate regression lines. The mean square corresponding to the sum of squares for adjusted means is tested over the pooled within-groups mean square and, as shown in step **8**, is highly significant. Thus we cannot assume a single regression line for these data. There is heterogeneity in membrane potential among the four cation systems when the log activity ratio is kept constant. The cation systems do affect membrane potential differentially.

It is customary in analysis of covariance to calculate adjusted means based on the pooled regression coefficient b_{within}. The formula for such an adjusted mean,

$$\bar{Y}_{i(\text{adj})} = \bar{Y}_i - b_{\text{within}}(\bar{X}_i - \bar{\bar{X}}) \tag{14.14}$$

is similar to that of Y_{adj} of Expression (14.11). We can calculate $\bar{Y}_{i(\text{adj})}$ using $b_{\text{within}} = 21.09292$ and $\bar{\bar{X}} = \Sigma^a\Sigma^n X/\Sigma^a n_i = -3.86/21 = -0.1838$ from Box 14.9. The calculations yield the following values for the four cation systems.

	$\bar{X}_i$	$\bar{Y}_i$	$\bar{Y}_{i(\text{adj})}$
Ca–Li	0.3125	10.05	−0.4184
Ca–Na	−0.1120	14.44	12.9255
Ca–K	−0.3550	17.167	20.7781
Sr–Na	−0.4033	22.283	26.9129

When the membrane potential of each cation system is adjusted for log activity ratio, the differentiation among the four becomes much more marked. Note that the activity ratios chosen for each cation system are such that their means decrease in the array, creating the mistaken impression that there might be a negative regression of membrane potential on log activity ratio. This is only an artifact, however, the slope b_{among} is negative in this example solely because of the choice of ranges of activity ratios. The standard error of an adjusted mean is

$$s_{\bar{Y}_{i(\text{adj})}} = \left[s^2_{Y \cdot X(\text{within})} \left(\frac{1}{n_i} + \frac{(\bar{X}_i - \bar{\bar{X}})^2}{\sum x^2_{\text{within}}} \right) \right]^{1/2}$$

It is used with t at $\Sigma^a n_i - a - 1$ degrees of freedom.

We may wish to test the significance of the differences among the adjusted means. Fundamentally this procedure is not different from the multiple-comparison techniques discussed in Sections 9.6 and 9.7. The choice between the T′- and GT2-methods depends on relative equalities of the quantities

$$\left(\frac{1}{n_i} + \frac{(\bar{X}_i - \bar{\bar{X}})^2}{\sum x^2_{\text{within}}} \right)$$

rather than of n_i as before. Since it would be quite unusual for these quantities to be equal or nearly equal, the GT2-method is likely to be chosen most often. Because computation of the *MSD* by the GT2-method is tedious, in this case we recommend the approximate comparison intervals due to Gabriel (see Box 9.11). They are computed here as

$$\bar{Y}_{i(\text{adj})} \pm \sqrt{\tfrac{1}{2}}\, m_{\alpha[k^*,\nu]} s_{\bar{Y}_{i(\text{adj})}} \tag{14.15}$$

where $m_{\alpha[k^*,\nu]}$ is the studentized maximum modulus from Table **M** with $k^* = k(k-1)/2 = a(a-1)/2$, and the appropriate degrees of freedom ν are those pertaining to $s^2_{Y \cdot X(\text{within})}$, namely $\Sigma^a n_i - a - 1$. The formula for $s_{\bar{Y}_{i(\text{adj})}}$, the standard error of an adjusted mean, is as given above.

As an example let us compute the 95% comparison limits for the adjusted mean of Ca–Li. Substituting into Expression (14.15), we obtain

$$\begin{aligned}
&-0.4184 \pm \sqrt{\tfrac{1}{2}}\, m_{.05[6,16]} s_{\bar{Y}_{i(\text{adj})}} \\
&= -0.4184 \pm \sqrt{\tfrac{1}{2}}\,(2.969) \left\{ s^2_{Y \cdot X(\text{within})} \left[\frac{1}{n_i} + \frac{(\bar{X}_1 - \bar{\bar{X}})^2}{\sum x^2_{\text{within}}} \right] \right\}^{1/2} \\
&= -0.4184 \pm (2.0994) \left\{ 3.00806 \left[\frac{1}{4} + \frac{(0.3125 - (-0.1838))^2}{12.53044} \right] \right\}^{1/2} \\
&= -0.4184 \pm 1.8908
\end{aligned}$$

$l_1 = -2.3092 \quad u_1 = 1.4724$

Computing similar intervals for the other adjusted means would reveal that the comparison intervals of all four adjusted means do not overlap, confirming the significance of the differences between them on the basis of unplanned multiple comparisons as well as for the planned overall test of Box 14.9.

Several uses of the analysis of covariance should have occurred to you by now. A common application is to increase precision. For example, in a study of the final weight of an animal, the initial weight is often also recorded. By adjusting final weight for initial weight, we can increase the precision of the analysis. Experimenters make such adjustments often simply by manipulating the variables—in the preceding example, for instance by employing weight gain, the difference between final and initial weights. These simple adjustments, however, are based on the assumption that β, the regression of one on the other, equals exactly 1. Analysis of covariance has the advantage of being able to produce a better adjustment, since it fits regression lines to the actual data instead of employing a hypothetical slope of unity.

An ancova may help to remove the effects of disturbing variables from an anova. For example, when students of unequal backgrounds are subjected to different tests, ancova can adjust for the differences in background. The example in Box 14.9 also falls into this category.

A third purpose of ancova is to reveal causal relations. Suppose, in each of four different countries, we group males over 50 years old into five groups of alcohol consumers ranging from no consumption to high consumption, and we record the frequency of heart attacks in these populations for all 20 groups (five each in four countries). If large initial differences in heart attack rates among the countries vanish after the means are adjusted by an ancova, we have strong evidence that alcohol consumption affects the frequency of heart attacks. An ancova also permits a test of regression slopes of means against regression slopes of individuals. Finally, analysis of covariance can be used in an anova to estimate missing values. Refer to Steel and Torrie (1980, section 17.11) for an application of this technique.

Turning next to a brief discussion of the assumptions of analysis of covariance, we would first like to reiterate the assumption of homogeneity of the slopes of the regression lines. If there is serious discrepancy between the slopes for the various groups, the test for the homogeneity of adjusted means is meaningless. Two nonparallel lines can, however, be compared over a particular range of interest by a test proposed by Tsutakawa and Hewett (1978). A second assumption of an ancova is linearity of the response. The heterogeneity of adjusted means, which we have assumed to be due to chemical difference, could also be the result of nonlinearity of the responses of the dependent on the independent variable. An ancova is based on a linear model. When this model fails, inferences about the null hypothesis will be in error. We could investigate nonlinearity of response before carrying out the ancova by employing the methods of Section 16.6.

Other assumptions of analysis of covariance include those of the appropriate designs of analysis of variance as well as of linear regression analysis. Thus covariance analysis is based on Model I regression. The covariate is to be measured without error and under the control of the investigator. In many applications of analysis of covariance, however, this is not strictly correct. For example, consider again the study of final weight in animals as the result of a treatment. Although we attempt to obtain greater precision in the results by recording initial weights of the animals as covariates and carrying out an ancova, the initial weights of animals are rarely, if ever, measured without error and are not likely to be under the control of the investigator. Such instances and others in which the investigator aims to obtain further insight into causal relationships in a given system by keeping a covariate constant are more appropriately studied by the method of partial correlations and other multivariate techniques (see Chapter 15). Although the analysis of covariance seems to be a useful technique for gaining insights into the interrelation of variables, we must wait until Model II regression techniques in ancova are developed before we can recommend its application to such cases.

Analysis of covariance is employed not only for single-classification designs, but also for more-complex designs. This subject is treated in greater depth in Snedecor and Cochran (1989, chapter 18) and in Bliss (1970, chapter 20). The discussion that follows will help you understand how to carry out an ancova in one of these designs. Single-classification analysis of covariance for one covariate is performed by our package of computer programs.

It will be useful to review the new concepts in this section by considering the detailed partitioning of the sums of squares undertaken in the computational steps of the analysis of covariance of Box 14.9. For pedagogical reasons the partitioning is shown in greater detail in Table 14.2 than is necessary for the usual ancova. We start out with a single-classification design in which samples of the dependent variate Y were divided into groups. The variation among groups is summarized by the sum of squares among groups in the first line of the table. Whenever there are differences in the means of the covariate X, we can regress the group means $\bar{Y}_i$ on the group means $\bar{X}_i$ in a manner similar to that of Box 14.4. In an idealized ancova example, one would expect the means of the covariates to be the same for all treatment groups; in reality, however, this is rarely the case, so the investigator can carry out a regression of the means of Y on the means of X. The sum of squares explained by such a linear regression is shown in the second line of Table 14.2, and the deviations from the computed regression are given in the third line. The mean square for linear regression is tested over that for deviations, and for the MS deviations is tested over MS_{within} in the fourth line of Table 14.2.

The variation of Y pooled within groups can be explained in large part by the common slope of Y and X computed from the within-groups sums of squares and sums of products. A principal reason for the variation of Y's within groups is that their associated covariates (the X's) vary. When the explained sum of squares, due to regression on the common slope within groups (shown in line 5), is subtracted from the within-groups sum of squares, we are left with the unexplained sum of squares representing deviations from the common slope within

Table 14.2 A DETAILED PARTITIONING OF THE SUMS OF SQUARES OF THE EXAMPLE FROM BOX 14.9.

Final ancova table

Source of variation	*df*	*SS*	*MS*	F_s
Among	3	390.65443	130.21814	0.394 *ns*
Linear regression	1	323.34568	323.34568	9.608 *ns*
Deviations from regression	2	67.30874	33.65437	
Within	17	5623.06362	330.76845	
Common slope within groups	1	5574.93465	5574.93465	1853.332***
Deviations from common slope within groups	16	48.12896	3.00806	
Differences among separate slopes	3	0.79640	0.26547	0.073 *ns*
Deviations in each group from its separate slope	13	47.33257	3.64097	
Among + within (= total)	20	6013.71805		
Common slope within total study	1	4197.00484	4197.00484	
Deviations from common slope within total study	19	1816.71162	95.61640	
Among a_i's	3	1768.58266	589.52755	195.983***
Deviations from common slope within groups	16	48.12896	3.00806	

groups. This value is the basic error term in the analysis of covariance. The next two sums of squares in the table represent a partitioning of the error sum of squares estimating deviation from a common slope within groups. The first sum of squares is due to the differences among the slopes of the several groups; the second is the sum of the deviations in each group from its own slope. Recall that the sum of squares for differences among the separate slopes, called "among *b*'s" in Box 14.9, was obtained by subtracting the sum of groups *SS* (= deviation in each group from its own slope) from the error *SS* (= deviations from a common slope within groups). The values provided in the upper portion of the ancova table, as just described, permit us to test for linearity of regressions among groups ($F_s = 9.608$) and within groups ($F_s = 1853.332$) and for homogeneity of slopes of the regression coefficients in the several groups ($F_s = 0.073$).

The lower part of Table 14.2 features the decomposition leading to the principal test of analysis of covariance—the test of the homogeneity of adjusted means. The reasons for labeling the first line of this section "Among + within (= total)" will be discussed presently. In this example this level is the total level of variation of the variable Y and represents the entire study (all a samples) with the distinctions between the groups removed. A common single regression line can be fitted to the entire data set. The sum of squares explained by the slope of such a line is shown in the second line of this part of the table and the sum of squares of deviations from the common slope in the third line. The latter (unexplained) sum of squares summarizes all the variation around the common regression line, representing unexplained variation around regression as well as differences among the treatment means. The unexplained sum of squares is again partitioned into that among Y-intercepts (adjusted means or treatment effects) and that representing deviation from the common slope within groups—that is, the error term in ancova that we encountered already in the upper part of the table. Recall that the sum of squares among the Y-intercepts is found by subtracting the unexplained sum of squares within groups (deviations from the common slope within groups) from the total unexplained SS (deviations from the common slope within the total study).

Now we will explain why the first line in the lower part of Table 14.2 is labeled "Among + within (= total)." In this single-classification design, the total level is identical to the sum of SS_{among} and SS_{within}, no matter if SS_{total} is obtained by summing these two sums of squares or by independent calculation. With a more complicated design, however, such as a two-way analysis of covariance, we are faced with the problem of having to study adjusted means separately for rows, columns, and even the interaction. In more-complex designs obviously there are more main effects and interactions to test. By computing a sum of squares that is the sum of the SS of the particular effect of interest plus the error SS, calculating an unexplained sum of squares for this summed level, and then subtracting from this unexplained SS the unexplained sum of squares of the error level, we will obtain the correct contrast—that is, the sum of squares due to differences among adjusted means for the particular effect that is being considered. Thus in an ancova for a randomized-complete-blocks designs, we could sum the sums of squares (and products) for treatments and discrepance (remainder) and compute an unexplained sum of squares for this level; we could also independently compute the unexplained sum of squares for the discrepance. We could then obtain the difference between these two unexplained sums of squares. This difference would be the appropriate sum of squares for testing the differences among the adjusted treatment means. Knowledge of this particular relationship should make it possible to construct and test balanced ancova designs similar to the ones encountered in the analyses of variance in earlier chapters.

14.10 Linear Comparisons in Anovas

In previous sections we noted the close relationship between regression and analysis of variance. The closeness of this relationship will become even more obvious in the present section. Our discussion here relates to planned tests in analysis of variance, which were introduced in Section 9.6. Recall that we were able to subdivide the sum of squares and the degrees of freedom among groups into separate sums of squares, each based on a single degree of freedom and furnishing certain desired information about the overall differences in the analysis of variance. The sum of squares among groups (treatments) could be decomposed into a set of orthogonal single-degree-of-freedom comparisons—that is, their *SS* and *df* were independent of each other and additive (added up to the *SS* and *df* among groups). An important idea put forth by R. A. Fisher is that all degrees of freedom and sums of squares in an anova eventually can be partitioned into single degrees of freedom with corresponding sums of squares that can be identified with certain contrasts. Not all of these contrasts will be scientifically meaningful, and it is generally not practical to carry out so complete a decomposition of the sums of squares; yet the general idea has considerable attraction and, as we will see in this section and in Section 16.6, in many cases we can probe deeply into the nature of the treatment and interaction sums of squares by means of individual-degree-of-freedom comparisons.

We will now provide a more general method, related to regression, for performing such individual-degree-of-freedom comparisons, at the same time laying the groundwork for the study of orthogonal polynomials (see Section 16.6). If the subject matter of Section 9.6 has by now become vague, we urge you to reread it before proceeding.

In Section 14.5 (replicated Y's for each value of X) we studied a single, classification anova in which survival of *Tribolium* beetles was regressed on density. The density classes were the groups of the anova, but at the same time these classes had numerical values of their own. Thus one density was 5/g, another 20/g, and so forth. Quite natural, therefore, we considered density an independent variable X, and we regressed the survival values Y on density X. However, we can also carry out a regression of response variable Y in an anova in which the groups are not expressly quantified. We do this by employing dummy variables to symbolize the groups and then regressing the response means $\bar{Y}_i$ on these dummy variables. This sounds like a futile exercise, but establishing a significant regression of Y on a carefully chosen dummy variable can yield important analytical insights. An example will clarify this point.

Refer to the pea section data that was introduced in Box 9.4. Table 9.1 featured the means for these data together with coefficients of linear comparison for two orthogonal contrasts, sugars versus control and mixed versus pure sugars. We will employ these coefficients as dummy variables. The significance tests for these contrasts, which in Section 9.6 were carried out as conventional analyses of variance, are now carried out as regressions on these dummy variables. Instead of the usual formula for the explained sum of squares, $(\Sigma xy)^2/\Sigma x^2$, we can write an analogous formula for an explained *SS* for regression of $\bar{Y}$ on the dummy variable c

$$\frac{\left(\sum c\bar{y}\right)^2}{\sum^{a} c^2}$$

omitting the subscripts from c_{ij} for simplicity. Note that the dummy variable c is already written in lower case; it is a deviate as well as a variate because its sum and mean are zero. Since $\Sigma^a c = 0$, $(\Sigma^a c\bar{y})^2/\Sigma^a c^2$ can be written as

$$\frac{\left[\sum^{a} c\,(\bar{Y} - \bar{\bar{Y}})\right]^2}{\sum^{a} c^2} = \frac{\left[\sum^{a} c\bar{Y} - \bar{\bar{Y}}\sum^{a} c\right]^2}{\sum^{a} c^2} = \frac{\left(\sum^{a} c\bar{Y}\right)^2}{\sum^{a} c^2}$$

The final step is to multiply the sum of squares by n to make it equivalent to a sum of squares of *groups,* not of *means,* thereby making it compatible with the rest of the anova to which it belongs. Therefore the final formula for the explained SS based on a linear comparison is

$$\frac{n\left(\sum^{a} c\bar{Y}\right)^2}{\sum^{a} c^2} \tag{14.16}$$

Now we apply Expression (14.16) to the two planned comparisons of Table 9.1. We compute

$$\begin{aligned} &10[(4 \times 70.1) + (-1 \times 59.3) + (-1 \times 58.2) + (-1 \times 58.0) \\ &\quad + (-1 \times 64.1)]^2 \\ &\quad \div [4^2 + (-1)^2 + (-1)^2 + (-1)^2 + (-1)^2] \\ &\quad = \frac{10\,[40.8]^2}{20} = 832.32 \end{aligned}$$

and

$$\begin{aligned} &10[(0 \times 70.1) + (-1 \times 59.3) + (-1 \times 58.2) + (3 \times 58.0) \\ &\quad + (-1 \times 64.1)]^2 \\ &\quad \div [0^2 + (-1)^2 + (-1)^2 + 3^2 + (-1)^2] \\ &\quad = \frac{10\,[-7.6]^2}{12} = 48.13 \end{aligned}$$

If you check back, you will find that these sums of squares for the difference between treatments and the control and between the pure and mixed sugars are the same as were found in Section 9.6. These values are incorporated into the anova as was shown there and prove to be significant.

Why do we go to all this trouble when we were able to carry out such an analysis by the more conventional method of obtaining sums of squares learned in Section 9.6? Although explaining how to evaluate a sum of squares from a linear comparison of treatments as shown here took some time, the setup of the data and the computation are generally simpler than the corresponding techniques of Box 9.8, once the idea has been mastered. A major advantage of the new technique is that it permits a comprehensive view of all comparisons through the table of coefficients, which often helps avoid inconsistencies. Furthermore, the coefficients enable us to test the orthogonality of the comparisons, as illustrated in Section 9.6. Finally, understanding the use of these coefficients will facilitate part of our study of curvilinear regression in Chapter 16.

Box 14.10 shows some examples of individual-degree-of-freedom comparisons that have been completely worked out. Let us emphasize again that these are all planned comparisons. The first example, the per diem fecundity in three selected lines of drosophila, comes from Box 9.8. First we test selected lines against nonselected lines. The coefficients are 1 for both RS and SS because there is one line in the contrasted set. The NS line however, has a coefficient of 2, since there are two lines, RS and SS, in its contrasted set. We arbitrarily assign a plus sign to one of the contrasted sets and a minus sign to the other. In testing resistant lines against susceptible lines, we set RS = +1 and SS = −1 and assign a coefficient of zero to the nonselected line, which is not involved in this test. The computations are straightforward and are outlined in Box 14.10. The results are the same as in Box 9.8. We find a significant difference in fecundity due to selection, but no difference in fecundity between the two selected lines.

The next analysis in Box 14.10 is a more extensive study in which the number of abdominal bristles in eight strains of houseflies differing in DDT resistance is analyzed. The strains are arranged in four groups according to their degree of resistance to the toxicant. This study was undertaken in the hope of finding a morphological correlate for DDT resistance, such as bristle number, that could be used (by inverse prediction) to indicate the resistance status of a housefly population. This example is a nested analysis of variance in which differences among jars within strains, as well as differences among strains were tested. The overall analysis of variance shows, however, that only differences among strains are highly significant; the number of bristles was not affected by environmental variations in the culture medium among different jars. Following the rules in Box 10.3, we do not pool the two mean squares but test the linear comparisons over the mean square of jars within strains. We now wish to partition the *SS* among strains based on seven degrees of freedom into separate single-degree-of-freedom comparisons to learn more about the differences in this study. For example, the first comparison suggested by the design of the study is resistant versus nonresistant strains—that is, groups 1 and 2 versus groups 3 and 4. The coefficients for this comparison are 5 for members of groups 1 and 2 because there are

Box 14.10 PLANNED INDIVIDUAL-DEGREE-OF-FREEDOM COMPARISONS OF MEANS IN ANALYSIS OF VARIANCE.

Per diem fecundity in *Drosophila melanogaster*. Data from anova of Box 9.8 (equal n): $a = 3$; $n = 25$.

		Resistant (RS) Line	Susceptible (SS) Line	Nonselected (NS) Line	(1)	(2)	(3)
Comparisons	$\bar{Y}$	25.256	23.628	33.372	$\sum^{a} c\bar{Y}$	$\sum c^2$	$\sum \hat{y}^2 = n(\mathbf{1})^2/\mathbf{2}$
Selected vs. nonselected	c_{i1}	+1	+1	−2	−17.860	6	1329.08
Resistant vs. susceptible	c_{i2}	+1	−1	0	1.628	2	33.13

Anova table

Source of variation	*df*	*SS*	*MS*
Strains	2	1362.21	681.10***
Selected vs. nonselected	1	1329.08	1329.08***
Resistant vs. susceptible	1	33.13	33.13 *ns*
Error	72	5659.02	78.60

Computation

See text for method of assigning correct coefficients of linear comparison.

1. $\sum^{a} c\bar{Y} = (1 \times 25.256) + (1 \times 23.628) + (-2 \times 33.372) = -17.860$ and $(1 \times 25.256) + (-1 \times 23.628) = 1.628$

Box 14.10 Continued

2. $\sum c^2 = (1)^2 + (1)^2 + (-2)^2 = 6$ and $(1)^2 + (-1)^2 = 2$

3. $\sum \hat{y}^2 = n(\text{quantity } \mathbf{1})^2/\text{quantity } \mathbf{2} = 25(-17.860)^2/6 = 1329.08$ and $25(1.628)^2/2 = 33.13$

Partitioning degrees of freedom in a larger example (degrees of freedom for strains not completely subdivided)

Abdominal bristle number in housefly strains. Strains are subdivided on the basis of resistance to DDT: $a = 8$ strains; $n = 24$ flies per strain.

		Strains										
		Group 1 Strongly resistant	Group 2 Slightly resistant		Group 3 Normal			Group 4 Susceptible		(1)	(2)	(3)
Meaningful comparisons		OL	LDD	RKS	RH	LC	BS	NKS	NH	$\sum^{a} c\bar{Y}$	$\sum c^2$	$\sum \hat{y}^2 = n(\mathbf{1})^2/(\mathbf{2})$
	$\bar{Y}$	34.2500	27.1250	32.9167	29.3750	27.4583	27.0000	30.50000	27.5417			
Resistant vs. nonresistant strains	c_{i1}	+5	+5	+5	−3	−3	−3	−3	−3	45.8335	120	420.14
Group 1 vs. group 2	c_{i2}	+2	−1	−1	0	0	0	0	0	8.4583	6	286.17
LDD vs. RKS (within group 2)	c_{i3}	0	+1	−1	0	0	0	0	0	−5.7917	2	402.53
Group 3 vs. group 4	c_{i4}	0	0	0	+2	+2	+2	−3	−3	−6.4585	30	33.37
Among strains of group 3	(*SS* for 2 *df*; compute in conventional manner as shown below)											
NKS vs. NH (within group 4)	c_{i5}	0	0	0	0	0	0	+1	−1	2.9583	2	105.02

Among strains of group 3 $SS = \sum^{k} n_i(\bar{Y}_i - \bar{\bar{Y}})^2 = 24(29.3750 - 27.9444)^2 + 24(27.4583 - 27.9444)^2$
$+ 24(27.0000 - 27.9444)^2 = 76.20$ (MS significant at $P < 0.05$)
Note that 27.9444 is the mean of group 3.

SOURCE: Data from Sokal and Hunter (1955).

Anova table

Source of variation	df	SS	MS
Strains	7	1323.42	189.06***
Resistant vs. nonresistant	1	420.14	420.14***
Group 1 vs. group 2	1	286.17	286.17***
LLD vs. RKS (within group 2)	1	402.52	402.53***
Group 3 vs. group 4	1	33.37	33.37 *ns*
Among strains of group 3	2	76.20	38.10 *ns*
NKS vs. NH (within group 4)	1	105.02	105.02 *ns*
Jars within strains	16	357.25	22.33 *ns*
Within jars	168	4663.25	27.76

Strains and all linear contrasts are tested over the mean square for jars within strains. Although this MS is not significant, following the rules of Box 10.3 we decided not to pool it with the mean square within jars.

Individual-degree-of-freedom comparisons in an anova with unequal n

Length of larval period (in hours) in lines of *Drosophila melanogaster* selected for short (SL) and long (LL) larval period compared with the control strain (CS).

BOX 14.10 CONTINUED

Comparisons		SL	CS	LL	(1) $\sum^{a} n_i c_i \bar{Y}_i$	(2) $\sum^{a} n_i c_i^2$	(3) $\sum \hat{y}^2 = \frac{(\mathbf{1})^2}{(\mathbf{2})}$
	n_i	80	69	33			
	$\bar{Y}_i$	100.9750	105.6667	110.3030			
Effect of selection (shortest vs. longest; linear)	Unweighted c_{i1}	(−1	0	+1)			
	Weighted $n_i c_{i1}$	−33	0	+80	24,625.9200	298,320	2032.84
Symmetry of selection (ends vs. middle; quadratic)	Unweighted c_{i2}	(−1	+2	−1)			
	Weighted $n_i c_{i2}$	−69	+113	−69	15,341.3289	1,419,054	165.85

Source: Data from Hunter (1959).

See text for assignment of correct coefficients of linear comparison. Computation as before except that in step **3** it is not necessary to multiply (quantity **1**)^2 by n because quantity **1,** $\Sigma^a n_i c_i \bar{Y}_i$, is already weighted by n_i.

Anova table

Source of variation	*df*	*SS*	*MS*
Among lines	2	2198.70	1099.35***
Effect of selection	1	2032.84	2034.84***
Symmetry of selection	1	165.85	165.85*
Error	179	7055.25	39.41

five strains comprising the contrasted set, groups 3 and 4; and, conversely, 3 for members of groups 3 and 4 because there are three strains in groups 1 and 2. Similar considerations govern the choice of coefficients for the other comparisons. The computations at the right side of the table in Box 14.10, which are no different from the previous calculations, are not discussed in detail.

We find that resistant flies differ from nonresistant ones in bristle number, that the very resistant group 1 differs from the slightly resistant group 2, and that group 2 is heterogeneous, since strain LDD is significantly different from strain RKS. The difference between group 3 and group 4 is not significant. The investigator had no logical basis for designing contrasts within group 3, consisting of three strains with two degrees of freedom among them. He therefore calculated the sum of squares pertaining to the two degrees of freedom among the three strains in the conventional manner, as shown at the bottom of the table in Box 14.10. The mean square among the three strains of group 3 is not significant. Finally, we find no difference between the two strains of group 4.

We may conclude that selection for resistance has affected abdominal bristle number variously in different strains, apparently raising it the most in the highly resistant strain, whereas among the nonresistant strains there are no differences for this variable. We might wish to consider one further contrast, between groups 2 and 3, which cannot be orthogonal to the previous contrasts, since they already used up the available 7 degrees of freedom. Because the data have already been divided into groups 1 and 2 versus groups 3 and 4, a contrast between groups 2 and 3 can no longer be orthogonal. We can easily convince ourselves of this fact by accumulating the products of the coefficients of the first row with those for the new contrast (let us call them c_{i6}): $\Sigma\, c_{i1}c_{i6} = (5 \times 0) + (5 \times 3) + (5 \times 3) + (-3 \times -2) + (-3 \times -2) + (-3 \times -2) + (-3 \times 0) + (-3 \times 0) = 48$. Since these products do not sum to zero, they are not orthogonal. A test of a set of such nonorthogonal contrasts can be carried out by the procedures described in Box 9.12.

The last example in Box 14.10 shows the minor complications that arise when sample sizes are unequal. The data are length of larval period in three strains of *Drosophila melanogaster,* one selected for short larval period (SL), one for long larval period (LL), and a control strain (CS). The sample sizes and sums are given in the first table. The simplest way of carrying out the computations for such a case is first to write down the "unweighted" coefficients as if the sample sizes were the same to give us a general idea of the contrasts we would like to compute. Note that the first contrast, strain SL versus strain LL, tests the overall effect of selection. If the strains selected for short and long larval period were not different, selection clearly would have been ineffective. The second comparison tests the symmetry of the results of selection. If selection for short larval period was as effective as selection for long larval period, then the control (unselected) strain should be exactly halfway between the two. This symmetry can be tested by comparing the control strain with the average of the short and long strains. Having established the contrasts, we now replace the unweighted coefficients by others weighted according to sample sizes. The weighted coefficient for each member of one set is the *total sample size* of the contrasted set. Thus, for the first comparison, the weighted coefficient for SL is the sample size of the contrasted LL; conversely, for the LL it is the sample size of the contrasted SL. In the second comparison, the weighted coefficient for both SL and LL is 69, the sample size of the contrasted CS, and the weighted coefficient for CS is the sum of the sample sizes of the contrasted lines, SL and LL, $80 + 33 = 113$. The signs of these coefficients are the same as those of the unweighted coefficients. The rest of the computations are as before, except that in step 3 we need not multiply the numerator by n because the coefficients are already weighted by sample size. The resulting anova table shows that there is a clear difference in length of larval period between the strains selected in opposite directions for this variable, and also that their distance from the nonselected control strain is not symmetrical.

We can estimate multiple confidence limits for means and contrasts by means of coefficients of linear comparison. Such a set of confidence limits is computed with an experimentwise error rate α so that all k confidence intervals that the investigator intends to construct contain the parameter with an overall probability of $1 - \alpha$. Note that these are confidence limits. They are thus not intended for significance tests based on testing overlaps between intervals, as are the comparison limits of Section 9.7. To set these multiple confidence limits for contrasts, employ Expression (14.17) for equal sample sizes:

$$\sum_{i=1}^{a} c_i \bar{Y}_i \pm t_{\alpha'[\nu]} \sqrt{\left(\frac{1}{n} \sum_{i=1}^{a} c_i^2\right) MS_{\text{within}}} \tag{14.17}$$

and Expression (14.18) for unequal sample sizes:

$$\sum_{i=1}^{a} c_i \bar{Y}_i \pm t_{\alpha'[\nu]} \sqrt{\left(\sum_{i=1}^{a} \frac{c_i^2}{n_i}\right) MS_{\text{within}}} \tag{14.18}$$

Multiple confidence limits for means are set as

$$\bar{Y}_i \pm t_{\alpha'[\nu]} \sqrt{\frac{MS_{\text{within}}}{n_i}} \tag{14.18m}$$

For the special case of the contrast between two means based on equal sample sizes Expression (14.17) simplifies to

$$(\bar{Y}_1 - \bar{Y}_2) \pm t_{\alpha'[a(n-1)]} \sqrt{\frac{2}{n} MS_{\text{within}}} \tag{14.19}$$

In all four cases $\alpha' = 1 - (1 - \alpha)^{1/k}$ where k is the number of contrasts intended. We do not have to compute α' but can look up $t_{\alpha'}$ directly in Table **C** as $t_{\alpha[k,\nu]}$.

Note that the length of the confidence interval differs for different contrasts and sample sizes. As an example, let us set confidence limits to the difference between the selected and nonselected lines (the first contrast in the first example of Box 14.10). We assume that we had planned to set two confidence intervals at a confidence level of 95%. From Statistical Table **C** we find that $t_{.05[2,72]} = 2.284$. Following Expression (14.17), we compute

$$[(+1)(25.256) + (+1)(23.628) + (-2)(33.372)] \pm 2.284 \sqrt{\left[\frac{1}{25}(1^2 + 1^2 + (-2)^2)\right] 78.60} = -17.860 \pm 9.920$$

14.11 Examining Residuals and Transformations in Regression

In Section 14.5 we stressed that the nature of the departure of points from linear regression gives us a clue to interpeting the goodness of fit. The modern tendency has been to carry out some examination of residuals in addition to the significance tests that we have already learned. Such an examination may detect outliers in a sample. Removal of such outliers may improve the regression fit considerably. One might also detect systematic departures from regression that can be adjusted by transformation of scale or by fitting a curvilinear regression line.

Two common procedures are used for such an examination. The first is to compute **leverage coefficients** (also known as the diagonal elements of the hat matrix; Hoaglin and Welsch, 1978) to match each dependent variate Y_i. These coefficients are computed as

$$h_i = \frac{1}{n} + \frac{(X_i - \bar{X})^2}{\sum x^2}$$

for $i = 1 \ldots n$. The values of h_i, which range from near 0 to 1, give an indication of the *leverage* of a given value of X_i—that is, the influence of the corresponding response value Y_i on the estimated value $\hat{Y}_i$. Note that the quantities h_i are the same quantities as those in square brackets in the formula for $s_{\hat{Y}}$ in Box 14.2. They are functions of the magnitude of the departure for any given value X_i for its mean; thus values of Y_i corresponding to deviant variates X_i (deviant from their mean $\bar{X}$, that is) are given greater weight in determining the fit of the line. For more than one value of Y per X, a single h_i is computed per X_i, and $1/\Sigma n_i$ replaces $1/n$ in the formula for h_i.

Examining the h_i values, however, is not sufficient. The second procedure is to compute and plot the **standardized residuals** from the regression line, which are the unexplained deviations $d_{Y \cdot X}$ divided by $s_{Y \cdot X}(1 - h_i)^{1/2}$, in order to put them into a standard deviation scale. Both sets of quantities have been computed for the weight loss data of Box 14.1 and are shown in Figure 14.18. Although several of the residuals seem suspiciously large, suggesting that a parabola might have fit the data better, by the criteria that we will describe shortly none of the h_i

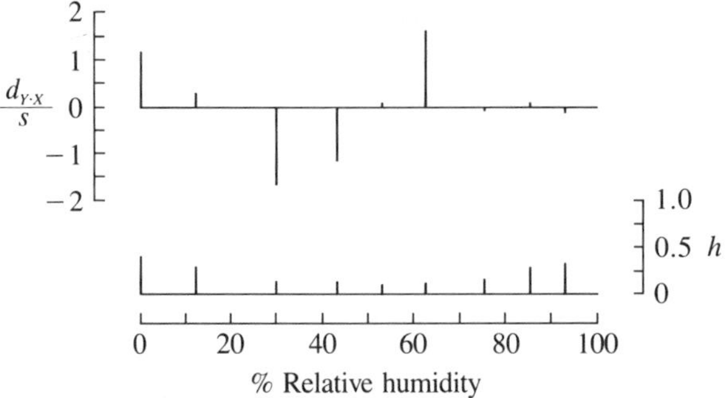

FIGURE 14.18 *Examination of residuals in regression. Weight loss data of Box 14.1 and Figure 14.8. The ordinate of the upper graph represents deviation from regression in standard deviation scale, that of the lower graph* h, *the leverage coefficients (diagonal elements of the hat matrix).*

or $d_{Y \cdot X}$ are high or outliers, respectively. In examining such a figure we look first at high values of h_i, which indicate points with potential high leverage. If such points also show high residuals, they may be affecting the slope of the line unduly. High values of h_i with low residuals are not a problem; they simply indicate consistency of the observations with the regression model. Large residuals for a point with a low leverage value are relatively unimportant, since they do not influence the regression line very much. When outliers are detected by this method, the regression may be recomputed after the outliers have been omitted. Hoaglin and Welsch (1978) recommend that values of h_i greater than $4/n$ be considered high. The standardized residuals can be tested by comparing them to $t_{\alpha[n-2]}$, although the nominal probability levels in such an exhaustive data analysis are only suggestive. Hoaglin and Welsch also refine the computation of the standardized residual for the ith observation by recomputing $s_{Y \cdot X}$, leaving that observation out of the computation, so that the observation being tested does not contribute to the unexplained error estimate. In practice, such a strategy is carried out only by computer.

In transforming either or both variables in regression, we aim at achieving a normal and homoscedastic distribution of points around the regression line. As a by-product of such a procedure we simplify a curvilinear relationship to a linear one, thereby usually increasing the proportion of the variance of the dependent variable explained by the independent variable. Rather than fit a curvilinear regression to points plotted on an arithmetic scale (see Section 16.6, it may be more expedient to compute a simple linear regression for variates plotted on a transformed scale. A general test of whether transformation will improve linear regression is to graph the points to be fitted on an arithmetic scale as well as on the scale suspected to improve the relationship. These operations can be carried out either by computer graphics or with suitably scaled graph paper. If the function straightens out and the systematic deviation of points around a visually fitted line is reduced, the transformation is worthwhile. If a suitable transformation cannot be found, methods of nonlinear least squares analysis may have to be employed. Such procedures are beyond the scope of this book. See Seber and Wild (1989) for a recent account. Programs for such analyses are found in Dixon (1985).

We will discuss briefly a few of the transformations common in regression analysis. Square root and arcsine transformations (see Sections 13.8 and 13.10, respectively) are not mentioned here, but they are also effective in regression cases involving data suited to such transformations (as in Box 14.7).

The **logarithmic transformation** is the most frequently used. Usually we transform the dependent variable Y. This transformation is indicated when percentage changes in the dependent variable vary directly with changes in the independent variable. Such a relationship is represented by the equation $\hat{Y} = ae^{bX}$, where a and b are constants and e is the base of the natural logarithm. Transformation results in the equation $\log \hat{Y} = \log a + b(\log e)X$. In this expression $\log e$ is a constant that, when multiplied by b, yields a new constant factor b', which is a regression coefficient in the new scale. Similarly, $\log a$ is a new Y-intercept, a'. We can then simply regress $\log Y$ on X to obtain the function $\log \hat{Y} = a' + b'X$ and obtain all our prediction equations and confidence intervals in this form. Figure 14.19 shows an example of transforming the dependent variate to logarithmic form, which results in considerable straightening of the response curve.

A logarithmic transformation of the independent variable X in regression is effective when proportional changes in the independent variable produce linear responses in the dependent variable. An example is the decline in weight of an organism as density increases, where the successive increases in density need to be in a constant ratio to effect equal decreases in weight. This example belongs to a well-known class of biological phenomena, another example of which is the Weber–Fechner law in physiology and psychology, which states that a stimulus has to be increased by a constant proportion to produce a constant increment in response. Figure 14.20 illustrates how logarithmic transformation of the independent variable results in the straightening of the regression line. For computation we transform X into logarithms.

Logarithmic transformation for both variables applies to situations in which the true relationship can be described by the formula $\hat{Y} = aX^b$. The curve described by this relationship becomes straight when both variables are transformed to the logarithmic scale, as Figure 14.21 shows. The regression equation is $\widehat{\log Y} = \log a + b \log X$, and the computation is carried out in the conventional manner. This equation is the well-known **allometric growth curve,** applicable in many organisms where the ratio between increments in structures of different size remains roughly constant, yielding a relatively great increase of

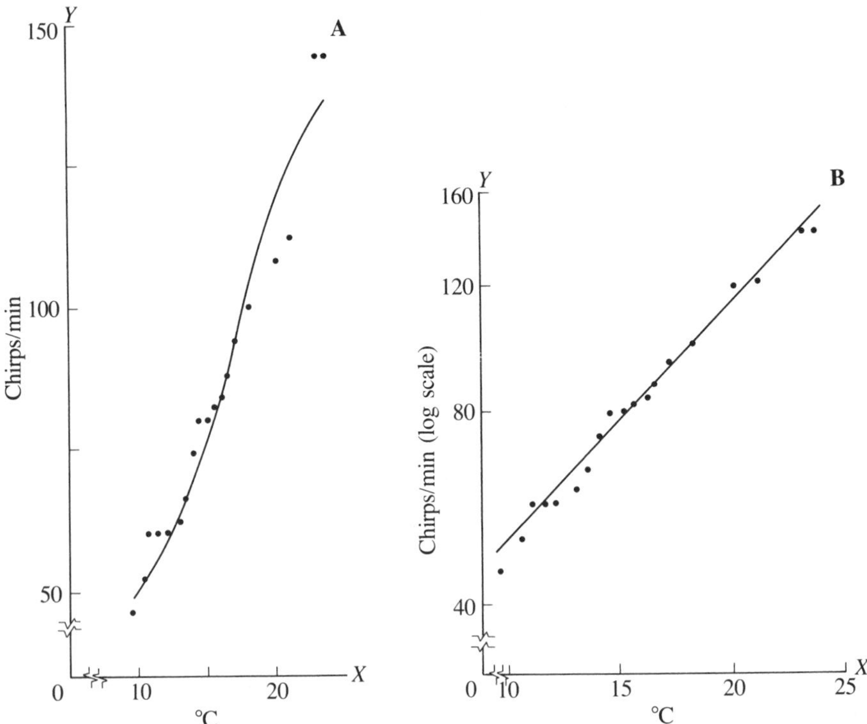

FIGURE 14.19 *Logarithmic transformation of a dependent variable in regression. Chirp rate as a function of temperature in males of the tree cricket* Oecanthus fultoni. *Each point represents the mean chirp rate per minute for all group observations at a given temperature in °C.* ***A.*** *Original data.* ***B.*** Y*'s plotted on logarithmic scale. (Data from Block, 1966.)*

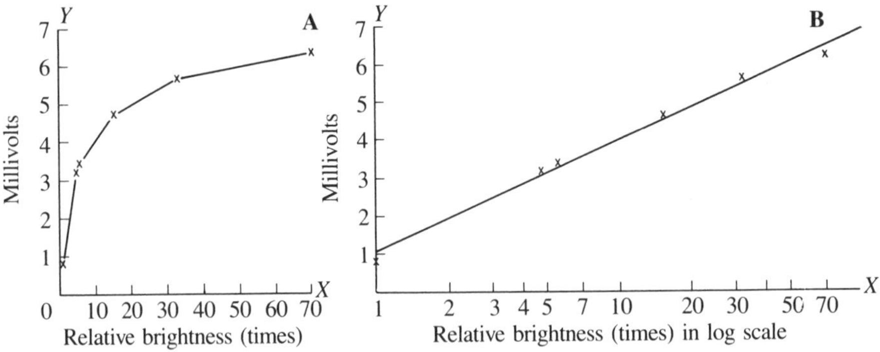

FIGURE 14.20 *Logarithmic transformation of the independent variable in regression. Electrical response as a function of illumination in cephalopod eyes.* ***A.*** *Original data.* ***B.*** Y*'s plotted on logarithmic scale. A proportional increase in* X *(relative brightness) produces a linear electrical response* Y. *(Data from Fröhlich, 1921.)*

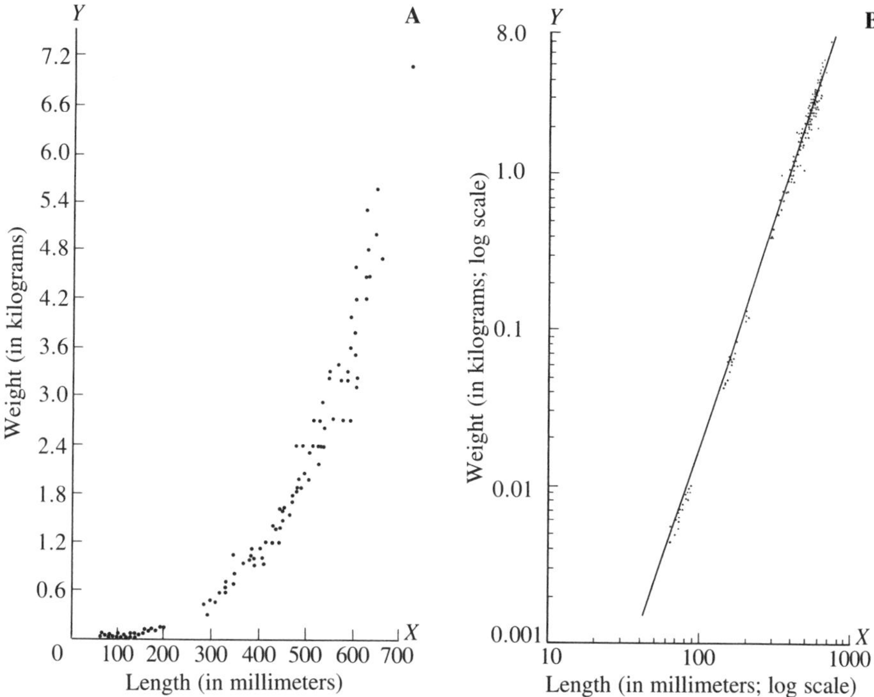

FIGURE 14.21 *Logarithmic transformation of both variables in regression. Weight versus length in the cabezon* Scorpaenichthys marmoratus. ***A.*** *Original data.* ***B.*** X*'s and* Y*'s plotted on logarithmic scale. (Data from O'Connell, 1953.)*

one variable with respect to the other on a linear scale. Examples are the greatly disproportionate growth of organs in some organisms, such as the antlers of deer or horns of stage beetles, with respect to their general body sizes. Note that the symbolism we use here is the converse of the conventional symbolism for allometric growth because we think it is important to retain the symbol a for Y-intercept and b for the regression coefficient. For comparison of structures subject to allometric growth, it is best to compare adjusted and transformed means after the allometric organ has been regressed on some measure of general body size. The analysis of allometric growth relationships is a subject of considerable depth that we cannot pursue here. Simpson et al. (1960) present an elementary discussion of this topic; Teissier (1960) and Gould (1966, 1975) give more advanced reviews of the field.

With models that are more complicated one may transform the variables, as well as rearrange the equation slightly to put it into a form amenable to ordinary linear regression analysis. For example, studies of population growth sometimes involve a slightly more complicated functional relationship that requires a double logarithmic transformation. The Ricker function $\hat{Y} = aXe^{-bX}$ (Ricker, 1954) relates the expected number of recruits Y into a population to the size of the parental population X. Taking logarithms of both sides of this equation results in $\ln \hat{Y} = \ln a + \ln X - bX$. If we move $\ln X$ to the left side, we can rewrite the equation as $\ln \hat{Y} - \ln X = \ln a - bX$. Since in regression analysis we assume that the independent variable is measured without error, we may consider regressing a new artificial variable $Y' = \ln Y - \ln X$ on the independent variable X, using simple linear regression techniques to estimate the slope, $-b$, and the intercept, $\ln a$. A complication sometimes overlooked is that the proportion of the variance of Y' as explained by regression on X should not be considered a measure of fit of this model. Because Y' is simply the difference between $\ln Y$ and $\ln X$, it is expected to be rather highly correlated with X even if there is no relationship between Y and X. To investigate the degree of fit of this model, one must see how well the predicted values $\widehat{\ln Y}$ match the logarithms of the observed value, $\ln Y$.

Many rate phenomena (a given performance per unit of time or per unit of population), such as wing beats per second or number of eggs laid per female, yield hyperbolic curves when plotted in original measurement scale. Thus, they form curves described by the general mathematical equations $bXY = 1$ or $(a + bX)Y = 1$. From these we can derive $1/Y = bX$ or $1/Y = a + bX$. Transforming the dependent variable into its reciprocal frequently results in a straight-line regression. The **reciprocal transformation** is illustrated in Figure 14.22.

We may also apply the Box–Cox transformation to either the dependent or the independent variables or to both of them simultaneously. In these cases the

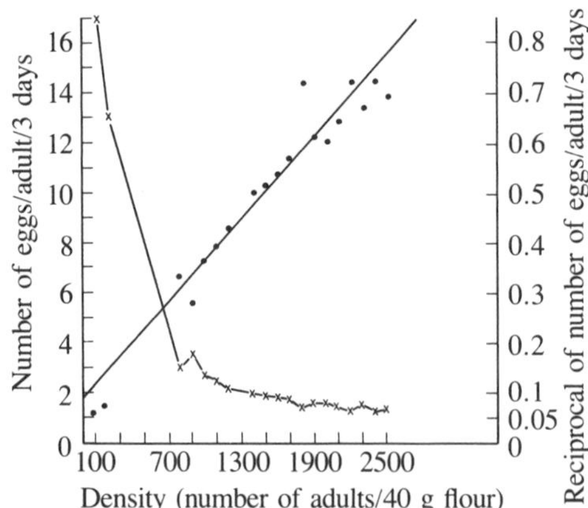

FIGURE 14.22 *The reciprocal transformation applied to the dependent variable. Fecundity in* Tribolium *beetles expressed as number of eggs per adult per three days graphed as a function of density. Transforming the dependent variable into reciprocal scale changes the relationship from a hyperbolic to a linear one. X-marks are plots in original scale, dots in reciprocal scale. (Data from R. R. Sokal, unpublished results.)*

log-likelihood function to be maximized is as given earlier (see Section 13.9). In the present case, s_T^2 is the variance of the deviations from regression, $s_{Y \cdot X}^2$, from the regression analysis carried out on the transformed data. Estimating $\hat{\lambda}$ and setting confidence limits are as described earlier. If we have more than one Y for each X and we are concerned with heteroscedasticity, we can maximize L' using Expression (13.4).

Some cumulative curves can be straightened by the **probit transformation.** Refresh your memory on the cumulative normal curve shown in Figure 6.5 and discussed in Section 6.7. Remember that by changing the ordinate into normal equivalent deviates (NEDs), we were able to make the cumulative distribution function of a normally distributed variable linear. **Probits** are simply normal equivalent deviates coded by the addition of 5.0, which avoids negative values for most deviates. The probit value 5.0 corresponds to a cumulative frequency of 50%, probit value 6.0 corresponds to a cumulative frequency of 84.13%, and probit value 3.0 corresponds to a cumulative frequency of 2.27%. Probit tables, giving the probit equivalents of cumulative percentages, are available in Fisher and Yates (1963) and Pearson and Hartley (1958).

Figure 14.23 shows mortality percentages for increasing doses of an insecticide. These percentages represent differing points of a cumulative frequency distribution. With increasing dosages an ever greater proportion of the sample dies, until at a high enough dose the entire sample is killed. If the doses of toxicants are transformed into logarithms, often the tolerances of many organisms to these poisons are distributed approximately normally. These transformed doses are often called **dosages**. Increasing dosages lead to a cumulative normal distribution of mortalities, often called **dosage—mortality curves.**

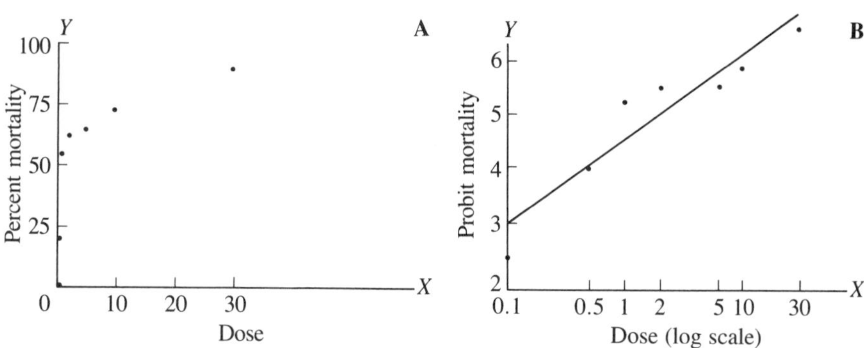

FIGURE 14.23 *Dosage-mortality data illustrating an application of the probit transformation. Data are mean mortalities for two replicates. Twenty* Drosophila melanogaster *per replicate were subjected to seven doses of an* "unknown" *insecticide in a class experiment.* ***A.*** *Original data.* ***B.*** *Probit transformation. The point at dose 0.1 that yielded 0% mortality has been assigned a probit value of 2.5 in lieu of $-\infty$, which cannot be plotted.*

These curves are the subject of an entire field of biometric analysis, **bioassay,** to which we can refer only in passing here. The most common technique in this field is **probit analysis**. Graphic approximations can be carried out on *probit paper,* which is probability graph paper in which the abscissa has been transformed into logarithmic scale. A regression line is fitted by eye to dosage–mortality data graphed on probit paper (see Figure 14.23). From the line fitted by eye the 50% kill is estimated by inverse prediction (see Section 14.7). Goulden (1952, chapter 18) and Goldstein (1964, latter half of chapter 4) present good basic discussions of probit analysis. For a discussion of greater depth, refer to Finney (1971).

Another cumulative distribution that superficially resembles the cumulative normal curve is the **logistic equation**. This distribution ranges from a proportion of zero to 1 and is especially suitable for proportions near 0 and 1. By transforming proportions p to **logits**—that is, $\ln(p/q)$, where $q = 1 - p$ —we can make the logistic function linear and compute regressions of data for proportions or occurrences. This method, which has become quite popular in recent years, may be the optimal technique for regressing proportions. Since they can be expressed as proportions, percentages can, of course, also be transformed to logits. Because this method is intimately related to the analysis of frequencies, we feature it in Chapter 17.

Finally, there is an iterative program that transforms both dependent and independent variables to yield regressions with minimal unexplained sums of squares. Known as the ACE algorithm (Breiman and Friedman, 1985), this program finds optimal nonlinear transformations. The user can restrict these to monotone transformations if so desired. The data may be continuous, ordered, or categorical, making this a very powerful transforming instrument. Since the ACE algorithm fits the dependent variable of a sample to its independent variable(s) empirically, it does not yield a general transformation formula, as do the other transformations we have considered. Adding to or subtracting from the data points in the sample would yield a different transformation. The ACE program is widely available. Its output can be fed to a graphics program, which reveals both the nature of the transformation and the improvement in the association of the transformed variables over their relation when untransformed.

When transforming variables, we should avoid the pitfall of comparing the goodness of fit of regressions based on untransformed and transformed variables by means of the coefficient of determination R^2. In Section 14.4 we defined this statistic only as $\Sigma\hat{y}^2/\Sigma y^2$, but Chapter 15 will give other formulations that are mathematically identical in the case of linear regressions with an intercept a. When the model is nonlinear or lacks a Y-intercept, however, these mathematical identities break down. The R^2 computed from transformed variables holds only for the linearized space resulting from the transformation and cannot be compared with an R^2 computed from the untransformed data. Kvålseth (1985) suggests computing R^2 as $1 - \Sigma(Y - \hat{Y}')^2/\Sigma(Y - \bar{Y}')$,^{2} where $\hat{Y}'$ is the backtransformed value of the estimated Y in the transformed space and $\bar{Y}'$ is the backtransformed mean. Kvålseth (1985) and Scott and Wild (1991) discuss these problems in some detail.

14.12 Nonparametric Tests for Regression

When transformations are unable to make the relationship between the dependent and independent variables linear, the investigator may wish to carry out a simpler nonparametric test in lieu of regression analysis. We show **Kendall's robust line-fit method** in Box 14.11 (Kendall and Gibbons, 1990, Section 13.4). We order the X and Y-values by the magnitude of X. Then we compute a slope S_{ji} for every pair of X-values, X_i and X_j, by the following formula:

$$S_{ji} = \frac{Y_j - Y_i}{X_j - X_i}$$

where i ranges from 1 to $n - 1$ and $j > i$. There will be $n(n - 1)/2$ such slope estimates S_{ji}. The nonparametric estimate b of the slope β is the median of the S_{ji}-values. To estimate the Y-intercept, compute the n values of $Y_i - bX_i$ and again choose the median. Using this approach, we find a regression equation of $Y = 8.78382 - 0.05436X$ for the weight loss/relative humidity data of Box 14.1. This equation is close to the estimate given in that box.

An appropriate significance test for b is the **ordering test** (Quenouille, 1952), which is equivalent to Kendall's rank correlation coefficient (see Box 15.7) and can be carried out most easily as such. According to this procedure we first rank variates X and Y. Then we arrange the independent variable X in increasing order of ranks and calculate the Kendall rank correlation of Y with X. The computational steps are shown in Box 15.6. If we carry out this computation for the weight loss data of Box 14.1 (reversing the order of percent relative humidity, X, which is negatively related to weight loss, Y), we obtain a quantity $N = 72$, which is significant at $P < 0.01$ (see Table **S**).

We conclude that there is a significant trend of weight loss as a function of relative humidity. The ranks of the weight losses are a perfect monotonic function on the ranks of the relative humidities. Consequently, b is significant.

Noether (1985) described a direct significance test for b obtained by Kendall's robust line-fit method, but since it too is related to Kendall's rank correlation, we limit ourselves to presenting the latter method. If we do not need either a prediction equation or a functional relationship, but simply want to know whether the dependent variable Y is a monotonically increasing (or decreasing) function of the independent variable X, we can simply carry out the rank correlation test. Such a test could also be applied to means in a single-classification analysis of variance that are assumed to represent a trend on X, such as the data in Box 14.4, where survival was considered a function of density. Note, however, that this

Box 14.11 KENDALL'S ROBUST LINE-FIT METHOD FOR NONPARAMETRIC REGRESSION.

Weight loss data of Box 14.1.

Weight loss in mg *(Y)*	8.98	8.14	6.67	6.08	5.90	5.83	4.68	4.20	3.72
Percent relative humidity *(X)*	0.0	12.0	29.5	43.0	53.0	62.5	75.5	85.0	93.0

The X-values are already arranged in ascending order of magnitude. If they had not been, then one should order them together with their paired Y-values.

Computation

1. Calculate $S_{ji} = (Y_j - Y_i)/(X_j - X_i)$ for all pairs X_i, X_j, where $j > i$.

$$S_{21} = (8.14 - 8.98)/(12.0 - 0.0) = -0.07000$$
$$S_{32} = (6.67 - 8.14)/(29.5 - 12.0) = -0.08400$$
$$\vdots$$
$$S_{31} = (6.67 - 8.98)/(29.5 - 0.0) = -0.07831$$
$$\vdots$$
$$S_{91} = (3.72 - 8.98)/(93.0 - 0.0) = -0.05656$$

The median of the $n(n - 1)/2$ slopes is $b = -0.05436$.

2. Estimate the Y-intercept: $Y_i - bX_i$ for $i = 1$ is 8.98000, for $i = 2$ is 8.79226, and so on

The median of the n intercepts is 8.78382.

3. The nonparametric regression equation is $Y = 8.78382 - 0.05436X$.

particular example cannot be tested by this method because there are only four densities and the minimum number of points required for significance testing by the rank correlation method is five.

When the data are presented as ranks in a randomized-complete-blocks design, such as the data analyzed by Friedman's test in Box 13.10, a monotonic trend among the treatment effects can be tested by **Page's L-test for ordered alternatives** (Page, 1963). This test can be performed easily using the layout of Box 13.10 for Friedman's method. Order the treatments in such a way as to produce a monotonically increasing response according to your alternative hypothesis (the null hypothesis, of course, is no trend). Then multiply the treatment rank sums $\Sigma^b R_{ij}$ by the rank order of the particular treatment. In the Rot Lake temperature data in Box 13.10, for example, the alternative hypothesis is that increases in depth produce decreases in temperature. It is advisable therefore to rank the greatest depths (15.5 m) as rank 1, with the other depths being ranked successively until rank 10, which is shallowest depth (0 m). Page's L-statistic is computed as

$$L = \sum_{i=1}^{a} i\left(\sum R_{ij}\right) \tag{14.20}$$

where i refers to the rank number of the treatment level. Applying this formula to the data in Box 13.10, we obtain $L = 1(4) + 2(8) + 3(12) + \cdots + 10(40) = 1540$. The critical value of L is given in Table **Z,** where a and b refer to the number of treatments and of blocks as before. The table furnishes critical values of L for a from 3 to 10 and for b from 2 to 24. For values beyond these arguments use the following expression:

$$L_\alpha = \frac{b(a^3 - a)}{12}\left[\frac{t_{(2\alpha)[\infty]}}{\sqrt{b(a-1)}} + \frac{3(a+1)}{a-1}\right] \tag{14.21}$$

Looking up L_α for these data, we obtain 1382 for $L_{.001}$. Since our observed L is much greater than this value, there is little doubt that there is a monotonic trend of temperature in relation to depth.

An alternative nonparametric test for trend in ordered variates is the runs-up-and-down test discussed in Section 18.2 and Box 18.3.

14.13 Model II Regression

In Model I regression, both estimation of functional relationships and prediction are carried out best by means of simple linear regression, but when both variables are subject to error, the appropriate method depends on the nature of the data. We therefore need to examine in some detail the possible types of situations subsumed under Model II regression.

The typical case involves two continuous variables distributed according to the bivariate normal distribution. This distribution will be discussed in Section 15.2. It will suffice for now to state that in a bivariate normal distribution both variables are normally distributed and the correlation or association between the two variables is determined by the magnitude of the parameter ρ (see Figures 15.1 and 15.2). Examples of such data include the following measurements: the length of both left and right wings in each individual of a sample of bees; the height of each person, as well as his score on an achievement test, in a sample of men; or the strength of muscle contraction and the naturally varying concentration of a particular cation in the blood of individuals in a sample of frogs. In all these cases, the variables vary naturally (individual differences are genetic and/or are environmentally caused), but there will likely be additional error due to measurement. The proportion of measurement error varies with the variable and depends on the circumstances of the study. Rarely do we encounter no measurement error. Fortunately, analysis of cases with error due only to inherent variability proceeds in the same manner as that of cases with error due to inherent variability *plus* measurement error, so we need not be concerned with measurement error as a distinct source of variation.

A second case of Model II regression arises when the sample does not follow a bivariate normal distribution for either or both of two reasons. First, approximately equal frequencies may be sampled along one of the axes. Thus, in a study relating chest width of men to their height, for example, we may set up classes for the height distribution and sample approximately equal frequencies of men for each height class in an attempt to have equal replication over the range of the observed data. Second, if the underlying parent population is not a bivariate normal distribution, the sample will reflect this departure from normality. Again the proportion of measurement error varies for different variables and studies. Many biological examples fall into this class.

A third case is quite rare in the biological sciences. Both variables lack natural variability and are subject to measurement error only. Most such instances come from the physical sciences; an example is measurements of the same piece of metal for conductivity at different uncontrolled temperatures in an attempt to establish a functional relation between these two variables. In such cases prediction could be carried out by simple linear regression, but the functional relationship would be biased (b would be lower in absolute value). Methods for estimating the functional relationship depend on whether the ratio between the measurement errors of Y and X is known. This problem is discussed by Mandel (1964, Section 12.5) and by Ricker (1973). Since its relevance to biometry is slight, we will not pursue it further here.

There is one special case of apparent Model II regression that permits us to apply Model I methods for tests of significance. This is the so-called *Berkson case,* in which independent variables are subject to error but are controlled by the experimenter. This situation occurs frequently in experimental work. You may have been uneasy earlier about applying Model I regression to cases in which different doses of a hormone or different densities are applied to organisms. How can we be sure that such independent variables are applied without error? Not only will we have error in administering the dose of hormone or reading the density, but surely the effective dose or density (the dose or density directly interacting with the organism) will not necessarily be the ones that we intend. Thus, there is an error δ that attaches itself to each of our intended independent variates X. We can now write $X = \zeta + \delta$, where X is the intended or nominal value of the independent variate, ζ is the actual or effective value, and δ is the error term making the difference between the intended and actual variates. In this model, X and δ are not expected to be correlated, since there is no reason to suppose, according to our model, that the magnitude of the intended variate and its error of application should be correlated. We may, therefore, use the ordinary Model I regression procedures to estimate slope without bias, and in such cases prediction and functional relationship can be obtained using simple linear regression. A readable discussion of these problems is found in Mandel (1964, Section 12.5).

Research on and controversy over Model II regression continues, and definitive recommendations are difficult to make. Much depends on the intentions of the investigator. If the regression line is being fitted mainly for purposes of prediction, then simple linear regression techniques (the Model I design already studied) are generally applied. Some further remarks on the distinction between prediction and functional relationship are, however, in order. Because in Model I regression the least squares simple regression equation $\hat{Y} = a + bX$ is appropriate for both purposes (prediction and determining functional relationship), investigators are tempted to use the line obtained in Model II cases interchangeably as well. If one has an equation relating Y to X, why not predict Y for a given value of X? Unbiased estimates of Y given X, and minimal-width confidence belts around the estimates of Y, however, are not generally given by functional equations that best describe the joint variation of two random variables. Thus investigators who wish to determine the mutual slope of two random variables and the Y-intercept of that slope (i.e., their functional relation), should not attempt to use the resulting equation to predict values of Y given X.

The reason that special techniques are necessary for Model II regression is that estimates of slopes of Y on X when both are subject to error are biased. Several approaches have been suggested to overcome the problem. When both variables are in the same units of measurement, determining the **slope of the major axis** (or *principal axis*) of the bivariate sample has been suggested. This method requires a knowledge of correlation techniques, so we postpone its presentation to Section 15.7, where it is employed to compute confidence ellipses for bivariate normal samples. We also illustrate in Section 15.7 how the confidence limits of the slope of the major axis can be computed. When the two variables have different units of measurement, the slope of the major axis is generally meaningless and another technique for Model II regression should be employed. When the units of measurement are predetermined and nonarbitrary, however, this criticism does not apply and the slope of the major axis might be a suitable measure of functional relationship among the variables. In morphometric work logarithmically transformed variables are often employed, and their functional relationship is estimated by the slope of the major axis.

Another way to overcome the scale dependence of the major axis is to standardize variables before the slope is computed—that is, to transform each of the two variables so that it has a mean of zero and a standard deviation of one. The principal axis of these standardized variables is known as the *reduced major axis* (Kermack and Haldane, 1950), *relation d'allométrie* (Teissier, 1948), **geometric mean (G M) regression** (Ricker, 1973), or *standard major axis* (Jolicoeur, 1975). The slope of this line is simply

$$v_{Y \cdot X} = \pm \sqrt{\frac{\sum y^2}{\sum x^2}} = \pm \frac{s_Y}{s_X} \qquad (14.22)$$

This slope can be shown to be the geometric mean of the linear regression coefficient of Y on X, and of the reciprocal of the regression coefficient of X on Y:

$$\pm \sqrt{b_{Y \cdot X} \frac{1}{b_{X \cdot Y}}} = \pm \sqrt{\frac{\sum xy}{\sum x^2} \cdot \frac{\sum y^2}{\sum xy}} = \pm \sqrt{\frac{\sum y^2}{\sum x^2}} = \pm \frac{s_y}{s_x}$$

which explains the name *geometric mean regression,* coined by Ricker (1973).

Note that $v_{Y \cdot X}$, unlike $b_{Y \cdot X}$, is independent of any joint function of X and Y. It is merely the ratio of two standard deviations. This property of the reduced major axis and other aspects of its behavior have led to serious criticisms of the method by Jolicoeur (1975) and Kuhry and Marcus (1977). The *GM* regression of X on Y, $v_{X \cdot Y}$, is obviously the same line, but because the axes are reversed, its slope is $1/v_{Y \cdot X}$. The *GM* regression line is computed as in linear regression, by forcing the line to pass through the bivariate mean $\bar{X}$, $\bar{Y}$. Thus the Y-intercept a_v can be obtained from $\bar{Y} - v\bar{X}$ by analogy with the formula for a. Note that the sign of v cannot be obtained from its formula; it is the sign of the sum of products Σxy. When the sum of products equals zero, there is no association between X and Y, the sign of v is undefined, and there is little point in fitting a regression line. The slope of the *GM* regression line will always be greater than that of the usual regression line, since $v_{Y \cdot X} = b_{Y \cdot X}/r_{XY}$. (See Chapter 15 for r_{XY}.)

The standard error of the *GM* regression slope, s_v, can be approximated by s_b, the standard error of the linear regression slope, and can be used to set confidence limits to v in the same way as for b. There is one difference, however. Whereas a test of the null hypothesis $H_0\!: \beta = 0$ was appropriate for linear regression and tested for the existence of any association between Y and X, a similar test is inappropriate for v, since this is a ratio of the two standard deviations and will not be zero or undefined unless one of the standard deviations equals zero (an unsuitable case for a regression problem).

A third approach to estimating functional relationship in Model II is to estimate the slope from segments of the sample arrayed by magnitude of the X variate. By the method illustrated here, **Bartlett's three-group method,** the array of pairs is divided into three equal groups, if possible, or at least into two equally sized end groups. Details of the computation follow. This method does not yield a conventional least squares regression line; consequently special methods must be used for significance testing. In an extensive review of regression analysis, Ricker (1973) argues that Bartlett's method and related techniques yield biased estimates of functional relationships and that the *GM* regression should be preferred on theoretical and empirical grounds. Kuhry and Marcus (1977) also criticize Bartlett's procedure as being unduly sensitive to errors in the ranked variable *X*, resulting in faulty rankings, which in turn produce a negative bias in the slope estimate.

Alternatively, one could use Kendall's robust line-fit method (see Section 14.12). Rayner (1985) reviewed several methods of Model II regression and presented two others, the **general structural relation** and the **structural relation.** Refer to Rayner (1985) for the formulas for these methods, which are complex. Rayner shows that simple linear regression, major axis, reduced major axis, and structural relation methods all are special cases of the general structural relation. This knowledge is helpful conceptually, but not computationally. Rayner's article concludes with some recommendations, which we have cast in the form of a taxonomic key and present as Table 14.3.

We illustrate the computation of Model II regression in Box 14.12. The data are from a sample of 11 fish whose weights before spawning and subsequent egg production were recorded. We want to estimate the functional relationship between number of eggs (Y) and weight of unspawned fish (X). These variables

Table 14.3 RULES FOR CHOOSING A MODEL II REGRESSION METHOD.

(Adapted from Rayner, 1985).

Polychotomous key		
1.(a)	Definite causality relation between variables	Employ simple linear regression (if the X variable is also known to be free of error, you can use standard error and confidence bands)
(b)	Causality relation between variables unknown or nonexistent	2
2.(a)	Error variances are known and > 0; covariance of errors known and $\neq 0$	Employ general structural relation (Rayner, 1985)
(b)	Magnitude of error variances not known but their ratio is; errors uncorrelated	Employ structural relation (Rayner, 1985)
(c)	Error variation not known and cannot be estimated	3
3.(a)	Dimensions and sizes differ	Employ reduced major axis
(b)	Similar dimensions and sizes	Employ major axis

Box 14.12 MODEL II REGRESSION.

Weights (X) of unspawned female cabezon (a California fish, *Scorpaenichthys marmoratus*) and the number of eggs they subsequently produced (Y). Sample size, $n = 11$.

(1) weight (to nearest 100 g) X	(2) eggs (in thousands) Y
14	61
17	37
24	65
25	69
27	54
33	93
34	87
37	89
40	100
41	90
42	97

Both variables are subject to error. It is desired to establish a functional relationship between number of eggs produced and weight before spawning.

I. *Simple linear regression*

The following quantities are used for comparison; some are also necessary for the computations illustrated in part **II.**

$$n = 11$$

$$\sum x^2 = 932.54545 \qquad \sum y^2 = 4188.72727 \qquad \sum xy = 1743.81818$$

$$\bar{X} = 30.36364 \qquad \bar{Y} = 76.54545 \qquad b_{Y \cdot X} = 1.86996$$

$$a = 19.76832 \qquad s_{Y \cdot X} = 10.15363 \qquad s_b = 0.33250$$

95% confidence limits for β: $L_1 = 1.118$, $L_2 = 2.622$.

II. *Reduced major axis (geometric mean regression)*

$$v_{Y \cdot X} = \sqrt{\Sigma y^2/\Sigma x^2} = \sqrt{4188.72727/932.54545} = 2.11937$$

$$a_v = 12.19366 \qquad s_v = s_b = 0.33250$$

95% confidence limits for v: $L_1 = 1.36726$, $L_2 = 2.871485$.

BOX 14.12 CONTINUED

III. *Bartlett's three-group method*

Computation

1. Arrange data by magnitude of X.
2. Divide the data pairs into thirds: $k = n/3 = 11/3 = 3.66 \approx 4$. If, as in this case, n is not evenly divisible by 3, the first and the last third should have the same sample size k. Find the means $\bar{X}_1$ and $\bar{Y}_1$ for the first third, $\bar{X}_3$ and $\bar{Y}_3$ for the last third, and $\bar{X}$ and $\bar{Y}$ for the total data.

	First third		Middle third		Last third	
	X_1	Y_1	X_2	Y_2	X_3	Y_3
	14	61	27	54	37	89
	17	37	33	93	40	100
	24	65	34	87	41	90
	25	69	31.3333	78.0	42	97
Means	20.0	58.0			40.0	94.0

3. The slope of the regression line is

$$b' = \frac{\bar{Y}_3 - \bar{Y}_1}{\bar{X}_3 - \bar{X}_1} = \frac{94.0 - 58.0}{40.0 - 20.0} = \frac{36.0}{20.0} = 1.80$$

4. The Y-intercept is

$$a' = \bar{Y} - b'\bar{X} = 76.54545 - (1.80)30.36364 = 21.89091$$

Hence, the functional relationship is

$$\hat{Y} = a' + b'X = 21.89091 + 1.80X$$

Figure 14.24 shows this line, fitted to the 11 data points. Had the data been inappropriately analyzed as a Model I regression, the equation would have been rather similar:

$$\hat{Y} = 19.768 + 1.870X$$

5. For confidence limits of β', the following statistics, which are the pooled sums of squares and sum of products of the three groups, must be computed:

$$\sum y^2 = \sum^{3} \sum^{k_i} (Y - \bar{Y})^2 = 1{,}588.00$$

$$\sum x^2 = \sum^{3} \sum^{k_i} (X - \bar{X})^2 = 128.6667$$

$$\sum xy = \sum^{3} \sum^{k_i} (X - \bar{X})(Y - \bar{Y}) = 298.00$$

Box 14.12 Continued

6. The confidence limits for β' are

$$\frac{\left(b'C - \sum xy\right) \pm \sqrt{C\left[b'^2\sum x^2 + \sum y^2 - 2b'\sum xy\right] - \sum x^2 \sum y^2 - \left(\sum xy\right)^2}}{C - \sum x^2}$$

where C is $k(\bar{X}_3 - \bar{X}_1)^2(n - 3)/2t^2_{\alpha[n-3]}$. The $+$ term yields the upper limit L_2, and the $-$ term, the lower limit L_1. In the present example, if we wish 95% confidence limits to β',

$$t_{.05[11-3]} = 2.306 \qquad k = 4 \qquad C = \frac{4(20)^2(11 - 3)}{2(2.306)^2} = 1203.542$$

Therefore,

$$L_1 = \left[\frac{1}{(1203.542 - 128.6667)}\right]\left\{[1.80(1203.542) - 298.0]\right.$$
$$\left. - \sqrt{\begin{matrix}1203.542[1.80^2(128.6667) + 1588.00 - 2(1.80)298.0] \\ - 128.6667(1588.00) - (298.0)^2\end{matrix}}\right\}$$

$$= \frac{1868.376 - \sqrt{828{,}678.466}}{1074.876} = \frac{1868.376 - 910.3137}{1074.876} = 0.891324$$

$$L_2 = \frac{1868.376 + 910.318}{1074.872} = 2.585127$$

are both measured with error, but they also vary naturally among individuals. Box 14.12 shows the several possible computations, whose results we will compare shortly. In view of the difference in measurement scales of the two variables, finding the slope of the major axis would be inappropriate in this case. The data in Box 14.12 are already arrayed by magnitude of the X variates, an arrangement necessary for carrying out Bartlett's three-group method. The computation for the *GM* regression is very simple. For Bartlett's procedure we divide the array of 11 pairs into two groups of 4 and one middle group of 3 pairs. The remaining computations are quite simple.

The Y-intercept is computed in the conventional manner, except that we call it a' to distinguish it from a in Model I linear regression. The regression line is fitted to the data plotted in Figure 14.24. Methods for setting confidence limits to the regression coefficient are also shown in Box 14.12. These are tedious to compute but contain no new complexities.

We can now compare the regression equations obtained by the four methods discussed here and by major axis regression, which we cover in Chapter 15.

Simple linear regression	$Y = 19.76682 + 1.86996X$
Reduced major axis regression	$Y = 12.19366 + 2.11937X$

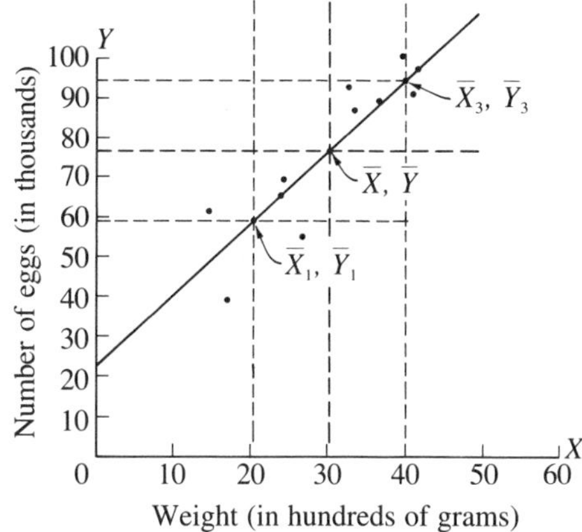

FIGURE 14.24 *Egg production and weight of 11 cabezon (data from Box 14.12) with equation* $\hat{Y} = 21.89091 + 1.80X$ *fitted by Bartlett's three-group method. The dashed lines indicate means of* X *and* Y *for the first and last third in the array of the data and for the total data.*

Major axis regression	$Y = 6.65668 + 2.30173X$
Bartlett's three-group method	$Y = 21.89091 + 1.80000X$
Kendall's robust line-fit method	$Y = 26.68421 + 1.68421X$

Since the correlation between the two variables is so high ($r = 0.88$; see Chapter 15), the slopes computed by the five methods are fairly similar. As mentioned earlier the problem of an optimal Model II strategy is still an open question. The proper criterion for showing an optimal functional relationship is unclear.

EXERCISES 14

14.1 The following temperatures (Y) were recorded in a rabbit at various times (X) after being inoculated with rinderpest virus. (Data from Carter and Mitchell, 1958.)

Time after injection (hours)	Temperature (°F)
24	102.8
32	104.5
48	106.5
56	107.0
72	103.9
80	103.2
96	103.1

Graph the data. Clearly the last three data points represent a different phenomenon from that of the first four pairs. *For the first four points:* (a) Calculate b. (b) Calculate the regression equation and draw the regression line on your graph. (c) Test the hypothesis that $\beta = 0$ and set 95% confidence limits. (d) Set 95% confidence limits to your estimate of the rabbit's temperature 50 hours after the injection. *Answers:* $b = 0.1300$, $\hat{Y}_{50} = 106.5$.

14.2 The following table is extracted from data by Sokoloff (1955). Adult weights of female *Drosophila persimilis* reared at 24°C are affected by their density as larvae. Carry out an anova among densities. Then calculate the regression of weight on density and partition the sums of squares among groups into that explained and that unexplained by linear regression. Graph the data with the regression line fitted to the means. Interpret your results.

Larval density	Mean weight of adults (in mg)	s of weights (not $s_{\bar{Y}}$)	n
1	1.356	0.180	9
3	1.356	0.133	34
5	1.284	0.130	50
6	1.252	0.105	63
10	0.989	0.130	83
20	0.664	0.141	144
40	0.475	0.083	24

14.3 Davis (1955) reported the following results in a study of the amount of energy metabolized by the English sparrow, *Passer domesticus,* under various constant temperature conditions and a 10-hour photoperiod.

Temperature (°C)	Calories $\bar{Y}$	n	s
0	24.9	6	1.77
4	23.4	4	1.99
10	24.2	4	2.07
18	18.7	5	1.43
26	15.2	7	1.52
34	13.7	7	2.70

Analyze and interpret. *Answer:* $b_{Y \cdot X} = -0.35558$, $F_s = 80.002$.

14.4 Using the complete data given in Exercise 14.1, calculate the regression equation and compare it with the one obtained in that exercise. Discuss the effect of including the last three points in the analysis.

14.5 The following results were obtained in a study of oxygen consumption (microliters/milligrams dry weight/hour) in the corn earworm *Heliothis zea* under controlled temperatures and photoperiods. (Data from Phillips and Newsom, 1966.)

Temperature (°C)	Photoperiod (hours)	
	10	14
18	0.51	1.61
21	0.53	1.64
24	0.89	1.73

Compute regression for each photoperiod separately and test for homogeneity of slopes. *Answer:* $s^2_{Y \cdot X} = 0.019267$ and 0.00060 for 10- and 14-hour photoperiods, respectively.

14.6 Length of developmental period (in days) of the potato leafhopper, *Empoasca fabae,* from egg to adult at various constant temperatures. (Data from Kouskolekas and Decker, 1966.) The original data were weighted means, but for this analysis we will consider them ordinary means.

Temperature °F	Mean length of developmental period (in days) $\bar{Y}$	Number reared n
59.8	58.1	10
67.6	27.3	20
70.0	26.8	10
70.4	26.3	9
74.0	19.1	43
75.3	19.0	4
78.0	16.5	20
80.4	15.9	21
81.4	14.8	16
83.2	14.2	14
88.4	14.4	27
91.4	14.6	7
92.5	15.3	4

Analyze and interpret. Compute deviations from the regression line $(\bar{Y}_i - \hat{\bar{Y}}_i)$ and plot against temperature.

14.7 Webber (1955) studied the relation between pupal weight (in milligrams) and the number of ovarioles in females of the blowfly *Lucilia cuprina* and obtained the following data.

Pupal weight	Number of ovarioles
6.30	62
7.91	76
8.20	81
8.25	80
8.65	83
9.48	105
16.08	142
18.15	152
21.70	184
23.12	186
21.88	182
26.20	230
28.82	193
32.00	260

Both variables are subject to error, but the count of the number of ovarioles is probably fairly accurate and could be used as the independent variable in a Model I regression. The equation needed, however, is a prediction equation for the number of ovarioles, given the pupal weight, because it is technically far easier to weigh a pupa than to dissect an adult and count the number of ovarioles. Estimate the number of ovarioles expected in a female pupa weighing 25 mg and set 95% confidence limits to your estimate. *Answer:* 95% confidence limits are 173.2 and 236.0.

14.8 Given the lower face widths (skeletal bigonial diameter in centimeters) for the 15 North American white females at ages 5 and 6 (see Box 11.5), evaluate their functional relationships by a Model II regression analysis. Also compute a prediction equation for face width at age 6 in terms of face width at age 5.

14.9 The experiment in Exercise 14.3 was repeated using a 15-hour photoperiod, yielding the following results.

Temperature (°C)	Calories $\bar{Y}$	n	s
0	24.3	6	1.93
10	25.1	7	1.98
18	22.2	8	3.67
26	13.8	10	4.01
34	16.4	6	2.92

Test for the equality of slopes of the regression lines for the 10- and 15-hour photo-periods. Comment on the results. *Answer:* $F_s = 0.003$.

14.10 Test whether the regression slopes of temperature on depth for the Rot Lake data in Box 11.3 are equal over the four days. Graph comparison limits for $b_{Y \cdot X}$ for the four days. Compute multiple confidence limits for the four slopes. Compute adjusted means for each day at $X = \bar{\bar{X}}$ and set multiple confidence limits to these values.

14.11 The following data were sampled at random from a larger study by Prange et al. (1979). They reinvestigated the question of whether birds have proportionately lighter skeletons than mammals because of their adaptation for flight. Compare the slopes for the estimated relationship between log skeletal mass and log body mass for the two samples. All weights are in kilograms. Compare the results one obtains using different methods for estimating structural relationships. Plot the data and superimpose the usual Model I regression line and the various Model II regression lines.

Birds		Mammals	
Skeletal Mass Y	Body Mass X	Skeletal Mass Y	Body Mass X
1.995	40.667	0.193	3.35
0.072	1.225	0.227	3.915
0.0054	0.163	0.0003	0.0063
0.203	2.504	0.039	0.790
0.043	0.701	0.027	0.820
0.027	0.416	0.244	4.836
0.186	2.379	0.002	0.030
0.0058	0.124	0.015	0.275
0.028	0.427	0.020	0.365
0.00174	0.031	0.0025	0.030
0.00182	0.029	0.0076	0.115
0.00102	0.020	1.146	22.7
0.024	0.383	0.748	11.950
0.00618	0.144	0.250	3.395
0.00184	0.038	0.107	2.460
0.00297	0.069	0.224	4.260
0.00183	0.045	0.233	4.210
0.00076	0.013	0.0173	0.350
0.00128	0.023	0.270	4.450
0.00049	0.0087	0.448	6.725
0.00062	0.0126	0.135	1.560
0.00061	0.0092	0.342	2.931
0.00117	0.019	4.120	20.000
0.00051	0.013	12.161	67.310
0.0011	0.032	1782.	6000.

Answer: For the bird data $b_{Y \cdot X} = 1.03059$ and $v_{Y \cdot X} = 1.07284$; for mammals the corresponding values are 1.10295 and 1.11272. The differences are not significant at the 5% level.

14.12 The following data were extracted from a larger study by Castillo (1938) concerned with determining the optimum amount of fish meal to be used as a supplement in rations for young growing ducklings.

Amount of fish meal							
5%		10%		15%		20%	
$\bar{Y}$	X	$\bar{Y}$	X	$\bar{Y}$	X	$\bar{Y}$	X
46.5	1	67.8	2	52.2	1	100.0	2
151.6	5	142.8	4	190.8	4	249.2	4
287.1	8	365.2	7	367.2	6	525.0	6
545.9	11	637.7	10	547.8	7	942.8	9
634.1	12	855.9	12	1012.8	11	1210.2	12

X = age in weeks.
$\bar{Y}$ = mean weight in grams (the original variates are not available, n varied from 11 to 46).

Does the amount of fish meal affect the rate of growth? Since the relationship between X and Y is obviously not linear, the Y variable should be transformed (try $Y' = Y^{1/2}$). Test for differences among adjusted means. If the overall test is significant, then use the GT2-method to test differences between pairs of adjusted means.

14.13 Carry out a nonparametric test for regression on the data in Exercise 14.7.

14.14 J. Calaprice (unpublished results) studied the effect of crowding on weight (in grams) of *Salmo trutta,* the brown trout. The fish were subjected to 4 densities: 175, 350, 525, and 700 per container. From each container 25 fish were sampled at random and weighed. The mean weights at the four densities are given below. A preliminary analysis of variance, also given below, showed that the densities differed in mean weight. Calculate the regression of weight on density for these trout and test its significance.

Density	175	350	525	700
Mean weight	3.51224	3.71020	2.92680	2.53960

Anova table

Source of variation	*df*	*SS*	*MS*	F_s
Among densities	3	21.63688	7.21229	6.944**
Within densities	96	99.71534	1.03870	
Total	99	121.35222		

15 CORRELATION

In this chapter we continue our discussion of bivariate statistics. Chapter 14 dealt with the functional relation of one variable to the other; here we discuss the measurement of the amount of association between two variables. This general topic is called correlation analysis.

It is not always obvious which type of analysis—regression or correlation—one should employ in a given problem. This topic has produced considerable confusion in the minds of investigators and also in the literature. We will try to make the distinction between these two approaches clear at the outset, in Section 15.1. In Section 15.2 you will be introduced to the product–moment correlation coefficient, the common correlation coefficient of the literature. We will derive a formula for this coefficient and present some of its theoretical background. The close mathematical relationship between regression and correlation analysis, as well as the algebraic relations between the t-test for paired comparisons and correlation, will be examined in Section 15.3. A fair amount of algebra is involved, but none of it is difficult, and its mastery will improve your overall understanding of several topics in this and previous chapters.

The computation of a product–moment correlation coefficient is the subject of Section 15.4. Section 15.5 presents tests of significance involving correlation coefficients. After this introduction to correlation coefficients, we discuss their applications in Section 15.6.

Section 15.7 examines some special computations with bivariate scattergrams, yielding the so-called principal axes of the ellipse of scattered points and the construction of concentric ellipses based on an observed sample of points. In Section 15.8 we offer several nonparametric methods for association to use when the necessary assumptions for tests involving correlation coefficients do not hold, or when quick but less than fully efficient tests are preferred for reasons of convenience.

When variables are nominal or categorical, the measures of association discussed in this chapter cannot be applied. Near the end of Section 17.4 we show how to compute a coefficient of association, the phi coefficient, between two categorical variables, each with two classes.

15.1 CORRELATION AND REGRESSION

Correlation and regression are often confused. Quite frequently correlation problems are treated as regression in the scientific literature, and the converse is equally true. There are several reasons for this confusion. First, the mathematical relations between the two methods of analysis are quite close. Algebraically, one can easily move from the formula for one to the formula for the other; hence the temptation to do so is great. Second, earlier texts did not make the distinction between the two approaches sufficiently clear, and the resulting confusion persists. Finally, although the approach chosen by an investigator may be correct for his/her intentions, the data available for analysis may make one or the other of the techniques inappropriate.

Let us examine these points at length. The many and close mathematical relations between regression and correlation will be detailed in Section 15.2. It suffices for now to state that for any given problem, most of the computational steps are the same, whether one carries out a regression or a correlation analysis. Recall that the fundamental quantity required for regression analysis is the sum of products. This same quantity is the base for computing the correlation coefficient. Because there are some simple mathematical relations between regression coefficients and their corresponding correlation coefficients, the temptation exists to compute a correlation coefficient corresponding to a given regression coefficient. As we will see later, this approach would be wrong unless our intention at the outset was to study association and the data were appropriate for such a computation.

An analogy from analysis of variance comes to mind. Recall the fundamental distinctions in purpose of Model I and Model II anovas (see Chapter 8). The computations for the two models, however, were almost the same, in some cases up to and including the initial significance tests. Only subsequent to these preliminary tests did the procedural paths separate. In Model I we estimated means, subdivided the differences among treatments into separate comparisons, and established confidence limits. In Model II we estimated variance components. Although it would be perfectly possible to carry out the computation for estimating a variance component in a Model I anova, the quantity obtained would not be an added variance component, but rather a fixed sum of squares whose magnitude is determined by treatment effects. Calling it a variance component would serve no useful purpose. Conversely, testing the significance between the means of random samples in a Model II anova by individual degree-of-freedom comparisons might yield some significant pairs, but establishing such differences would be meaningless in a Model II because of the random sampling of the groups.

In a similar manner we can easily compute a correlation coefficient from data that were properly analyzed by Model I regression, but such a coefficient is meaningless as an estimate of any population correlation coefficient. Conversely, data for which correlations could have been properly computed easily yield a Model I regression equation of one variable on the other, but such a slope would be biased; the appropriate equation depends on the purpose of the investigator and the nature of the variables.

Let us then look at the intentions or purposes behind the two types of analyses. In regression we intend to describe the dependence of a variable Y on an independent variable X. As we have seen, we employ regression equations to support hypotheses regarding the possible causation of changes in Y by changes in X, to predict Y in terms of X, and to explain some of the variation of Y by X, by using the latter variable as a statistical control. Studies of the effects of temperature on heart rate, nitrogen content of soil on growth rate in a plant, age of an animal on blood pressure, or dose of an insecticide on mortality of an insect population are typical examples of regression for these purposes.

In correlation, by contrast, we generally want to determine whether two variables are interdependent, or **covary**—that is, vary together. We do not express one as a function of the other. There is no distinction between independent and dependent variables. With the pair of variables whose correlation is studied, one may well be the cause of the other, but we neither know nor assume this. A more typical (but not essential) assumption is that both variables are effects of a common cause. What we wish to estimate is the degree to which these variables vary together. For example, we might be interested in the correlation between arm length and leg length in a population of mammals or between body weight and egg production in female blowflies or between days to maturity and number of seeds in a weed. Reasons for demonstrating and measuring association between pairs of variables need not concern us yet. We will discuss this topic in Section 15.6. All we need to know now is that when we wish to establish the degree of association between pairs of variables in a population sample, correlation analysis is the proper approach.

Even if we attempt the correct method for our purposes, we may run afoul of the nature of the data. For example, in an attempt to establish cholesterol content of blood as a function of weight, we may take a random sample of men of the same age group, obtain each individual's cholesterol content and weight, and regress the former on the latter. However, both these variables will have been measured with error. Individual variates of the supposedly independent variable X were not deliberately chosen or controlled by the experimenter. The underlying assumptions of Model I regression do not hold, and fitting a Model I regression to the data is not legitimate, although you will have no difficulty finding instances of such improper practices in the published research literature. If what we want is an equation describing the dependence of Y on X, we should carry out a Model II regression. If it is the degree of association between the variables (interdependence) that is of interest, however, we should carry out a correlation analysis, for which the data in this example are suitable. The converse difficulty is trying to obtain a correlation coefficient from data that are properly computed as a regression—that is, when X is fixed. An example is heart rate of a poikilotherm as a function of temperature, where several temperatures have been applied in an experiment. Such a correlation coefficient is easily obtained mathematically but would simply be a numerical value, not an estimate of a parametric measure of correlation. The square of the correlation coefficient can be interpreted in a way that is relevant to a regression problem, but it is not an estimate of a squared parametric correlation.

This discussion is summarized in Table 15.1, which shows the relations between correlation and regression. The two columns of the table indicate the two conditions of the pair of variables: in one case, one random and measured with error, the other fixed; in the other case, both variables random. We have departed from the usual convention of labeling the pair of variables Y and X or X_1, X_2 for both correlation and regression analyses. In regression we continue the use of Y for the dependent variable and X for the independent variable, but in correlation both of the variables are random variables, which we have throughout the text designated as Y. We therefore refer to the two variables as Y_1 and Y_2. The rows of the table indicate the intention of the investigator in carrying out the analysis, and the four quadrants of the table indicate the appropriate procedures for a given combination of intention of investigator and nature of the pair of variables.

Table 15.1 **THE RELATIONS BETWEEN CORRELATION AND REGRESSION.**

The correct computation for any combination of purposes and variables is shown.

Purpose of investigator	Nature of the two variables	
	Y random, X fixed	Y_1, Y_2 both random
Establish and estimate dependence of one variable upon another (describe functional relationship and/or predict one in terms of the other).	Model I regression.	Model II regression. (Model I generally inappropriate. For exceptions, the Berkson case and some instances of prediction, see Section 14.13.)
Establish and estimate association (interdependence) between two variables.	Meaningless for this case. If desired, an estimate of the proportion of the variation of Y explained by X can be obtained as the square of the correlation coefficient between X and Y.	Correlation coefficient. (Significance tests entirely appropriate only if Y_1, Y_2 are distributed as bivariate normal variables.)

15.2 The Product–Moment Correlation Coefficient

There are many correlation coefficients in statistics. The most common is the **product–moment correlation coefficient,** the current formulation of which was worked out by Karl Pearson. We will derive its formula using an intuitive approach similar to the one we used to introduce the regression coefficient (see Section 14.3).

Because it is a measure of covariation, the sum of products is likely to be the basic quantity from which the formula for the correlation coefficient is derived. We will label the variables whose correlation is to be estimated as Y_1 and Y_2. Their sum of products, therefore, will be $\Sigma y_1 y_2$ and their covariance $\Sigma y_1 y_2/(n - 1) = s_{12}$. The latter quantity is analogous to a variance, that is, a sum of squares divided by its degrees of freedom.

A standard deviation is expressed in original measurement units, such as inches, grams, or cubic centimeters. Similarly, a regression coefficient is expressed as units of Y per unit of X, such as 5.2 grams per day. A measure of association, however, should be independent of the original scale of measurement so that we can compare the degree of association in one pair of variables with that in another. One way to eliminate units is to divide the covariance by the standard deviations of variables Y_1 and Y_2:

$$r_{Y_1Y_2} = \frac{\sum y_1 y_2}{(n - 1)s_{Y_1}s_{Y_2}} \tag{15.1}$$

This expression is the formula for the product–moment correlation coefficient $r_{Y_1Y_2}$ between variables Y_1 and Y_2. Simplifying the symbolism, we obtain

$$r_{12} = \frac{\sum y_1 y_2}{(n - 1)s_1 s_2} = \frac{s_{12}}{s_1 s_2} \tag{15.2}$$

This notation is adequate unless you have two-digit variable numbers, in which case commas are necessary. For example, $r_{12,14}$ is the correlation between variables Y_{12} and Y_{14}. Expression (15.2) can be rewritten in another common form. Since

$$s\sqrt{n - 1} = \sqrt{s^2(n - 1)} = \sqrt{\frac{\sum y^2}{n - 1}(n - 1)} = \sqrt{\sum y^2}$$

we can rewrite Expression (15.2) as

$$r_{12} = \frac{\sum y_1 y_2}{\sqrt{\sum y_1^2 \sum y_2^2}} \tag{15.3}$$

To state Expression (15.2) more generally for variables Y_j and Y_k, we can write it as

$$r_{jk} = \frac{s_{jk}}{s_j s_k} \tag{15.4}$$

The correlation coefficient r_{jk} can range from $+1$ for perfect positive association to -1 for perfect negative association. The meaning of these values is intuitively obvious when we consider the correlation of a variable Y_j with itself. Expression (15.3) would then yield

$$r_{jj} = \frac{\sum y_j y_j}{\sqrt{\sum y_j^2 \sum y_j^2}} = \frac{\sum y_j^2}{\sum y_j^2} = 1$$

a perfect correlation of $+1$. If deviations in one variable were paired with opposite but equal deviations representing another variable, the result would be a correlation of -1 because the sum of products in the numerator would be negative. Proof that the correlation coefficient is bounded by $+1$ and -1 will be given shortly.

We can compute r_{jk} for any set of paired values of Y_j and Y_k, no matter what their underlying distribution is and even if they properly should be analyzed by regression. In cases in which regression would be more appropriate, however, the correlation coefficient is simply a mathematical index, not a sample statistic estimating an unknown parameter, and would be of little interest. If, however, the variates follow a specified distribution, the **bivariate normal distribution** (Figure 15.1), the correlation coefficient r_{jk} will estimate a parameter of that distribution symbolized by ρ_{jk}. Actually, r_{jk} is slightly biased and underestimates ρ_{jk} in small samples. Kendall and Stuart (1961, section 26.17) suggest the following corrected formula:

$$r_{jk}^* = r_{jk}\left[1 + \frac{1 - r_{jk}^2}{2(n-4)}\right]$$

where r_{jk}^* is the unbiased estimate of ρ_{jk}, and r_{jk} and n have their previous meanings. This bias can be ignored when sample size $n > 42$, since in these cases r and r^* are identical to two decimal places. The probability density function for the bivariate normal distribution has a formidable-looking formula, which we will not reproduce here.

Let us approach the distribution empirically. Suppose you have sampled 100 items and measured two variables on each item, obtaining two samples of 100 variates in this manner. If you plot these 100 items on a graph in which the variables Y_1 and Y_2 are the coordinates, you will obtain a scattergram of points (as in Figure 15.3A). Let us assume that both variables, Y_1 and Y_2, are normally distributed and that they are independent of each other; the fact that one individual is greater than the mean in character Y_1 thus has no effect on its value for

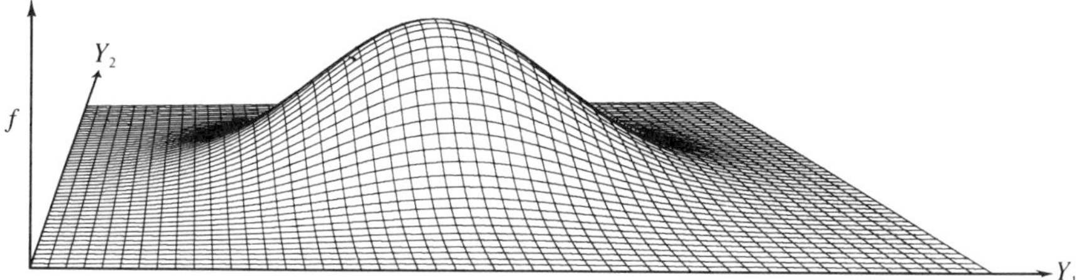

FIGURE 15.1 *Bivariate normal frequency distribution. The parametric correlation ρ between variables* Y_1 *and* Y_2 *equals zero. The frequency distribution may be visualized as a bell-shaped mound.*

variable Y_2. This same individual may therefore be greater or less than the mean for variable Y_2. If there is absolutely no relation between Y_1 and Y_2 and if the two variables are standardized to make their scales comparable, the outline of the scattergram is roughly circular. Of course, for a sample of 100 items, the circle would be only imperfectly outlined; the larger the sample, the more clearly could one discern a circle with the central area around the intersection $\bar{Y}_1$, $\bar{Y}_2$ heavily darkened because of the aggregation there of many points. Continued sampling requires superimposing new points upon previous points. If you visualize these points in a physical sense, such as grains of sand, a mound peaked in the shape of a bell would gradually form. This three-dimensional realization of a normal distribution is shown in perspective in Figure 15.1. Regarded from either coordinate axis, the mound has a two-dimensional appearance, with the outline of a normal distribution curve, the two views giving the distributions of Y_1 and Y_2, respectively.

If we assume that the two variables Y_1 and Y_2 are not independent but are correlated positively to some degree, then an individual that has a large value of Y_1 is likely to have a large value of Y_2 as well. Similarly, a small value of Y_1 will likely be associated with a small value of Y_2. Were you to sample items from such a population, the resulting scattergram (shown in Figure 15.3D) would be elongated in the form of an ellipse because the parts of the circle that formerly included individuals high for one variable and low for the other (and vice versa) would now be scarcely represented. Continued sampling (with the sand grain model) yields a three-dimensional elliptical mound (Figure 15.2). If correlation is perfect, all the data would fall along a single regression line (the identical line would describe the regression of Y_1 on Y_2 and of Y_2 on Y_1). In a physical model they would be represented by a flat, essentially two-dimensional normal curve lying on this regression line.

The shape of the outline of the scattergram and of the resulting mound is a function of the degree of correlation between the two variables, which is the parameter ρ_{jk} of the bivariate normal distribution. By analogy with Expression

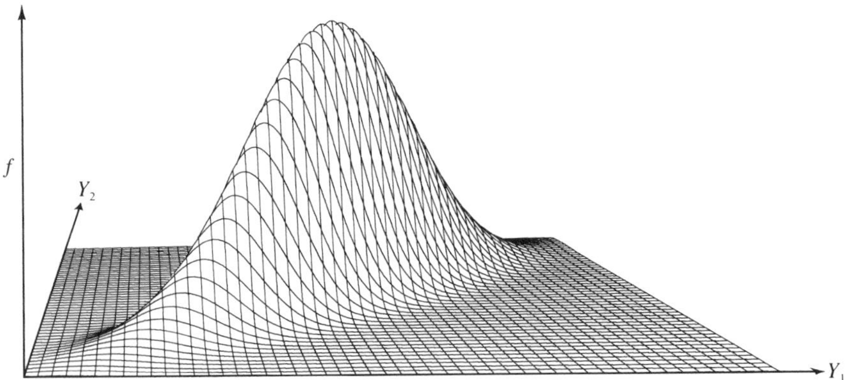

FIGURE 15.2 *Bivariate normal frequency distribution. The parametric correlation* ρ *between variables* Y_1 *and* Y_2 *equals 0.9. The bell-shaped mound of Figure 15.1 has become elongated.*

(15.2), the parameter ρ_{jk} can be defined as

$$\rho_{jk} = \frac{\sigma_{jk}}{\sigma_j \sigma_k} \tag{15.5}$$

where σ_{jk} is the parametric covariance of variables Y_j and Y_k, and σ_j and σ_k are the parametric standard deviations of variables Y_j and Y_k, respectively, as before. When two variables are distributed according to the bivariate normal distribution, a sample correlation coefficient r_{jk} estimates the parametric correlation coefficient ρ_{jk}. We can make some statements about the sampling distribution of ρ_{jk} and set confidence limits to it.

Regrettably, the elliptical shape of scattergrams of correlated variables is not usually very clear unless either large samples have been taken or the parametric correlation ρ_{jk} is high. To illustrate this point, we show in Figure 15.3 several graphs illustrating scattergrams resulting from samples of 100 items from bivariate normal populations with differing values of ρ_{jk}. Note that when $\rho_{jk} = 0$ (Figure 15.3A), the circular distribution is only very vaguely outlined. A much larger sample is required to demonstrate the circular shape of the distribution more clearly. No substantial difference is visible in Figure 15.3B, where $\rho_{jk} =$ 0.3. Because we know that this graph depicts a positive correlation, we can visualize a positive slope in the scattergram, but without this knowledge the positive slope would be difficult to detect visually. The shape in Figure 15.3C, where $\rho_{jk} = 0.5$, is clearer but still does not exhibit an unequivocal trend. In general, correlation cannot be inferred from inspecting scattergrams based on samples from populations with ρ_{jk} between -0.5 and $+0.5$ unless the sample is very large, say, 500. The graph in Figure 15.3G, also sampled from a population with $\rho_{jk} = 0.5$ but based on a sample of 500, illustrates this point. Here the positive slope and elliptical outline of the scattergram are quite evident. Figure

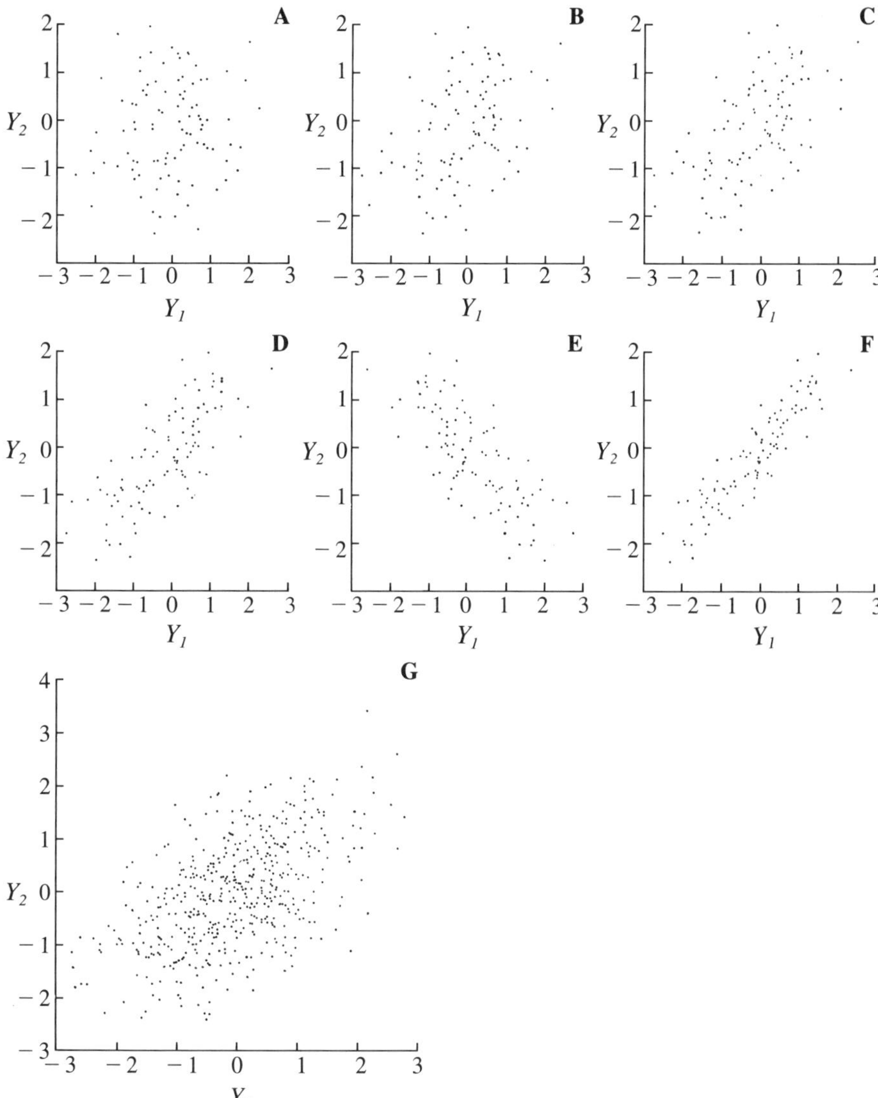

FIGURE 15.3 *Random samples from bivariate normal distributions with varying values of the parametric correlation coefficient ρ. Sample size* n = *100 in all graphs except* ***G****, where* n = *500.* ***A****. ρ = 0.* ***B****. ρ = 0.3.* ***C****. ρ = 0.5.* ***D****. ρ = 0.7.* ***E****. ρ = −0.7.* ***F****. ρ = 0.9.* ***G****. ρ = 0.5.*

15.3D, based on $\rho_{jk} = 0.7$ and $n = 100$, shows the trend. (Figure 15.3E, based on the same magnitude of ρ_{jk}, represents the equivalent negative correlation.) Finally, Figure 15.3F, representing a correlation of $\rho_{jk} = 0.9$, shows a tight association between the variables, reasonably approximating an ellipse of points.

Now let us return to the equation for the sample correlation coefficient shown in Expression (15.3). Squaring both sides results in

$$r_{12}^2 = \frac{\left(\sum y_1 y_2\right)^2}{\sum y_1^2 \sum y_2^2} = \frac{\left(\sum y_1 y_2\right)^2}{\sum y_1^2} \cdot \frac{1}{\sum y_2^2}$$

Look at the left term of the last expression. It is the square of the sum of products of variables Y_1 and Y_2, divided by the sum of squares of Y_1. If this were a regression problem, this term would be the formula for the explained sum of squares of variable Y_2 on variable Y_1, $\Sigma\hat{y}_2^2$. Using the symbolism that we used for regression (see Chapter 14), we would get $\Sigma\hat{y}^2 = (\Sigma xy)^2/\Sigma x^2$. Thus, we can write

$$r_{12}^2 = \frac{\sum \hat{y}_2^2}{\sum y_2^2} \tag{15.6}$$

The square of the correlation coefficient, therefore, is the ratio of the explained sum of squares of variable Y_2 divided by the total sum of squares of variable Y_2 or, equivalently,

$$r_{12}^2 = \frac{\sum \hat{y}_1^2}{\sum y_1^2} \tag{15.6a}$$

which could be derived as easily as Expression (15.6) was. Remember that since we are not really regressing one variable on the other, explaining Y_1 by Y_2 is just as legitimate as the other way around. This ratio is a proportion between zero and one. This range becomes obvious after a little contemplation of the meaning of this formula. The explained sum of squares of any variable must be smaller than its total sum of squares. In other words, if all the variation of a variable has been explained, its explained sum of squares can be as great as the total sum of squares, but no greater. When none of the variance can be explained by the other variable with which the covariance has been computed, the explained sum of squares is at its lowest value, zero. Thus, we obtain an important measure of the proportion of the variation of one variable determined by the variation of the other. This quantity, the square of the correlation coefficient, r_{12}^2, is called the **coefficient of determination.**

We encountered the coefficient of determination in Section 14.4. It ranges from zero to 1 and must be positive regardless of whether the correlation coefficient is negative or positive. Incidentally, here is proof that the correlation coefficient cannot vary beyond -1 and $+1$. Since its square is the coefficient of determination, and we have just shown that the bounds of the latter are zero to 1, obviously the bounds of its square root are ± 1.

The coefficient of determination is also useful when one is considering the relative importance of correlations of different magnitudes. As Figure 15.3 shows, the rate at which the distributions of the scatter diagrams vary from a circular outline to an elliptical outline seems to be more directly proportional to r^2 than to r. Thus, in Figure 15.3B, with $\rho^2 = 0.09$, the correlation is difficult to detect visually. By the time we reach Figure 15.3D, with $\rho^2 = 0.49$, however, the presence of correlation is very apparent.

The coefficient of determination may be useful in regression analysis also. Recall that in a regression we used an anova to partition the total sum of squares into explained and unexplained sums of squares. After such an analysis of variance has been carried out, one can obtain the ratio of the explained sums of squares over the total SS as a measure of the proportion of the total variation that has been explained by the regression. As we explained in Section 15.1, however, the square root of such a coefficient of determination would not be meaningful as an estimate of the parametric correlation of these variables.

We can invert Expression (15.6) and use it to compute the explained and unexplained sum of squares in cases where we know the coefficients of correlation or determination. Thus, given a sample in which you know that the correlation between two variables is r_{12}, and given the total sum of squares of variable Y_1 as Σy_1^2, you can compute the explained SS as $\Sigma \hat{y}_1^2 = r_{12}^2 \Sigma y_1^2$ and the unexplained SS as $\Sigma d_{1\cdot 2}^2 = (1 - r_{12}^2)\Sigma y_1^2$ and subject these to an analysis of variance. The quantity $1 - r_{12}^2$ is sometimes known as the **coefficient of nondetermination** because it expresses the proportion of variance of a variable that has *not* been explained by another variable. The square root of this coefficient, $\sqrt{1 - r_{12}^2}$, is known as the **coefficient of alienation,** measuring the lack of association between variables Y_1 and Y_2.

We will now examine some mathematical relations between the coefficients of correlation and regression. At the risk of sounding redundant, we stress again that the ease of converting one coefficient into the the other does not mean that the two types of coefficients can be used interchangeably on the same type of data. An important relationship between the correlation coefficient and the regression coefficient can be derived as follows from Expression (15.3):

$$r_{12} = \frac{\sum y_1 y_2}{\sqrt{\sum y_1^2 \sum y_2^2}} = \frac{\sum y_1 y_2}{\sqrt{\sum y_1^2}} \cdot \frac{1}{\sqrt{\sum y_2^2}}$$

Multiplying the numerator and the denominator of this expression by $\sqrt{\Sigma y_1^2}$, we obtain

$$r_{12} = \frac{\sum y_1 y_2}{\sqrt{\sum y_1^2}\sqrt{\sum y_1^2}} \cdot \frac{\sqrt{\sum y_1^2}}{\sqrt{\sum y_2^2}} = \frac{\sum y_1 y_2}{\sum y_1^2} \cdot \frac{\sqrt{\sum y_1^2}}{\sqrt{\sum y_2^2}}$$

Dividing the numerator and denominator of the right-hand term of this expression by $\sqrt{n-1}$, we obtain

$$r_{12} = \frac{\sum y_1 y_2}{\sum y_1^2} \cdot \frac{\sqrt{\dfrac{\sum y_1^2}{n-1}}}{\sqrt{\dfrac{\sum y_2^2}{n-1}}} = b_{2\cdot 1}\frac{s_1}{s_2} \tag{15.7}$$

Similarly, we could demonstrate that

$$r_{12} = b_{1\cdot 2}\frac{s_2}{s_1} \tag{15.7a}$$

and hence

$$b_{2\cdot 1} = r_{12}\frac{s_2}{s_1} \quad \text{and} \quad b_{1\cdot 2} = r_{12}\frac{s_1}{s_2} \tag{15.7b}$$

In these expressions $b_{2\cdot 1}$ is the regression coefficient for variable Y_2 and Y_1. The correlation coefficient is thus the regression slope multiplied by the ratio of the standard deviations of the variables. The correlation coefficient may therefore be regarded as a standardized regression coefficient. If the two standard deviations are identical, the regression and correlation coefficients will also be identical.

A second relation between the two coefficients is the following: multiplying together the $b_{2\cdot 1}$ and $b_{1\cdot 2}$ terms of Expression (15.7b), we obtain

$$b_{2\cdot 1}b_{1\cdot 2} = r_{12}\frac{s_2}{s_1}r_{12}\frac{s_1}{s_2} = r_{12}^2$$

Therefore,

$$r_{12} = \pm\sqrt{b_{2\cdot 1}b_{1\cdot 2}} \tag{15.8}$$

Thus, if we know the two regression coefficients, we can easily compute the correlation coefficient as the geometric mean (see Section 4.2) of the two regression coefficients. The sign of the correlation coefficient will be that of the regression coefficients and is, of course, determined by the sign of the sum of products $\Sigma y_1 y_2$.

If the two regression lines are drawn on the same graph, correlation is a measure of the angle between them. If this angle is a right angle, there can be no correlation in the data; the result is a circular scattergram. If this angle is very small, then correlation is very high. In the case of perfect correlation, the two regression lines coincide, and the angle between them becomes zero.

In Section 15.4 we will learn efficient methods of computing correlation coefficients.

15.3 THE VARIANCE OF SUMS AND DIFFERENCES

Having engaged in Section 15.2 in more than the usual amount of algebra for this book, we hope the reader is in a frame of mind to tolerate a little more. Now that we know about the coefficient of correlation, some of the earlier work on paired comparisons (see Section 11.5) can be put into proper perspective. In Section A.13 of the appendix we show that the variance of a sum of two variables is

$$s^2_{(Y_1+Y_2)} = s_1^2 + s_2^2 + 2r_{12}s_1s_2 \tag{15.9}$$

where s_1 and s_2 are standard deviations of Y_1 and Y_2, respectively, and r_{12} is the correlation coefficient between these variables. Similarly, for a difference between two variables, we obtain the following equation for variance:

$$s^2_{(Y_1-Y_2)} = s_1^2 + s_2^2 - 2r_{12}s_1s_2 \tag{15.9a}$$

Expression (15.9) indicates that if we make a new composite variable that is the sum of two other variables, the variance of this new variable will be the sum of the variances of the variables of which it is composed plus an added term, which is a function of the standard deviations of these two variables and of the correlation between them. Section A.13 shows that this added term is twice the covariance of Y_1 and Y_2. When the two variables being summed are uncorrelated, this added covariance term will also be zero, and the variance of the sum will simply be the sum of the variances of the two variables. This is why, in an anova or in a t-test of the difference between the two means, we had to assume the independence of the two variables to permit us to add their variances. Otherwise we would have had to allow for a covariance term.

By contrast, in the paired-comparisons technique we expect correlation between the variables, since each pair shares a common experience. The guinea pig data in Box 15.1, which we analyzed by a nonparametric technique in Box 13.11, make this point clearer. First we calculate the correlation coefficient between variables Y_1 and Y_2 and find it to be quite high, 0.882. If these data were (improperly) analyzed by a t-test for difference of means (see Section 9.4), the result, as shown in step **2** of Box 15.1, would not be significant because the standard error for the t-test is too large, being based on the sum of the variances of the two variables without subtracting a term to account for the correlation. The paired-comparison test (see Section 11.5 and Box 11.5) automatically subtracts such a term, resulting in a smaller standard error and consequently in a larger value of t_s, since the numerator of the ratio remains the same. Thus, whenever correlation between two variables is positive, the variance of their differences will be considerably smaller than the sum of their variances; this is why the paired-comparison test must be used in place of the t-test for difference of means. These considerations are equally true for the corresponding analyses of variance: single-classification and two-way anova.

At the bottom of Box 15.1 we show how to estimate the correlation coefficient from the variances of the variables and of their differences. Although this

Box 15.1 RELATIONS BETWEEN THE CORRELATION COEFFICIENT AND THE PAIRED-COMPARISONS TEST.

Litter size in guinea pigs. (Data from Box 13.11.)

Year	Strain B Y_1	Strain 13 Y_2	$D = Y_1 - Y_2$
1916	2.68	2.36	+0.32
1917	2.60	2.41	+0.19
1918	2.43	2.39	+0.04
1919	2.90	2.85	+0.05
1920	2.94	2.82	+0.12
1921	2.70	2.73	−0.03
1922	2.66	2.58	+0.08
1923	2.98	2.89	+0.09
1924	2.85	2.78	+0.07
$\bar{Y}$	2.749	2.646	+0.1033

Computation

1. Computing the correlation coefficient between Y_1 and Y_2:

$$n = 9$$

$$\sum y_1^2 = 0.2619 \qquad \sum y_2^2 = 0.3638$$

$$s_1^2 = 0.03274 \qquad s_2^2 = 0.04548$$

$$\sum y_1 y_2 = 0.2723 \qquad \sqrt{\sum y_1^2 \sum y_2^2} = 0.3087$$

$$r_{12} = 0.882$$

2. t-test for difference of means (in the manner of Box 9.6):

$$t_s = \frac{(\bar{Y}_1 - \bar{Y}_2) - (\mu_1 - \mu_2)}{\sqrt{\frac{1}{n}(s_1^2 + s_2^2)}} = \frac{0.1033}{0.093226} = 1.108\ ns$$

3. t-test for paired comparisons (in the manner of Box 11.5):

$$s_D^2 = 0.01015 \qquad s_{\bar{D}} = 0.03358$$

$$t_s = \frac{\bar{D} - 0}{s_{\bar{D}}} = \frac{0.1033}{0.03358} = 3.077 > t_{.02[8]}$$

The difference in outcome of the two tests in steps **2** and **3** shows presence of correlation. We can calculate r_{12} in the usual way (step **1**) or as follows:

BOX 15.1 CONTINUED

$$s_D^2 = s_{(Y_1 - Y_2)}^2 = s_1^2 + s_2^2 - 2r_{12}s_1s_2 \text{ [from Expression (15.9a)]}$$

$$2r_{12}s_1s_2 = s_1^2 + s_2^2 - s_D^2$$

$$r_{12} = \frac{s_1^2 + s_2^2 - s_D^2}{2s_1s_2}$$

$$= \frac{0.03274 + 0.04548 - 0.01015}{2(0.1809)(0.2132)}$$

$$= \frac{0.06807}{0.077136}$$

$$= 0.882$$

method is not recommended, it does provide further insight into the relationships that we have just discussed.

Finally, you might ask how negative correlation would affect the considerations of the last four paragraphs. Such a situation might arise with a compensatory phenomenon. For instance, if two animals share a cage and the total amount of food available is constant, the more one animals eats, the less the other animal will get. If the two animals can be differentiated—for example, by sex—then over a series of cages there will be a negative correlation between the amount of food eaten by males and females. In such a case the standard error of the paired-comparison test will be greater than that of the difference-of-means test because the covariance term is now added to the sum of the two variances (the negative sign before the covariance term and the negative sign of the covariance itself together yield a plus sign). Therefore, with negative correlations a paired-comparisons test is less likely to result in rejection of the null hypothesis than is the test for difference of means. This is another instance where two-way analysis of variance would be more useful than a t-test. In this case, an overall test for difference of means of the variables (treatments) has little meaning. Also differences due to the classification factor grouping them into pairs will not mean much in view of the negative correlation. The important aspect of the analysis is the negative interaction between classification factor and food intake. To establish the significance of such an interaction, however, one needs to have replication of each subgroup. Otherwise the only way to demonstrate such a relationship is by computing a correlation coefficient between the variables.

15.4 COMPUTING THE PRODUCT–MOMENT CORRELATION COEFFICIENT

The computation of a product-moment correlation coefficient is quite simple. The basic quantities needed are the same as those required for computing the regression coefficient (see Section 14.3). Box 15.2 illustrates the computation. The example is based on a sample of 12 crabs in which gill weight, Y_1, and body weight, Y_2, have been recorded. We wish to know whether there is a correlation between the weight of the gill and that of the body, the latter representing a measure of overall size. The existence of a positive correlation might lead you to conclude that a bigger-bodied crab, with its resulting greater amount of metabolism, would require larger gills to provide the necessary oxygen. The correlation coefficient computed in Box 15.2 (0.87) agrees with the clear slope and narrow elliptical outline of the scattergram for these data (Figure 15.4).

This example has another interesting feature: It is an example of a **part-whole correlation.** Although only a small fraction of the total, gill weight is a part of body weight, and the correlation is therefore between two variables, one of which is a part of another. We can rewrite these variables as follows: Y_1 is gill weight as before, but Y_2, the body weight, is made up of Y_0, body weight *minus* gill weight, and Y_1, gill weight. Since $Y_2 = Y_0 + Y_1$, Y_1 and Y_2 should be correlated positively to some degree because Y_1 is part of Y_2 and will clearly be correlated positively with itself. The correlation between Y_1 and Y_2 depends on the degree of correlation between Y_0 and Y_1 and can be predicted from the formula

$$r_{12} = \frac{s_1 + r_{01}s_0}{\sqrt{s_0^2 + 2r_{01}s_0s_1 + s_1^2}} \qquad (15.10)$$

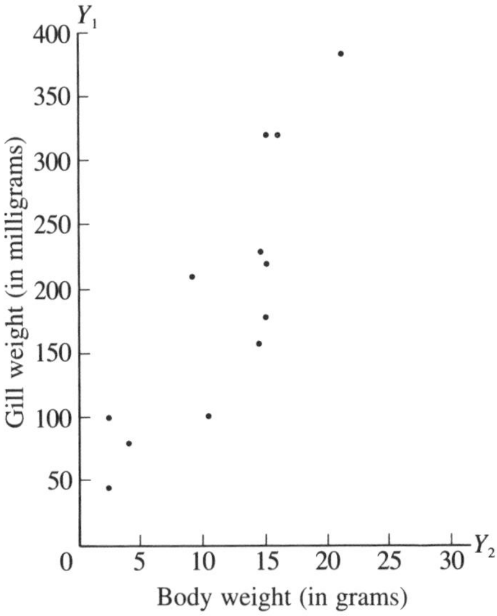

FIGURE 15.4 *Scattergram for the crab data of Box 15.2.*

Box 15.2 COMPUTING THE PRODUCT-MOMENT CORRELATION COEFFICIENT FOR SMALL AND MEDIUM-SIZED SAMPLES.

Relationship between gill weight and body weight in the crab *Pachygrapsus crassipes; n* = 12.

(1) Y_1 Gill weight in milligrams	(2) Y_2 Body weight in grams
159	14.40
179	15.20
100	11.30
45	2.50
384	22.70
230	14.90
100	1.41
320	15.81
80	4.19
220	15.39
320	17.25
210	9.52

SOURCE: Data from L. Miller (unpublished results).

Computation

1. $\bar{Y}_1 = 195.58333$, $\bar{Y}_2 = 12.04750$
2. Sum of squares of $Y_1 = \sum y_1^2 = \sum (Y_1 - \bar{Y}_1)^2 = 124{,}368.9111$
3. Sum of squares of $Y_2 = \sum y_2^2 = \sum (Y_2 - \bar{Y}_2)^2 = 462.47817$
4. Sum of products $= \sum y_1 y_2 = \sum (Y_1 - \bar{Y}_1)(Y_2 - \bar{Y}_2) = 6561.61737$
5. Product-moment correlation coefficient [by Expression (15.3)]:

$$r_{12} = \frac{\sum y_1 y_2}{\sqrt{\sum y_1^2 \sum y_2^2}} = \frac{\text{quantity } \mathbf{4}}{\sqrt{\text{quantity } \mathbf{2} \times \text{quantity } \mathbf{3}}}$$

$$= \frac{6561.61737}{\sqrt{(124{,}368.9111)(462.47817)}}$$

$$= \frac{6561.61737}{7584.05607} = 0.86519 \approx 0.87$$

Box 15.3 COMPUTING THE PRODUCT-MOMENT CORRELATION COEFFICIENT FROM A TWO-WAY FREQUENCY DISTRIBUTION.

Lengths of the median cross vein in the right (Y_1) and left (Y_2) forewings of 500 worker bees. The variates are coded (mm × 50). Data by Phillips (1929).

Length of vein m, left forewing (Y_2)

Length of vein m, right forewing (Y_1)

															f_1	Y_1
														1	1	86
									1	1	2	2	3		9	85
									3	7	2	1	1		14	84
							1	4	6	16	3	1			31	83
						3	13	19	12	5	2				54	82
						8	18	22	19	3	1				71	81
				2	6	23	46	23	1	1					102	80
				10	17	34	25	8							94	79
			2	14	19	12	11	1							59	78
		1	4	19	14	6	1								45	77
	1	5	2	4	3	1									16	76
			1	1		1									3	75
		1													1	74
f_2	1	7	9	50	59	88	115	77	42	33	10	4	4	1	500	
Y_2	74	75	76	77	78	79	80	81	82	83	84	85	86	87		

In this table the frequencies and class marks of variables Y_1 and Y_2 are given at the right and at the bottom, respectively.

Computation

In our sample, $n = \Sigma f_1 = \Sigma f_2 = \Sigma f_{12} = 500$. We use f_{12} for the frequencies inside the two-way table. The following quantities may now be calculated:

1. $\sum f_1 Y_1 = 39{,}947$
2. $\sum f_2 Y_2 = 39{,}918$
3. $\bar{Y}_1 = \dfrac{\sum f_1 Y_1}{\sum f_1} = \dfrac{39{,}947}{500} = 79.8940$
4. $\bar{Y}_2 = \dfrac{\sum f_2 Y_2}{\sum f_2} = \dfrac{39{,}918}{500} = 79.836$
5. Sum of squares of $Y_1 = \sum y_1^2 = \sum f_1(Y_1 - \bar{Y}_1)^2 = 2147.382$

BOX 15.3 CONTINUED

6. Sum of squares of $Y_2 = \sum y_2^2 = \sum f_2(Y_2 - \bar{Y}_2)^2 = 2110.552$

7. Sum of products of Y_1 and $Y_2 = \sum y_1 y_2 = \sum f_{12}(Y_1 - \bar{Y}_1)(Y_2 - \bar{Y}_2) = 1782.308$

8. Correlation coefficient $= r_{12} = \dfrac{\sum y_1 y_2}{\sqrt{\sum y_1^2 \sum y_2^2}}$

$$= \frac{\text{quantity } \mathbf{7}}{\sqrt{\text{quantity } \mathbf{5} \times \text{quantity } \mathbf{6}}}$$

$$= \frac{1782.308}{\sqrt{(2147.382)(2110.552)}} = 0.837202 \approx 0.837$$

Thus, the correlation of a part with a whole is a function of s_1 and s_0, the standard deviations of the part and of the whole minus that part, as well as of r_{01}, the correlation between these variables. When the correlation $r_{01} = 0$, the formula for a part-whole correlation simplifies to

$$r_{12} = \frac{1}{\sqrt{1 + (s_0^2/s_1^2)}} \tag{15.10a}$$

and the correlation is then simply a function of the ratio of the variances. If, in addition to no correlation between the part and its complement, the variances were equal, then the formula for the part-whole correlation would simplify to $r_{12} = 1/\sqrt{1 + 1} = 1/\sqrt{2} = 0.707$, which shows that when two parts are equally variable, the correlation of one part with their sum will be quite high. In the past, such correlations were considered spurious. They are not spurious, however, but rather are logical consequences of the way in which the variables were defined. As long as we keep this in mind, there is nothing necessarily wrong with calculating such correlations, except that one should not be surprised to find a sizable correlation of a part with a whole. When the part has a much smaller variance than the whole, Expression (15.10) shows that the expected value of the part-whole correlation is not likely to be great unless r_{01} is substantial.

Sometimes data for correlation analysis are given in the form of **two-way frequency distributions** for reasons very much the same as those enumerated in Section 2.5 for simple frequency distributions. The calculation of the sums of squares and the sum of products must then be redesigned so that the sums of squares are computed from the marginal totals and the sum of products is computed from the cells of the two-way table.

The data in Box 15.3 are wing vein lengths from the right and left forewings of 500 worker bees. Obviously, we would expect a correlation between these variables. Larger bees are expected to have larger wings on both sides of the body, and wing vein length reflects wing size. The degree to which the right and left sides are correlated is a measure of developmental homeostasis. Although these particular data are quite old, measuring the degree of symmetry of organisms by correlating bilaterally paired organs has received considerable attention in recent years. The computational steps for obtaining the basic quantities from the two-way frequency distribution are illustrated in Box 15.3. Once the basic quantities have been obtained, the rest of the computation of the correlation coefficient is the same as in Box 15.2. We find the correlation to be quite high; yet remember that a correlation of 0.837202 yields a lower coefficient of determination (0.7009). Thus, only 70% of the variance of wing vein length on one side is accounted for by variation of the wing vein length on the opposite side.

Many correlation programs exist. The BIOM-pc computer program performs this operation. We should stress, however, that it is important to examine a scattergram of the points as part of the analysis. Inspection may reveal that the scatter of points, rather than circular or elliptical, shows a peculiar shape that might be improved by transformation. Most computer programs display a bivariate scattergram. In addition, the discussion of leverage coefficients in Section 14.11 in conjunction with regression applies also to correlation. Outliers seriously affect values of a sample correlation coefficient.

The advantage of the computer becomes especially noticeable when more than two variables have been recorded for each item. For p variables the outcome of the study will be a $p \times p$ **correlation matrix**—that is, a table of correlation coefficients of variable 1 against variable 2, variable 1 against variable 3, and so on until variable 1 is correlated with variable p. This process is followed by the correlation of 2 against 3, 2 against 4, and so on up to 2 against p. The final value in the correlation matrix is the correlation of variable $p - 1$ against variable p. Computing such correlation matrices in the precomputer days was a very difficult and tedious procedure, often taking several months for sizable studies. The ease with which correlation matrices can now be obtained has given impetus to considerable research employing multivariate analysis, some of which is discussed in Chapter 16.

15.5 Significance Tests in Correlation

The most common significance test determines whether a sample correlation coefficient could have come from a population with a parametric correlation coefficient of zero. The null hypothesis is, therefore, $H_0: \rho = 0$, implying that the two variables are uncorrelated. If the sample comes from a bivariate normal distribution and $\rho = 0$, the standard error of the correlation coefficient is

$$s_r = \sqrt{\frac{1 - r^2}{n - 2}}$$

The null hypothesis is tested as a t-test with $n - 2$ degrees of freedom:

$$t_s = \frac{r - 0}{\sqrt{\dfrac{1 - r^2}{n - 2}}} = r\sqrt{\frac{n - 2}{1 - r^2}}$$

Note that this standard error applies only when $\rho = 0$, so it cannot be applied to testing a hypothesis that ρ is a specific value other than zero. The t-test for the significance of r is mathematically equivalent to the t-test for the significance of b. This similarity is analogous to the situation in Model I and Model II single-classification anovas, where the same F-test establishes the significance regardless of the model.

Significance tests following this formula have been carried out systematically and are tabulated in Statistical Table **R,** which permits a sample correlation coefficient to be tested for significance without further computation. Box 15.4 illustrates tests of the hypothesis $H_0 : \rho = 0$, using Table **R,** as well as the t-test discussed above.

For ρ close to ± 1.0, the distribution of sample values of r is markedly asymmetrical, and, although a standard error has been found for r in such cases, it should not be applied unless the sample is very large ($n > 500$). To overcome this difficulty, we transform r to a function z, developed by R. A. Fisher. The formula for z is

$$z = \frac{1}{2}\ln\left(\frac{1 + r}{1 - r}\right) \tag{15.11}$$

You may recognize this expression as $z = \tanh^{-1} r$, the formula for the inverse hyperbolic tangent of r. This function is found on most statistical pocket calculators. Inspection of Expression (15.11) will show that when $r = 0$, z also equals zero, since $\frac{1}{2}\ln 1$ equals zero. As r approaches 1, however, $(1 + r)/(1 - r)$ approaches infinity; consequently, z approaches infinity. Therefore, substantial differences between r and z occur at the higher values for r. Thus, when $r = 0.115$, $z = 0.11551$; for $r = -0.531$ we obtain $z = -0.59154$; and when $r = 0.972$, $z = 2.12730$. Note by how much z exceeds r in this last pair of values. The inverse function, $r = \tanh z$, yields values of r given a value of z. Thus, $z = 0.70$ corresponds to $r = 0.60437$, and $z = -2.76$ corresponds to $r = -0.99202$.

The advantage of the **z-transformation** is that whereas correlation coefficients are distributed in skewed fashion for values of $\rho \neq 0$, the values of z are approximately normally distributed for any value of its parameter, which we call ζ (zeta), following the usual convention. The expected variance of z is

$$\sigma_z^2 = \frac{1}{n - 3} \tag{15.12}$$

Box 15.4 TESTS OF SIGNIFICANCE, CONFIDENCE LIMITS, AND REQUIRED SAMPLE SIZES FOR CORRELATION COEFFICIENTS.

Test of the null hypothesis $H_0: \rho = 0$ *versus* $H_1: \rho \neq 0$

The simplest procedure is to consult Statistical Table **R,** where the critical values of r are tabulated for $df = n - 2$ from 1 to 1000. If the absolute value of the observed r is greater than the tabulated value in the column for two variables, we reject the null hypothesis.

Examples:—In Box 15.2 we found the correlation between body weight and gill weight to be 0.86519, based on a sample of $n = 12$. For 10 degrees of freedom the critical values are 0.576 at the 5% level and 0.708 at the 1% level of significance. Since the observed correlation is greater than both of these, we can reject the null hypothesis, $H_0: \rho = 0$, at $P < 0.01$.

In Box 15.3 we obtained the correlation $r = 0.837202$, based upon $n = 500$ observations. This value is far greater than the critical values (0.088 and 0.115) given in Table **R.** Again we reject the null hypothesis that $\rho = 0$ at $P \ll 0.01$.

Table **R** is based upon the following test, which may be carried out when the table is not available or when an exact test is needed at significance levels or at degrees of freedom other than those furnished in the table. The null hypothesis is tested by means of the t-distribution (with $n - 2$ df) by using the standard error of r. When $\rho = 0$,

$$s_r = \sqrt{(1 - r^2)/(n - 2)}$$

Therefore,

$$t_s = \frac{(r - 0)}{\sqrt{(1 - r^2)/(n - 2)}} = r\sqrt{(n - 2)/(1 - r^2)}$$

For the data of Box 15.2, this value would be

$$t_s = 0.86519 \sqrt{(12 - 2)/(1 - 0.86519^2)} = 0.86519 \sqrt{10/0.25145}$$
$$= 0.86519 \sqrt{39.76993} = 0.86519\ (6.30634) = 5.45618 > t_{.001[10]}$$

For a one-tailed test the 0.10 and 0.02 values of t should be used for 5% and 1% significance tests, respectively. Such tests would apply if the alternative hypothesis were $H_1: \rho > 0$ or $H_1: \rho < 0$, rather than $H_1: \rho \neq 0$.

When n is greater than 50, we can also make use of the z-transformation described in the text. For the data of Box 15.3, since $\sigma_z = 1/\sqrt{n - 3}$, we test

$$t_s = \frac{z - 0}{1/\sqrt{n - 3}} = z\sqrt{n - 3}$$

Since z is distributed normally and we are using a parametric standard deviation, we compare t_s with $t_{\alpha[\infty]}$ or employ Statistical Table **A,** areas of the normal curve. For

Box 15.4 Continued

$r = 0.837202$, $z = 1.21174$ and $t_s = 1.21174\sqrt{497} = 27.01402$. In the table of areas of a normal curve (Table **A**), this value yields a very small probability ($<10^{-6}$).

For $n < 50$ (or in critical cases when $n < 100$), more exact procedures using z are given in the last section of this box.

Test of the null hypothesis $H_0: \rho = \rho_1$, where $\rho_1 \neq 0$

To test this hypothesis we cannot use Table **R** or the t-test given above, but must make use of the z-transformation.

Suppose we wish to test the null hypothesis $H_0: \rho = +0.50$ versus $H_1: \rho \neq +0.50$ for the data of Box 15.3. We would use the following expression:

$$t_s = \frac{z - \zeta}{1/\sqrt{n-3}} = (z - \zeta)\sqrt{n-3}$$

where z and ζ are the z-transformations of r and ρ, respectively. Again we compare t_s with $t_{\alpha[\infty]}$ or look it up in Table **A.** Using the $\tanh^{-1}$ function, we find

For $r = 0.837202$ $\quad z = 1.21174$

For $\rho = 0.500$ $\quad \zeta = 0.54931$

Therefore

$$t_s = (1.21174 - 0.54931)(\sqrt{497}) = 14.76807$$

The probability of obtaining such a value of t_s by random sampling is $P < 10^{-6}$ (see Table **A**). It is most unlikely that the parametric correlation between right and left wing veins is 0.5.

Confidence limits

If $n > 50$, we can set confidence limits to r using the z-transformation. First we convert the sample r to z, set confidence limits to z, and then transform these limits back to the r-scale. The following calculations determine 95% confidence limits for the data of Box 15.3 ($r = 0.837202$; $z = 1.21174$; $\alpha = 0.05$):

$$L_1 = z - t_{\alpha[\infty]}\sigma_z = z - \frac{t_{.05[\infty]}}{\sqrt{n-3}} = 1.21174 - \frac{1.960}{\sqrt{497}}$$

$$= 1.21174 - 0.08792 = 1.12383$$

$$L_2 = z + \frac{t_{.05[\infty]}}{\sqrt{n-3}} = 1.21174 + 0.08792 = 1.29966$$

We retransform these z-values to the r-scale by means of the tanh function:

$$L_1 \approx 0.80890 \quad \text{and} \quad L_2 \approx 0.86164$$

are the 95% confidence limits around $r = 0.837$.

BOX 15.4 CONTINUED

Sample size needed to detect a true difference between an observed and an expected correlation

To be 90% certain that the null hypothesis, $H_0: \rho = 0$, will be rejected at the 1% level of significance when $|\rho_1| \geq 0.6$, we should employ sample size n. This sample size can be found by the following formula, which is Expression (10.3.5) in Cohen (1977):

$$n = \left(\frac{t_{\alpha[\infty]} + t_{2(1-P)[\infty]}}{\zeta_1}\right)^2 + 3$$

where n = desired sample size, α = intended significance level, P = desired probability that the null hypothesis will be rejected when $|\rho_1| \geq$ a set value, and ζ_1 = the z-transformation of the expected value ρ_1.

For the present example, we find by consulting Table **B** and computing $\zeta_1 = \tanh^{-1} \rho_1 = 0.6931$ that

$$n = \left(\frac{2.576 + 1.282}{0.6931}\right)^2 + 3 = 33.98$$

We therefore need a sample of 34 paired observations to meet the specified conditions.

The use of the z*-transformation when sample sizes* n < *50*

For small samples, calculating exact probabilities is difficult. The following modified z-transformation has been suggested by Hotelling (1953) for use in small samples:

$$z^* = z - \frac{3z + r}{4(n-1)}$$

$$\sigma_{z^*} = \frac{1}{\sqrt{n-1}}$$

The distribution of z^* is closer to a normal distribution than is z. We do not know for how small a sample size this transformation is adequate, but most likely it should not be used for $n < 10$. As an example, z^* and σ_{z^*} are computed for the data of Box 15.2:

$$r = 0.86519 \qquad n = 12 \qquad z = 1.31363$$

$$z^* = z - \frac{(3z + r)}{4(n-1)} = 1.31363 - \frac{3(1.31363) + 0.86519}{4(11)}$$

$$= 1.31363 - 0.10923 = 1.20440$$

$$\sigma_{z^*} = \frac{1}{\sqrt{n-1}} = \frac{1}{\sqrt{11}} = \frac{1}{3.31662} = 0.30151$$

The differences between z and z^* and between their standard errors are slight and are important only for critical tests near the borderline of significance. For example, to test the null hypothesis $H_0: \rho = 0$ versus $H_1: \rho \neq 0$,

$$t_s = (z^* - \zeta^*)\sqrt{n-1}$$

Box 15.4 Continued

where ζ^* is the z^*-transformation applied to the ρ of the null hypothesis:

$$\zeta^* = \zeta - \frac{(3\zeta + \rho)}{4n}$$

which equals zero since ζ and ρ are both zero. Again we compare t_s with $t_{\alpha[\infty]}$ or look it up in Table **A.** Applying this formula to the gill-weight–body-weight correlation of Box 15.2, we obtain

$$t_s = (1.20440)\sqrt{11} = 1.20440\,(3.31662) = 3.99454 \qquad P < 0.001$$

If we had used the ordinary z-transformation, our result would have been

$$t_s = (z - \zeta)\sqrt{n - 3} = 1.31363\sqrt{9} = 3.94088$$

leading to the same conclusion as before.

We can use Expressions (15.13) and (15.14) to set confidence limits to ρ in small samples. These limits can be backtransformed from limits expressed as z^* by employing Expression (15.15).

This approximation is adequate for sample sizes $n \geq 50$ and tolerable even when $n \geq 25$. An interesting aspect of the variance of z evident from Expression (15.12) is that it is independent of the magnitude of r and is simply a function of sample size n. Where the significance tests or confidence limits are critical and a better approximation is desired, Hotelling (1953) recommends another transformation, z^*, which is computed as follows:

$$z^* = z - \frac{3z + r}{4(n - 1)} \tag{15.13}$$

Its variance is given as

$$\sigma^2_{z^*} = \frac{1}{n - 1} \tag{15.14}$$

This expression is actually another approximation for a more involved expression but should be satisfactory for sample sizes $n \geq 10$. The back transformation is carried out by

$$r = 4(n - 1)(z - z^*) - 3z \tag{15.15}$$

As Box 15.4 shows, for sample sizes greater than 50 we can also use the z-transformation to test the significance of a sample r employing the hypothesis $H_0: \rho = 0$. In the second section of Box 15.4 we show the test of a null hypothesis that $\rho \neq 0$. We may hypothesize that the true correlation between two variables is a given value ρ different from zero. Hypothesis about the expected correlation between two variables are frequent in genetic work, and we may wish to test observed data against such a hypothesis. Although there is no a-priori reason to assume that the true correlation between vein lengths of the right and left wings of the bees is 0.5, we show the test of such a hypothesis to illustrate the method. Corresponding to $\rho = 0.5$ is ζ, the parametric value of z—that is, the z-transformation of ρ. Note that the probability that the sample r of 0.837 could have resulted from a population with $\rho = 0.5$ is vanishingly small.

Next in Box 15.4 we show how to assign confidence limits to a sample correlation coefficient r. We use the z-transformation, which will result in asymmetrical confidence limits when these are retransformed to r-scale, as when setting confidence limits with variables subjected to square root or logarithmic transformations. We also show how to determine the sample size needed to detect that a sample correlation is different from a given expected value. The approach is quite similar to that in Box 9.14, except that the solution is found directly, not iteratively, since the appropriate degrees of freedom for the z-transformation are invariably infinity.

Occasionally we measure two variables in several samples and obtain correlation coefficients for each sample. It may be of interest to know whether these correlation coefficients can be considered samples from a population exhibiting a common correlation among the variables. One way of stating the null hypothesis is to say that the k sampled correlation coefficients are homogeneous and estimate a common parametric value of ρ. Such problems arise occasionally in systematics, where taxonomic differentiation results in differences in correlation between body structures. Thus, not only do characters vary among taxa, or geographically within a taxon, but the correlation between pairs of characters also changes in different populations. The example in Box 15.5 illustrates the test of homogeneity among correlation coefficients. In this case the variables are wing length and width of a band on the wing in females of 10 populations of a species of butterfly. These 10 populations were taken from different islands and localities in the Caribbean. Note that the sample correlation coefficients differ considerably, from a low of 0.29 to a high of 0.70. The computations are quite simple and in effect consist of calculating a weighted sum of squares of the z-values corresponding to the correlation coefficients. This sum of squares is called X^2 and is distributed as χ^2 with $k - 1$ degrees of freedom.

The reason that the weighted sum of squares of z should be distributed as χ^2 may not be immediately obvious. This sum of squares is the quantity

$$X^2 = \sum^k \left(\frac{z - \bar{z}}{1/\sqrt{n_i - 3}}\right)^2 = \sum^k (n_i - 3)(z - \bar{z})^2$$

The denominator in the parentheses is the parametric standard deviation of z, σ_z. If we had equal sample sizes, and thus equal standard deviations, we could rewrite this expression as $(1/\sigma_z^2)\Sigma^k(z - \bar{z})^2$. In this form the expression is analogous to Expression (7.7), $(1/\sigma^2)\Sigma(Y_i - \bar{Y})^2$, which, as you may recall, followed the χ^2-distribution with $n - 1$ degrees of freedom. Thus, if the z's represent a

Box 15.5 TEXT OF HOMOGENEITY AMONG TWO OR MORE CORRELATION COEFFICIENTS.

Correlation coefficients between length of wing and width of band on wing in females of 10 populations of the butterfly *Heliconius charitonius based on varying sample sizes.*

(1) n_i	(2) $n_i - 3$	(3) r_i	(4) z_i	(5) Weighted $z_i = (n_i - 3)z_i$
100	97	0.29	0.2986	28.96093
46	43	0.70	0.8673	37.29392
28	25	0.58	0.6625	16.56157
74	71	0.56	0.6328	44.93116
33	30	0.55	0.6184	18.55144
27	24	0.67	0.8107	19.45786
52	49	0.65	0.7753	37.98964
26	23	0.61	0.7089	16.30519
20	17	0.64	0.7582	12.88895
17	14	0.56	0.6328	8.85966
423	393			241.8003

SOURCE: Data from Brown and Comstock (1952).

We wish to test whether the $k = 10$ sample r's could have been taken from the same population; if so, we will combine them into an estimate of ρ.

Computation

1. Compute average z:

$$\bar{z} = \frac{\sum^{k} \text{weighted } z_i}{\sum (n_i - 3)} = \frac{\text{sum of column (5)}}{\text{sum of column (2)}} = \frac{241.8003}{393} = 0.615268$$

2. To test that all r's are from the same population compute

$$X^2 = \sum^{k} (n_i - 3)(z_i - \bar{z})^2 = 15.26352$$

X^2 is to be compared with values in the χ^2-table for $k - 1$ *df* (Statistical Table **D**):

$$\chi^2_{.10[9]} = 14.684 \qquad \chi^2_{.05[9]} = 16.919$$

We do not quite have sufficient evidence to reject the null hypothesis of no heterogeneity among the correlation coefficients.

BOX 15.5 CONTINUED

3. Estimate common ρ: The r-value corresponding to $\bar{z} = 0.615268$ is $r = 0.547825 \approx 0.55$.

The special case of two correlation coefficients

When we have only two correlation coefficients we may test $H_0: \rho_1 = \rho_2$ versus $H_1: \rho_1 \neq \rho_2$ as follows:

$$t_s = \frac{z_1 - z_2}{\sqrt{\dfrac{1}{n_1 - 3} + \dfrac{1}{n_2 - 3}}}$$

Since $z_1 - z_2$ is normally distributed and we are using a parametric standard deviation, we compare t_s with $t_{\alpha[\infty]}$ or employ Statistical Table **A,** areas of the normal curve.

For example, the correlation between body weight and wing length in *Drosophila pseudoobscura* was found by Sokoloff (1966) to be 0.552 in a sample of $n_1 = 39$ at the Grand Canyon and 0.665 in a sample of $n_2 = 20$ at Flagstaff, Arizona.

Grand Canyon: $z_1 = 0.6213$ Flagstaff: $z_2 = 0.8017$

$$t_s = \frac{0.6213 - 0.8017}{\sqrt{\frac{1}{36} + \frac{1}{17}}} = \frac{-0.1804}{\sqrt{0.086601}} = \frac{-0.1804}{0.29428} = -0.6130$$

By linear interpolation in Table **A,** we find the probability of a t_s being between ± 0.6130 to be about $2(0.22941) = 0.45882$, so we clearly have no evidence on which to reject the null hypothesis.

Procedures to follow for small sample sizes are suggested in the text (Section 15.5).

random sample from a single normally distributed population, the value of X^2 obtained as shown here should follow the χ^2-distribution with $k - 1$ degrees of freedom.

In the example in Box 15.5, we find that X^2 is smaller than the 5% critical value of χ^2. We therefore do not reject the null hypothesis. The correlation coefficients in the 10 populations should thus be considered homogeneous (even though the $r = 0.29$ appears quite deviant from the rest). This result is rather typical; the r-values have to be very different or based on larger sample sizes before this test can detect differences. If the sample correlations are adjudged homogeneous, we can use them to estimate an overall average to obtain a common estimate of ρ. We estimate ρ by backtransforming the value of $\bar{z}$ to r. The estimated common correlation is 0.547825.

This average is weighted by the reciprocals of the variances, which are a function of sample size. If some samples are small and an especially precise estimate of ρ is desired, we must allow for a bias introduced in our sample estimate ζ. An unbiased estimate, z_u, is given by

$$z_u = z - \frac{\rho}{2(n-1)}$$

Note that to obtain such an estimate we need to know ρ, which is usually impossible; if we did know ρ, then we would, of course, not be interested in z_u. We therefore obtain a more approximate estimate of ζ from the sample r by the following formula:

$$z_u = z - \frac{r}{(2n-5)}$$

These biases are of the same general type as the bias in estimating the population variance as $\Sigma y^2/n$, first encountered in Section 4.7. In practice we would correct the estimate $\bar{z}$ by subtracting from it

$$\frac{\bar{r}\sum^{k}\left(\frac{n_i - 3}{2n_i - 5}\right)}{\sum^{k}(n_i - 3)}$$

For the example in Box 15.5 this correction amounts to 0.006848, which results in a new estimate of $\bar{z} = 0.615268 - 0.006848 = 0.60842$. The new estimate of $\bar{r}$ is, therefore, 0.543014, which is only slightly different from the original $\bar{r}$ of 0.547825. Although the literature cautions against applying the homogeneity test (and the two-sample test discussed next) to data with small samples, it gives no clear instructions on what constitutes a small sample or on how to proceed in such cases.

A test for the significance of the difference between two sample correlation coefficients is a special case of the test for homogeneity among r's and can be carried out in the manner just described. A simpler procedure is shown at the end of Box 15.5. A standard error for the difference is computed and tested against a table of areas of the normal curve (Table **A**). In the example in Box 15.5 the correlation between body weight and wing length in two *Drosophila* populations was tested, and the difference in correlation coefficients between the two populations was found not significant. The formula given is an acceptable approximation when the smaller of the two samples is greater than 25. It is also frequently used with even smaller sample size, as this example shows.

15.6 Applications of Correlation

The purpose of correlation analysis is to measure the intensity of association observed between two variables and to test whether this correlation greater than could be expected by chance alone. Once established, such an association is likely to lead to reasoning about causal relationships between the variables. Students of statistics are taught early not to confuse significant correlation with causation. We are also warned about so-called nonsense correlations, of which a well-known case is the positive correlation between the number of Baptist ministers and the total liquor consumption in cities with populations of over 10,000 in the United States. Individual cases of correlation must be carefully analyzed before inferences are drawn from them. It is useful to distinguish correlations in which one variable is the entire, or more likely the partial, cause of another, from others in which the two correlated variables have a common cause and from more complicated situations involving both direct influence and common causes.

Almost all observed correlations can be described in terms of one of the structural designs in Figure 15.5 or in terms of a more complicated structure composed of these elements. There are probably few situations in nature in which two variables are totally correlated, either because one is the sole cause of the other (Figure 15.5A) or because both arise from a common sole cause (Figure 15.5B). Exceptions include trivial cases such as age of tree and number of growth rings (an example of Figure 15.5A). The model described by Figure 15.5C is more common. The correlation between age (Y_2) and weight (Y_1) in an animal is likely to be of this type. Factors other than age (Y_3, Y_4) also affect the animal, so the correlation between Y_1 and Y_2 is unlikely to be perfect. Weight (Y_1) and size (Y_2) of an animal usually have age (Y_4) as a common factor, leading to a

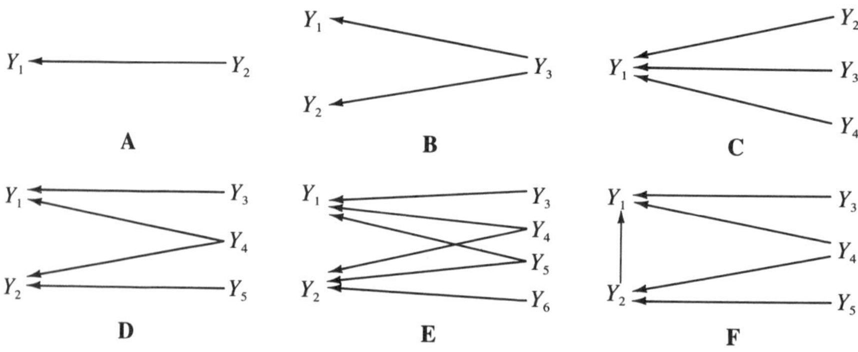

FIGURE 15.5 *Different structural explanations for the observed correlation* r_{12} *between* Y_1 *and* Y_2 *(assuming only linear relations between variables). **A**.* Y_2 *is the entire cause of the variation of* Y_1. *In such a case* r_{12} *must equal 1, and* r_{12}^2 *must also equal 1. **B**. The common cause* Y_3 *totally determines variables* Y_1 *and* Y_2. *Again,* r_{12} *must be 1. **C**. In this case,* Y_2 *is one of several causes of* Y_1, *and* r_{12} *will be less than 1. **D**. The correlation between variables* Y_1 *and* Y_2 *is due to a common cause* Y_4. *Since other causes,* Y_3 *and* Y_5, *also determine* Y_1 *and* Y_2, *respectively, the correlation between these variables is not perfect. **E**. The correlation between variables* Y_1 *and* Y_2 *is due to two common causes,* Y_4 *and* Y_5. ***F**. The correlation between variables* Y_1 *and* Y_2 *is due to the direct effect of* Y_2 *on* Y_1, *as well as to a common cause,* Y_4.

model such as that in Figure 15.5D. Most correlations in nature are likely to be of the type shown in Figure 15.5E, in which two or more common causes (Y_4 and Y_5) lead to correlation of variables Y_1 and Y_2. Correlation between morphological parts in organisms is surely of this type, with many morphogenetic forces exerting an influence on the correlated variables. The next degree of complexity is shown in Figure 15.5F, where the correlation between Y_1 and Y_2 is due both to the direct effect of one of the variables (Y_2) as well as to a common cause (Y_4).

These relationships can be even more complicated. Adult weight and length of developmental period, for example, are correlated in many insects. There is a direct effect of length of developmental period on adult weight: The longer the immature insect feeds, the heavier the adult will be. The density at which the insects were reared, however, affects both larval period and weight. A high density reduces the weight because the food supply decreases, but generally increases the developmental period. Thus, the same common cause increases one but lowers the other of the two correlated variables, hence decreasing the correlation caused by the direct effect of density on weight. Models of this type can rapidly become quite complex, including instances of mutual causation between two variables.

The establishment of a significant correlation does not tell us which of these structural models is appropriate. Further analysis is needed to distinguish between models. Path analysis (see Section 16.3) is especially well suited to the study of models such as those of Figure 15.5.

The traditional distinction of real versus nonsense or illusory correlation is of little use. In the supposedly legitimate correlations, the relations were of the types shown in Figure 15.5 and causal connections were known or at least believed to be clearly understood. In so-called illusory correlations, no reasonable connection between the variables can be found; or if one is demonstrated, it is of no real interest or may be shown to be an artifact of the sampling procedure. For example, the correlation between Baptist ministers and total liquor consumption is simply a consequence of city size. The larger the city, the more Baptist ministers it will contain on the average and the greater the liquor consumption will be. Because city size determines both variables, this case is really an example of Figure 15.5D, correlation due to a common cause. The correlation is of little interest to anyone studying either the distribution of Baptist ministers or the consumption of alcohol. Some correlations have time as the common factor, and processes that change with time are frequently likely to be correlated, not because of any functional biological reasons but simply because the change with time in the two variables under consideration happens to be in the same direction. Thus, size of an insect population increasing through the summer may be correlated with the height of some weeds, but this association may simply be a function of the passage of time. There may be no ecological relation between the plant and the insects.

Perhaps the only correlations properly called nonsense or illusory are those assumed by popular belief or scientific intuition that, when tested by proper statistical methodology using adequate sample sizes, are found to be not significant. Thus, if we can show that there is no significant correlation between the amount of saturated fats eaten and the degree of atherosclerosis, we can consider this correlation to be illusory. Remember also that when testing significance of correlations at conventional levels of significance, you must allow for type I error (see Section 7.8), which will otherwise cause a certain percentage of correlations to be judged significant when in fact the parametric value of $\rho = 0$.

Correlation coefficients have a history of extensive use and application dating back to the English biometric school at the turn of the century. Recently this technique has been applied somewhat less frequently as biological research has become more experimental. In experiments in which one factor is varied and the response of another variable to the deliberate variation of the first is examined, the method of regression is more appropriate, as has already been discussed. Large areas of biology and of other sciences remain, however, where the experimental method is not suitable because variables cannot be brought under control of the investigator. There are many areas of ecology, systematics, evolution, and other fields in which experimental methods are difficult to apply. As yet the weather cannot be controlled, nor can historical evolutionary factors be altered. Nevertheless, we need to understand the scientific mechanisms underlying these phenomena as much as we understand those in biochemistry or experimental embryology. In such cases, correlation analysis is a preliminary descriptive technique to estimate the degrees of association among the variables involved. Other techniques such as factor analysis, briefly discussed in Section 16.7, lead to additional analytical insights into the mechanisms of such phenomena even without applying the experimental method.

15.7 Principal Axes and Confidence Regions

We have used the correlation coefficient as a measure of intensity of association between two variables in a bivariate scattergram of the type in Figure 15.3. Occasionally it is of interest to characterize the trend of the scattered points by a line. A measure of the reliability of the bivariate mean is also desirable. Simple confidence limits are not adequate, since points on the scattergram can vary in two dimensions. Instead we refer to **confidence regions** around the sample mean $(\bar{Y}_1, \bar{Y}_2)$ and around individual observations in the scattergram. In the past, research workers often used regression lines to characterize two-way scattergrams. With correlation analysis, however, in which both variables are random, regression lines are not appropriate. Moreover, deciding which of the two variables to regress on the other would be difficult. Sometimes both regression lines are shown cutting through the scattergram, and the viewer can visualize a trend line intermediate between them. The method of principal axes described in this section provides a single axis to represent the trend expressed by the scattergram and is preferred for other reasons as well, several of which we will mention. We noted in Section 14.13 that the slope of this axis has been suggested as one of the ways of estimating the slope of the functional relationship in Model II regression when both variables are in the same units of measurement.

We saw in Section 15.2 that the bivariate normal distribution can be represented by concentric ellipses describing the topography of a bell-shaped mound. In the univariate work of Chapter 6 we assumed that observed samples had been taken from a normal distribution and, consequently, calculated expected frequencies (areas under the normal curve). Analogously, we will calculate ellipses in these data representing various volumes under the bell-shaped surface of the bivariate normal distribution. In effect, the ellipses are the contour lines of the three-dimensional model of Figure 15.2. In the univariate case, we assigned confidence limits to the mean and calculated expected frequencies for normal distributions whose mean and standard deviation were identical to the sample statistics. These calculations required two parameters: a fixed point, the mean μ for which our best estimate is $\bar{Y}$, and a measure of variation, the standard deviation σ, which we estimate as s. The corresponding values necessary in the bivariate case are the sample mean $(\bar{Y}_1, \bar{Y}_2)$, the standard deviations s_1 and s_2, and the covariance s_{12}. Recall from analytical geometry that an ellipse can be described by two **principal axes,** the major axis and the minor axis at right angles to each other, the major axis being the longest possible axis of the ellipse. Once the required parameters are estimated, we can draw ellipses around the mean.

Our first task, therefore, is to find the slope and equation of the major axis from a sample. A criterion of fitting a regression line through bivariate data is that it must pass through $\bar{X}, \bar{Y})$ and that the SS of the deviations of the observed Y-values from regression is minimal (see Section 14.3). These deviations are parallel to the Y-axis (see Figure 14.7). A corresponding criterion for locating the principal axis is that the SS of deviations, which represent perpendiculars to the principal axes, be at a minimum. These perpendiculars take into consideration the deviation of a given point with respect to both $\bar{Y}_1$ and $\bar{Y}_2$ (Figure 15.6). By

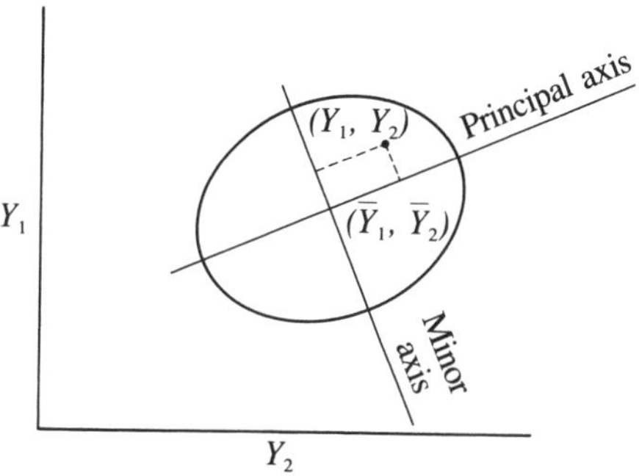

FIGURE 15.6 *Equal-frequency ellipse in which principal and minor axes and deviations of an observation* (Y_1, Y_2) *from these axes are shown. Note that the deviations are measured in a direction perpendicular to the principal and minor axes rather than perpendicular to the abscissa, as was done in regression analysis.*

contrast, with regression the deviation was measured with respect only to Y, but not to X.

The equation of the principal axis defined in this way is $Y_1 = \bar{Y}_1 + b_1(Y_2 - \bar{Y}_2)$. The expression involves the means of the two variables and b_1, the slope of the principal axis. This slope is computed from $b_1 = s_{12}/(\lambda_1 - s_1^2)$, where s_{12} is the covariance $\Sigma y_1 y_2/(n - 1)$, s_1^2 is the variance of Y_1, and λ_1 is a new quantity defined as follows in terms of variances and covariance of Y_1 and Y_2:

$$\lambda_1 = \tfrac{1}{2}[s_1^2 + s_2^2 + \sqrt{(s_1^2 + s_2^2)^2 - 4(s_1^2 s_2^2 - s_{12}^2)}]$$

The slope, b_2, of the minor axis, which is at right angles to the slope of the major axis, can be found easily as $b_2 = -1/b_1$. Details of the computation are given in Box 15.6.

To describe the shape of an ellipse, we must also know the ratio of the lengths of its major and minor axes. This ratio is $\sqrt{\lambda_1/\lambda_2}$, where $\lambda_2 = s_1^2 + s_2^2 - \lambda_1$. In mathematics and statistics λ_1 and λ_2 are important quantities, known by many names. Among other terms, they are called **eigenvalues,** *latent roots,* or *characteristic roots* of the variance–covariance matrix of Y_1 and Y_2. In the type of problem we are discussing here, the eigenvalues are quantities analogous to variances; λ_1 and λ_2 measure variability along the major and minor axes, respectively. The major and minor axes are known as eigenvectors, latent vectors, or characteristic vectors. In data that are highly correlated and consequently represented by a very narrow, elongated ellipse, most of the variance can be accounted for by the major axis; in this case the value of λ_1 would be very great with respect to the magnitude of λ_2. If the quantities λ_1 and λ_2 are equal, the major and minor axes would be of equal length, the data would be represented by a circle, and hence there would be no correlation between the variables. In Figure 15.7 the major and minor axes do not appear to be orthogonal (at right angles) because Y_1 and Y_2 are not drawn to the same scale. In Figure 15.6, however, where both axes are of identical scale, the orthogonality is evident.

Step **5** of Box 15.6 illustrates the computation of approximate confidence limits to the slope of the principal axis. This calculation is based on the approximate procedure for small samples proposed by Jolicoeur (1968), exact only for large n. With small n and low r, H may be > 1 and as a result the limits will not exist.

Step **6** of Box 15.6 shows a method for finding confidence ellipses for the bivariate mean (μ_1, μ_2). Since two variables are involved, we cannot state univariate confidence intervals, but need to identify a confidence region describing an ellipse around the point $(\bar{Y}_1, \bar{Y}_2)$. The shape of the ellipse is a function of the correlation between the variables, and the size (area) of the ellipse is a function of the confidence coefficient $1 - \alpha$. The equation describing the ellipse is shown in Box 15.6. We expect $100(1 - \alpha)\%$ of such ellipses to contain the true mean (μ_1, μ_2). Drawing an ellipse is a tedious procedure because it requires plotting a sufficient number of points to be able to draw a curve connecting them. By substituting our sample estimates of $\bar{Y}_1$ and μ_1 and $\bar{Y}_2$ for μ_2, and evaluating C_α

Box 15.6 PRINCIPAL AXES AND CONFIDENCE REGIONS FOR A BIVARIATE SCATTERGRAM.

Data from Box 15.2.

1. Compute $s_1^2, s_2^2, s_{12}, \overline{Y}_1$, and $\overline{Y}_2$. Using the sums of squares and cross products given in Box 15.2, we find

$$s_1^2 = \frac{\sum y_1^2}{n-1} = \frac{124{,}368.9167}{11} = 11{,}306.26465$$

$$s_2^2 = \frac{\sum y_2^2}{n-1} = \frac{462.4782}{11} = 42.04347$$

$$s_{12} = \frac{\sum y_1 y_2}{n-1} = \frac{6561.6175}{11} = 596.51067$$

$$\overline{Y}_1 = \frac{\sum Y_1}{n} = \frac{2347}{12} = 195.58333$$

$$\overline{Y}_2 = \frac{\sum Y_2}{n} = \frac{144.57}{12} = 12.04750$$

2. The eigenvalues (see text) are found as follows (note that s_{12}^2 is the square of the covariance s_{12}):

Let $D = \sqrt{(s_1^2 + s_2^2)^2 - 4(s_1^2 s_2^2 - s_{12}^2)} = 11{,}327.22290$. Then

$$\lambda_1 = \frac{s_1^2 + s_2^2 + D}{2} = 11{,}337.76550$$

and

$$\lambda_2 = \frac{s_1^2 + s_2^2 - D}{2} = 10.54261$$

or

$$\lambda_2 = s_1^2 + s_2^2 - \lambda_1$$

3. The slope of the principal axis is

$$b_1 = \frac{s_{12}}{\lambda_1 - s_1^2} = 18.93633$$

The equation of this axis is

$$Y_1 = \overline{Y}_1 + b_1(Y_2 - \overline{Y}_2) = -32.55208 + 18.93633\, Y_2$$

and it is plotted in Figure 15.7.

4. The slope of the minor axis (at right angles to the principal axis) is

$$b_2 = \frac{-1}{b_1} = -0.0528$$

Box 15.6 Continued

The equation of the minor axis thus is

$$Y_1 = \bar{Y}_1 + b_2(Y_2 - \bar{Y}_2) = 196.21954 - 0.05281\ Y_2$$

5. 95% confidence limits for β_1, the slope of the principal axis:

$$\text{Let } H = \frac{F_{.05[1,n-2]}}{[(\lambda_1/\lambda_2) + (\lambda_2/\lambda_1) - 2](n-2)}$$

$$= \frac{4.96}{[(11{,}337.76550/10.54261) + (10.54261/11{,}337.76550) - 2]10}$$

$$= 0.000{,}462{,}073$$

and

$$A = \sqrt{\frac{H}{(1-H)}}$$

$$= 0.021{,}500{,}846$$

The 95% confidence limits to the slope β_1 are

$$L_1 = \frac{b_1 - A}{1 + b_1 A} = \frac{18.936 - 0.0215}{1 + 18.936(0.0215)} = \frac{18.9145}{1.407} = 13.44197$$

$$L_2 = \frac{b_1 + A}{1 - b_1 A} = \frac{18.936 + 0.0215}{1 - 18.936(0.0215)} = \frac{18.9575}{0.59287} = 31.97729$$

Shortcut using trigonometric functions:

$$L_1 = \tan\left(\arctan b_1 - \tfrac{1}{2}\arcsin 2\sqrt{H}\right)$$

$$L_2 = \tan\left(\arctan b_1 + \tfrac{1}{2}\arcsin 2\sqrt{H}\right)$$

6. Confidence ellipses for the bivariate mean (μ_1, μ_2): The confidence region describes an ellipse around the point ($\bar{Y}_1$, $\bar{Y}_2$). The equation for a $100(1 - \alpha)$% confidence region is

$$P[s_1^2(Y_2 - \mu_2)^2 - 2s_{12}(Y_1 - \mu_1)(Y_2 - \mu_2) + s_2^2(Y_1 - \mu_1)^2 \le C_\alpha] = 1 - \alpha$$

where μ_1 is the true mean for Y_1, μ_2 is the true mean for Y_2, and

$$C_\alpha = \frac{\lambda_1\lambda_2(n-1)2}{n(n-2)} F_{\alpha[2,n-2]}$$

For a 95% confidence region applied to our example, $C_{.05}$ would be computed as follows:

$$F_{.05[2,10]} = 4.10$$

$$C_{.05} = \frac{(11{,}337.76550)(10.54261)(12-1)2}{12(12-2)} \times 4.10$$

$$= 89{,}846.44602$$

The simplest way to make use of the confidence region is to graph its boundary. Several shortcuts are possible by computing the coordinates of particular points for which most of the equation above reduces to zero.

BOX 15.6 CONTINUED

The following are the coordinates of points a to h on Figure 15.7:

a. $Y_2 = \bar{Y}_2 = 12.0475$

$$Y_1 = \bar{Y}_1 + \sqrt{\frac{C_\alpha}{s_2^2}} = 241.8109$$

b. $Y_2 = \bar{Y}_2 = 12.0475$

$$Y_1 = \bar{Y}_1 - \sqrt{\frac{C_\alpha}{s_2^2}} = 149.3557$$

c. $Y_2 = \bar{Y}_2 + \sqrt{\frac{C_\alpha}{s_1^2}} = 14.8665$

$$Y_1 = \bar{Y}_1 = 195.5833$$

d. $Y_2 = \bar{Y}_2 - \sqrt{\frac{C_\alpha}{s_1^2}} = 9.2285$

$$Y_1 = \bar{Y}_1 = 195.5833$$

e. $Y_2 = \bar{Y}_2 + \sqrt{\frac{C_\alpha}{\lambda_2(1 + b_1^2)}} = 16.9158$

$$Y_1 = \bar{Y}_1 + b_1(Y_2 - \bar{Y}_2) = 287.7707$$

f. $Y_2 = \bar{Y}_2 - \sqrt{\frac{C_\alpha}{\lambda_2(1 + b_1^2)}} = 7.1792$

$$Y_1 = \bar{Y}_1 + b_1(Y_2 - \bar{Y}_2) = 103.3959$$

g. $Y_2 = \bar{Y}_2 = \sqrt{\frac{C_\alpha}{\lambda_1(1 + b_2^2)}} = 14.8586$

$$Y_1 = \bar{Y}_1 + b_2(Y_2 - \bar{Y}_2) = 195.4349$$

h. $Y_2 = \bar{Y}_2 - \sqrt{\frac{C_\alpha}{\lambda_1(1 + b_2^2)}} = 9.2364$

$$Y_1 = \bar{Y}_1 + b_2(Y_2 - \bar{Y}_2) = 195.7318$$

When these points are plotted and the axes drawn, it becomes apparent that the two axes will not appear to be at right angles unless Y_1 and Y_2 are drawn to the same scale. This was not practical for Figure 15.7, since the units of measurement were very different.

It is sometimes of interest to draw ''equal-frequency ellipses'' to describe the covariation found in a sample. Such an ellipse is the bivariate analogue of the $100(1 - \alpha)\%$ expected range of items in a univariate distribution. One can construct an ellipse so that it encloses about $100(1 - \alpha)\%$ of the observations in the given sample. To do this, set

$$C_\alpha = \frac{\lambda_1\lambda_2(n - 1)2}{(n - 2)} F_{\alpha[2,n-2]}$$

and use the procedures described in step **6** for confidence regions.

as shown in the box, we can solve for various sample values of Y_1, given possible values of Y_2. These equations are tedious to work out, so we try to plot only points for which most of the equation can be simplified considerably. We have labeled such points with lower case letters in the accompanying Figure 15.7. At points a and b, $Y_2 = \bar{Y}_2$, at c and d the ellipse is cut at $Y_1 = \bar{Y}_1$, at e and f the ellipse is cut by the major axis, and at g and h it is cut by the minor axis. These points are evaluated for the example of Box 15.6 and plotted in Figure 15.7.

This plot is an equal-frequency ellipse of means, corresponding to a confidence interval in the univariate case. If we desire an equal-frequency ellipse for observations analogous to a range of $\pm 1.960s$ around a sample mean, which would yield the expected 95% range of items, we use a different formula for C_α, which has the same relation to the previous C_α as the variance of items has to the variance of means (see Box 15.6 for the formula). All of these computations are performed by the correlation routine in the BIOM-pc computer program. Altman (1978) presents another, slightly better technique for plotting equal-frequency ellipses by hand.

The technique of principal axes is important in multivariate analysis, where instead of ellipses we encounter clouds of observations describing hyperellipsoids in a multidimensional space. To simplify the description of these clouds of points, we calculate principal axes through the hyperellipsoids. An important property that we have not yet emphasized is that the eigenvalues, which represent the variance along the principal axes, are such that as we determine successive principal axes representing the major axis, the second major axis, and so

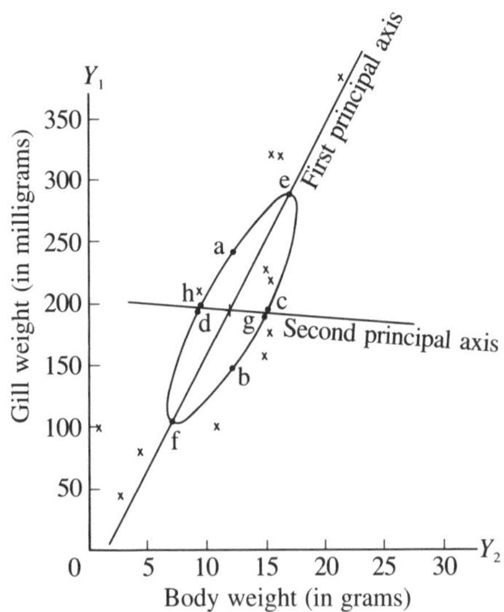

FIGURE 15.7 *95% confidence region for the bivariate mean (μ_1, μ_2) using the crab data of Box 15.1. Axes do not appear at right angles since the ordinate and abscissa are not drawn to the same scale. (See text for explanation.)*

forth of the hyperellipsoid, we are successively finding the orthogonal axis that accounts for the greatest, second greatest, and successively smaller amounts of variation. This technique is called principal component analysis, a branch of multivariate analysis, which is discussed in Section 16.7.

15.8 Nonparametric Tests for Association

Remember that the product–moment correlation coefficient describes the linear component of relations between two variables. Depending on the nonlinearity and the range of values used in computing the coefficient, even a strictly deterministic phenomenon may have a nonsignificant linear correlation. For example, the correlation between X and Y for the relation $Y = (X - 5)^2$ in the range $1 \leq X \leq 10$ is $r = 0.37$ (not significant at 5% for a sample size of 10). For the range $11 \leq X \leq 20$, however, $r = 0.99$, which is highly significant.

When we know that our data do not conform to a bivariate normal distribution, but we wish to test for the significance of association between the two variables, one method of analysis is to rank the variates and calculate a coefficient of rank correlation. This approach belongs to the general family of nonparametric methods we encountered in Chapter 13, where we learned methods for analyses of ranked variates paralleling anovas. In other cases especially suited to ranking methods, we cannot measure the variable on an absolute scale, but only on an ordinal scale. This limitation is typical of data in which we estimate relative performance, as in assigning positions in a class. For example, we can say that A is the best student, B the second best student, C and D are both equal to each other and next best, and so on. Two instructors may independently rank a group of students, and we can then test whether these two sets of rankings are independent, which one would not expect if the judgments of the instructors are based on objective evidence. Of greater biological interest are the following examples. We might wish to correlate order of emergence in a sample of insects with a ranking in size, or order of germination in a sample of plants with rank order of flowering. A geneticist might predict the rank order of performance of a series of n genotypes he synthesizes and wants to show the correlation of his prediction with the rank orders of the realized performance of these genotypes. A taxonomist might wish to array n organisms from those most like form X to those least like it. Will a similar array prepared by a second taxonomist be significantly correlated with the first—that is, are the taxonomic judgments of the two observers correlated?

There are several rank correlation coefficients. In Box 15.7 we present **Kendall's coefficient of rank correlation,** generally symbolized by τ (the Greek letter tau), although it is a sample statistic, not a parameter. The formula for Kendall's coefficient of rank correlation is $\tau = N/n(n - 1)$, where n, as usual, is the sample size and N is a count of ranks that can be obtained in a variety of ways (we show three simple approaches in Box 15.7). If a second variable, Y_2, is perfectly correlated with the first variable, Y_1, then the Y_2-variates should be in

Box 15.7 KENDALL'S COEFFICIENT OF RANK CORRELATION, τ.

Computation of a rank correlation coefficient between the total length (Y_1) of 15 aphid stem mothers and the mean thorax length (Y_2) of their parthenogenetic offspring (based on measurement of four alates, or winged forms); $n = 15$ pairs of observations. Measurements are in micrometer units.

(1) Stem mother	(2) Y_1	(3) R_1	(4) Y_2	(5) R_2	(1) Stem mother	(2) Y_1	(3) R_1	(4) Y_2	(5) R_2
1	8.7	8	5.95	9	8	6.5	2	4.18	1
2	8.5	6	5.65	4	9	6.6	3	6.15	13
3	9.4	9	6.00	10	10	10.6	12	5.93	8
4	10.0	10	5.70	6.5	11	10.2	11	5.70	6.5
5	6.3	1	4.70	2	12	7.2	4	5.68	5
6	7.8	5	5.53	3	13	8.6	7	6.13	12
7	11.9	15	6.40	15	14	11.1	13	6.30	14
					15	11.6	14	6.03	11

SOURCE: Data from a more extensive study by R. R. Sokal (unpublished results).

Computation

1. Rank variables Y_1 and Y_2 separately and then replace the original variates with the ranks (assign average ranks in the case of ties). These ranks are listed in columns (3) and (5) above. Because they were tied, the 4th and 11th variates of variable Y_2 were assigned an average rank of 6.5.

2. Write down the n ranks of one of the two variables in order, paired with the rank values assigned for the other variable (as shown in the next table). If only one variable has ties, order the pairs by the variable without ties (as in the present example). If both variables have ties, it does not matter which of the variables is ordered. There are three equivalent methods for computing the rank correlation coefficient.

3a. The conventional method is to obtain a sum of the counts C_i, as follows. Examine the first value in the column of ranks R_2 paired with the ordered column. In our case, this is rank 2. Count all ranks subsequent to it that are higher than the rank being considered. In this example, count all ranks greater than 2. There are 14 ranks following the 2, and all of them except rank 1 are greater than 2. Therefore, we count a score of $C_1 = 13$. Now we look at the next rank (rank 1) and find that all 13 subsequent ranks are greater than it; therefore, C_2 is also equal to 13. C_3, however, is equal to 2 because only ranks 14 and 15 are higher than rank 13. Continue in this manner, taking each rank of the variable in turn and counting the number of higher ranks subsequent to it. This procedure can usually be done in one's head, but we

BOX 15.7 CONTINUED

show it here explicitly so that the method will be entirely clear. In the case of ties, count a $\frac{1}{2}$. Thus, for C_{10}, there are $4\frac{1}{2}$ ranks greater than the first rank of 6.5.

R_1	R_2	Subsequent ranks greater than pivotal rank R_2	Counts C_i
1	2	13, 5, 3, 4, 12, 9, 10, 6.5, 6.5, 8, 14, 11, 15	13
2	1	13, 5, 3, 4, 12, 9, 10, 6.5, 6.5, 8, 14, 11, 15	13
3	13	14, 15	2
4	5	12, 9, 10, 6.5, 6.5, 8, 14, 11, 15	9
5	3	4, 12, 9, 10, 6.5, 6.5, 8, 14, 11, 15	10
6	4	12, 9, 10, 6.5, 6.5, 8, 14, 11, 15	9
7	12	14, 15	2
8	9	10, 14, 11, 15	4
9	10	14, 11, 15	3
10	6.5	(6.5), 8, 14, 11, 15	$4\frac{1}{2}$
11	6.5	8, 14, 11, 15	4
12	8	14, 11, 15	3
13	14	15	1
14	11	15	1
15	15		0
			$78\frac{1}{2} = \sum^{n} C_i$

We then compute the following quantity:

$$N = 4 \sum^{n} C_i - n(n-1) = 4(78\tfrac{1}{2}) - 15(14) = 314 - 210 = 104$$

3b. An equivalent procedure, which is simpler when there are no ties in the R_1 ranks, is to look at each rank in the R_2 column not in the order in which they appear in the column, but in numerical order, crossing out the rank after we are done with it. With each rank in succession, we count the number of ranks (k_i) that are not crossed out, *no matter of what magnitude,* that appear to the right of (or below) rank i. In this example, we start with rank 1 (the second rank in the R_2 column). Since 13 ranks appear below rank 1 and none of these ranks have been crossed out yet, $k_1 = 13$. We cross out rank 1. Next we look at rank 2. Since rank 1 has been crossed out, k_2 also equals 13. We cross out rank 2. Since 10 ranks follow rank 3, $k_3 = 10$, and so on. We compute the sum of all the counts:

$$\sum^{n} k_i = 13 + 13 + 10 + 9 + 9 + 5 + 4 + 3 + 4 + 3 + 1 + 2 + 2 + 1 + 0$$
$$= 79$$

The quantity N is now defined as

$$N = 4 \sum^{n} k_i - \sum^{m} T_2 - n(n-1)$$

BOX 15.7 CONTINUED

The new quantity $\Sigma^m T_2$ is a correction term needed to correct for the presence of ties in the R_2 ranks (it is zero if there are no ties). A T-value equal to $t(t - 1)$ is computed for each group of t tied variates and summed over m such groups. In our case, there is $m = 1$ group of $t = 2$ tied variates; therefore

$$\sum^m T_2 = T = 2(1) = 2,$$

$$N = 4(79) - 2 - 15(14) = 316 - 2 - 210 = 104$$

Note that $\sum^n C_i = \sum^n k_i - \dfrac{\sum^m T_2}{4}$.

3c. The following semigraphic procedure is often very convenient. Draw lines connecting identical ranks. In the case of ties, draw the lines from the tied ranks in such a way that they do not intersect. Count the number of intersections formed by the lines ($X = 26$ in the present case). The lines need not be straight but they should be drawn so as to avoid ambiguous multiple intersections.

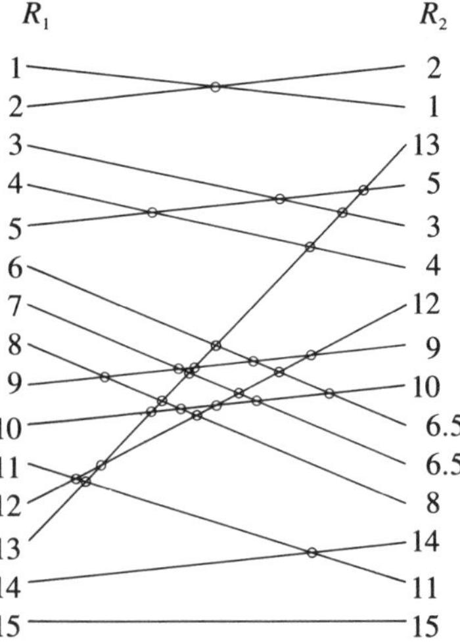

The quantity N previously found by counting ranks can now be computed as follows:

$$N = n(n - 1) - 4X - \sum^m T_2$$

where $\Sigma^m T_2$ is the correction term for ties defined above. As we saw in step **3b,** in the present example $\Sigma^m T_2 = 2$. Therefore

$$N = 15(14) - 4(26) - 2 = 210 - 104 - 2 = 104$$

If there is a negative correlation between the ranks, ranking one of the variables in reverse order will simplify the counting of intersections. The sign of N will then have to be reversed.

BOX 15.7 CONTINUED

4. The Kendall coefficient of rank correlation, τ, can be found as follows:

$$\tau = \frac{N}{\sqrt{\left[n(n-1) - \sum^{m} T_1\right]\left[n(n-1) - \sum^{m} T_2\right]}}$$

where $\Sigma^m T_1$ and $\Sigma^m T_2$ are the sums of correction terms for ties in the ranks of variable Y_1 and Y_2, respectively, defined as in step **3b.** In our example $\Sigma^m T_1 = 0$, since there were no ties in the ranks R_1; $\Sigma^m T_2 = 2$ because of one group of $t = 2$ tied ranks R_2. $T = t(t-1) = 2(2-1) = 2$. Had there been more groups of ties, we would have summed the T's. Thus,

$$\tau = \frac{104}{\sqrt{[15(14)][15(14) - 2]}} = 0.4976$$

If there are no ties, the equation can be simplified to

$$\tau = \frac{N}{n(n-1)}$$

J. S. Smart (personal communication) suggests the following treatment for ties. When it is believed that the ties are real, as for example scores of error-free integers, then the procedure described above is appropriate. When it is believed that the ties are only apparent, however, because differences in items are being obscured by limitations on precision of measurement or judgment, then it would be appropriate to average the correlation coefficient over all possible ways of assigning untied ranks to the tied observations. This result can be obtained directly by using formula $N/n(n-1)$, where the numerator is the N corrected for ties as shown above and the denominator is the one appropriate for the case without ties. It could be argued that the present example falls into the second category. The aphids are probably not truly tied in length, but the measurement device was unable to discriminate at a more precise level. Thus $T = 104/15(14) = 0.4952$, a very slight decrease in the strength of the association.

5. To test significance for sample sizes > 40, we can use a normal approximation to test the null hypothesis that the true value of $\tau = 0$:

$$t_s = \frac{\tau}{\sqrt{2(2n+5)/9n(n-1)}} = \frac{0.4976}{\sqrt{2[2(15)+5]/9(15)(14)}}$$

$$= 2.59 \qquad \text{compared with } t_{\alpha[\infty]}$$

When we look up this value in Statistical Table **A** (areas of the normal curve), we find the probability of such a t_s arising by chance to be 0.0096 (both tails).

When $n \leq 40$, the approximation given here is not adequate, and Statistical Table **S** must be consulted. The table gives various (two-tailed) critical values for $n = 4$ to 40. These are exact only if there are no ties. If there are ties, a special table must be consulted (see Burr, 1960).

the same order as the Y_1-variates. If the correlation is less than perfect, however, the order of the Y_2-variates will not entirely correspond to that of the Y_1. The quantity N measures how well the second variable corresponds to the order of the first. It has a maximal value of $n(n - 1)$ and a minimal value of $-n(n - 1)$. The following sample will make this clear. Suppose we have a sample of five individuals that have been arrayed by rank of variable Y_1:

Y_1	1 2 3 4 5
Y_2	1 3 2 5 4

Note that the ranking by variable Y_2 is not totally concordant with that by Y_1. One of the techniques in Box 15.7 (step **3a**) is to count the number of higher ranks following any given rank, sum this quantity for all ranks, multiply the sum, $\Sigma^n C_i$, by 4, and subtract from it a correction factor $n(n - 1)$ to obtain a statistic N. For variable Y_1 we find $\Sigma^n C_i = 4 + 3 + 2 + 1 + 0 = 10$; then we compute $N = 4\Sigma^n C_i - n(n - 1) = 40 - 5(4) = 20$ to obtain the maximum possible score $N = n(n - 1) = 20$. Obviously, Y_1 being ordered is always perfectly concordant with itself. For Y_2, however, we obtain $\Sigma^n C_i = 4 + 2 + 2 + 0 + 0 = 8$, and $N = 4(8) - 5(4) = 12$. Since the maximum score of N is $n(n - 1) = 20$ and the observed score 12, an obvious coefficient suggests itself as

$$\tau = \frac{N}{n(n-1)} = \frac{4\sum^{n} C_i - n(n-1)}{n(n-1)} = \frac{12}{20} = 0.6$$

Kendall's τ is the converse of another coefficient called the **coefficient of disarray,** equal to $-N/n(n - 1)$, which measures the degree to which two sets of rankings are in disarray, or not ordered in the same way. Frequently, ties present minor computational complications, which are dealt with in Box 15.7, where the correlation is between total body size of aphid stem mothers and mean thorax length of their offspring. In this case, there was no special need to turn to rank correlation except that there is some evidence that these data are bimodal and not normally distributed. The significance of τ for sample sizes greater than 40 can easily be tested by a standard error shown in Box 15.7. For sample sizes up to 40 look up critical values of N in Table **S.** This table is exact for samples without ties only. A table featuring ties is given in Burr (1960). The normal approximation given in step **5** of Box 15.7 is not conservative and should be used with some caution.

Another rank correlation coefficient was devised by Charles E. Spearman. It is computed for data arranged in a manner similar to that described in Box 15.7. There is no simple mathematical relation between the two coefficients of rank correlation. Spearman's coefficient, r_s, can be computed directly from the differences between the ranks R_1 and R_2 of paired variables 1 and 2 as follows:

$$r_s = 1 - \frac{6\sum^{n} (R_1 - R_2)^2}{n(n^2 - 1)}$$

Box 15.8 OLMSTEAD AND TUKEY'S CORNER TEST FOR ASSOCIATION.

A graphic "quick and dirty" method for determining the presence but not the magnitude of correlation. Data from Box 15.7: $n = 15$.

1. Prepare a scattergram of the data to be tested. The scattergram in Figure 15.8 shows total length of stem mother along the ordinate, Y_1, and mean thorax length of alates along the abscissa, Y_2.
2. Draw the medians of Y_1 and Y_2 into the scattergram. When n is an odd number, the median lines will run across the median item of each array. When n is an even number, the median lines will pass between the pair of central variates of each array. Label the upper right and lower left quadrants + and the upper left and lower right quadrants −, respectively.
3. Apply a ruler to the left side of the scattergram. Note the leftmost point and the sign of the quadrant in which it is located. Now slowly move the rule to the right. Count 1 for each point, including the first, preceded by the sign of the quadrant in which it is located. Stop counting as soon as you reach a point that crosses the median perpendicular to your rule (the Y_1 median in this case). In other words, if your first point is +1, stop counting when you reach a negative point or vice versa. Enter the count ("quadrant count") of your points preceded by the appropriate sign at the left side of a square table.

   Repeat this procedure starting at the bottom of the scattergram. Enter the quadrant count at the bottom of the table. Continue with the right side and top of the scattergram and enter quadrant counts at the right and top sides of the table, respectively.

   Following this procedure on the scattergram in Figure 15.8, we obtain +5 for the left quadrant, +2 for the bottom, and +2 for the right. Obtaining the top quadrant count involves some complications. Since the sample size is odd in this example, two points will lie on the medians (unless the same point is the median observation for both variables). The fourth point from the top lies on the Y_2 median at $Y_1 = 10.6$. In such a case, we are told to ignore this point (as well as the point on the Y_1 median at $Y_2 = 5.95$) and replace them by a single point at $Y_1 = 10.6$ and $Y_2 = 5.95$. This new point is marked by a triangle in the figure. The top quadrant count becomes +4.

   The broken lines in Figure 15.8 have been drawn to indicate the limits across which points change sign.
4. Sum the four quadrant counts, take the absolute value of this "quadrant sum," and read the probability from Statistical Table **T.** In the present example the absolute value of the quadrant sum $|S| = 13$, which is significant at 0.02. Thus, it is quite improbable that the association observed is due to sampling error.

NOTE: Another complication, not encountered in our example, occurs when on moving the rule inward, we are faced with a tie—two or more points on the same level and carrying opposite signs. In such a case, count the number of points in the tied group with the same sign as the quadrant counts and divide by (1 + the number of points with the opposite sign). In the case of two tied points, this will give $\frac{1}{2}$.

Alternatively, r_s can be computed as the product–moment correlation coefficient of the two columns of ranks.

If the two variables Y_1 and Y_2 are independent, then the numerical value of Spearman's coefficient of rank correlation is highly correlated with that of Kendall's coefficient ($\rho_{r_s\tau} = 1$ for $n = 2$, decreases to 0.98 for $n = 5$, and then increases to 1.0 again as $n \rightarrow \infty$). If the true correlation is not zero, the two coefficients are sensitive to different types of departures from independence. Specifically, r_s gives greater weight to pairs of ranks that are further apart, while τ weights each disagreement in rank equally. Therefore r_s is more appropriate when there is less certainty about the reliability of close ranks. When $n \leq 10$, special tables are needed for significance testing. Table P in Siegel (1956) furnishes critical values of r_s up to $n = 30$ (for even n only above $n = 10$); table I in Bradley (1968) provides critical values of $D = \Sigma(R_1 - R_2)^2$ for all samples sizes up to $n = 30$. For $n > 10$, one can test r_s as an ordinary product–moment correlation coefficient with relatively little error. However, the normal approximation for τ is considered more accurate than that for r_s.

When ranks of more than two variables are to be compared, one can obtain an overall measure of agreement among the rankings of the k variables. This value is W, **Kendall's coefficient of concordance.** The data would be set up as in the

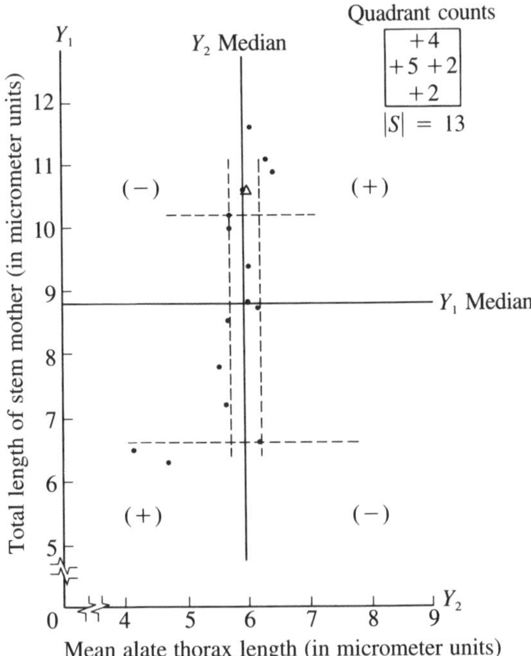

FIGURE 15.8 *Scattergram of aphid morphological data from Box 15.7, illustrating the procedure for Olmstead and Tukey's corner test for association. (See Box 15.8 for explanation.)*

second table of Box 13.10 (Friedman's method for randomized blocks). We could ask the following question for that example: Do the temperature rankings at the 10 depths agree for the four days of the study? The formula of the coefficient of concordance is

$$W = \frac{X^2}{k(n-1)} \tag{15.16}$$

where X^2 is as defined in Box 13.10, k is the number of variables, and n is the number of items per variable. Note that in the symbolism of Box 13.10, $k = b$ and $n = a$. Thus for the example of the temperatures of Rot Lake, we find $W = 36/4(9) = 1.000$. Values of W can range from 0 to 1. It should not surprise us that there is perfect concordance among the temperature readings, since this fact is obvious on visual inspection of the data. The significance test for W is the same as that for Friedman's method: X^2 is compared with $\chi^2_{[n-1]}$.

It is sometimes desirable to establish the significance of the association between two variables not by computation, but by inspection of a scattergram. **Olmstead and Tukey's corner test for association** is designed for this purpose and permits a significance test for plotted points, regardless of whether the exact numerical values are known to the investigator. The method is described in Box 15.8 and illustrated in Figure 15.8.

EXERCISES 15

15.1 Graph the following data in the form of a bivariate scatter diagram. Compute the correlation coefficient and set 95% confidence intervals to ρ for these data from a study of geographic variation in the aphid *Pemphigus populitransversus*. The values in the data table represent locality means based on equal sample sizes for 23 localities in eastern North America. The variables, extracted from Sokal and Thomas (1965), are expressed in millimeters. Y_1 = tibia length; Y_2 = tarsus length. The correlation coefficient will estimate correlation of these two variables over localities. *Answer:* $r = 0.910$.

Locality code number	Y_1	Y_2
1	0.631	0.140
2	0.644	0.139
3	0.612	0.140
4	0.632	0.141
5	0.675	0.155
6	0.653	0.148
7	0.655	0.146
8	0.615	0.136
9	0.712	0.159
10	0.626	0.140

Locality code number	Y_1	Y_2
11	0.597	0.133
12	0.625	0.144
13	0.657	0.147
14	0.586	0.134
15	0.574	0.134
16	0.551	0.127
17	0.556	0.130
18	0.665	0.147
19	0.585	0.138
20	0.629	0.150
21	0.671	0.148
22	0.703	0.151
23	0.662	0.142

IMPORTANT NOTE: The significance of this correlation, calculated in the usual way, is likely to be overstated because these data are spatially autocorrelated (Sokal and Oden, 1978a, b). That is, individual localities are not truly independent observations, but nearby localities tend to be similar. Clifford, Richardson, and Hémon (1989) present a method of correcting the statistical significance of correlation coefficients based on spatially autocorrelated data.

15.2 This two-way frequency distribution was presented by Band and Ives (1963) in a study of viability of different combinations of chromosomes in drosophila. Each variate represents means of paired cultures run at two temperatures.

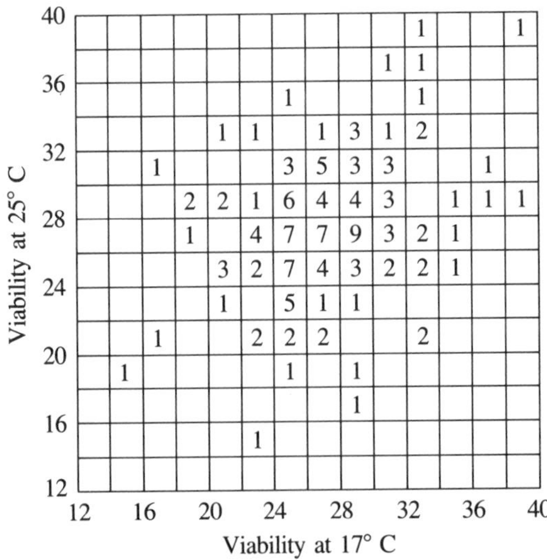

Note that the figures given in the margin are class limits, not class marks (see Section 2.5). Compute the correlation coefficient. Test the null hypothesis $\rho = 0$ against the alternative hypothesis $\rho > 0$.

15.3 The following data were extracted from a larger study of Brower (1959) on speciation in a group of swallowtail butterflies. Morphological measurements are in millimeters coded $\times$ 8.

Species	Specimen number	Y_1 Length of 8th tergite	Y_2 Length of superuncus
Papilio multicaudatus	1	24.0	14.0
	2	21.0	15.0
	3	20.0	17.5
	4	21.5	16.5
	5	21.5	16.0
	6	25.5	16.0
	7	25.5	17.5
	8	28.5	16.5
	9	23.5	15.0
	10	22.0	15.5
	11	22.5	17.5
	12	20.5	19.0
	13	21.0	13.5
	14	19.5	19.0
	15	26.0	18.0
	16	23.0	17.0
	17	21.0	18.0
	18	21.0	17.0
	19	20.5	16.0
	20	22.5	15.5
Papilio rutulus	21	20.0	11.5
	22	21.5	11.0
	23	18.5	10.0
	24	20.0	11.0
	25	19.0	11.0
	26	20.5	11.0
	27	19.5	11.0
	28	19.0	10.5
	29	21.5	11.0
	30	20.0	11.5
	31	21.5	10.0
	32	20.5	12.0

Species	Specimen number	Y_1 Length of 8th tergite	Y_2 Length of superuncus
Papilio rutulus (continued)	33	20.0	10.5
	34	21.5	12.5
	35	17.5	12.0
	36	21.0	12.5
	37	21.0	11.5
	38	21.0	12.0
	39	19.5	10.5
	40	19.0	11.0
	41	18.0	11.5
	42	21.5	10.5
	43	23.0	11.0
	44	22.5	11.5
	45	19.0	13.0
	47	22.5	14.0
	48	21.0	12.5
Papilio glaucus	49	17.5	9.0
	50	19.0	9.0
	51	18.0	10.0
	52	19.0	8.0
	53	19.0	8.5
	54	21.0	8.5
	55	19.0	9.0
	56	21.0	9.5
	57	19.0	9.5
	58	17.5	9.0
	64	23.5	10.5
	65	16.0	9.5
	66	19.5	10.0
	67	17.0	7.0
	68	16.0	8.0
	69	18.0	9.0
	70	18.5	8.5
	71	19.0	8.5
	72	16.5	8.5
	73	18.5	9.5
	74	17.5	9.5
	75	19.5	8.0
	76	18.5	8.0
	77	19.0	9.0
	78	20.5	9.0
	79	17.0	9.5

Species	Specimen number	Y_1 Length of 8th tergite	Y_2 Length of superuncus
Papilio glaucus (continued)	80	18.5	9.0
	81	16.5	9.0
	82	17.5	9.0
	83	19.0	9.0
	84	18.5	7.5
	85	18.5	8.5
	86	17.5	9.0
	87	18.5	9.5
	88	16.5	11.0
	89	23.0	9.0
	90	22.0	8.0
	91	18.0	10.0
	92	21.5	9.0
	93	18.5	9.0
	94	20.5	8.0
	95	21.0	10.0
	96	21.0	8.0
	97	22.5	9.5
	98	18.0	10.0
Papilio eurymedon	99	17.5	11.5
	100	17.5	11.5
	101	17.0	11.0
	102	18.5	11.5
	103	19.5	10.0
	104	16.0	11.5
	105	17.0	10.5
	106	17.5	10.5
	107	19.0	10.5
	108	19.0	12.0
	109	19.0	11.5
	110	21.0	11.0
	111	19.5	10.5
	112	19.5	10.5
	113	19.5	9.5
	114	19.0	10.5
	115	19.5	12.0
	116	21.5	11.5
	117	16.5	11.0
	118	21.5	13.5

Compute the correlation coefficient separately for each species. Test for homogeneity of the four correlation coefficients. *Answer:* For *Papilio rutulus* $r = 0.1958$.

15.4 Graph the data for *Papilio glaucus* in Exercise 15.3 in the form of a scatter diagram. (a) Compute the principal and minor axes. (b) Draw the 95% equal frequency ellipse for the scatter diagram. (c) Draw the 95% confidence ellipse for the bivariate mean (μ_1, μ_2).

15.5 Test for the presence of association between tibia length and tarsus length in the data of Exercise 15.1 using (a) Olmstead and Tukey's corner test of association and (b) Kendall's coefficient of rank correlation. *Answer:* $|S| = 27$.

15.6 The following four correlation coefficients are extracted from a paper by Clark (1941). They represent correlations between skull width and skull length in four samples of the deer mouse, *Peromyscus maniculatus sonoriensis,* from different localities. Can we consider the four sample r's to have come from a single population with parametric correlation ρ? If so, what is your estimate of ρ?

	n	r
Miller Canyon, Ariz.	71	0.68
Winslow, Ariz.	54	0.54
San Felipe, Calif.	83	0.56
Victorville, Calif.	57	0.43

15.7 The following table of data is from an unpublished morphometric study of the cottonwood, *Populus deltoides,* by T. J. Crovello. One hundred leaves from one tree were measured when fresh and after drying. The variables shown are fresh leaf width (Y_1) and dry leaf width (Y_2), both in millimeters. Group the data in a two-way frequency distribution and calculate r. *Answer:* For the ungrouped data $r = 0.974$.

Y_1	Y_2	Y_1	Y_2	Y_1	Y_2
90	88	94	87	86	80
88	87	83	81	98	90
55	52	90	88	100	98
100	95	90	84	70	65
86	83	91	90	100	96
90	88	88	86	105	98
82	77	98	94	95	88
78	75	98	94	95	90
115	109	89	85	93	89
100	95	90	86	95	97
110	105	105	102	95	90
84	78	95	90	104	99
76	71	100	93	81	78
100	97	104	99	85	83

Y_1	Y_2	Y_1	Y_2	Y_1	Y_2
110	105	84	78	108	103
95	90	85	81	112	106
99	98	89	92	96	90
104	100	118	110	98	93
92	92	104	97	83	78
80	82	105	100	103	98
110	106	113	108	107	104
105	97	106	100	72	67
101	98	87	83	105	100
95	91	100	96	80	75
80	76	108	103	99	95
103	97	101	97	108	102
108	103	92	90	86	84
113	110	84	85	94	92
90	85	89	85	106	104
97	93	120	111	98	104
107	106	96	86	100	96
112	115	97	94	95	90
101	98	86	83	108	102
95	91				

15.8 Buley (1936) studied the relative consumption of diatoms and dinoflagellates by the California sea mussel, *Mytilus californianus.* Test whether the four variables listed are concordant in their pattern of variation during the course of the study.

	Average numbers of diatoms and dinoflagellates			
	In plankton hauls		In mussel stomachs	
Date	Diatoms/ℓ seawater	Dinofl./ℓ seawater	Diatoms per mussel	Dinofl. per mussel
9/5	7,340	860	3,525	20,200
9/7	6,960	480	13,000	37,750
9/13	1,160	480	2,450	32,700
9/15	1,380	820	897	37,790
9/19	1,480	340	2,500	53,125
10/5	3,960	1,300	2,000	100,350
10/7	3,080	1,860	850	38,940
10/11	80	6,500	525	691,650
10/13	240	5,300	1,575	363,860
10/19	100	2,040	4,900	260,725

	Average numbers of diatoms and dinoflagellates			
	In plankton hauls		In mussel stomachs	
Date	Diatoms/ℓ seawater	Dinofl./ℓ seawater	Diatoms per mussel	Dinofl. per mussel
10/21	480	1,820	500	350,950
10/26	220	4,300	1,860	241,350
11/3	640	2,160	300	137,000
11/11	40	3,260	130	780,018
11/19	0	1,380	625	59,437
11/22	220	740	175	26,370
12/2	180	260	1,500	5,000
12/9	1,020	340	700	1,566
12/16	120	160	3,333	11,000
12/23	820	260	1,300	2,066
12/29	20	40	1,400	930
1/6	100	300	4,000	430
1/13	8,800	300	1,300	1,150
1/20	940	300	550	800
1/27	240	0	1,100	750
2/3	1,100	80	850	500
2/11	27,400	20	2,500	850
2/18	11,240	60	650	150
2/25	103,040	40	2,600	250
3/3	179,760	80	1,550	200
3/10	7,940	40	2,400	900
3/16	20	280	400	2,000
3/23	7,720	100	400	1,550
3/27	55,920	420	50	50
3/28	601,980	300	1,150	3,400
3/31	628,000	3,200	16,800	5,800
Total	1,663,740	40,220	80,345	3,271,557

Counts are averages for 25 liters of seawater and 5 mussel stomachs for each date.

16 Multiple and Curvilinear Regression

A corollary to the uses of regression for studying causation and prediction (as described in Section 14.6) is its application to reduce the unexplained error variance of a variable. Prior to regression on X, all of s_Y^2, the variance of Y, is "unexplained." After regression, the unexplained or residual variance, $s_{Y \cdot X}^2$, is smaller. We could visualize regressing the unexplained deviations $d_{Y \cdot X}$ on a third variable Z in order to remove more of the variance of Y. We would then obtain a new residual variance of Y, $s_{Y \cdot XZ}^2$, in which the portion of the fluctuations of Y determined by X and by Z has been removed. The new residual deviations $d_{Y \cdot XZ}$ would be those that neither X nor Z could explain. This process could go on indefinitely. In fact, this process of successively regressing a dependent variable Y on a series of independent variables, X, Z, W, . . . , or more conventionally, $X_1, X_2, X_3, \ldots$, is analogous to the fundamental process of science. Scientists examine a phenomenon, study its variation, and successively reduce its unexplained variation as more and more of its causes are understood. We might state tongue-in-cheek that the aim of science is to reduce residual variances $\sigma_{Y \cdot X_{i \ldots}}^2$ to zero.

Regressing a variable Y on a series of independent variables could be done by successively regressing deviations in the manner we just described. Generally, however, if we suspect several variables of being functionally related to Y, we try to regress Y on all of them simultaneously. This technique is called **multiple regression.** It computes the least squares best-fitting linear function of two or more independent variables. Multiple regression is explained in Section 16.1, and computations are illustrated for two and three independent variables. Tests of significance in multiple regression are discussed and illustrated in Section 16.2.

In Section 16.3 we discuss path analysis, which is an important technique in its own right and which will help your understanding of various methods employed in this chapter. Path analysis is a method for studying the direct and indirect effects of one set of variables taken as causes on another set taken as effects.

Just as we were able to extend the subject of regression to more than two independent variables, so are we able to study correlations between pairs of variables, other variables being held constant. This topic, partial correlation, and a related topic, multiple correlation, are covered in Section 16.4. In Section 16.5 we discuss how to choose predictor variables in multiple regression studies.

Section 16.6 introduces curvilinear regression for relationships between variables that cannot be expressed as straight lines. Finally, in Section 16.7 we mention briefly some advanced topics in the general area of multivariate analysis, referring to sources where these topics can be pursued in depth.

Although the distinction between Model I and II regression (see Section 14.2) also applies to the methods discussed in this chapter, techniques for handling Model II cases have not been developed. Therefore, exercise caution in interpreting the results obtained by the Model I techniques of this chapter when the assumptions of that model have not been met.

16.1 Multiple Regression: Computation

There are two main purposes of multiple regression analysis. One is to establish a linear prediction equation that will enable a better prediction of a dependent variable Y than would be possible by any single independent variable X_j. Typically, one aims for that subset of all possible predictor variables that predicts a significant and appreciable proportion of the variance of Y, trading off adequacy of prediction against the cost of measuring more predictor variables. The second purpose of multiple regression is to estimate and fit a structural model to "explain" variation in the observations of Y in terms of the independent variables X_j. Among a set of putative causal variables, which ones affect the dependent variable significantly and appreciably, and what are the estimates of the relative magnitudes of the contributions of the independent variables? To examine these goals, consider the following example.

Table 16.1 shows data on air pollution in 41 American cities. The dependent variable is the annual arithmetic mean concentration of sulfur dioxide expressed in micrograms per cubic meter. When data were available, means for three years, 1969, 1970, and 1971, were averaged. The predictor variables are two human ecological variables—number of manufacturing enterprises employing 20 or more workers (X_2), and population size (X_3)—and four climatic averages for weather stations at these cities: average annual temperature (X_1), average annual wind speed (X_4), average annual precipitation (X_5), and average number of days with precipitation per year (X_6).

The multiple regression equation can be stated in two distinct but interrelated ways, the conventional and the standardized form. The conventional equation is

$$\hat{Y} = a + b_{Y1\cdot}X_1 + b_{Y2\cdot}X_2 + \cdots + b_{Yk\cdot}X_k \tag{16.1}$$

Table 16.1 AIR POLLUTION (SO_2 CONTENT OF AIR) IN 41 U.S. CITIES ASSOCIATED WITH SIX ENVIRONMENTAL VARIABLES.

Cities	Y	X_1	X_2	X_3	X_4	X_5	X_6
Phoenix	10	70.3	213	582	6.0	7.05	36
Little Rock	13	61.0	91	132	8.2	48.52	100
San Francisco	12	56.7	453	716	8.7	20.66	67
Denver	17	51.9	454	515	9.0	12.95	86
Hartford	56	49.1	412	158	9.0	43.37	127
Wilmington	36	54.0	80	80	9.0	40.25	114
Washington	29	57.3	434	757	9.3	38.89	111
Jacksonville	14	68.4	136	529	8.8	54.47	116
Miami	10	75.5	207	335	9.0	59.80	128
Atlanta	24	61.5	368	497	9.1	48.34	115
Chicago	110	50.6	3344	3369	10.4	34.44	122
Indianapolis	28	52.3	361	746	9.7	38.74	121
Des Moines	17	49.0	104	201	11.2	30.85	103
Wichita	8	56.6	125	277	12.7	30.58	82
Louisville	30	55.6	291	593	8.3	43.11	123
New Orleans	9	68.3	204	361	8.4	56.77	113
Baltimore	47	55.0	625	905	9.6	41.31	111
Detroit	35	49.9	1064	1513	10.1	30.96	129
Minneapolis–St. Paul	29	43.5	699	744	10.6	25.94	137
Kansas City	14	54.5	381	507	10.0	37.00	99
St. Louis	56	55.9	775	622	9.5	35.89	105
Omaha	14	51.5	181	347	10.9	30.18	98
Albuquerque	11	56.8	46	244	8.9	7.77	58
Albany	46	47.6	44	116	8.8	33.36	135
Buffalo	11	47.1	391	463	12.4	36.11	166
Cincinnati	23	54.0	462	453	7.1	39.04	132

Notes:
Y—SO_2 content of air in micrograms per cubic meter.
X_1—Average annual temperature in °F.
X_2—Number of manufacturing enterprises employing 20 or more workers.
X_3—Population size (1970 census); in thousands.
X_4—Average annual wind speed in miles per hour.
X_5—Average annual precipitation in inches.
X_6—Average number of days with precipitation per year.
Data compiled from several government publications. Cities are in alphabetical order by state.

TABLE 16.1 CONTINUED

Cities	Y	X_1	X_2	X_3	X_4	X_5	X_6
Cleveland	65	49.7	1007	751	10.9	34.99	155
Columbus	26	51.5	266	540	8.6	37.01	134
Philadelphia	69	54.6	1692	1950	9.6	39.93	115
Pittsburgh	61	50.4	347	520	9.4	36.22	147
Providence	94	50.0	343	179	10.6	42.75	125
Memphis	10	61.6	337	624	9.2	49.10	105
Nashville	18	59.4	275	448	7.9	46.00	119
Dallas	9	66.2	641	844	10.9	35.94	78
Houston	10	68.9	721	1233	10.8	48.19	103
Salt Lake City	28	51.0	137	176	8.7	15.17	89
Norfolk	31	59.3	96	308	10.6	44.68	116
Richmond	26	57.8	197	299	7.6	42.59	115
Seattle	29	51.1	379	531	9.4	38.79	164
Charleston	31	55.2	35	71	6.5	40.75	148
Milwaukee	16	45.7	569	717	11.8	29.07	123
Mean	30.049	55.763	463.10	608.61	9.4439	36.769	113.90
s	23.472	7.2277	563.47	579.11	1.4286	11.772	26.506

where the estimate of the dependent variable $\hat{Y}$ is a function of k independent variables $X_1, X_2, \ldots, X_k$. A coefficient such as $b_{Yj\cdot}$ denotes the regression coefficient of Y on variable X_j that one would expect if all the other variables in the regression equation had been held constant experimentally. It is called a **partial regression coefficient.** The period in the subscript of such coefficients separates the two labeled variables (dependent and independent variables) from other independent variables that are held constant. In those cases where the period terminates the subscript, all other independent variables in the study are held constant. Figure 16.1 is a diagrammatic representation of a multiple regression of Y on two independent variables, X_1 and X_2. The shaded area represents the regression plane, which is a generalization of the regression line from Chapter 14. The plane is situated so as to minimize the sum of squared vertical distances from the points to the plane. The slope or tilt of the plane is determined by the partial regression coefficients that correspond to the slope of the plane in the X_1 and in the X_2 directions. Such coefficients express the rate of change of variable Y per unit of variable j with all other variables in the study held constant. Thus, if for the example in Table 16.1 we compute $b_{Y1\cdot 2\ldots 6}$, the partial regression coefficient yields the rate of change of SO_2 concentration as a function of

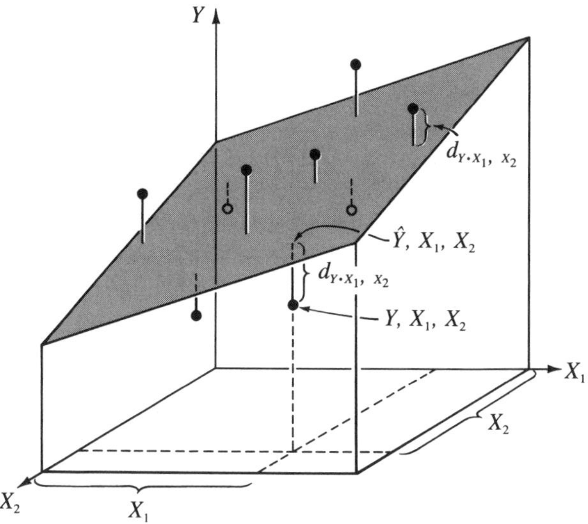

FIGURE 16.1 *Multiple regression of* Y *on* X_1 *and* X_2. *Note that the regression line has become a plane.*

average temperature, with variables X_2 through X_6 kept constant. The coefficient would be expressed as μg/m^3/°F. Expression (16.1) is a direct extension of the by now familiar regression equation $\hat{Y} = a + bX$ from Chapter 14.

In the standardized form of the equation the variables are transformed to standard deviates by subtracting means and dividing by the standard deviation

$$y' = \frac{Y - \bar{Y}}{s_Y}$$

and

$$x_j' = \frac{X_j - \bar{X}_j}{s_{X_j}}$$

In standardized form the multiple regression equation is thus

$$\hat{y}' = b'_{Y1\,.}x_1' + b'_{Y2\,.}x_2' + \cdots + b'_{Yk\,.}x_k' \tag{16.2}$$

Here the coefficients $b'_{Yj\,.}$ are **standard partial regression coefficients** also known as *b-primes, beta coefficients,* or *beta weights,* and are in a simple relation to the conventional partial regression coefficients $b_{Yj\,.}$

$$b'_{Yj\,.} = b_{Yj\,.}\frac{s_{X_j}}{s_Y} \tag{16.3}$$

Instead of expressing a rate of change in the original measurement units as in $b_{Yj\,.}$, the standard partial regression gives the rate of change in standard deviation units of Y per one standard deviation unit of X_j (all other X variables kept constant, of course). One advantage of standard partial regression coefficients is that their magnitudes can be compared directly to show the relative standardized strengths of the effects of several independent variables on the same dependent variable. This property eliminates the effect of differences in measurement scale for different independent variables. Standard partial regression implies that the range of variation in the independent variables is comparable to that expected to be encountered in future studies.

In standard form, the familiar simple linear regression equation would be

$$\hat{y}' = b'_{Y \cdot X} x' \tag{16.4}$$

We saw in Section 15.2 that the standard regression coefficient $b'_{Y \cdot X}$ equals the correlation coefficient r_{XY} [see Expression (15.7)]. Thus another way of looking at a correlation coefficient is as the slope in standard deviation units of Y on X or X on Y. Standard partial regression coefficients, however, are not equal to the partial *correlation* coefficients, which we will discuss in Section 16.4.

Many multiple regression analyses require computation of both conventional and standard partial regression coefficients, and most computer programs routinely furnish both types of coefficients. Once the standard deviations of the variables are known, one type of coefficient is easily transformed into the other. Computing conventional and standard partial regression coefficients directly, however, requires different equations.

The computation of the $b_{Yj \cdot}$ is based on the same principles as those in ordinary regression. As in simple linear regression, we want the coefficients that minimize the sum of squared errors $Y - \hat{Y}$. The derivation of these coefficients involves the calculus of vector variables and results in a series of simultaneous linear equations. These equations are called **normal equations,** and they can be written as follows for the general case:

$$\begin{array}{ccccccccccc}
 & X_1 & & X_2 & & \cdots & & X_k & & Y \\
X_1 & s_1^2\, b_{Y1 \cdot} & + & s_{12} b_{Y2 \cdot} & + & \cdots & + & s_{1k} b_{Yk \cdot} & = & s_{1Y} \\
X_2 & s_{12} b_{Y1 \cdot} & + & s_2^2\, b_{Y2 \cdot} & + & \cdots & + & s_{2k} b_{Yk \cdot} & = & s_{2Y} \\
\vdots & \vdots & & \vdots & & \vdots & & \vdots & & \vdots \\
X_k & s_{1k} b_{Y1 \cdot} & + & s_{2k} b_{Y2 \cdot} & + & \cdots & + & s_k^2\, b_{Yk \cdot} & = & s_{kY}
\end{array} \tag{16.5}$$

Note the regularities in these equations, which are presented so that the variables and coefficients line up as in a table or matrix. The rows of the table are labeled for the independent variables, as are the first k columns of the table. The last column is labeled for the dependent variable Y. The partial regression coefficients are lined up in columns, and there are k such values of $b_{Yj \cdot}$, which are the unknowns in a normal equation. The coefficients in the equations are variances, s_i^2, and covariances, s_{ij}, of the independent variables $X_1, \ldots, X_k$. The right-hand sides of the equations are covariances between variables X_i and Y. The variances s_i^2 along the diagonal correspond to the covariances of each variable with itself. Because there are k equations for k unknowns, we can solve for the regression coefficients using a variety of standard procedures.

When we want standardized partial regression coefficients, we can set up an analogous set of normal equations by replacing the $b_{Yj\,\cdot}$ by $b'_{Yj\,\cdot}$, setting all variances to unity, and replacing all covariances by correlation coefficients. We will show here the method used to determine standard partial regression coefficients to enable the reader to solve small problems with a desk or pocket calculator.

Although eventually we will employ all six predictor variables in our example, we shall first try to account for the amount of pollution Y in terms of two variables—average temperature (X_1) and number of manufacturing enterprises (X_2)—then in terms of three variables, the two already mentioned and population size (X_3). We will proceed in this fashion to explain the method for simple cases, where the amount of computation is moderate. Problems involving four or more independent variables nowadays are almost never carried out without a computer, and the reader usually need not be concerned whether the program solves the normal equations for variances and covariances or for correlations.

Computation of the multiple regression on two (and three) independent variables is explained in Box 16.1. The normal equations for the two variables in standardized form are:

$$\begin{aligned} b'_{Y1\,\cdot} + r_{12}b'_{Y2\,\cdot} &= r_{1Y} \\ r_{12}b'_{Y1\,\cdot} + \quad b'_{Y2\,\cdot} &= r_{2Y} \end{aligned} \tag{16.6}$$

We can solve for the b-primes by the customary algebraic techniques for solving simultaneous equations. Box 16.1 gives explicit solutions for the standard partial regression coefficients in terms of the known correlation for two and three independent variables. These solutions are based on the determinantal method for solving simultaneous equations. No solution is possible if $r_{12} = \pm 1$. In step **2** of part **I** we obtain $b'_{Y1\,\cdot\,2} = -0.3227$ and $b'_{Y2\,\cdot\,1} = 0.5835$. Since these are standard partial regression coefficients they express the average change in standard deviation units of the dependent variable Y for one standard deviation unit of each independent variable, the other one being kept constant. Thus if we consider towns with the same number of manufacturing enterprises (variable X_2 kept constant), $b'_{Y1\,\cdot\,2} = -0.3227$ means that for an increase of one standard deviation of annual temperature, there will be a decrease in the SO_2 pollution by roughly one-third of its standard deviation. By contrast, the number of factories increases pollution; when cities are compared at the same temperature (average annual temperature, X_1, kept constant), an increase of one standard deviation in number of enterprises will result in a corresponding increase of 0.5835 standard deviations of SO_2. The positive effect of manufacturing is greater in absolute magnitude than the negative effect of temperature.

Because the standard partial regression coefficients are free of the original measurement scale, they are generally used to estimate the relative importance of the influence of each of the set of independent variables being considered in the

Box 16.1 MULTIPLE REGRESSION FOR TWO AND THREE INDEPENDENT VARIABLES.

Data from Table 16.1, where means and standard deviations are also furnished.

I. *Two independent variables*

We propose to regress SO_2 content of air, Y, on average temperature, X_1, and number of manufacturing enterprises, X_2.

1. Compute the correlations among the three variables concerned following the procedures of Box 15.2. We obtain

$$r_{12} = -0.1900$$
$$r_{1Y} = -0.4336$$
$$r_{2Y} = 0.6448$$

2. Formulas for the standard partial regression coefficients are as shown below. These are solutions to the normal equations [Expression (16.6)]. As r_{12} approaches unity the denominator must be computed with extra precision.

$$b'_{Y1\cdot 2} = \frac{(r_{1Y} - r_{2Y}r_{12})}{(1 - r_{12}^2)} = \frac{[-0.4336 - (0.6448)(-0.1900)]}{[1 - (-0.1900)^2]}$$

$$= \frac{-0.3111}{0.9639} = -0.3227$$

$$b'_{Y2\cdot 1} = \frac{(r_{2Y} - r_{1Y}r_{12})}{(1 - r_{12}^2)} = \frac{[0.6448 - (-0.4336)(-0.1900)]}{0.9639}$$

$$= \frac{0.5624}{0.9639} = 0.5835$$

3. The multiple regression equation in standard format is

$$\hat{y}' = -0.3227x'_1 + 0.5835x'_2$$

4. Conventional partial regression coefficients, based on Expression (16.3):

$$b_{Y1\cdot 2} = b'_{Y1\cdot 2}\frac{s_Y}{s_{X_1}} = -0.3227\left(\frac{23.472}{7.2277}\right) = -1.0480$$

$$b_{Y2\cdot 1} = b'_{Y2\cdot 1}\frac{s_Y}{s_{X_2}} = 0.5835\left(\frac{23.472}{563.47}\right) = 0.02431$$

5. The Y-intercept is

$$a = \bar{Y} - b_{Y1\cdot 2}\bar{X}_1 - b_{Y2\cdot 1}\bar{X}_2$$
$$= 30.049 - (-1.0480)(55.763) - (0.02431)(463.10)$$
$$= 77.231$$

Box 16.1 Continued

The prediction equation is

$$\hat{Y} = 77.231 - 1.0480X_1 + 0.02431X_2$$

II. *Three independent variables*

Multiple regression of Y on X_1, and X_2 explained in part **I,** as well as on X_3, population size.

1. The necessary correlations among the four variables are

	Y	X_1	X_2	X_3
Y	1			
X_1	−0.4336	1		
X_2	0.6448	−0.1900	1	
X_3	0.4938	−0.0627	0.9553	1

2. Formulas for the standard partial regression coefficients. Compute

$$\begin{aligned} D &= 1 + 2(r_{12}r_{13}r_{23}) - r_{12}^2 - r_{13}^2 - r_{23}^2 \\ &= 1 + 2[(-0.1900)(-0.0627)(0.9553)] - (-0.1900)^2 \\ &\quad - (-0.0627)^2 - (0.9553)^2 \\ &= 0.07013 \end{aligned}$$

$$\begin{aligned} b'_{Y1\cdot 2,3} &= [(1 - r_{23}^2)r_{1Y} + (r_{23}r_{3Y} - r_{2Y})r_{12} + (r_{23}r_{2Y} - r_{3Y})r_{13}]/D \\ &= \{[1 - (0.9553)^2](-0.4336) \\ &\quad + [(0.9553)(0.4938) - 0.6448](-0.1900) \\ &\quad + [(0.9553)(0.6448) - 0.4938](-0.0627)\}/0.07013 \\ &= -0.01267/0.07013 = -0.1807 \end{aligned}$$

$$\begin{aligned} b'_{Y2\cdot 1,3} &= [(1 - r_{13}^2)r_{2Y} + (r_{13}r_{3Y} - r_{1Y})r_{12} + (r_{13}r_{1Y} - r_{3Y})r_{23}]/D \\ &= \{[1 - (-0.0627)^2](0.6448) \\ &\quad + [(-0.0627)(0.4938) - (-0.4336)](-0.1900) \\ &\quad + [(-0.0627)(-0.4336) - (0.4938)](0.9553)\}/0.07013 \\ &= 0.1200/0.07013 = 1.7111 \end{aligned}$$

$$\begin{aligned} b'_{Y3\cdot 1,2} &= [(1 - r_{12}^2)r_{3Y} + (r_{12}r_{2Y} - r_{1Y})r_{13} + (r_{12}r_{1Y} - r_{2Y})r_{23}]/D \\ &= \{[1 - (-0.1900)^2](0.4938) \\ &\quad + [(-0.1900)(0.6448) - (-0.4336)](-0.0627) \\ &\quad + [(-0.1900)(-0.4336) - (0.6448)](0.9553)\}/0.07013 \\ &= -0.08081/0.07013 = -1.1522 \end{aligned}$$

BOX 16.1 CONTINUED

3. The multiple regression equation in standard format is

$$\hat{y}' = -0.1807x'_1 + 1.7111x'_2 - 1.1522x'_3$$

4. Conventional partial regression coefficients, based on Expression (16.3):

$$b_{Y1\cdot 2,3} = b'_{Y1\cdot 2,3}\frac{s_Y}{s_{X_1}} = -0.1807\left(\frac{23.472}{7.2277}\right) = -0.5868$$

$$b_{Y2\cdot 1,3} = b'_{Y2\cdot 1,3}\frac{s_Y}{s_{X_2}} = 1.7111\left(\frac{23.472}{563.47}\right) = 0.07128$$

$$b_{Y3\cdot 1,2} = b'_{Y3\cdot 1,2}\frac{s_Y}{s_{X_3}} = -1.1522\left(\frac{23.472}{579.11}\right) = -0.04670$$

5. The Y-intercept is

$$\begin{aligned} a &= \bar{Y} - b_{Y1\cdot 2,3}\bar{X}_1 - b_{Y2\cdot 1,3}\bar{X}_2 - b_{Y3\cdot 1,2}\bar{X}_3 \\ &= 30.049 - (-0.5868)(55.763) - (0.07128)(463.10) - (-0.04670)(608.61) \\ &= 58.183 \end{aligned}$$

The prediction equation is

$$\hat{Y} = 58.183 - 0.5868X_1 + 0.07128X_2 - 0.04670X_3$$

equation. In this example, manufacturing appears to affect SO_2 pollution more than average temperature does. In the next section we will present a test for the significance of the difference between two b-primes. In Box 16.1 we write the multiple regression equation in step **3** of part **I** in standardized form. This is the equation that would have been obtained if all three variables had been transformed to standard deviates before the computation. Workers concerned with best estimates of the population standard partial regression coefficients, $\beta'_{Yj\,\cdot}$, may wish to apply a correction suggested by Mayer and Younger (1976).

Next we obtain the conventional partial regression coefficients by inverting Expression (16.3), shown as step **4** in part **I** of Box 16.1, which yields $b_{Y1\cdot 2} = -1.0480$ and $b_{Y2\cdot 1} = 0.02431$. Note that $b_{Y1\cdot}$ and $b_{Y2\cdot}$ are in original measurement units. Thus $b_{Y1\cdot 2} = -1.0480$ μg/m^3/°F. To complete the multiple regression equation we replace Y, X_1, and X_2 by $\bar{Y}$, $\bar{X}_1$, and $\bar{X}_2$, and solve for a, the Y-intercept. We obtain $a = 77.231$ in step **5** and can now write out the entire multiple regression equation in terms of original measurement units as

$$\hat{Y} = 77.231 - 1.0480X_1 + 0.02431X_2$$

Note that unless the independent variables are measured in comparable units, you cannot infer the importance of an effect from conventional partial regression coefficients; numerically X_1 now seems to have a greater effect than X_2 does, but we have seen that exactly the converse is true when b-primes are considered.

Next we turn to a computation involving three independent variables. In addition to X_1 and X_2, we will include the variable X_3 from Table 16.1. This variable is population size (to the nearest thousand) in the 1970 census. How can we predict SO_2 content of the air in terms of this variable and the previous two? The problem is worked out in part **II** of Box 16.1. The explicit solutions for the three standard partial regression coefficients are given in step **2;** they are far more complicated than those for only two independent variables. Although formulas for multiple regression coefficients involving more independent variables could be given, they are much too tedious to evaluate without a computer program. The formulas involve division by a quantity D, which is the determinant of the correlation matrix of the independent variables. By referring to the two-variable case you will see that there, too, a constant divisor (also the determinant) was employed for each coefficient. This is because the solution of the normal equations was carried out by means of determinants. The method of determinants is not recommended for solving multiple regression equations involving more than three independent variables because this approach is cumbersome and very sensitive to rounding errors. We will discuss the recommended procedure later in this section.

Inspection of the standard multiple regression equation (step **3** of part **I** or **II** in Box 16.1) reveals three points of interest. Standard partial regression coefficients can be greater than one (one standard deviation change in the independent variable may effect more than one standard deviation change in the dependent variable). More important, the value of $b_{Yj\,\cdot}$ for Y and X_j depends on which other independent variables have been kept constant. Introducing population size X_3 into the equation not only changed the partial regression coefficients on temperature and number of manufacturing establishments, but also altered their relative magnitudes. The reason for this effect is that the independent variables are correlated with each other. Later in this chapter we will see exactly how this correlation affects the standard partial regression coefficients. For now it is important to remember that as we add or drop independent variables, we have to recompute the entire problem with a new set of normal equations, a tedious task impractical without a computer when there are four or more independent variables. The third point is that the sign of a partial regression coefficient need not be the same as that of the correlation between the two variables concerned. Note that $r_{3Y} = 0.4938$, while $b'_{Y3\,\cdot\,1,2} = -1.1522$. Such a disparity is not possible in simple linear regression, where the sign of the covariance in the numerator determines the sign of both the correlation and the regression coefficients. Again, this phenomenon is due to correlation among independent variables and will be discussed later in this chapter.

A corollary to the last two points is that partial regression coefficients may differ in sign and magnitude depending on which independent variables are kept constant. Readers with an appreciation for graphic rather than arithmetic relations may find the illustration in Mullett (1972) instructive. One can plot the three variables

Y	X_1	X_2
−5	−4	3
−7	−2	3
−1	−2	1
−3	0	1
3	0	−1
1	2	−1
7	2	−3
5	4	−3

as simple linear regressions of Y on X_1 to obtain the graph of Figure 16.2A, which shows the least squares equation $\hat{Y} = 1.5X_1$ fitted to these data. If we compute the partial regression coefficient of Y on X_1 with X_2 fixed, we obtain a slope of $b_{Y1 \cdot 2} = -1.0$. We can easily compute the four regression lines of Y on X_1, corresponding to these four values of X_2. Line segments representing these regression lines are shown in Figure 16.2B. Note that the slopes of the partial regression coefficients are now negative, while the overall slope of Y on X_1 is positive. The least squares fit of Y on X_1 given X_2 constant is thus quite different from the least squares fit of Y on X_1 without any second independent variable kept constant.

The new equation based on three variables shows that number of factories still affects SO_2 content of air most strongly, but that temperature is no longer important when factories *and* population size are kept constant. However, the new variable, population size, now affects SO_2 importantly *in a negative direction.* This seems counterintuitive at first. Should we not expect more pollution in more populous cities? Remember, however, that this is a *partial* regression. In cities with the same average temperature and number of factories (i.e., with these variables kept constant), an increase in population size probably implies a larger residential area, hence less SO_2 emissions per unit volume of air. We will return to the interpretation of this case later. The conventional partial regression coefficients and the multiple regression equation are found in steps **4** and **5,** respectively, in part **II** of Box 16.1. The equation $\hat{Y} = 58.183 - 0.5868X_1 + 0.07128X_2 - 0.04670X_3$ permits prediction of the SO_2 content of air for a city with given values for the three variables. Thus in Phoenix, where the average annual temperature is 70.3°F and there are 213 manufacturing establishments with 582,000 inhabitants (see Table 16.1), we would expect

$$\hat{Y} = 58.183 - (0.5868)(70.3) + (0.07128)(213) - (0.04670)(582)$$
$$= 4.9\ \mu\text{g } SO_2/\text{m}^3$$

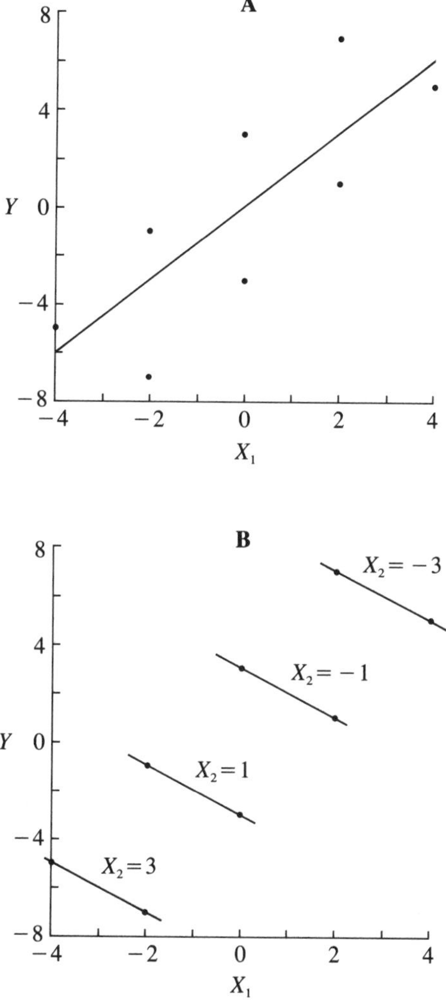

FIGURE 16.2 *Comparison of simple linear and partial regression in an artificial data set.* **A.** *Simple linear regression of* Y *on* X_1. **B.** *The partial regression of* Y *on* X_1, *with variable* X_2 *held constant for the four distinct values of* X_2. *The variates upon which this figure is based are given in the text.*

The observed value is 10 μg SO_2/m^3. As in simple linear regression, some observations are closer to the predicted values, others farther away from them. The goodness of fit will be estimated by means of the unexplained sum of squares, as shown in Section 16.2.

The BIOM-pc package of statistical computer programs solves the normal equations by matrix methods. The program computes standard and conventional partial regression coefficients, the Y-intercept, and other statistics necessary for interpreting the multiple regression equation. Most computer programs solve the normal equations, either in the form already familiar (i.e., solving for b-primes when correlation coefficients are given), or in another form in which conventional partial regression coefficients $b_{Yj\cdot}$ replace the standard ones, sums of products $\Sigma x_i x_j$ and $\Sigma x_i y$ replace the correlations r_{ij} and r_{iY}, respectively, and sums of squares Σx_i^2 replace r_{ii} (equals unity) in the diagonal entries. Thus

$$
\begin{array}{cccccccccc}
 & X_1 & & X_2 & & \cdots & & X_k & & Y \\
X_1 & b_{Y1\cdot}\sum x_1^2 & + & b_{Y2\cdot}\sum x_1x_2 & + & \cdots & + & b_{Yk\cdot}\sum x_1x_k & = & \sum x_1y \\
X_2 & b_{Y1\cdot}\sum x_1x_2 & + & b_{Y2\cdot}\sum x_2^2 & + & \cdots & + & b_{Yk\cdot}\sum x_2x_k & = & \sum x_2y \\
\vdots & \vdots & & \vdots & & \vdots & & \vdots & & \vdots \\
X_k & b_{Y1\cdot}\sum x_1x_k & + & b_{Y2\cdot}\sum x_2x_k & + & \cdots & + & b_{Yk\cdot}\sum x_k^2 & = & \sum x_ky
\end{array}
\tag{16.7}
$$

The only difference between this expression and Expression (16.5) is that the latter is divided by $n - 1$ to change sums of squares and products into variances and covariances. Those of you familiar with elementary matrix algebra (Green and Carrol, 1978, is a good introduction oriented toward statistical applications) will be able to see that the equations in Expression (16.7) can be expressed in matrix form as

$$\mathbf{Ab} = \mathbf{c} \tag{16.8}$$

where **A** is the sum of squares–sum of products matrix (a square, symmetrical matrix of dimension k), **b** is the column vector of unknown partial regression coefficients, and **c** is the column vector of sums of products of the k independent variables with the dependent variable Y. To solve for **b,** we write the equation as

$$\mathbf{b} = \mathbf{A}^{-1}\mathbf{c} \tag{16.9}$$

Thus we can find the regression coefficients by premultiplying **c** by the inverse of the sum of squares–sum of products matrix. The solution by Expression (16.9) is useful because, as you will see in the next section, the elements of the inverse matrix $\mathbf{A}^{-1}$ are needed to compute standard errors for most statistics in multiple regression. These elements of $\mathbf{A}^{-1}$ are usually called **Gaussian multipliers** and are symbolized by c_{ii} and c_{ij} for diagonal and off-diagonal elements, respectively. Since $\mathbf{A}^{-1}$ is also a square, symmetrical matrix of dimension k, there will be k diagonal elements, $c_{11}, \ldots, c_{kk}$, and $k(k - 1)/2$ different off-diagonal elements, $c_{ij} = c_{ji}$. Because calculating the inverse of a sizable matrix is practical only by computer, we will not concern ourselves here with the details of the computation.

16.2 Multiple Regression: Significance Tests

In Section 14.4 we learned how to compute a sum of squares for the deviations $\hat{y} = \hat{Y} - \bar{Y}$ of the predicted values from the mean in a linear regression. This explained sum of squares, $\Sigma\hat{y}^2$, was also used to test significance of the regression by analysis of variance, and in Section 15.2 we saw that the ratio $\Sigma\hat{y}^2/\Sigma y^2$ expresses the proportion of the variation of Y due to variation in X. This proportion was called the coefficient of determination, which in simple linear regression was shown to be r^2_{XY}, the square of the correlation coefficient between the two variables.

We could predict a value of $\hat{Y}$ for each triplet of values of X_1, X_2, and X_3 of the example in Section 16.1, evaluate the deviation $\hat{y} = \hat{Y} - \bar{Y}$, and compute $\Sigma\hat{y}^2_{\cdot 1,2,3}$, the sum of squares of Y explained by variables X_1, X_2, and X_3. The ratio $\Sigma\hat{y}^2_{\cdot 1 \ldots k}/\Sigma y^2$ is known as the **coefficient of multiple determination.** It is generally symbolized by $R^2_{Y\cdot 1 \ldots k}$ and is an estimate of the proportion of the variation of Y jointly explained by variables X_1 through X_k. A more practical way of computing this coefficient is by the following formula, which we state without proof:

$$R^2_{Y\cdot 1 \ldots k} = r_{1Y}b'_{Y1\cdot} + r_{2Y}b'_{Y2\cdot} + \cdots + r_{kY}b'_{Yk\cdot} \tag{16.10}$$

Thus, for the two independent variables in part **I** of Box 16.1, we get

$$\begin{aligned} R^2_{Y\cdot 1,2} &= r_{1Y}b'_{Y1\cdot 2} + r_{2Y}b'_{Y2\cdot 1} \\ &= (-0.4336)(-0.3227) + (0.6448)(0.5835) \\ &= 0.5162 \end{aligned}$$

This value of R^2 shows that almost 52% of the total variation of SO_2 in the air is explained by average annual temperature and number of manufacturing establishments (in the population of cities from which our 41 cities are a sample). Remember that the significance test for linear regression was carried out by an anova and that in Section 15.2 we learned one of several ways to compute the explained and unexplained sums of squares. We showed there that $\Sigma\hat{y}^2 = r^2_{XY}\Sigma y^2$ and that $\Sigma d^2_{Y\cdot X} = (1 - r^2_{XY})\Sigma y^2$. These formulas can easily be derived from Expression (15.6), $r^2_{12} = \Sigma\hat{y}^2_2/\Sigma y^2_2 = \Sigma\hat{y}^2/\Sigma y^2$. In a similar manner we can compute the sum of squares of Y due to regression on $X_1, \ldots, X_k$:

$$\sum \hat{y}^2_{\cdot 1 \ldots k} = R^2_{Y\cdot 1 \ldots k} \sum y^2 \tag{16.11}$$

as well as the sum of squares unexplained by these variables:

$$\sum d^2_{Y\cdot 1 \ldots k} = (1 - R^2_{Y\cdot 1 \ldots k}) \sum y^2 \tag{16.12}$$

The square root of the coefficient of multiple determination is the coefficient of multiple correlation (see Section 16.4).

Now let us use these coefficients in an **overall significance test of the multiple regression.** We proceed as in the analysis of variance for linear regression in Section 14.4. What will be the degrees of freedom for $\Sigma \hat{y}^2_{\cdot 1 \ldots k}$, the sum of squares of Y explained by variables X_1 through X_k? In the case of simple linear regression we counted one degree of freedom for $\Sigma \hat{y}^2$ due to regression on X. Now that we have k separate X-variables, we allocate k degrees of freedom to $\Sigma \hat{y}^2_{\cdot 1 \ldots k}$. What remains is $\Sigma d^2_{Y \cdot 1 \ldots k}$ with $n - k - 1$ *df*, which agrees with the earlier instance in which the unexplained *SS* had $n - 2$ *df*, since in that case $k = 1$ (see Section 14.4). We will use these facts to test for the significance of the multiple regression of SO_2 on temperature and manufacturing enterprises in Box 16.1. Earlier in this section we computed $R^2_{Y \cdot 1,2} = 0.5162$. Hence $1 - R^2_{Y \cdot 1,2} = 0.4838$. The total sum of squares for SO_2 is 22,038 (computation not shown). We can therefore obtain the explained and unexplained sums of squares by simple multiplication using Expressions (16.11) and (16.12): $\Sigma \hat{y}^2_{\cdot 1,k} = 0.5162 \times 22{,}038 = 11{,}376$ and $\Sigma d^2_{Y \cdot 1,2} = 0.4838 \times 22{,}038 = 10{,}662$. We display the computation in conventional anova format:

Source of variation	*df*	*SS*	*MS*	F_s
Explained—due to regression on X_1 and X_2	2	11,376	5688	20.27***
Unexplained—error around regression plane	38	10,662	280.58	

The value of F_s tells us that we can safely reject the null hypothesis.

In multiple regression analysis we will want to carry out repeatedly tests such as the one just completed. Even for one and the same example we will want to test the effects of a single independent variable, of two variables, and of k variables. A look at the F_s value just obtained will show us how we can simplify the computation further. We computed

$$F_s = \frac{\sum \hat{y}^2_{\cdot 1 \ldots k}/k}{\sum d^2_{Y \cdot 1 \ldots k}/(n - k - 1)}$$

From Expressions (16.11) and (16.12), however,

$$F_s = \frac{R^2_{Y \cdot 1 \ldots k} \sum y^2/k}{(1 - R^2_{Y \cdot 1 \ldots k}) \sum y^2/(n - k - 1)}$$

$$= \frac{R^2_{Y \cdot 1 \ldots k}/k}{(1 - R^2_{Y \cdot 1 \ldots k})/(n - k - 1)} \qquad (16.13)$$

Thus for significance tests we need only compute the coefficients of multiple determination and their complements, the **coefficients of multiple nondetermination,** and compare their ratio (properly weighted by degrees of freedom) to $F_{\alpha[k, n-k-1]}$.

These calculations are shown in step **1** of Box 16.2. Note that the values of the coefficients of determination in Box 16.2 were obtained by computer and therefore differ slightly from values that would be calculated from Box 16.1 because of rounding errors. Variable X_2, number of manufacturing enterprises, has the highest correlation of any independent variable with Y, the SO_2 content of air ($r_{2Y} = 0.6448$). We might try to explain Y entirely in terms of X_2. The coefficient of determination $r_{2Y}^2 = 0.41573$ and is highly significant by the F-test shown in Box 16.2. When we regress Y on X_1 as well as X_2 we obtain $R_{Y \cdot 1,2}^2 = 0.51611$, an appreciable increase over the 0.41573 due to X_2 alone. Again, this coefficient is highly significant. In Box 16.2 we also show that including X_3 among the independent variables raises R^2 to 0.61255; including all available independent variables ($X_1 - X_6$) further raises the coefficient of determination to 0.66951. All regressions tested are highly significant.

What hypotheses are we testing with these tests? There are two mathematically equivalent ways of looking at this matter. We test $H_0: \rho_{Y \cdot 1 \ldots k}^2 = 0$, which is to say that the parametric R^2 equals zero and none of the variation of Y can be explained by $X_1, \ldots, X_k$. Alternatively, we test $H_0: \beta_{Y1\cdot} = \beta_{Y2\cdot} = \cdots = \beta_{Yk\cdot} = 0$, that is, that all of the partial regression coefficients are equal to zero.

The tests carried out in step **1** of Box 16.2 raise new questions that are almost self-evident and lead to *significance tests for additional independent variables.* Since X_2 determined a significant proportion of the variation of Y, it is not surprising that the variation explained by X_1 and X_2 together should also be significant, since the explained sum of squares increases as additional independent variables are included in the regression equation. But is the observed increase in the explained sum of squares itself significant? If we compute an explained sum of squares based on a larger number, k_2, of independent variables and subtract from it an explained sum of squares based on a smaller number of variables, k_1, the resulting difference, $\Sigma \hat{y}_{\cdot 1 \ldots k_2}^2 - \Sigma \hat{y}_{\cdot 1 \ldots k_1}^2$, is a sum of squares pertaining to the added explanation due to variables $X_{k_1+1}, \ldots, X_{k_2}$. This sum of squares possesses $k_2 - k_1$ degrees of freedom and can be tested against the unexplained sum of squares $\Sigma d_{Y \cdot 1 \ldots k_2}^2$ with $n - k_2 - 1$ *df*. We are thus partitioning the explained *SS* with the greater number of variables into two sums of squares—one due to the effect of the first k_1 variables and one due to the effect of the additional $k_2 - k_1$ variables, with the k_1 variables held constant. Rather than set the tests up as formal analyses of variance, we can again use the shortcut employed in Expression (16.13) and write

$$F_s = \frac{(R_{Y \cdot 1 \ldots k_2}^2 - R_{Y \cdot 1 \ldots k_1}^2)/(k_2 - k_1)}{(1 - R_{Y \cdot 1 \ldots k_2}^2)/(n - k_2 - 1)} \tag{16.14}$$

You should be able to derive this formula easily from the facts already given. These tests are carried out on the air pollution data from Table 16.1 in step **2** of Box 16.2. The increases in determination are significant for the additions of X_1 and then of X_3 to X_2. Adding variables X_4, X_5, and X_6, however, increases the

Box 16.2 TESTS OF SIGNIFICANCE IN MULTIPLE REGRESSION.

Based on data in Table 16.1, analyses in Box 16.1, and standard errors in Box 16.3.

$$n = 41 \quad \sum y^2 = 22038$$
$$r^2_{2Y} = 0.41573$$
$$R^2_{Y \cdot 1,2} = 0.51611$$
$$R^2_{Y \cdot 1,2,3} = 0.61255$$
$$R^2_{Y \cdot 1 \ldots 6} = 0.66951$$

The coefficients of multiple determination for two and three independent variables ($R^2_{Y \cdot 1,2}$ and $R^2_{Y \cdot 1,2,3}$) could be obtained from Box 16.1 by means of Expression (16.10). Their values here, as well as the value of the coefficient of all six independent variables ($R_{Y \cdot 1 \ldots 6}$), were calculated by computer, which explains any discrepancies in terminal digits between these values and those calculated earlier in the text.

1. Testing the significance of regression on k independent variables. Compare

$$F_s = \frac{R^2_{Y \cdot 1 \ldots k}/k}{(1 - R^2_{Y \cdot 1 \ldots k})/(n - k - 1)}$$

with $F_{\alpha[k, n-k-1]}$.

$k = 1$; regression of Y on X_2:

$$F_s = \frac{r^2_{2Y}/1}{(1 - r^2_{2Y})/(41 - 1 - 1)} = \frac{0.41573}{0.58427/39} = 27.750^{***}$$

$k = 2$; regression of Y on X_1 and X_2:

$$F_s = \frac{R^2_{Y \cdot 1,2}/2}{(1 - R^2_{Y \cdot 1,2})/(41 - 2 - 1)} = \frac{0.51611/2}{0.48389/38} = 20.265^{***}$$

$k = 3$; regression of Y on X_1, X_2, and X_3:

$$F_s = \frac{R^2_{Y \cdot 1,2,3}/3}{(1 - R^2_{Y \cdot 1,2,3})/(41 - 3 - 1)} = \frac{0.61255/3}{0.38745/37} = 19.498^{***}$$

$k = 6$; regression of Y on $X_1 - X_6$:

$$F_s = \frac{R^2_{Y \cdot 1 \ldots 6}/6}{(1 - R^2_{Y \cdot 1 \ldots 6})/(41 - 6 - 1)} = \frac{0.66951/6}{0.33049/34} = 11.480^{***}$$

All proportions of the variation of Y determined by variable X_2, and by X_2 in combination with other variables as shown, are highly significant at $P < 0.001$. We conclude that number of manufacturing concerns, X_2, can significantly predict SO_2 content in air singly and even better when in combinations with other variables such as average temperature, X_1, and population size, X_3.

BOX 16.2 CONTINUED

2. Testing the significance of the increase in proportion of variance determined as the number of dependent variables is augmented from k_1 to k_2. Compare

$$F_s = \frac{(R^2_{Y \cdot 1 \ldots k_2} - R^2_{Y \cdot 1 \ldots k_1})/(k_2 - k_1)}{(1 - R^2_{Y \cdot 1 \ldots k_2})/(n - k_2 - 1)}$$

with $F_{\alpha[k_2 - k_1, n - k_2 - 1]}$.

Testing the increment in determination due to addition of X_1 over determination by X_2 alone; $k_2 = 2$, $k_1 = 1$:

$$F_s = \frac{(R^2_{Y \cdot 1,2} - r^2_{2Y})/(2 - 1)}{(1 - R^2_{Y \cdot 1,2})/(41 - 2 - 1)} = \frac{(0.51611 - 0.41573)/1}{(1 - 0.51611)/38}$$

$$= \frac{0.10038}{0.48389/38} = 7.883^{**}$$

Testing the increment in determination due to addition of X_3 over joint determination by X_1 and X_2; $k_2 = 3$, $k_1 = 2$:

$$F_s = \frac{(R^2_{Y \cdot 1,2,3} - R^2_{Y \cdot 1,2})/(3 - 2)}{(1 - R^2_{Y \cdot 1,2,3})/(41 - 3 - 1)} = \frac{(0.61255 - 0.51611)/1}{(1 - 0.61255)/37}$$

$$= \frac{0.09644}{0.38745/37} = 9.210^{**}$$

Testing the increment in determination due to addition of X_4, X_5, and X_6 over joint determination by X_1, X_2, and X_3; $k_2 = 6$, $k_1 = 3$:

$$F_s = \frac{(R^2_{Y \cdot 1 \ldots 6} - R^2_{Y \cdot 1,2,3})/(6 - 3)}{(1 - R^2_{Y \cdot 1 \ldots 6})/(41 - 6 - 1)} = \frac{(0.66951 - 0.61255)/3}{(1 - 0.66951)/34}$$

$$= \frac{0.05696/3}{0.33049/34} = 1.953\ ns$$

We conclude that the increased determination by first adding X_1, then X_3, is highly significant in each case. The subsequent addition of variables X_4, X_5, and X_6, however, does not add further significant explanation of the variation of Y.

For the following tests of significance, additional quantities are needed. All the examples are from the regression of Y on X_1 and X_2 as shown in part **I** of Box 16.1. The unexplained standard deviation

$$s_{Y \cdot 1,2} = \left[\frac{(1 - R^2_{Y \cdot 1,2}) \sum y^2}{(n - k - 1)}\right]^{1/2} = \left[\frac{(1 - 0.51611)22038}{38}\right]^{1/2} = 16.7520$$

Also needed are the Gaussian multipliers. Their computation for this simple case with two independent variables is shown in the text. For larger problems one needs to invert the sum of squares–sum of products matrix of the independent variables. In this case $c_{11} = 4.964{,}929{,}584 \times 10^{-4}$, $c_{22} = 8.168{,}974{,}959 \times 10^{-8}$, and $c_{12} = 1.210{,}292{,}356 \times 10^{-6}$.

Box 16.2 Continued

3. Standard error of the partial regression coefficient:

$$s_{b_{Y1\cdot}} = s_{Y\cdot}\sqrt{c_{11}} = 16.7520\sqrt{4.964,929,584 \times 10^{-4}} = 0.37327$$

$$s_{b_{Y2\cdot}} = s_{Y\cdot}\sqrt{c_{22}} = 16.7520\sqrt{8.168,974,959 \times 10^{-8}} = 0.004,788,0$$

4. Testing the significance of the partial regression coefficients:

$$t_s = \frac{(b_{Y1\cdot} - 0)}{s_{b_{Y1\cdot}}} = \frac{-1.0480}{0.37327} = -2.808^{**}$$

$$t_s = \frac{(b_{Y2\cdot} - 0)}{s_{b_{Y2\cdot}}} = \frac{0.02431}{0.004,788,0} = 5.077^{**}$$

$$t_{.01[38]} = 2.712$$

5. 95% confidence limits for partial regression coefficients:

$$t_{.05[38]}s_{b_{Y1\cdot}} = 2.025(0.37327) = 0.75587$$

$$L_1 = b_{Y1\cdot} - t_{.05[38]}s_{b_{Y1\cdot}} = -1.0480 - 0.75587 = -1.8039$$

$$L_2 = b_{Y1\cdot} + t_{.05[38]}s_{b_{Y1\cdot}} = -1.0480 + 0.75587 = -0.2981$$

$$t_{.05[38]}s_{b_{Y2\cdot}} = 2.025(0.004,788,0) = 0.009,695,7$$

$$L_1 = b_{Y2\cdot} - t_{.05[38]}s_{b_{Y2\cdot}} = 0.02431 - 0.009,695,7 = 0.01461$$

$$L_2 = b_{Y2\cdot} + t_{.05[38]}s_{b_{Y2\cdot}} = 0.02431 + 0.009,695,7 = 0.03401$$

6. Standard error of standard partial regression coefficients:

$$s_{b'_{Y1\cdot}} = \sqrt{\frac{(1 - R^2_{Y\cdot 1,2})}{n - k - 1} c_{11} \sum x_1^2}$$

$$= \sqrt{\frac{(1 - 0.51611)}{38} (4.964,929,584 \times 10^{-4})(2089.595122)}$$

$$= \sqrt{(0.012,733,95)(1.037,469,26)}$$

$$= 0.11494$$

$$s_{b'_{Y2\cdot}} = \sqrt{\frac{(1 - R^2_{Y\cdot 1,2})}{n - k - 1} c_{22} \sum x_2^2}$$

$$= \sqrt{\frac{(1 - 0.51611)}{38} (8.168,974,959 \times 10^{-8})(12,700,115.61)}$$

$$= \sqrt{(0.012,733,95)(1.037,469,26)}$$

$$= 0.11494$$

Significance tests and confidence limit calculations for standard partial regression coefficients are analogous to those for partial regression coefficients, shown in steps **4** and **5,** respectively.

BOX 16.2 CONTINUED

7. Testing the significance of the difference between two partial regression coefficients:

$$s_{b_{Y1\cdot}-b_{Y2\cdot}} = s_{Y\cdot}\sqrt{c_{11}+c_{22}-2c_{12}}$$
$$= 16.7520[4.964{,}929{,}584\times10^{-4}+8.168{,}974{,}959\times10^{-8} -2(1.210{,}292{,}356\times10^{-6})]^{1/2}$$
$$= 0.37285$$

$$t_s = \frac{b_{Y1\cdot}-b_{Y2\cdot}}{s_{Y\cdot}\sqrt{c_{11}+c_{22}-2c_{12}}} = \frac{-1.048-0.02431}{0.37285}$$
$$= 2.876^{**}$$

$$t_{.01[38]} = 2.712$$

This test is shown here for illustrative purposes. It makes sense only when the regression coefficients being compared are in identical measurement units, which is not the case in this example.

8. Standard error of the sampled mean $\bar{Y}$ (at $\bar{X}_1, \bar{X}_2$):

$$s_{\bar{Y}} = \sqrt{\frac{s^2_{Y\cdot}}{n}} = \frac{16.7520}{\sqrt{41}} = 2.6162$$

9. 95% confidence limits for the mean μ_Y corresponding to $\bar{X}_1, \bar{X}_2$. ($\bar{Y} = 30.049$):

$$t_{.05[38]}s_{\bar{Y}} = 2.025(2.6162) = 5.29785$$
$$L_1 = \bar{Y} - t_{.05[38]}s_{\bar{Y}} = 30.049 - 5.298 = 24.751$$
$$L_2 = \bar{Y} + t_{.05[38]}s_{\bar{Y}} = 30.049 + 5.298 = 35.347$$

10. Standard error of $\hat{Y}$, an estimated Y for specified values of X_1 and X_2. We propose to estimate SO_2 content of air (Y) for a town with an annual average temperature of 50°F (X_1) and 1000 manufacturing enterprises (X_2).

$$s_{\hat{Y}} = s_{Y\cdot}\sqrt{\frac{1}{n}+\sum_{i=1}^{k}c_{ii}x_i^2+2\sum_{ij}c_{ij}x_ix_j}$$
$$= s_{Y\cdot}\left[\frac{1}{n}+c_{11}(X_1-\bar{X}_1)^2+c_{22}(X_2-\bar{X}_2)^2+2c_{12}(X_1-\bar{X}_1)(X_2-\bar{X}_2)\right]^{1/2}$$
$$= 16.7520\left[\frac{1}{41}+4.964{,}929{,}584\times10^{-4}(50-7.2277)^2 +8.168{,}974{,}959\times10^{-8}(1000-563.47)^2 +2(1.210{,}292{,}356\times10^{-6})(50-7.2277)(1000-563.47)\right]^{1/2}$$
$$= 16.6972$$

Box 16.2 Continued

11. 95% confidence limits for μ_Y corresponding to the estimate $\hat{Y}$ for $X_1 = 50$ and $X_2 = 1000$. Employing the prediction equation from part **I**, step **5** of Box 16.1,

$$\hat{Y} = 77.231 - 1.0480(50) + 0.0243(1000) = 49.141$$

$$t_{.05[38]}s_{\hat{Y}} = 2.025(16.6972) = 33.81183$$

$$L_1 = \hat{Y} - t_{.05[38]}s_{\hat{Y}} = 49.141 - 33.812 = 15.329$$

$$L_2 = \hat{Y} + t_{.05[38]}s_{\hat{Y}} = 49.141 + 33.812 = 82.953$$

12. Standard error for a predicted mean $\bar{Y}$ to be obtained in a new study of towns with $X_1 = 50°\text{F}$ and $X_2 = 1000$ manufacturing enterprises. Our best prediction for this mean would be $\bar{Y} = \hat{Y} = 49.141\ \mu\text{g/m}^3$. If the new study were based on a sample size of $m = 5$, the standard error of the predicted mean would be

$$\hat{s}_{\bar{Y}} = s_{Y\cdot}\sqrt{\frac{1}{m} + \frac{1}{n} + \sum_{i=1}^{k} c_{ii}x_i^2 + 2\sum_{ij} c_{ij}x_ix_j}$$

$$= 16.7520\left(\frac{1}{5} + 0.99347\right)^{1/2} \text{(using numerical values obtained in step } \mathbf{10}\text{)}$$

$$= 16.7520(1.09246) = 18.3009$$

13. 95% prediction limits for a sample mean of 5 towns at $X_1 = 50°\text{F}$ and $X_2 = 1000$ manufacturing enterprises:

$$t_{.05[38]}\hat{s}_{\bar{Y}} = 2.025(18.3009) = 37.05934$$

$$L_1 = \hat{Y} - t_{.05[38]}\hat{s}_{\bar{Y}} = 49.141 - 37.059 = 12.082$$

$$L_2 = \hat{Y} + t_{.05[38]}\hat{s}_{\bar{Y}} = 49.141 + 37.059 = 86.200$$

14. Testing the significance of $R_{Y\cdot 1\ldots k}$ for $k \leq 4$. Use Statistical Table **R** and look up critical values for $\nu = n - k - 1$ and k ($k + 1$ is called m in Table **R**).
Example: Test the significance of $R_{Y\cdot 1,2}$. $R^2_{Y\cdot 1,2} = 0.51611$. Therefore $R_{Y\cdot 1,2} = 0.7184$, which is greater than the 0.01 critical value for 35 *df* and $k = 2$, which is 0.481. The actual $\nu = n - k - 1 = 38$. This is not listed in Table **R** but there is no need to interpolate in view of the magnitude of the coefficient. We reject the null hypothesis that $\rho_{Y\cdot 1,2} = 0$.

coefficient of multiple determination only a little; with X_1, X_2, and X_3 only, $R^2 = 0.61255$; with all six X-variables, $R^2 = 0.66951$. This increase is not significant. Thus we cannot reject the hypothesis that $\beta_{y4\cdot} = \beta_{Y5\cdot} = \beta_{Y6\cdot} = 0$, and we would be tempted to omit X_4, X_5, and X_6 from the prediction equation (although the best fit for the observed data is obtained by using all available variables).

We might formalize the tests just carried out by stating their null hypothesis as follows: $H_0\colon \beta_{Yk_1+1\cdot} = \beta_{Yk_1+2\cdot} = \cdots = \beta_{Yk_2\cdot} = 0$. This statement says that the subset of partial regression coefficients due to variables X_{k_1+1} through X_{k_2} does not differ from zero.

All the other tests of significance familiar from simple linear regression apply to multiple regression as well. Remember that the key variance necessary in linear regression is $s^2_{Y \cdot X}$, the unexplained mean square with $n - 2$ degrees of freedom. The analogous quantity in multiple regression is the unexplained mean square $s^2_{Y \cdot 1 \ldots k}$, computed as follows:

$$\frac{(1 - R^2_{Y \cdot 1 \ldots k}) \sum y^2}{n - k - 1}$$

In addition to $s^2_{Y \cdot}$ we need the Gaussian multipliers, the elements of the inverse of the *SS–SP* (sum of squares–sum of products) matrix (see Section 16.1). Larger matrices, for cases with more than two independent variables, should be inverted by computer. Many computer programs for multiple regression analysis automatically calculate various standard errors of interest, and the user may not see or need to employ the Gaussian multipliers explicitly. However, you should at least know of their existence and how they might be employed if the standard error you wish to use is not included in the program available to you.

As an example, we will employ the regression of Y on X_1 and X_2 from Box 16.1. At the beginning of Box 16.2 we computed $r^2_{Y \cdot 1,2} = 0.51611$. From this value we can obtain

$$s_{Y \cdot 1,2} = \left[\frac{(1 - R^2_{Y \cdot 1,2}) \sum y^2}{n - k - 1}\right]^{1/2} = \left[\frac{(1 - 0.51611)22{,}038}{38}\right]^{1/2} = 16.7520$$

In this simple example with only two independent variables one can easily obtain the Gaussian multipliers without a computer. These quantities are the elements of the inverse of the *SS–SP* matrix $\mathbf{S}$. In the present example

$$\mathbf{S} = \begin{bmatrix} \sum x_1^2 & \sum x_1 x_2 \\ \sum x_1 x_2 & \sum x_2^2 \end{bmatrix} = \begin{bmatrix} 2089.595122 & -30{,}958.85366 \\ -30{,}958.85366 & 12{,}700{,}115.61 \end{bmatrix}$$

These values were obtained by computation from the data in Table 16.1. In this 2×2 matrix, computing an inverse is a trivial task. First we compute the determinant, $|\mathbf{S}| = \Sigma x_1^2 \Sigma x_2^2 - (\Sigma x_1 x_2)^2 = 2.557{,}964{,}901 \times 10^{10}$, and its reciprocal, $1/|\mathbf{S}| = 3.909{,}357{,}785 \times 10^{-11}$. Then we compute the elements of the inverse of $\mathbf{S}$ as follows:

$$\frac{1}{|\mathbf{S}|}\begin{bmatrix} \sum x_2^2 & -\sum x_1 x_2 \\ -\sum x_1 x_2 & \sum x_1^2 \end{bmatrix} = \begin{bmatrix} c_{11} & c_{12} \\ c_{12} & c_{22} \end{bmatrix}$$

$$3.909{,}357{,}785 \times 10^{-11}\begin{bmatrix} 12{,}700{,}115.61 & 30{,}958.85366 \\ 30{,}958.85366 & 2089.595122 \end{bmatrix}$$

$$= \begin{bmatrix} 4.964{,}929{,}584 \times 10^{-4} & 1.210{,}292{,}356 \times 10^{-6} \\ 1.210{,}292{,}356 \times 10^{-6} & 8.168{,}974{,}959 \times 10^{-8} \end{bmatrix}$$

Using the variance $s^2_{Y\cdot}$ and the Gaussian multipliers, we can set standard errors and confidence limits to various statistics related to multiple regression. The necessary standard errors are listed in Box 16.3.

The first row of Box 16.3 shows the *standard error of the partial regression coefficient* for the regression of Y on X_j, all other independent variables being kept constant. Its computation is illustrated in step **3** of Box 16.2, while in step **4** of the same box we show the by now familiar t-test for *significance of the partial regression coefficients.* Both coefficients are highly significant: Temperature and manufacturing enterprises both explain SO_2 content of air, each significantly affecting the dependent variable when the other is kept constant. Note that in contrast to simple linear regression, the significance test of the regression coefficients does not follow from that of the overall regression (the anova test in step **1** of Box 16.2). The anova tests whether both X_1 and X_2 explain a significant proportion of the variance of Y. If this were the case, we still would not know whether X_1 or X_2 or both are significantly related to Y. For example, we saw in step **1** of Box 16.2 that all six independent variables significantly explained Y. If we tested the partial regression coefficients, however, we would find that only those on X_1, X_2, and X_3 are significant. The standard errors of the partial regression coefficients also permit *setting of confidence limits for partial regression coefficients* in a manner by now familiar. The procedure is illustrated in step **5** of Box 16.2.

The *standard error of standard partial regression coefficients* is computed by the formula in the second row of Box 16.3. Sample computations are shown in step **6** of Box 16.2. Tests of significance (not shown) given identical results to those on partial regression coefficients. Thus the standard errors of b-primes simply allow for the standardizations. The fact that the standard errors of $b'_{Y2\cdot1}$ in this case have identical numerical values (within rounding error) is a peculiarity of the case with two independent variables. When there are more than two such variables, the standard errors differ. We can set confidence limits to b-primes in the customary manner.

The factors $c_{jj}\Sigma x_j^2$ in the equations for $s_{b'_{Yj}}$ (see Box 16.3) are called the **variance inflation factors** (VIF) (Marquardt, 1970), since they represent the factor by which the standardized unexplained variance $(1 - R^2)/(n - k - 1)$ is inflated because of the pattern of intercorrelation among the independent variables. If the independent variables are uncorrelated, the VIF will equal unity. In the present case (with $r_{12} = -0.19$), they are equal to 1.037,469,26. As the independent variables become more highly correlated, the factors can become much larger. When the factors are greater than 100, the ordinary single precision arithmetic on most computers will not permit a sufficiently accurate solution of the normal equations (Snee, 1973).

Although the test makes sense only when the independent variables concerned are in the same units of measurement, we may wish to *test differences between pairs of partial regression coefficients.* A standard error for such

Box 16.3 STANDARD ERRORS OF MULTIPLE REGRESSION STATISTICS.

For explanation of this box, see Section 16.2; the degrees of freedom appropriate to these tests are $\nu = n - k - 1$, where n = sample size and k = number of independent variables in the study.

Statistic	s	Formula for standard error
$b_{Yj\cdot}$ (partial regression coefficient of Y on X_j)	$s_{b_{Yj\cdot}}$	$s_{Y\cdot}\sqrt{c_{jj}}$
$b'_{Yj\cdot}$ (standard partial regression coefficient of Y on X_j)	$s_{b'_{Yj\cdot}}$	$\sqrt{\dfrac{(1 - R^2_{Y\cdot 1\ldots k})}{n - k - 1} c_{jj}\sum x_j^2}$
$b_{Yi\cdot} - b_{Yj\cdot}$ (difference between two partial regression coefficients)	$s_{b_{Yi\cdot} - b_{Yj\cdot}}$	$s_{Y\cdot}\sqrt{c_{ii} + c_{jj} - 2c_{ij}}$
$\bar{Y}$ (sample mean)	$s_{\bar{Y}}$	$\sqrt{\dfrac{s^2_{Y\cdot}}{n}}$
$\hat{Y}$ (estimated Y for given values of $X_1, X_2, \ldots, X_k$)	$s_{\hat{Y}}$	$s_{Y\cdot}\sqrt{\dfrac{1}{n} + \sum_{i=1}^{k} c_{ii}x_i^2 + 2\sum_{ij} c_{ij}x_ix_j}$
Predicted Y for given values of $X_1, X_2, \ldots, X_k$	$\hat{s}_Y$	$s_{Y\cdot}\sqrt{1 + \dfrac{1}{n} + \sum_{i=1}^{k} c_{ii}x_i^2 + 2\sum_{ij} c_{ij}x_ix_j}$
Predicted mean $\bar{Y}$ of m items for given values of $X_1, X_2, \ldots, X_k$	$\hat{s}_{\bar{Y}}$	$s_{Y\cdot}\sqrt{\dfrac{1}{m} + \dfrac{1}{n} + \sum_{i=1}^{k} c_{ii}x_i^2 + 2\sum_{ij} c_{ij}x_ix_j}$

NOTES: All standard errors except that for $b'_{Yj\cdot}$ contain the unexplained variance $s^2_{Y\cdot 1\ldots k}$ (shown in simplified symbolism as $s^2_{Y\cdot}$). This can be obtained from the anova testing overall regression. The terms c_{ii}, c_{jj}, and c_{ij} are the Gaussian multipliers, the elements of the inverse of the sum of squares–sum of products matrix.

In the last three rows the estimated $\hat{Y}$, Y, and $\bar{Y}$ are for a specified set of values of $X_1, X_2, \ldots, X_k$. The deviations under the radicals on the right are for these specified values. Thus $x_1 = X_1 - \bar{X}_1$, where X_1 is the specified variate of variable X_1. The summation symbolized by Σ_{ij} is over all pairs of variables X_i, X_j, where $i < j$, i ranges from 1 to $k - 1$, and j ranges from $i + 1$ to k.

differences is shown in the third row of Box. 16.3, and its use is illustrated in step **7** of Box 16.2, although its employment in this particular case is not of interest. Change in SO_2 per degree Fahrenheit cannot readily be compared with change per number of manufacturing enterprises.

The other standard errors in Box 16.3 correspond to those for simple linear regression in Box 14.2, which are illustrated in Box 14.3. A new, smaller *standard error of* $\overline{Y}$ can be computed, leading to narrower confidence limits (see steps **8** and **9** of Box 16.2). When we wish to construct confidence "belts" around the plane, we need to take the deviation of the specified values of X_1 and X_2 from their respective means into consideration, as was done in simple linear regression. The formulas are more complicated, however, because the partial regression coefficients are not independent of each other and because in addition to the Gaussian multipliers that furnish their variance (remember that $s^2_{Y\cdot}c_{jj}$ is the variance for $b_{Yj\cdot}$), we need those that provide their covariance ($s^2_{Y\cdot}c_{ij}$). The confidence surfaces are closest to the plane at $\overline{X}_1, \overline{X}_2$ and curve away from it as X_1 and X_2 depart from their respective means. The computations for the *standard error of an estimated Y for specified values of* X_j and those for the *standard error of predicted sample mean based on m items* are shown in steps **10** and **12,** respectively, and computations of confidence or prediction limits based on the standard errors in steps **11** and **13,** respectively, of Box 16.2.

16.3 Path Analysis

For the remainder of our discussion on multiple regression a different point of view will be useful. In this section we will introduce a powerful analytical tool known as path analysis, which is well worth studying for its own sake as well as for the light it sheds on multiple regression.

We could represent the relations between SO_2 content of air and the three independent variables, X_1, X_2, X_3, as shown in Figure 16.3. This representation implies that these independent variables are immediate causes of the dependent variable Y. We know that this is not true. Complicated demographic, environmental, and physicochemical factors determine the sulfur dioxide concentration in any given city, and all the cautions about cause-and-effect relationships voiced in Sections 14.6 and 15.6 apply here as well. Nevertheless, it is convenient to consider the model in this simplistic manner for the insights it gives us about the mathematical relationships between the variables. In addition to causal relationship, we make another assumption, namely that the independent and dependent variables are related by a linear and additive model such as the multiple regression equations in Expressions (16.1) and (16.2). Rather than use the terms *independent* and *dependent* variables, which carry with them definite assumptions relating to Model I regression that are generally not met in path analysis, we will frequently employ the more neutral terms **predictor** and **criterion** variables,

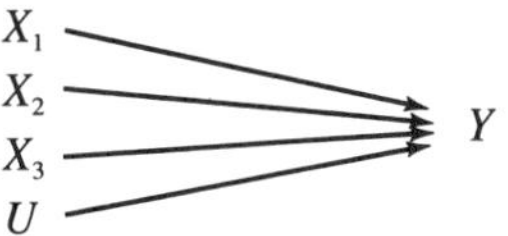

FIGURE 16.3 *Path diagram showing three uncorrelated predictor (independent) variables,* X_1, X_2, *and* X_3, *and a residual variable,* U, *affecting one criterion (dependent) variable,* Y.

respectively—terminology that is common in multiple regression and path analysis in the social sciences.

Diagrams such as Figure 16.3 are known as **path diagrams** and obey the following conventions. Cause-and-effect relationships are depicted by one-headed arrows. The criterion variables are generally indicated as completely determined—that is, all the factors contributing to the total variation of variable Y are drawn in. Since in most instances one does not know, and has no hope of ever knowing, *all* the factors that explain the variation of a given dependent (criterion) variable, these unknown factors are lumped together as a single, unknown, residual variable (shown as U in Figure 16.3). The effect of this pooled variable on Y cannot be interpreted in terms of direction, but we can ascribe to it the proportion of the total variance of Y not determined by the other causal variables. In fact, this is a familiar quantity. If we consider the total variance of Y to be 1, then the proportion of the variance due to U is r^2_{UY}, which is the coefficient of determination between Y and U. The proportion that is determined by U is the proportion not determined by X_1, X_2, and X_3. Therefore, $r^2_{UY} = 1 - R^2_{Y \cdot 1,2,3}$, the coefficient of nondetermination of Y by X_1, X_2, and X_3.

One other assumption is implied by the diagram in Figure 16.3. Because the bases of the arrows from variables X_1, X_2, X_3, and synthetic variable U are not connected, it is implied that they are uncorrelated—that they are **independent causes.** Such an assumption is plausible in certain situations; in this specific case, however, it is not. Furthermore, we know empirically that there are correlations among the three predictor variables—low correlation between temperature and the others but a very high correlation (0.9553) between number of manufacturing enterprises (X_2) and population size (X_3). Thus the model in Figure 16.3 cannot describe our data, but we will pursue this simplistic model for a little while longer to explore its consequences. Let us therefore begin by assuming that there are no correlations among X_1, X_2, X_3, and U.

In such a case, what is the independent effect of X_1 on Y? By independent effect, we mean the effect in the absence of the other predictor variables X_2 and X_3. This problem is by now familiar—such an effect can be quantified by a partial regression coefficient. Because we will want to compare the magnitudes of such effects for different factors, we will employ standard partial regression coefficients. The effect of X_1 on Y is thus expressed as $b'_{Y1 \cdot 2,3}$. The effects of the other two predictor variables can be quantified similarly by the appropriate b-

primes. How do we obtain these? Remembering the normal equations of Expression (16.6), we can write in standardized form

$$\begin{aligned} b'_{Y1\cdot2,3} + r_{12}b'_{Y2\cdot1,3} + r_{13}b'_{Y3\cdot1,2} &= r_{1Y} \\ r_{12}b'_{Y1\cdot2,3} + b'_{Y2\cdot1,3} + r_{23}b'_{Y3\cdot1,2} &= r_{2Y} \\ r_{13}b'_{Y1\cdot2,3} + r_{23}b'_{Y2\cdot1,3} + b'_{Y3\cdot1,2} &= r_{3Y} \end{aligned} \tag{16.15}$$

We do not have to use simultaneous equations or matrix methods, however, to solve these equations. We have assumed (incorrectly, but for purposes of completing our study of Figure 16.3) that the causal variables are not correlated. Lack of correlation would make $r_{12} = r_{13} = r_{23} = 0$, so all terms involving these correlation coefficients on the left-hand side of the equations vanish. We can therefore write down the solution of the equations by inspection: $b'_{Y1\cdot2,3} = r_{1Y}$, $b'_{Y2\cdot1,3} = r_{2Y}$, and $b'_{Y3\cdot2,3} = r_{3Y\cdot}$. Thus the standard partial regression coefficients in this special case are equal to the correlation coefficients of the two variables concerned.

In Expression (15.7) we showed that if the simple linear regression coefficient $b_{Y\cdot X}$ is standardized by dividing by the appropriate standard deviations, we obtain the correlation coefficient r_{XY}. Thus it follows that for the special case of Figure 16.3, $b'_{Y1\cdot2,3} = r_{1Y} = b'_{Y1}$; there are similar relations for the other two variables. When you think about it, this makes a good deal of sense. Since the causal variables X_1, X_2, and X_3 are here assumed to be independent, the effect of X_1 on Y with X_2 and X_3 kept constant ($b'_{Y1\cdot2,3}$) should be no different from the effect of X_1 on Y when X_2 and X_3 are not considered (b'_{Y1}), which, as we have seen, is the same as the correlation between X_1 and Y.

In path analysis, the standard partial regression coefficient that estimates the strength of the relationship between cause X_1 and effect Y is called a **path coefficient** and is symbolized by p_{Y1}. From what we have learned already about path analysis we can make the first useful conclusion: In the case of independent causes, a path coefficient is equal to the correlation between the predictor and the criterion variable.

Thus in Figure 16.3 $p_{Y1} = r_{1Y}$, $p_{Y2} = r_{2Y}$, and $p_{Y3} = r_{3Y}$. What proportion of the variance of Y is determined by X_1? Obviously, the answer is the proportion estimated by the coefficient of determination r^2_{1Y}, which equals p^2_{Y1}. Since the predictor variables are not correlated, the proportion of the total variance that each one determines is additive, and we can therefore write the following equation for the total determination of the variance of Y:

$$p^2_{Y1} + p^2_{Y2} + p^2_{Y3} + r^2_{UY} = 1 \tag{16.16}$$

Sometimes you may encounter path diagrams for which the strength of the effect of the unknown variable U on the criterion variable Y is given as $p_{YU} = r_{UY} = (1 - R^2_{Y\cdot})^{1/2}$. This expression is all right as long as we keep in mind that variable U does not exist as an entity but rather is a composite of all the unknown sources of unexplained variation, so any single effect ascribed to it is meaningless, except as a measure of the amount of unexplained variation.

Now we will make the model more realistic for our data—we will consider **correlated causes.** In Figure 16.4, variables X_1, X_2, and X_3 are connected by double-headed arrows. In path diagrams such arrows indicate correlation between predictor or causal variables. What is the correlation between X_1 and Y now? In the first line of the normal equations of Expression (16.15), the correlations r_{12} and r_{13} are now not equal to zero, and to obtain the standard partial regression coefficients we have to solve the normal equations. These equations also show us, however, how to evaluate the correlation between predictor and criterion variables by means of path coefficients. If we rewrite the first normal equation in terms of path coefficients, we obtain $r_{1Y} = p_{Y1} + r_{12}p_{Y2} + r_{13}p_{Y3}$. This expression leads directly to the general rule for finding the correlation between two variables in a path diagram: *The correlation between two variables is the sum of the products of the chains of path coefficients or correlations along all of the paths by which they are connected.*

In evaluating the correlation between X_1 and Y in Figure 16.4, then, the first path between them is the direct one, p_{Y1}, the second is the indirect one via X_2, which is $r_{12}p_{Y2}$, while the third is an indirect path via X_3, $r_{13}p_{Y3}$. Note that you cannot consider the chain $r_{12}r_{23}p_{Y3}$ to contribute to the correlation r_{1Y} via X_3. Only the common causes behind the correlation r_{13} contribute via X_3, not the common causes behind the correlation r_{23}. The correlation r_{13} is indicated in Figure 16.4 by the outer double-headed arrow between X_1 and X_3. The following simple rule prevents such errors in evaluating correlation coefficients: *In a chain connecting two variables, no more than one double-headed arrow (correlation between predictor variables) is permitted.* Knowing the values of the correlations represented by the double-headed arrows and of the path coefficients (the standard partial regression coefficients obtained in step **2** of Box 16.1), we can compute the correlation between any X and Y. Using our newly learned rules for obtaining correlations, we can write the equation for finding the correlation r_{2Y}:

$$\begin{aligned} r_{2Y} &= p_{Y2} + r_{12}p_{Y1} + r_{23}p_{Y3} \\ &= 1.7111 + (-0.1900)(-0.1807) + (0.9553)(-1.1522) \\ &= 0.6447 \end{aligned}$$

Within rounding error, this result is the same as the value of 0.6448 computed from the original data (see Box 16.1).

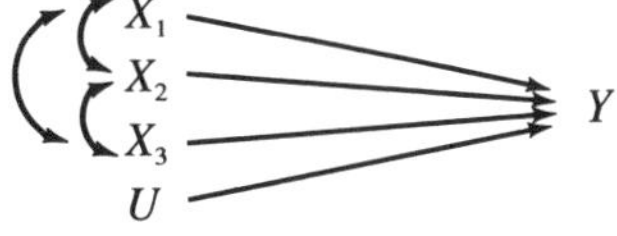

FIGURE 16.4 *Path diagram showing three correlated predictor variables,* X_1, X_2, *and* X_3, *and an independent variable,* U, *affecting one criterion variable,* Y.

The equation tells us something about the components of a correlation coefficient. In this particular system, the correlation between X_2 and Y is made up not only of the direct contribution of X_2 to Y (the path coefficient p_{Y2}), but also of common causes between X_2 and any other variables in the system that have a direct effect on Y. These additional contributions from common causes tend to increase the correlation between the two variables concerned, except in cases where either the paths or the correlation due to common causes are negative. In such cases, compensating effects tend to dampen the variation of the criterion variable and hence decrease correlation.

Note also that some of the path coefficients are larger than 1. This is so because, as noted in Section 16.1, no limitation was set on the magnitude of standard partial regression coefficients. If a coefficient is larger than 1, however, the predictor variable must be connected to the criterion variable through at least one other chain of opposite sign and of a magnitude that confines the total correlation between them to the limits of -1 and $+1$ for correlation coefficients.

Next we develop the formula for total determination of variable Y in the path diagram of Figure 16.4. Remembering the additive model for these paths, we can write $Y = X_1 + X_2 + X_3 + U$. Thus the variance of Y must be a function of the variances of these variables. We have already examined the variance of a sum (see Section 15.3), and the formula we need now is merely an extension of Expression (15.9). We can write

$$\sum_i p_{Yi}^2 + 2\sum_{ij} p_{Yi}p_{Yj}r_{ij} + r_{UY}^2 = 1 \tag{16.17}$$

where Σ_i is summation of all i $(1, \ldots, k)$ causal variables and Σ_{ij} is summation over all pairs of causal variables X_i, X_j, where $i < j$ and i ranges from 1 to $k - 1$, while j ranges from $i + 1$ to k. In words, this formula indicates that the total variance of Y (which in standardized scale equals unity) is the sum of the squares of the path coefficients along all paths leading to Y, added to twice the sum of the products of the path coefficients along all paths leading to Y multiplied by their correlations for all possible pairs of predictor variables. We will now compute the determination of variable Y for the example in Figure 16.4. Evaluating the formula just given, we obtain

$$\begin{aligned} &p_{Y1}^2 + p_{Y2}^2 + p_{Y3}^2 + 2(p_{Y1}p_{Y2}r_{12} + p_{Y1}p_{Y3}r_{13} + p_{Y2}p_{Y3}r_{23}) + r_{UY}^2 \\ &\quad = (-0.1807)^2 + (1.7111)^2 + (-1.1522)^2 \\ &\qquad + 2[(-0.1807)(1.7111)(-0.1900) + (-0.1807)(-1.1522)(-0.0627) \\ &\qquad + (1.7111)(-1.1522)(0.9553)] + 0.38745 \\ &\quad = 1.0001 \end{aligned}$$

The result agrees within rounding error with the expected value of 1. Note that in Expression (16.17) the terms expressing determination by known causes, $\Sigma_i\, p_{Yi}^2 + 2\, \Sigma_{ij}\, p_{Yi}p_{Yj}r_{ij}$, equal the coefficient of determination, R^2. This relationship is obvious, since r_{UY}^2 is the coefficient of nondetermination and $R^2 = 1 - r_{UY}^2$.

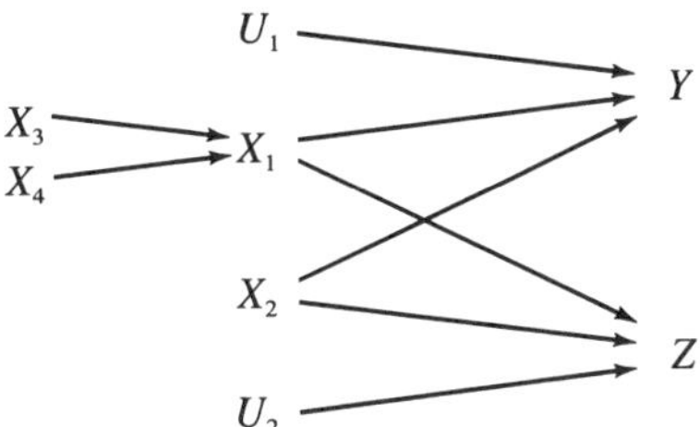

FIGURE 16.5 *Path diagram showing two criterion variables,* Y *and* Z, *which are correlated by having the common predictor variables* X_1 *and* X_2. *Variable* X_1 *is completely determined by uncorrelated predictor variables,* X_3 *and* X_4.

We must explore three more situations before we can apply our new knowledge of path analysis. We may wish to express the value of a path coefficient in terms of its constituent paths. An example is shown in Figure 16.5. The path from X_3 to Y goes via X_1. What is the value for p_{Y3}? We can write the prediction equation for Y as follows: $\hat{y}' = p_{YU_1}u_1' + p_{Y1}x_1' + p_{Y2}x_2'$, using notation for standardized variables. Similarly, $\hat{x}_1' = p_{13}x_3' + p_{14}x_4'$. Substituting $\hat{x}_1'$ into the equation for $\hat{y}'$, we obtain

$$\begin{aligned}\hat{y}' &= p_{YU_1}u_1' + p_{Y1}(p_{13}x_3' + p_{14}x_4') + p_{Y2}x_2' \\ &= p_{YU_1}u_1' + p_{Y1}p_{13}x_3' + p_{Y1}p_{14}x_4' + p_{Y2}x_2'\end{aligned}$$

Thus the standard partial regression coefficient of Y on X_3, which is the path coefficient p_{Y3}, is shown to be the product $p_{Y1}p_{13}$. Consequently the proportion of the variance of Y due to X_3, p_{Y3}^2, must be $p_{Y1}^2p_{13}^2$. Thus a **compound path** is the product of the constituent paths. This is true even when paths are bidirectional, as in Figure 16.6, where one of the paths from Y to Z is via X_1 and X_2. Note that this path, symbolized as $p_{Y(12)Z}$, is not the correlation r_{YZ}; it is just one constituent in that correlation, which we will discuss presently. In such a bidirectional path, at most one two-headed arrow representing a correlation coefficient may be included.

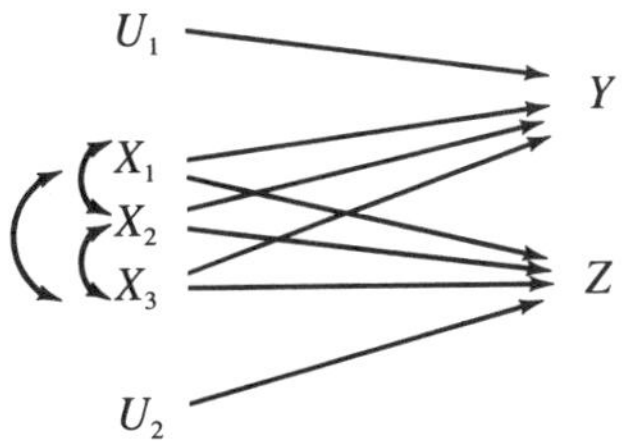

FIGURE 16.6 *Path diagram showing two criterion variables,* Y *and* Z, *which are correlated by having the common correlated causes,* X_1, X_2, *and* X_3.

Figure 16.5 also illustrates another important relationship—**common independent causes.** Two criterion variables are determined by two (or more) predictor variables, plus those representing unexplained variation. For the moment we need not be concerned that one of these variables, X_1, is determined by even more remote causes, X_3 and X_4. Both X_1 and X_2 affect Y and Z, and the correlation between Y and Z is due to these two common causes only. We can therefore employ the rules stated earlier that the correlation between two variables is the sum of all the paths connecting them, and we compute the two bidirectional paths $p_{Y(1)Z}$ and $p_{Y(2)Z}$ as the products of their constituent paths. Thus, $p_{Y(1)Z} = p_{Y1}p_{Z1}$ and $p_{Y(2)Z} = p_{Y2}p_{Z2}$, yielding $r_{YZ} = p_{Y1}p_{Z1} + p_{Y2}p_{Z2}$, and the general relation is

$$r_{YZ} = \sum_i p_{Yi}p_{Zi} \tag{16.18}$$

for all common independent causes, i. since the predictors in Figure 16.5 are not correlated, the total determination of Y or Z presents no new problems. Thus, to express the total determination of Y, we can write

$$p_{Y1}^2 + p_{Y2}^2 + r_{U_1Y}^2 = 1$$

or, if we wish to express X_1 in terms of X_3 and X_4, we obtain

$$p_{Y3}^2 + p_{Y4}^2 + p_{Y2}^2 + r_{U_1Y}^2 = 1$$

Finally, we evaluate a correlation in terms of **common correlated causes** as shown in Figure 16.6. The correlation between Y and Z is still the sum of all paths connecting them, but now there are many paths, nine to be specific. The formula for this case is

$$r_{YZ} = p_{Y1}p_{Z1} + p_{Y2}p_{Z2} + p_{Y3}p_{Z3} + p_{Y1}p_{Z2}r_{12} + p_{Y2}p_{Z1}r_{12} + p_{Y1}p_{Z3}r_{13} + p_{Y3}p_{Z1}r_{13} + p_{Y2}p_{Z3}r_{23} + p_{Y3}p_{Z2}r_{23}$$

The first three terms are bidirectional paths connecting Y and Z via X_1, X_2, and X_3, respectively. The next two involve paths from Y to Z via the correlation r_{12}. Note that there are two such paths: from Y through X_1 and X_2 to Z, or from Y to X_2 to X_1 and Z. Different path coefficients are involved each time; the only common element is the correlation r_{12}. Note, however, that in any one bidirectional path the correlation occurs only once. Confirm for yourself how the remaining four terms are obtained by tracing the paths in the figure. The formula can be generalized to

$$r_{YZ} = \sum_i p_{Yi}p_{Zi} + \sum_{ij} (p_{Yi}p_{Zj} + p_{Yj}p_{Zi})r_{ij} \tag{16.19}$$

The determination of the variance of either Y or Z presents no new features.

The rules for expressing correlations and determinations by path coefficients are summarized in Table 16.2, which gives general formulas that might serve as mnemonic aides.

Table 16.2 SUMMARY FORMULAS FOR PATH COEFFICIENTS.

	Correlation	Determination
One criterion variable		
Independent causes[a]	$r_{jY} = p_{Yj}$	$1 = \sum_i p_{Yi}^2 + r_{UY}^2$
Correlated causes[b]	$r_{jY} = p_{Yj} + \sum_{i \neq j} r_{ij} p_{Yi}$	$1 = \sum_i p_{Yi}^2 + 2 \sum_{ij} p_{Yi} p_{Yj} r_{ij} + r_{UY}^2$
Chains[c]	$r_{jY} = p_{Yj} = p_{Yi} p_{ij}$	
Two criterion variables		
Independent causes[c]	$r_{YZ} = \sum_i p_{Yi} p_{Zi}$	
Correlated causes[d]	$r_{YZ} = \sum_i p_{Yi} p_{Zi} + \sum_{ij} (p_{Yi} p_{Zj} + p_{Yj} p_{Zi}) r_{ij}$	

Notes:

$\sum_i$ is summation over all i $(1 \cdots k)$ causal variables.

$\sum_{i \neq j}$ is summation over all i $(1 \cdots k)$ causal variables except for j.

$\sum_{ij}$ is summation over all pairs of causal variables X_i, X_j where $i < j$, i ranges from 1 to $k - 1$, and j ranges from $i + 1$ to k.

[a] See Figure 16.3
[b] See Figure 16.4
[c] See Figure 16.5
[d] See Figure 16.6

You now know the essential features of path coefficients and path analysis. To understand the subject fully, however, we must review some special situations a few of which we will present later as the need arises. Now that we have achieved a reasonable mastery of the subject, though, what can we do with it? Path coefficients were first introduced by Sewall Wright in the early 1920s as a device for evaluating the contributions of various factors in complicated interaction systems in animal breeding, physiology, and related subjects. Path coefficients proved useful for predicting correlations in various fields, especially in population genetics, and during the next three decades they were employed largely for predicting the consequences of breeding patterns in population and quantitative genetics. Although this use of path coefficients has diminished somewhat in recent years in favor of newer techniques, there has been a marked resurgence in the application of path analysis to problems testing structural models. Social scientists have discovered path analysis and are now among its most active users. In biology, too, there is a growing renaissance of path analysis for model testing, its original purpose.

There are three major uses of path analysis. The first is to predict correlation in systems where the strengths of component parts are known. Most such situations occur in population genetics, and we will not discuss them here. For more information on this subject see Li (1975, 1976) and Wright (1968, 1969). We will illustrate a different application of path analysis. The example is artificial, but we chose it to illustrate an important theoretical point. Suppose you wish to predict the correlation between the sums and differences of two random variables. Assume that you have sampled n pairs of numbers from a table of random numbers (such as Statistical Table **FF**) and you have arrayed them into columns, X_1 and X_2. We now generate two new columns, $S = X_1 + X_2$ and $D = X_1 - X_2$. What is the correlation between variables S and D, which are both completely determined by variables X_1 and X_2? Although this problem can be solved algebraically, solving it by path coefficients is even simpler. The relationship is diagrammed in Figure 16.7. Since S and D are both completely determined by X_1 and X_2, there is no need for a residual path from unknown factors U for either variable, and we can write the following equations for the determination of the two variables, respectively: $p_{S1}^2 + p_{S2}^2 = 1$ and $p_{D1}^2 + p_{D2}^2 = 1$. In view of the equal determination by the two predictor variables, all squared coefficients equal 0.5, and the path coefficients are $\sqrt{\frac{1}{2}}$. We make p_{D2} negative, since it is a negative contribution to D. Because X_1 and X_2 are independent random variables, they are not correlated and we can predict the correlation between S and D, using the formula for independent common causes from Table 16.2. Thus we obtain

$$r_{SD} = p_{S1}p_{D1} + p_{S2}p_{D2} = \sqrt{\tfrac{1}{2}} \cdot \sqrt{\tfrac{1}{2}} + \sqrt{\tfrac{1}{2}} \cdot \left(-\sqrt{\tfrac{1}{2}}\right) = 0$$

The results may surprise you. The two variables S and D show no correlation. The positive contribution toward the correlation by X_1 is negated by the negative contribution due to X_2. Thus if you can specify a model exactly (and it need not be such a simple model as this) and if you are familiar with the rules for handling path coefficients, evaluating resulting correlations is a very simple task.

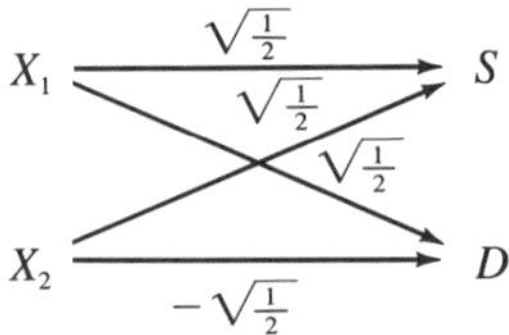

FIGURE 16.7 *Path diagram illustrating computation of the correlation between sums and differences.*

This example illustrates another important point. When you observe a low or zero correlation between two variables, you cannot really be certain that they have no common causes without further study of the underlying structure of the model. Here we have a completely determined model; yet if you had no insight into the structure, you would assume that S and D had no common predictors. This notion can be extended to high positive or negative correlations as well. We do not know how many underlying common causes contribute to establishing any observed correlation; in many cases there may be positive as well as negative contributions, some of which will cancel each other out.

A related example, presented by Hamilton (1987), is explained here in terms of path coefficients. We encounter two independent variables, X_1 and X_2, that are only weakly correlated with a dependent variable Y. The correlation of Y with X_2 is 0.4341, but that of Y with X_1 is essentially zero (0.0025). The correlation between the two independent variables, however, is very strong and negative (-0.8998). The relations are shown in Figure 16.8. We work out the path coefficients using by now familiar procedures and arrive at $p_{Y1} = 2.0650$ and $p_{Y2} = 2.2922$. When we compute the determination of Y, we find that $p_{Y1}^2 + p_{Y2}^2 + 2p_{Y1}p_{Y2}r_{12}$ already yields an R^2 of 1, leaving essentially nothing to be determined by p_{YU}^2, the remaining unknown factors. Thus, what appeared to be a case of weak correlation between the X-variables and Y when judged by zero-order correlation is actually a case of complete determination of the criterion variable by the predictor variables. As Hamilton (1987) points out, this example illustrates the often counterintuitive nature of multivariate relationships. In addition, such cases almost always have negative relations in a chain of their paths, in this case in the correlation between X_1 and X_2. In this system, both predictor variables, X_1 and X_2, strongly affect the criterion variable, but because of their tight negative linkage, they dampen each other's effects, producing weak or no pairwise correlations with the criterion variable. Nevertheless, they completely determine the variation of Y.

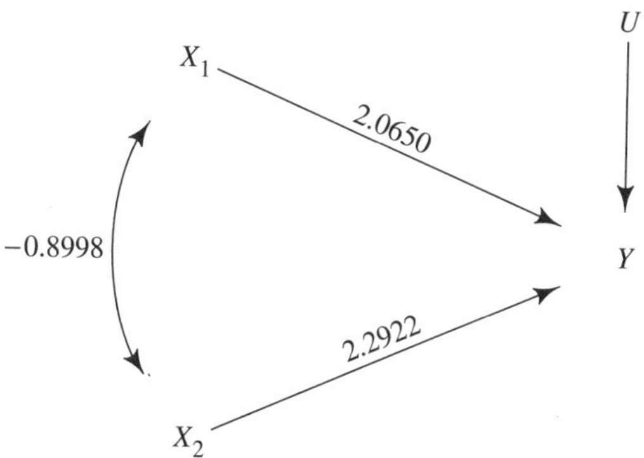

FIGURE 16.8 *Path coefficient diagram for the example presented by Hamilton (1987). For explanation see text.*

The second use of path analysis is as an aid for illuminating certain other statistical topics where its combination of graphic representation with simple algebraic manipulation is preferable over alternative, wholly algebraic formulations of these models. Examples are partial and multiple correlations, adding or deleting variables in multiple regression analysis, and factor analysis. The first two of these topics will be discussed in later sections of this chapter; we therefore defer discussion of the relation of path coefficients to these topics until then.

The third use of path analysis, and for our purposes here its major application, is the construction and evaluation of alternative structural models—theory testing. As an example, let us return to the sulfur dioxide pollution data analyzed in previous sections and earlier in this section in the context of learning path coefficients. In Figure 16.4 we illustrated the example using a path diagram characteristic of multiple regression analysis: All predictor variables affect the criterion variable Y directly but the predictors are also correlated among themselves. Is this really the most reasonable model for this example? Remember that X_1 is average annual temperature, X_2 is number of manufacturing enterprises with 20 or more employees, and X_3 is population size of the city. Should each of these variables be shown as directly affecting SO_2 content, or might one or two of them act indirectly by affecting the third variable? Temperature (X_1) is unlikely to affect population size appreciably in this data set. It is also unlikely that the number of manufacturing enterprises (X_2) is a function of temperature. Thus it seems that temperature should continue to affect SO_2 content directly. The climate undoubtedly affects the physicochemical processes determining the degree of pollution. However, population and the number of manufacturing enterprises might form a causal chain. Do manufacturing enterprises attract their work force (i.e., increase the population), or does a population base generate its number of manufacturing enterprises? Probably there is a complex interplay of cause and

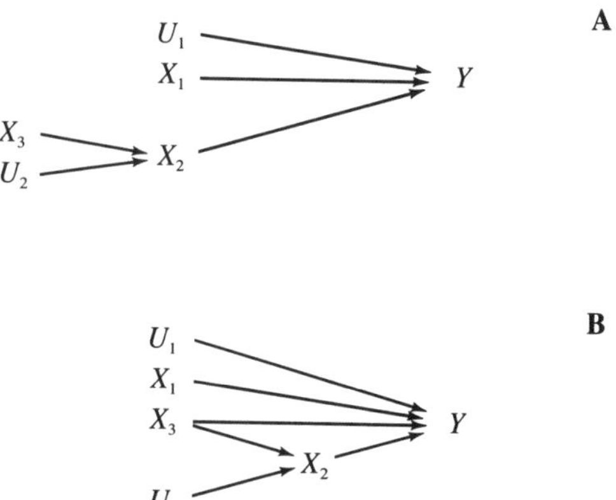

FIGURE 16.9 *Alternative interpretations of the pollution data. (For explanation see text.)*

effect, but a reasonable model worth investigating is one in which population size partly determines the number of manufacturing enterprises and enterprises partly determine SO_2 content.

The model in Figure 16.9A assumes that the population (X_3) does not directly engage in activities affecting the sulfur dioxide concentration in the air. Its influence is solely on increasing the number of manufacturing enterprises, which in turn release pollutants. The model as specified allows for correlation between X_3 and X_2 but not for correlation between either of these and X_1. We are encouraged in constructing this model by the observed low correlations r_{12} and r_{13}, which are -0.1900 and -0.0627, respectively. We have available to us (in part **II** of Box 16.1) the observed correlations among the three predictor variables and of the three with Y, the SO_2 concentration. We therefore wish to work out the path coefficients for the diagram in Figure 16.9A to see how well they reproduce the observed correlations. First we write specification equations for all the path coefficients as follows:

$$p_{Y1} = r_{1Y} = -0.4336$$

$$p_{Y2} = r_{2Y} = 0.6448$$

$$p_{23} = r_{23} = 0.9553$$

Of the correlations of the three predictor variables with Y, two are simply equal to their corresponding path coefficients and must be correct because we designated them by the observed values. The correlation $r_{3Y} = p_{Y3} = p_{23}p_{Y2}$, however, is a product of two of these designated path coefficients. When we estimate r_{3Y} in this manner, we obtain $(0.9553)(0.6448) = 0.6160$. This estimate of r_{3Y} is poor, since the observed correlation in Box 16.1 is 0.4938.

We can compute the coefficient of nondetermination $r^2_{U_1Y}$ from the formula for complete determination of Y, which is

$$p^2_{Y1} + p^2_{Y2} + r^2_{U_1Y} = 1$$

Since we know the path coefficients, $r^2_{U_1Y}$ is the only remaining unknown in the equation. We estimate $r^2_{U_1Y} = 0.3962$, slightly higher than the coefficient of nondetermination implied by the multiple regression model (in Box 16.2, $R^2_{Y\cdot 1,2,3} = 0.61255$; hence $1 - R^2_{Y\cdot 1,2,3} = 0.3874$). Thus on the basis of reproduction of the observed correlation and by the percent variation of Y that has been determined, we conclude that the model of Figure 16.9A is not satisfactory. Perhaps we should have allowed for correlations among predictor variables, or the paths in the diagram may have been linked incorrectly.

In Figure 16.9B we propose another model for the same example. Temperature (X_1) still independently affects SO_2 content. Population size, however, now affects pollution in two ways, indirectly through number of manufacturing enterprises, as before, but also directly. It is quite plausible that large numbers of people cause SO_2 to be liberated in the air through activities other than manufacturing enterprises. Note again that X_1 is not correlated with either X_3 or X_2. We obtain the following specification equations for the relations expressed in Figure 16.9B:

$$p_{Y1} = r_{1Y} = -0.4336$$

$$r_{3Y} = p_{23}p_{Y2} + p_{Y3}$$

$$r_{2Y} = p_{Y2} + p_{23}p_{Y3}$$

Note that p_{Y1} is the same as before, but the paths from X_2 and X_3 have changed because of the configuration of the path diagram. To solve for p_{Y2} and p_{Y3} we solve the two simultaneous equations shown here. The results are $p_{Y2} = 1.9802$ and $p_{Y3} = -1.3979$. These values differ appreciably from those obtained for the model in Figure 16.9A because the two models differ structurally. If we try now to reconstitute correlations r_{2Y} and r_{3Y}, we find that the predictions match the observations perfectly. This, however, is no cause for satisfaction. In a triangular relationship such as that among Y, X_2, and X_3 in Figure 16.9B, we can always solve exactly for the values of the path coefficients, but we have no independent test of these estimates. However, we can estimate the amount of variance of Y determined by the model.

The equation for the determination of Y is

$$p_{Y1}^2 + p_{Y3}^2 + p_{Y2}^2 + 2p_{Y3}p_{23}p_{Y2} + r_{U_1Y}^2 = 1$$

Substituting known values of path coefficients, we obtain the coefficient of nondetermination $r_{U_1Y}^2 = 0.2254$. This model thus explains even more of the variation than the standard multiple regression model of Figure 16.4. Note that one of our predictions is not fulfilled. We thought X_3 would have an independent positive effect on pollution; people by their numbers and activities would produce more SO_2. The very opposite seems to be the case. Although people contribute to pollution through increased number of manufacturing enterprises, which in turn pollute the air, the independent effect of X_3 on Y is strongly negative. Do people tend to clean up air? Paradoxically, yes. We are concerned with the effect of people when X_2, number of manufacturing enterprises, is held constant. Of the cities in Table 16.1, those with large numbers of people but with the number of manufacturing enterprises kept constant tend to be those with large "bedroom communities," so, all other things being equal, they contribute less to pollution than do nonresidential occupants. Note also that the effects of people and manufacturing enterprises as indicated by the absolute magnitudes of those path coefficients are greater than that of temperature (−0.4336). Apparently, the lower the temperature, the more sulfur dioxide is produced. This correlation may be related to increased use of fossil fuels. These two models are, of course, far from exhaustive, but comparing them should give you some feel for what we can do with such models. Remember that you have already observed correlations among all variables and that you are merely trying to reproduce these correlations by a heuristic path diagram. We now turn to another example to illustrate some different points.

In a laboratory study of fluctuations in housefly populations, biweekly censuses resulted in counts of live flies and counts of flies that had died during the previous two weeks. If, instead of the census of live adults, we record the change in adult number since the last census (and call this variable Y), the density of the population at the previous census (calling it X_1), and the number of flies that died during the two-week period (calling it X_2), we can represent the interrelations by the path diagram in Figure 16.10, which represents one hypothesis. The diagram implies that density (X_1) determines mortality (X_2) in the population and that this mortality contributes to the decrease in population size. Through its effect on fecundity, however, density also affects population size directly. The observed correlations for these data are as follows: $r_{1Y} = -0.134$, $r_{2Y} = -0.573$, and $r_{12} = 0.540$; from these values, by procedures that should by now be familiar, we obtain the path coefficients that are indicated in the diagram.

From the diagram, then, density appears to affect the number of flies dying (X_2), which in turn has a strong negative effect on the population size eventually recorded. At the same time, density has a mild positive effect on the change in population size, Y. We might wish to change the model by including a variable not directly observed: the number of adult flies emerging in the population cage during the two-week period. We can argue convincingly that, except for experimental error, the change during the two weeks must reflect the balance between recruitment (newly emerged adult flies) and adult death. If we designate recruitment by X_3, we can write $Y = X_3 - X_2$, not allowing for experimental error. On this assumption, birth and death make equal contributions to the variance of Y and we can write by definition that the paths from X_3 to X_2 and from X_3 to Y are $+\sqrt{\frac{1}{2}}$ and $-\sqrt{\frac{1}{2}}$, respectively, as they were in the path diagram of Figure 16.7. In the path diagram for this new model (Figure 16.11), the unexplained factor completing the determination of Y has been removed, since presumably X_2 and X_3 completely determine Y. Now X_2 and X_3 are determined by unexplained factors, together with their common cause X_1 (density). For this path diagram, we lack the value for path p_{31} from density to the unrecorded recruitment variable. We can obtain this value easily, however, from the equation for the correlation, $r_{1Y} = p_{31}p_{Y3} + p_{21}p_{Y2}$. In this equation all variables but p_{31} are known, and it is

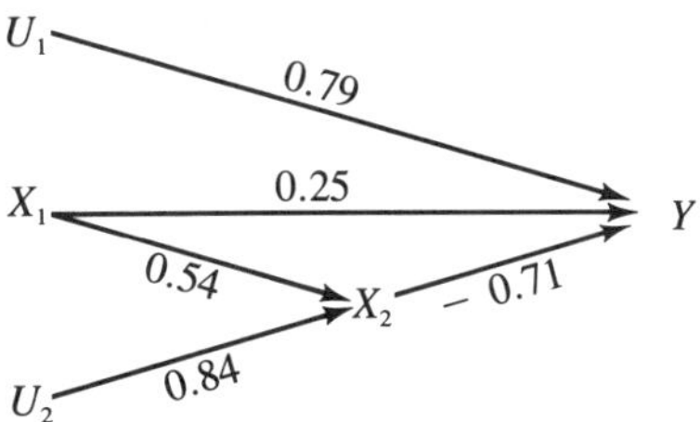

FIGURE 16.10 *Path coefficient diagram illustrating causal relationships in the housefly population study. (For explanation see text.)*

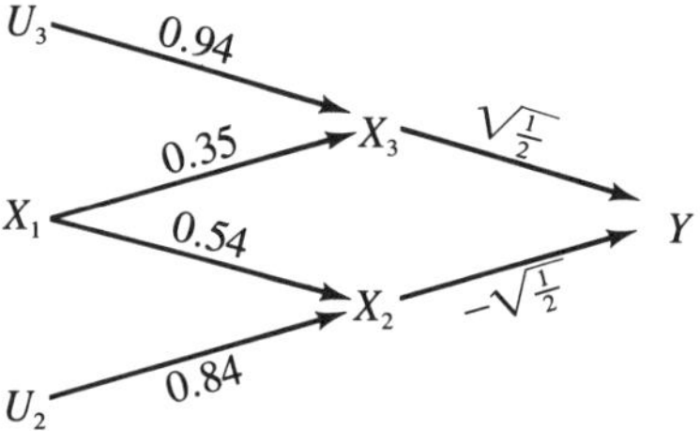

FIGURE 16.11 *Path coefficient diagram illustrating an alternative interpretation of the housefly population data. (For explanation see text.)*

easy to solve for the path coefficient, which turns out to be 0.351. Thus, we learn that density influences the number of flies added to the population in the next two weeks positively but weakly.

This influence may be weak because the effect of density on adult recruitment is complex. The more flies there are, the more eggs laid and the higher the input into the population. At the same time, the higher the density, the greater the crowding and the less adult input into the population (because of lower fecundities and increased egg and larval mortality). The relations may well not be linear, as assumed by the path coefficient model, in which case further elaboration of the model would be indicated. What is important about this example is that we were able to obtain path coefficients from an observed variable to one that was not even measured (number of new flies added to the population). Using path coefficients, one can frequently construct similar models in which the effects of hypothesized intermediate variables can be estimated if there are enough equations to determine the unknown coefficients.

Another approach is to assume a reasonable experimental error in this system so that the two variables X_2 and X_3 do not completely determine Y, thus requiring an unknown factor to complete the determination of Y. In such a case we might hesitate to postulate that the path coefficients from X_2 and X_3 were equal in magnitude. We could do a separate experiment in which the effect of density on adult production is investigated. From this experiment we could obtain an estimate of the correlation between X_1 and X_3. We could then employ this correlation, together with the other known values, to obtain estimates of p_{Y2} and p_{Y3}.

Occasionally situations arise in which a given unknown path coefficient is overdetermined; that is, there are more equations than unknowns. In such cases, one tries to adjust the values to effect a reasonable compromise. Estimating p_{31} from Figure 16.11 was one such example. When based on the equation for r_{1Y}, $p_{31} = 0.3504$, but when based on r_{2Y}, $p_{31} = 0.3512$.

We have not discussed the issue of significance of path coefficients. Although in some cases the significance tests for multiple regression analysis would apply, the main emphasis in path analysis has been on estimating magnitudes and directions of interactions rather than on the statistical significance of individual coefficients. Remember also that underlying path analysis is linearity of the regressions. Any nonlinear effects must be accounted for in the unexplained variation or by the transformation of variables. We will employ path coefficients again later in this chapter.

16.4 PARTIAL AND MULTIPLE CORRELATION

A **partial correlation coefficient** measures the correlation between any pair of variables when other, specified variables, have been held constant. For example, the correlation $r_{13 \cdot 245}$ is the correlation between variables Y_1 and Y_3 when variables Y_2, Y_4, and Y_5 are held constant. We can keep these other variables constant experimentally or statistically, and either method should give equivalent results. Again, we encounter the parallel approaches of experimental control and statistical control. It might not be immediately obvious why correlations should differ depending on whether other variables are free to vary or kept constant. If you refer back to Figure 15.5, however, the reasons will become obvious. In Figure 15.5D the correlation between Y_1 and Y_2 depends entirely on the common cause, Y_4. If we calculate the partial correlation $r_{12 \cdot 4}$, we should expect this to be zero because there is no reason why Y_1 and Y_2 should be correlated when Y_4 does not vary. Keeping variable Y_4 constant in Figure 15.5E would enhance the relative correlating influence of common cause Y_5 on Y_1 and Y_2, because a larger proportion of the variation of Y_1 and Y_2 would now be accounted for by the common element Y_5. If both Y_4 and Y_5 are kept constant, the partial correlation $r_{12 \cdot 45}$ would again be zero. Partial correlations are simple to compute when only three variables are involved. The formula for the partial correlation between variables Y_1 and Y_2, with variable Y_3 held constant, is

$$r_{12 \cdot 3} = \frac{r_{12} - r_{13}r_{23}}{\sqrt{(1 - r_{13}^2)(1 - r_{23}^2)}} \tag{16.20}$$

We will illustrate the computation of a partial correlation coefficient with an example from morphology based on a study of the intercorrelations among 18 morphological variables of alate aphids, *Pemphigus populitransversus* (Sokal, 1962). We compute the following three correlations for the characters forewing length (W), thorax length (T), and length of antennal segment 3 (A): $r_{TW} = 0.86$, $r_{AT} = 0.75$, $r_{AW} = 0.85$. There is a substantial correlation between wings and antennae. Some of this correlation is undoubtedly due to general size—that is, a bigger aphid will have longer wings and longer antennae. Is there also a special developmental factor, however, that tends to give aphids of the same general body length longer antennae if they have longer wings? We can investigate this question by means of partial correlations. Following Expression (16.20) we can write

$$r_{AW \cdot T} = \frac{r_{AW} - r_{AT}r_{TW}}{\sqrt{(1 - r_{AT}^2)(1 - r_{TW}^2)}} = \frac{0.85 - (0.75)(0.86)}{\sqrt{[1 - (0.75)^2][1 - (0.86)^2]}}$$
$$= 0.60736$$

The partial correlation between antennal segment 3 length and wing length, with thorax length kept constant, is 0.61. Thus, although thorax length is known to be an adequate predictor of general size, it is insufficient for explaining the correlation between wing length and antennal segment 3 length. Other common causes must contribute to the correlation between these two morphological variables.

The partial correlation just computed is a first-order partial correlation coefficient (the ordinary product–moment correlation r_{12} is known as a zero-order correlation). We can also compute higher-order partial correlation coefficients. The second order partial correlation between variables Y_1 and Y_2 with Y_3 and Y_4 kept constant is

$$r_{12 \cdot 34} = \frac{r_{12 \cdot 3} - r_{14 \cdot 3} r_{24 \cdot 3}}{\sqrt{(1 - r_{14 \cdot 3}^2)(1 - r_{24 \cdot 3}^2)}} \tag{16.21}$$

From this formula we should have no difficulty constructing formulas for other second-order partial correlations or even higher-order partial correlations. However, coefficients of partial correlation of higher orders are rarely calculated; the exception comes about when just one coefficient for two specific variables is the subject of interest. When a large matrix of correlation coefficients is available, methods for decomposing it—such as multiple regression analysis, path analysis, or principal-component or factor analysis—are preferable (see Section 16.7). Figure 16.12 illustrates the meaning of partial correlation graphically.

We can obtain further insight into the nature of partial correlation and the basis for Expression (16.20) by depicting the relationship by means of a path diagram. In Figure 16.13 we show the relations between the three morphological variables as if thorax and wing length were causes of antennal segment 3 length. This choice is arbitrary; we could have shown wing length as the criterion with antennal segment 3 and thorax length as predictors. Given the three known correlation coefficients, we can work out the path coefficients p_{AW} and p_{AT} by the methods of Section 16.3. The path coefficients are $p_{AW} = 0.78725$ and $p_{AT} = 0.072965$, while the coefficients of nondetermination are as follows: $r_{U_A A}^2 = 0.27611$, $r_{U_W W}^2 = r_{U_T T}^2 = 0.2604$.

We will now evaluate the partial correlation $r_{AW \cdot T}$. In this partial correlation T (thorax length) is considered constant. By making it constant we alter the pattern of action in the path diagram. We eliminate path p_{AT} and the resulting diagram is much simpler. The broken lines in Figure 16.13 show paths and correlations that vanish when thorax length is assumed to be constant.

To obtain the partial correlation of W with A, variable T kept constant, we recompute the path from W to A under the new conditions. Remember that a path coefficient is a standard partial regression coefficient. It thus measures the strength of a relationship as a proportion of the total standard deviation. In standardized variables, variances equal 1. With one variable kept constant, however, the variances of neither A nor W remain at unity. The new variance of W as a proportion of the old variance is $1 - r_{WT}^2 = r_{U_W W}^2 = 0.2604$. The old variance of A was unity, and by the formula for determining the variance of a variable in a

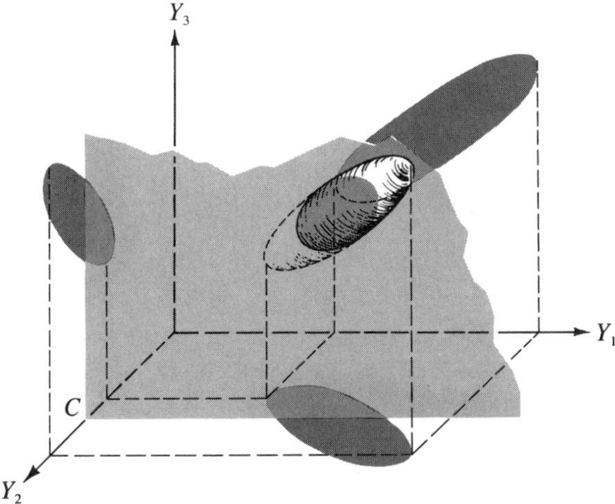

FIGURE 16.12 *Geometric interpretation of partial correlation. A three-dimensional equal-frequency ellipsoid is shown at center with its two-dimensional projections (shadows) shown on each wall. Cutting through the figure at the point* $Y_2 = C$ *(a constant) is a plane. The trace of the ellipsoid on this plane is an equal-frequency ellipse (the central shaded ellipse). This ellipse represents the covariation in variable* Y_1 *and* Y_3 *when variable* Y_2 *is held constant at* C. *The magnitude of the partial correlation between* Y_1 *and* Y_3 *is reflected in the shape of this ellipse. The more elongate this ellipse is, the higher the partial correlation is. The shapes of the three elliptical shadows on the walls represent the ordinary correlations between the pairs of variables defining these walls.*

path diagram (see Section 16.3) we write

$$1 = p_{AT}^2 + p_{AW}^2 + 2p_{AT}p_{AW}r_{TW} + r_{AU_A}^2$$

What remains of this variance when T is kept constant? All terms involving T vanish, leaving the unexplained variation $r_{AU_A}^2$ and the contribution due to the one remaining path coefficient, p_{AW}^2. But not all of p_{AW}^2 remains. Remember that we have just seen that the effect of W has been diminished by the disappearance of the correlation r_{TW}. Thus the remaining determination of A due to W is

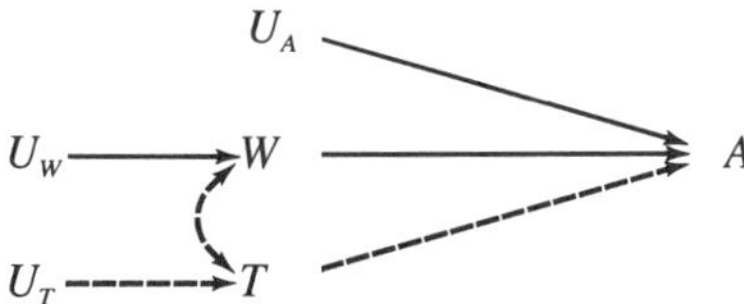

FIGURE 16.13 *Path diagram illustrating partial correlation of wing length and antenna length, with thorax length kept constant.*

$(1 - r_{TW}^2)p_{AW}^2$, and we can write the determination of A, T being kept constant, as follows:

$$1 - p_{AT}^2 - 2p_{AT}p_{AW}r_{TW} - r_{TW}^2 p_{AW}^2 = r_{AU_A}^2 + (1 - r_{TW}^2)p_{AW}^2$$

Referring back to Expression (16.3) will help you recall how standard partial regression coefficients were rescaled by dividing by the respective standard deviations of the variables. Here again we have to rescale the path coefficients in terms of the new standard deviations, which are less than unity. The appropriate scale factors are

$$p_{AW\cdot T} = \frac{p_{AW}\sqrt{1 - r_{WT}^2}}{\sqrt{r_{AU_A}^2 + (1 - r_{TW}^2)p_{AW}^2}} = \frac{0.78725\sqrt{0.2604}}{\sqrt{0.2761 + (0.2604)(0.78725)^2}} = 0.60736$$

Since the only path connecting W and A in the reduced diagram is this direct path, it is not surprising that the value of the path coefficient is equal to the value of the partial correlation coefficient $r_{AW\cdot T}$ as calculated earlier. Expressed in general terms, the adjusted path coefficient is

$$p_{Y1\cdot 2} = p_{Y1}\frac{\sqrt{1 - r_{12}^2}}{\sqrt{1 - r_{Y2}^2}} \tag{16.22}$$

or expressed in words, it is the old path coefficient multiplied by the square root of the ratio of the remaining variation of variable X_1 to that of variable Y when variable X_2 is kept constant. If we wish to express this relationship in terms of partial r's and b's, we can write

$$r_{Y1\cdot 2} = b'_{Y1\cdot 2}\frac{\sqrt{1 - r_{12}^2}}{\sqrt{1 - r_{Y2}^2}} \tag{16.22a}$$

When partial correlations are to be estimated between two criterion variables such as Y and Z in Figure 16.14, these path coefficient formulas must be modified. Thus to calculate the partial correlation $r_{YZ\cdot 1}$ we first calculate the correlation between Y and Z over the remaining paths, which is $p_{Y2}p_{Z2}$. This quantity must be multiplied by the square root of the proportion of the variance of X_2 remaining after X_1 has been kept constant. In the example in Figure 16.14 there is no change in the variance of X_2 when X_1 is kept constant. Therefore, the

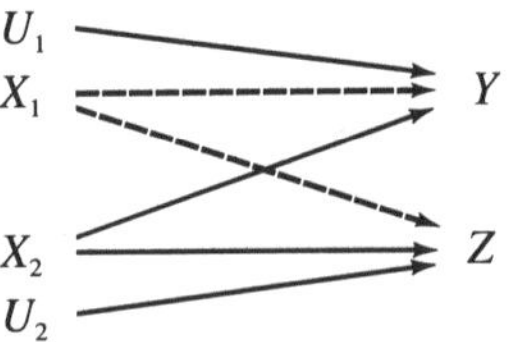

FIGURE 16.14 *Path diagram illustrating partial correlation of* Y *and* Z *with variable* X_1 *kept constant.*

proportion of X_2 remaining is 1, and $p_{Y2}p_{Z2}$ is multiplied by unity. Just as we had to adjust for the remaining proportion of variation of variable Y in Expression (16.22), we now have to adjust for the remaining variations of both Y and Z. This adjustment signifies that a given correlating force may be more or less important depending on how much of the variation of Y or Z remains. Keeping X_1 constant in this instance reduces the determination of Y to $1 - p_{Y1}^2$ and that of Z to $1 - p_{Z1}^2$. Thus we can write the formula for the desired partial correlation by inspection as follows:

$$r_{YZ \cdot 1} = \frac{p_{Y2}p_{Z2}(1)}{\sqrt{1 - p_{Y1}^2}\sqrt{1 - p_{Z1}^2}}$$

To test the significance of a partial correlation coefficient, one can proceed as in Box 15.4 for ordinary correlation coefficients, except that an adjustment must be made in the degrees of freedom for the number of variables held constant. To test the null hypothesis that $\rho_{YZ \cdot 1 \ldots m} = 0$ we compute

$$t_s = r_{YZ \cdot 1 \ldots m}\sqrt{\frac{n - 2 - m}{1 - r_{YZ \cdot 1 \ldots m}^2}}$$

and test against $t_{\alpha[n-2-m]}$, where m is the number of variables kept constant. We can also use the z-transformation (see Section 15.5), but its standard error is now modified to $1/\sqrt{n - 3 - m}$ where m is again the number of variables kept constant.

We should briefly mention $R_{Y \cdot 1 \ldots k}$, the **coefficient of multiple correlation,** which is the positive square root of the coefficient of multiple determination. Since the latter ranges from 0 to 1, the multiple correlation coefficient can assume only positive values from zero to unity. This range distinguishes the multiple correlation coefficient from the ordinary correlation coefficient, which ranges from -1 to $+1$. The coefficient of multiple correlation is a measure of how well the predicted values $\hat{Y}_i$ fit the observations Y_i. If we were to compute the entire array of values $\hat{Y}_i$ and calculate a correlation between $\hat{Y}_i$ and the observations Y_i, the resulting correlation coefficient $r_{\hat{Y}Y}$ would equal $R_{Y \cdot 1 \ldots k}$, which is indeed a correlation coefficient. This relation also holds for the bivariate case. Thus $r_{\hat{Y}Y} = r_{XY}$, where $\hat{Y}$ is the estimate due to regression on X.

Our discussion of the coefficients of multiple correlation and determination must include one additional point. When $R_{Y \cdot 1 \ldots k}^2$ is estimated in the conventional way, it tends to overestimate $\rho_{Y \cdot 1 \ldots k}^2$, the parametric coefficient of determination because the least squares estimates of the partial regression coefficients are fitted on the assumption that the variances and covariances in the normal equations [Expression (16.6)] are true values. In fact, they are only estimates. In the anova test of significance of the multiple regression (see Section 16.2), this fact is allowed for by dividing the explained and unexplained sums of squares by the appropriate degrees of freedom. $R_{Y \cdot 1 \ldots k}^2$, however, is computed from sums of squares and is not so corrected. It is customary therefore to compute

$$R^2_{Y \cdot 1 \ldots k}(\text{adj.}) = 1 - (1 - R^2_{Y \cdot 1 \ldots k})\left(\frac{n-1}{n-k-1}\right) \tag{16.23}$$

an adjusted coefficient of multiple determination. This value, corrected for the degrees of freedom, is always less than the $R^2_{Y \cdot 1 \ldots k}$ estimated from the sums of squares of the anova. The adjustment will be drastic when n is small and k is relatively large. By manipulating the terms in Expression (16.23) we can even show that for low values of ρ^2_Y and of n, the introduction of additional independent variables actually *decreases* $R^2_{Y\cdot}$ (adj.), although the values of estimated $R^2_{Y\cdot}$ tend to increase as the number of independent variables is augmented from k_1 to k_2. This phenomenon is the reason that many statisticians recommend that n equal or exceed $30k$, a condition not frequently met in biological research.

We can use Statistical Table **R** to look up the significance of an estimated $R_{Y \cdot 1 \ldots k}$ for $k \leq 4$, just as we used this table to test the significance of a simple correlation r_{12}. If we wish to test $R_{Y \cdot 1,2}$ from Box 16.2, for example, we proceed as follows. In the box we find $R^2_{Y \cdot 1,2} = 0.51611$; hence $R_{Y \cdot 1,2} = 0.7184$. This value is greater than the critical values in Table **R** (0.397 and 0.481) for two independent variables and $\nu = 35$; the actual $\nu = n - m = n - k - 1 = 41 - 2 - 1 = 38$, but there is no need to interpolate in the table in view of the magnitude of the multiple correlation coefficient. The null hypothesis that $\rho_{Y \cdot 1,2} = 0$ can be rejected.

In closing this section we would like to contrast the path coefficient approach with the partial correlation approach. With path coefficients we make an explicit a-priori model of the system and then estimate the strength of the links given that model. The advantage of path coefficients is that all of the assumptions are made explicitly, and it becomes clear that inferences are a function of the underlying model. When we make different path coefficient models of the same process, we come up with quite different coefficients. By contrast, when we compute partial correlations, we often assume that we are carrying out "objective" and model-free computations. Partial correlations are also not model-free, however, because certain variables are isolated as the ones to be kept constant. Partial correlations are a specific interpretation of relations among the variables concerned.

16.5 Choosing Predictor Variables

Multiple regression analysis, like simple linear regression, is used both for prediction and for establishing functional relations among the variables (structural models). Prediction is the more frequent application in multiple regression, and various techniques have been developed that attempt in a systematic manner to find the best set of predictor variables. By "best" is usually meant some combination of parsimony and adequacy of prediction. That is, one wishes to predict the value of variable Y using as few predictor variables as possible while at the same time making $\Sigma(Y - \hat{Y})^2$ as small as possible. These requirements conflict with each other: Increasing the number of variables will explain more of the variance of Y, but there is usually a point of diminishing returns beyond which additional variables are able to explain only a small amount of additional variation of Y. Thus the aim is to omit variables that do not make an appreciable contribution.

A complication in such endeavors is that every time a new independent (predictor) variable is added or removed from a given data set, another column of contrasts is added or removed from the set of normal equations. As we saw in Box 16.1, $b'_{Y1 \cdot 2} \neq b'_{Y1 \cdot 23}$, and the equations have to be recomputed. Fortunately, this is easily accomplished nowadays with fast computers.

There are two basic strategies for choosing a set of predictor variables. One approach is to start with a single variable and then add variables one at a time until adding variables does not produce an appreciable increase in the coefficient of determination. This process is known as **forward selection.** A criterion for the order in which to add variables is necessary. The obvious approach is to add the variable whose addition would produce the largest increment in R^2. We can accomplish this task in two ways without having to recompute the solutions to the normal equations for all possible combinations of variables. Given that k predictor variables have already been selected, one can add next the variable j (among those not yet part of the set of predictors) that has the highest partial correlation $r_{Yj \cdot 1 \ldots k}$. By calculating partial correlations for all possible candidates, we can determine which variable has the strongest partial correlation.

A second common approach is to compute the F_s-value corresponding to a test of the significance of the increment of R^2 due to the addition of a new variable. This quantity is known as the "F-to-enter" value. The variable with the largest F-to-enter value is added first, but only if that particular F-value is significant at a specified nominal level. The partial-r criterion and F_s criterion are the same in that they choose the same predictor—they are monotone functions of one another. After a given variable has been added, the normal equations are solved again; the new partial regression coefficients are computed, along with the new values of partial correlations and/or F-to-enter values for the remaining candidates for entry into the set of predictor variables. The next variable to be selected is chosen by the same criterion as before—that is, the highest partial correlation or highest F-to-enter value. How do we know when to stop adding variables? The general rule is to stop when the F-to-enter value becomes nonsignificant at a specified nominal probability level (the "P-to-enter" value).

The logical alternative to forward selection is to include initially all of the predictor variables in the multiple regression equation and then to drop those whose removal does not appreciably lower the determination of Y. This process is called **backward elimination.** Again, the partial correlation coefficient between each predictor variable and the criterion variable, with the other predictor variables held constant, is useful. The variable with the smallest partial correlation is removed first from the set of predictor variables. One can also compute an

"F-to-remove" value for each predictor variable. In this case, we stop removing variables when the F-to-remove value becomes significant at a specified nominal level (the "P-to-remove" level).

Both strategies—forward selection and backward elimination—suffer a major drawback. Once a variable has been added in forward selection it remains in the predictor set even though the added variable may contribute little to the determination of Y when other variables are taken into consideration. Similarly, once a variable has been deleted in backward elimination it stays out of the predictor set even though it may become a good predictor at a later stage when many other predictors have been eliminated. This possible instability depends upon the patterns of intercorrelations among the predictor variables. If we drew path diagrams for the different constellations of predictor variables as they relate to the criterion variable, we would notice that the set of paths that needs to be traced to evaluate the correlations between predictors and criterion variable would change materially depending on the subset of predictor variables chosen for a given solution. To overcome some of these difficulties, the **stepwise procedure** has been suggested.

The stepwise procedure is carried out either in the forward or in the backward direction, as explained above. After some variables have been added (or deleted, depending on the type of solution), however, the data set is scanned for any added variables that should now be deleted (or for any deleted variables that should be added again). Let us apply this method to the pollution data with six independent variables from Table 16.1. In Box 16.4 we show edited results of such an analysis. For the previous edition of this book, we computed the values with the stepwise multiple regression program of MIDAS, the statistical program package available at the University of Michigan. For the present edition, we confirmed these results by means of SAS, a widely used statistical package. Predictor variables are considered for addition (or deletion) in the order of highest (or lowest) partial correlations with the criterion variable. The decisions on whether to add or remove a variable are based on P-to-enter (or P-to-remove) values.

The usual significance levels in a stepwise multiple regression problem are approximate at best. In testing a large number of the possible combinations of predictor variables, we are carrying out unplanned tests (comparable to the unplanned multiple-comparison tests of Section 9.7), but the probabilities are conventionally determined as if we were performing planned tests. Thus the P-to-enter or P-to-remove values should not be interpreted as conventional levels of significance,since they may greatly underestimate the correct probabilities. Most computer programs allow the user to specify the P-to-enter or P-to-remove values. We have specified 0.05 for the P-to-enter value and 0.10 for the P-to-remove value. Since the method is only heuristic, the user can experiment freely with other values. We should also point out that with a large number of predictor variables only a relatively small number of combinations of variables out of the total possible are tested by the program. Thus it is possible that the program will not discover best solution.

Box 16.4 STEPWISE MULTIPLE REGRESSION.

Air pollution data from Table 16.1. Regression of Y (SO_2 content of air) on variables X_1 through X_6. (For explanation see text.)

Forward selection procedure ($P \leq 0.05$ to add, $P \geq 0.10$ to remove)

Step 1. Regression on X_2:

Anova table

Source of variation	df	SS	MS	F_s	P
Explained	1	9161.7	9161.7	27.750	<0.0001
Unexplained	39	12,876	330.16		
Total	40	22,038			

$R^2_{Y \cdot 2} = 0.41573 \qquad R_{Y \cdot 2} = 0.64477$

Table of coefficients

Predictor variables	r_{Yi}	a or b_{Y2}	SE (of a or b)	P	Decision
a		17.611	3.6916	<0.0001	
X_2	0.64477	0.026859	0.0050987	<0.0001	retain

Remaining variables	$r_{Yi \cdot 2}$	P	Decision
X_1	−0.41451	0.0078	hold
X_3	−0.54035	0.0003	add
X_4	−0.07911	0.6275	do not add
X_5	0.09843	0.5457	do not add
X_6	0.37556	0.0169	hold

Step 2. Regression on X_2 and X_3:

Anova table

Source of variation	df	SS	MS	F_s	P
Explained	2	12,921	6460.6	26.929	<0.0001
Unexplained	38	9116.6	239.91		
Total	40	22,038			

$R^2_{Y \cdot 2,3} = 0.58632 \qquad R_{Y \cdot 2,3} = 0.76572$

Table of coefficients

Predictor variables	$r_{Yi \cdot j}$	a or $b_{Yi \cdot j}$	SE (of a or b)	P (for a or b)	Decision
a		26.325	3.8404	<0.0001	
X_2	0.67301	0.082434	0.014697	<0.0001	retain
X_3	−0.54035	−0.056607	0.014300	0.0003	retain

Remaining variables	$r_{Yi \cdot 2,3}$	P	Decision
X_1	−0.25179	0.1220	do not add
X_4	−0.12695	0.4412	do not add
X_5	0.12752	0.4391	do not add
X_6	0.27411	0.0913	do not add

Program terminates.

Backward elimination procedure ($P \leq 0.05$ to add, $P \geq 0.10$ to remove)

Step 1. Regression on all predictor variables (X_1 to X_6):

Anova table

Source of variation	df	SS	MS	F_s	P
Explained	6	14,755	2459.1	11.480	<0.0001
Unexplained	34	7283.3	214.21		
Total	40	22,038			

$R^2_{Y \cdot 1 \ldots 6} = 0.66951 \qquad R_{Y \cdot 1 \ldots 6} = 0.81824$

Table of coefficients

Predictor variables	$r_{Yi \cdot}$	a or $b_{Yi \cdot}$	SE (of a or b)	P (for a or b)	Decision
a		111.73	47.318	0.0241	
X_1	−0.33040	−1.2679	0.62118	0.0491	do not remove
X_2	0.57727	0.064918	0.015748	0.0002	do not remove
X_3	−0.40665	−0.039277	0.015133	0.0138	do not remove
X_4	−0.28788	−3.1814	1.8150	0.0887	do not remove
X_5	0.23542	0.51236	0.36276	0.1669	hold
X_6	−0.05501	−0.052050	0.16201	0.7500	remove
Remaining variables			None		

BOX 16.4 CONTINUED

Step 2. Regression on X_1 to X_5:

Anova table

Source of variation	df	SS	MS	F_s	P
Explained	5	14,733	2946.5	14.117	<0.0001
Unexplained	35	7305.4	208.73		
Total	40	22,038			

$R^2_{Y \cdot 1 \ldots 5} = 0.66851 \qquad R_{Y \cdot 1 \ldots 5} = 0.81762$

Table of coefficients

Predictor variables	$r_{Yi \cdot}$	a or $b_{Yi \cdot}$	SE (of a or b)	P (for a or b)	Decision
a		100.15	30.175	0.0022	
X_1	−0.41472	−1.1213	0.41586	0.0107	do not remove
X_2	0.57651	0.064887	0.015545	0.0002	do not remove
X_3	−0.40666	−0.039335	0.014937	0.0125	do not remove
X_4	−0.28303	−3.0824	1.7656	0.0896	do not remove
X_5	0.31156	0.41947	0.21624	0.0605	do not remove

Remaining variables	$r_{Yi \cdot 1-5}$	P	Decision
X_6	−0.05501	0.7500	do not add

Program terminates.

NOTE: These results are edited from computer output.

An alternative exploratory strategy, available in many computer programs, is to request the subset of $m < k$ independent variables that best predict a given dependent variable by multiple linear regression. The program will try all possible combinations of m variables and select the one that yields the highest R^2 (or a specified number of combinations yielding the highest R^2-values). Note, however, that even the best-fitting model for a particular dataset may not be the "true" model. No statistical method can be relied upon to select the true model automatically. Statistical procedures such as multiple regression can be used to explore alternative models and to test consistency of a particular model with the data. Achieving good models requires building them from relevant theories that suggest appropriate predictor variables connected by plausible functional relationships.

In Box 16.4 a variable will be added to the set of predictors if adding it yields a partial regression corresponding to a nominal significance of $P < 0.05$. By contrast, a variable in the set of predictors will be removed only if the nominal significance of its partial regression coefficient is $P > 0.10$. The order in which acceptable variables are added or undesirable ones are dropped is by decreasing or increasing magnitude of the partial correlation coefficient (which is correlated with the nominal significance level of the partial b). Box 16.4 shows the forward solution first. In the analysis of variance of step **1,** variable X_2 has been chosen because its correlation with the variable Y is the highest of any predictor variable in the study (0.6448). Regressing Y on X_2 explains 0.41573 of the variance of Y, and the regression is nominally highly significant. Following the anova table is a table of partial correlation coefficients of the remaining variables with Y and of the nominal significance values of the partial regression coefficients that would be computed for each of these variables were it to be added to the present predictor set, which thus far includes only variable X_2. From this list variables X_1, X_3, and X_6 might be considered candidates for admission, and of these, variable X_3 should be considered first because its partial correlation is the strongest. Accordingly, the program recomputes the multiple regression equation for variables X_2 and X_3, both of which have significant partial regression coefficients, and which together explain 0.58632 of the total variance of Y. The values of the remaining partial correlations have changed appreciably, as have the nominal significance values of the standard partial regression coefficients. Since none of the latter have P-values less than 0.05, the program stops adding variables. A check of the variables in the set reveals that both variables have P-values less than 0.10, so none are removed. At this point the program stops and prints out a summary of the activity carried out during the stepwise regression program. In this instance only two variables were added and none was deleted.

The second half of Box 16.4 shows the backward elimination of variables by the stepwise procedure for the same example. We start with the regression of Y on all six variables. A significant proportion of the variation (0.66951) is explained by all six. The partial correlation coefficients are listed in the table following the anova table, as are the significance values of the partial regression coefficients. Only X_1, X_2, and X_3 appear to have significant partial regressions with Y when all six variables are considered. We remove the variable with the weakest partial correlation, namely X_6. Removing it lowers the proportion explained only trivially to 0.66851 but raises the partial correlations, lowering the probability values of the significance tests for the partial regression coefficients. Thus, after X_6 has been removed from the multiple regression, the partial regression coefficients of X_1 through X_5 all have associated P-values that are less than 0.10, although two (X_4 and X_5) have probability values between 0.05 and 0.10, making them not nominally significant by conventional standards. The program parameters as we have specified them, however, would not remove variables unless $P \geq 0.10$. The program also tests whether any of the removed variables should be readmitted, but since only variable X_6 has been removed and its partial correlation coefficient has a P-value >0.05, it remains outside and the program terminates.

We have thus obtained two different solutions by the forward and backward techniques using the parameters for adding and removing variables as stated earlier. We can easily see why the particular criteria for admission and removal have led to these disparate results. If we had made the criteria for admission and rejection more similar to each other—say, 0.05 to admit and 0.06 to remove—we would expect the backward selection to remove variables X_4 and X_5 as well, leaving regression on X_1, X_2, and X_3. In such a regression, the partial regression coefficient of Y on X_1 has a P-value of 0.122, which even by the old criterion would indicate that variable X_1 should be removed. This process would have ended in a solution identical to the one obtained by forward selection. The relative indeterminacy of these stepwise regression analyses should be obvious. In Boxes 16.1 and 16.2 we built up the regression by first regressing Y on X_1 and X_2 and were able to show that adding X_3 provided a significant increase in percent of determination. Had we started with X_2, however, then added X_3, and then X_1, the addition of X_1 would not have resulted in a nominally significant increment in determination. Indeed, the stepwise regression procedure (with our stated parameters) would remove X_1 from a set containing the predictor variables X_1, X_2, and X_3. Variable X_1 is average annual temperature. When some of the other weather variables, X_4, X_5, and X_6, are added to the set, as in the backward solution, then the partial effect of X_1 suddenly becomes quite large, and unless we remove the weather variables, X_1 continues to retain an important effect.

Thus one cannot assume that the subset of variables found by a stepwise regression program necessarily corresponds to the most important set of variables. The addition or deletion of a variable can sometimes cause a complete change in the membership of the predictor set. One should not lose sight of the purpose of stepwise multiple regression analysis, which is simply to find the smallest set of predictor variables that does an adequate job of prediction. Variables left out of the predictor set are not necessarily unimportant (they may simply be correlated with other variables in the predictor set).

To run a stepwise regression program, we must furnish the program with an upper limit on the number of iterations the program is permitted to take. A situation can easily develop where the program gets into a loop, adding one variable in one step and removing it in the next step. Limits for the criterion for addition and removal that are too close together may lead to such a situation.

Decisions to employ or reject variables need not be made only on the basis of nominal statistical significance and importance. Considerations such as ease of observation, cost of obtaining the variable, and so on also play an important role.

When an investigator is faced with a substantial number of putative predictor variables and has to decide which to include and which to omit, it is useful to know the consequences of the possible types of decisions. As we have seen, much will depend on the correlation structure of the variables, but some general principles have been summarized by Rao (1971). In a multiple regression situation in which the "truth" is known—that is, all the variables that contribute to an explanation of the dependent variable are known—the omission of a variable specified by the truth introduces bias to least squares estimates but reduces their variances. By contrast, including an irrelevant variable—that is, a predictor variable in addition to the truth—does not produce bias in the least squares estimates but does increase the variance and mean square error of all least squares estimates. Thus unless a complete prediction equation is desired, omission of a variable that contributes little may be a favored procedure, since the smallest estimates of mean square error would be preferable over those that are unbiased. By contrast, adding variables that are not in the true specifications may decrease efficiency because of the increase in mean square errors.

After you have decided which subset of independent variables to employ to predict the dependent variable Y, you may wish to rank the independent variables in the order of their importance for predicting Y. This problem does not have a unique solution. We have already seen one way of ranking variables: We can employ the absolute magnitude of the standard partial regression coefficients to indicate an order of importance. The complication is that when the independent variables are correlated, the relative magnitude of a coefficient depends on the other variables being considered at the same time. The discussion in Section 16.1 about the example in Figure 16.2 illustrates how the importance of a variable depends upon the other variables being held constant. Kruskal (1987) suggested using an average of squared partial correlation coefficients. He computes the average of the squared partial correlations between Y and X_i with all possible combinations of the other independent variables held constant, e.g., r^2_{Y1}, $r^2_{Y1 \cdot 2}$, $r^2_{Y1 \cdot 3}, \ldots, r^2_{Y1 \cdot k}, r^2_{Y1 \cdot 23}, \ldots, r^2_{Y1 \cdot 23 \ldots k}$. There are $2^{(k-1)}$ such combinations for each of the k independent variables. By this criterion, the variable with the highest average is considered the most important, that with the next highest is the second most important, and so on.

For the city pollution data of Table 16.1, we obtained the following averages (each based on 32 correlations) for the six variables: 0.16161, 0.40433, 0.25043, 0.03251, 0.04080, and 0.08159. The ranking is: 3, 1, 2, 6, 5, 4. The ranking based on the b_i values (computed from b_i values shown in the results of step **1** of the backward elimination procedure in Box 16.4) are 3, 1, 2, 5, 4, 6. The ranking based on the partial correlations shown in the same table are 3, 1, 2, 4, 5, 6. These rankings agree only for the three most important variables. Kruskal points out that his procedure is appropriate only when the independent variables are random, a condition which actually violates the Model I regression assumption.

In practice unbiased least squares estimates are not often used for the regression coefficients (β). Whenever we use a procedure to obtain a solution with a reduced number of variables, we compromise the least squares principle by setting some *coefficients equal to zero. If the variables eliminated correspond to variables that have no effect in the true model, then we may expect this solution to be a better predictor in* future *samples even though we obtain a worse fit in the observed sample. Other types of biased estimation methods, such as ridge regression and latent root regression analysis, have been proposed. These methods are designed to yield more satisfactory predictive models when one is analyzing data in which the independent variables are too highly intercorrelated (a problem referred to as the "multicollinearity problem"). The ridge regression method is easy to understand. A small constant,* k *(the ridge coefficient), is added to the diagonal elements of the coefficient matrix on the left-hand side of the normal equations, yielding* $(\mathbf{X}^t\mathbf{X} + \mathrm{k}\mathbf{I})\mathbf{b} = \mathbf{X}^t\mathbf{Y}$. *The next step is to investigate the effect of using small positive values for* k *(*$\mathrm{k} = 0$ *yields the ordinary least squares solution). Marquardt and Snee (1975) give an example in which the ridge regression technique yields a much better prediction equation than the one obtained by ordinary least squares. Latent root regression analysis is a more complex technique and involves the regression of the dependent variable onto the principal axes of the correlation matrix consisting of both the dependent and the independent variables. This method should result in an improved prediction equation when the independent variables are highly correlated (and this high correlation cannot be explained by the joint correlation of the independent variables with the dependent variable). The details of these biased regression techniques are beyond the scope of the present text. We should consider these more advanced techniques when analyzing data sets for which the variance inflation factors are large (*>10*). Marquardt and Snee (1975) present an introduction to ridge regression. Latent root regression is described by Webster, Gunst, and Mason (1974). Hocking (1976) and Draper and Smith (1981) give useful surveys of the entire field.*

Equally important is the relation between number of variables and number of observations. The number of observations must be at least one greater than the number of independent variables for a solution to be possible; much larger sample sizes are desirable in order to obtain good estimates of the population parameters. Far too many studies of multiple regression have been carried out in biology using few observations and many different predictor variables. The number of variables chosen and number of observations should be such that the adjustment to the explained coefficient of determination discussed in Section 16.4 results in a very small percentage decrease.

In carrying out ordinary or stepwise multiple regression analysis, we must consider the model being tested. As we saw in Section 16.2, the overall significance test of multiple regression involves the null hypothesis that $\rho^2_{Y \cdot 1 \ldots k} = 0$ and that $\beta_{Y1\cdot} = \beta_{Y2\cdot} = \cdots = \beta_{Yk\cdot} = 0$. What if the null hypothesis is rejected? Rejection does not imply that all partial regression coefficients are significant. When tested separately, some of them indeed may not be significantly different from zero. In fact, occasionally none of the $b_{Yi\cdot}$ are significant, even though the overall test of multiple regression is significant. In such a case each of the independent variables is making its own contribution, but the effect of each variable is so slight that it takes their combined effects to result in a significant regression equation. It might not be possible in such a case to eliminate variables and still obtain a significant regression. By contrast, the overall test may not be significant, even though all or some of the partial regression coefficients are significant.

It must be clearly understood in such a model what is being tested by the significance test of an individual partial regression coefficient. The null hypothesis is that the dependent variable Y can be explained by a multiple regression model in which the particular variable X_j being tested has a partial regression coefficient of value zero against the dependent variable. Looking simultaneously at two or more such tests in a single multiple regression table, however, is not legitimate because independent variables are typically correlated. Only when independent variables are uncorrelated (orthogonal) with each other are the tests of individual regression coefficients independent. We saw in Section 16.3 how inclusion or exclusion of a predictor variable affects the magnitude of the remaining path coefficients. After the significance test of one partial regression coefficient, the remaining partial regression coefficients are affected in a similar manner. Thus when evaluating the significance of single variables one should proceed in a stepwise fashion in order to be able to judge significance tests for single regression coefficients one variable at a time. Situations in which the overall test is not significant but some individual regression coefficients are significantly different from zero might lead us to conclude that the significant independent variables should be included in a prediction equation. The overall nonsignificance of the results, however, would indicate that there is no evidence that the chosen set of predictor variables has any predictability. This problem is discussed in greater detail by Cramer (1972).

IMPORTANT NOTE: Before accepting the results of a multiple regression analysis you should attempt to validate the adequacy of the solution. If you know enough about the phenomena under study, compare the predictions ($\hat{Y}$) and the coefficients ($b_{Yi\cdot}$) with biological theory. For example, ask whether the signs of the coefficients are reasonable. You can also check the reasonableness of extrapolations made using the model. Often theory does not provide enough information to verify the reasonableness of the solution. The solution should then be tested using additional data. These data could be a portion of the data already available (which has not been used in the computations), or additional data could be collected specifically for testing the model. Testing the multiple regression model using new data is very important because multiple regression analysis is so powerful that it is able to capitalize on chance patterns in the observed sample to achieve a small $s^2_{Y\cdot\cdot}$. Future predictions, however, may be quite poor if the original sample data are not representative of the population. Validation is especially useful when the stepwise multiple regression analysis yields very different estimates for the β's as one adds or deletes variables. Snee (1977) gives a general survey of methods for validating regression models.

16.6 Curvilinear Regression

In Figure 14.12 we saw that a linear regression is not always sufficient to account for differences among sample means. Often there are significant deviations around linear regression. You learned to test the significance of such deviations in Section 14.5. Figures 14.12E and 14.12F illustrate the difference in types of deviations around regression. The former appears to be a curvilinear trend; the latter seems to represent added heterogeneity without any apparent regularity, which might arise from a situation such as the following. Suppose we measure the weight of four samples of mice, each sample representing a different age group. We would expect the weights to be different. If we also fed the four samples different rations, however, we could expect the sample means to deviate significantly around the regression on age because of the differences in amount or quality of food eaten.

Instead of being linear, the relation between weight and age might best be described by fitting a **curvilinear regression,** which can be expressed as a polynomial function of the following general form:

$$\hat{Y} = a + bX + cX^2 + dX^3 + \cdots \tag{16.24}$$

Such an expression uses increasing powers of X, the independent variable, and a different regression coefficient preceding each power of X. Note that this form of the equation is formally identical with the multiple regression equation [Expression (16.1)]. The independent variables $X_1, X_2, \ldots, X_k$ are now replaced by powers of X, that is, $X^1, X^2, \ldots, X^k$. We will make use of this relationship in computing the coefficients of the powers of X.

A curvilinear regression equation that involves only terms of X and X^2 will yield a parabola. As the power of X increases, the curve becomes more and more complex and will be able to fit a given set of data increasingly well. With each added power of X, however, the mean square over which the regression MS must be tested loses another degree of freedom. Remember that the test for linear regression employs a mean square for deviations from regression based on $a - 2$ degrees of freedom. When there are only $a = 5$ means to be regressed, the highest-order polynomial we can fit is a cubic, since a cubic polynomial would leave us with an MS for deviations with a single degree of freedom ($5 - 2 - 1 - 1 = 1$). Even when there are many groups, fifth and even higher powers are fitted only occasionally. For most biological work, terms of X no higher than cubic are used.

Generally, such polynomial regressions are empirical fits. That is, in most cases we cannot read structural meaning into the terms of X^2 and X^3. We simply use them to obtain a better-fitting regression line to a set of points for any of the purposes mentioned in Section 14.6. Since our aim is to find the best-fitting line, we first fit a linear regression to a set of data, then see if we can remove a significant portion from the residual sum of squares by adding a quadratic term to the regression equation. We do the same for the cubic and possibly for higher terms. Curve fitting is often, therefore, a sequential addition algorithm with a significance test for each increase in powers of X to find out whether a significant improvement in fit has been made. Of course, if we had some a priori basis for expecting, say, a quadratic relationship, then we should proceed directly to test this. As we increase the number of terms in the polynomial, the coefficients of the powers of X also change, just as the partial regression coefficients changed as we added independent variables in stepwise multiple regression analysis. To be precise in symbolism, we should write the first three polynomial regressions as follows:

$$\hat{Y} = a_0 + b_0 X$$

$$\hat{Y} = a_1 + b_1 X + c_1 X^2$$

$$\hat{Y} = a_2 + b_2 X + c_2 X^2 + d_2 X^3$$

where in most cases $a_0 \neq a_1 \neq a_2$, $b_0 \neq b_1 \neq b_2$, and $c_1 \neq c_2$.

Computing a polynomial regression using multiple regression analysis is straightforward. If no special computer program for polynomial regression is available, then a multiple regression program can be employed with the independent variable X taking the place of X_1 in the multiple regression analysis and the successive powers $X^2, X^3, \ldots, X^k$ taking the place of $X_2, X_3, \ldots, X_k$, respectively. Serious numerical problems can arise, however, when one goes beyond cubic regression. One way to minimize these problems is to code the X-variable so that it has a mean of zero and its sum of squares equals 1.

Box 16.5 shows the result of a multiple regression analysis of gene frequencies in a mussel as a function of distance along the shore line. The data are given as arcsine-transformed allele frequencies (Y) regressed on distance of the sample from Southport, Connecticut. The program was instructed to furnish the first five powers of X. The analysis of variance and the coefficients for powers 1 through 5 are shown in Box 16.5. Note that none of the coefficients of the quintic equation is significant when judged by its standard error, but overall the quintic equation removes a significant amount of the variation of variable Y (93%). This result does not correspond to the results of the first anova table in Box 16.5, which tests the explained sums of squares in a successive, incremental manner. Will the addition of a quartic or quintic term explain a significant added proportion of the variance? By contrast, the coefficients indicate the partial contributions of the separate terms, given that a quintic equation has been fitted to the data.

It appears from the anova, however, that adding powers beyond the cubic does not significantly increase the coefficient of determination, which is 0.91119 for the cubic equation. The equation needed to be recomputed using only X, X^2, and X^3. The results are shown in the second half of Box 16.5. The polynomial regression equation is

$$\hat{Y} = 26.22325 - 0.94408X + 0.042145X^2 - 0.000350X^3$$

Now the separate partial regression coefficients are significant when judged by their standard errors. The cubic equation is illustrated in Figure 16.15. The line

Box 16.5 POLYNOMIAL REGRESSION.

Frequency of the *Lap*[94] allele (in samples of *Mytilus edulis*) regressed on the location of the population. Location of the population expressed in miles east of Southport, Connecticut. The allele frequencies have been transformed using the angular transformation.

Frequency of allele $Lap^{94}(Y)$	Miles east of Southport, Conn. (X)
0.155	1
0.15	9
0.165	11
0.115	17
0.11	18.5
0.16	25
0.17	36
0.355	40.5
0.43	42
0.305	44
0.315	45
0.46	51
0.545	54
0.47	55.5
0.45	57
0.51	61
0.525	67

SOURCE: Unpublished data by R. K. Koehn (graph shown in Koehn, 1978).

Testing the polynomials up to the fifth power

Anova table

Source of variation	*df*	*SS*	*MS*	F_s
Explained	5	1588.6071	317.7214	28.124***
Linear	1	1418.3746	1418.3746	125.553***
Quadratic	1	57.2773	57.2773	5.070*
Cubic	1	85.1075	85.1075	7.534*
Quartic	1	11.8538	11.8538	1.049
Quintic	1	15.9939	15.9939	1.416
Unexplained	11	124.2671	11.2970	
Total	16	1712.8742		

$R^2_{Y\cdot} = 0.92745$ $\quad R_{Y\cdot} = 0.96304$

Box 16.5 Continued

Table of coefficients

Powers of X	a or $b_{Yi\cdot}$	SE (of a or b)
a	22.2401	4.20617
X	1.049,167,3	1.36982
X^2	−0.151,721,35	0.13422
X^3	0.006,555,5	5.05596×10^{-3}
X^4	−0.000,103,313	8.08923×10^{-5}
X^5	0.000,000,551,848	4.63791×10^{-7}

Solution of cubic polynomial

Anova table

Source of variation	df	SS	MS	F_s
Explained	3	1560.7594	520.2531	44.462***
Unexplained	13	152.1148	11.7011	
Total	16	1712.8742		

$R^2_{Y\cdot} = 0.91119 \qquad R_{Y\cdot} = 0.95456$

Table of coefficients

Powers of X	a or $b_{Yi\cdot}$	SE (of a or b)
a	26.22325	3.07616
X	−0.94408	0.38040
X^2	0.042145	0.01244
X^3	−0.000350	0.0001706

fits the data fairly well (which we expected, since the coefficient of determination is so large, 0.91119). The considerations affecting the interpretation and choice of power of the polynomial regression are quite similar to those discussed near the end of the previous section.

Another useful way of examining the goodness of fit is to look at the residuals, a procedure introduced in conjunction with simple linear regression in

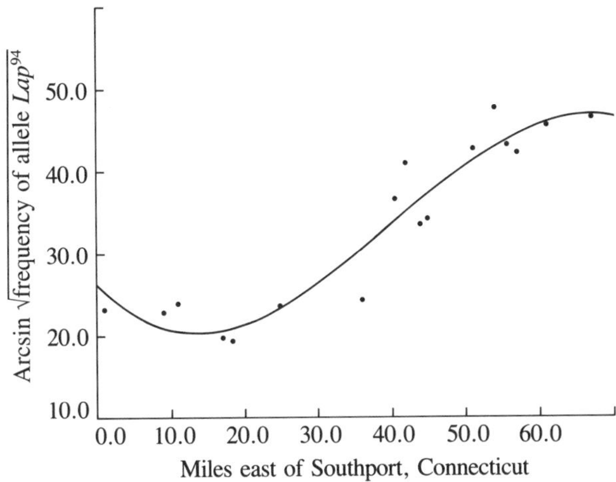

FIGURE 16.15 *A cubic polynomial fitted to the data from Box 16.5 on gene frequency in the blue mussel,* Mytilus edulis, *as a function of distance from Southport, Connecticut.*

Section 14.11. Figure 16.16 shows the standardized residuals to the linear and to the cubic fits for the example in Figure 16.15. Note that the residuals to the linear fit still show considerable trend, which is removed by fitting a polynomial to the third degree, for which the residuals are not patterned.

The coefficients of the powers of the predictor variable are not customarily tested for significance, but if desired, such a test can be carried out exactly as in multiple regression analysis. Typically, one would retain lower powers of X up to the highest power one wishes to test. If, on the other hand, there is evidence that only a higher power of X relates meaningfully to the dependent variable Y, while lower powers have no effect and no biological meaning, then the lower powers could be omitted. For such tests a stepwise multiple regression program would be entirely appropriate. Recall, however, that in Section 16.5 we cautioned against accepting the P-values employed in stepwise multiple regression as anything more than indicators of the importance of a given variable. For similar reasons we cannot easily ascribe significance to the coefficients for a given power of X. Furthermore, Hurvich and Tsai (1990) point out that if a given dataset is used to determine a specific model (i.e., the polynomial function to be fitted), one cannot also use the same dataset to establish significance and/or confidence bounds for the regression estimates. In a simulation study Hurvich and Tsai showed that actual coverage rates (confidence coefficients) were substantially below nominal rates. They suggested randomly splitting the dataset into two (not necessarily equally sized) portions, determining the model in one portion, and carrying out estimates and significance tests using the second portion. Before accepting a given polynomial solution as anything more than a

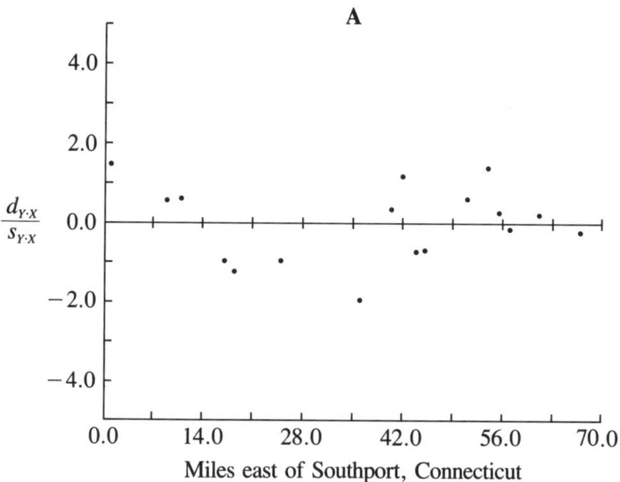

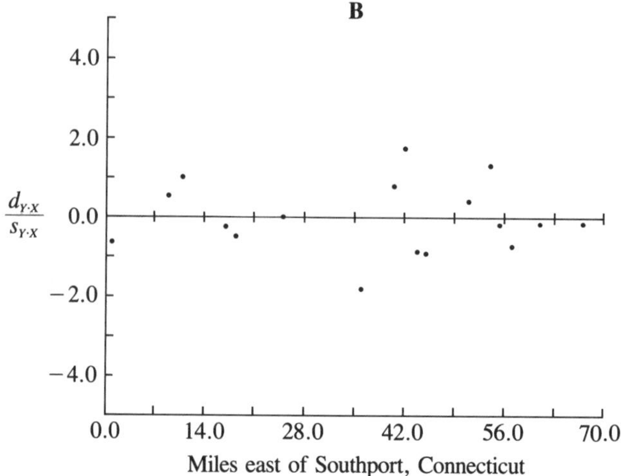

FIGURE 16.16 *Plots of standardized residuals for the* Mytilus edulis *data of Box 16.5.* ***A.*** *Residuals from linear regression.* ***B.*** *Residuals from cubic regression.*

description of the observed sample, empirical testing of the coefficients as described in the previous section is advisable.

Since high correlations among the independent variables can lead to considerable rounding errors when the calculations are done by computer, it is often useful to use powers of $X - \overline{X}$ rather than of X. This substitution reduces the correlations between odd and even powers to zero (but even powers will still be correlated). In any case, double or higher precision arithmetic is essential.

Curvilinear regression by fitting a polynomial in X can be carried out by the BIOM-pc package of computer programs.

By using **orthogonal polynomials,** we can circumvent the recomputation of coefficients as powers are added to a polynomial equation. Expression (16.24) is replaced by another polynomial regression equation,

$$\hat{Y} = A + B\xi_1 + C\xi_2 + D\xi_3 + \cdots \tag{16.25}$$

where the capital letters again are constant coefficients analogous to regression coefficients, and the ξ_j's are coefficients of special orthogonal polynomials of X; that is, the ξ_j's are complicated functions of the powers of X. These coefficients are uncorrelated with each other; by contrast, the powers of X in Expression (16.24) are clearly correlated. Thus $\Sigma \xi_j = 0$ and $\Sigma\, \xi_j \xi_k = 0$. As a result the **S** matrix is diagonal, which makes the normal equations trivial to solve. Since they are orthogonal, it is possible to add successive terms to the equation until no significant improvement in fit is obtained. Using the simple procedures presented in this book, orthogonal polynomials can be used only for data in which the values of the independent variable X are evenly spaced with an equal number of replicates.

There are separate sets of these orthogonal polynomials for each sample size n and for increasing powers of X. In Statistical Table **EE** we list orthogonal polynomials that permit the fitting of maximally third-degree regression equations to sample sizes up to $n = 12$. Coefficients for sample sizes up to 75 are found in Fisher and Yates (1963, table XXIII). Pearson and Hartley (1958, table 47) furnish coefficients for fitting sixth-degree equations up to sample size 52. Since the polynomial coefficients are frequently fractional, they are multiplied by the coefficients λ_j furnished underneath each column of polynomials in Table **EE.** The sum of squares of the polynomials, a convenient quantity for computation, is also provided. Remember that since the sum and hence the mean of each column of polynomial coefficients is zero, their sum of squares $\Sigma\,(\xi_j - \bar{\xi})^2$ is just the sum of their squares $\Sigma\xi_j^2$. The coefficients that have been multiplied by λ_j are generally symbolized ξ_j'. The coefficients listed in Table **EE** are not the only sets of orthogonal polynomials, but they are those commonly employed for curvilinear regression.

These orthogonal polynomials are used as dummy variables on which Y is regressed in the manner presented in Section 14.10. They are one type of coefficient of linear comparisons, whose values for different powers of X are mutually orthogonal for any given sample size n. The unweighted coefficients in the example on length of larval period in *Drosophila melanogaster* (see Box 14.10) are actually orthogonal polynomial coefficients. By successively regressing the Y-values on linear, quadratic, cubic, and higher coefficients, we learn whether the data show significant curvilinear regression and to what degree.

Box. 16.6 illustrates such a computation for the trout data from Exercise 14.14, which are set up in the manner of the linear comparisons of Box 14.10. As coefficients of linear comparison, we use the coefficients of the orthogonal polynomials given in Table **EE** for $n = 4$ points. The computation of the explained

Box 16.6 TESTING FOR CURVILINEAR REGRESSION BY ORTHOGONAL POLYNOMIALS.

This method applies to equally spaced X-values with equal sample sizes at each X. (Trout data from Exercise 14.14; length of trout regressed on density; $a = 4$, $n = 25$). The coefficients for the orthogonal polynomials are found in Statistical Table **EE** under sample size 4, since there are four means being regressed. These coefficients are used in linear comparisons of the type carried out in Box 14.10.

		Density X						
Regression		175	350	525	700	(1)	(2)	(3)
	$\bar{Y}$	3.51224	3.71020	2.92680	2.53960	$\sum^{a} c\bar{Y}$	$\sum c^2$	$\sum \hat{y}^2 = n(1)^2/(2)$
Linear	c_{iL}	−3	−1	1	3	−3.70132	20	17.12471
Quadratic	c_{iQ}	1	−1	−1	1	−0.58516	4	2.14008
Cubic	c_{iC}	−1	3	−3	1	1.37756	20	2.37209

Completed anova

Source of variation	*df*	*SS*	*MS*	F_s
Among densities	3	21.63688	7.212	6.941**
Linear regression	1	17.12471	17.124	16.481***
Quadratic regression	1	2.14008	2.140	2.060 *ns*
Cubic regression	1	2.37209	2.372	2.283 *ns*
Within densities	96	99.71535	1.039	

sum of squares due to the linear, quadratic, and cubic components is as shown in Box 14.10 and is not discussed further here. The completed anova is shown at the bottom of Box 16.6. It is obvious that there is a significant difference among densities, and there is a highly significant mean square for linear regression. If you carried out Exercise 14.14, however, you would have found the linear regression to be nonsignificant. How can we explain this discrepancy?

If we test for successively higher orders of polynomials we use the residual *MS* as a denominator mean square. Since the residual *MS* is large in this example, we find linear regression not significant. If, on the other hand, we decide to test a-priori for several individual-degree-of-freedom comparisons, we assume by implication that there is no random heterogeneity around linear regression and that the departures from linearity are due to the curvilinear functional relationships. We then test over the error *MS*. The model changes, and the interpretation changes accordingly. Decomposing the *SS* for deviations around linear regression into quadratic and cubic terms showed neither term to be significant, which is not surprising in view of the lack of significance of the *MS* for deviations. Thus, a linear regression seems adequate to describe the relation between length of trout and density at which it has been reared.

We restrict our discussion of orthogonal polynomials in relation to curvilinear regression to testing for degree of curvilinearity. Although it is possible to obtain a curvilinear regression equation expressed in terms of orthogonal polynomials as given by Expression (16.25), there seems little point in doing so since the polynomial can be evaluated by a multiple regression program. Furthermore, converting the orthogonal polynomial equation into the usual powers of X is complex and subject to rounding errors.

We can also use orthogonal polynomials to decompose an interaction sum of squares into single-degree-of-freedom contrasts. An example of this application is shown in Box 16.7, which is largely self-explanatory. The analysis is a typical two-way anova with replication (see Chapter 11) with significant interaction between the two factors (presence or absence of auxin and number of hours in sucrose) affecting growth of pea sections. There is no overall linear regression of length of pea sections on number of hours in sucrose, but when we partition the interaction and test for the difference in linear regression on hours in sucrose between those samples run with auxin and those without auxin, we find a highly significant difference of the linear component. This difference is clear from Figure 16.17, which shows that number of hours in sucrose increased the length

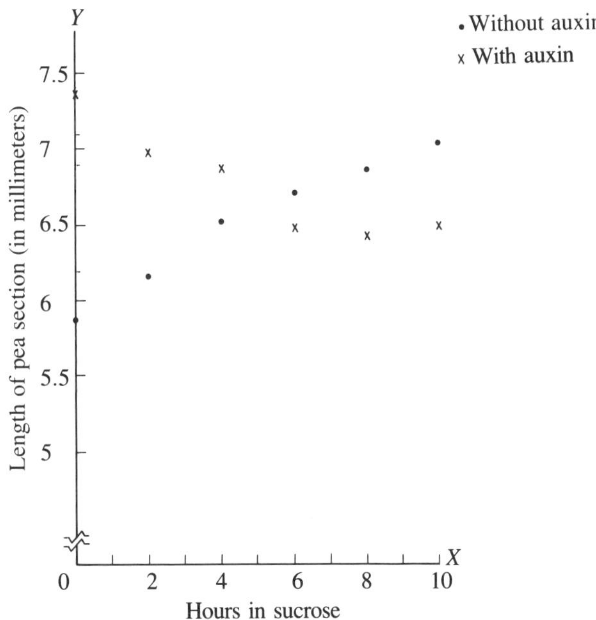

FIGURE 16.17 *Mean lengths of pea sections grown with or without auxin and treated with sugar for varying periods. Data from Box 16.7.*

Box 16.7 PARTITIONING INTERACTION IN A TWO-WAY ANOVA BY MEANS OF ORTHOGONAL POLYNOMIALS.

Lengths of pea sections in millimeters after 20 hours of growth in tissue culture. The sections had been subjected to two treatments: (1) presence or absence of auxin and (2) various periods of time in the presence of sucrose. Ten sections were run per treatment combination.

Subgroup means for treatment combinations based on 10 items

	Hours in sucrose					
Auxin	0	2	4	6	8	10
Absent	58.9	61.7	65.2	67.1	68.7	70.7
Present	73.5	69.8	67.4	64.7	64.2	65.0
$\bar{Y}$	66.20	65.75	66.30	65.90	66.45	67.85

SOURCE: Data from W. Purves (unpublished results).

The means are plotted in Figure 16.17. A two-way anova yielded the following results:

Anova table

Source of variation	df	SS	MS	F_s
Time in sucrose	5	56.5417	11.3083	<1 *ns*
Auxin	1	126.0750	126.0750	4.120*
Time × Auxin interaction	5	1584.4750	316.8950	10.355**
Within subgroups	108	3304.9917	30.6018	

In view of the significant interaction, it is not especially interesting to partition the sums of squares for the effect of time in sucrose, but this procedure is done here because the sums of squares explained by linear, quadratic, and cubic regression are needed for subsequent computations.

Computation proceeds in the manner of Box 14.10: $n = 10$ observations per time period; $a = 6$ time periods; $b = 2$ auxin treatments.

	X = hours in sucrose						(1)	(2)	(3)
Comparisons	0	2	4	6	8	10	$\sum^{a} c\bar{Y}$	$\sum c^2$	$\sum \hat{y}^2 = n(1)^2/(2)$
$\bar{Y}$	66.20	65.75	66.30	65.90	66.45	67.85			
Linear (c_{iL})	−5	−3	−1	1	3	5	9.95	70	28.2864
Quadratic (c_{iQ})	5	−1	−4	−4	−1	5	9.25	84	20.3720
Cubic (c_{iC})	−5	7	4	−4	−7	5	4.95	180	2.7225
									$51.3809 = \sum$

BOX 16.7 CONTINUED

The coefficients of linear comparisons are the orthogonal polynomial coefficients from Statistical Table **EE** for sample size 6.

SS for deviations from cubic regression

$$= SS\text{ (time in sucrose)} - SS\text{ (linear + quadratic + cubic)}$$
$$= 56.5417 - 51.3809 = 5.1608$$

None of the regression terms is significant. Since the sums of squares unexplained after regression was quite small, higher polynomials were not fitted.

There is only one degree of freedom for the effects of auxin; therefore no further partitions are possible. The interaction *SS* has 5 *df*, however, so it is possible to subdivide this *SS* further to see whether a particular aspect of the interaction is more important than others.

There are many logical ways to partition the interaction, depending upon the nature of the factors. One way would be to study the differences in the linear, quadratic, cubic, and possibly higher regressions in the two treatments. To do this, perform the analysis carried out in the first part of this box on the column means separately for each row treatment as follows.

Without auxin

	X = hours in sucrose						(1)	(2)	(3)
Comparisons	0	2	4	6	8	10	$\sum^{a} c\bar{Y}$	$\sum c^2$	$\sum \hat{y}^2 = n(1)^2/(2)$
$\bar{Y}$	58.9	61.7	65.2	67.1	68.7	70.7			
Linear (c_{iL})	−5	−3	−1	1	3	5	81.9	70	958.2300
Quadratic (c_{iQ})	5	−1	−4	−4	−1	5	−11.6	84	16.0190
Cubic (c_{iC})	−5	7	4	−4	−7	5	2.4	180	0.3200

With auxin

	X = hours in sucrose						(1)	(2)	(3)
Comparisons	0	2	4	6	8	10	$\sum^{a} c\bar{Y}$	$\sum c^2$	$\sum \hat{y}^2 = n(1)^2/(2)$
$\bar{Y}$	73.5	69.8	67.4	64.7	64.2	65.0			
Linear (c_{iL})	−5	−3	−1	1	3	5	−62.0	70	549.1429
Quadratic (c_{iQ})	5	−1	−4	−4	−1	5	30.1	84	107.8583
Cubic (c_{iC})	−5	7	4	−4	−7	5	7.5	180	3.1250

BOX 16.7 CONTINUED

To test if the two treatments differ in the slope of their regression lines, add together the sums of squares explained by the separate regressions for each treatment row and subtract the sums of squares explained by the common regression line (computed from the column totals). Calculated in this manner, each SS has $b - 1$ degrees of freedom.

$$SS_{\text{time}_L \times \text{auxin}} = 958.2300 + 549.1429 - 28.2864 = 1479.0865$$

$$F_s = 1479.0865/30.6018 = 48.333^{**}$$

The slope of the regression of pea section length on hours in sucrose (linear regression) differs, depending on the presence or absence of auxin. To test if the shapes of the two lines also differ, we perform similar operations on the sums of squares explained by higher-order regression terms (in this case quadratic and cubic terms).

$$SS_{\text{time}_Q \times \text{auxin}} = 16.0190 + 107.8583 - 20.3720 = 103.5053$$

$$F_s = 103.5053/30.6018 = 3.382 (0.1 > P > 0.05)$$

$$SS_{\text{time}_C \times \text{auxin}} = 0.3200 + 3.1250 - 2.7225 = 0.7225$$

$$F_s < 1 \; ns$$

Since the SS explained by the remaining terms is so small ($1584.4750 - 1479.0865 - 103.5053 - 0.7225 = 1.1607$), it is not worthwhile to compute the individual sums of squares for higher-order components of the time × auxin interaction ($F_s < 1$). If any of these sums of squares were significant, we could conclude that the lines differed in shape.

Complete anova table

Source of variation	df	SS	MS	F_s
Subgroups	11	1767.0917		
Time in sucrose	5	56.5417	11.3083	<1 *ns*
Time_L	1	28.2864	28.2864	<1 *ns*
Time_Q	1	20.3720	20.3720	<1 *ns*
Time_C	1	2.7225	2.7225	<1 *ns*
Residual	2	5.1608	2.5804	<1 *ns*
Auxin	1	126.0750	126.0750	4.120*
Time × Auxin	5	1584.4750	316.8950	10.355**
Time_L × Auxin	1	1479.0865	1479.0865	48.333**
Time_Q × Auxin	1	103.5053	103.5053	3.382 ($0.1 > P > 0.05$)
Time_C × Auxin	1	0.7225	0.7225	<1 *ns*
Residual	2	1.1607	0.5804	<1 *ns*
Within subgroups	108	3304.9917	30.6018	
Total	119			

of pea sections without auxin but decreased the length of those with auxin. When the data are considered together, as they would be in an overall analysis of variance, time in sucrose does not seem to affect the length of the pea sections, which is an entirely misleading finding. Thus the careful study of interactions is frequently repaid with considerable insight into the mechanism of a given scientific process.

Remember that the method we have presented using tables of orthogonal polynomials can be applied only with equally spaced values of X and equally replicated samples of Y. Unequal sample sizes and unequally spaced values of X require more complicated procedures if orthogonal polynomials are to be fitted. Remember also that one common way of circumventing the computations of curvilinear regression is to transform the data to linearity, as discussed in Section 14.11.

Although the functions described in this section have been curvilinear, i.e., not straight lines, they have all been of the form

$$\mathrm{Y} = \mathrm{a} + \mathrm{bX} + \mathrm{cX}^2 + \mathrm{dX}^3 + \cdots + \epsilon \qquad (16.26)$$

where the coefficients a, b, c, d, . . . *estimate parameters* $\alpha, \beta, \gamma, \delta, \ldots$ *and* ϵ *is the error term. Although the curve described by this function is not linear, the parameters of the function are. The unknown parameters to be estimated are in a simple additive relationship. There are other equations, however, whose parameters are* nonlinear. *Examples are*

$$\mathrm{Y} = e^{(\mathrm{a} + \mathrm{bX}^2 + \epsilon)} \qquad (16.27)$$

and

$$\mathrm{Y} = \mathrm{a} - \mathrm{bc}^{\mathrm{X}} + \epsilon \qquad (16.28)$$

We can handle Expression (16.27) by transforming to natural logarithms, yielding

$$\ln \mathrm{Y} = \mathrm{a} + \mathrm{bX}^2 + \epsilon \qquad (16.27a)$$

and solving the new equation by ordinary least squares methods. This function is therefore called an ***intrinsically linear*** *equation. There is no transformation, however, that will make Expression (16.28) linear. If we wish to fit this equation to a dataset by least squares for the residual terms* ϵ*, we must carry out complex computational procedures known as* **nonlinear regression.** *Why would one wish to do so? In some cases the investigator may have direct knowledge of the true (nonlinear) structure of the process, or the theory governing it may require a nonlinear equation. In some cases, although the equation may be intrinsically linear, i.e., transformable, the error terms* ϵ *may be distributed normally only in the original scale. Gene exchange between populations as a function of distance, size of organisms as a function of time, or crop yield as a function of amount of fertilizer may be functions of this type. Alternatively, in many studies one would* *like to interpret the results in terms of the original, untransformed variables. In such cases, too, nonlinear regression is a feasible approach.*

There is no direct least squares solution to such equations. Rather, an iterative approach (available in many program packages) is required. There are several methods for approaching the least squares solution—the method of steepest descent and Marquardt's method, among others—which the user of the program selects. Problems that arise not infrequently with this approach include the iterative process reaching local minima and not yielding the global least squares solution, or the parameter estimates "blowing up" unless new, reduced estimates are fed to the program. An example of fitting parameters to an asymptotic regression, such as that described by Expression (16.28) is given in Snedecor and Cochran (1989, sections 19.7 and 19.8). For a more extensive discussion, see Draper and Smith (1981, chapter 10). An advanced treatment of nonlinear regression is found in Seber and Wild (1989).

16.7 Advanced Topics in Regression and Correlation

As you may have gathered from the three chapters devoted to them regression and correlation analysis are important subjects in statistics and invaluable tools in biometric work. Many other techniques had to be omitted to keep this book a reasonable length. In this section we briefly describe several other important methods to give you some idea of their applications and refer you to suitable references where you may learn more about the techniques if you find that your work requires them.

Although developed by R. A. Fisher in the 1930s, only recently, as computers have become more available has ***discriminant function analysis*** *been applied much in biology, especially in systematics. The basic problem that this technique solves is easy to understand. Suppose we have samples from two populations representing, for example, different sexes, species, or disease states. We measure a character and find that although the means of the two distributions are not identical, the distributions overlap considerably. On the basis of this one character, then, we could not, with any degree of certainty, identify an unknown specimen as belonging to one or the other of the two populations. If a second character were measured, it might also differentiate the populations somewhat, but not absolutely. A discriminant function is shown in Figure 16.18, in which samples of two species of limpets are plotted against two characters,* X_1 *and* X_2*, commonly used by taxonomists. On the basis of either* X_1 *or* X_2*, assigning an unknown specimen to either of the two species with any reasonable degree of certainty would be impossible. The degree of overlap of the histograms along the two coordinate axes makes this abundantly clear.*

Discriminant function analysis constructs a new variable Z, which is a linear function of variables X_1 *and* X_2*. This function is of the form* $Z = \Sigma \lambda_i X_i$*, which is the equation of a line cutting across the intermixed cluster of points representing*

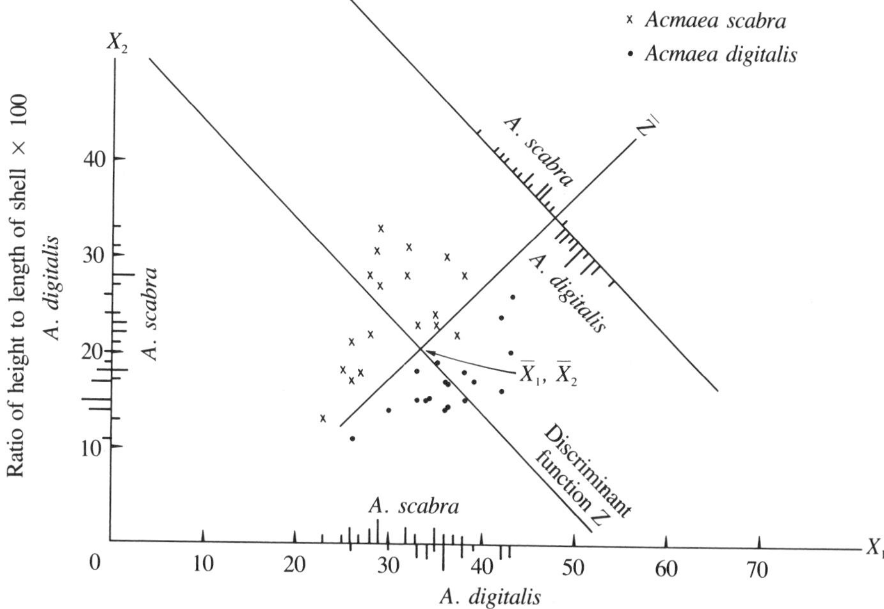

FIGURE 16.18 *A discriminant function. Histograms for* Acmeae digitalis *below abscissa and to left of ordinate; those for* Acmea scabra *above abscissa and to right of ordinate. (For explanation see text.)*

the two species. This function is defined in such a way that members of one population will have high values for Z *and members of the other will have low values. Figure 16.18 shows a line drawn parallel to the discriminant function on which a histogram of the* Z*-values (discriminant function scores) has been graphed. Note that the histograms do not overlap except for a single individual of* Acmaea scabra. *The discriminant function has served admirably to separate the two groups. Given a new, unknown specimen belonging to one of these two species, you could measure its* X_1 *and* X_2, *calculate a new value of* Z *employing the previously calculated coefficients* λ_1 *and* λ_2, *and assign the new individual to one of the two groups. There are also techniques for estimating the expected accuracy of such assignments.*

One way of computing a discriminant function is to use multiple regression. If all members of the first group are coded $Y_i = 1$, *and all members of the second group are coded* $Y_i = -1$, *we can compute a multiple regression equation of* Y *on all independent variables* X_j. *The multiple regression equation then yields the least squares best predictor of group membership, here symbolized by variable* Y. *As you can see, this procedure is very useful when we need to assign unknown specimens to previously recognized groups. Such situations arise in taxonomic identification where taxa have been previously established. If taxa have not been established, discriminant functions are of no assistance; other techniques, collectively called numerical taxonomy (Sneath and Sokal, 1973), have to be employed (see the end of this section).*

Discriminant function analysis can save time and money. Using a complicated or expensive technique, a researcher may be able to determine definitively the sex of an immature form of an insect. Discriminant function analysis, however, enables the allocation of an immature to one sex or the other simply by the measurement of several easily and inexpensively obtained external characteristics. Descriptions of discriminant functions are found in most books on multivariate analysis, including Hand (1981) and Morrison (1990). Computer programs for discriminant functions are widely available (e.g., in SAS, BMDP, SPSS, SYSTAT, NTSYSpc, and many other packages).

The discriminant function is a topic in the general area of ***multivariate analysis,*** *the branch of statistics dealing with the simultaneous variation of two or more dependent variables. As computer facilities have become widely available and microcomputers more powerful, these methods are ever more widely applied, and by now various types of multivariate analyses are among the arsenal of most research biologists. The degree of statistical sophistication required for work of this type is still considered relatively high, however, and is beyond the scope of this text. One technique of note is* ***multivariate analysis of variance,*** *manova for short, which carries out a generalization of an anova for cases in which several dependent variables have been measured for two or more samples. Introductions to multivariate analysis with applications in biology appear in Pielou (1984), Morrison (1990), and Reyment (1991). Everitt and Dunn (1992) is a good introduction to multivariate methods. More-comprehensive coverage is provided by Green (1978), Hand and Taylor (1987), Tatsuoka (1988), Krzanowski (1988), and Jackson (1991).*

Exploratory studies looking for structure in multivariate data often start with a ***correlation matrix,*** *which is a symmetrical table of correlation coefficients for each variable with all others (see Section 15.4). Such a matrix serves as a point of departure for many statistical techniques. Given the standard deviations of each variable, we can obtain covariances.*

From a matrix of correlations may emerge patterns of structure. Various methods of grouping the variables according to the magnitudes and interrelationships among their correlations have been developed. These methods are known generally as ***cluster analysis.*** *Early attempts at cluster analysis date back to the turn of the century, but recently there have been extensive developments in this field. Everitt (1980) gives a good introduction to cluster analysis. Comprehensive reviews are provided by Anderberg (1973), Sneath and Sokal (1973), Hartigan (1975), and Späth (1975).*

Whereas cluster analysis groups variables that are highly correlated with each other and excludes from a cluster those that are not, ***factor analysis*** *aims to* *express covariation in terms of* k *underlying factors that explain a large part of the variance and covariance of the original variables. In this model, the variables are considered linear combinations of the underlying factors. The number of factors considered is usually much lower than the number of variables in the study. There are several methods of factor analysis, most of them quite complex and time-consuming. The advent of computers, however, has made the wide use of factor analysis feasible. Introductory texts include Mulaik (1972), Harman (1976), Kim and Mueller (1978), and Gorsuch (1983). In biology, the goals of factor analysis are (1) resolving complex relationships into the interaction of fewer and simpler factors, whether they are physiological and environmental variables underlying behavioral correlations, or morphometric trends underlying morphological correlations, and (2) isolating and identifying causal factors behind biological correlations. For an example of the second see Sokal et al. (1980). A discussion of the application of factor analysis to work in biological systematics appears in Sokal (1965). A comprehensive discussion of factor analysis as a scientific tool is given by Cattell (1978).*

The analysis of matrices representing similarities or dissimilarities among taxonomic units has, in recent years, developed into an extensive field of research called **numerical taxonomy.** *This quantitative method is defined as the* numerical evaluation of the affinity or similarity between taxonomic units and the ordering of these units into taxa on the basis of their affinities. *One procedure in numerical taxonomy consists of computing similarity or dissimilarity matrices, sometimes composed of correlation coefficients, and clustering these matrices into taxonomic systems based on the values they contain. Another procedure estimates evolutionary branching sequences from data matrices using principles such as parsimony, maximum likelihood, or arbitrary heuristics. All these procedures require a considerable amount of computation. This is a rapidly moving field. Interested readers should consult Sneath and Sokal (1973), Clifford and Stephenson (1975), Gordon (1981), Dunn and Everitt (1982), and Swofford and Olsen (1990), as well as current literature in the journals* Systematic Biology, Evolution, Journal of Molecular Evolution, *and* Cladistics. *The statistical background required for numerical taxonomy is such that readers of this book should have little difficulty in applying the methods.*

EXERCISES 16

16.1 The following correlation matrix was taken from a study of the effect of growth rates, timing of breeding, and territory size on the survival of chicks of the glaucous-winged gull (Hunt and Hunt, 1976). Survival was estimated as percent chick survival to 500 g; growth rate is expressed as grams gained per day; and territory size is in square meters.

		Survival	Growth rate	Hatching date	Territory size
Y	Survival	1.00			
X_1	Growth rate	0.42	1.00		
X_2	Hatching date	−0.16	−0.22	1.00	
X_3	Territory size	0.10	0.19	−0.44	1.00
		$n = 104$			

Which variable is the single best predictor of survival? Which variable has the largest standardized partial regression coefficient? *Answer:* $b'_{Y1\cdot 23} = 0.40536$.

16.2 Use polynomial regression to predict SO_2, given population size, using the data from Table 16.1 (Y_1 and X_3). What degree polynomial should be used? How well does it fit? Are the deviations from the regression line what one would expect?

16.3 The following data are from an unpublished study by J. Levinton. This experiment measured sediment ingestion, estimated by the cumulative number (Y) of fecal pellets made of sedimentary particles by the mud snail *Hydrobia minuta.* As all of the available sedimentary grains are exhausted, the rate of accumulation slowed.

	Time in hours X	Cumulative number of pellets Y		Time in hours X	Cumulative number of pellets Y
1	6	119	10	24	774
2	6	111	11	24	896
3	6	87	12	24	670
4	12	348	13	42	896
5	12	210	14	42	729
6	12	272	15	42	886
7	18	438	16	78	763
8	18	778	17	78	954
9	18	548	18	78	954

Use polynomial regression to predict Y as a function of X. Plot the data showing the fitted linear, quadratic, cubic, and quartic curves. Plot the standardized residuals from these curves. Comment on the fit of the quartic curve to these data. *Answer:* $R^2_{\hat{Y}\cdot 1} = 0.5636$, $R^2_{\hat{Y}\cdot 123} = 0.8961$, $\hat{Y} = -338.9084 + 75.7313X - 1.5446X^2 + 0.0099X^3$.

16.4 The following data on growth of Hudson River striped bass are from Exhibits UT-49 and UT-50 of the 1978 Hudson River adjudicatory hearings. Perform a multiple regression analysis of incremental growth (change in mean length from last half of July to last half of August) using the three independent variables. ''Degree rise'' is 4°C divided by the number of days it took the river temperature to rise from 16°C to 20°C. The variable F_7 is a standardized measure of flow in the river.

	Incremental growth	Population density	Degree rise	F_7
Year	Y	X_1	X_2	X_3
1965	23.7	1.2	0.3077	−3.2356
1966	18.3	10.0	0.4444	−1.3405
1967	27.8	4.2	0.4000	−1.5533
1968	18.5	1.3	0.1818	0.0070
1969	7.9	32.1	0.2105	−0.1987
1970	15.3	16.7	0.1820	−1.1177
1972	21.2	9.7	0.2667	4.3428
1973	21.8	28.6	0.5000	1.8497
1974	22.1	9.5	0.1540	0.8988
1975	25.2	18.3	0.4444	1.0978
1976	25.1	11.3	0.4444	5.1556

16.5 Mark et al. (1958) present the following data on the passage of thiobarbiturates from the blood plasma into the brain.

Time (in minutes)	Log plasma concentration	Log brain concentration
X	Y_1	Y_2
3	2.3324	1.8808
3	2.2406	1.8633
30	2.1399	1.9912
30	2.1553	2.0128
60	2.0864	2.1173
60	2.0531	2.1335

Assuming the path model $X \rightarrow Y_1 \rightarrow Y_2$, estimate the path coefficients and percent determination of Y_2 by this model. *Answer:* $p_{21} = -1.0605$.

16.6 Represent the data given in Exercise 16.1 in terms of a path coefficient model. Draw the diagram and show all path coefficients. Express the correlation between survival and territory size in terms of path coefficients.

16.7 Asmundson (1931) presents the following statistics based on 707 chicken eggs.

Variable	Mean	Standard deviation
1. Length (mm)	56.93	3.02
2. Breadth (mm)	41.00	1.51
3. Maximum-minimum diameter (mm)	0.17	0.14
4. Length × 100/breadth	71.83	4.05
5. Weight of yolk (g)	16.10	2.11
6. Weight of albumen (g)	31.52	3.98
7. Weight of shell (g)	5.63	0.83

	Correlations						
	1	2	3	4	5	6	7
1	X						
2	0.208	X					
3	−0.092	0.116	X				
4	−0.758	0.431	0.166	X			
5	0.428	0.476	−0.131	−0.103	X		
6	0.582	0.726	0.105	−0.079	0.197	X	
7	0.279	0.570	0.008	0.118	0.091	0.611	X

Which variable is the best single predictor of the weight of the egg yolk (variable 5)? Develop a prediction equation for the weight of the egg yolk. Consider various subsets of the variables in order to find the smallest set of variables that enables a good prediction. Give the final prediction equation in both the standardized and the conventional form. Set 95% confidence limits to your estimates. Compute the variance inflation factors for the various models considered. *Answer:* $r_{52}^2 = 0.2266$, $R_{5 \cdot 1267}^2 = 0.6734$, $b'_{51 \cdot 267} = 0.81754$, VIF for variable 4 when using all 6 predictors is 16.1485.

16.8 Evaluate r_{XY} in the following path diagram.

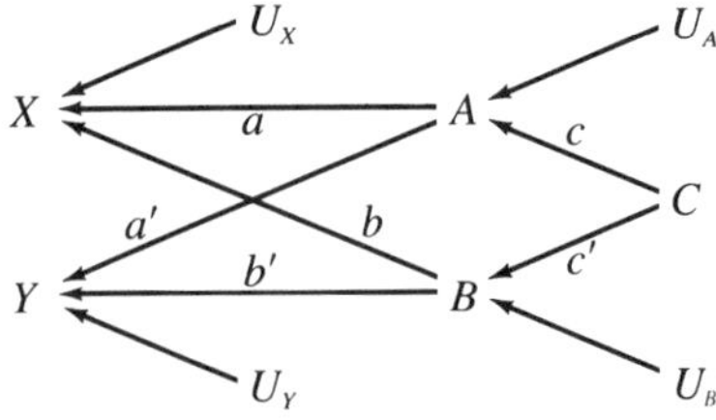

16.9 Recalculate the multiple regression of the air pollution data of Table 16.1 after recoding X_1 as $X_1' = 10(X_1 - 40)$, X_4 as $X_4' = 10(X_4 - 5)$, and X_5 as $X_5' = 100X_5$. List the statistics that need uncoding and furnish formulas for doing so.

17 ANALYSIS OF FREQUENCIES

Almost all our work so far has dealt with estimating parameters and testing hypotheses for continuous variables. This chapter focuses on tests of hypotheses about frequencies. Biological variables may be distributed into two or more classes, depending on a criterion such as arbitrary class limits in a continuous variable or a set of mutually exclusive attributes. An example of the former is a frequency distribution of birth weights (a continuous variable that is divided into an arbitrary number of contiguous classes); an example of the latter is a qualitative frequency distribution such as the frequency of individuals of 10 different species obtained from a soil sample. For any such distribution we may hypothesize that it was sampled from a population in which the frequencies of the various classes represent certain parametric proportions of the total frequency.

We need a test of goodness of fit for our observed frequency distribution to the expected frequency distribution representing our hypothesis. We first realized the need for such a test in Chapters 5 and 6, where we calculated expected binomial, Poisson, and normal frequency distributions but were unable to decide whether an observed sample distribution departed significantly from the theoretical one. In Section 17.1 we introduce the idea of goodness of fit, discuss the types of significance tests that are appropriate and the basic rationale behind them, and develop general computational formulas for these tests.

Section 17.2 illustrates computations for goodness of fit when the data are arranged by a single criterion of classification, as in a one-way quantitative or qualitative frequency distribution. This test applies to cases expected to follow one of the well-known frequency distributions such as the binomial, Poisson, or normal distribution. It also applies to expected distributions following other laws suggested by the scientific subject matter under investigation, such as, for example, tests of goodness of fit of observed genetic ratios against expected Mendelian frequencies. The goodness-of-fit tests discussed in this section are the G, chi-square, and Kolmogorov-Smirnov tests.

An interesting property of the G-test statistic is its additivity: The results of several G-tests can be summed to yield meaningful results. Also, overall G-tests can be partitioned into separate G-tests representing individual degrees of freedom in a manner analogous to that practiced in anova. Such procedures are discussed in Section 17.3, which deals with pooled, total, and interaction G.

We proceed to significance tests of frequencies in two-way classifications—called tests of independence. Section 17.4 discusses various two-way classifications, including the common tests of 2×2 tables in which each of two criteria of classification are used to divide the frequencies into two classes, yielding a four-cell table. Procedures for analyzing such data, including the G-test, Fisher's exact test, and the traditional chi-square test are presented. We also show how to compute a measure of association (the phi coefficient) from a 2×2 table. Section 17.5 discusses procedures for testing frequency distributions classified on the basis of more than two criteria.

Section 17.6 reexamines 2×2 and $R \times C$ contingency tables from the point of view of proportions. The logit transformation of such proportions is stressed, and odds ratios and their logarithms as a way of expressing the relative success rates of different criteria of classification are introduced. This section also covers the relation of these quantities to logistic regression, which is a useful regression technique for proportions. The Mantel-Haenszel procedure for obtaining overall odds ratios from replicated 2×2 contingency tables is shown. Section 17.6 also presents methods for testing the difference between two percentages and for estimating sample sizes when planning experiments for such a test. Section 17.7 deals with goodness-of-fit tests in two-way classifications in which successive columns of the two-way table represent repeated measurements of the same individuals rather than different individuals, as in the examples from Section 17.4. This design is analogous to the randomized-blocks design of Section 11.4.

Although we mention and illustrate chi-square tests in several sections of this chapter, we emphasize the G-statistic throughout. The former is a traditional way of analyzing such cases, but as we will explain, G has theoretical advantages over X^2 in addition to being computationally simpler, not only by computer but also on most calculators.

17.1 Introduction to Tests for Goodness of Fit

Your by now extensive experience with statistical hypothesis testing should make the basic idea of a goodness-of-fit test easy to understand. Assume that a geneticist has carried out a crossing experiment between two F_1 hybrids and obtains an F_2 progeny of 90 offspring, 80 of which appear to be wild-type and 10 of which are the mutant phenotype. The geneticist assumed dominance and expected a 3:1 ratio of the phenotypes, but the actual ratio is 80/10 = 8:1. Expected values for p and q are $\hat{p} = 0.75$ and $\hat{q} = 0.25$ for the wild-type and mutant, respectively. Note that we use the caret (generally called "hat" in statistics) to indicate hypothetical or expected values of the binomial proportions.

We have not used Greek letters here to symbolize parameters, because they would be awkward. The *observed* proportions of these two classes are $p = 0.89$ and $q = 0.11$, respectively. Another way of noting the contrast between observation and expectation is to state it in frequencies: The observed frequencies are $f_1 = 80$ and $f_2 = 10$ for the two phenotypes; expected frequencies are $\hat{f}_1 = \hat{p}n = 0.75(90) = 67.5$ and $\hat{f}_2 = \hat{q}n = 0.25(90) = 22.5$, respectively, where n refers to the sample size of offspring from the cross. Note that when we sum the expected frequencies, they yield $67.5 + 22.5 = 90 = n$, as they should.

The obvious question is whether the deviation from the 3 : 1 hypothesis observed in our sample is of such a magnitude as to be improbable. In other words, do the observed data differ enough from the expected values to cause us to reject the null hypothesis? For this example, you already know two methods for coming to a decision about the null hypothesis. This is a binomial distribution in which p is the probability of being a wild type and q is the probability of being a mutant. We can work out the probability of obtaining an outcome of 80 wild-type phenotypes and 10 mutants, as well as all "worse" cases for $\hat{p} = 0.75$ and $\hat{q} = 0.25$, in a sample of $n = 90$ offspring. We use the conventional binomial expression, $(\hat{p} + \hat{q})^n$, except that p and q are hypothesized, and we replace the symbol k by n, which we adopted in Chapter 5 as the appropriate symbol for the sum of all the frequencies in a frequency distribution. In this example, we have only one sample, so what would ordinarily be labeled k in the binomial is, at the same time, n. Such a problem was illustrated in Table 5.3 and Section 5.2. We can compute the cumulative probability of the tail of the binomial distribution by the method of Box 5.1. The result is a probability of 0.000849 for all outcomes as deviant or more deviant from the hypothesis. Note that this is a one-tailed test, the alternative hypothesis being that there are more wild-type offspring than the Mendelian hypothesis would postulate. Assuming $\hat{p} = 0.75$ and $\hat{q} = 0.25$, the observed sample is, consequently, a very unusual outcome, and we conclude that there is a significant deviation from expectation.

A less time-consuming approach based on the same principle is to look up confidence limits for the binomial proportions as we did for the sign test in Section 13.12. Interpolation in Statistical Table **P** shows that for a sample of $n = 90$, an observed percentage of 89% would yield approximate 99% confidence limits of 78 and 96 for the true percentage of wild-type individuals. Clearly, the hypothesized value of $\hat{p} = 0.75$ is beyond the 99% confidence bounds.

Now, let us develop a third approach to evaluating the null hypothesis by a goodness-of-fit test. Table 17.1 illustrates how we might proceed. In the first column are the observed frequencies f representing the outcome of the experiment. Column (2) shows the observed frequencies as (observed) proportions p and q computed as f_1/n and f_2/n, respectively. Column (3) lists the expected proportions for the particular null hypothesis being tested. In this case, the hypothesis is a 3 : 1 ratio, corresponding to expected proportions $\hat{p} = 0.75$ and

Table 17.1 DEVELOPING THE *G*-TEST (LIKELIHOOD RATIO TEST) FOR GOODNESS OF FIT.

Observed and expected frequencies from the outcome of a genetic cross, assuming a 3 : 1 ratio of phenotypes among the offspring.

Phenotypes	(1) Observed frequencies f	(2) Observed proportions $\frac{f}{n}$	(3) Expected proportions $\hat{p}$ and $\hat{q}$	(4) Expected frequencies $\hat{f}$	(5) Ratio $\frac{f}{\hat{f}}$	(6) $f \ln\left(\frac{f}{\hat{f}}\right)$
Wild type	80	$p = \frac{8}{9}$	$\hat{p} = 0.75$	$\hat{p}n = 67.5$	1.185185	13.59192
Mutant	10	$q = \frac{1}{9}$	$\hat{q} = 0.25$	$\hat{q}n = 22.5$	0.444444	−8.10930
Sum	90	1.0	1.0	90.0		$\ln L = 5.48262$

$\hat{q} = 0.25$, as we have seen. Column (4) gives the expected frequencies, which we have already calculated for these proportions as $\hat{f}_1 = \hat{p}n = 0.75(90) = 67.5$ and $\hat{f}_2 = \hat{q}n = 0.25(90) = 22.5$.

The log-likelihood ratio test for goodness of fit may be developed as follows. Using Expression (5.9) for the expected relative frequencies in a binomial distribution, we compute two quantities of interest to us here:

$$\binom{90}{80}\left(\frac{80}{90}\right)^{80}\left(\frac{10}{90}\right)^{10} = 0.132{,}683{,}8$$

$$\binom{90}{80}\left(\frac{3}{4}\right)^{80}\left(\frac{1}{4}\right)^{10} = 0.000{,}551{,}754{,}9$$

The first quantity is the probability of observing the sampled results (80 wild types and 10 mutants) on the hypothesis that $\hat{p} = p$—that is, that the population parameter equals the observed sample proportion. The second quantity is the probability of observing the sampled results assuming that $\hat{p} = \frac{3}{4}$, as per the Mendelian null hypothesis. Note that these expressions yield the probabilities for the observed outcomes only, *not for observed and all worse outcomes.* Thus, $P = 0.000{,}551{,}8$ is less than the earlier computed $P = 0.000{,}849$, which is the probability of 10 *and fewer* mutants, assuming $\hat{p} = \frac{3}{4}$ and $\hat{q} = \frac{1}{4}$.

The first probability (0.132,683,8) is greater than the second (0.000,551,754,9), since the hypothesis is based on the observed data. If the observed proportion p is equal to the proportion $\hat{p}$ postulated under the null hypothesis, then the two computed probabilities will be equal and their ratio, L, will equal 1. The greater the difference between p and $\hat{p}$ (the expected proportion under the null hypothesis), the higher the ratio will be (the probability based on p is divided by the probability that is based on $\hat{p}$ or that is defined by the null hypothesis). This relationship indicates that the ratio of these two probabilities or *likelihoods* can be used as a statistic to measure the degree of agreement between sampled and expected frequencies. A test based on such a ratio is called a **likelihood ratio test.** In our case, $L = 0.132{,}683{,}8/0.000{,}551{,}754{,}9 = 240.4761$. The theoretical distribution of this ratio is, in general, complex and poorly known. It has been shown, however, that the distribution of

$$G = 2 \ln L \tag{17.1}$$

can be approximated by the χ^2-distribution when sample sizes are large (for a definition of "large" in this case, see Section 17.2). The appropriate number of degrees of freedom is 1. Why? As Table 17.1 shows, there are two frequency cells for these data. Together they have to add to 90. The outcome of the sampling experiment could have been any number of mutants from 0 to 90, but the number of wild types consequently would have to be constrained so that the total would add up to 90. One of the cells in the table is free to vary, the other is constrained. Hence, there is one degree of freedom.

In our case,

$$G = 2 \ln L = 2(5.48262) = 10.96524$$

If we compare this observed value with a χ^2-distribution with one degree of freedom, we find that the result is significant ($P < 0.001$). We thus reject the 3:1 hypothesis and conclude that the proportion of wild types is greater than 0.75. The geneticist must, consequently, look for a mechanism explaining this departure from expectation.

Notation for the log-likelihood ratio test is not yet standardized. The symbol $2I$ is sometimes used for G. The letter I is used because G can be interpreted as twice the amount of information present in the sample for distinguishing between the $\hat{p}$ of the null hypothesis and that of the alternative hypothesis. The term "information" is used here in the sense defined in the field of information theory. Interested readers with a knowledge of mathematical statistics may find Kullback (1959) interesting in this regard. Employing the **G-test** (= $2I$-test)—as the log-likelihood ratio test is also called—does not require any special mathematical sophistication, however, and as we shall see, this method has some considerable advantages over the traditional chi-square tests.

We will now develop a simple computational formula for G. Referring to Expression (5.9), we can rewrite the two probabilities computed earlier as

$$\binom{n}{f_1} p^{f_1} q^{f_2} \tag{17.2}$$

and

$$\binom{n}{f_1} \hat{p}^{f_1} \hat{q}^{f_2} \tag{17.2a}$$

but

$$L = \frac{\binom{n}{f_1} p^{f_1} q^{f_2}}{\binom{n}{f_1} \hat{p}^{f_1} \hat{q}^{f_2}} = \left(\frac{p}{\hat{p}}\right)^{f_1} \left(\frac{q}{\hat{q}}\right)^{f_2}$$

Since $f_1 = np$ and $\hat{f}_1 = n\hat{p}$, and similarly $f_2 = nq$ and $\hat{f}_2 = n\hat{q}$,

$$L = \left(\frac{f_1}{\hat{f}_1}\right)^{f_1} \left(\frac{f_2}{\hat{f}_2}\right)^{f_2}$$

and

$$\ln L = f_1 \ln \left(\frac{f_1}{\hat{f}_1}\right) + f_2 \ln \left(\frac{f_2}{\hat{f}_1}\right) \tag{17.3}$$

The computational steps implied by Expression (17.3) are shown in columns (5) and (6) of Table 17.1 Column (5) contains the ratios of observed over expected frequencies. These ratios would be 1 in the unlikely case of a perfect fit of observations to the hypothesis. In such a case, the logarithms of these ratios, in column (6), would be 0, as would their sum. Consequently G, which is twice the natural logarithm of L, would be 0, indicating a perfect fit of the observations to the expectations.

Next, let us study the distribution of G. Suppose our expectation of a 3 : 1 ratio is correct. What would be the distribution of the sample statistics G if we took a series of samples of 90 items from a population in proportions 3 : 1 and tested them against the 3 : 1 hypothesis? If the sample came out exactly in a 3 : 1 ratio (mathematically impossible for 90 items), then our value of G would be zero. The greater the departure from expectation, the greater the value of G. Figure 17.1A shows the expected distribution of G's for all possible samples of 90 items from a binomial population specified as $\hat{p} = 0.75$ and $\hat{q} = 0.25$. On the distribution of G we have superimposed the probability density function of the χ^2-distribution for one degree of freedom, broken up into corresponding classes for ease of comparison. There is general agreement between the two distributions. The G-values appear to be distributed approximately as χ^2 with one degree of freedom.

The test for goodness of fit can be applied to a distribution with more than two classes. Table 17.2 shows the outcome of a dihybrid cross in tomato genetics in which the expected ratio of phenotypes is 9 : 3 : 3 : 1. The table is set up as before, with expected frequencies calculated as $\hat{f}_i = \hat{p}_i n$, where the values of $\hat{p}_i$ are the expected probabilities for the $a = 4$ classes. Since the probabilities of all possible outcomes sum to 1, $\Sigma^a \hat{p}_i = 1$. The values of $\hat{p}_i$ for this example are $\hat{p}_1 = \frac{9}{16}$,

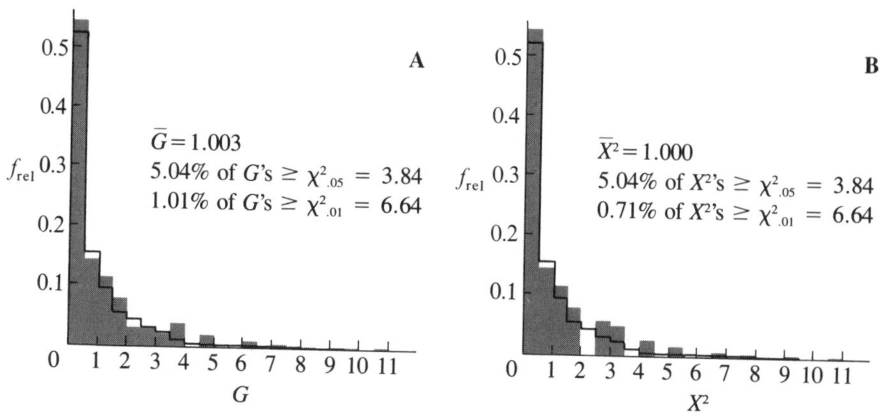

FIGURE 17.1 *Expected distribution of* G *and* X^2 *for all possible samples of size* $n = 90$ *drawn from a binomial population with* $\hat{p} = 0.75$. *The shaded histograms show the distribution of the resulting values of* G (**A**) *and* X^2 (**B**) *computed for each sample on the hypothesis that* $\hat{p} = 0.75$ *and* $\hat{q} = 0.25$. *The expected* $\chi^2_{[1]}$*-distribution is superimposed for comparison.*

Table 17.2 THE G-TEST FOR GOODNESS OF FIT FOR MORE THAN TWO CLASSES.

Observed and expected frequencies in a dihybrid cross between tall, potato-leaf tomatoes and dwarf, cut-leaf tomatoes.

Phenotypes	(1) Observed frequencies f	(2) Expected frequencies $\hat{f}$	(3) Ratio $\frac{f}{\hat{f}}$	(4) $f \ln\left(\frac{f}{\hat{f}}\right)$
Tall, cut-leaf	926	906.1875	1.021864	20.02752
Tall, potato-leaf	288	302.0625	0.953445	−13.72996
Dwarf, cut-leaf	293	302.0625	0.969998	−8.92517
Dwarf, potato-leaf	104	100.6875	1.032899	3.36640
Sum	1611	1611.000		$\ln L = 0.73879$

SOURCE: Data from MacArthur (1931).

$\hat{p}_2 = \frac{3}{16}$, $\hat{p}_3 = \frac{3}{16}$, and $\hat{p}_4 = \frac{1}{16}$. Again, we calculate ratios of observed over expected frequencies in column (3) and $f \ln(f/\hat{f})$ in column (4). The sum of the values in column (4) yields $\ln L = 0.73879$; hence $G = 2 \ln L = 1.478$. This operation can be expressed by the following general computational formula, whose derivation, based on the multinomial expectations (for more than two classes), is shown in Section A.14 in the appendix:

$$G = 2 \sum^{a} f_i \ln\left(\frac{f_i}{\hat{f}_i}\right) \tag{17.4}$$

Thus the formula can be seen as the sum of the independent contributions of departures from expectation $[\ln(f_i/\hat{f}_i)]$ weighted by the frequency of the particular class (f_i).

How can we evaluate the outcome of our test for goodness of fit? We need to know how many degrees of freedom there are in this example to be able to compare the observed value of G with the appropriate χ^2-distribution. We now have four classes constrained by a sum. Thus, any three of them can vary freely, but the fourth class must constitute the difference between the total sum and the sum of the first three. Thus, in a case with four classes we have three degrees of freedom, and in general when we have a classes, we have $a - 1$ degrees of freedom.

In Statistical Table **D** (chi-square), under three degrees of freedom we find the probability of obtaining a $\chi^2_{[3]}$-value > 1.478 to be slightly more than 50%. We therefore have no reason for rejecting the null hypothesis. Insofar as we can tell, the dihybrid cross of Table 17.2 follows the expected 9 : 3 : 3 : 1 ratio.

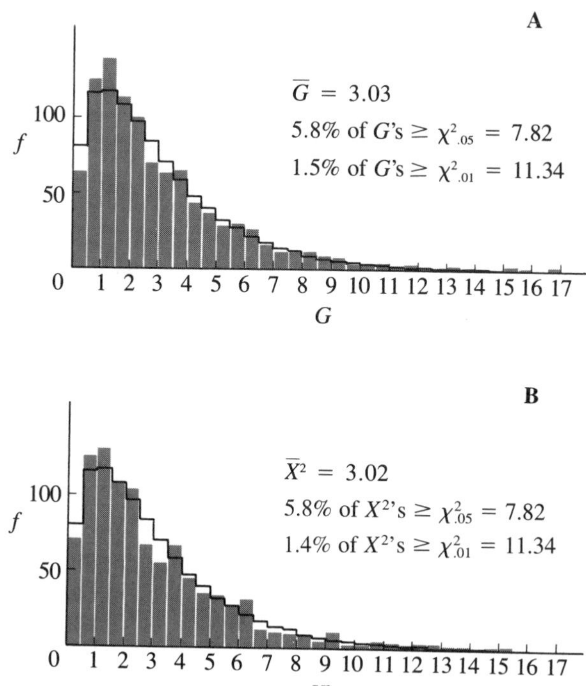

FIGURE 17.2 *Results of a sampling experiment in which 1000 samples of size n = 165 were drawn from a population divided into four classes in the ratio 9:3:3:1. The shaded histograms show the distribution of the resulting values of G* (**A**) *and* X^2(**B**) *computed for each sample on the hypothesis that the parametric ratio is 9:3:3:1. The expected* $\chi^2_{[3]}$*-distribution is superimposed for comparison.*

Figure 17.2 shows the results of a sampling experiment. One thousand samples of size $n = 165$ were drawn from a population with a 9:3:3:1 ratio. (A sample size of $n = 1611$, as in the actual example of Table 17.2, was not used because sampling would have required an excessive amount of computer time.) The 1000 G-values are shown in the form of a histogram with the probability density function of the $\chi^2_{[3]}$-distribution superimposed for ease of comparison. The fit appears better than in Figure 17.1 because a larger sample size was used ($n = 165$ instead of $n = 90$).

Specifying the alternative hypothesis is more difficult than is specifying the null hypothesis. In some genetics experiments, there are only a limited number of possible alternative hypotheses. Thus, if the 3 : 1 ratio is rejected, a 1 : 1 ratio or a more-complicated ratio might apply in a given case. In other examples, however, there may be a variety of complex alternative hypotheses that are difficult to specify. The main consideration in such cases is the acceptance or rejection of the null hypothesis. If we accept a 3 : 1 ratio, we still do not know absolutely that

3 : 1 is the true ratio. All we know is that, given our sample, we cannot show that it is *not* the true ratio.

In some goodness-of-fit tests, we subtract more than one degree of freedom from the number of classes, a. These are instances in which the parameters for the null hypothesis have been estimated from the sample data themselves, in contrast with the null hypotheses tested in Tables 17.1 and 17.2. In these two cases, the hypothesis to be tested was based on the investigator's general knowledge of the specific problem and of Mendelian genetics. The values of $\hat{p} = 0.75$ and $\hat{q} = 0.25$ were dictated by the 3 : 1 hypothesis and were not estimated from the sampled data. Similarly, the 9 : 3 : 3 : 1 hypothesis was also based on general genetic theory. For this reason, the expected frequencies are said to have been based on an **extrinsic hypothesis,** a hypothesis external to the data. By contrast, consider the expected frequencies of birth weights under the assumption of normality (see Box 6.1). To compute these frequencies we needed values for μ and σ, which we estimated from the sample mean $\bar{Y}$ and the sample standard deviation s of the birth weights. Therefore the two parameters of the computed normal distribution, the mean and the standard deviation, came from the sampled observations themselves. The expected normal frequencies thus represent an **intrinsic hypothesis.** In such a case, to obtain the correct number of degrees of freedom for the G-test of goodness of fit (and for the chi-square test described later in this section) we would subtract from a, the number of classes into which the data had been grouped, not only one degree of freedom for n, the sum of the frequencies, but also two additional degrees of freedom—one for the estimate of the mean and the other for the estimate of the standard deviation. Thus, in such a case a sample statistic G would be compared with chi-square for $a - 3$ degrees of freedom. Exact rules on how many degrees of freedom to subtract and computations for such tests are given in the next section.

Now let us introduce you to an alternative technique. We must acquaint you with this method because it is the traditional approach, which you will see applied in the earlier literature and in a substantial number of current research publications. We turn once more to the genetic cross with 80 wild-type and 10 mutant individuals, this time laid out in Table 17.3.

First we calculate $f - \hat{f}$, the deviation of observed from expected frequencies. Note that the sum of these deviations equals zero, for reasons very similar to those causing the sum of deviations from a mean to add to zero. In an example with two classes, then, the deviations are always equal and opposite in sign (proof of this property is developed in Section A.15 of the appendix). Following our previous approach of making all deviations positive by squaring them, we square $(f - \hat{f})$ in column (4) to yield a measure of the magnitude of the deviation from expectation. This quantity is expressed as a proportion of the expected frequency. After all, if the expected frequency were 13.0, a deviation of 12.5 would be extremely large, comprising almost 100% of $\hat{f}$, but such a deviation would represent only 10% of an expected frequency of 125.0. Each value in column (5) is the quotient of the quantity in column (4) divided by that in column

Table 17.3 DEVELOPING THE CHI-SQUARE TEST FOR GOODNESS OF FIT.

Observed and expected frequencies from the outcome of a genetic cross, assuming a 3 : 1 ratio of phenotypes among the offspring.

Phenotypes	(1) Observed frequencies f	(2) Expected frequencies $\hat{f}$	(3) Deviations from expectation $f - \hat{f}$	(4) Deviations squared $(f - \hat{f})^2$	(5) $\frac{(f - \hat{f})^2}{\hat{f}}$
Wild type	80	$\hat{p}n = 67.5$	12.5	156.25	2.31481
Mutant	10	$\hat{q}n = 22.5$	−12.5	156.25	6.94444
Sum	90	90.0	0		$X^2 = 9.25926$

(2). Note that the magnitude of the quotient is greater for the second row (mutant phenotypes), in which the $\hat{f}$ is smaller. The next step in developing the test statistic is to sum these quotients, which yields a value of 9.25926, shown at the foot of column (5).

What shall we call this new statistic? We have some nomenclatural problems here. Many of you will have recognized this procedure as the so-called **chi-square test,** regularly taught to beginning genetics classes. The name of the test is too well established to make a change practical, but in fact, the quantity just computed, the sum of column (5), could not possibly be a χ^2-value. The χ^2-distribution is a continuous and theoretical frequency distribution, whereas our quantity, 9.25926, is a sample statistic based on discrete frequencies. The latter point is easily seen if you visualize other possible outcomes. For instance, we could have had as few as zero mutants matched by 90 wild-type individuals (assuming that the total number of offspring $n = 90$ remains constant), or we could have had 1, 2, 3, or more mutants, in each case balanced by the correct number of wild-type offspring to yield a total of 90. Observed frequencies change in unit increments, and since the expected frequencies remain constant, deviations, their squares, and the quotients are not continuous variables but can assume only certain values.

The reason that this test has been called the chi-square test and that many persons inappropriately call the statistic obtained as the sum of column (5) a chi-square is that the sampling distribution of this sum approximates that of a χ^2-distribution with one degree of freedom. We have previously encountered cases of this sort. The G-statistic we just studied is distributed as approximately chi-square. Similarly, Bartlett's test for homogeneity of variances (see Section 13.3) and the test for homogeneity of correlation coefficients (see Section 15.5) both employ a sample statistic distributed approximately as a chi-square. Since the sample statistic is not a chi-square, however, in these two tests we followed the prevalent convention of labeling the sample statistic X^2 rather than χ^2. We will follow the same practice here, although in one sense this is misleading. The three tests just mentioned, the homogeneity tests for variances and correlation coefficients and the goodness-of-fit test shown in Table 17.3, have only one thing in common: Their sample statistic is approximately distributed as chi-square. For this reason, we give the sample statistic a Latin letter resembling χ^2. Structurally, however, the three tests are quite different, and to be strictly correct we should provide each of these with a separate symbol to emphasize that the three sample statistics are not the same. This approach was taken in the case of the *G*-test, where the sample statistic is called *G*, although it is approximately chi-square distributed. Since we have followed a policy of conservatism on symbolism, we retain the X^2, but the reader should be clear that the sample statistic X^2 in this chapter is for the test of goodness of fit and is not the same as the X^2 for homogeneity of either variances or correlations. Some authors have called X^2 (for goodness of fit) the *Pearson statistic.*

Why should the distribution of X^2 as computed in Table 17.3 approximate that of χ^2? Rigorous proof requires a knowledge of mathematical statistics beyond the scope of this book, but we will provide two simple illustrations—one through an inspection of the formula for X^2 and the other empirically by means of a sampling experiment. Let us express the computational steps of Table 17.3 as a formula:

$$X^2 = \frac{(f_1 - \hat{f}_1)^2}{\hat{f}_1} + \frac{(f_2 - \hat{f}_2)^2}{\hat{f}_2}$$

$$= \sum^{2} \frac{(f_i - \hat{f}_i)^2}{\hat{f}_i} \tag{17.5}$$

Expression (17.5) is the summation of the two quotients of deviations squared over expected frequencies. Using simple algebra, as shown in Section A.16 of the appendix we can change Expression (17.5) into

$$X^2 = \frac{(f_1 - \hat{p}n)^2}{\hat{p}\hat{q}n} \tag{17.6}$$

or

$$X^2 = \frac{(f_2 - \hat{q}n)^2}{\hat{p}\hat{q}n} \tag{17.7}$$

The numerator of either of these expressions is the square of the deviation between observed and expected frequencies in one class; the denominator is the product of the expected probabilities $\hat{p}$ and $\hat{q}$ multiplied by sample size n. Recall Expression (7.6)

$$\frac{1}{\sigma^2} \sum (Y_i - \mu)^2$$

which describes the χ^2-distribution. Its numerator is a sum of squares, and its denominator is a parametric variance. Now look at Expression (17.6) again. Its numerator is a deviation from expectation squared, which is a sum of squares with one degree of freedom. The denominator is the variance of the binomial distribution, a parametric variance. It therefore appears quite reasonable from Expression (17.6) that X^2 as computed in Table 17.3 is distributed as chi-square with one degree of freedom.

Empirically, we can show that the distribution of the X^2-statistic is indeed close to a chi-square distribution with one degree of freedom as we did for the G-statistic. The results are graphed in Figure 17.1B. Note that discontinuity is more apparent in X^2 than in G. Also, G appears to follow the chi-square distribution a bit more closely. The value of $X^2 = 9.25926$ from Table 17.3, when compared with the critical value of χ^2 (Statistical Table **D**), is highly significant ($P < 0.005$). The chi-square test is always one-tailed. Since the deviations are squared, negative and positive deviations both result in positive values of X^2. Our conclusions are the same as with the G-test. In general, X^2 will be numerically similar to G.

We can apply the chi-square test for goodness of fit to a distribution with more than two classes as well. We will practice this application on the dihybrid cross in Table 17.2. Squared deviations and their quotients over expected frequencies must now be summed for four classes. The operation can be described by the formula

$$X^2 = \sum^{a} \frac{(f_i - \hat{f}_i)^2}{\hat{f}_i} \tag{17.8}$$

which is a generalization of Expression (17.5). The sum of these quotients yields a value of $X^2 = 1.46872$. Again, this four-class example contains three degrees of freedom, so we compare our observed X^2 against $\chi^2_{[3]}$.

In Table **D** (chi-square), under three degrees of freedom, we find the probability of $\chi^2_{[3]} > 1.469$ to be between 90% and 50%. We therefore have no reason to reject the null hypothesis. As far as we can tell, the dihybrid cross of Table 17.2 follows the 9:3:3:1 ratio. In addition to the distribution of G-values, Figure 17.2 shows the results of computing X^2 for the sampling experiment with four classes. Both distributions follow the χ^2-distribution closely.

Expression (17.8) can be applied to any chi-square test of goodness of fit, although its use is somewhat cumbersome and requires the computation of deviations, their squaring, and their division by expected frequencies over all of the classes.

17.2 Single-Classification Tests for Goodness of Fit

Before discussing in detail the computational steps involved in tests of goodness of fit of single-classification frequency distributions, some remarks on the choice of a test statistic are in order. We have already stated that the traditional method for such a test is the chi-square test for goodness of fit. The new approach by the G-test, however, has been recommended on theoretical grounds. The G-test is computationally simpler, especially in complicated designs. The BIOM-pc computer program uses G exclusively for goodness-of-fit tests. Box 17.1 demonstrates how to carry out a chi-square test so that you can understand published results using this statistic. The more complex designs featured in later sections employ the G-test exclusively.

The G- and chi-square tests can be applied to frequency distributions of nominal variables as well as to those of continuous variables grouped suitably. For continuous sample distributions, however, the preferred method is the Kolmogorov–Smirnov test. This simpler and more powerful procedure is described at the end of this section.

The G-tests for goodness of fit for single-classification frequency distributions are given in Box 17.1, which treats only cases in which the expected frequencies are based on a hypothesis extrinsic to the data: There are a classes, and the expected proportions in each class are assumed on the basis of outside knowledge and are not functions of parameters estimated from the sample. We start with the general case for any number of classes, where the number of classes is symbolized by a, to emphasize the analogy with analysis of variance. The data are the results of a complicated genetic cross expected to result in an 18:6:6:2:12:4:12:4 ratio. In all, 241 progeny were obtained and classified into the eight phenotypic classes. The expected frequencies can be computed simply by multiplying the total sample size n by the expected probabilities of occurrence. The overall fit to expectation is quite good. This is a test with an extrinsic hypothesis because the genetic ratio tested is based on considerations prior and external to the sample observations tested in the example.

We learned in the previous section that the value of G is to be compared with a critical value of χ^2 for $a - 1$ degrees of freedom. Because the actual type I error of G-tests tends to be higher than the intended level, a correction for G to obtain a better approximation to the chi-square distribution has been suggested by Williams (1976). He divides G by a correction factor q (not to be confused with a proportion), which is computed as $q = 1 + (a^2 - 1)/6n\nu$. Note that for the extrinsic hypothesis ν will always be $a - 1$. The effect of this correction is to reduce the observed value of G slightly. When we compare our result with the χ^2-distribution, we find that the value of G obtained from our sample is not significant. We do not have sufficient evidence to reject the null hypothesis and are led to conclude that the sample is consistent with the specified genetic ratio. Especially in examples such as the one just analyzed, however, remember that we have not specified an alternative hypothesis; there are numerous alternative hypotheses that also could not be excluded if we were to carry out a significance test for them. We have not really proven that the data are distributed as specified; there are various other genetic ratio hypotheses that could also be plausible.

No comprehensive study has determined how small a sample can be and still be suitable for the G-test of goodness of fit. The usual rule of thumb is that the

Box 17.1 TESTS FOR GOODNESS OF FIT. SINGLE-CLASSIFICATION, EXPECTED FREQUENCIES BASED ON HYPOTHESIS EXTRINSIC TO THE SAMPLED DATA.

G-test

1. Frequencies divided into $a \geq 2$ classes

In a genetic experiment involving a cross between two varieties of the bean *Phaseolus vulgaris,* Smith (1939) obtained the following results:

Phenotypes ($a = 8$)	Observed frequencies f	Expected frequencies $\hat{f}$
Purple/buff	63	67.78125
Purple/testaceous	31	22.59375
Red/buff	28	22.59375
Red/testaceous	12	7.53125
Purple	39	45.18750
Oxblood red	16	15.06250
Buff	40	45.18750
Testaceous	12	15.06250
Total	241	241.00000

The expected frequencies $\hat{f}_i$ were computed on the basis of the expected ratio of 18 : 6 : 6 : 2 : 12 : 4 : 12 : 4. Compute $\hat{f}_i$ as $\hat{p}_i n$. Thus, $\hat{f}_1 = \hat{p}_1 n = (\frac{18}{64}) \times 241 = 67.78125$.

Employing Expression (17.4) we obtain:

$$G = 2\sum^{a} f_i \ln\left(\frac{f_i}{\hat{f}_i}\right)$$

$$= 2\left[63 \ln\left(\frac{63}{67.78125}\right) + \cdots + 12 \ln\left(\frac{12}{15.06250}\right)\right]$$

$$= 8.82396$$

Since our observed $G = 8.824 < \chi^2_{.05[7]} = 14.067$, we may consider the data consistent with Smith's null hypothesis.

Applying Williams's correction to G, to obtain a better approximation to χ^2, we compute

$$q = 1 + \frac{a^2 - 1}{6n(a-1)} = 1 + \frac{a+1}{6n}$$

$$= 1 + \frac{9}{6(241)}$$

$$= 1.006224$$

$$G_{\text{adj}} = \frac{G}{q} = \frac{8.82396}{1.006224} = 8.76938$$

BOX 17.1 CONTINUED

2. Special case of frequencies divided into $a = 2$ classes

For the G-test there is no special shortcut equation for the case in which $a = 2$.

In an F_2 cross in drosophila the following 176 progeny were obtained, of which 130 were wild-type flies and 46 were ebony mutants. Assuming that the mutant is an autosomal recessive, one would expect a ratio of 3 wild-type flies to each mutant fly. To test whether the observed results are consistent with this 3:1 hypothesis, we set up the data as follows.

Flies	f	Hypothesis	$\hat{f}$	Adjusted f
Wild type	$f_1 = 130$	$\hat{p} = 0.75$	$\hat{p}n = 132.0$	130.5
Ebony mutant	$f_2 = 46$	$\hat{q} = 0.25$	$\hat{q}n = 44.0$	45.5
	$n = 176$		176.0	176.0

Computing G from Expression (17.4), we obtain

$$G = 2\sum^{a} f_i \ln\left(\frac{f_i}{\hat{f}_i}\right)$$
$$= 2\left[130 \ln\left(\frac{130}{132}\right) + 46 \ln\left(\frac{46}{44}\right)\right]$$
$$= 0.12002$$

Williams's correction for the two-cell case is $q = 1 + \dfrac{1}{2n}$, which is

$$1 + \frac{1}{2(176)} = 1.00284$$

in this example.

$$G_{\text{adj}} = G/q = 0.12002/1.00284 = 0.11968$$

Since $G_{\text{adj}} \ll \chi^2_{.05[1]} = 3.841$ (Statistical Table **D**), we clearly do not have sufficient evidence to reject our null hypothesis.

An alternative adjustment discussed in Section 17.2 is the correction for continuity. We calculate this adjustment by computing an adjusted observed frequency f_i, which is changed to reduce the difference between the observed and the corresponding expected frequencies by $\frac{1}{2}$, as shown in the table above. Then we employ Expression (17.4) as before to obtain G_{adj}. This correction may be applied when n is less than about 200. The computations are carried out as described above, but using adjusted f's instead of the original f's (see text).

$$G_{\text{adj}} = 2\sum^{a} f_i \ln\left(\frac{f_i}{\hat{f}_i}\right)$$
$$= 2\left[130.5 \ln\left(\frac{130.5}{132.0}\right) + 45.5 \ln\left(\frac{45.5}{44.0}\right)\right]$$
$$= 0.06768$$

BOX 17.1 CONTINUED

Note that G_{adj} is now smaller because the continuity correction results in a more conservative test.

Chi-square test

1. Frequencies divided into $a \geq 2$ classes

Employ Expression (17.8):

$$X^2 = \sum^{a} \frac{(f_i - \hat{f}_i)^2}{\hat{f}_i}$$

for the bean example featured earlier in this box:

$$X^2 = \frac{(63 - 67.78125)^2}{67.78125} + \frac{(31 - 22.59375)^2}{22.59375} + \cdots + \frac{(12 - 15.06250)^2}{15.06250} = 9.53389$$

This value is to be compared to a χ^2-distribution (Statistical Table **D**) with $a - 1$ *df* ($df = 8 - 1 = 7$ in this example). Since $X^2 < \chi^2_{.05[7]} = 14.067$, the data may be considered consistent with the null hypothesis (as was concluded by the G-test).

2. Special case of frequencies divided into $a = 2$ classes.

Employ Expression (17.6):

$$X^2 = \frac{(f_1 - \hat{p}n)^2}{\hat{p}\hat{q}n}$$

or Expression (17.7):

$$X^2 = \frac{(f_2 - \hat{q}n)^2}{\hat{p}\hat{q}n}$$

Applying the chi-square test to the drosophila data tabled in this box, by Expression (17.6),

$$X^2 = \frac{(130 - 132.0)^2}{0.75 \times 0.25 \times 176} = \frac{4.0}{33.0} = 0.12121$$

Since $X^2 < \chi^2_{.05[1]} = 3.841$ (Table **D**), we do not have sufficient evidence to reject our null hypothesis.

Adjustment for small sample sizes in cases with a = 2 classes:

If	Compute
$n > 200$	X^2 by Expressions (17.5), (17.6), or (17.8)
$200 \geq n > 25$	X^2_{adj} (continuity correction) by Expressions (17.5C) or (17.6C). Note that the adjustment will result in a conservative test.
$n < 25$	Exact probabilities of the binomial as shown in Table 5.3 (Section 5.2), employing the method of Box 5.1

smallest $\hat{f}$ should be five or more. Whenever classes with expected frequencies of less than five occur, expected and observed frequencies for those classes are generally pooled with an adjacent class to obtain a joint class with an expected frequency $\hat{f} > 5$ (such a case is illustrated later in Box 17.2). Conahan (1970) found (on the basis of an analysis of 126 equiprobable multinomial distributions) that if $\hat{f}$ is > 10, G gives a good approximation to the exact multinomial probability. She found that G was satisfactory (and better than X^2) for $a \geq 5$ and $\hat{f} \geq 3$, and she recommended the exact test when $a \geq 5$ and $\hat{f} < 3$ and for $a < 5$ and $\hat{f} < 5$.

An exact test could be carried out in a manner quite similar to working out the exact probabilities for the binomial distribution, as shown in Table 7.2. One first calculates the probability of obtaining the observed outcome on the null hypothesis. The general term for the multinomial distribution (for four classes) is given by Expression (17.11) in Section 17.4. This expression could readily be extended to any number of classes as required by the distribution to be tested. Unless the probability of the observed outcome under the null hypothesis is already greater than α, the specified type I error, the probabilities of all worse cases, must be evaluated as in the binomial case discussed in Section 7.8. A definition of ''all worse cases'' is not so simple in the multinomial distribution. The conventional procedure is to sum all probabilities $P_i \leq P_0$, where P_i is the probability of other outcomes under the null hypothesis and P_0 is the probability of the observed outcome under the null hypothesis. Such a computation could involve a prodigious amount of calculation in all but trivial cases. The StatXact (1991) program evaluates the exact probability of the null hypothesis in such cases.

When expected frequencies are unequal, Conahan's recommendations do not directly apply but could be interpreted conservatively by applying them to the smallest expected frequencies. Larntz (1978) compared X^2 and G at just the 5% level of significance and found that when $1.5 < \hat{f} < 4$, G rejects the null hypothesis much too often. Also when $f = 0$ or 1, G is not a close approximation to χ^2. Unfortunately this study did not consider the correction suggested by Williams (1976). Monte Carlo simulations (see Section 18.3) that we performed show that with this correction, the G-statistic approximates the χ^2-distribution more closely than it does without it.

In a goodness-of-fit analysis involving more than two classes, as in the bean example of Box 17.1, with eight classes, we may wish to probe deeper into the nature and, ultimately, reasons for departure of some classes from expectation. Such schemes are comparable to single-degree-of-freedom decompositions in analysis of variance (see Sections 9.6 and 9.7). As we did earlier, we must distinguish between planned and unplanned tests. Certain planned tests—e.g., computing the goodness of fit for all single phenotypes (such as buff) together versus mixed phenotypes (such as purple/buff) also placed in one group—can be carried out simply. We evaluate their significance with one degree of freedom in the ordinary manner. By contrast, when we carry out tests suggested by the outcome of the overall analysis, we have to use the Bonferroni or Dunn–Šidák approach, and to obtain an experimentwise error rate α, we must carry out each individual test at a critical probability of $\alpha' = \alpha/k$ (Bonferroni) or $\alpha' = 1 - (1 - \alpha)^{1/k}$ (Dunn–Šidák), where k is the number of intended tests. If we wish to test all possible sets, we undertake a simultaneous test procedure using $\chi^2_{\alpha'[a-1]}$ as the critical value for all tests carried out. This test will ensure an α experimentwise error rate.

Next we consider the case for $a = 2$ cells. There is no special simplified equation for the G-test. One employs the same general equation [Expression (17.4)]. When Expression (17.4) is written explicitly for the two-cell case, it corresponds to Expression (17.3) multiplied by 2.

In tests of goodness of fit involving only two classes, the values of G and X^2, as computed from Expressions (17.4) and (17.5), (17.6), or (17.7), typically result in type I errors at a level higher than the intended one. Williams's correction reduces the value of G and results in a more conservative test (see Box 17.1). An alternative correction that has been used widely is the **correction for continuity,** usually applied to make the value of G or X^2 approximate the χ^2-distribution more closely. This correction consists of adding or subtracting 0.5 from the observed frequencies in such a way as to minimize the value of G or X^2. In the case of the G-test we simply adjust the f_i, changing them to reduce the difference between them and the corresponding expected frequencies by one half. Then we employ Expression (17.4) for G to obtain G_{adj}. In our experience, the correction for continuity results in excessively conservative tests.

The two-cell example in Box 17.1 is a genetic cross with an expected 3:1 ratio. The G-test is adjusted by either Williams's correction or by the continuity correction, but not by both. The latter yields the more conservative results. The expected frequencies differ very little from the observed frequencies; it is no surprise, therefore, that the resulting value of G_{adj} is far less than the critical value of χ^2 at one degree of freedom. Inspection of the chi-square table reveals that roughly 80% of all samples from a population with the expected ratio would show greater deviations than the sample at hand. When we apply the continuity correction to the computation of G for the genetic cross of Table 17.1, based on a sample size of $n = 90$, we obtain $G_{adj} = 10.01214$. After Williams's correction $G_{adj} = 10.82987$. Note that though the G_{adj} values are lower than the unadjusted G, 10.96524, the significance level does not change, remaining at $P < 0.005$.

These corrections are recommended as routine by some, while others suggest that they are necessary only for sample sizes $n < 200$. Actually, the corrections make only a small difference in G or X^2 even when sample sizes are below 200. We have found the continuity correction too conservative and therefore recommend that the Williams correction be applied routinely, although it will have little effect when sample sizes are large. For sample sizes of 25 or less, work out the exact probabilities as shown in Table 5.3 (see Section 5.2), employing the method of Box 5.1.

The conceptual alternative hypothesis in this genetic cross with two progeny classes was $H_1{:}p \neq \hat{p} = \frac{3}{4}$; that is, the test was two-sided, since we assumed that the proportion p of wild-type individuals could be less than or greater than the expected proportion $\hat{p} = \frac{3}{4}$. If our alternative hypothesis is that the proportion of wild types could be only less than the expected $\frac{3}{4}$, we should carry out a test of a one-sided conceptual hypothesis, and the chi-square table should be entered in the column for 2α if a type I error of α is desired. For $\alpha > 2$ classes a simple one-sided alternative hypothesis cannot be constructed. These same considerations apply to the chi-square test with two classes, which we discuss next.

Having dealt with the problems in Box 17.1 by means of the G-test, we now turn to the *chi-square test,* which appears in the second half of that box. We employ Expression (17.8) to obtain $X^2 = 9.53389$ for the bean example with $a = 8$ classes. Again our test statistic is less than the critical value of χ^2.

The correction for continuity in the two-cell case with small sample sizes is made for X^2 by subtracting 0.5 from the absolute values of the deviation $f - \hat{f}$, yielding the following adjusted versions of the corresponding formulas, (the letter C follows the expression numbers to indicate this correction).

$$X^2_{\text{adj}} = \sum^{2} \frac{(|f_i - \hat{f}_i| - \frac{1}{2})^2}{\hat{f}_i} \tag{17.5C}$$

$$X^2_{\text{adj}} = \frac{(|f_1 - \hat{p}_1 n| - \frac{1}{2})^2}{\hat{p}\hat{q}n} \tag{17.6C}$$

Knowledge of the size and direction of deviations from expectation aids the formulation of new hypotheses and experiments; inspection of $\hat{f}$ and of deviations $f - \hat{f}$ is therefore recommended.

Let us compare the results of all these methods for a test for goodness of fit in a small sample. We will use the example of Table 5.3: a litter of 17 offspring, of which 14 were females and 3 were males. We wish to test these data against a hypothetical sex ratio of 1 : 1. By the rules we have enumerated here, we should use the exact probability, which was worked out in Table 5.3 as 0.006,363,42, the probability of a deviation as great or greater in one direction from the 1 : 1 hypothesis. We approximate the probability of a two-tailed test by doubling this probability to yield $P = 0.01273$.

Applying the G-test to the example of Table 5.3, we obtain the following results:

G [by Expression (17.4)]

$$= 2\left[14 \ln\left(\frac{14}{8.5}\right) + 3 \ln\left(\frac{3}{8.5}\right)\right]$$

$$= 7.72303 \qquad P = 0.00545$$

G_{adj} (by Williams's correction)

$$= \frac{7.72303}{1.02941}$$

$$= 7.50237 \qquad P = 0.00620$$

G_{adj} (by the continuity correction)

$$= 2\left[13.5 \ln\left(\frac{13.5}{8.5}\right) + 3.5 \ln\left(\frac{3.5}{8.5}\right)\right]$$

$$= 6.27971 \qquad P = 0.01221$$

X^2 [by Expression (17.6)]

$$= \frac{(14 - 8.5)^2}{0.5 \times 0.5 \times 17}$$

$$= 7.11765 \qquad P = 0.00763$$

X^2_{adj} [by Expression (17.6C)]

$$= \frac{(14 - 8.5 - \frac{1}{2})^2}{0.5 \times 0.5 \times 17}$$

$$= 5.88235 \qquad P = 0.01529$$

We looked up these probabilities in Statistical Table **A,** areas of the normal curve, by entering the table with argument $\sqrt{G}$ or $\sqrt{X^2}$ (since $\sqrt{\chi^2_{\alpha[1]}} = t_{\alpha[\infty]}$). In this instance, there is no question by any of the tests that the results deviate significantly from the 1 : 1 hypothesis. Although the probability resulting from the G-test adjusted by the continuity correction is closest to the exact probability in this case, we would expect Williams's correction to yield probabilities closer to the exact values on the average. For both G and X^2 the adjusted value yields more nearly correct probabilities. In another experiment, carried out as described earlier for Figure 17.1 but with the continuity correction applied to both G and X^2, the percentages of type I error for the G_{adj} sampling experiment were smaller than the intended 5% and 1% values (3.84% and 0.77%, respectively); the percentages for X^2_{adj} were even smaller (2.76% and 0.71%, respectively).

Next we test the goodness of fit of frequencies arrayed in a single classification where the expected frequencies are based on a hypothesis intrinsic to the sampled data. Box 17.2 illustrates the computation using the sex ratio data in sibships of 12 introduced in Table 5.4. Only the G-test is illustrated, but the chi-square test could have been employed as well, using Expression (17.8). Recall that the expected frequencies in these data are based on the binomial distribution, with the parametric proportion of males $\hat{p}_{♂}$ estimated from the observed frequencies of the sample ($p_{♂}$ = 0.519,215). In Box 17.2 we employ the simple computational formula [Expression (17.4)] for efficient computation. The signs of the deviations are shown in column (5) of the table to indicate the pattern of departures from expectation.

Two aspects make this computation different from that in part **1** of Box 17.1. As we have explained, the G-test does not yield very accurate probabilities for small $\hat{f}_i$. The cells with $\hat{f}_i < 3$ (when $a \geq 5$) or $\hat{f}_i < 5$ (when $a < 5$) are generally grouped with adjacent classes so that the new $\hat{f}_i$ are large enough. The grouping of classes results in a less powerful test with respect to alternative hypotheses

Box 17.2 G-TEST FOR GOODNESS OF FIT. SINGLE-CLASSIFICATION, EXPECTED FREQUENCIES BASED ON HYPOTHESIS INTRINSIC TO THE SAMPLED DATA.

Sex ratio in 6115 sibships of 12 in Saxony. (Data from Table 5.4.) Column (4) gives the expected frequencies, assuming a binomial distribution. Computed originally to one decimal place, they are given here to five decimal places to give sufficient accuracy to the computation of G.

(1) ♂♂	(2) ♀♀	(3) f		(4) $\hat{f}$		(5) Deviation from expectation
12	0	7	52	2.34727	28.42973	+
11	1	45		26.08246		
10	2	181		132.83570		+
9	3	478		410.01256		+
8	4	829		854.24665		−
7	5	1112		1265.63031		−
6	6	1343		1367.27936		−
5	7	1033		1085.21070		−
4	8	670		628.05501		+
3	9	286		258.47513		+
2	10	104		71.80317		+
1	11	24	27	12.08884	13.02168	+
0	12	3		0.93284		
		6115 = n		6115.00000		

Since expected frequencies $\hat{f}_i < 3$ for $a = 13$ classes should be avoided, we group the classes at both tails with the adjacent classes to create classes of adequate size. Corresponding classes of observed frequencies f_i should be grouped to match. The number of classes after grouping is $a = 11$.

Compute G by Expression (17.4):

$$G = 2\sum^{a} f_i \ln\left(\frac{f_i}{\hat{f}_i}\right) = 2\left[52 \ln\left(\frac{52}{28.42973}\right) + 181 \ln\left(\frac{181}{132.83570}\right) + \cdots + 27 \ln\left(\frac{27}{13.02168}\right)\right]$$

$$= 94.87155$$

Since there are $a = 11$ classes, the degrees of freedom are maximally $a - 1 = 10$. Because the mean $\hat{p}_♂$ was estimated from the $p_♂$ of the sample however, another degree of freedom is removed, and the sample G is compared to a χ^2-distribution with $a - 2$

Box 17.2 Continued

$= 11 - 2 = 9$ degrees of freedom. Applying Williams's correction to the observed G, we obtain

$$q = 1 + \frac{(a^2 - 1)}{6nv}$$

$$= 1 + \frac{(11^2 - 1)}{6(6115)(9)} = 1.000{,}363{,}4$$

$$G_{adj} = \frac{G}{q} = \frac{94.87155}{1.000{,}363{,}4} = 94.83709$$

$$G_{adj} = 94.83709 > \chi^2_{.001[9]} = 27.877$$

The null hypothesis—that the sample data follow a binomial distribution—is therefore rejected decisively.

Typically, the following degrees of freedom pertain to G-tests for goodness of fit with expected frequencies based on a hypothesis *intrinsic* to the sample data (a is the number of classes after grouping, if any):

Distribution	Parameters estimated from sample	*df*
Binomial	$\hat{p}$	$a - 2$
Normal	μ, σ	$a - 3$
Poisson	μ	$a - 2$

When the parameters for such distributions are estimated from hypotheses *extrinsic* to the sampled data, the degrees of freedom are uniformly $a - 1$.

These tests could also be carried out by the chi-square test, employing Expression (17.8) and the same critical value of χ^2 as for the G-test.

that differ mostly in terms of expected frequencies in the tails of the distribution. By these criteria the classes of $\hat{f}_i$ at both tails of the distribution are too small. We group them by adding their frequencies to those in contiguous classes as shown in Box 17.2. Clearly, the observed frequencies must also be grouped to match the grouping of expected frequencies. The number of classes a is the number *after* grouping has taken place. In this case, $a = 11$.

The other new feature is the number of degrees of freedom considered for the significance test. We always subtract one degree of freedom for the fixed sum (here $n = 6115$). In addition, we subtract one degree of freedom for every parameter of the expected frequency distribution estimated from the sampled distribution. In this case, we estimated $\hat{p}_{♂}$ from the sample; therefore, a second degree of freedom is subtracted from a, making the final number of degrees of freedom $a - 2 = 11 - 2 = 9$. Comparing the corrected sample value of $G_{adj} = 94.83709$ with the critical value of χ^2 at 9 degrees of freedom, we find it highly significant ($P \ll 0.001$, assuming that the null hypothesis is correct). We therefore reject this hypothesis and conclude that the sex ratios are not binomially distributed. As is evident from the pattern of deviations, there is an excess of sibships in which one sex or the other predominates. Had we applied the chi-square test to these data, the critical value would have been the same ($\chi^2_{\alpha[9]}$).

The G-test (and the chi-square test) for testing the goodness of fit of a set of data to an expected frequency distribution can be applied not only to the binomial distribution, but also to the normal, Poisson, and other distributions. For a normal distribution we customarily estimate the two parameters μ and σ from the sampled data. Hence the appropriate degrees of freedom are $a - 3$. In the Poisson distribution, only one parameter, μ must be estimated, so the appropriate degrees of freedom are $a - 2$. Remember that the critical issue is whether the expected frequencies are based on an extrinsic or intrinsic hypothesis, not on whether the test is against a normal distribution or some other distribution. An investigator may have a hypothesis about the mean and standard deviation of a distribution and make the expected normal frequencies conform to this hypothesis. In such a case, the two extra degrees of freedom are not subtracted, so the degrees of freedom would be only $a - 1$, since estimates of the parameters were not obtained from the sample.

Conventional tests of goodness of fit specify the alternative hypothesis simply as departures from expectation, hence they are one-sided. Occasionally one may wish to test the alternative hypothesis that the observed frequencies are too deviant or too close to the expected frequencies. In such a case, critical values for chi-square should be looked up in columns $1 - \alpha/2$ and $\alpha/2$ of the chi-square table when a type I error of α is desired.

A nonparametric test that is applicable to continuous frequency distributions, where it has greater power than the G- or chi-square tests for goodness of fit, is the **Kolmogorov–Smirnov test.** The Kolmogorov–Smirnov test for goodness of fit is especially useful with small samples. In such cases it is not necessary (actually it is inadvisable) to group classes. We have already learned the general approach of this test in the Kolmogorov–Smirnov two-sample test (see Section 13.11). As in its earlier application, this test is based on differences between two cumulative relative frequency distributions, this time between an observed (F) and an expected ($\hat{F}$) distribution. An important difference is that since $\hat{F}$ is a continuous function, the largest difference between $\hat{F}$ and F is found by computing differences both before and after each time F steps up. Harter, Khamis, and Lamb (1984) showed that one can allow for both differences by using $(i - 0.5)/n$ as the observed cumulative relative frequencies. Then $d_{max} = g_{max} + 1/2n$, where g_{max} is the largest absolute difference between $\hat{F}$ and F.

We can look up tabled critical values and decide whether the maximum difference between the observed and expected cumulative frequency distributions is significant. An example will make this clear. Consider the body weights of crabs from Box 15.2. In Table 17.4 we have arrayed the 12 measurements in order of magnitude in Column (2). The mean of these readings is 12.0475, and the standard deviation is 6.484094. Column (3) shows the standardized deviation of each of these readings from its mean. We use the symbols $\hat{F}$ for cumulative expected frequencies and F for cumulative observed frequencies, either relative or absolute (it will be obvious in any given example which one it is). The cumulative relative expected frequencies in column (4) of Table 17.4 are based on the areas of the normal curve corresponding to the deviations in column (3). Column (5) lists the adjusted cumulative relative observed frequency of the variates using the relation $F_{0.5} = (i - 0.5)/n$. Since we have a sample of 12 items, each successive item increases the cumulative frequency by $\frac{1}{12}$.

If we were to graph these cumulative distribution functions, we would obtain the two curves of Figure 17.3. Notice that the curve representing the observed frequencies is a stepped function because the sample contains a finite number of observations. The relative deviations $g_{0.5}$, shown in column (6) of Table 17.4, between the observed and expected cumulative frequencies differ for

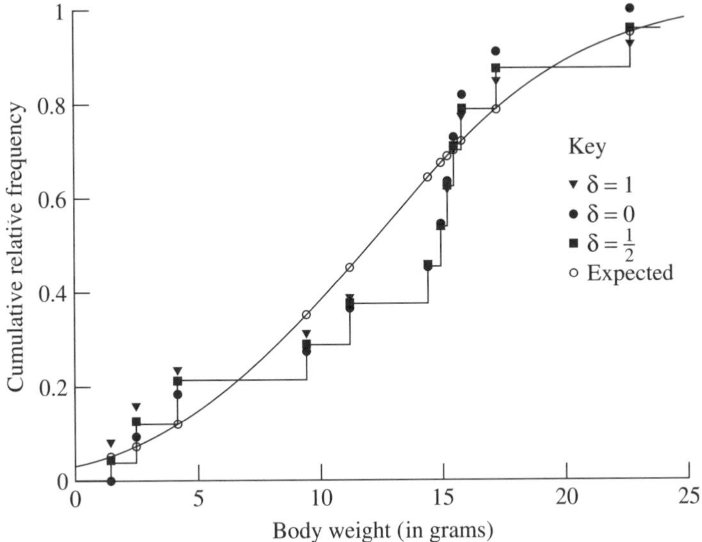

FIGURE 17.3 *Observed cumulative frequency distribution of body weights of 12 crabs (stepped line). The expected cumulative distribution for these data, assuming normality, is superimposed (smooth curve). Data from Box 15.2 and Table 17.4. Deviation for the sixth variate is $g_{\max,0.5}$, the largest absolute deviation in this example. The largest difference, $d_{\max}$, between the two distributions is $g_{\max,0.5} + 1/2n$.*

Table 17.4 KOLMOGOROV–SMIRNOV TEST FOR GOODNESS OF FIT USING DELTA ADJUSTMENT.

Cumulative expected frequencies ($\hat{F}$) *of body weight in grams of 12 crabs. (Data from Box 15.2.)* $\bar{Y} = 12.0475$; $s = 6.484094$.

(1) i	(2) Y	(3) $(Y - \bar{Y})/s$	(4) $\hat{F}$	(5) $F_{0.5}$	(6) $g_{0.5}$	(7) F_0	(8) g_0	(9) F_1	(10) g_1
1	1.41	−1.64055	0.05045	0.04167	0.00878	0.0769231	0.0264731	0.00000	0.05045
2	2.50	−1.47245	0.07045	0.12500	0.05455	0.1538462	0.0833962	0.09091	0.02046
3	4.19	−1.21181	0.11279	0.20833	0.09554	0.2307692	0.1179792	0.18182	0.06903
4	9.52	−0.38980	0.34834	0.29167	0.05667	0.3076923	0.0406477	0.27273	0.07561
5	11.30	−0.11528	0.45411	0.37500	0.07911	0.3846154	0.0694946	0.36364	0.09047
6	14.40	0.36281	0.64163	0.45833	0.18330	0.4615385	0.1800915	0.45455	0.18708
7	14.90	0.43992	0.67000	0.54167	0.12833	0.5384615	0.1315385	0.54545	0.12455
8	15.20	0.48619	0.68658	0.62500	0.06158	0.6153846	0.0711954	0.63636	0.05022
9	15.39	0.51549	0.69690	0.70833	0.01143	0.6923077	0.0045923	0.72727	0.03037
10	15.81	0.58027	0.71913	0.79167	0.07254	0.7692308	0.0501008	0.81818	0.09905
11	17.25	0.80235	0.78882	0.87500	0.08618	0.8461538	0.0573338	0.90909	0.12027
12	22.70	1.64287	0.94979	0.95833	0.00854	0.9230769	0.0267131	1.00000	0.05021

$$D_{0.5} = d_{\max,0.5} = g_{\max,0.5} + 1/2n = 0.18330 + 1/24 = 0.224963$$

$$D_0 = d_{\max,0} = g_{\max,0} = 0.180092$$

$$D_1 = d_{\max,1} = g_{\max,1} + 1/(n-1) = 0.277994 + 1/11 = 0.277994$$

different values of the abscissa (body weight). The maximum unsigned difference, $g_{\max 0.5} = 0.18330$, is at the sixth variate, $Y = 14.40$. Had the difference been in absolute rather than relative frequencies, it would have been necessary to divide it by n to obtain $g_{\max}$. The test statistic is computed as $D_{0.5} = g_{\max 0.5} + 1/2n = 0.224963$. If the expected distribution had been based on an extrinsic hypothesis, the appropriate test would have been to compare D to its critical value in Statistical Table **X.** In this table of the critical values of the Kolmogorov–Smirnov function, we would find that the 5% critical value for $n = 12$ and $\delta = 0.5$ (see the next paragraph for an explanation of δ) is 0.33376. Table **X** furnishes critical values of D for several values of α and for n up to 100. A formula for computing critical values for $n > 100$ is given in Table **X** and in Box 17.3.

Khamis (1992) proposed a simple adjustment to the Kolmogorov–Smirnov test that in some cases greatly increases its power to detect departures from the expected distribution. His δ-corrected Kolmogorov–Smirnov test is performed by using $F_\delta = (i - \delta)/(n - 2\delta + 1)$ as the observed cumulative relative frequency distribution and $D_\delta = g_\delta + \delta/(n - 2\delta + 1)$ as the test statistic. This test is performed twice: once for $\delta = 0$ and once for $\delta = 1$. The results are shown in columns (7) through (10) in Table 17.4. Note that the conventional test is equivalent to using $\delta = 0.5$. These values are looked up in Table **X,** and the level of significance of the test is taken as the smaller of the two probabilities. In our example, D_0 corresponds to a probability greater than 0.2 and D_1 corresponds to a probability slightly less than 0.2, so the observed distribution would be considered not significantly different (at an α level slightly less than 0.2) if this were a test of an extrinsic hypothesis.

In a Monte Carlo study (see Section 18.3) Lilliefors (1967) pointed out that the standard Kolmogorov–Smirnov tables, such as Table **X,** are unduly conservative for tests in which the expected normal frequencies were computed using $\bar{Y}$ and s^2 (i.e., fitted to an intrinsic hypothesis). In other words, the probability of rejecting the null hypothesis is much smaller than the indicated value of α using the critical values furnished in Table **X.** Thus, we have furnished new critical values for the test, from Khamis (1992), in Statistical Table **Y**. The example in Table 17.4 analyzing body weights of crabs is a case in point. The 5% critical values of D for $n = 12$ and $\delta = 0.5$ is 0.24497, lower than before, but the observed statistic, although significant at the 10% level, is not significant at the 5% level. Using the δ-correction, we compare $D_0 = 0.18009$ to 0.1932 and $D_1 = 0.27799$ to 0.30901 and find them still not significant at the 5% level. For larger samples Gonzalez, Sahni, and Franta (1977) describe an efficient algorithm for the Kolmogorov–Smirnov test that eliminates the need to sort the data.

Although the Kolmogorov–Smirnov test as just carried out is properly applied only to continuous functions without tied values, in a large sample one can make an approximate test using data grouped into a frequency distribution. In large samples, the effect of different values of δ can be ignored. An example of the Kolmogorov–Smirnov test applied to a large sample (the birth weights of Chinese children from Box 6.1) is shown in Box 17.3. The computation is

Box 17.3 KOLMOGOROV–SMIRNOV TEST FOR GOODNESS OF FIT. FREQUENCY DISTRIBUTION, HYPOTHESIS EXTRINSIC OR INTRINSIC TO SAMPLED DATA.

In Box 6.1 (birth weights of Chinese children) the observed frequencies were qualitatively compared to the frequencies expected on the basis of a normal distribution with $\mu = \bar{Y}$ and $\sigma = s$. The following procedure tests the significance of this agreement.

1. Form cumulative frequency distributions for the observed frequencies and for the expected frequencies (based on Boxes 6.1 and 6.3). For observed cumulative frequencies, compute $F_{0.5} = (F - 0.5)/n$ for relative observed frequencies or $F_{0.5} = F - 0.5$ for absolute observed frequencies. Using absolute frequencies we obtain the following:

(1) Y (oz)	(2) $F_{0.5}$	(3) $\hat{F}$	(4) $g_{0.5} = \lvert F_{0.5} - \hat{F}\rvert$
59.5	1.5	2.8	1.3
67.5	7.5	22.7	15.2
75.5	46.5	118.3	71.8
83.5	431.5	468.5	37.0
91.5	1319.5	1368.6	49.1
99.5	3048.5	3021.2	27.3
107.5	5288.5	5184.9	103.6 ← largest difference
115.5	7295.5	7204.7	90.8
123.5	8528.5	8532.6	4.1
131.5	9169.5	9180.0	10.5
139.5	9370.5	9400.5	30.0
147.5	9444.5	9454.5	10.0
155.5	9458.5	9464.0	5.5
163.5	9463.5	9464.9	1.4
171.5	9464.5	9464.9	0.4

2. Compute $g_{0.5}$, the absolute value of the differences between the cumulative frequencies [these values are given in column (4)], and locate $g_{\max,0.5}$, the largest difference—in this case 103.6.

For absolute frequencies, the Kolmogorov–Smirnov test statistic is

$$D_{0.5} = \frac{\text{largest difference} + 0.5}{n} = \frac{103.6 + 0.5}{9465} = 0.0110$$

Had we used relative frequencies, we would not have divided by n. In such a case, the test statistic is $D_{0.5} = g_{0.5} + 1/2n =$ largest difference.

BOX 17.3 CONTINUED

These data have been fitted to a cumulative normal distribution based on the sample mean and standard deviation (an intrinsic hypothesis). The sample D should be tested against the adjusted critical D in Statistical Table **Y.** However, tabulated values are furnished only for $n \leq 30$, an unlikely case for a frequency distribution. For larger samples, an asymptotic approximation is given in the table. For $\alpha = 0.05$ it is $0.886/\sqrt{n}$, and for $\alpha = 0.01$ it is $1.031/\sqrt{n}$. In the present case,

$$D_{0.5(\text{adj})} = \frac{0.886}{\sqrt{9465}} = 0.00911$$

and

$$D_{.01(\text{adj})} = \frac{1.031}{\sqrt{9465}} = 0.0106$$

Since our observed D is greater than the 1% critical value, we reject the null hypothesis that the birth weights are normally distributed.

In the rarer cases of fitting observations to an extrinsic hypothesis we use the standard Kolmogorov–Smirnov critical values in Statistical Table **X** for samples up to $n \leq 100$. We would compare the value of D with the entries in Table **X.** We reject the null hypothesis if the observed D is greater than the critical value of D_α.

For larger samples, we can calculate critical values as follows:

$$D_\alpha = \sqrt{\frac{-\ln\left(\frac{1}{2}\alpha\right)}{2n}}$$

For $\alpha = 0.05$ D_α is $1.358/\sqrt{n}$, and for $\alpha = 0.01$ it is $1.628/\sqrt{n}$.

The Kolmogorov–Smirnov test is properly applied to continuous frequency distributions only. It is conservative in the case of grouped data.

straightforward, using a frequency distribution instead of individual observations. In this instance, we find differences between absolute cumulative frequencies, which must be divided by n before they can be compared with the critical values. Again, corrected critical values of D should be employed because the data are fitted to an intrinsic hypothesis. Since Table **Y** furnishes values for samples sizes up to $n = 100$ only, we use the asymptotic approximations given in the table and in Box 17.3.

In practice it is not necessary to calculate all differences and find the largest one; any difference greater than the critical value will make the test significant. In this example we find that the observed maximum absolute difference is greater than the 1% critical value and reject the null hypothesis. The birth weights clearly do not follow the normal distribution. These conclusions agree with our earlier findings that these data are skewed significantly (see Box 7.5) and with the graphic test for departure from normality (see Box 6.3). Had the expected frequencies been based on an extrinsic hypothesis, we would have employed the ordinary Kolmogorov–Smirnov critical values in Table **X.** When applied to grouped data in a frequency distribution, the Kolmogorov–Smirnov test tends to be conservative; that is, the null hypothesis is rejected at a frequency less than α (D'Agostino and Noether, 1973).

Although it assumes a continuously distributed variable, the Kolmogorov-Smirnov test is often used for a discrete variable because using standard table results in a conservative test. A procedure for determining exact significance levels when the Kolmogorov–Smirnov test is applied to discrete distributions is explained by Conover (1972). Problems with computating the exact significance levels are addressed by Bartels et al. (1977). For the relatively constrained case in which the total number of discrete variates $n \leq 30$, the number of discrete classes $k \leq 10$, and the expected frequencies $\hat{f}_i$ are equal, Pettit and Stephens (1977) furnish a table that gives critical values for $S = \max |F - \hat{F}|$. These authors also report that their table provides a good approximation to the distribution of S with unequal $\hat{f}_i$, as long as the inequalities are not excessive. For small samples with few classes, consulting this table might be the preferred way of testing discrete frequency distributions.

The Kolmogorov–Smirnov statistic can also be used for setting confidence limits to an entire cumulative frequency distribution. By computing $F_i \pm D_\alpha$ for all values or classes of Y_i, we obtain $100(1 - \alpha)\%$ confidence limits. The 95% confidence limits of the cumulative distribution function of the crab body weight data are shown in Figure 17.4. We expect 95% of such sets of confidence limits to completely enclose the true cumulative frequency distribution.

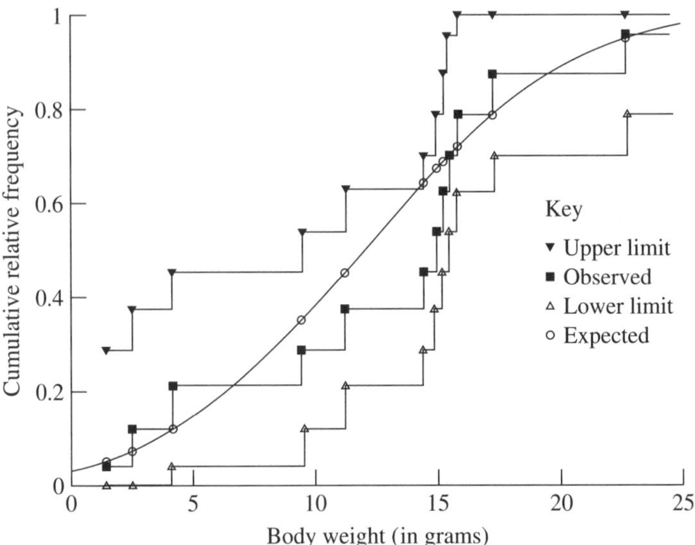

FIGURE 17.4 *The 95% confidence interval for the observed cumulative frequency distribution of Figure 17.3.*

The BIOM-pc package of computer programs performs the computations described in this section. Goodness-of-fit tests of continuous variables for normality are performed using g_1, g_2, and the Kolmogorov–Smirnov statistic. Tests on frequency data are performed using the G-test for goodness of fit to binomial, Poisson, or arbitrary expected distributions or to simple ratios.

17.3 Replicated Tests of Goodness of Fit

Frequently experiments leading to tests of goodness of fit are replicated. For example, the data in Box 17.4 are from a genetic experiment on houseflies in which eight replicate crosses were carried out, with the offspring expected in a ratio of 9 wild-type to 7 pale-eyed flies. Although we could simply pool the results from the eight progenies and test the hypothesis of a 9 : 7 ratio on the grand totals, in doing so we lose information. The tests that can be performed on such data are illustrated here in Box 17.4 (and later in Box 17.5), using the G-test. Such a test is called an *interaction* or **heterogeneity G-test.** Corresponding chi-square tests can be performed, but these have the disadvantage that the X^2 values are not completely additive, so the anovalike table given in step **6** of Box 17.4 would be only approximately correct.

The order of computations depends in part on the design of the experiment. If the replicates are obtained in sequence, with an appreciable time interval between each, the first logical step is likely to be a goodness-of-fit test of each replicate. Thus, if the progenies were obtained a week apart, the investigator usually would wish to carry out a significance test for each of the progenies as it is obtained, reserving the overall analysis for the end of the experiment. On the other hand, if the results are obtained simultaneously, the overall tests of goodness of fit may be carried out first, with individual replicates analyzed if warranted by the data and the interest of the investigator. In the example of Box 17.4, the data were collected essentially simultaneously, so we leave the analysis of the separate progenies until later in the box (step **5**). The final outcome of the analysis is identical regardless of which way the problem is approached.

First we test whether the outcomes of all the replicates (progenies in this case) are homogeneous—that is, whether the sampled proportions (of wild types) in the eight progenies could have come from a single population. If they are not uniform among replicates, the proportions are said to be heterogeneous. Five hypothetical heterogeneous replicates might have the following relations between the observed and expected proportions p_1 and $\hat{p}_1$ of wild type flies, respectively:

$$p_1 > \hat{p}_1$$

$$p_1 > \hat{p}_1$$

$$p_1 = \hat{p}_1$$

$$p_1 < \hat{p}_1$$

$$p_1 > \hat{p}_1$$

Box 17.4 REPLICATED GOODNESS-OF-FIT TESTS (*G*-STATISTIC):

Numbers of wild-type (N^+) and pale-eyed (ge) flies in $b = 8$ progenies of the housefly, *Musca domestica*. The expected ratio of the $a = 2$ phenotypic classes N^+ and ge is 9:7.

Progeny	N^+	ge	$\sum$
1	83	47	130
2	77	43	120
3	110	96	206
4	92	58	150
5	51	31	82
6	48	61	109
7	70	42	112
8	85	66	151
$\sum$	616	444	1060 = n

SOURCE: Data from R. L. Sullivan (unpublished results).

The order of computations may differ somewhat, depending on whether all replicates are obtained simultaneously or over a period of time (see text for further discussion of this issue). The resulting final analysis will be the same in any case. In this example the progenies were obtained simultaneously; we therefore proceed immediately to an overall analysis of the data, reserving tests of goodness of fit of individual progenies for step **5**.

Computation

1. The following quantities are necessary for the intended computations. They can be obtained with the help of a natural logarithm function key on a calculator.

a. The sum of the $f \ln f$ transforms of all frequencies in the table:

$$\sum^{a}\sum^{b} f \ln f = 83 \ln 83 + 77 \ln 77 + \cdots + 42 \ln 42 + 66 \ln 66$$
$$= 4502.74287$$

b. The sum of the $f \ln f$ transforms of the column sums of the table:

$$\sum^{a}\left(\sum^{b} f\right) \ln \left(\sum^{b} f\right) = 616 \ln 616 + 444 \ln 444$$
$$= 6663.26624$$

c. The sum of the $f \ln f$ transforms of the row sums of the table:

$$\sum^{b}\left(\sum^{a} f\right) \ln \left(\sum^{a} f\right) = 130 \ln 130 + \cdots + 151 \ln 151$$
$$= 5215.20530$$

Box 17.4 Continued

d. $n \ln n = \left(\sum^{a}\sum^{b} f\right) \ln \left(\sum^{a}\sum^{b} f\right) = 1060 \ln 1060 = 7383.98564$

2. Are the ratios of the progenies (replicates) homogeneous? Compute G_H (G for heterogeneity).

$$\begin{aligned} G_H &= 2[\text{quantity } \mathbf{a} - \text{quantity } \mathbf{b} - \text{quantity } \mathbf{c} + \text{quantity } \mathbf{d}] \\ &= 2[4502.74287 - 6663.26624 - 5215.20530 + 7383.98564] \\ &= 16.51394 \end{aligned}$$

The value is compared to a χ^2-distribution with $(a - 1)(b - 1) = (2 - 1)(8 - 1) = 7$ degrees of freedom. Since $\chi^2_{.05[7]} = 14.067$, there is evidence that the ratios are not all homogeneous among the progenies.

3. A second test is for goodness of fit to the 9 : 7 hypothesis for the pooled data, or all eight replicates pooled together. These frequencies are $\Sigma^b f_1 = 616$ and $\Sigma^b f_2 = 444$. The test is carried out conveniently by Expression (17.4):

$$\begin{aligned} G_P &= 2\left[\sum^{b} f_1 \ln\left(\frac{\sum^{b} f_1}{n\hat{p}}\right) + \sum^{b} f_2 \ln\left(\frac{\sum^{b} f_2}{n\hat{q}}\right)\right] \\ &= 2\left[616 \ln\left(616 \div \frac{1060 \times 9}{16}\right) + 444 \ln\left(444 \div \frac{1060 \times 7}{16}\right)\right] \\ &= 1.50039 \end{aligned}$$

This value is not significant when compared to a χ^2-distribution with $a - 1 -$ (number of parameters estimated from the sample) $= 2 - 1 - 0 = 1$ degree of freedom.

4. Adding the G for heterogeneity to the pooled G, we obtain a total G:

$$G_H + G_P = G_T = 16.51394 + 1.50039 = 18.01433$$

with $b(a - 1) = 8(2 - 1) = 8$ degrees of freedom. This quantity can be used to test whether the data as a whole fit the expected ratio. Since $\chi^2_{.05[8]} = 15.507$, we conclude that they do not. Since G_H was significant and G_P was small, we know that the significance of G_T is due to the significant heterogeneity. One of two analyses might next be done, depending on the interests of the investigator. One is described in step **5**; the other, an unplanned test of differences among the progenies using an STP procedure, is described in Box 17.5.

5. Partitioning the total G into contributions due to individual samples. For each sample compute G by Expression (17.4). For example, for progeny 1

$$\begin{aligned} G &= 2\left[83 \ln\left(83 \div \frac{130 \times 9}{16}\right) + 47 \ln\left(47 \div \frac{130 \times 7}{16}\right)\right] \\ &= 3.10069 \end{aligned}$$

This value has $a - 1 -$ (number of parameters estimated from sample) $= 2 - 1 - 0 = 1$ degrees of freedom. It is not significant.

BOX 17.4 CONTINUED

We can continue testing each progeny in this fashion. In doing so, we find only progeny 6 significant, as shown here:

$$G = 2\left[48 \ln\left(48 \div \frac{109 \times 9}{16}\right) + 61 \ln\left(61 \div \frac{109 \times 7}{16}\right)\right]$$

$$= 6.53781^*$$

Since the proportion of wild-type flies in progeny 6 equals 48/109 = 0.440, we see that the discrepancy was in the direction of having too few wild-type flies ($\hat{p}_1 = \frac{9}{16} = 0.5625$).

As a computational check, we add the eight individual G's and note that their sum equals the total G ($G_T = 18.01433$) within rounding error.

6. We can summarize our results in the following tables (* = $0.01 < P \leq 0.05$).

Tests	*df*	*G*
Pooled	1	1.50039
Heterogeneity	7	16.51394*
Total	8	18.01433*

Progeny	*df*	*G*
1	1	3.10069
2	1	3.11166
3	1	0.67795
4	1	1.59162
5	1	1.19268
6	1	6.53781*
7	1	1.80182
8	1	0.00011
Total	8	18.01434*

NOTE: If individual values of G have been computed first, the heterogeneity G could have been obtained simply by subtracting pooled G from the total G (the sum of the individual G's) and would not have needed to be computed independently as in step **2.**

By analogy with analysis of variance (see Chapter 11), we see immediately why such heterogeneity could also be called *interaction.*

The test is shown in step **2** of Box 17.4. G_H (heterogeneity G) is computed by a formula analogous to the interaction formula in a two-way orthogonal anova: $\bar{Y} - \bar{R} - \bar{C} + \bar{\bar{Y}}$ (see Section 11.2):

$$\begin{aligned} G_H = 2[&\text{sum of } f \ln f \text{ of frequencies in each cell of the table} \\ &- \text{sum of } f \ln f \text{ of column sums of frequencies} \\ &- \text{sum of } f \ln f \text{ of row sums of frequencies} \\ &+ f \ln f \text{ of total number of items in the table}] \end{aligned}$$

We find that we must reject the null hypothesis for the data of Box 17.4 because one or more of the progenies differ in outcome from the others. If we had simply pooled all the data and carried out an overall test, we could not, of course, have detected this heterogeneity.

The next test is carried out on the pooled frequencies, which means the eight progenies are treated as though they were a single large progeny. The test is exactly the same as that for a single classification, as shown in Section 17.2 using Expression (17.4). We find that the pooled data do not deviate significantly from expectation. This lack of deviation might surprise you in view of the significant heterogeneity of the data, but it occurs when replicates are heterogeneous in such a way that they tend to compensate for their respective deviations from expectation.

We can illustrate this phenomenon with an artificial, deliberately exaggerated example. Suppose we test sex ratios in six bottles of drosophila, three of which have only male offspring and three only female offspring. If we pooled these data (assuming that the number of male and female offspring were approximately the same), we would be unable to reject the 50 : 50 sex ratio hypothesis for these drosophila, although there clearly is no such sex ratio in these flies. Heterogeneity in this example, would, of course, be highly significant.

The scientific relations expressed by these tests for heterogeneity and of pooled data are of a fundamental nature and must be clearly understood. To deepen and confirm your understanding of them, we furnish other hypothetical examples on sex ratios in drosophila in which the differences are more subtle. Four examples are given in Table 17.5.

In example A we have four progenies, each with a slight excess of females, but each not significant by individual G-tests using Expression (17.4). Although each individual G is suspiciously high, none is significant. When we add these values of G together to obtain a total G, however, the test is significant at the 2.5% level. When considered together, suspicious results for individual progenies yield an improbable total result. We conclude from the total G that obtaining four sample progenies, each of which is as far from expectation as these are, is unlikely. This finding is reinforced when we examine the pooled G. Since all progenies deviate in the same direction, the pooled frequencies show this deviation clearly and, being based on a larger sample size, yield a highly significant value of G. To obtain the heterogeneity G, we subtract pooled from total G. The resulting heterogeneity G of 0.07363 with 3 degrees of freedom is clearly not significant, indicating that the deviations from expectation of the progenies are in the same direction and are not significantly different from each other.

Example B has the same total frequencies as example A, but in progenies 2 and 4 the frequencies of males and females have been reversed. The individual values of G are necessarily the same as before, suspicious but not significant. The total G, being the sum of the individual ones, must also be the same as before, indicating that these progenies do not conform to expectation. The important difference between examples A and B lies in the pooled and

Table 17.5 HYPOTHETICAL REPLICATED TESTS OF GOODNESS OF FIT ILLUSTRATING MEANING OF TOTAL, POOLED, AND HETEROGENEITY *G*.

Computation of G is illustrated in Box 17.4, using formulas from Box 17.1. The null hypothesis being tested is a 50 : 50 sex ratio.

	Progenies	♀	♂	n		df	G
A	1	59	41	100		1	3.25773 *ns*
	2	58	42	100		1	2.57104 *ns*
	3	57	42	99		1	2.28150 *ns*
	4	58	40	98		1	3.32497 *ns*
					Total	4	11.43523 $P < 0.025$
	$\sum$	232	165	397	Pooled	1	11.36160 $P < 0.001$
					Heterogeneity	3	0.07363 *ns*
B	1	59	41	100		1	3.25773 *ns*
	2	42	58	100		1	2.57104 *ns*
	3	57	42	99		1	2.28150 *ns*
	4	40	58	98		1	3.32497 *ns*
					Total	4	11.43523 $P < 0.025$
	$\sum$	198	199	397	Pooled	1	0.00252 *ns*
					Heterogeneity	3	11.43272 $P < 0.01$
C	1	59	41	100		1	3.25773 *ns*
	2	58	42	100		1	2.57104 *ns*
	3	72	26	98		1	22.46416 $P < 0.001$
	4	73	26	99		1	23.23751 $P < 0.001$
					Total	4	51.53044 $P < 0.001$
	$\sum$	262	135	397	Pooled	1	41.35016 $P < 0.001$
					Heterogeneity	3	10.18027 $P < 0.025$
D	1	52	48	100		1	0.16004 *ns*
	2	36	64	100		1	7.94580 $P < 0.005$
	3	52	47	99		1	0.25263 *ns*
	4	52	46	98		1	0.36758 *ns*
					Total	4	8.72605 *ns*
	$\sum$	192	205	397	Pooled	1	0.42577 *ns*
					Heterogeneity	3	8.30028 $P < 0.05$

heterogeneity G. Since two of the progenies in B favor females and the other two favor males, the pooled frequencies of males and females are almost identical. The pooled G is therefore negligible, 0.00252, a suspiciously good fit, unlikely to occur in actual, experimental data. Subtracting such a low pooled G from the total G leaves a large heterogeneity G, indicating that the suspicious departures from expectation are not in a uniform direction, but are in different directions, as we know them to be.

Example C illustrates a case in which all three tests are significant. All four progenies have an excess of females. The first two have only a small (nonsignificant) excess, and the last two deviate highly significantly from expectation. The total G is consequently highly significant. The pooled G, in view of the consistent trend in favor of females, is also highly significant; and the heterogeneity G is significant as well, showing that the trend, although in all cases favoring females, is not uniform in magnitude.

Finally, in example D we show that the total G is an overall measure of the departure from expectation of the several progenies. Whereas progeny 2 has a significantly larger number of males, progenies 1, 3, and 4 have slight, nonsignificant excesses in favor of females. The total G, although suspiciously high, is not significant; the fit of the other progenies was good enough that the departure from expectation by progeny 2 was not sufficient to affect our overall judgment based on the total G. The frequencies in the progenies tended to compensate for each other, making the pooled G not significant, but on subtraction, the heterogeneity G is significant, indicating correctly that the data are heterogeneous, although neither total nor pooled G had been significant.

When G-values are to be summed or partitioned, neither the Williams nor the continuity correction is used because adjusted G-values are not additive.

Returning to the example in Box 17.4, we now add heterogeneity G to pooled G to obtain total G (step **4**). We find that the data as a whole do not fit the expected ratio. Occasionally G_T is significant, even though G_H and G_P are nonsignificant but appreciable. In such a case we would be led to conclude that the overall fit of the data to the hypothesis is poor, but we would be unable to pinpoint the exact nature of the departure from expectation.

When we compute G separately for each replicate of the study, we find only progeny 6 departing significantly from expectation, by having fewer wild-type and more mutant flies. The analysis can be summarized by a table similar to an anova table (step **6**).

Having determined that the replicates are not homogeneous, we now want to know the reason for this heterogeneity. Precisely which samples are homogeneous and which samples are different from the rest significantly enough to cause the heterogeneity? Although sometimes we can locate sources of heterogeneity by simple inspection, the routine procedures for carrying out unplanned tests to accomplish this task, outlined in Box 17.5, are usually necessary. For all pairwise (or fewer) unplanned comparisons we can employ an experimentwise error rate α and adjust the probability per comparison (as proposed by Šidák; see

Section 9.6) to $\alpha' = 1 - (1 - \alpha)^{1/k}$, making use of Statistical Table **E** and setting $k < b(b - 1)/2$. Alternatively, when comparisons will be made for all possible sets of replicates, use the simultaneous test procedure shown in Box 17.5. In carrying out the computations we find that the significance of the heterogeneity appears to be due to differences between progenies 1 and 2 on one hand and 6 on the other. The next step might be to try to determine why these progenies gave different results. Perhaps the media in which they were reared were different, or perhaps there were biological differences among their parents.

Although the replicated goodness-of-fit tests in Boxes 17.4 and 17.5 tested a simple 9:7 ratio, these techniques can also be used for more complicated

Box 17.5 UNPLANNED TESTS OF THE HOMOGENEITY OF REPLICATES TESTED FOR GOODNESS OF FIT.

When there is significant heterogeneity among replicates, the following techniques can be used to test whether all replicates differ from one another or whether there are homogeneous sets of replicates differing from other such sets or from single replicates. We will use the housefly progenies of Box 17.4 to illustrate the technique.

When testing fewer than all pairwise unplanned comparisons, we can employ an experimentwise error rate α and adjust the probability per comparison to $\alpha' = 1 - (1 - \alpha)^{1/k}$ with $k < b(b - 1)/2$. Thus let us assume we wish to make only four nonorthogonal comparisons. Then we employ $\alpha' = 1 - 1(1 - \alpha)^{1/k} = 1 - (0.95)^{1/4} = 0.01274$. We need not be concerned with evaluating χ^2 for this probability because we can employ Statistical Table **E,** which gives us the critical value directly. We look up $\chi^2_{\alpha[k,\nu]}$, where α is the experimentwise type I error rate, k is the number of intended comparisons, and $\nu = (a - 1)(b - 1)$, the degrees of freedom of the test (usually this will be $a - 1$). Thus $\chi^2_{.05[4,1]} = 6.205$, and any of the four intended tests whose $G_H \geq 6.205$ will be considered significant.

Alternatively, we may wish to test all pairs of replicates or the pairs that differ the most (which amounts to the same thing). In such a case consult Table **E** for $\chi^2_{\alpha[k,\nu]}$, where $k = b(b - 1)/2$. Thus to compare all pairs of replicates in the eight housefly progenies, employ $\chi^2_{.05[28,1]} = 9.712$ as a critical value. All values of G_H comparing pairs of replicates equal to or greater than this critical value will be considered significant.

Finally, when we wish to test all possible sets (which requires a still more conservative procedure), we carry out a simultaneous test procedure, based on the following critical value: For the heterogeneity test to be just significant at the α level, G_H must equal $\chi^2_{\alpha[(a-1)(b-1)]}$. This value is used as the constant critical value for G in the STP tests. In our case, $a = 2$ and $b = 8$; therefore $\chi^2_{.05[7]} = 14.067$ can be used as a critical value if a 5% experimentwise error rate is accepted. The test procedure consists of a heterogeneity test for all sets of replicates (progenies in this example) taken 2, 3, 4, . . . at a time. If there are many replicates a computer program is useful. Usually not all of these tests need to be carried out, however, as we will show.

BOX 17.5 CONTINUED

The simultaneous test procedure

1. Compute the proportion of wild-type flies p_i for each progeny. Reorder progenies by magnitude of p_i.

(1) Progeny	(2) N^+	(3) ge	(4) $\sum$	(5) p_i
2	77	43	120	0.642
1	83	47	130	0.638
7	70	42	112	0.625
5	51	31	82	0.622
4	92	58	150	0.613
8	85	66	151	0.563
3	110	96	206	0.534
6	48	61	109	0.440

2. Since progenies 2 through 4 (i.e., 1, 2, 4, 5, and 7) have similar p_i's, we will test this set first. Compute G_H using the formula in step **2** of Box 17.4. The general layout of the formula for this and subsequent values of G_H will be $G_H = 2$[(sum of $f \ln f$ for frequencies in the set) $-$ (sum of $f \ln f$ for column sums of the set) $-$ (sum of $f \ln f$ for row sums of the set) + ($f \ln f$ for total sum of the set)].

$$\begin{aligned} G_H &= 2[(77 \ln 77 + 43 \ln 43 + \cdots + 58 \ln 58) - (373 \ln 373 + 221 \ln 221) \\ &\quad -(120 \ln 120 + 130 \ln 130 + \cdots + 150 \ln 150) + (594 \ln 594)] \\ &= 2[2456.78908 - 3401.74271 - 2848.69663 + 3793.80632] \\ &= 0.31210 \end{aligned}$$

which is far less than our critical $G_{.05} = 14.067$; this set of replicates therefore is considered homogeneous.

3. We now try adding progeny 8 to our set (this progeny has the next lower value of p_i). In computing a new G_H we can make use of some of the intermediate computations already carried out. The new column totals are 373 + 85 = 458 and 221 + 66 = 287, and their sum is 745. Also we can recompute the new sum of $f \ln f$ for the frequencies in the set by adding the $f \ln f$ for the new member to the sum for the former set. Similarly the former sum of the row sums is incremented by the $f \ln f$ of the new row.

$$\begin{aligned} G_H &= 2[(2456.78908 + 85 \ln 85 + 66 \ln 66) - (458 \ln 458 + 287 \ln 287) \\ &\quad -(2848.69663 + 151 \ln 151) + 745 \ln 745] \\ &= 2[3110.93165 - 4430.37748 - 3606.30588 + 4926.97124] \\ &= 2.43905 \end{aligned}$$

This set is also homogeneous.

Box 17.5 Continued

4. Progeny 3 has the next lower value of p_i. We add it to the set and recompute G_H.

$$\begin{aligned} G_H &= 2[(3110.93165 + 110 \ln 110 + 96 \ln 96) - (568 \ln 568 \\ &\quad +383 \ln 383) - (3606.30588 + 206 \ln 206) + 951 \ln 951] \\ &= 2[4066.16192 - 5880.42237 - 4703.84837 + 6521.49587] \\ &= 6.77410 \end{aligned}$$

Again, the set is homogeneous. It is also a maximally nonsignificant set, since the test of the complete set yields a significant G_H of 16.51394, as shown in Box 17.4. The results there show that adding progeny 6 to set {1, 2, 3, 4, 5, 7, 8} would yield a significantly heterogeneous set.

5. We could now try to find a maximally nonsignificant set involving progeny 6. Starting with progenies 6 and 3, we obtain

$$\begin{aligned} G_H &= 2[(110 \ln 110 + \cdots 61 \ln 61) - (158 \ln 158 + 157 \ln 157) \\ &\quad -(206 \ln 206 + 109 \ln 109) + 315 \ln 315] \\ &= 2[1391.8112 - 1593.72061 - 1608.8994 + 1812.06038] \\ &= 2.50317 \end{aligned}$$

This value of G_H is not significant. Continuing in this fashion, we find that the set {3, 4, 5, 6, 7, and 8} is also maximally nonsignificant, since its G_H (11.80127) would increase to 14.44762 if we added progeny 1.

Thus, we may consider progenies {1, 2, 3, 4, 5, 7, and 8} and {3, 4, 5, 6, 7, and 8} to be homogeneous sets. Arraying the progenies by proportion of wild types (high to low), we can summarize their relations as follows:

Progeny	2	1	7	5	4	8	3	6
	———	———	———	———	———	———	———	
			———	———	———	———	———	———

The next question to address is why progenies 1 and 2 (high proportion of wild types) were different from progeny 6 (low proportion of wild types).

situations, such as single-classification data with more than two classes, including replicated goodness-of-fit tests to binomial or Poisson distributions.

Single-classification and replicated goodness-of-fit tests are performed by the BIOM-pc package of computer programs. The unplanned simultaneous test described in Box 17.5 is also performed.

17.4 Tests of Independence: Two-way Tables

The notion of statistical or probabilistic independence was introduced in Section 5.1, where we showed that if two events are independent, the probability of their occurring together can be computed as the product of their separate probabilities. For example, suppose that among the progeny of a certain genetic cross, the probability of a corn kernel's being red is $\frac{1}{2}$ and the probability of a kernel's being dented is $\frac{1}{3}$. Assuming statistical independence of these two characteristics, the probability of obtaining a dented red kernel would be $\frac{1}{2} \times \frac{1}{3} = \frac{1}{6}$.

The appropriate statistical test for the genetic problem is to test the frequencies for goodness of fit to the expected ratios of 2 (red, not dented) : 2 (not red, not dented) : 1 (red, dented) : 1 (not red, dented). This would be a simultaneous test of two null hypotheses: that the expected proportions are $\frac{1}{2}$ and $\frac{1}{3}$ for red and dented, respectively, and that these two properties are independent. The first null hypothesis tests the Mendelian model in general. The second tests whether these characters assort independently—that is, whether they are determined by genes located in different linkage groups. Rejection of the second hypothesis is taken as evidence that the characters are linked—that is, located nearby on the same chromosome.

In many instances in biology the second hypothesis, concerning the independence of two properties, is of great interest and the first hypothesis, regarding the true proportion of one or both properties, is of little interest. In fact, often no hypothesis regarding the parametric values $\hat{p}_i$ can be formulated by the investigator. We will cite several examples of such situations which lead to the test of independence that we will present in this section. We employ this test whenever we wish to test whether two different properties, each occurring in two states, are dependent on each other. For instance, specimens of a certain moth may occur in two color phases—light and dark. Fifty specimens of each phase may be exposed in the open, subject to predation by birds. The number of surviving moths is counted after a fixed interval of time. The proportion preyed upon may differ in the two color phases. The two properties in this example are color and survival. We can divide our sample into four classes: light-colored survivors, light-colored prey, dark survivors, and dark prey. If the probability of being preyed upon is independent of the color of the moth, the expected frequencies of these four classes can be computed simply as independent products of the proportion of each color (in our experiment, $\frac{1}{2}$) and the overall proportion preyed upon in the entire sample. If the statistical test of independence that we explain in this section shows that the two properties are not independent, we will be led to conclude that one of the color phases is more susceptible to predation than the other. In this example, this issue is the one of biological importance; the exact proportions of the two properties are of little interest. The proportion of the color phases is arbitrary, and the production of survival is of interest here only insofar as it differs for the two phases.

Another example relates to a sampling experiment carried out by a plant ecologist. The first step is to obtain a random sample of 100 individuals of a fairly rare species of tree distributed over an area of 400 square miles. For each tree the ecologist notes whether it is rooted in a serpentine soil or not, and whether the leaves are pubescent or smooth. Thus the sample of $n = 100$ trees can be divided into four groups: serpentine–pubescent, serpentine–smooth, nonserpentine–pubescent, and nonserpentine–smooth. If the probability of a tree's being pubescent is independent of its location, our null hypothesis of the independence of these properties is true. If, on the other hand, the proportion of pubescence differs for the two types of soils, our statistical test is expected to result in rejection of the null hypothesis of independence. Again, the expected frequencies are products of the independent proportions of the two properties—serpentine versus nonserpentine, and pubescent versus smooth. In this instance the proportions themselves may be of interest to the investigator.

The example we will work out here in detail is from immunology. A sample of 111 mice was divided into two groups, 57 that received a standard dose of pathogenic bacteria followed by an antiserum and a control group of 54 that received the bacteria but no antiserum. After sufficient time had elapsed for an incubation period and for the disease to run its course, 38 dead mice and 73 survivors were counted. Of those that died, 13 had received bacteria *and* antiserum, whereas 25 had received bacteria only. A question of interest is whether the antiserum protected the mice who received it, leading to a proportionally higher number of survivors in that group. Here again the proportions of these properties are of no more interest than they were in the first example (predation on moths).

Such data can be displayed conveniently in the form of a **two-way table** as shown below. Two-way and multiway tables (more than two criteria) are often referred to as **contingency tables.** This type of two-way table, in which each of the two criteria is divided into two classes, is known as a **2 × 2 table.**

		Dead	Alive	$\sum$
Bacteria and antiserum		13	44	57
Bacteria only		25	29	54
	$\sum$	38	73	111

We can confirm from the table that 13 mice received bacteria and antiserum but died. The totals in the margins give the number of mice exhibiting any one property: 57 mice received bacteria and antiserum; 73 mice survived the experiment. Altogether 111 mice were involved in the experiment; they constitute the total sample. For discussion, we label the cells of the table and the row and column sums as follows:

a	b	$a + b$
c	d	$c + d$
$a + c$	$b + d$	n

From a two-way table one can systematically compute the expected frequencies (based on the null hypothesis of independence) and compare them with the observed frequencies. For example, the expected frequency for cell d (bacteria, alive) would be

$$\hat{f}_{\text{bact,alive}} = n\hat{p}_{\text{bact,alive}} = n\hat{p}_{\text{bact}} \times \hat{p}_{\text{alive}} = n\left(\frac{c + d}{n}\right)\left(\frac{b + d}{n}\right)$$

$$= \frac{(c + d)(b + d)}{n}$$

which in our case would be (54)(73)/111 = 35.5, a value higher than the observed frequency, 29. In general, *to compute the expected frequencies for a cell in the table, multiply its row total by its column total and divide the product by the grand total.* The expected frequencies can also be displayed conveniently in the form of a two-way table:

		Dead	Alive	$\sum$
Bacteria and antiserum		19.514	37.486	57.000
Bacteria only		18.486	35.514	54.000
	$\sum$	38.000	73.000	111.000

The row and column sums of this table are identical to those in the table of observed frequencies, because the expected frequencies were computed on the basis of these same row and column totals. It should therefore be clear that a test of independence will not test whether a property occurs at a given proportion but can test only whether the two properties are manifested independently.

Before we turn to the actual tests of significance, we need to consider different models based on the design of the experiment or sampling procedure. These models relate to whether the totals in the margins of the two-way table are fixed by the investigator or are free to vary and reflect population parameters. In the example of the trees growing in two types of soils and exhibiting two types of leaves, only the total sample size n was fixed; 100 trees had been sampled. The proportion of trees growing in serpentine soils was an outcome of the sampling experiment and was not under the control of the investigator. Similarly, the proportion of leaves that were pubescent was not fixed by the investigator but was an outcome of the experiment. For a test of the null hypothesis one should compute the probability of obtaining the frequencies observed in the 2×2 table or worse departures from independence out of all possible 2×2 tables with the same total sample size n. We will call this design the Model I for a two-way table.

There is another type of Model I, in which the same individuals are subjected to two successive tests. Again only the total sample size is fixed, but the hypotheses to be tested are different. This type of Model I is discussed in Section 17.7.

A second design, Model II, would have the margin totals for one of the two criteria fixed. An example is the immunology experiment that we just discussed in detail. The number of mice given bacteria plus antiserum and the number given only bacteria were fixed by the experimenter at 57 and 54 animals, respectively. The proportion that survived the experiment was free to vary, depending on the effect of the treatment. In testing the null hypothesis of independence one should compute the probability of getting the observed results or worse departures from independence out of all possible two-way tables with the same marginal totals for the two treatment classes.

The moth example cited at the beginning of this section is another instance of a Model II test for independence. The investigator released 50 specimens of each color phase; hence the marginal totals for color phase were fixed by the experiment. The proportion that were prey depended on the natural forces involved and was free to vary.

In a third model for the design of a 2×2 table, both margin totals are fixed by the experiment. We have not yet encountered an example of this model. Assume that we wish to investigate the preference of weevil larvae for two types of beans differing in seed coats. The investigator places a fixed number of beans (100) in a jar—50 of type A and 50 of type B. Then 70 first-instar larvae are added to the jar. At these densities no more than one larva will attack any one bean. After enough time has elapsed to permit each larva to enter a bean, a count is made of how many beans of each type have been attacked by a weevil larva. Again, we have four frequency classes of beans: A–attacked, B–attacked, A–not attacked, and B–not attacked. The number of beans of both types and the total number of beans that were attacked were both fixed by the experimenter. (The number attacked was equal to the number of larvae released in the jar.) When testing the null hypothesis of independence in this model, which we call Model III, we should compute the probability of obtaining the observed frequencies and all worse departures from independence out of all possible 2×2 tables with the same margin totals for both criteria—bean type and presence or absence of attacking larvae.

The tests of independence proposed for these 2×2 tables are of three kinds: the G-test, an exact probability test known as Fisher's exact test, and the chi-square test. Conventional formulations of the first two tests were designed originally for specific models: The G-test, based on Expression (17.13) (and illustrated later in Box 17.6), was intended for Model I; Fisher's exact test [see Expression (17.14) and Box 17.7] was intended for Model III. The chi-square test, which has been the most common test in the past, was not designed for any particular model. The designs represented by the three models have not been clearly distinguished in textbooks of statistics, and research workers have been applying each of the tests to all three models. The criteria for choosing among the tests have not been the model of the design, but rather the sample size and the computational effort involved. These criteria are no longer relevant. Fortunately, the different techniques seem to provide similar results even when applied to the inappropriate model.

Our recommendations for analyzing 2 × 2 tables are summarized in the following table:

Model	Totals	Recommended test
I	not fixed	*G*-test of independence
II	fixed for one criterion	*G*-test of independence
III	fixed for both criteria	Fisher's exact test

The **G-test of independence** tests the goodness of fit of the observed cell frequencies to their expected frequencies. The equations for the standard *G*-test of independence were designed for Model I and are based on a **multinomial distribution,** which is a discrete probability distribution that is a generalization of the binomial distribution to the case where an attribute has more than two classes. We do not discuss the multinomial distribution in detail; interested readers should refer to Ostle (1963; Section 5.12). The probability of observing the cell frequencies a, b, c, and d, assuming a multinomial distribution, is computed as follows:

$$\frac{n!}{a!b!c!d!}\left(\frac{a}{n}\right)^a\left(\frac{b}{n}\right)^b\left(\frac{c}{n}\right)^c\left(\frac{d}{n}\right)^d \qquad (17.9)$$

If we assume that the row and column classifications are independent, this expression changes to

$$\left[\frac{n!}{a!b!c!d!}\left(\frac{(a+b)(a+c)}{n^2}\right)^a\left(\frac{(a+b)(b+d)}{n^2}\right)^b\right] \times \left[\left(\frac{(a+c)(c+d)}{n^2}\right)^c\left(\frac{(b+d)(c+d)}{n^2}\right)^d\right] \qquad (17.10)$$

If we compute the natural logarithm of the ratio of these two probabilities or likelihoods, the result is

$$\ln\left[\frac{a^a b^b c^c d^d n^n}{(a+b)^{a+b}(a+c)^{a+c}(b+d)^{b+d}(c+d)^{c+d}}\right] = a\ln a + b\ln b + c\ln c + d\ln d + n\ln n - (a+b)\ln(a+b) - (a+c)\ln(a+c) - (b+d)\ln(b+d) - (c+d)\ln(c+d) \qquad (17.11)$$

Multiplying Expression (17.11) by 2 produces G, which is approximately distributed as χ^2 with one degree of freedom. In this manner we compute G directly without having to calculate expected frequencies ($\hat{f}$) first.

In this case also the actual type I error of the *G*-test tends to be higher than the intended level. For this reason the Williams correction is recommended (see

Section 17.2). (Its formula for a 2×2 table is given in Box 17.6.) An alternative correction, frequently recommended with small samples is Yates's correction for continuity, which, as we saw in Section 17.2, adjusts the observed frequencies by adding or subtracting 0.5 in such a way as to reduce each $|f_i - \hat{f}_i|$. In a 2×2 test of independence we add 0.5 to cells a and d and subtract 0.5 from cells b and c when the quantity $ad - bc$ is negative. If this quantity is positive we subtract 0.5 from cells a and d and add 0.5 to cells b and c. Because this correction results in a very conservative test, however, we prefer Williams's correction.

The observed G should be compared with χ^2 for one degree of freedom for reasons that we will examine shortly.

Box 17.6 illustrates the G-test of independence (using the Williams and the Yates corrections) applied to the plant ecology sampling experiment (discussed earlier in this section) in which trees with one of two types of leaves were rooted in one of two different soils. The result of the analysis shows clearly that we cannot reject the null hypothesis of independence between soil type and leaf type. The presence of pubescent leaves is independent of whether or not the tree is rooted in serpentine soils.

The advent of easy, inexpensive computing has made it possible to solve Model I problems exactly rather than by the G-test, which is only an approximation. The program StatXact (1991) evaluates the probability of the observed frequencies a, b, c, and d and all less probable ones occurring if the row and column classifications are independent, using the multinomial distribution. When we subject the plant ecology data to this program, we obtain a probability of $P = 0.3471$ almost instantly. This exact procedure is, of course, preferable to the G-test, which yields a slightly more liberal result ($P = 0.2484$ in this case).

The method designed especially for a Model III 2×2 table is **Fisher's exact test.** The computation is based on the hypergeometric distribution (see Section 5.4) with four classes. These probabilities are computed assuming that the row and column classifications are independent (the null hypothesis) and that the row and column totals are fixed. Fisher's exact test answers the following question: Given two-way tables with the same fixed margin totals as the observed one, what is the chance of obtaining the observed cell frequencies a, b, c, and d and all cell frequencies that represent a greater deviation from expectation?

The total number of ways in which a two-way table with fixed margin totals can be obtained is

$$\binom{n}{a+b}\binom{n}{a+c} = \frac{n!}{(a+b)!(c+d)!} \times \frac{n!}{(a+c)!(b+d)!}$$

This total is simply the product of the number of ways of taking $(a + b)$ items from n, multiplied by the number of combinations of $(a + c)$ items taken from n. From the coefficients of the multinomial distribution it can be shown that there are $n!/a!b!c!d!$ ways of obtaining the cell frequencies that we observed. Therefore the probability of obtaining a 2×2 table with cell frequencies a, b, c, and d

Box 17.6 *G*-TEST OF INDEPENDENCE.

A plant ecologist samples 100 trees of a rare species from a 400-square-mile area. He records for each tree whether or not it is rooted in serpentine soils and whether its leaves are pubescent or smooth. This represents a Model I design (only n is fixed).

Soil	Pubescent	Smooth	Totals
Serpentine	12	22	34
Not Serpentine	16	50	66
Totals	28	72	$100 = n$

The conventional algebraic representation of this table is as follows:

			$\sum$
	a	b	$a + b$
	b	d	$c + d$
$\sum$	$a + c$	$b + d$	$a + b + c + d = n$

Compute the following quantities:

1. $\sum f \ln f$ for the cell frequencies

$$= 12 \ln 12 + 22 \ln 22 + 16 \ln 16 + 50 \ln 50$$
$$= 337.78438$$

2. $\sum f \ln f$ for the row and column totals

$$= 34 \ln 34 + 66 \ln 66 + 28 \ln 28 + 72 \ln 72$$
$$= 797.63516$$

3. $n \ln n = 100 \ln 100 = 460.51702$

4. Compute G by multiplying Expression (17.11) by 2:

$$G = 2[\text{quantity } \mathbf{1} - \text{quantity } \mathbf{2} + \text{quantity } \mathbf{3}]$$
$$= 2[337.78438 - 797.63516 + 460.51702]$$
$$= 2[0.66624] = 1.33249$$

Williams's correction

Williams's correction for a 2×2 table is

$$q = 1 + \frac{\left(\frac{n}{a + b} + \frac{n}{c + d} - 1\right)\left(\frac{n}{a + c} + \frac{n}{b + d} - 1\right)}{6n}$$

BOX 17.6 CONTINUED

For these data we obtain

$$q = 1 + \frac{\left(\frac{100}{34} + \frac{100}{66} - 1\right)\left(\frac{100}{28} + \frac{100}{72} - 1\right)}{6(100)}$$

$$= 1.02281$$

$$G_{adj} = G/q = 1.33249/1.02281 = 1.30277$$

Compare G_{adj} with the critical value of χ^2 for one degree of freedom. Since our observed G_{adj} is much less than $\chi^2_{.05[1]} = 3.841$, we accept the null hypothesis that the leaf type is independent of the type of soil in which the tree is rooted.

Yates's correction for continuity (an alternative correction)

Adjust the frequencies in the table as follows. If $ad - bc$ is positive, as it is in our example, since $(12 \times 50) - (16 \times 22) = 248$, subtract $\frac{1}{2}$ from a and d and add $\frac{1}{2}$ to b and c. If $ad - bc$ were negative, we would add $\frac{1}{2}$ to a and d and subtract $\frac{1}{2}$ from b and c. The new 2×2 table is

Soil	Pubescent	Smooth	Totals
Serpentine	$11\frac{1}{2}$	$22\frac{1}{2}$	34
Not serpentine	$16\frac{1}{2}$	$49\frac{1}{2}$	66
Totals	28	72	100

Compute the following quantities:

1. $\sum f \ln f$ for the cell frequencies

$$= 11\tfrac{1}{2} \ln 11\tfrac{1}{2} + 22\tfrac{1}{2} \ln 22\tfrac{1}{2} + 16\tfrac{1}{2} \ln 16\tfrac{1}{2} + 49\tfrac{1}{2} \ln 49\tfrac{1}{2}$$
$$= 337.54418$$

Quantities **2** and **3** are the same as those earlier in this box.

4. Compute G_{adj} by multiplying Expression (17.11) by 2:

$$G_{adj} = 2[\text{quantity } \mathbf{1} - \text{quantity } \mathbf{2} + \text{quantity } \mathbf{3}]$$
$$= 2[337.54418 - 797.63516 = 460.51702] = 2[0.42604] = 0.85208$$

Compared with the critical value of $\chi^2_{.05[1]}$, the observed G_{adj} is not significant.

is computed as the ratio of the two quantities given above. This expression can be simplified to

$$P = \frac{(a + b)!(c + d)!(a + c)!(b + d)!}{a!b!c!d!n!} \tag{17.12}$$

We must then compute the probability of all "worse" cases. The simplest method is to list systematically all the worse cases, compute the probability of each, and sum these probabilities. Since usually we are interested in a two-tailed test, we should consider all worse cases in both tails, which is what the BIOM-pc computer program does. When the computations are carried out by pocket calculator, it may be convenient to compute the probability for the tail that contains the observed result and double this probability. Computation using a calculator can be simplified somewhat by noting that the numerator (products of factorials of the marginal totals) and $n!$ in the denominator are constant for all cases. Only the frequencies a, b, c, and d change in each table.

Fisher's exact test is illustrated in Box 17.7, which shows the results of an experiment with acacia trees and ants. All but 28 trees of two species of acacia were cleared from an area in Central America. There were 15 trees of species A and 13 of species B. These trees had been freed from ants by insecticide. Next 16 separate colonies of ant species X, obtained from cut-down trees of species A in an area nearby, were brought in. The colonies were situated roughly equidistant to the 28 trees and were permitted to invade them. The number of trees of each species that was reinfested by ants was recorded. This experiment is clearly a Model III problem, since the total number of trees of the two species and the number of infested trees are both fixed. (The number of infested trees is fixed because there are only 16 ant colonies and each colony can infest only one tree; hence 12 trees must remain free of ants. No more than one colony will invade a single tree.) From the results it appears that the invasion rate is much higher for the acacia species A, in which these ants had lived before; there were $(13)(100)/15 = 86.7\%$ invasions of species A, contrasted with $(3)(100)/13 = 23.1\%$ invasions of species B. We conclude in Box 17.7 that the probability of obtaining as bad a fit to expectation as shown in these data (or worse) is approximately 0.002. We therefore reject the null hypothesis of independence and conclude that the difference in invasion rate is related to the host species, acacia species A being more attractive than species B.

The computations can be long if the sample is large. When the lowest cell frequency in the table is greater than 10, the method is impractical without a computer. When no margin total is greater than 15 (and consequently the total sample size is no greater than 30), we do not have to carry out the computations but can look up the probabilities in published tables. Table 38 of the *Biometrika Tables* (Pearson and Hartley, 1958) serves this purpose; it appears in slightly simpler form as Table I of Siegel (1956). More extensive tables were published by Finney et al. (1963). Instructions for the use of these tables, which are somewhat complicated, can be found in the cited sources.

Box 17.7 FISHER'S EXACT TEST OF INDEPENDENCE.

Invasion rate of two species of acacia trees by ant colonies. For details of this experiment see text.

Acacia species	Not invaded	Invaded by ant colony	Totals
A	2	13	15
B	10	3	13
Totals	12	16	28

The conventional algebraic representation of this table is as follows:

			$\sum$
	a	b	$a + b$
	c	d	$c + d$
$\sum$	$a + c$	$b + d$	$a + b + c + d = n$

Computation

We employ Expression (17.12). Many pocket and tabletop calculators now have factorial keys that compute factorials up to the capacity of the machine, usually $\leq 69!$ When the numbers in a 2×2 table are in excess of this value or when a factorial key is not available, employ a table of common logarithms of factorials, such as Table **1** in the 2nd edition of the Rohlf and Sokal *Statistical Tables*.

1. Direct computation of Expression (17.12) yields $P = 0.000{,}987{,}12$. Assuming independence of the invasion rate and the proportion of the two species of acacia trees, as well as fixed margin totals, this value is the probability of obtaining the results of the experiment by chance.

Because factorials yield such larger numbers, however, it is often more convenient to evaluate Expression (17.12) in terms of logarithms. Most computer programs employ this device transparent to the user. Here we show how to evaluate the probability using common logarithms.

Compute the sum of transforms of the margin totals and subtract from it the transform of the total frequency:

$$
\begin{aligned}
\log (a+b)! &= \log 15! = 12.11650\\
\log (c+d)! &= \log 13! = 9.79428\\
\log (b+d)! &= \log 12! = 8.68034\\
\log (a+c)! &= \log 16! = 13.32062\\
-\log n! &= -\log 28! = -29.48414\\
& \qquad\qquad\quad 14.42760
\end{aligned}
$$

BOX 17.7 CONTINUED

2. In a manner similar to step **1** compute the sum of the transforms of the observed cell frequencies:

$$\log a! = \log\ 2! = 0.30103$$
$$\log b! = \log 13! = 9.79428$$
$$\log c! = \log 10! = 6.55976$$
$$\log d! = \log\ 3! = \underline{0.77815}$$
$$17.43322$$

3. Subtract quantity **2** from quantity **1** and compute the antilogarithm of the result. The end result will be a probability (which must be ≤ 1).

$$14.42760 - 17.43322 = -3.00563$$

$$\text{antilog}\ -3.00563 = 0.000{,}987{,}12$$

Next we compute the probability of all worse cases.

4. If $ad - bc$ is negative (as in the present example), decrease cells a and d by one and add unity to cells b and c. In this new 2×2 table the cell frequencies depart even more from expected frequencies (on the assumption of independence) than do the experimental results.

Acacia species	Not invaded	Invaded by ant colony	Σ
A	1	14	15
B	11	2	13
Σ	12	16	28

Note that the margin totals have remained the same. Thus the numerator of Expression (17.12), quantity **1,** will remain unchanged and we need only compute a new value corresponding to the denominator, quantity **2.**

If $ad - bc$ had been positive, we would have decreased cells b and c by unity, and increased cells a and d by the same amount.

By direct computation we obtained $P = 0.000{,}038{,}46$. To save space we do not show steps **2** and **3** worked out in terms of logarithms.

5. Step **4** is repeated until one of the cell frequencies reaches zero. In the present example this happens on the very next iteration.

Acacia species	Not invaded	Invaded by ant colony	Σ
A	0	15	15
B	12	1	13
Σ	12	16	28

By direct computation, $P = 0.000{,}000{,}4273$.

BOX 17.7 CONTINUED

6. Assuming independence of the invasion rate and the proportions of the two species of acacia trees, as well as fixed margin totals, the total probability of obtaining the results of our experiment and all less likely outcomes is obtained by summing the probabilities computed in steps **3, 4,** and **5** and multiplying this sum by 2 (for a two-tailed test).

$$P = 2(0.000{,}987{,}12 + 0.000{,}038{,}46 + 0.000{,}000{,}43)$$
$$= 2(0.001{,}026{,}01) = 0.002{,}052{,}02$$

The probability computed in this manner is still not "exact" because the distribution of the probabilities associated with the possible outcomes is not symmetrical. The correct, but more tedious procedure is to calculate the cumulative probability for the second tail of the distribution in a manner analogous to the computation shown in this box.

We start with the most extreme case by setting to zero the smaller of the two frequencies b or c (with $ad - bc$ negative) or a or d (with $ad - bc$ positive). The other cell frequencies are adjusted to preserve the margin totals. Compute the probability of the second tail as in steps **2** and **3.** Continue the calculation by repeating step **4** until the probability of a resulting outcome exceeds that obtained in step **3** for the first tail. The probability of obtaining the observed results assuming independence (for a two-tailed test) is the sum of the probabilities obtained from steps **3** through **5** earlier, plus the sum of the newly computed probabilities for the second tail, *excluding the last one,* which exceeded the P-value for step **3** in the first tail. In this example the cumulative probability for the second tail is 0.000,598,3 and the overall probability for the two-tailed test thus is 0.000,598,3 + 0.001,026 = 0.001,624,3. The discrepancy between the exact solution and the conservative value obtained by doubling the one-tailed probability increases with the difference between the two row or column totals.

Based on these results, we reject the null hypothesis that invasion rate is independent of the host species (at the 0.2% level of significance).

If no computer is accessible the G-test can be used to yield approximate probabilities when sample sizes are so large that calculator operation is impractical. When n is smaller, as in Box 17.7, the approximation is not so good. In this example G_{adj} yields a probability of 0.000,698,68, whereas Fisher's exact test yields a P-value of 0.001,624,3.

The traditional method for testing independence in a 2 × 2 table is the **chi-square test of independence.** To carry out such a test we could compute expected frequencies as shown earlier and apply Expression (17.8), but there is a much simpler way. The equation

$$X^2 = \frac{(ad - bc)^2 n}{(a + b)(c + d)(a + c)(b + d)} \tag{17.13}$$

can be derived from Expression (17.6). For the immunology experiment with which we introduced this section, this expression would yield

$$X^2 = \frac{[(13 \times 29) - (44 \times 25)]^2 \times 111}{57 \times 54 \times 38 \times 73} = \frac{58{,}022{,}919}{8{,}538{,}372} = 6.79555$$

Yates's adjustment for continuity is contained in the following expression:

$$X^2_{\text{adj}} = \frac{\left(|ad - bc| - \frac{n}{2}\right)^2 n}{(a + b)(c + d)(a + c)(b + d)} \qquad (17.13\text{Y})$$

Grizzle (1967) has shown that applying the Yates's correction to chi-square tests with Models I and II almost always results in an unduly conservative test (the type I error is much lower than desired). Yates's correction therefore seems unnecessary even with quite small samples, such as $n = 20$. Applying Expression (17.13Y) to the immunology experiment yields $X^2_{\text{adj}} = 5.79229$, substantially lower than 6.7955. Nevertheless, since $X^2_{\text{adj}} > X^2_{.025[1]}$, we conclude that mortality in these mice is not independent of the presence of antiserum. Note that the percentage mortality among animals given bacteria *and* antiserum is (13)(100)/57 = 22.8%, considerably lower than the mortality of (25)(100)/54 = 46.3% among mice to whom only bacteria were administered. Clearly, the antiserum has been effective in reducing mortality.

Tests of independence need not be restricted to 2 × 2 tables. In the two-way cases considered in this section, we are concerned with only two properties, but each of these properties may be divided into any number of classes. For example, organisms may occur in four color classes and be sampled from three localities, requiring a 4 × 3 test of independence. Such a test would examine whether the color proportions exhibited by the margin totals are independent of the localities at which the individuals were sampled. Such tests are often called **R × C tests of independence,** *R* and *C* standing for the number of rows and columns in the frequency table, respectively. Another example, examined in detail in Box 17.8, concerns bright red color patterns found in samples of a species of tiger beetle on four occasions during spring and summer. Of interest was whether the percentage of bright red individuals changed significantly during the time of observation. We therefore test whether the proportion of beetles colored bright red (55.7% for the entire study) is independent of the time of collection.

As shown in step **5** of Box 17.8, the following is a simple general rule for computing the *G*-test of independence:

$$\begin{aligned} G = 2[&(\textstyle\sum f \ln f \text{ for the cell frequencies}) \\ &- (\textstyle\sum f \ln f \text{ for the row and column totals}) + n \ln n] \end{aligned}$$

You can compute these quantities directly. We can also apply the Williams correction to estimates of *G* from *R* × *C* tables. The formula, now more complicated, is given in Box 17.8. The adjustment in *G* is minor when the sample is

Box 17.8 $R \times C$ TEST OF INDEPENDENCE USING G-TEST.

Frequencies of color patterns of a species of tiger beetle *(Cicindela fulgida)* in different seasons.

Season	Color pattern ($a = 2$)			
($b = 4$)	Bright red	Not bright red	Totals	% Bright red
Early spring	29	11	40	72.5
Late spring	273	191	464	58.8
Early summer	8	31	39	20.5
Late summer	64	64	128	50.0
Totals	374	297	$671 = n$	55.7

SOURCE: (Data from H. L. Willis unpublished results).

Computation

1. Sum of transforms of the frequencies in the body of the contingency table

$$= \sum^{b}\sum^{a} f \ln f = 29 \ln 29 + 11 \ln 11 + \cdots + 64 \ln 64 = 3314.02462$$

2. Sum of transforms of the row totals $= \sum^{b}\left(\sum^{a} f\right) \ln\left(\sum^{a} f\right)$

$$= 40 \ln 40 + \cdots + 128 \ln 128 = 3760.40039$$

3. Sum of the transforms of the column totals $= \sum^{a}\left(\sum^{b} f\right) \ln\left(\sum^{b} f\right)$

$$= 374 \ln 374 + 297 \ln 297 = 3906.71011$$

4. Transform of the grand total $= n \ln n = 671 \ln 671 = 4367.38409$

5. $G = 2[\text{quantity } \mathbf{1} - \text{quantity } \mathbf{2} - \text{quantity } \mathbf{3} + \text{quantity } \mathbf{4}]$

$$= 2[3314.02462 - 3760.40039 - 3906.71011 + 4367.38409] = 2[14.29821]$$

$$= 28.59642$$

6. Williams's correction for an $R \times C$ table is

$$q = 1 + \frac{\left(n \sum^{b} \frac{1}{\sum^{a} f} - 1\right)\left(n \sum^{a} \frac{1}{\sum^{b} f} - 1\right)}{6n(a-1)(b-1)}$$

$$= 1 + \frac{\left[671\left(\frac{1}{40} + \cdots + \frac{1}{128}\right) - 1\right]\left[671\left(\frac{1}{374} + \frac{1}{297}\right) - 1\right]}{6(671)(1)(3)}$$

$$= 1.01003$$

BOX 17.8 CONTINUED

A simple (lower-bound) estimate of q is given by

$$q_{\text{min}} = 1 + \frac{(a+1)(b+1)}{6n}$$

$$= 1 + \frac{(2+1)(4+1)}{6(671)}$$

$$= 1.00373$$

Using the more exact value for q,

$$G_{\text{adj}} = G/q = 28.59642/1.01003 = 28.312$$

We compare this value with a χ^2-distribution with $(a - 1)(b - 1)$ degrees of freedom, where a is the number of columns and b the number of rows in the table. In our case, $df = (2 - 1)(4 - 1) = 3$.

Since $\chi^2_{.001[3]} = 16.266$, our G-value is significant at $P \ll 0.001$, and we must reject the null hypothesis that frequency of color pattern is independent of season.

large, as in this example; when computing by hand, the correction is necessary only when total sample size is small and the observed G-value is of marginal significance.

The results in Box 17.8 show clearly that the frequency of bright red color patterns in these tiger beetles is dependent on the season. We note a decrease of bright red beetles in late spring and early summer, followed by an increase again in late summer. To find out whether the differences between successive points in time are significant, we could subject these data to unplanned tests similar to those in Box 17.5. In such an analysis we would test the independence of selected subsets of the data.

You may have wondered about the relationship between the replicated goodness-of-fit tests in Box 17.4 and the $R \times C$ tests discussed here. The computational outline for the heterogeneity statistic, G_H, shown in Step **2** of Box 17.4, is the same as that for the test statistic G in the $R \times C$ test (step **5** of Box 17.8). Are these quantities identical? Computationally speaking, they are. In fact, we could obtain the G-value of an $R \times C$ test of independence by running the data as though each row or column were one replicate of a series of replicated goodness-of-fit tests and obtaining the value of G_H. In Box 17.4, however, we were testing an *extrinsic hypothesis,* a 9 : 7 ratio that was applied to all eight replicates (progenies) of the experiment. In the $R \times C$ test of independence (Box 17.8), *the hypothesis is based on the totals in the margin of the table.* Thus we are testing each row of the table as though it were a replicate against the hypothesis that 55.7% of the beetles have the bright red color pattern (and consequently 44.3% do not have it).

The three models defined earlier apply to the $R \times C$ test of independence as well. Thus the example in Box 17.8 is Model I, since the investigator did not limit the number of beetles to be collected at any one season and had no control over the number of beetles in the two color classes. No methods have been devised especially for Models II and III; such cases are analyzed in the manner described above. Note, however, that the StatXact (1991) program evaluates the exact probability, or a Monte Carlo approximation thereof, for a Model I or Model III $R \times C$ test in a reasonably short time.

The *degrees of freedom for G-tests and chi-square tests of independence* are always the same and can be computed using the rules given earlier (see Section 17.2). There are k cells in the table, but we must subtract one degree of freedom for each independent parameter we have estimated from the data, and we must, of course, subtract one degree of freedom for the observed total sample size, n. We have also estimated $a - 1$ row probabilities and $b - 1$ column probabilities, where a and b are the number of rows and columns in the table, respectively. Thus, there are $k - (a - 1) - (b - 1) - 1 = k - a - b + 1$ degrees of freedom for the test. Since $k = a \times b$, however, by substitution this expression becomes $(a \times b) - a - b + 1 = (a - 1) \times (b - 1)$, the conventional expression for the degrees of freedom in a two-way test of independence. Thus, the degrees of freedom in the example of Box 17.8, a 4×2 case, was $(4 - 1) \times (2 - 1) = 3$. In all 2×2 cases there is clearly only $(2 - 1) \times (2 - 1) = 1$ degree of freedom.

Another name for test of independence is **test of association**. If two properties are not independent of each other they are *associated*. Thus, in the example testing relative frequency of two leaf types on two different soils we can speak of an association between leaf types and soils. In the immunology experiment there is a negative association between presence of antiserum and mortality. **Association** is thus similar to correlation, but it is a more general term, applying to attributes as well as to continuous variables.

The computations described in this section can be performed by the BIOM-pc package of computer programs—both the Fisher's exact test, and the $R \times C$ test for independence using the G-test.

Suggestions for graphic representation of two-way contingency tables have been made by Snee (1974). He depicts $R \times 2$ tables by bars representing the proportions of one of the binary properties with confidence limits superimposed on the top of each bar. The confidence limits of the proportions are evaluated as discussed at the end of Section 13.12. For $R \times 3$ tables an approximate $100(1 - \alpha)\%$ confidence region for the parameters of the multinominal distribution of any given row is the set of points $\hat{p}_1, \hat{p}_2, \hat{p}_3$ satisfying the equation

$$\frac{p_1^2}{\hat{p}_1} + \frac{p_2^2}{\hat{p}_2} + \frac{p_3^2}{\hat{p}_3} \leq \frac{\chi^2_{\alpha[2]}}{n} + 1$$

This region can be represented on triangular graph paper.

It may be of interest to test subsets of an $R \times C$ table for homogeneity. These tests can be carried out in a manner entirely analogous to the unplanned tests in Box 17.5. We try to find the set of largest homogeneous tables, starting with various (all, if possible) smallest tables, which of course are 2×2 tables. Those that prove homogeneous are then augmented by another row or column, and testing is continued until addition of another row or column makes the table significant. From the final listing of maximal homogeneous subsets we can construct a scheme of relations among the rows and columns of the overall table. Such decomposition of a sizable $R \times C$ table requires extensive computation and should be done by computer. The BIOM-pc program includes an optional feature for an unplanned test of all subsets of rows and columns in an $R \times C$ contingency table.

In the tests of independence that we have discussed in this section, one way of looking for suspected lack of independence is to examine the percentage occurrence of one of the properties in the two classes based on the other property. For example, we compared the percentage of smooth leaves on the two types of soils, and we studied the percentage mortality with or without antiserum. This way of looking at a test of independence suggests another test of such data. Can we develop a test for the significance of differences between two percentages? Several such tests are found in the statistical literature. In our experience, however, none of these tests can match the 2×2 G-test of independence in ease of computation and/or closeness of the observed type I error to the intended one. For this reason we recommend the 2×2 G-test of independence to test the significance of the difference between two percentages or proportions.

For a divergent opinion, see D'Agostino, Chase, and Belanger (1988). This topic has received considerable attention from statisticians in recent years (Little, 1989; Greenland, 1991), but no consensus on how to proceed has been reached. The argument pivots on the correct model of the 2×2 table to be analyzed—i.e., which margin totals are fixed. With intermediate to large sample sizes, this issue is probably of only academic interest, but if n is small (say, $n < 25$), the selection of the correct model may matter.

A 2×2 contingency table can be thought of as expressing the association between the two nominal criteria that define the margins. In the immunology example with bacteria and antiserum, the table expresses the association between survival and the administration of antiserum. It is not surprising that statisticians have developed a variety of **coefficients of association** for such data. We feature only one of them, the **phi coefficient,** which is defined as

$$\phi = \sqrt{\frac{X^2}{n}} \tag{17.14}$$

where X^2 is the test statistic from a chi-square test of the 2×2 table, unadjusted for continuity, and n is the total sample size. The phi coefficient ranges between 0 and 1, and the positive or negative direction of the association can

Box 17.9 TESTING FOR EQUALITY OF MEDIANS BY TWO-WAY TESTS OF INDEPENDENCE.

Two samples

Length of cheliceral base in chigger nymphs. (Data from Box 13.7.) Overall median of 26 items = 118 or rank 13.5.

	Sample A	Sample B	$\sum$
Below median	5	7	12
Above median	9	3	12
$\sum$	14	10	24

A little experimentation will show that the marginal totals of this table are fixed, so we proceed to Fisher's exact test (see Box 17.7), which yields a P-value of 0.1069. The two medians cannot be shown to differ.

More than two samples

Length of pea sections. (Data from Box 13.6.) Overall median of 50 items = 60.5 or rank 25.5.

	Treatments					
	Control	2% glucose	2% fructose	1% glucose + 1% fructose	2% sucrose	$\sum$
Below median	0	8	8	9	0	25
Above median	10	2	2	1	10	25
$\sum$	10	10	10	10	10	

Although the margin totals of this table are fixed (it is a Model III $R \times C$ test), we have not learned an exact test of the probability of independence. We therefore evaluate $G = 42.797$ as an approximation. Since there are $(5 - 1)(2 - 1) = 4$ degrees of freedom, $P \ll 0.0001$, and we reject the null hypothesis of independence. We conclude that the medians of the five samples are not the same.

There now exist computer programs that evaluate the exact probability of an $R \times C$ table. When we subjected the pea section table to the FIsher option of the StatXact (1991) program, we obtained $P = 0.1602 \times 10^{-9}$. The conclusion remains the same.

be ascertained from the sign of the determinant of the table, $ad - bc$. By substituting Expression (17.13) into Expression (17.14), we obtain a direct computational formula for phi:

$$\phi = \frac{(ad - bc)}{\sqrt{(a + b)(c + d)(a + c)(b + d)}} \tag{17.15}$$

This formulation ranges from -1 to $+1$, automatically attaching a sign to the index. Evaluated for the bacteria and antiserum example, Expression (17.15) yields

$$\phi = \frac{(13 \times 29 - 25 \times 44)}{\sqrt{57 \times 54 \times 38 \times 73}} = -0.247$$

Because of the arrangement of the table, the association is negative. The phi coefficient is only moderately strong. Specifically, the coefficient says that the administration of only bacteria is negatively related to survival in the experiment. If all antiserum-treated mice had survived and none of the bacteria-only group were left alive, there would be perfect association. Substitution of $a = d = 0$ into Expression (17.15) would yield -1.

Before we proceed to three-way tables, we should mention one more application of the two-way table. We showed in Chapter 13 how, if the assumptions of the analysis of variance are not met, we can carry out tests on ranks—the Mann–Whitney U-test for two samples and the Kruskal–Wallis test for several samples. As we mentioned earlier, there are tests of location applied to ranked data. A (weaker) test only of medians can be constructed as follows. For two or more samples, evaluate the overall median of all samples. Then record for each sample the number of observations below and above this overall median. Omit observations lying exactly on the median. Finally, test these tables for independence by the G-test. Rejection of the null hypothesis implies that the proportions to each side of the overall median differ among the samples; hence the medians of the samples also differ. Box 17.9 shows this test applied to two examples.

17.5 Analysis of Three-way and Multiway Tables

As with anovas where we proceeded from two-way classifications to three- and four-way tables, in analysis of frequencies we can turn from a test of independence in a two-way table to tests involving tables of three, four, or more dimensions. Unfortunately, the analysis of higher-dimensional tables is much more complicated than that of two-way tables, since we must test for more-complex patterns of associations among various combinations of the factors.

In recent years the analysis of contingency tables has been based upon **log-linear models,** which are the counterpart for attribute data to the linear model for continuous variables in analysis of variance and multiple regression. For a two-way contingency table we have (by analogy to a two-way anova) the model

$$\ln \hat{f}_{ij} = \mu + \alpha_i + \beta_j + \alpha\beta_{ij} \tag{17.16}$$

where $\hat{f}_{ij}$ is the expected frequency in row i, column j of the two-way contingency table, μ is the mean of the logarithms of the expected frequencies, α_i and β_j are the effects of categories i and j of factors A and B, respectively, and the $\alpha\beta_{ij}$ interaction term expresses the dependence of category i of factor A on category j of factor B. For a three-way table the log-linear model is

$$\ln \hat{f}_{ijk} = \mu + \alpha_i + \beta_j + \gamma_k + \alpha\beta_{ij} + \alpha\gamma_{ik} + \beta\gamma_{jk} + \alpha\beta\gamma_{ijk} \tag{17.17}$$

In contrast to the analysis of variance (where the main effects are of the most interest), we are interested primarily in testing for the presence of interactions. Thus, in the two-way contingency tables of the previous section we tested the hypothesis that interaction is not present ($H_0: \alpha\beta_{ij} = 0$ for all ij), since this is a test of the independence of factors A and B. Testing for the significance of a term in the log-linear model requires, in theory, fitting two models, one with the term present and one with it omitted. The G-statistic for goodness of fit is computed for each of the two models, and the difference between the G-values is used to test the significance of the term being left out.

In many cases we can evaluate the G-statistics directly without having to compute the expected frequencies explicitly. It is a good practice, however, to compute the expected frequencies routinely, since this process permits examination of the deviations between the observed and the expected frequencies for each cell. The pattern of deviations may suggest additional models that should be tested (as Box 17.10 later shows).

The expected frequencies for any model are functions of the smallest set of marginal tables (one-way, two-way, and higher-dimensional tables of marginal totals) sufficient to compute these frequencies. The marginal tables in this minimal set are called configurations by Bishop, Fienberg, and Holland (1975). Thus, for example, the expected frequencies when testing for independence in a two-way table are a function of the products of the row and column sums. The test for the three-way interaction in a three-way table is a function of the AB, BC, and AC two-way tables. To find the minimal set of marginal totals needed, we examine the log-linear model to be fitted. Starting with the highest-order terms in the model, we form a set consisting of marginal tables corresponding to the terms in the model. If, for example, the term $\alpha\beta_{ij}$ is present, we include the AB two-way table in the minimal set. Such a table would contain observed frequencies f_{ij+}. This symbol stands for the observed frequencies in row i and column j summed over all the depths. If the marginal tables already in the minimal set correspond to all the terms in the model, no additional tables are needed. If there are still terms in the model that lack corresponding margin tables, however, we next consider lower-order terms and add marginal tables to the set until all terms in the model are accounted for. Note that marginal tables need not be added if they are redundant (i.e., the marginal totals can be obtained from marginal tables already in the set). For the model $\ln \hat{f}_{ijk} = \mu + \alpha_i + \beta_j + \gamma_k + \alpha\beta_{ij} + \beta\gamma_{jk}$ the AB and BC margin tables would suffice because from these we can obtain the $\alpha\beta$ and $\beta\gamma$ totals, as well as the α, β, and γ totals. If the $\beta\gamma$ term were not present in the model, the marginal tables AB and C would suffice to estimate it.

An iterative proportional-fitting algorithm can always be used to compute the $\hat{f}$. This algorithm consists simply of an iterative scaling of an initial estimate of $\hat{f}$ until its marginal totals agree with those in the minimal set of marginal totals used for the model. For some models $\hat{f}$ can be estimated directly. In such cases, when the iterative procedure is used, it will converge after only a single iteration. In the analysis of three-way tables (the only design taken up in detail in this section) only the test for the ABC interaction requires the iterative procedure. All other estimated frequencies in a three-way table can be estimated directly. Analyses of four-way and higher-dimensional tables require iterative solutions for many of the tests. These analyses are not practical without a computer.

The example in Box 17.10 is an analysis of an experiment in which adult drosophila were classified according to three factors: sex, whether they were healthy or poisoned, and the location in which they had pupated. In this experiment drosophila larvae were reared in a medium containing DDT. It was of interest to learn whether the two sexes had different tolerances to the poison and also whether the site where the fly had pupated (in the medium, at the margin of the medium, on top of the medium, or away from the medium on the wall of the vial) affected its chance for subsequent survival.

In the analysis of a multiway table we test a hierarchy of models, starting with the most complex. For example, as in a Model I three-way anova, in a three-way table (see Box 17.10) we first test for the three-factor interaction. This test is done by fitting a model with the $\alpha\beta\gamma$ terms deleted. This is a "partial association" model between each pair of variables, implying that there are pairwise associations between all of the factors but no joint associations between the three factors considered simultaneously. Computing $\hat{f}$ directly is not possible; an iterative proportional-fitting algorithm must be used, as shown in Box 17.10. Only five iterations were needed for that example. The expected frequencies are then compared with observed frequencies using the G-test. The very small G-value obtained (1.365) indicates that we have little evidence for the presence of three-factor interaction, so the $\alpha\beta\gamma$ term can safely be dropped from the model.

If the three-factor interaction term were significant, the degree of association between any pair of variables would depend upon the different levels of a third variable. For example, the significance could mean that the association between factors A and B differs depending on the level of factor C (or that the association between A and C depends upon B or that the association between B and C depends upon A). In such a case, we would not attempt to fit any simpler models, but would make separate two-way tests of independence within each level of one of the factors.

Since the three-factor interaction is not significant for the example of Box 17.10, we test for the significance of the two-factor effects. There are three possible models with one two-factor effect absent. The expected frequencies can

Box 17.10 ANALYSIS OF A THREE-WAY TABLE USING LOG-LINEAR MODELS.

Emerged drosophila are classified according to three factors: pupation site, *A* (in the medium, IM; at the margin of the medium, AM; on the wall of the vial, OW; and on top of the medium, OM); sex, *B*; and mortality, *C* (healthy, H; poisoned, P).

Pupation site ($a = 4$) *A*	Sex ($b = 2$) *B*	Mortality ($c = 2$) *C*: Healthy (H)	Poisoned (P)	Totals
IM	♀	55	6	61
	♂	34	17	51
		89	23	112
AM	♀	23	1	24
	♂	15	5	20
		38	6	44
OW	♀	7	4	11
	♂	3	5	8
		10	9	19
OM	♀	8	3	11
	♂	5	3	8
		13	6	19
Totals		150	44	194

SOURCE: Data from R. R. Sokal (unpublished results).

Sex *B*	Mortality *C*: Healthy	Poisoned	Totals
♀	93	14	107
♂	57	30	87
Totals	150	44	194

The second table is the $B \times C$ two-way table needed for the computations below: the observed frequencies of sex against mortality summed over the four pupation sites. The other two-way tables ($A \times B$ and $A \times C$) are shown as subtotals in the first table. Numerous analyses can be performed on data such as these. Following the suggestion of Bishop, Fienberg, and Holland (1975), we will test a series of hierarchical models

BOX 17.10 CONTINUED

based on the log-linear model:

$$\ln \hat{f}_{ijk} = \mu + \alpha_i + \beta_j + \gamma_k + \alpha\beta_{ij} + \alpha\gamma_{ik} + \beta\gamma_{jk} + \alpha\beta\gamma_{ijk}$$

I. *Test for three-factor interaction (test* $\alpha\beta\gamma = 0$)

This is the only test in a three-way table for which the expected frequencies cannot be calculated directly. One therefore employs the iterative proportional-fitting algorithm to compute the $\hat{f}_{ijk}$, which are to be compared to the f_{ijk} using the G-test.

Computation

The procedure consists of adjusting an initial estimate of the $\hat{f}_{ijk}$ until their two-way margin sums $\hat{f}_{AB+}$, $\hat{f}_{A+C}$, and $\hat{f}_{+BC}$ agree with the observed two-way margin sums f_{AB+}, f_{A+C}, and f_{+BC}. The symbol f_{AB+} stands for the margin totals of an $A \times B$ table whose cell frequencies f_{ij+} for row i and column j are summed over all the depths. An $A \times B$ table can be abstracted from the three-way table given at the beginning of this box. Thus f_{12+}, the observed frequency for pupation site IM and sex ♂ summed over the two mortalities, equals 51. The margin totals for this table are 112, 44, 19, and 19 for pupation site, f_{A++}, and $(61 + 24 + 11 + 11 = 107)$ and $(51 + 20 + 8 + 8 = 87)$ for sex, f_{+B+}. The latter margin frequencies, as well as those for mortality (f_{++C}), are obtained more readily from the $B \times C$ two-way table shown below the three-way table. Here the frequencies f_{+jk} are the observed values of column j and depth k summed over the rows, and the margin totals f_{+BC} represent the totals for factors B and C, respectively. Since the computations are practical only when done by computer, we give merely a general outline of the steps.

1. Computation of the initial estimates for $\hat{f}_{ijk}$. A satisfactory initial estimate is simply

$$\hat{f}^{(0)}_{ijk} = \frac{n}{abc} = \frac{194}{4 \times 2 \times 2} = 12.125$$

for all i, j, and k. The zero superscript refers to thse original estimates.

2. AB adjustments. Form an AB two-way table of $\hat{f}_{ij+}$ values (based on the current estimate of $\hat{f}_{ijk}$) by pooling the $\hat{f}^{(0)}_{ijk}$ values over factor C. Since each cell is the sum of two values of $\hat{f}^{(0)}_{ijk}$ (for the healthy and poisoned categories of Factor C), pooling yields $\hat{f}^{(0)}_{ij+} = 2 \times 12.125 = 24.25$ in each cell.

	B	
A	♀	♂
IM	24.25	24.25
AM	24.25	24.25
OW	24.25	24.25
OM	24.25	24.25

BOX 17.10 CONTINUED

New estimates, $\hat{f}^{(0)'}_{ijk}$, are then computed using the following relationship:

$$\hat{f}^{(0)'}_{ijk} = \hat{f}^{(0)}_{ijk} \frac{f_{ij+}}{\hat{f}^{(0)}_{ij+}}$$

For example,

$$\hat{f}^{(0)'}_{111} = 12.125 \frac{61}{24.25} = 30.5$$

and

$$\hat{f}^{(0)'}_{422} = 12.125 \frac{8}{24.25} = 4.0$$

These computations result in the following three-way table of estimated frequencies.

		C	
A	*B*	H	P
IM	♀	30.5	30.5
	♂	25.5	25.5
AM	♀	12.0	12.0
	♂	10.0	10.0
OW	♀	5.5	5.5
	♂	4.0	4.0
OM	♀	5.5	5.5
	♂	4.0	4.0

3. Similar adjustments are made for the *AC* and then for the *BC* two-way tables.

The resulting three-way table of estimated frequencies at the end of the first iteration is

		C	
A	*B*	H	P
IM	♀	54.632	7.163
	♂	34.231	16.099
AM	♀	23.361	1.871
	♂	14.589	4.192
OW	♀	6.525	2.979
	♂	3.556	5.825
OM	♀	8.483	1.986
	♂	4.623	3.883

Box 17.10 Continued

4. Test for convergence. Next the current $\hat{f}^{(x)}_{ijk}$ values are compared to those at the end of the previous iteration (or the initial estimates $\hat{f}^{(0)}_{ijk}$ the first time we pass through step **4**). If they agree to sufficient accuracy, we can stop; otherwise we must repeat the AB, AC, and BC adjustments described in steps **2** and **3,** using the current $\hat{f}^{(x)}_{ijk}$ values as input to step **2.** Note that if the estimates themselves are of special interest (rather than simply the tests of significance), the iterations should be continued until the maximum $|\hat{f}^{(x)}_{ijk} - \hat{f}^{(x-1)}_{ijk}|$ is less than a tolerance, say 0.001.

Another convenient measure of convergence is the change in the G-statistic computed at the end of each cycle of adjustments. We compute G in the conventional manner:

$$2\sum^{a}\sum^{b}\sum^{c} f_{ijk} \ln\left(\frac{f_{ijk}}{\hat{f}^{(x)}_{ijk}}\right)$$

as shown below. When $|\Delta G|$ is less than, say 0.001, one can stop. For the present data the G and ΔG values and the maximum absolute difference between expected frequencies obtained for five iterations were as follows:

Iteration	G	ΔG	$\max|\hat{f}^{(x)}_{ijk} - \hat{f}^{(x-1)}_{ijk}|$
1	2.193	—	42.507
2	1.392	−0.802	0.884
3	1.366	−0.025	0.120
4	1.365	−0.001	0.023
5	1.365	−0.000	0.005

The convergence was quite rapid (as is typical). The ΔG criterion often requires fewer iterations than does the deviation between expected frequencies. Since G always decreases monotonically, we can also stop when it falls below its critical value. In the present case we could have stopped at iteration 1 because 2.193 is less than $\chi^2_{.05[3]} = 7.815$.

The final table of estimates $\hat{f}^{(5)}_{ijk}$ is shown here at the left, observed frequencies are in the middle, and Freeman–Tukey deviates, described below, are on the right.

		Expected C		Observed C		Freeman–Tukey deviates	
A	B	H	P	H	P	H	P
IM	♀	54.418	6.582	55	6	0.112	−0.132
	♂	34.584	16.417	34	17	−0.057	0.201
AM	♀	22.394	1.605	23	1	0.178	−0.310
	♂	15.607	4.393	15	5	−0.091	0.376
OW	♀	7.313	3.688	7	4	−0.026	0.267
	♂	2.685	5.314	3	5	0.306	−0.032
OM	♀	8.875	2.125	8	3	−0.213	0.650
	♂	4.125	3.875	5	3	0.502	−0.330

Box 17.10 Continued

The G-test for goodness of fit yields

$$G_{ABC} = 2\sum^{a}\sum^{b}\sum^{c} f_{ijk} \ln\left(\frac{f_{ijk}}{\hat{f}_{ijk}}\right) = 1.365$$

which has $(a - 1)(b - 1)(c - 1) = (4 - 1)(2 - 1)(2 - 1) = 3$ degrees of freedom. It is therefore not significant at the $\alpha = 0.05$ level. If it had been significant, we would have used the Williams (1976) correction by dividing G by

$$q_{\min} = 1 + \frac{(a^2 - 1)(b^2 - 1)(c^2 - 1)}{6(a - 1)(b - 1)(c - 1)n} = 1 + \frac{(a + 1)(b + 1)(c + 1)}{6n} = 1.03866$$

The more exact q-value is more complex in this case. We are also reassured about the lack of significance, since none of the individual Freeman–Tukey deviates, $\sqrt{f_{ijk}} + \sqrt{f_{ijk} + 1} - \sqrt{4\hat{f}_{ijk} + 1}$, is large (although there does seem to be a pattern). The deviates can be compared with

$$\sqrt{\frac{\nu\chi^2_{.05[1]}}{abc}} = \sqrt{\frac{3(3.841)}{16}} = 0.849$$

as an approximate criterion for being "large."

A significant G would indicate the presence of a three-way interaction (e.g., the degree of association between sex and mortality differing for different pupation sites). As in a three-way Model I anova, we would then have little interest in testing the two-way interactions and the main effects.

II. *Test for two-factor interactions (conditional independence)*

Since the $\alpha\beta\gamma$ term is not significant, it is of interest to test the $\alpha\beta$, $\alpha\gamma$, and $\beta\gamma$ terms. Simple direct estimates exist for these tests.

1. Test for $\alpha\beta_{ij} = 0$, for all ij (test of independence of A and B, given the level of C). The estimates are given by

$$\hat{f}_{ijk} = \frac{f_{i+k}f_{+jk}}{f_{++k}}$$

Thus, for example,

$$\hat{f}_{111} = \frac{89 \times 93}{150} = 55.180$$

and

$$\hat{f}_{422} = \frac{6 \times 30}{44} = 4.091$$

Box 17.10 Continued

		Expected C		Freeman–Tukey deviates	
A	*B*	H	P	H	P
IM	♀	55.180	7.318	0.009	−0.407
	♂	33.820	15.682	0.073	0.383
AM	♀	23.560	1.909	−0.064	−0.524
	♂	14.440	4.091	0.207	0.519
OW	♀	6.200	2.864	0.395	0.707
	♂	3.800	6.136	−0.293	0.369
OM	♀	8.060	1.909	0.063	0.793
	♂	4.940	4.091	0.129	−0.435

The G-test for goodness of fit to the model is

$$G_{AB(C)} = 2\sum^{a}\sum^{b}\sum^{c} f_{ijk} \ln\left(\frac{f_{ijk}}{\hat{f}_{ijk}}\right) = 2.869 \; ns$$

This value has $(a-1)(b-1)c = (4-1)(2-1)2 = 6$ degrees of freedom. The decrease in fit due to dropping the $\alpha\beta$ terms from the model is $G_{AB(C)} - G_{ABC} = 2.869 - 1.365 = 1.504$, which has $6 - 3 = 3$ degrees of freedom and is clearly not significant. We conclude that A and B are independent at each level of factor C. As before, the Freeman–Tukey deviates are also quite small.

Since direct estimates exist, we can also compute the G-values directly from observed frequencies and sums by using a natural logarithm key.

$$G = 2\left[\sum^{a}\sum^{b}\sum^{c} f_{ijk}\ln f_{ijk} - \sum^{a}\sum^{c} f_{i+k}\ln f_{i+k} - \sum^{b}\sum^{c} f_{+jk}\ln f_{+jk} + \sum^{c} f_{++k}\ln f_{++k}\right]$$

$$= 2[581.783 - 707.480 - 790.969 + 918.099]$$

$$= 2.866 \; ns$$

2. Test for $\alpha\gamma_{ik} = 0$, for all ik (test of independence of A and C, given the level of B). The estimates are given by

$$\hat{f}_{ijk} = \frac{f_{ij+} f_{+jk}}{f_{+j+}}$$

Thus, for example,

$$\hat{f}_{111} = \frac{61 \times 93}{107} = 53.019$$

and

$$\hat{f}_{422} = \frac{8 \times 30}{87} = 2.759$$

Box 17.10 Continued

		Expected C		Freeman–Tukey deviates	
A	B	H	P	H	P
IM	♀	53.019	7.981	0.302	−0.643
	♂	33.414	17.586	0.143	−0.081
AM	♀	20.860	3.140	0.506	−1.268*
	♂	13.103	6.897	0.565	−0.661
OW	♀	9.561	1.439	−0.790	1.637*
	♂	5.241	2.759	−0.955	1.216*
OM	♀	9.561	1.439	−0.436	1.133
	♂	5.241	2.759	−0.001	0.263

We can compute G by the test for goodness of fit or by direct computation as shown:

$$G_{AC(B)} = 2\left[\sum^{a}\sum^{b}\sum^{c} f_{ijk}\ln f_{ijk} - \sum^{a}\sum^{b} f_{ij+}\ln f_{ij+} - \sum^{b}\sum^{c} f_{+jk}\ln f_{+jk} + \sum^{b} f_{+j+}\ln f_{+j+}\right]$$

$$= 2[581.783 - 673.499 - 790.968 + 888.527]$$
$$= 11.684$$

This quantity has $(a-1)b(c-1) = (4-1)2(2-1) = 6$ degrees of freedom and is not quite significant at the 0.05 level.

The Williams (1976) correction is

$$q_{\min} = 1 + \frac{(a^2-1)b^2(c^2-1)}{6(a-1)b(c-1)n} = 1.02577$$

Hence $G = 11.684/1.02577 = 11.390$. The decrease in fit due to dropping $\alpha\gamma$ from the model is $G_{AC(B)} - G_{ABC} = 11.684 - 1.365 = 10.319^*$ with $6 - 3 = 3$ degrees of freedom and is significant at the 5% level. Note that for successive fitting, we do not use the Williams correction.

The pattern (IM plus AM versus OW plus OM) of the residuals from the model expressed as Freeman–Tukey deviates) is suspicious. To detect "large" deviates we can compare their absolute values with the rough criterion

$$\sqrt{\frac{\nu\chi^2_{.05[1]}}{abc}} = \sqrt{\frac{6(3.841)}{16}} = 1.200$$

Three of the deviates thus seem large, so it is of interest to partition $G_{AC(B)}$ even though the G-value is not quite significant.

Three planned orthogonal comparisons can be made; they are shown in Box 17.11.

3. Test for $\beta\gamma_{jk} = 0$ for all jk (test of independence of B and C, given the level of A).

BOX 17.10 CONTINUED

The estimates are given by

$$\hat{f}_{ijk} = \frac{f_{ij+}f_{i+k}}{f_{i++}}$$

but to conserve space are not shown here.

The G-statistic can be computed directly using the equation

$$G_{BC(A)} = 2\left[\sum^{a}\sum^{b}\sum^{c} f_{ijk} \ln f_{ijk} - \sum^{a}\sum^{b} f_{ij+} \ln f_{ij+} - \sum^{a}\sum^{c} f_{i+k} \ln f_{i+k} + \sum^{a} f_{i++} \ln f_{i++}\right]$$

$$= 2[581.783 - 673.500 - 707.480 + 806.864]$$

$$= 15.334^{**}$$

This quantity has $a(b - 1)(c - 1) = 4(2 - 1)(2 - 1) = 4$ degrees of freedom and is significant at the 0.01 level. The decrease in fit due to dropping $\beta\gamma$ is $15.338 - 1.365 = 13.973^{**}$, which has $4 - 3 = 1$ degree of freedom. We conclude that factors B and C are not independent at each level of factor A.

III. *Test for* $\alpha\beta_{ij} = \alpha\gamma_{ik} = 0$ *for all ij and ik*

If two of the two-way independence terms (e.g., $\alpha\beta_{ij}$ and $\alpha\gamma_{ik}$) were deleted from the log-linear model, it would be of interest to test for the complete independence of factor A versus factors B and C; i.e., we would wish to test the model

$$\ln \hat{f}_{ijk} = \mu + \alpha_i + \beta_j + \gamma_k + \beta\gamma_{jk}$$

The estimates are given by

$$\hat{f}_{ijk} = \frac{f_{i++}f_{+jk}}{f_{+++}}$$

where $f_{+++} = n$.

		Expected C		Freeman–Tukey deviates	
A	B	H	P	H	P
IM	♀	53.691	8.082	0.211	−0.678
	♂	32.907	17.320	0.231	−0.017
AM	♀	21.093	3.175	0.455	−1.287
	♂	12.928	6.804	0.613	−0.626
OW	♀	9.108	1.371	−0.644	1.690*
	♂	5.582	2.938	−1.098	1.114
OM	♀	9.108	1.371	−0.290	1.186
	♂	5.582	2.938	−0.145	0.161

Box 17.10 Continued

G can be computed directly by the equation

$$G_{A,BC} = 2\left[\sum^{a}\sum^{b}\sum^{c} f_{ijk}\ln f_{ijk} - \sum^{a} f_{i++}\ln f_{i++} - \sum^{b}\sum^{c} f_{+jk}\ln f_{+jk} + n\ln n\right]$$
$$= 2[581.783 - 806.864 - 790.969 + 1021.964]$$
$$= 11.828\ ns$$

This quantity has $(a-1)(bc-1) = (4-1)[(2\times 2)-1] = 9$ degrees of freedom and is not significant. We would therefore conclude that factor A is completely independent of factors B and C. An examination of the residuals from the model expressed as Freeman–Tukey deviates, however, indicates that the fit is not good and one of the deviates is greater than the criterion:

$$\sqrt{\frac{\nu\chi^2_{.05[1]}}{abc}} = \sqrt{\frac{9(3.841)}{16}} = 1.470$$

The pattern of the deviates is the same as that in part **II,** step **2** of this box (apparently due to the lack of independence between the IM-plus-AM-versus-OW-plus-OM contrast and mortality).

IV. *Test for* $\alpha\beta_{ij} = \alpha\gamma_{ik} = \beta\gamma_{jk} = 0$ *for all ij, ik, and jk. (Test of independence of A, B, and C).*

If $\beta\gamma_{jk}$ had been found not significant in part **II,** step **3,** we would next have tested for the complete independence of all three factors. That is, we would have tested the model

$$\ln \hat{f}_{ijk} = \mu + \alpha_i + \beta_j + \gamma_k$$

The direct estimates for this model are given by

$$\hat{f}_{ijk} = \frac{f_{i++}f_{+j+}f_{++k}}{n^2}$$

G can be computed directly by the equation

$$G_{A,B,C} = 2\left[\sum^{a}\sum^{b}\sum^{c} f_{ijk}\ln f_{ijk} - \sum^{a} f_{i++}\ln f_{i++} - \sum^{b} f_{+j+}\ln f_{+j+} - \sum^{c} f_{++k}\ln f_{++k} + 2n\ln n\right]$$
$$= 2[581.783 - 806.864 - 918.099 - 888.527 + 2(1021.964)]$$
$$= 24.442$$

This quantity has $abc - a - b - c + 2 = 4(2)2 - 4 - 2 - 2 + 2 = 10$ degrees of freedom. We do not test it here because we already rejected this model in part **II,** step **3.** It is given here for completeness. The Williams correction factor is

$$q_{\min} = 1 + \frac{a^2b^2c^2 - a^2 - b^2 - c^2 + 2}{6(abc - a - b - c + 2)n} = 1.02010$$

Box 17.10 Continued

V. *Final log-linear model*

The final log-liner model is

$$\ln \hat{f}_{ijk} = \mu + \alpha_i + \beta_j + \gamma_k + \alpha\gamma_{ik} + \beta\gamma_{jk}$$

The estimated values, $\hat{f}_{ijk}$, for this model are given in part **II,** step **1,** where the $\alpha\beta$ had been omitted.

We conclude that variable A (pupation site) and B (sex) are conditionally independent; that is, they are independent for each level of variable C (mortality). Both A and B are associated with variable C, however. We also found that we could simplify the model by condensing categories IM with AM and OW with OM, since the dependence of mortality on pupation site was *between* these categories rather than within them. Table 17.6 gives the estimates for the parameters in this final model. Their computation and interpretation are explained in the text.

be estimated directly for these models, so the G-test for goodness of fit can also be computed directly. We find that dropping the $\alpha\beta$ term results in a nonsignificant $G_{AB(C)} = 2.869$. The increase in G ($2.869 - 1.365 = 1.504$ with $6 - 3 = 3$ degrees of freedom) is also nonsignificant. The G-value from the model with the $\alpha\gamma$ term deleted is not quite significant at the 5% level; however, the increase ($11.684 - 1.365 = 10.319$ with $6 - 3 = 3$ degrees of freedom) is significant at the 5% level. An examination of the deviations of observed from expected frequencies reveals a suspicious pattern that supports the rejection of this model even though the G-value is not quite significant. The critical value used to test the deviates is a reasonable approximation. The value

$$\sqrt{\frac{\chi^2_{[\nu]}}{abc}}$$

suggested by Bishop, Fienberg, and Holland (1975) often results in a type I error much greater than the nominal α. The $\beta\gamma$ term is clearly significant. The increase in G due to the deletion of $\beta\gamma$ from the model is $15.338 - 1.365 = 13.973$ with $4 - 3 = 1$ degree of freedom. Because of the suspicious pattern of deviates found in the test for $\alpha\gamma = 0$ and because more than one degree of freedom was involved, we partition the G-test into three orthogonal comparisons as shown in Box 17.11. We find that most of the deviation from the model is due to the differences in factor C for the first two and the last two levels of factor A.

There are also three models with two of the two-factor effects absent—for example, the model $\ln \hat{f}_{ijk} = \mu + \alpha_i + \beta_j + \gamma_k + \beta\gamma_{jk}$, which implies that variable A is completely independent of B and C, but that variables B and C are associated (i.e., not independent). Although an example of such a test is shown in Box 17.10 to illustrate the method, it is not appropriate for the present data, since we found $\alpha\gamma$ as well as $\beta\gamma$ to have significant effects.

Box 17.11 DECOMPOSITION OF A TEST FOR CONDITIONAL INDEPENDENCE IN A THREE-WAY TABLE.

The drosophila data in Box 17.10 show a significant deviation from independence of pupation site (A) and mortality (C) from sex (B). Since there are large residuals, we want to partition the $G_{AC(B)}$ into its logical components. For these data, three comparisons (which happen to be orthogonal) are relevant.

I. *IM + AM versus OW + OM*

The expected frequencies are computed as before, but using the condensed table of observed frequencies given in the middle.

		Expected C		Observed C		Freeman–Tukey deviates	
A	*B*	H	P	H	P	H	P
IM + AM	♀	73.879	11.121	78	7	0.500	−1.270
	♂	46.517	24.483	49	22	0.394	−0.460
OW + OM	♀	19.121	2.879	15	7	−0.930	1.937
	♂	10.483	5.517	8	8	−0.724	1.025

The G-value for AC, given B, is computed as before, but using only the frequencies in the condensed table above.

$$G_{AC(B)} = 2(699.660 - 792.640 - 790.968 + 888.527)$$
$$= 9.157^*$$

The degrees of freedom are also computed as before but equal $(2 - 1)2(2 - 1) = 2$ because there are only two levels for factor A in the condensed table.

II. *IM versus AM*

		Expected C		Observed C		Freeman–Tukey deviates	
A	*B*	H	P	H	P	H	P
IM	♀	55.976	5.024	55	6	−0.097	0.502
	♂	35.197	15.803	34	17	−0.160	0.353
AM	♀	22.024	1.976	23	1	0.256	−0.570
	♂	13.803	6.197	15	5	0.376	−0.393

$G_{AC(B)} = 1.307\ ns \qquad df = 2$

Box 17.11 Continued

III. *OW versus OM*

		Expected C		Observed C		Freeman–Tukey deviates	
A	B	H	P	H	P	H	P
OW	♀	7.500	3.500	7	4	−0.094	0.363
	♂	4.000	4.000	3	5	−0.391	0.562
OM	♀	7.500	3.500	8	3	0.261	−0.141
	♂	4.000	4.000	5	3	0.562	−0.391

$G_{AC(B)} = 1.221$ *ns* $\quad df = 2$

Since the comparisons are orthogonal, we can confirm that both G and df have been partitioned correctly: $9.157 + 1.307 + 1.221 = 11.685$ and $2 + 2 + 2 = 6$.

In the final model that we can test, all two-factor effects are absent. If this model fits, the implication is that all three variables are completely independent of one another.

Figure 17.5 shows the relationships among the tests of the hierarchical models in a three-way table (using the data given in Box 17.10). The entries at the nodes of the diagram correspond to the G-values for the goodness-of-fit tests for the various models. The values along the arrows represent the increase in the G-value due to the parameter being deleted when one passes from the model at the base of the arrow to the model at the head of the arrow. In practice, one would not routinely make all of these tests but would proceed sequentially until reaching the simplest model that adequately fit the data.

If any of the marginal totals are fixed by the design of the experiment (e.g., the experimenter determine how many males and females are to be exposed to each treatment combination), then the corresponding term must be present in all models tested. If, for example, the AB table is fixed, then the $\alpha\beta_{ij}$ term must be present in all of the models—that is, it must not be dropped. The effect of $\alpha\beta$ itself is therefore never tested, since it is under the investigator's control.

An additional complication in the analysis of a multiway table is that we have tested a large number of models and, furthermore, some models have been tested only as a consequence of our having observed nonsignificant results in tests of other models. Unless one can argue that the tests are all of interest a priori, one needs to assess the overall probability of making a type I error. A simple approximate procedure has been described by Whittaker and Aitkin (1978). In Box 17.10 we tested for the presence of the $\alpha\beta\gamma$ interaction at the nominal level of $\alpha_1 = 0.05$. Since $\alpha\beta\gamma$ was not significantly different from zero, we then tested

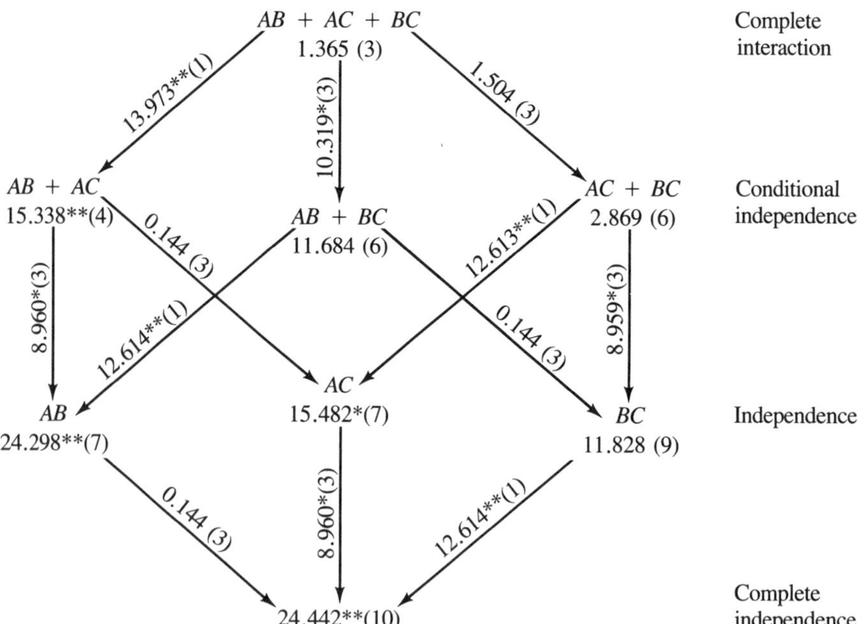

FIGURE 17.5 *Diagrammatic representation of the hierarchical models tested in Box 17.10. The quantities at the nodes are* G-*values corresponding to the goodness-of-fit tests from the various models tested. The quantities along the arrows correspond to the differences in* G-*values between the model at the base of the arrow and the model at the head of the arrow. These differences correspond to the* G-*value explained by the parameters being deleted from the model. The values in parentheses are the degrees of freedom associated with the corresponding* G-*values.*

three models at a nominal level of $\alpha_2 = 0.05$. At this point the overall level of significance was $\alpha = \alpha_1 + (1 - \alpha_1)\alpha_2 = 0.0975$.

If more than two of the conditional independence models had been non-significant, we would have tested for complete independence at the nominal level of significance $\alpha_3 = 0.05$. In this case the overall level of significance would have been raised to $\alpha = \alpha_1 + (1 - \alpha_1)\alpha_2 + (1 - \alpha_1)(1 - \alpha_2)\alpha_3 = 0.14263$. In order to avoid having such a large probability of a type I error, one may wish to use smaller values for α_1, α_2, and α_3 so that the overall level of significance is closer to 0.05.

Up to now we have dealt only with expected frequencies and goodness-of-fit tests for the various models. It is also possible (and often of interest) to estimate the parameters themselves in the log-linear model (i.e., the α_i, β_j . . . values). For any given model these values can easily be computed in the same way as in a corresponding anova model without replication.

For the final log-linear model in Box 17.10 let

$$l_{ijk} = \ln \hat{f}_{ijk} = \mu + \alpha_i + \beta_j + \gamma_k + \alpha\gamma_{ik} + \beta\gamma_{jk}$$

Then $\hat{\mu}$ is simply the mean of $l_{ijk}(\Sigma\Sigma\Sigma l_{ijk}/abc)$. The main effects are simply deviates from the mean: $\alpha_i = \bar{l}_i - \hat{\mu}$, $\beta_j = \bar{l}_j - \hat{\mu}$, and $\gamma_k = \bar{l}_k - \hat{\mu}$. The interaction terms are $\alpha\gamma_{ik} = \bar{l}_{ik} - \bar{l}_i - \bar{l}_k + \hat{\mu}$ and $\beta\gamma_{jk} = \bar{l}_{jk} - \bar{l}_j - \bar{l}_k + \hat{\mu}$. The numerical values of the estimates are given in Table 17.6. As a check one may verify that $\hat{f}_{111}$ can be computed as $\exp[1.99531 + 1.07331 + (-0.06815) + 0.53032 + 0.16688 + 0.31292] = \exp(4.01059) = 55.179$. Compare this value with 55.180 (identical within rounding error), given in the table of expected frequencies in part **II**, step **1** of Box 17.10.

When one of the variables is of special interest and has only two categories (such as mortality: healthy versus poisoned, in the present example), it is convenient to express the results in terms of the log odds (see Section 17.6) of the estimated incidences as a function of the other variables. For example, the logarithmic estimated frequencies of the healthy flies are

$$\ln \hat{f}_{ij1} = \gamma_1 + \alpha\gamma_{i1} + \beta\gamma_{j1}$$

Table 17.6 ESTIMATES OF THE PARAMETERS FOR THE LOG-LINEAR MODEL $\ln \hat{f}_{ijk} = \mu + \alpha_I + \beta_J + \gamma_K + \alpha\gamma_{IK} + \beta\gamma_{JK}$ FOR THE DATA OF BOX 17.10.

$\hat{\mu} = 1.99531$

$\hat{\alpha}_1 = 1.07331$, $\hat{\alpha}_2 = -0.02408$, $\hat{\alpha}_3 = -0.48883$, $\hat{\alpha}_4 = -0.56040$

$\hat{\beta}_1 = -0.06815$, $\hat{\beta}_2 = 0.06815$

$\hat{\gamma}_1 = 0.53032$, $\hat{\gamma}_2 = -0.53032$

Table of $\widehat{\alpha\gamma}$

	k	
i	1	2
1	0.16688	−0.16689
2	0.41323	−0.41324
3	−0.47503	0.45702
4	−0.12309	0.12308

Table of $\widehat{\beta\gamma}$

	k	
j	1	2
1	0.31292	−0.13292
2	−0.31292	0.31292

Those of the poisoned flies are

$$\ln \hat{f}_{ij2} = \gamma_2 + \alpha\gamma_{i2} + \beta\gamma_{j2}$$

Therefore, the log odds of being healthy versus poisoned are

$$\ln \frac{\hat{f}_{ij1}}{\hat{f}_{ij2}} = (\gamma_1 - \gamma_2) + (\alpha\gamma_{i1} - \alpha\gamma_{i2}) + (\beta\gamma_{j1} - \beta\gamma_{j2})$$

where the term $(\gamma_1 - \gamma_2)$ is equal to 1.06064. This value indicates that the overall odds ratio (or risk) of being in the healthy versus poisoned state is $\exp(1.06064) = 2.888$. The second term, $(\alpha\gamma_{i1} - \alpha\gamma_{i2})$, is a function of pupation site and takes on the following values: 0.33376 for IM, 0.82646 for AM, −0.95006 for OW, and −0.24618 for OM. Thus the odds ratio is increased if the position is IM or AM and is decreased if the position is OW or OM. The last term, $(\beta\gamma_{j1} - \beta\gamma_{j2})$, gives the contribution of sex to the log odds: 0.62584 for ♀, −0.62584 for ♂ (indicating that being female increases the chances of a fly being in the healthy state). The relative risk of being healthy versus poisoned for a female pupating in the medium is then $\exp(1.06064 + 0.33376 + 0.62584) = \exp(2.02024) = 7.540$ (which agrees with the direct computation $\hat{f}_{111}/\hat{f}_{112} = 55.180/7.318 = 7.540$ using the $\hat{f}$ given in part **II**, step **1** of Box 17.10).

See Bishop, Fienberg, and Holland (1975) for methods to compute approximate standard errors, as well as confidence limits of the parameters. Also described is the testing of linear comparisons of the parameters in order to test specific hypotheses. Ku and Kullback (1974) show the analysis of a three-way and a four-way table.

The BIOM-pc program carries out an analysis of all models specified in Figure 17.5 for a three-way contingency table.

17.6 Analysis of Proportions

In Section 17.4, which dealt with two-way tables, our major emphasis was on testing whether the two variables are independent. We will now present an alternative, equally useful approach. We can focus on the proportions observed for one class of the first variable in the classes of the second. Let us return to the immunology example of Section 17.4 to clarify this point. Recall that 44 of 57 mice who received bacteria *and* antiserum survived, whereas the survival rate for mice who received only bacteria was only 29 out of 54. We can express these results as proportions and arrange them in a table, as shown here:

	Dead	Alive	$\sum$
Bacteria and antiserum	$p_1 = 0.22807$	$q_1 = 0.77193$	1.0
Bacteria only	$p_2 = 0.46296$	$q_2 = 0.53704$	1.0

The survival proportion under the treatment criterion is roughly 0.77; under the controls it is 0.54. How should we express this difference? We could describe it as a simple difference of $0.77 - 0.54 = 0.23$, but that might not be meaningful without referring to the magnitude of the proportions of occurrences being compared. An example will make this clear. Suppose the difference were only 0.01. If the survival proportion with bacteria alone were only 0.02, then an increase of 0.01 would be a change of 50%. If the survival proportion were 0.50, however, the change would be only 2%. An alternative way of expressing the difference in proportions is by stating them as **odds ratios.** If both bacteria and antiserum are administered, the odds that the mouse will survive is the ratio q_1/p_1 to 1. In our example this is $0.77193/0.22807 = 3.38462:1$. A mouse is thus more than three times as likely to survive than to die if it receives antiserum along with its bacterial injection. The survival odds for the treatment with only bacteria are $q_2/p_2 = 0.53704/0.46296 = 1.16000:1$. Thus without antiserum, the odds on survival are only slightly greater than 1.

The ratio of these two odds is a useful way to compare the difference in outcomes. We evaluate the odds ratio, ω, as

$$\omega = \frac{q_1/p_1}{q_2/p_2} = \frac{3.38462}{1.16000} = 2.91778 \qquad (17.18)$$

Thus, the odds on survival are almost three times better when antiserum is given than when it is withheld. When the odds ratio is one, obviously there is no difference in the outcomes of the two treatments. Can we test the odds ratio for a departure from the null case of 1.0? Such as test is equivalent to a test of independence for a 2×2 table, so we can use the methods of Section 17.4. The natural log transform of the odds ratio, commonly called the log-odds ratio, is an important quantity that we will discuss shortly. Odds ratios are perhaps most meaningful in Model II two-way tables where we compare the outcomes for the two levels of the fixed treatment, but they have been used widely for the other models as well. In many applications, e.g., epidemiology, both q_1 and q_2 are usually small. In such cases, $p_2/p_1 \cong 1$ and the quantity q_1/q_2, called the **relative risk,** approximates the odds ratio ω.

In Section 14.11 we introduced the **logit transformation** of a proportion p, which is $\ln(p/q)$. This transformation stretches out very small and very large proportions and makes the relations between proportions and some causative variables linear. We now realize that a logit is the natural logarithm of the *odds* expressed by a given proportion. Logit $q = \ln(q/p)$. Let us now examine the difference between the logit transforms of two proportions:

$$\text{logit } q_1 - \text{logit } q_2 = \ln\frac{q_1}{p_1} - \ln\frac{q_2}{p_2} = \ln\frac{q_1/p_1}{q_2/p_2} = \ln\omega$$

Note that the difference between the logits equals the logarithm of the odds ratio of the corresponding 2×2 table [see Expression (17.18)]. This relationship has made the **log-odds ratio** a very popular statistic expressing the difference between two proportions. The odds ratio ω can be obtained from the log-odds ratio as $e^{\text{log-odds}}$. In the example of the mice, the odds ratio is 2.91778. The corresponding log-odds ratio is ln 2.91778 = 1.07082. This quantity is also the difference between the logits of the two proportions $\ln q_1/p_1 - \ln q_2/p_2 = \ln 3.38462 - \ln 1.16000 = 1.21924 - 0.14842 = 1.07082$.

The odds ratio is easy to understand because it is in a "natural" scale. An odds ratio of 2.91778, for example, says that the odds that a mouse will survive are 2.91778 greater if antiserum is included in their injection. By transforming the odds ratio into its logarithm, this simple interpretation is lost, but at the same time we obtain a quantity that is approximately normally distributed. The larger the log odds, the greater the difference between proportions. When the proportions are equal, the odds ratio is 1 and the log-odds ratio is 0.

Note that when $p = 0$ or 1 the logit cannot be computed. One method to avoid this problem is to use what are called **empirical logits,** $\ln((a + \frac{1}{2})/(b + \frac{1}{2}))$, where a and b are observed frequencies that add up to n, the sample size. Another method is simply to replace the zeros by a small constant (such as $\frac{1}{2}$) and to decrease corresponding frequencies equal to the marginal totals by the same amount to maintain the original margin totals.

For large samples, a good estimate of the standard error of the log-odds ratio can be obtained from the formula

$$s_{\ln \omega} = \left(\frac{1}{a} + \frac{1}{b} + \frac{1}{c} + \frac{1}{d}\right)^{1/2} \tag{17.19}$$

For the mice data this value is 0.41729. Some authors prefer to add a continuity correction by adding $\frac{1}{2}$ to each of the four denominators in Expression (17.19). When we evaluate this corrected standard error for the mice, we obtain the slightly smaller value 0.41190. Approximate 95% confidence limits for the log odds therefore are $1.07082 \pm 1.96(0.41190)$, or 0.26350 and 1.87814. In odds-ratio scale these limits correspond to 1.30147 and 6.54135. The antiserum has clearly helped improve the odds for survival in the mice.

Another application of this standard error is to test the null hypothesis of no difference between the observed odds ratio and an established standard. Suppose a veterinarian is deciding whether to inoculate the mouse colony against the bacterium in question by injecting antiserum but does not consider inoculation with antiserum worthwhile unless the odds for survival upon infection with the bacterium are at least eight times those for mice not inoculated with antiserum. We compute

$$\frac{(\ln \omega - \ln \omega_{\text{st}})^2}{s^2_{\ln \omega}} = \frac{(1.07082 - \ln 8)^2}{0.41190^2} = 5.996$$

This ratio of a single squared difference over an estimated variance is distributed as chi-square with one degree of freedom. Since the true odds ratio is less than 8, it may not be worthwhile to inoculate the mice.

For small samples, the standard error given by Expression (17.19) may not be very accurate. Various techniques exist for obtaining better estimates. All are computationally intensive and are used only with computer programs. Interested readers should consult the review by Fleiss (1979).

We now turn to odds ratios computed from replicated 2×2 tables, which may arise from a variety of designs. Here we will mention only two. In the first design we consider the separate tables to be replicates sampled from the same underlying population. For example, we may wish to test the incidence of heart disease in samples from various cities. For each sample we have also recorded whether the person was a smoker. Each 2×2 table then is smoker (yes/no) versus heart disease (yes/no) for a given city. We would like to test the homogeneity of the relation between the two variables over the cities (this is a test of conditional independence, as discussed in Section 17.5), to construct a joint estimate of the odds ratio from all the 2×2 tables (cities), and to test this estimate against the null hypothesis of an odds ratio of 1.

In the second design we suspect that the odds ratio is affected by a covariate, and we wish to control it by constructing separate 2×2 tables for different classes of the covariate. For example, the immunologist who has been inoculating mice with bacteria with or without antiserum suspects that the age of the mice plays a role in the outcome of the test. The experimenter consequently divides the test mice into three groups by age—immature, mature, and old—and divides each group roughly into two subgroups, experimentals and controls. The questions posed to the results are the same as in the first design. We will analyze an artificial example that extends the immunology experiment. The data and analysis are featured in Box 17.12.

Let us first establish the symbolism we will be using. We will designate the cells and margins in each table using the now familiar a, b, c, d, and n notation. Since there now are $k = 3$ tables, the frequencies of each table are subscripted to indicate the number $1, \ldots, k$ of the table or sampling stratum concerned. If we examine the proportion alive after treatment in the three experiments in Box 17.12, we note an apparent trend toward increasing survival in the older animals. Is this real? We first test whether there is homogeneity in the odds ratios of the three tables by calculating the odds ratio ω_i of each table using Expression (17.19) and then computing the natural logarithm of the ω_i. Heterogeneity of the odds ratios, symbolized by X_H^2, is tested with the formula

$$X_H^2 = \sum_i [w_i(\ln\omega_i - \overline{\ln\omega_i})^2] \tag{17.20}$$

This quantity is a weighted sum of squares that is distributed as chi-square with $k - 1$ (in this case, 2) degree of freedom. The weights w_i are inversely proportional to the variance of the log odds. In our example, $X_H^2 = 2.118$, which is not significant despite the apparent trend toward higher survival with age. This test does not take trend into account. To do so, logistic regression, discussed later in this section, must be used. The test of homogeneity of odds ratios is equivalent to

Box 17.12 TEST OF HOMOGENEITY OF ODDS RATIOS FOR REPLICATED 2 × 2 TABLES: THE MANTEL–HAENSZEL PROCEDURE

Three experiments on immature, mature, and old mice, testing survival of injection with bacteria only versus injection with bacteria plus antiserum. There are $i = 1, \ldots, k$ strata. In this problem $k = 3$.

	Immature			Mature			Old		
	Dead	Alive	$\sum$	Dead	Alive	$\sum$	Dead	Alive	$\sum$
Bacteria only	26	29	55	43	41	84	15	23	38
Bacteria and antiserum	21	34	55	21	58	79	8	37	45
$\sum$	47	63	110	64	99	163	23	60	83

We examine the proportion alive in the 6 experimental groups (3 ages × 2 treatments) to obtain an overview of the results.

	Immature proportion alive	Mature proportion alive	Old proportion alive
Bacteria only	0.527	0.488	0.605
Bacteria and antiserum	0.618	0.734	0.822

Visual inspection indicates a trend toward higher survival in older mice, both with and without antiserum.

Mantel–Haenszel procedure to test odds ratios for homogeneity.

1. Compute odds ratio ω for each table. Rather than employ a formula based on proportions, such as Expression (17.18), these data can be computed most easily by the following equation:

$$\omega' = \frac{(a + 0.5)(d + 0.5)}{(b + 0.5)(c + 0.5)}$$

which is expressed in terms of cell frequencies. This formula includes continuity corrections of 0.5, which are indicated by the prime following the symbol for odds ratio. Thus,

$$\omega_1' = \frac{(26.5)(34.5)}{(29.5)(21.5)} = 1.441466$$

BOX 17.12 CONTINUED

2. Transform the values of ω' to their natural logarithms, $\ln \omega'$. The results are the adjusted log-odds ratios. For example, $\ln \omega'_1 = \ln 1.441466 = 0.365661$.
3. Compute weights, w, for the log odds. These weights are the reciprocals of the variances of the log odds, again corrected for continuity, as indicated by the prime:

$$w' = \frac{1}{\left(\dfrac{1}{a+0.5} + \dfrac{1}{b+0.5} + \dfrac{1}{c+0.5} + \dfrac{1}{d+0.5}\right)}$$

Thus

$$w'_1 = \frac{1}{\left(\dfrac{1}{26.5}\right) + \left(\dfrac{1}{29.5}\right) + \left(\dfrac{1}{21.5}\right) + \left(\dfrac{1}{34.5}\right)} = 6.796651$$

We summarize the quantities obtained in steps **1** through **3** in the following table.

	ω'	$\ln \omega'$	w'
Immature	1.441466	0.365661	6.796651
Mature	2.852059	1.048041	9.034196
Old	2.909887	1.068114	3.977993

4. Next compute the weighted mean of the adjusted log odds, $\overline{\ln \omega'}$ by the following formula:

$$\overline{\ln \omega'} = \frac{\sum_i w'_i \ln \omega_i}{\sum_i w'_i}$$

5. Finally, compute Expression (17.20), the weighted sum of squares of the adjusted log odds:

$$X_H^2 = \sum_i \left[w'_i(\ln \omega'_i - \overline{\ln \omega'})^2\right]$$

This quantity is distributed as chi-square with $k - 1$ degrees of freedom. When we compute this value for the data in this example we obtain $X_H^2 = 2.118$, which is not significant. We conclude that the log-odds ratios, and hence the odds ratios, are homogeneous, despite what appears to be a trend toward higher survival in older mice, and we are therefore encouraged to estimate an overall average odds ratio for our experiments. If the odds ratios are heterogeneous, separate estimates for each stratum (or at least for some strata) are indicated. To test for a trend we can use logistic regression analysis (see text for explanation).

Mantel–Haenszel estimator and its significance.

6. First compute Expression (17.21):

$$\omega_{MH} = \frac{\sum_i a_i d_i / n_i}{\sum_i b_i c_i / n_i}$$

Box 17.12 Continued

which is the Mantel–Haenszel estimator of the odds ratios for the three strata. Thus,

$$\omega_{MH} = \frac{(26)(34)/110 + \cdots + (15)(37)/83}{(29)(21)/110 + \cdots + (23)(8)/83} = 2.303$$

We conclude that over the three experiments, the odds of survival for a mouse with antiserum were 2.3 times those for mice with bacteria alone.

Finally, we test this ratio for significance. That is, we ask, Is the odds ratio greater than 1? The formula for the Mantel–Haenszel statistic is cumbersome but straightforward:

$$X^2_{MH} = \frac{\left[\left|\sum_i a_i - \sum_i (a+b)_i(a+c)_i/n_i\right| - 0.5\right]^2}{\sum_i (a+b)_i(a+c)_i(b+d)_i(c+d)_i/(n^3-n^2)_i}$$

This statistic is a function of the sums of one cell and of the margin totals of the k tables. It also contains a continuity correction. It is distributed as chi-square with one degree of freedom and tests the null hypothesis that there is no relationship between the presence of the antiserum and survival. When evaluated for these data, $X^2_{MH} = 13.110$, which yields $P < 0.001$. We conclude that the two criteria are related in these three studies, and that in consequence, the joint odds ratio is greater than 1.

the test of no three-factor interaction in a $2 \times 2 \times k$ three-way contingency table. For the data in Box 17.12, the methods of Box 17.10 yield $G_{ABC} = 2.161$ (2.125 with Williams's correction).

The next step is to estimate the joint odds ratio for the three experiments, using the formula

$$\omega_{MH} = \frac{\sum_i a_i d_i / n_i}{\sum_i b_i c_i / n_i} \tag{17.21}$$

The subscript *MH* refers to Mantel and Haenszel (1959), who in a seminal paper proposed this method of joint estimation, as well as a significance test for the estimate. For the three 2×2 tables in Box 17.12, the Mantel–Haenszel estimate is $\omega_{MH} = 2.303$. On the average, then, antiserum increases the odds for survival appreciably (2.303 times).

This estimate can be tested for difference from 1 (which would imply that on average factors A and B are independent within strata) by the following equation:

$$X^2_{MH} = \frac{\left(\left|\sum_i a_i - \sum_i (a+b)_i\,(a+c)_i/n_i\right| - 0.5\right)^2}{\sum_i (a+b)_i\,(a+c)_i\,(b+d)_i\,(c+d)_i/(n^3-n^2)_i} \tag{17.22}$$

which is distributed as chi-square with one degree of freedom. For the data in Box 17.12 we obtain $X^2_{MH} = 13.110$, which is significant at $P < 0.001$. We conclude that there is a highly significant increase in survival upon administration of the antiserum at all ages. This hypothesis can also be tested by computing $G_{AB(C)} - G_{ABC}$, which also has just 1 degree of freedom. For the present example, this quantity is equal to $16.300 - 2.161 = 14.139$. One advantage of the Mantel–Haenszel test is that it can be carried out on 2×2 tables that contain very few individual observations, as long as the overall sample size, $\Sigma_i n_i$, of all strata is substantial.

It may be necessary to determine the sample size required for a test of equality of proportions. The method described in Box 17.13 is based on principles similar to those given in Box 9.14 and discussed in Section 9.8. The formula for n was developed by Casagrande, Pike, and Smith (1978). It is an approximation that enables an explicit expression for the sample size, so iteration is not necessary to obtain a solution, as it was in Section 9.8. Box 17.13 also illustrates how to estimate the required sample size for a test of the difference between two proportions when one of them is theoretical rather than a sample value. This formula would apply in a single-classification goodness-of-fit test with only two classes. Finding sample sizes required for one-way classifications with more than two classes, such as those discussed in Section 17.2, is difficult because of the great variety of possible deviations from expectation.

Our final topic for this section is **logistic regression.** Many situations arise in which researchers must investigate relations between a proportion and a continuous variable (with perhaps other variables taken into account). Examples are mortality as a function of dose of toxicant, morbidity as related to amount of exposure, parasitism as a function of density of intermediate hosts of the parasites, gene frequencies as affected by climatic variables, or survival as a function of age (as in the mice example). Logistic regression has become popular in recent years because it generalizes analyses of frequency data from the anovalike log-linear models to the regressionlike models, often permitting additional insights into data, as we will see.

Logistic regression relates the proportions p_i of a dependent variable to an independent variable X according to the following model:

$$\hat{p} = \frac{e^{a + bX}}{1 + e^{a + bX}} \tag{17.23}$$

To see what this function will look like for the logit transform of $\hat{p}$, first evaluate $1 - \hat{p}$ from Expression (17.23):

$$1 - \hat{p} = \frac{1 + e^{a + bX} - e^{a + bX}}{1 + e^{a + bX}} = \frac{1}{1 + e^{a + bX}} \tag{17.24}$$

Then

$$\frac{\hat{p}}{1 - \hat{p}} = \frac{e^{a + bX}/(1 + e^{a + bX})}{1/(1 + e^{a + bX})} = e^{a + bX} \tag{17.25}$$

Box 17.13 SAMPLE SIZE REQUIRED TO DETECT A GIVEN "TRUE" DIFFERENCE BETWEEN TWO PROPORTIONS.

The rationale behind this method is discussed in Section 17.6. Given the significance level α, we want to estimate the sample size, n, needed for a $1 - \beta$ chance of detecting a difference between two true proportions p_1 and p_2 ($p_1 > p_2$). Let

$$\bar{p} = (p_1 + p_2)/2$$

$$A = [t_{\alpha[\infty]}\sqrt{2\bar{p}(1 - \bar{p})} + t_{2\beta[\infty]}\sqrt{p_1(1 - p_1) + p_2(1 - p_2)}]^2$$

where $t_{\alpha[\infty]}$ and $t_{2\beta[\infty]}$ are critical values from a two-tailed t-table. Then

$$n = \frac{A[1 + \sqrt{1 + 4(p_1 - p_2)/A}]^2}{4(p_1 - p_2)^2}$$

Let us apply this method to the following example: What sample size is required in each of two samples if one wishes to be $(1 - \beta) = 0.80$ (or 80%) certain of detecting a true difference between two proportions, $p_1 = 0.65$ and $p_2 = 0.55$, at the $\alpha = 0.05$ level of significance?

$$\bar{p} = \frac{(0.65 + 0.55)}{2} = 0.60$$

$$t_{[.05]\infty} = 1.960$$

$$2\beta = 2(1 - 0.80) = 0.40$$

$$t_{.40[\infty]} = 0.842$$

$$\begin{aligned} A &= [1.960\sqrt{2(0.6)(0.4)} + 0.842\sqrt{0.65(0.35) + 0.55(0.45)}]^2 \\ &= [1.960(0.69282) + 0.842(0.68920)]^2 \\ &= 3.75676 \end{aligned}$$

Then

$$\begin{aligned} n &= \frac{3.75676(1 + \sqrt{1 + 4(0.65 - 0.55)/3.75676})^2}{4(0.65 - 0.55)^2} \\ &= \frac{15.81692}{0.04} \\ &= 395.42310 \end{aligned}$$

Thus 396 observations would be needed in each sample.

If we wish to determine the sample size when one of the proportions is a theoretical rather than a sample value (as in genetic research), we simply redefine the quantity A as follows:

$$A = [t_{\alpha[\infty]}\sqrt{p_1(1 - p_1)} + t_{2\beta[\infty]}\sqrt{p_2(1 - p_2)}]^2$$

For example, suppose we want to plan an experiment in which the expected ratio is 3 : 1 (thus $p_1 = 0.75$). What sample size should we use to be $1 - \beta = 0.80$ (or 80%) sure of

BOX 17.13 CONTINUED

detecting another ratio, such as a 1 : 1 ratio ($p_2 = 0.50$), at the $\alpha = 0.05$ level of significance?

$$A = [1.960\sqrt{0.75(0.25)} + 0.842\sqrt{0.5(0.5)}]^2$$
$$= [1.960(0.43301) + 0.842(0.5)]^2$$
$$= 1.61215$$

Then

$$n = \frac{1.61215[1 + \sqrt{1 + 4(0.75 - 0.50)/1.61215}]^2}{4(0.75 - 0.50)^2}$$
$$= \frac{8.32853}{0.25}$$
$$= 33.31413$$

Therefore 34 observations are required.

Consequently

$$\ln\left(\frac{\hat{p}}{1 - \hat{p}}\right) = a + bX \qquad (17.26)$$

Note that $\text{logit}(\hat{p}) = \ln[\hat{p}/(1 - \hat{p})]$ is related to X by simple linear regression (see Section 14.11). Thus the logit is linear in its parameters, it may be continuous, and it ranges from $-\infty$ to $+\infty$, rather than from 0 to 1. The error around the line, however, is not distributed normally.

Since the dependent variable is binomially rather than normally distributed, the **maximum-likelihood method** is used to fit a regression line to logit-transformed data rather than the least-squares method as described in Chapters 14 and 16. Recall the notion of likelihood we encountered in Section 17.1. The maximum-likelihood method will result in the values for the parameters (a and b) that make the observed values in our data set seem most probable. Unfortunately, the solution to the nonlinear maximum-likelihood equations is complex and in practice can be carried out iteratively only by computer. We therefore do not furnish computational formulas in this book. We should point out that the maximum-likelihood approach of fitting a regression line to bivariate or multivariate data is the general and correct method. It just so happens that the least squares method that we employed in Chapter 14 and 16 yields results that are mathematically identical to the maximum-likelihood solution when the dependent variable is distributed normally.

A typical logistic regression program, such as in the BIOM-pc package, yields estimates for a and b, as well as for their standard errors, which are also tedious to compute and therefore not formulated here. The program usually shows the ratio of the coefficients over the standard errors to give an indication of the significance of the results. Next, it shows the log-likelihood of the chosen fit, given the estimates for parameters a and b. Finally, the program shows the G-value of $-2n$ times the ratio of the log-likelihood of the observed data, given their observed values as parameters, over the log-likelihood of the model with the fitted parameters. This G-value is approximately distributed as chi-square with one degree of freedom and tests the hypothesis that there is no linear relationship between the logit and the independent variable. Note that logistic regression need not be restricted to a single independent variable, as shown in Expression (17.26), but frequently is carried out as a multiple regression on several independent variables.

An alternative significance test for a logistic regression is to examine the ratios of the fitted coefficients over their standard errors and to evaluate these ratios as normal deviates (i.e., as $t_{[\infty]}$). This test, also known as the *Wald test*, usually gives results that are similar to those obtained with the G-test.

Before we present an example of logistic regression, we will discuss the meaning of a logistic regression coefficient. Note that the independent variable X need not be continuous. It could also be a multistate ordinal or nominal variable, or even a binary variable. It is the latter case that yields the simplest relationship. Suppose we have a sample of n individuals, of whom a given number $(a + b)$ have lung cancer and a given number $(c + d)$ do not; $(a + c)$ individuals had smoked, and $(b + d)$ had not. Note that we have employed the familiar notation of a 2×2 contingency table. We show such a table here:

		Smoking $X = 1$	Smoking $X = 0$	
Lung cancer	$Y = 1$	$\hat{p}(1) = \dfrac{e^{a+b}}{1 + e^{a+b}}$	$\hat{p}(0) = \dfrac{e^{a}}{1 + e^{a}}$	$a + b$
	$Y = 0$	$1 - \hat{p}(1) = \dfrac{1}{1 + e^{a+b}}$	$1 - \hat{p}(0) = \dfrac{1}{1 + e^{a}}$	$c + d$
		$a + c$	$b + d$	n

In this table, 1 stands for presence and 0 stands for absence of the X or Y property. The values of $\hat{p}(1)$ and $\hat{p}(0)$ are the estimated proportions of lung cancer cases in smokers and nonsmokers, respectively. Recalling relations presented earlier in this section, we note that the odds of lung cancer are $\hat{p}(1)/(1 - \hat{p}(1))$ for smokers and $\hat{p}(0)/(1 - \hat{p}(0))$ for nonsmokers. The formulas in the cells are the logistic regressions, derived from Expressions (17.25) and (17.26), when X in turn assumes values of 1 and 0, respectively. Now recall that the odds ratio for a 2×2

table is $(q_1/p_1)/(q_2/p_2)$, which in this case could be rewritten as

$$\frac{\hat{p}(1)/(1 - \hat{p}(1))}{\hat{p}(0)/(1 - \hat{p}(0))}$$

When we substitute the logistic regression terms for the $\hat{p}$'s and cancel, we obtain $\omega = e^b$; thus,

$$\ln \omega = b$$

We have obtained the important result that for a dichotomous independent variable the logistic regression coefficient is the log-odds ratio, which is the difference between the two logits. It follows that when we exponentiate the logistic regression coefficient, we obtain an equation for the odds ratio for the data set in question. Knowing that the logistic regression coefficient yields the log-odds ratio for a difference of 1 in the units of the independent variable is useful for an interpretation of the results, even though it is not a practical method for computation.

When the independent variable is multistate categorical, with each of k states representing a separate group of the independent variable, we can obtain a solution by recoding it to $k - 1$ design variables for the k groups. Each design variable is binary and assigns 1 to one of $k - 1$ groups and zero to all others. That is, one group is considered a referent group, and all design variables code zero for it. Consider the example in Box 17.14, which shows frequencies of the ACP1 genotypes A + BA at four Sardinian villages. We consider the first village, Fonni, to be the referent sample, and the other three samples are evaluated with respect to it. The odds ratios are computed in a straightforward manner, following the established formula. We find that Aritzu-Belvi and Lode have substantially lower odds of possessing an A + BA genotype, but that Oristano has a genotype pattern very similar to that of Fonni. Box 17.14 also shows the design variables and the results of a multiple logistic regression of the genotype frequencies on the design variables. Computationally, the first approach is simpler, but the logistic regression enables us to test significance and set confidence limits.

Finally, in Box 17.15 we examine a data set that has a continuous independent variable. Sardinian villages were analyzed for genotype frequencies at the ACP1 locus (this is the complete data set from which the data in Box 17.14 were extracted). An attempt was made to relate the frequency of this genotype to the elevation of each village and to the genotype at the ADA locus, since interaction was suspected. A multiple logistic regression program is essential for obtaining results. Since an interaction between the effects of ADA and elevation was found, separate logistic regressions were computed for each ADA class. The regression coefficient for the ADA effect, taking elevation into account, 0.57409, is a log-odds ratio and indicates that the odds for ADA11 at an elevation of 0 are larger than for the ADA12 and ADA22 genotypes by a factor of

Box 17.14 LOGISTIC REGRESSION WITH A MULTISTATE CATEGORICAL INDEPENDENT VARIABLE.

Data (extracted from the data set in Box 17.15) are ACP1 genotype frequencies in four Sardinian villages, using only individuals with ADA11 genotype.

	Village				
	Fonni	Aritzu-Belvi	Lode	Oristano	$\sum$
A + BA	34	13	34	62	143
Other	66	94	85	123	368
$\sum$	100	107	119	185	511
Odds ratio, ω	1.0	0.26846	0.77647	0.97848	
Log-odds ratio, $\ln \omega$	0	−1.31505	−0.25300	−0.02176	

Computation

1. Since this table contains counts rather than proportions, we use the former to obtain odds ratios and log-odds ratios. For example, for Aritzu-Belvi versus Fonni

$$\omega = \frac{13 \times 66}{34 \times 94} = 0.26846$$

We proceed similarly for each of the other two villages versus Fonni.

2. An alternative procedure is to subject the data to a multiple logistic regression program, by constructing three design variables as follows.

	Design variables		
Village	1	2	3
Fonni	0	0	0
Aritzu-Belvi	1	0	0
Lode	0	1	0
Oristano	0	0	1

The variables, serving as regressors of the proportion A + BA genotypes, yield the following results when subjected to a logistic regression program.

BOX 17.14 CONTINUED

Village	$b_{Yi \cdot jk}$	SE_b	*Coeff.* / SE	e^b
Aritzu-Belvi	−1.31505	0.36349	−3.61785	0.26846
Lode	−0.25300	0.29281	−0.86402	0.77647
Oristano	−0.02176	0.26234	−0.08293	0.97848

The partial regression coefficients are the log-odds ratios, and their exponents are the odds ratios. The standard errors enable us to estimate confidence limits for the log-odds ratios, and, by extension, for the odds ratios. The 95% confidence limits for log-odds ratios are $b_{Yi \cdot jk} \pm t_{.05[\infty]}\mathrm{SE}_b$; 95% confidence limits for odds ratios are $e^{(b_{Yi \cdot jk} \pm t_{.05[\infty]}\mathrm{SE}_b)}$.

$e^{0.57409} = 1.77551$. (This difference is not, however, significant.) The regression coefficient of log odds on altitude (in meters), which is shown in Box 17.15, is also a log-odds ratio, but for a difference in elevation of 1 meter, which is too small to be of interest. We therefore multiply the estimates by a scale factor, say c. The log-odds ratio then is cb, and the odds ratio is e^{cb} for units c times larger than the original units. Units of 100 m seem much more convenient for these data. We obtain a regression coefficient of 0.00090653 in Box 17.15 for the ADA11 genotype and multiply this by 100 to yield the scaled log-odds ratio of 0.090653. When we exponentiate this value we obtain a scaled odds ratio of 1.09489. Thus for every increase in elevation of 100 m, the odds for the designated ACP1 genotypes increase by a factor of 1.09489. For the ADA12 and ADA22 genotypes we have a decrease in the odds for the designated ACP1 genotypes of $e^{-0.052426} = 0.94892$ per 100 meters.

Plots of the original data with the logistic regression lines are shown in Figures 17.6 and 17.7 for logit and proportion scales, respectively. As in the regression analyses described in Chapters 14 and 16, analyzing the residuals from the fitted model is important. Removal of bad data points may greatly improve the fit, and the pattern of deviations may suggest other explanatory variables that can be added to the model. Note that the point for ADA for the village of Lode seems to be an outlier. This point is based on a sample size of only seven, however, and thus is not the most deviant point statistically.

It is interesting to compare these results with an analysis of these same data (columns 4 and 5 of the first table in Box 17.15) as a $2 \times 2 \times 13$ three-way contingency table. If we use the methods of Box 17.10, we find that the three-way interaction is not significant, nor is ACP $\times$ ADA independent within vil-

Box 17.15 LOGISTIC REGRESSION ANALYSIS.

Data are ACP1 genotype frequencies in 13 Sardinian villages, collected to determine the relationship, if any, between genotype and elevation. Samples were separated on the basis of the ADA locus because it was believed there could be an interaction between it and the ACP1 locus.

(1)	(2)	(3)	(4)	(5)	(6)
			ACP1		
Village	Elevation (m)	ADA	A + BA	Other	*n*
Fonni	1000	ADA12 & ADA22	9	6	15
		ADA11	34	66	100
Seulo	797	ADA12 & ADA22	3	10	13
		ADA11	27	85	112
Aritzu-Belvi	796	ADA12 & ADA22	3	10	13
		ADA11	13	94	107
Burcei	648	ADA12 & ADA22	9	9	18
		ADA11	26	41	67
Lanusei	590	ADA12 & ADA22	9	16	25
		ADA11	43	69	112
Bitti	550	ADA12 & ADA22	4	16	20
		ADA11	26	93	119
Jerzu	442	ADA12 & ADA22	3	7	10
		ADA11	16	29	45
Lode	345	ADA12 & ADA22	5	2	7
		ADA11	34	85	119
Sedilo	228	ADA12 & ADA22	1	8	9
		ADA11	29	75	104
Ottana	185	ADA12 & ADA22	2	6	8
		ADA11	38	56	94
Villasimius	45	ADA12 & ADA22	1	3	4
		ADA11	25	31	56
Tortoli	15	ADA12 & ADA22	1	14	15
		ADA11	38	69	107
Oristano	10	ADA12 & ADA22	9	17	26
		ADA11	62	123	185

SOURCE: Data from E. Bottini (unpublished results).

Analysis

1. To perform a logistic regression, one of the variables must be selected as the dependent variable, Y. We select as Y the state of the ACP1 locus and consider the presence

Box 17.15 Continued

of the A or BA genotypes a ''success'' ($Y = 1$) and the presence of any other genotypes a ''failure'' ($Y = 0$).

2. A design variable corresponding to the two states of the ADA variable, column (3), is coded 0 for the ADA12 and ADA22 genotypes and 1 for the ADA11 genotype. Elevation, column (2), is used directly as an independent variable.
3. Design variables to allow testing for the presence of interaction between ADA and elevation can be constructed in more than one way. The simplest method is to multiply the independent variables for ADA and elevation together element by element. The result is a variable that for samples with the ADA12 and ADA22 genotypes is 0 and for those with the ADA11 genotype is equal to the village elevation.
4. The input to a logistic regression program thus consists of columns (4) and (5) as the dependent variable and various combinations of independent variables corresponding to the models of interest. The results are shown in the following table. (Intermediate computations are not shown, since they involve an iterative solution of the maximum-likelihood equations.)

(1) Model	(2) Log likelihood	(3) G	(4) df
Constant	−936.35665	67.74416***	25
Constant + ADA	−936.29664	67.62415***	24
Constant + Elev	−934.36246	63.75579***	24
Constant + ADA + Elev	−934.27078	63.57243***	23
Constant + ADA + Elev + ADA × Elev	−930.46984	55.97055***	22

For all models the G-values from the goodness-of-fit test are significant, indicating that additional independent variables are needed to fit the data completely.

5. Next we must compute the G-values corresponding to each effect we wish to test. First we test for the presence of an interaction (as we would in a two-way anova). The G-value corresponding to the effect of an ADA × Elev interaction is obtained by computing the difference between the goodness-of-fit G-values for a model with just the main effects (ADA + Elev) and a model with the main effects plus interaction (ADA + Elev + ADA × Elev). In our case, $G_{\text{ADA} \times \text{Elev}} = 63.57243 - 55.97055 = 7.60188$. This G-value has just 1 degree of freedom because we are subtracting a G-value with 22 degrees of freedom from one with 23 degrees of freedom. This G-value can also be computed as twice the difference between the log likelihoods of the two models we are comparing.

   The effect of ADA is computed by comparing the fit of the model with only ADA against that of the model with ADA + Elev. In our example, $G_{\text{ADA}} = 67.62415 - 63.57243 = 4.05172$. Similarly, the effect of elevation is computed by comparing the fit of the model with only elevation against that of the model with ADA + Elev.

BOX 17.15 CONTINUED

$G_{\text{Elev}} = 63.75579 - 63.57243 = 0.18336$. The degrees of freedom are equal to 1 in each case because we are subtracting a G-value with 23 degrees of freedom from one with 24 degrees of freedom. These results are shown in the following table.

(1) Source	(2) G	(3) df
ADA (adj. for Elev)	4.05172*	1
Elev (adj. for ADA)	0.18336	1
ADA × Elev	7.60188**	1

Thus there is significant interaction between the ADA locus and elevation in their effect on the frequency of the ACP1 genotypes. As in a Model I two-way anova, there is usually little interest in the tests for the main effects when an interaction effect is present.

6. Since it is significant we want to investigate further this interaction between the ADA locus and elevation. One approach is to regress elevation on ACP1 separately within each of the two ADA genotype classes. This can be done as two separate regressions or simultaneously by replacing the independent variables used above for elevation and for interaction by variables that are equal to elevation for samples corresponding to a particular ADA class and zero otherwise. The log likelihood and G-values will be the same as before, but now there will be separate parameters for the logistic regression of elevation on ACP1 within each ADA genotype class. The resulting parameter estimates, their standard errors, and a significance test are shown in the following table.

(1) Parameter	(2) b	(3) s_b	(4) t_s
Constant	−1.16084	0.29901	−3.88249**
ADA	0.57409	0.31371	1.829992
Elev (for ADA = 0)	0.00090653	0.00050988	1.777909
Elev (for ADA = 1)	−0.00052426	0.00018528	−2.829569**

These results indicate that the odds for the ACP1 A + BA genotypes tend to increase with elevation for individuals with the ADA12 and ADA22 genotypes (but not significantly) with a factor of $e^{0.00090653} = 1.000907$ per meter increase in elevation. On the other hand, the odds decrease significantly among individuals with the ADA11 genotype. The odds for ACP1 for the ADA11 genotype change by the factor $e^{-0.00052426} = 0.99948$ per meter increase in elevation.

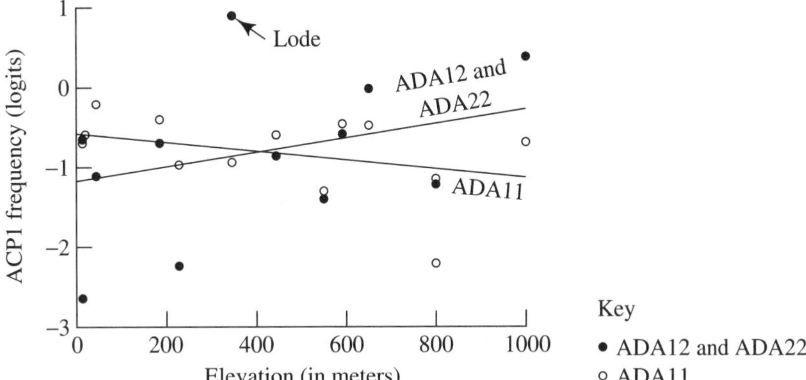

FIGURE 17.6 *Logistic regression of frequencies of ACP1 against elevation, separately for two ADA classes. Data from Box 17.15. Ordinate in logit scale.*

lages. The ACP × ADA tables are not homogeneous *across* villages, however, nor are the ACP × Village tables homogeneous for the two ADA genotypes, which is consistent with our results, since the villages differ in elevation. When an independent variable is continuous or when it is categorical but has more than two classes, the logistic regression approach is usually more informative than analysis as a three-way table is because it expresses the results directly in terms of the continuous variable. It also results in more powerful tests than do log-linear models because the effect being tested has fewer degrees of freedom.

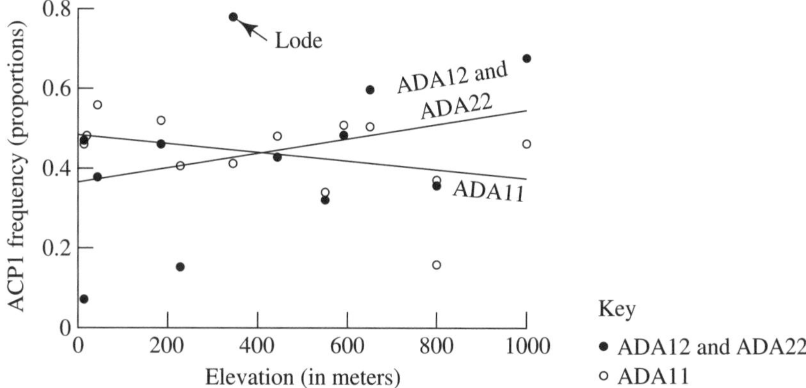

FIGURE 17.7 *Logistic regression of frequencies of ACP1 against elevation, separately for two ADA classes. Data from Box 17.15. Ordinate in proportion scale. Although the regression lines appear linear over the range of the independent variable shown, they actually are curvilinear functions.*

For more information on the application of logistic regression analysis to more-complex designs (including generalizations to the case of dependent variables with more than two response categories), refer to Agresti (1990). For interpreting the results of logistic regression analyses, Hosmer and Lemeshow (1989) may be useful.

In closing this section we would like to caution readers against complications that may arise when some of the assumptions underlying tests of proportions are violated. For example, if it is suspected that a factor X *predisposes toward disease* A, *and a retrospective study is carried out of patients with* A *in a hospital, investigators frequently will assemble a control group manifesting disease* B. *If the hospital admission rates differ for* A *and* B, *an apparent association may result between* A *and* X, *even if they are independent. This phenomenon is known as* ***Berkson's fallacy.*** *Another possible complication is that prospective studies (those that follow a cohort of individuals forward through time) and retrospective studies (those that examine events that occurred prior to a study) may yield quite different results as a function of differential duration and mortalities from disease. This case is known as* ***Neyman's example.*** *Finally, when several* 2×2 *tables are aggregated, an odds ratio that is similar in the* 2×2 *tables may become quite different when computed overall for a pooled contingency table. This case is known as* ***Simpson's paradox.*** *Workers should be aware of these (only apparent) anomalies and guard against them when designing their experiments. Goddard (1991) presents a readable discussion of these phenomena.*

17.7 RANDOMIZED BLOCKS FOR FREQUENCY DATA

The tests of independence of Section 17.4 can be regarded as tests of the effect of treatments on the percentage of an attribute. In that section, each treatment was applied to independently and randomly selected individuals that were different for the separate treatments. Thus, in the immunology example, two samples of mice were taken and subjected to different treatments—bacteria alone, or bacteria and antiserum. These mice had presumably been selected randomly from the pool of animals available to the investigator. In the example dealing with the two leaf types found on two different types of soils, the samples were obviously separate, consisting of different groups of trees on the two soils.

Sometimes collecting data in this manner is neither possible nor desirable. For example, consider an experiment that compares the phototactic response of a species of ticks under various experimental conditions. We could take a sample of ticks, divide it into several groups, and test each group under a separate set of conditions. What if we knew, however, that the behavior of the organisms varies depending on factors such as age or condition of the animal? In such a case, by chance more older individuals may be placed into one group than into the other, or some other sampling accident may take place. Thus treatment effects might be confounded with heterogeneity of the experimental subjects. There are three possible solutions to such a problem. If the factors causing heterogeneity (such as age or prior condition) are known, we can try to keep them constant. Using this approach, to obtain animals of uniform age we may have to select from a large base population of animals of various ages, which could be difficult and/or expensive.

If the factors causing heterogeneity of response are unknown, there are two possible approaches. One is simply to take large enough samples to overcome the heterogeneity. For obvious reasons, this solution is often not feasible nor desirable.

The alternative procedure, which is the subject of this section, is to use the same individuals for both treatments. Using this design, even if there is some heterogeneity in the inherent responses to the organisms, one can still look for and test for a significant *change* in their behavior due to the treatment. This same experimental design was discussed in Section 11.4 (randomized blocks). Whereas that section dealt with continuous variables that can be analyzed by analysis of variance, however, here we are considering attributes for which we can record only the frequency or percentage of individuals possessing a certain characteristic. As in its previous application, the method here is based on the correlation between treatments over blocks (which correspond to individuals in the example above) and assumes the absence of treatment × block interaction. Because frequently the same sample of individuals is exposed to two or more treatments, this method is also known as *repeated testing of the same individuals.* Another appropriate name for this technique is *testing of correlated proportions.*

This design is also indicated when there is some reason to believe that a prior treatment will affect the responses in a subsequent treatment. In such cases, clearly the same sample must be used because the effects of one treatment have to be compared with the effects of the prior treatment in the same individuals. An example is conditioning or learning in animals. Given a certain stimulus, individuals in a group of animals may or may not respond. If the same or a different stimulus is given at a later time, the presence or absence of a response in these animals may be conditioned by their previous response.

Our first example of repeated testing of the same individuals is the case of two treatments (only one repetition of the test). For continuous variables, this design is one of paired comparisons (see Section 11.5). This is a special case of randomized blocks in which there are only two treatments; hence each individual or block is tested twice. By analogy, for attributes we have the **McNemar test for significance of changes.** Box 17.16 shows an example in which 30 ticks were each tested twice for their phototactic response—first in a 1-inch arena and then in a 2-inch arena. The data consist of 30 paired scores. A plus was recorded if the tick left the arena toward the light, a minus if it moved away from the light. The

Box 17.16 RANDOMIZED BLOCKS FOR FREQUENCY DATA. (REPEATED TESTING OF THE SAME INDIVIDUALS OR TESTING OF CORRELATED PROPORTIONS.)

Individuals tested twice. McNemar test for significance of changes

The response of the rabbit tick *(Haemaphysalis leporispalustris)* to light was measured by placing individual ticks in arenas 1 or 2 inches in diameter and recording whether the tick left the arena on the side toward (+) or away from (−) the light. Each tick was scored first in the 1-inch arena and then in the 2-inch arena. Of interest was whether the response of an individual to light would change depending on the distance a tick had to crawl before its response was recorded.

The results can be summarized as follows:

	1-inch arena		
2-inch arena	−	+	Totals
−	8	5	13
+	9	8	17
Totals	17	13	30

SOURCE: Data from V. E. Nelson (unpublished results).

Note that the pattern of response appeared to reverse itself. In the 1-inch arena 13 out of 30 ticks turned toward the light, whereas 17 of the 30 did so in the 2-inch arena.

We label the cells of this table in the conventional manner:

	−	+	$\sum$
−	a	b	$a + b$
+	c	d	$c + d$
$\sum$	$a + c$	$b + d$	n

We wish to test whether the change in proportion between the two trials is significant. The null hypothesis is that the frequencies of the two types of "changers" are equal—that is, that the number of ticks that followed a (+) response with a (−) equaled those following a (−) response with a (+). If this condition is satisfied, the proportions in the

BOX 17.16 CONTINUED

two trials must be the same. Concisely stated, the null hypothesis is $H_0:(b - c) = 0$. The appropriate test statistic is computed by Expression (17.27):

$$G = 2\left[b \ln\left(\frac{2b}{b+c}\right) + c \ln\left(\frac{2c}{b+c}\right)\right]$$

$$= 2\left[5 \ln\left(\frac{2(5)}{5+9}\right) + 9 \ln\left(\frac{2(9)}{5+9}\right)\right]$$

$$= 2\left[5 \ln\left(\frac{10}{14}\right) + 9 \ln\left(\frac{18}{14}\right)\right]$$

$$= 2[-1.68236 + 2.26183] = 1.15894$$

It is advisable to apply Williams's correction, which in this case is $q = 1 - 1/2n$. Thus

$$q = 1 + \frac{1}{2(14)} = 1.03571$$

$$G_{\text{adj}} = G/q = 1.118$$

These G-values approximate the distribution of χ^2 with one degree of freedom. Clearly, there is little evidence of a significant change in behavioral response.

Individuals tested three or more times. Cochran's Q-test

The data are scores expressing the condition of 24 plants *(Acacia cornigera)* observed in three different months. All of these plants were from an experimental plot in which the growing shoots had been deprived of the ant *Pseudomyrmex ferruginea,* which normally lives within the stems. A positive rating represents a plant that has all or most of its uppermost shoot apices still intact and in which most of the foliage does not show the feeding damage of phytophagous insects. A negative rating represents a plant that has all of its growing shoot apices eaten off and in which more than half of the foliage shows severe damage. In the complete study (not shown here) there was also a control plot.

1. The data are arranged as a two-way table, using 0's for negative ratings and 1's for positive ratings. The $b = 24$ rows (blocks) of this table correspond to the individual plants scored, and the $a = 3$ columns (treatments) correspond to the repeated scoring of these plants in different months.

BOX 17.16 CONTINUED

Individual plants (blocks)	Month in which scored (treatments)			Totals
	March	June	August	
1	0	1	0	1
2	0	0	0	0
3	0	1	0	1
4	1	0	0	1
5	1	1	0	2
6	1	0	0	1
7	0	0	0	0
8	1	1	0	2
9	1	1	0	2
10	1	0	0	1
11	1	0	0	1
12	1	1	0	2
13	0	1	1	2
14	0	1	1	2
15	0	0	0	0
16	0	1	0	1
17	0	1	0	1
18	1	0	0	1
19	1	0	0	1
20	1	0	0	1
21	1	0	1	2
22	1	1	1	3
23	1	0	0	1
24	1	1	0	2
Totals	15	12	4	31

SOURCE: Data from D. H. Janzen (unpublished results).

2. Compute the row and column sums and record them along the margins of the table.

3. Grand total of the frequencies in the table $= \sum^{a} \sum^{b} Y = 31$

4. Sum of the row sums squared $= \sum^{b} \left(\sum^{a} Y \right)^2$

$$= 1^2 + 0^2 + 1^2 + \cdots + 2^2 = 53$$

Box 17.16 Continued

5. Sum of the column sums squared $= \sum^{a}\left(\sum^{b} Y\right)^2 = 15^2 + 12^2 + 4^2 = 385$

6. Test statistic [Expression (17.28)] $= Q$

$$= \frac{(a-1)\left[a\sum^{a}\left(\sum^{b} Y\right)^2 - \left(\sum^{a}\sum^{b} Y\right)^2\right]}{a\sum^{a}\sum^{b} Y - \sum^{b}\left(\sum^{a} Y\right)^2}$$

$$= \frac{(a-1)[a \times \text{quantity } \mathbf{5} - (\text{quantity } \mathbf{3})^2]}{(a \times \text{quantity } \mathbf{3}) - \text{quantity } \mathbf{4}}$$

$$= \frac{(3-1)[3 \times 385 - 31^2]}{(3 \times 31) - 53} = \frac{2(194)}{40}$$

$$= 9.700^{**}$$

This value is to be compared with a χ^2-distribution with $a - 1 = 2$ degrees of freedom. Since $Q = 9.700 > \chi^2_{.01[2]} = 9.210$, we reject the null hypothesis that the proportion of plants in good condition remained the same as the season progressed. It is reasonable to conclude, therefore, that when the plants are deprived of their protecting ants, their condition gradually deteriorates. Many of the plants in the treatment plot died a few months later, but the control plot showed little change.

30 paired observations can be conveniently recorded in a 2×2 table as shown in the box and repeated here:

	1-inch arena		
2-inch arena	−	+	$\sum$
−	8	5	13
+	9	8	17
$\sum$	17	13	30

Unlike the situation in Section 17.4, where we were interested in the proportions implied by the cells of the table, we are now more interested in the proportions in the margins. For example, it is no longer of particular interest that 8 of the 17 ticks that had a negative response in the 1-inch arena also responded negatively in the 2-inch arena. We are more interested in the fact that in the 1-inch arena 17 out of 30 ticks showed a negative reaction, whereas in the 2-inch arena 13 out of 30 ticks showed a negative reaction. We could rearrange the table as follows:

	−	+	$\sum$
1-inch arena	17	13	30
2-inch arena	13	17	30
$\sum$	30	30	60

This arrangement would imply, however, that our sample size is $n = 60$ ticks, twice the actual number of ticks involved. For this reason we cannot use the methods of Section 17.4 on such data. The proportions within this second table are not independent of one another; they are all based upon the same 30 ticks.

Inspecting the earlier table, we note that the difference between the proportions in the margins is a function only of the number of ticks that are ''changers''—that responded differently to the second test (the 2-inch arena) than they did to the first test. If there are no changers, or if the two types of changers are equally frequent, the marginal frequencies will be the same for both sets. In our example, 5 ticks changed from a positive to a negative reaction in the 2-inch arena, and 9 changed from a negative to a positive response. We may state as our null hypothesis that the proportion of individuals reacting negatively will be the same in both sizes of arenas. This statement implies that we expect the frequencies of the two types of changers to be equal. We can easily construct a G-test of this hypothesis. If we label the four cells of the table in the conventional manner (a, b, c, d), we can construct the following simple formula for G (approximating χ^2 with one degree of freedom):

$$G = 2\left[b \ln\left(\frac{2b}{b + c}\right) + c \ln\left(\frac{2c}{b + c}\right)\right] \tag{17.27}$$

This formula is derived easily from Expression (17.3). For the data in Box 17.16 $G = 1.15894$. Following our recommendation in Section 17.2 and Box 17.1, we apply Williams's correction to G, although in this case, with a nonsignificant G-value, the correction cannot affect the outcome. In the box we show that $G_{\text{adj}} = 1.118$. In view of the nonsignificance of G we accept the null hypothesis that there is no significant change in response of the ticks between the 1-inch and the 2-inch arenas.

When samples are small $[(b + c) < 25]$, an ''exact'' binomial test is recommended (note, however, that we have not followed this advice in Box 17.16). Such a test is carried out as follows. If we let k equal $b + c$, and let Y equal b or c, whichever is greater, then the desired probability is twice the probability of observing Y or more events of one type out of k from a binomial distribution with $p = q = 0.5$. We can employ the methods of Section 13.12 to evaluate this probability, either computing the probability or inferring it from the confidence limits for $\hat{p}$ in Statistical Table **P.**

In the tick experiment described in Box 17.16, all ticks were tested first in the 1-inch arena and then in the 2-inch arena. The experimenter has therefore confounded two possible effects, size of arena and order of test. The results suggest that more ticks were light-positive in the 2-inch arena (but recall that we cannot show this trend to be statistically significant). However, this potential effect could be due to the fact that by the time ticks were placed into a 2-inch arena, they had already been tested once before. If we wish to control for the possibility of such an effect, we can subject half the ticks to the 1-inch arena first, followed by the 2-inch arena, and reverse the order of arenas for the other half. Even though we employ equal frequencies for the two orders of treatment, however, we cannot be certain that these occur in equal frequency in the two types of changers. How can we analyze such a situation? The following table is an artificial data set intended to describe an imagined repetition of the tick experiment with a larger number of ticks, 100:

	1-inch arena		
2-inch arena	−	+	Totals
−	81	9	90
+	1	9	10
Totals	82	18	100

The numbers in this example are based on an experiment by Gart (1969), who developed the technique we are about to describe. In this data set there are only a few light-positive responses among the ticks. Of 10 changers in this table, 9 are positive in a 1-inch arena but negative in a 2-inch arena, and only 1 is negative, then positive. When we carry out McNemar's test and calculate the exact one-tailed probability of this outcome on the null hypothesis that the number of changers in each direction is equal, we obtain $P = 0.011$. Thus there is evidence that more ticks are light-positive in the 1-inch arena than in the 2-inch arena. Until now we have not considered the order of presentation of the arenas. The following table describes the minor diagonal of the 2×2 table above, i.e., the changers by tabulating the responses to first and second stimuli (treatments) against the order in which the two arenas were presented.

	Order of administration		
	First 1-inch then 2-inch	First 2-inch then 1-inch	Totals
Positive response to first treatment	5	1	6
Positive response to second treatment	0	4	4
Totals	5	5	10

Note that the total n equals 10, the number of changers. If we carry out Fisher's exact test (one-tailed) on this table (see Box 17.7), we obtain $P = 0.0238$, which suggests that the response of the first versus the second treatment is not independent of what the treatment is (i.e., 1-inch or 2-inch arena). From inspection of the table we conclude that the 1-inch arena induces significantly more light-positive responses. Now let us rearrange this table to indicate responses to the two arenas as these relate to which one was offered first.

	Order of administration		
	First 1-inch then 2-inch	First 2-inch then 1-inch	Totals
Positive response to 1-inch arena	5	4	9
Positive response to 2-inch arena	0	1	1
Totals	5	5	10

We obtain a table with the same total counts and column totals, but the row totals have changed. A one-tailed Fisher's exact test yields $P = 0.5$. We conclude that order of administration is independent of type of treatment and plays no role in the response of the ticks.

A method for comparing the changes in two groups, each tested twice, is illustrated by Feuer and Kessler (1989). When the responses of the individuals can occur in $k = 3$ *or more classes, rather than the two that we have considered, and the individuals are tested twice, the data can be arranged as a* $k \times k$ *square table. The diagonal frequencies,* f_{ii}, *of this table would list the frequencies of individuals that responded twice identically, and the off-diagonal frequencies,* f_{ij}, *would be those in class* i *for the first test and class* j *for the second test (or vice versa for frequencies* f_{ji}*). Such a design could also arise when two judges classify the same group of individuals into* k *classes, or when* n *males divided into* k *morphs mate with an equal number of females divided into the same* k *morphs. Maxwell (1970) shows how the margin frequencies of such tables can be compared and how to test the null hypothesis* $H_0: f_{ij} = f_{ji}$ *for all* i *and* j.

When individuals have been tested three or more times for the presence or absence of an attribute, a test analogous to the randomized-blocks design is possible. This test is called **Cochran's *Q*-test.** Its test statistic is based on a model that differs from those used for previous *G*-tests. The original observations are replaced by ones and zeros ($Y = 1$ corresponds to the presence of the attribute; $Y = 0$ corresponds to the absence of the attribute). The resulting data are entered in a two-way table and analyzed according to the standard equations for a randomized-blocks design. For tests of significance, however, the usual test statistic is modified so that the results are expected to follow the chi-square rather than the F-distribution. If we let A stand for the columns (or the treatment effects) and B for the rows (or blocks) of the randomized-blocks design (the individuals), the test statistic Q is computed as follows:

$$Q = \frac{SS_A}{(SS_{\text{total}} - SS_B)/(df_{\text{total}} - df_B)}$$

Since the data consist only of ones and zeros, a great deal of simplification is possible, resulting in the following equation:

$$Q = \frac{(a-1)\left[a \sum^{a}\left(\sum^{b} Y\right)^2 - \left(\sum^{a}\sum^{b} Y\right)^2\right]}{a \sum^{a}\sum^{b} Y - \sum^{b}\left(\sum^{a} Y\right)^2} \qquad (17.28)$$

Q is expected to follow the chi-square distribution with $a - 1$ degrees of freedom (where a is the number of times each of the b individuals has been tested). There does not appear to be a G-test equivalent to the Cochran Q-test.

In the second half of Box 17.16, Cochran's Q-test is applied to data showing that acacias gradually deteriorate when the ants of a species that normally lives within its stems are removed. The test shows that there is a significant change in the frequency of plants in "good condition" when the plants are observed in successive months after having been deprived of the ants which normally prevent herbivores from feeding on the tender growing shoots. See Box 17.16 for details.

EXERCISES 17

17.1 In an experiment to determine the mode of inheritance of the *green* mutant, 146 wild-type and 30 mutant offspring were obtained when F_1 generation houseflies were crossed. Test whether the data agree with the hypothesis that the ratio of wild types to mutants is 3 : 1. *Answer:* $G_{\text{adj}} = 6.43715$, $X^2_{\text{adj}} = 5.523$.

17.2 Locality A has been sampled extensively for snakes of species S. An examination of the 167 adult males that have been collected reveals that 35 of these have pale-colored bands around their necks. From locality B, 90 miles away, we obtain a sample of 27 adult males of the same species, 6 of which show the bands. What is the chance that both samples are from the same statistical population with respect to frequency of bands?

17.3 Of 445 specimens of the butterfly *Erebia epipsodea* from mountainous areas, 2.5% have light color patches on their wings. Of 65 specimens from the prairie, 70.8% have such patches (unpublished data from P. R. Ehrlich). Is this difference significant? Calculate the odds ratio. If you have access to a logistic regression program, assign confidence limits to the odds ratio. *Answer:* $G = 175.516$, $G_{adj} = 171.453$.

17.4 The following data on the sex ratios of adult long-billed marsh wrens in Seattle were presented by Verner (1964).

	Date	Males	Females
1961	May 6	14	11
	May 11	14	12
	May 25	14	12
	June 13	13	11
	June 19	13	9
	June 25	12	8
	July 8	9	6
1962	April 18	8	8
	May 22	6	8
	May 31	6	7
	June 2	6	7
	June 20	8	8
	July 23	7	5

Test whether there is a 1 : 1 sex ratio for each date and whether the observations on different dates are homogeneous. If heterogeneity is indicated, perform an unplanned test to obtain sets of dates homogeneous for sex ratios. Is there any evidence of heterogeneity between years?

17.5 In a study of polymorphism of chromosomal inversions in the grasshopper *Moraba scurra,* Lewontin and White (1960) gave the following results for the composition of a population at Royalla "B," Australia, in 1958.

		Chromosome CD		
		St/St	St/Bl	Bl/Bl
Chromosome EF	Td/Td	22	96	75
	St/Td	8	56	64
	St/St	0	6	6

Are the frequencies of the three different combinations of chromosome EF independent of those of the three combinations of chromosome CD? *Answer:* $G = 7.396$, $q_{min} = 1.008008$.

17.6 In this experiment, stem mothers of the aphid *Myzus persicae* were placed on one of three diets one day before giving birth to nymphs (data from Mittler and Dadd, 1966). Test whether the percentage of nymphs that developed into winged forms depends on the diet of the mother.

Type of diet	% Winged forms	n
Synthetic diet	100	216
Cotyledon ''sandwich''	92	230
Free cotyledon	36	75

17.7 Test agreement of observed frequencies to those expected on the basis of a Poisson distribution for the data given in (a) Table 5.5, (b) Table 5.6, (c) Table 5.7, (d) Table 5.9, and (e) Table 5.10. *Answer:* (a) $G = 40.152$. When using more precise values of $\hat{f}$ than are given in Table 5.5, one obtains $G = 40.408$.

17.8 Test agreement of observed frequencies to those expected on the basis of a binomial distribution for the data given in Tables 5.1 and 5.2.

17.9 For the pigeon data of Exercise 2.4 test the agreement of the observed distribution to a normal distribution by (a) using the Komolgorov–Smirnov test on the ungrouped data and (b) using G-tests and chi-square tests for goodness of fit to the expected frequencies calculated in Exercise 6.1. *Answer:* (a) $D = 0.09386$.

17.10 Repeat Exercise 17.9 with the pigeon data as transformed in Exercise 2.6. Use the expected frequencies computed in Exercise 6.2.

17.11 Calculate expected normal frequencies for the butterfat data of Exercise 4.3. Test the goodness of fit of the observed to the expected frequencies by means of the Kolmogorov–Smirnov test. *Answer:* $D = 0.06150$.

17.12 If a standard drug cures disease X 60% of the time, what sample size would be required in an experiment designed to have an 80% chance of detecting (at the 5% level of significance) a new drug that would cure disease X (a) at least 65% of the time, (b) at least 70% of the time, and (c) at least 80% of the time?

17.13 The following table of data is from a study of caste differentiation in the ant *Myrmica rubra* by Brian and Hibble (1964).

Frequencies of larvae of two sizes in three colonies at weekly intervals

	Week	Small larvae	Large larvae
Colony 1	2	12	47
	3	19	29
	4	9	52
	5	5	42
	6	8	58
	7	2	43
	8	0	35
	9	7	50
	10	9	38
Colony 2	2	0	47
	3	28	36
	4	2	44
	5	1	45
	6	4	33
	7	2	46
	8	3	38
	9	3	33
	10	5	50
Colony 3	2	55	23
	3	71	4
	4	40	21
	5	29	9
	6	42	17
	7	26	8
	8	28	21
	9	29	13
	10	57	7

Are the three factors (A = larval size, B = colonies, and C = weeks) jointly independent? If not, test if the factors taken two at a time are independent. Is the degree of association between weeks and larval size homogeneous over the three colonies—that is, is there any interaction? *Answer:* $G_{A.B.C.} = 724.750$, 42 df; $G_{ABC} = 30.174$, 16 df (using the Williams correction; see Section 17.2).

17.14 In studying the effect of 1% indoleacetic acid in lanolin on fruit setting in muskmelon, Burrell and Whitaker (1939) obtained the results in the following three-way table.

		Number of flowers that did or did not set fruit			
		Treatment		Control	
		Fruit set	Fruit not set	Fruit set	Fruit not set
	1	13	10	5	23
	2	2	10	1	7
Replication	3	13	1	6	10
	4	5	6	5	9
	5	13	5	6	14

Is the effect of the treatment on fruit setting independent of the effect of different replications? Within each treatment are the replicates independent of fruit setting? Examine the Freeman–Tukey deviates for the model that tests for conditional independence of treatment versus fruit setting, given the replicates.

17.15 A doctor had 25 patients who suffered from disease X during a recent epidemic. Although antibiotic B appeared to have no curative value against X, only 6 out of 11 patients administered B developed secondary infections, but 11 out of 14 patients not given B developed such infections. Would the doctor be justified in believing that the administration of B was beneficial to his patients? Compare the results of the *G*-test (using Williams's correction) with those of the chi-square test using the continuity correction. Calculate the odds ratio. If you have access to a logistic regression program, assign confidence limits to the odds ratio. *Answer:* $X^2_{\text{adj}} = 0.716$.

17.16 In F_2 crosses between progeny of yellow-green and chlorophyll-deficient strains of lettuce, Whitaker (1944) obtained the following results.

Family number	Yellow-green	Chlorophyll-deficient
15487	482	156
15395	92	40
15488	28	12
15396	256	96
15397	158	47

Test agreement to a 3 : 1 ratio. Are the proportions of the two color types homogeneous over the families?

17.17 Perform a logistic regression of the log odds of being male on date, using the data from Exercise 17.4. For the regression, replace the calendar dates with their equivalent number of Julian days of the year. Test whether the relationship is the same for both years.

17.18 In a study of the relationship between mammary tumors and dietary fiber and fat in rats, Cohen et al. (1991) obtained the following data. (Each treatment class contained 30 rats.)

Diet	No. rats with tumors
High fat	27
High fat + fiber	20
Low fat	19
Low fat + fiber	14

Compute the odds ratios of interest and their confidence limits. Test whether there is evidence of differences among the treatments. Also test whether the effect of fat content of the diet is independent of the effect of fiber in the diet.

17.19 The following results from a logistic regression of the effects of sex, race, and tar on the odds ratio for Kreyberg I lung cancer were extracted from the study by Harris et al. 1993. The logistic regression coefficients have been adjusted for age, year of diagnosis, alcohol consumption, body size, education, and other variables in the original study. The regression coefficients for tar exposure are relative to zero exposure.

Variable	Regression
Sex (F = 0, M = 1)	0.05
Race (black = 0, white = 1)	0.48
Cumulative (lifetime) tar exposure (in kg):	
1–4	1.94
5–8	3.02
≥ 8	3.76
Tar × Sex:	
1–4, M	0.23
5–8, M	0.58
≥ 8, M	0.58

Estimate the odds ratio for a white male who has accumulated 5 to 8 kg of tar in his lifetime. Compare this value with that of a black male who has accumulated the same amount of tar.

17.20 The following data were taken from a case-control study by Zang and Wynder (1992) on the risk of lung cancer among cigarette smokers. Control individuals were patients admitted to the same hospital on the same day, matched as far as possible with respect to age and gender, but with different diagnoses. Adjustments for age are not included here (as they were in the published study) in order to simplify the computations.

Most recent number of cigarettes per day	Males Kreyberg I Lung cancer	Males Control	Females Kreyberg I Lung cancer	Females Control
Never smoked	15	822	22	941
1–10	36	136	20	118
11–20	133	328	122	193
21–40	226	311	156	115
≥41	127	91	40	17

Compute the odds ratios and log-odds ratios for having lung cancer relative to those who have never smoked (separately for each sex). Illustrate graphically. Analyze using the current level of smoking as a multistate categorical variable. Do the two sexes show the same relationships?

17.21 Reanalyze Exercise 17.20 using logistic regression analysis with the current level of smoking as an independent variable. Use the following arbitrary numerical values for this variable for the classes: 0, 5.5, 15.5, 31.5, and 45.5. Do the two sexes show the same relationship? Superimpose the logistic regression function on a plot of the data, as in Figures 17.6 and 17.7.

18 Miscellaneous Methods

We conclude our introduction to biometry by presenting seven methods that do not easily fit into any of the previous chapters. Five of them are nonparametric tests of significance and all are useful in a variety of situations.

A method for combining the outcome of several experiments to obtain an overall significance test of a given hypothesis is the topic of Section 18.1. In Section 18.2 we discuss runs tests, which test for randomness of nominal data in a sequence of events. Section 18.3 introduces randomization tests, which are extremely versatile for testing the significance of a variety of statistics about whose distribution we can make no assumptions for a randomization. The only requirement test is random sampling of the data. This method is generally very heavy on computation and usually requires a digital computer. Often, however, it is the technique of choice. After showing some applications of randomization tests, we illustrate two special methods, isotonic regression and the Mantel test, that require randomization tests for an evaluation of their significance.

The topic of randomization tests leads us naturally to Tukey's jackknife and Efron's bootstrap (Section 18.4), procedures that correct for bias in the estimate of a statistic and also furnish an approximate standard error. Finally, in Section 18.5 we discuss the outlook for future developments in biometry and introduce the concept of data analysis, which promises considerable progress in biometric research.

18.1 Combining Probabilities from Tests of Significance

Combining probabilities from significance tests is a very useful technique, which we personally have employed repeatedly. A series of separate significance tests on different sets of data may test the same scientific (but not statistical) hypothesis. Each test produces a probability value for the particular outcome, assuming the null hypothesis to be correct. These probabilities may be low enough to suggest but not establish statistical significance. Fisher (1954, Section 21.1) developed a technique to combine these probabilities to create an overall test for significance.

Although we have used this method repeatedly with actual data, we will illustrate it with a hypothetical example to show the full range of its applicability. Box 18.1 describes experiments in which a physiologist is studying oxygen consumption in two species of crayfish. Experiment 1 is a careful study of oxygen consumption of individuals of species A and B, with separate readings obtained for each of 14 individuals of A and 12 of B. In this and all subsequent experiments, species A appears to have the higher oxygen consumption. A t-test results in a value of $t_s = 1.90$, and since the degrees of freedom are 24, $P = 0.073$. The null hypothesis cannot be rejected, yet the probability is suggestively low.

The investigator has recourse to four other experiments. In a published study by another author, equal numbers of the two species had been left in separate

Box 18.1 **COMBINING PROBABILITIES FROM INDEPENDENT TESTS OF SIGNIFICANCE.**

Five experiments testing oxygen consumption in two species of crayfish. See Section 18.1 for detailed account of each experiment and statistical test.

Test	P	$\ln P$
1. t-test		
$n_1 = 14, n_2 = 12$ 24 df $t_s = 1.90$	0.073	−2.61730
2. Chi-square test for independence		
1 df $X^2 = 2.95$	0.086	−2.45341
3. Mann–Whitney U-test		
$n_1 = 10, n_2 = 10$ $U_s = 72$	0.10	−2.30259
4. Anova, single-degree-of-freedom comparison		
$\nu_1 = 1, \nu_2 = 20$ $F_s = 3.522$	0.080	−2.52573
5. t' (t-test with unequal variances)		
16.2 df $t'_s = 2.040$	0.060	−2.81341
Total		−12.71243

$-2\sum \ln P = -2(-12.71243) = 25.42486$

If all the null hypotheses in the experiments were true, this quantity would be distributed as χ^2 with $2k$ degrees of freedom (k = the number of separate tests and probabilities).

Since our value of $-2\sum \ln P$ is greater than $\chi^2_{.01[10]} = 23.209$, we reject the null hypothesis that there was no difference between the oxygen consumption of the two species in the five experiments.

containers with the same volume of water, and the proportion of survivors had been recorded after a fixed time interval. Fewer of species A survived, and the significance of this result was tested by a chi-square test for independence with one degree of freedom, yielding $X^2 = 2.95$, corresponding to a probability of 0.086. The investigator undertakes a similar test; however, he puts only 10 individuals of each species into a flask, limits oxygen supply by pouring paraffin over the surface, and studies time to death of each crayfish, assuming that those with the greater oxygen consumption will die first. Although he has fairly exact times of death for all crayfish, because of the uncertainty of the distribution of this variable he prefers to analyze it by a nonparametric technique involving ranking and employs the Mann–Whitney statistic. Although interpolation is difficult in Statistical Table **U,** he obtains an approximate probability of 0.10 for $U_s = 72$ for this experiment.

A fourth study involving an analysis of separate batches of four species of crayfish (including species A and B) leads to an analysis of variance. A single-degree-of-freedom comparison between A and B yields a suggestive F_s-value with a corresponding P-value of 0.080. Finally, intrigued by all these results, the physiologist undertakes yet another experiment involving measurements of individual crayfish. Because he entrusts part of this experiment to an inaccurate technician, however, the result is a much greater variance for the oxygen readings for species A than those for species B, so a regular t-test cannot be applied. Using the interpolated value (see Section 13.3 and Box 13.3), the investigator obtains a value of $t' = 2.040$ at 16.2 degrees of freedom, corresponding to a probability of 0.060. What can he conclude as a result of all these findings?

The computation is based on the fact that ln P is distributed as $-\frac{1}{2}\chi^2_{[2]}$ or -2 ln P is distributed as $\chi^2_{[2]}$. By evaluating twice the negative natural logarithm of each of the five probabilities, considering each to be a $\chi^2_{[2]}$, and summing these values in the manner of Section 17.3, we obtain a total that can be looked up under 10 degrees of freedom in this example, since there are two degrees of freedom for each probability value. In Box 18.1, the resulting $-2 \sum \ln P$, when compared with $\chi^2_{[10]}$, yields a probability between 0.01 and 0.001, leading to an overall rejection of the null hypothesis. On the basis of all these experiments together, then, we conclude that species A has a higher oxygen consumption than does species B. Details of the computation are furnished in Box 18.1

The usefulness of combining probabilities in this instance should be apparent. This method can also be used when the separate significance tests to be combined are all of the same type but a joint overall statistical analysis is not possible. For example, suppose we measure the differences in production of a metabolite by two groups of mice, each with different diets. This experiment is repeated three times for one-week intervals. The analysis could be set up as a two-way anova with weeks as blocks and the two diets as the columns of the table. It may be, however, that the replication of this experiment is unequal, presenting the more complex problem of trying to solve a two-way anova with unequal and disproportional subclass numbers. Rather than do this, we may wish to carry out separate anovas or t-tests on each week, obtain the probability value for each of the tests, and combine it in the manner just shown (of course, one would not be able to test for interaction by this technique).

Fisher's method for combining probabilities has also been used by one of us to show that patterns of genetic variation in human populations match patterns of linguistic variation. These analyses were based on different sets of samples for different genetic loci, so they could not be combined into a single overall analysis. By obtaining a P-value for the test on each locus and then combining these P-values by Fisher's method, we were able to conclude that overall genetic variation is related to linguistic variation.

A limitation of this method is that the exact probability is required for each significance test. Because the conventional statistical tables furnish only selected probabilities, either inverse interpolation must be used in these tables or more-comprehensive tables of the probability distribution functions must be consulted. Alternatively, computer programs that calculate the tail areas are widely available; they are included in the BIOM-pc package. Ling (1978) reviews procedures for the t-, χ^2-, and F-distributions. When exact probability values are not available, a conservative approach that avoids interpolation is to employ the next higher tabled P. If $-2\Sigma \ln P$ nevertheless emerges significant, we can be quite firm in rejecting the null hypothesis.

Combining probabilities is a simple example from a wider array of techniques that have become popular in recent years under the name of ***meta-analysis.*** *These techniques, which have been applied extensively in medical research and in ecology, are used to reach overall conclusions about the variables of interest from numerous results published in the relevant literature. Often such meta-analytic studies include as many as 200 or more published papers. Interested readers should refer to Hedges and Olkin (1985).*

18.2 Tests for Randomness of Nominal Data: Runs Tests

In this section we discuss ways of testing whether events occur in a random sequence or whether the probability of a given event is a function of the outcome of a previous event. These tests, which are of great general usefulness, are known as **runs tests** (for reasons that will become obvious shortly). Runs tests are best introduced by an example. Consider the following familiar hypothetical case, which has undisputable biological implications. Imagine a line of 20 youngsters in front of a cinema box office, with an equal number of boys and girls. They could be lined up in many ways based on the permutations of the 10 boys and 10 girls. We will not calculate exactly how many arrangements are possible, but will concentrate on extreme departures from random arrangement. If these youngsters were eight-year-olds, a natural arrangement would be 10 boys followed by

10 girls, shown symbolically as

BBBBBBBBBBGGGGGGGGGG

Clearly, this arrangement is not random and would be quite unlikely to occur by chance. The other extreme is an arrangement that would be likely in the same group of youngsters 10 years later:

BGBGBGBGBGBGBGBGBGBG

Note the regular alternation between the sexes. How can we devise a test for these departures from randomness?

Let us call a sequence of one or more like elements preceded and followed by unlike elements a **run.** The initial and terminal sequences, however, can only be followed or preceded, respectively, by unlike elements. In our case the eight-year-olds comprise two runs, one of 10 boys, followed by one of 10 girls. On the other hand, the eighteen-year-olds comprise 20 runs, since each sequence consists of a single individual preceded and/or followed by a member of the opposite sex.

Statisticians have worked out the expected mean and distribution of the number of runs in dichotomized samples containing n_1 individuals of one type and n_2 individuals of the other. The critical numbers of runs at both tails of the distribution of runs have been tabulated for unequal sample sizes between 2 and 20 for n_1 and n_2 and for equal sample sizes from n between 10 and 100 (see Statistical Table **AA**). Any number of runs that is *equal to or less than* the desired critical value at the left half of any row in the table or is *equal to or greater than* the desired critical value in the right half of any row leads to a rejection of the hypothesis of random arrangement. For larger samples, we use a normal approximation to the expected value and standard deviation of the distribution of runs, as shown in Box 18.2. We can then test whether the observed number of runs deviates significantly from the expected value, basing our conclusion on the table of areas of the normal curve (Statistical Table **A**). For small samples, there are few computations, and even for large samples the computation is quite simple.

Box 18.2 shows a runs test for dichotomized data, such as the sexual composition of the box office line we examined earlier. There are three major applications of the runs test for dichotomized data in biometric work.

The first of these is for a dichotomy that occurs naturally, as in differences between two sexes, two color phases, two different species, and so on. The example in Box 18.2 is such a case. Among bees and wasps, fertilized eggs give rise to females, unfertilized eggs to males. Females fertilize some of these eggs by releasing sperm stored in the spermatheca during copulation, while laying others unfertilized, which result in males. Are fertilized eggs laid randomly or do they occur in batches? In Box 18.2 the sequence of male and female eggs is shown, with runs underlined. There are four runs. Consulting Table **AA,** we find that if the sequence is random, the probability of as few as 4 runs or as many as

Box 18.2 A RUNS TEST FOR DICHOTOMIZED DATA.

A wasp produced 18 offspring in the following sequence, where F stands for females (fertilized eggs) and M for males (unfertilized eggs); n_1 (females) $= 12$; n_2 (males) $= 6$; r (number of runs) $= 4$ (sequences of one or more like elements preceded and/or followed by unlike elements). The runs have been underlined.

F F F F F M M M F F F F F F F M M M

Critical values ($P = 0.05$) for r from Statistical Table **AA:**
Lower bound (from column 0.025) $= r = 4$
Upper bound (from column 0.975) $= r = 13$

We conclude that the two types of eggs were not laid in random sequence. The eggs are laid in sequences of one sex. A sample value of $r = 13$ would have indicated an alternating sequence of male and female eggs.

When one of two unequal sample sizes is greater than 20, we use a normal approximation and test

$$t_s = \frac{r - \mu_r}{\sigma_r} = \frac{r - [2n_1n_2/(n_1 + n_2)] - 1}{\sqrt{[2n_1n_2(2n_1n_2 - n_1 - n_2)]/[(n_1 + n_2)^2(n_1 + n_2 - 1)]}}$$

where μ_r is the expected number of runs, and σ_r is its standard deviation. For $n_1 = n_2 = n$ this formula simplifies to

$$t_s = \frac{r - \mu_r}{\sigma_r} = \frac{[r - n - 1]}{\sqrt{n(n-1)/(2n-1)}}$$

We reject the null hypothesis at the 5% level if t_s is greater than 1.960 (using Statistical Table **A**). For equal sample sizes, however, Table **AA** furnishes critical values for r up to $n = 100$.

13 runs is only 0.05. We may therefore conclude that the two types of eggs were not laid in a random sequence, but were laid in sequences of one sex. The boys and girls lined up in front of the box office, discussed earlier, are another instance of naturally occurring dichotomies.

A second application of the runs test is for studying sequences of dichotomies defined by the investigator—for example, the plus and minus signs representing deviations from expectation in Table 5.4, the sex ratios in 6115 sibships of 12. Recall that we fitted expected binomial frequencies to these data, discovering that observed frequencies were higher than expected at the tails and deficient at the center of the distribution. The G-test for goodness of fit discussed in Section 17.2 substantiated the departure from expectation by a significant value of G. By looking only at the signs of the deviations, we are ignoring their magnitude. Thus it is conceivable (but improbable) that we might have a sequence of positive deviations for the first half of the distribution, followed by a sequence of negative deviations, with all of these deviations being quite minute and the observed data fitting their expectations quite faithfully. In cases where the fit to expectation is satisfactory, however, pattern of pluses and minuses among the deviations is much more likely to be random, and any departure from randomness, especially in the direction of long sequences, will be matched by large deviations from expectation.

Let us look at the pattern in column (6) of Table 5.4. There are 9 plus signs and 4 minus signs, and only 3 runs in the data. This result is significant at $P = 0.025$. We conclude that the sequence of signs of the deviations is not random. Of course, this test does not prove that these data do not fit the binomial expectations. This was not the hypothesis being tested. The test does, however, by an extremely simple method, suggest that all is not as expected in this instance and leads us to further tests of various hypotheses.

The third major application of the runs test is for dichotomized data, the so-called **runs test above and below the median.** This test is especially useful in testing the random sequence of a series of observations. As you will recall, we have stressed that independence of the observations is a fundamental assumption of analysis of variance, without which the value of the significance test is put into question (see Section 13.2). In Section 13.2 you were referred to a significance test to be introduced later. This is the runs test above and below the median. We find the median item in a given sample to be used for a t-test or an anova and label all items above the median item + and all those below it − (any variates equal to the median must be ignored). Then we array these items in their natural order in the sample and undertake a runs test of the pluses and minuses. If individual variates are independent of their predecessors—that is, the data are in random order—the values above and below the median (the plus and minus signs) should be in random sequence. If, on the other hand, the data lack independence, there may be either more or fewer runs than expected.

It is obvious how a sequence over an area as, for example, a transect across a field, might lack independence. Several plots might occur on a particularly rich patch of ground followed by other plots on relatively infertile soil. A similar lack of independence can occur in a temporal sequence. For example, a technician weighing specimens for a given analysis might commit various types of errors. He or she might have a bias for high readings, and would periodically overcorrect these by deliberate low readings, creating an unusual alternation of high and low readings—that is, a greater number of runs than expected. Or the balance used for weighing may gradually go out of adjustment during the experiment, yielding a trend in which all the later-weighed specimens would tend to be lighter than the earlier-weighed specimens. This phenomenon would tend to place earlier-weighed specimens above the median, and later-weighed ones below the median. A similar trend would occur if the specimens had been dried in an oven from which they were all removed before weighing, permitting them to absorb water from the atmosphere before being weighed. Specimens weighed last would have had the greatest opportunity to absorb water and would likely be heaviest. In a perfect linear trend, the plus or minus deviations from the median would be in two groups to each side of the median. In such a case the runs test provides a simple, distribution-free test for the significance of regression with a null hypothesis simpler than that of the ordering test discussed in Section 14.12. Neither test can substitute for regression analysis, of course, since it does not fit a line or allow one to predict the dependent variable.

We will briefly consider a second type of runs test called **runs up and down,** that is especially suited to trend data. If n items are ordered in their natural sequence of occurrence and the sign of the difference from the previous value is recorded for each item except the first, we can carry out a runs test on the resulting sequence of $n - 1$ signs. These signs would all be alike if the data were monotonically increasing or decreasing. Cyclical data would show more than the number of runs expected in a random sequence of values. The expected distribution of runs has been tabulated for sample sizes up to $n = 25$ (see Statistical Table **BB**). The normal approximation shown in Box 18.3 is adequate for larger samples.

Box 18.3 illustrates a test for runs up and down. The data are from a selection experiment in which survival to the pupal stage was a measure of fitness, possibly correlated with the progress of selection. According to the simple analysis of this test, the results show no significant trend in successive differences. Visual inspection of the data, however, indicates an apparent downward trend followed by a recovery after generation 20 as Figure 18.1 shows. This discrepancy illustrates two points: the need for careful statistical testing of what may appear obvious, and that all runs tests apply to attributes only. Runs tests work with dichotomized attributes, such as pluses and minuses or males and females, but they take neither absolute magnitude nor order of magnitude into consideration.

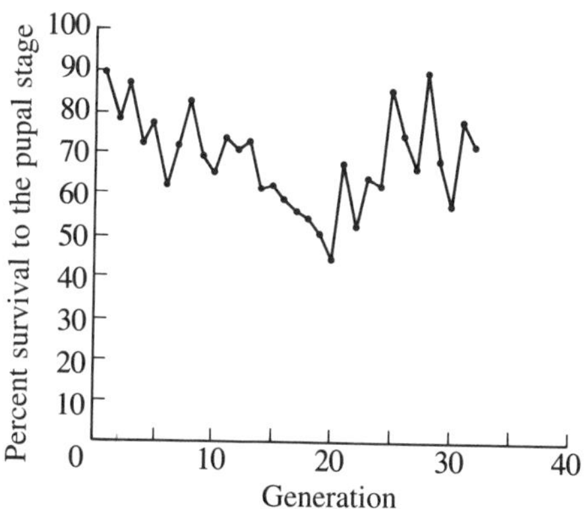

FIGURE 18.1 *Percent survival to pupal stage in the CP line of* Drosophila melanogaster *selected for peripheral pupation site. (Data from Sokal, 1966.)*

Box 18.3 A RUNS TEST FOR TREND DATA (RUNS UP AND DOWN).

Percent survival to the pupal stage in 32 generations of a line of *Drosophila melanogaster* selected for central pupation site. The data are graphed in Figure 18.1. From the differences between successive generations, the following set of signs is obtained (the runs have been underlined): (+) and (−) indicate an increase or a decrease, respectively, from the previous generation.

Generation	1	2	3	4	5	6	7	8	9	10	11
Increase or decrease		−	+	−	+	−	+	+	−	−	+

Generation	12	13	14	15	16	17	18	19	20	21	22
Increase or decrease	−	+	−	+	−	−	−	−	−	+	−

Generation	23	24	25	26	27	28	29	30	31	32
Increase or decrease	+	−	+	−	−	+	−	−	+	−

SOURCE: Data from Sokal (1966).

Number of points or elements, $n = 31$. Number of runs (sets of like signs preceded or followed by unlike signs), $r = 23$. Since $n > 25$, we test r by means of a normal approximation:

$$t_s = \frac{r - \mu_r}{\sigma_r} = \frac{r - [(2n - 1)/3]}{\sqrt{(16n - 29)/90}}$$

where μ_r is the expected number of runs and σ_r is its standard deviation. The numbers in this expression are constants independent of the specific problem. In this example,

$$t_s = \frac{23 - [61/3]}{\sqrt{[16(31) - 29]/90}} = \frac{23 - 20.333}{\sqrt{5.1889}} = \frac{2.667}{2.278} = 1.171$$

Clearly, the number of runs is not significantly different from expectation (since t_s is less than 1.960). We cannot show any departure from a random trend in survival over the 32 generations tested.

Obtaining 5% significance in this sample would require $\mu_r \pm 1.960\sigma_r$ runs, which works out to $20.333 \pm 1.960(2.278) = 15.87$ or 24.80 runs. Thus, with 15 or fewer runs we would have concluded that one or more systematic trends were exhibited by the data. With 25 or more runs we would be led to suspect a more or less regular alternation of differences, as might be caused by a cyclical phenomenon.

For sample sizes $n \leq 25$ the significance of an observed number of runs can be looked up directly in Statistical Table **BB**.

Therefore, a small positive deviation counts as much as a large positive deviation. Had the experiment of Box 18.3 and Figure 18.1 been terminated at generation 20, we could have shown a significant downward trend using ordinary regression analysis, even though this trend cannot be shown by the runs test, which uses less information from the data. With this caution, we readily recommend runs tests as ''quick and dirty'' methods for assessing trends and randomness. Runs tests are also useful when the investigator is unwilling to make assumptions necessary for more elaborate tests, such as a regression analysis. When ranks or actual measurements are not at hand, runs tests are often the only practical alternatives.

18.3 RANDOMIZATION TESTS

As the quantitative analysis of biological data becomes ever more common, we are frequently required to test for significance in situations where we know little or nothing about the expected distribution of the variables or statistics being tested, or where we know that the data to be analyzed do not meet the assumptions required for the customary statistical tests. In such instances, the **randomization test,** a nonparametric approach, has been found to be a powerful tool, and because of its structural simplicity it has considerable appeal. A randomization test involves three steps: (1) Consider an observed sample of variates or frequencies as one of many possible but equally likely different outcomes that could have arisen by chance. (2) Enumerate the possible outcomes that could be obtained by randomly rearranging the variates or frequencies. (3) On the basis of the resulting distribution of outcomes, decide whether the single outcome observed is deviant (= improbable) enough to warrant rejection of the null hypothesis. In other words, is the probability of obtaining an observation as deviant as or more deviant than the single outcome less than the desired significance level α? Although many randomization tests involve heavy computation, the general availability of computers has made them feasible, so they are applied frequently. For a comprehensive treatment of the subject, see Manly (1991).

There are two types of randomization tests, although the terminological distinctions between them are not always made. Sometimes determining all the possible outcomes in a given problem is feasible (either by probability theory or by enumeration). We can then perform an **exact randomization test** and base our decision on the proportion of outcomes that are as deviant as or more deviant than the observed sample in one or both tails. If this proportion were less than, say, $\alpha = 0.05$, we would reject the null hypothesis. If the structure of the problem is not too complicated, we can compute the number of possible outcomes using probability theory. You are already familiar with two such examples. In Section 5.2 we worked out the probability of sibships with 14 females and 3 males on the assumption that all simple events were equally likely (since the assumed sex ratio was 50:50). In Section 17.4 we used Fisher's exact test to evaluate the probability of a given outcome (and of more deviant outcomes), assuming independence of two criteria of classification and the equal likelihood of all possible distributions of the n observations into the four cells, given fixed margin totals.

Let us consider an example of an exact randomization test in which we can use probability theory. In a selection experiment in the flour beetle, *Tribolium castaneum,* Sokal and Sonleitner (1968) studied average deviations of zygotic frequencies of eggs as compared with equilibrium expectations based on adults of the parent generation. In one series, deviations were recorded for generations 2 through 8, as well as for generations 18 through 28; no observations were made for generations 9 through 17. Preliminary results suggested that the eggs showed an increase of mutant homozygotes and a decrease of wild-type homozygotes. Was this trend uniform through the experiment or was it altered during the later generations? The results showed that the deviation pattern, ‘‘decrease of wild type—increase of mutants,’’ included the seven early generations, as well as one of the late generations. All other patterns of deviations, considered as a single group, contained only late generations. These data suggest that there was a distinctive pattern of deviations of zygotic frequencies during the early generations. We can represent the results as follows:

Pattern of deviations	Generations	
Decrease of wild type—increase of mutants	2, 3, 4, 5, 6, 7, 8, 20	$n_1 = 8$
Other types of deviations	18, 19, 21, 22, 23, 24, 25, 26, 27, 28	$n_2 = 10$

Let us assume that the arrangement of the generations in the two deviation groups of size 8 and 10 is due entirely to chance. In how many different ways can we arrange the total of 18 generations into two groups of 8 and 10 generations, respectively? This problem is simple if we remember some of what we learned in Section 5.2. We can evaluate the number of different arrangements by computing the number of ways in which 8 items can be selected from 18 to form the first deviation group (of necessity leaving behind the second group of 10 items). There are $\binom{n_1 + n_2}{n_1} = \binom{18}{8} = 18!/8!10! = 43{,}758$ ways of setting up the groups as specified. Out of this very large number of possible arrangements, how many could have resulted in the distribution of the items as shown in the original sample (that is, 7 early and 1 late generation in the first deviation group)? Since there are 7 early generations and 11 late ones, there are 11 ways in which 1 late generation can combine with the 7 early generations to form the sample of 8 making up the first group of deviations. Thus, the probability of obtaining a result as unusual as the one observed is 11 out of 43,758 outcomes, i.e., only 0.000,251,4.

Since this is a two-tailed test (the late generations could favor deviations of the first type), we must consider the probability of obtaining such deviant samples in the other direction. In this case the other extreme is 0 early and 8 late generations, with a probability of $\binom{11}{8}\bigg/\binom{18}{8} = 165/43758 = 0.003{,}770{,}7$. As mentioned in Chapter 17, an outcome is "worse" than another only if its probability is less than or equal to the probability of the first outcome. Since the event "0 early and 8 late" has a probability $P = 0.003{,}770{,}7$, which is greater than $P = 0.000{,}251{,}4$ for the event "7 early and 1 late," we do not combine these two probabilities. The probability for the two-tailed test is therefore 0.000,251,4. We must clearly reject our null hypothesis that the arrangement of the early and late generations in the two groups of deviations is entirely at random.

Suppose the outcome had not been so clear-cut and there had been 6 early generations and 2 late ones in the sample of eight deviations of the first type. There are $\binom{7}{6} = 7$ ways of obtaining 6 early generations out of the 7 available, and these have to be matched with 2 late generations, which can be taken in $\binom{11}{2} = 55$ ways out of the 11 late ones. Therefore, there are $7 \times 55 = 385$ ways of obtaining 6 early and 2 late generations for a sample of 8 deviations. Even this outcome would have represented only 385 out of 43,758 possible outcomes, or a probability of 0.008,798. Considering the other tail, we must add the probability 0.003,770,7 of 0 early and 8 late generations to give us a two-tailed probability of 0.012,820,5. We still conclude that the arrangement is unlikely to have arisen randomly.

A group of 8 deviations composed of 5 early and 3 late generations would have a probability of 0.079,185,5, since there are 3465 ways of obtaining such an arrangement. The cumulative two-tailed probability, including all worse outcomes, would thus be 0.144,796,4, and we would be unable to reject the null hypothesis of random allocation of the generations to the two types of deviations.

For the experiment we described, we therefore conclude that during the early generations eggs were produced representing fewer wild-type and more mutant genotypes than their parents, but that during later generations other types of deviations prevailed.

In this example we were able to take advantage of our knowledge of probability theory to forecast the outcome of repeated random selection of 8 deviations out of the universe of 18. We were not required to list all possible types of outcomes of the sampling procedure, since we could directly calculate the probability of the outcomes that interested us. In many randomization tests, however, direct calculation is not possible because the statistic studied is a function of the variates such as the mean, standard deviation, g_1, or even an unconventional statistic.

Table 18.1 SIMILARITY COEFFICIENTS (CORRELATION COEFFICIENTS) AMONG 10 SPECIES SELECTED FROM THE *HOPLITIS* COMPLEX OF BEES.

Species code numbers are shown at the margins of the half-matrix.

	Species									
Species	4	5	8	26	35	36	40	50	67	68
4	—									
5	.65	—								
8	.70	.84	—							
26	.40	.43	.49	—						
35	.41	.45	.49	.60	—					
36	.41	.42	.48	.52	.94	—				
40	.17	.39	.44	.57	.42	.29	—			
50	.35	.48	.52	.45	.41	.33	.51	—		
67	.36	.37	.41	.20	.25	.29	.20	.49	—	
68	.33	.36	.43	.18	.25	.29	.19	.46	.96	—

SOURCE: Data from Sokal (1958).

To illustrate an exact randomization test based on enumeration, we have taken a problem from numerical taxonomy. In this example we compute the significance of an unconventional statistic. Table 18.1 is a half-matrix of correlation coefficients indicating the degree of similarity among 10 species of bees whose numerical code names are shown at the margins of the matrix. The correlation coefficients are computed from a set of variates representing the characters of these species. The matrix enables us to look up the similarity between any two species. Thus, the similarity between species 26 and 36 is 0.52.

Conventional taxonomic work has resulted in species 4, 5, and 8 being assigned to a single genus, *Proteriades.* We will now justify this judgment by developing a criterion for distinctness of a group. By distinctness we mean a measure of homogeneity or cohesion of the members of a group relative to their similarity with other species. A very simple and intuitively obvious method for determining distinctness is to take average similarity among the members of the group (that is, the average similarity between species 4 and 5, 4 and 8, and 5 and 8) and subtract from it the average similarity of each of the species 4, 5, and 8 with all the other species that are not in this group. This calculation yields a positive or negative value that indicates the distinctness of the original group. The higher the positive value, the more distinct the group; that is, similarity is high within the group and low with species outside the group. Conversely, if the distinctness value is negative, the average similarity with species outside of the group is higher than that within the group; hence there would be little justification for its formal recognition. When we calculate the distinctness value for the group containing species 4, 5, and 8, we obtain $\frac{1}{3}(0.65 + 0.70 + 0.84) - \frac{1}{21}(0.40 + 0.41 + 0.41 + \cdots + 0.43 + 0.45 + \cdots + 0.49 + \cdots + 0.41 + 0.43) = 0.730 - 0.409 = 0.321$, indicating that this group is markedly distinct from the other species.

How can we compute the probability of obtaining such a value, or a more extreme one, by chance alone? The similarity coefficients (in this case correlation coefficients) in the half-matrix of Table 18.1 have an unknown distribution, and the distribution of the distinctness values is even less evident. The problem was solved by computing distinctness values for all possible samples of 3 out of the 10 species. There are $\binom{10}{3} = 120$ different ways of making sets of 3 out of 10 species. The distribution of the 120 distinctness values is shown in Figure 18.2. Five percent of the distinctness values lie beyond −0.148 and 0.210 at the

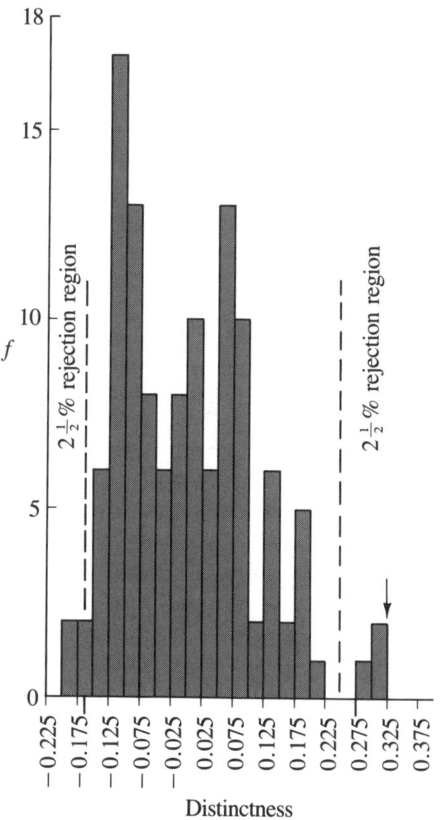

FIGURE 18.2 *Frequency distribution of distinctness values for all sets of 3 species from 10 species of bees (see Table 18.1). The broken lines mark off $2\frac{1}{2}\%$ rejection regions. The arrow identifies the distinctness value of the group being tested, that containing species 4, 5, and 8.*

two tails of the distribution. Thus the distinctness value 0.321 of the group containing species 4, 5, and 8 is significantly deviant. On the other hand, an artificial group made up of species 4, 26, and 50 (which have been placed in different genera) forms a group with little cohesion (distinctness = −0.039, not significant according to Figure 18.2). Had this problem been larger (i.e., with more possible outcomes), we might have decided to carry out a sampled randomization test (which we will discuss shortly) by taking several hundred samples and computing distinctness values for these, rather than for all outcomes. In a similarity matrix among 100 species there are 161,700 possible sets of 3; clearly we would not wish to compute a distribution of distinctness values for all of them.

The second type of randomization test is a **sampled randomization** test. It is used whenever the number of possible sampling outcomes is so large that total enumeration is impractical. We take random samples, compute our statistic, empirically find a distribution of this statistic, and decide on the significance of our original observation. The following example illustrates this procedure. In an experiment studying the effects of selection, 25 larval ticks were exposed to cold shock treatment. Only 9 larvae survived. Measurements of the scutum (in micrometers) were made for both dead and surviving larvae. The results are as follows:

Surviving larvae	$n_1 = 9$, $\bar{Y}_1 = 210.63$, $s_1^2 = 16.4025$		
211.3	218.5	211.2	205.1
211.9	204.9	211.4	211.9
209.5			

Killed larvae	$n_2 = 16$, $\bar{Y}_2 = 214.43$, $s_2^2 = 51.0036$		
219.2	211.1	210.4	219.5
205.1	222.8	210.1	218.4
213.4	210.2	213.1	204.6
206.7	212.7	224.4	229.2

The investigator hypothesized that variation in scutum length reflects general genetic variability and that only ticks from a restricted portion of the genetic spectrum would survive. This hypothesis would be supported if the variance of scutum length of the sample of ticks surviving the cold shock is lower than that for the ticks that were killed.

How can we test this hypothesis? We can calculate a variance, s_2^2, of the scutum length of the 16 nonsurviving ticks and divide it by the variance, s_1^2, based on the 9 survivors. In the observed data, $s_1^2 = 16.4025$ and $s_2^2 = 51.0036$, so $s_2^2/s_1^2 = 3.1095$. This value suggests that the hypothesis is justified. However, this ratio does not follow the F-distribution. The two variances are not obtained from independent, normally distributed samples, but from a (nonrandom) sorting of individuals within a circumscribed universe, the 25 experimental organisms. Also, on inspection, the scutum lengths do not appear to be distributed normally. The investigator in this example, who was unaware of any appropriate statistical technique for such a test, proceeded with a randomization test. There are $\binom{n_1 + n_2}{n_1} = \binom{25}{9} = 25!/9!16! = 2{,}042{,}975$ ways of taking 9 ticks out of the sample of 25. For each of these we can compute the two variances and their ratio, s_2^2/s_1^2, and then determine whether the ratio observed in the sample is sufficiently deviant in terms of the total distribution of the sampled ratios to merit rejection of the null hypothesis of random assortment. Such an undertaking, however, would involve an astronomical number of computations. The investigator therefore obtained 500 random partitions of the data into samples of 9 and 16 from the population of the 25 scutum lengths and computed s_1^2, s_2^2, and s_2^2/s_1^2 for each case. Figure 18.3 illustrates the distribution of ratios obtained by running the problem on a computer. The ratio 3.1095 obviously is not in the upper 5% tail of the distribution of 500 empirically obtained ratios. Thus we cannot support our hypothesis with the evidence from this experiment. Of the randomized variance ratios, 8% are equal to or greater than the observed value 3.1095. Based on a sample of 500, the 99% confidence limits of this percentage are 5.23 and 11.60 (see Statistical Table **P**). Thus it is improbable that the true level of significance for the entire population of 2,042,975 partitions is less than 5%.

The BIOM-pc computer program carries out a randomization test for selected statistics. Green (1977) describes a general randomization test program.

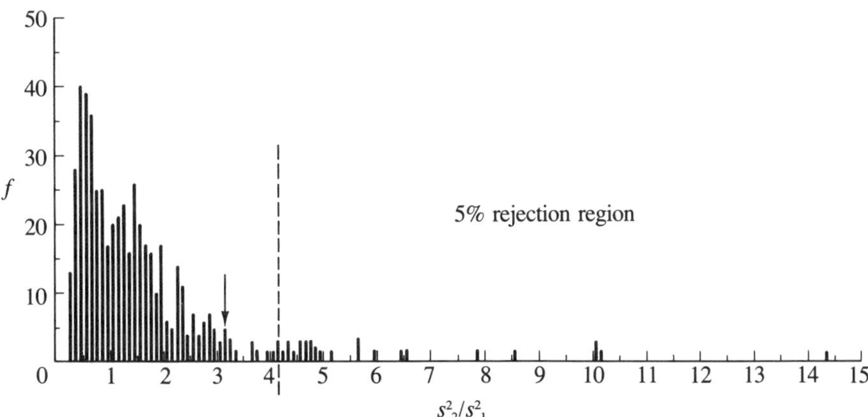

FIGURE 18.3 *Distribution of ratios s_2^2/s_1^2 obtained from 500 random partitions of the 25 scutum lengths into sets of 9 and 16 each. The one-tailed 5% rejection region is marked off by a broken line. The arrow identifies the observed ratio being tested.*

Randomization tests are most useful when the statistic to be computed has an unknown distribution or a distribution other than that required for the customary tests of significance. The variance ratio and distinctness index discussed in this section fall into this category. With the help of a computer we can apply this randomization test to any statistic of interest—perhaps one generated especially for the problem at hand, which measures some aspect of the data of particular interest to us. In the past, the development of such a new statistic often required considerable mathematical skill to find its expected distribution and critical values. In some instances these problems proved intractable even for experienced mathematical statisticians. By randomization tests and with the help of a digital computer, a nonmathematician can empirically obtain a sampling distribution of any statistic he or she cares to define and, as a result of this empirical distribution, decide whether a given observation is significantly deviant from expectation.

A randomization test can involve complicated functions and require the processing of large amounts of data, which would be impossible without powerful computational facilities. Without such approaches, there would be no hope of establishing significance levels for these statistics. The tremendous power and versatility of some of these methods, such as the Mantel test described later in this section, are undeniable.

Sampled randomization tests belong to the general category of **Monte Carlo methods,** in which complex mathematical and statistical problems are solved by random sampling from a simulated population on a computer. We will present one more example to illustrate the general Monte Carlo approach. Lewontin and Felsenstein (1965) addressed the problem of chi-square tests of independence in $R \times C$ tables in which $R = 2$ and the expected frequencies in one of the rows are very small (considerably below the minimum of 5 that is generally recommended). They set up 54 different types of tables, varying the number of rows and the marginal totals. For each type of table they randomly allocated frequencies limited only by the marginal totals. From the table they computed X^2 in the conventional manner. Each randomization and X^2-computation was repeated 5000 times for each of the 54 types of tables. The resulting distributions of X^2 were compared with chi-square distributions for the appropriate degrees of freedom. From these experiments and comparisons, Lewontin and Felsenstein concluded that the probability of type I error given by the conventional chi-square test is, in general, conservative for 5 or more degrees of freedom, even when expectations are very small in each cell. This important reassurance to the research worker carrying out a test of independence with low expectations in one row of the table could not easily have been obtained without computers and the Monte Carlo method. Readers may recognize that the sampling experiments on confidence limits and the G- and X^2-statistics calculated by computer in Chapters 7 and 17 also fall into this general category. We conclude this section with accounts of two special applications of randomization tests that are useful in biology: isotonic regression and the Mantel test.

In Section 14.5 we learned that a single-classification anova, whose groups correspond to a measurement variable X, can be tested against a stricter alternative hypothesis than that applied in the anova. The conventional alternative hypothesis of anova is that at least one group mean differs from the others. The stricter hypothesis relates to regression and states that the group means are a linear function of the variable X. Many situations arise in science, however, where the investigator is not prepared to posit such a strong alternative hypothesis. He or she may be prepared to say only that the means are in monotonic order, or weaker yet, that they are in isotonic order, meaning that designated means are greater than or equal to, or less than or equal to each other. These situations include not only cases in which the category differentiating the groups is a quantitative variable, but also those in which it is a qualitative variable—for example, the effects of different drugs for which the experimenter has a clear notion of the ordering of their anticipated effects. In Sections 7.8 and 9.4 we mentioned that when there are only two groups, you can increase the power of the test considerably if you anticipate the direction of the differences and can carry out a one-tailed significance test. There is no direct analogue of a one-tailed test in an analysis of variance with many groups, but by specifying an alternative hypothesis with ordered expectations, one can increase the power of the analysis considerably. Gaines and Rice (1990), who have brought this approach to the attention of biologists, have referred to it as **isotonic regression.**

In Box 18.4 we analyze the data set from Exercise 14.14, which gives the weight of brown trout at four densities in increasing order. When we analyze these data by analysis of variance, we detect differences among densities that are statistically significant, but if you carried out Exercise 14.14, you found the linear regression to be not significant. The data do not conform very well to a linear regression model. In the second density, 350 fish/container, the weight of the fish was higher than at the preceding lower density and at the succeeding higher density. Could we have obtained significance if we had relaxed our alternative hypothesis, asking only for isotonicity of the means? Note that for analysis of variance, any inequality among the means, such as the ones observed in this data set, enlarges the mean square among groups, thereby increasing its significance. But, if the hypothesis that the investigator wishes to test is either linear regression or isotonic regression, the excessive magnitude of the second group should not contribute to the rejection of the null hypothesis because it counters the expectations of the alternative hypothesis.

Box 18.4 illustrates isotonic regression analysis. In step **1** we specify an alternative hypothesis. We describe only the simplest isotonic regression model; other models are given in Gaines and Rice (1990). In step **2** we make the group means conform to the expectations of the alternative hypothesis that the means stay the same or decrease as density increases. Since $\mu_{350} > \mu_{175}$, we average μ_{350} with its predecessor, making the remaining three groups $\mu_{175+350}$, μ_{525}, and μ_{700}. If this were insufficient to achieve the expected relationship, further merging would be done until the array of means was isotonic. We now arrive at

Box 18.4 ISOTONIC REGRESSION.

Effect of crowding on weight of *Salmo trutta,* the brown trout; X = number of fish per container; Y = weight in grams; $n = 25$ (observations per group); $a = 4$ groups. Original data omitted to conserve space.

	Density = X			
	175	350	525	700
$\bar{Y}_i$	3.51224	3.71020	2.92680	2.53960

SOURCE: Unpublished data of J. Calaprice.

Completed anova table with regression

	Source of variation	*df*	*SS*	*MS*	F_s
$\bar{Y} - \bar{\bar{Y}}$	Among densities	3	21.63688	7.21229	6.944**
$\hat{Y} - \bar{\bar{Y}}$	Linear regression	1	17.12471	17.12471	7.590 *ns*
$\bar{Y} - \hat{Y}$	Deviations from regression	2	4.51218	2.25609	2.172 *ns*
$Y - \bar{Y}$	Within groups	96	99.71534	1.03870	
$Y - \bar{\bar{Y}}$	Total	99	121.35222		

Computation

1. The null hypothesis of the anova is $H_0: \mu_1 = \mu_2 = \cdots = \mu_a$. The alternative hypothesis of the ordinary least squares regression as computed in the anova is $H_1: \mu_Y = a + bX$. The simplest alternative hypothesis of isotonic regression as employed here is $H_1: \mu_1 \geq \mu_2 \geq \cdots \geq \mu_a$, with at least one inequality retained. This alternative hypothesis is appropriate for this example of density effects with a negative regression slope. In the case of positive regression, the inequality signs would be reversed.
2. We make the data conform to the specifications of the alternative hypothesis. Since $\mu_{175} < \mu_{350}$, we amalgamate these two groups and compute their weighted mean:

$$\frac{1}{50}[(25 \times 3.51224) + (25 \times 3.71020)] = 3.61122$$

This procedure is repeated (always starting at the same end of the array of means) until the specifications are met. In our example, a single amalgamation, as shown, is sufficient. We call the means of the new amalgamated sample $\bar{Y}_i^*$.

BOX 18.4 CONTINUED

3. Compute a new test statistic, $E^2 = SS_{\text{among}}/SS_{\text{total}}$, where SS_{among} is based on the $m \leq a$ amalgamated means $\bar{Y}_i^*$:

$$E^2 = \frac{\sum^{m} n_i(\bar{Y}_i^* - \bar{\bar{Y}})^2}{\sum^{m}\sum^{n_i} (Y - \bar{\bar{Y}})^2} = \frac{21.14703}{121.35222} = 0.17426$$

4. The general method for evaluating the probability of obtaining an E^2 value equal to or greater than the observed value is to perform a sampling experiment, since available tables are limited to equal sample sizes and only a few groups. For the present problem we drew 4 random samples of size 25 from the same normal distribution 10,000 times and then counted how often we obtained an E^2 value that was equal to or greater than the observed value. Thus we assume normality and homoscedasticity of the data. None of the samples from these data resulted in an E^2 value larger than the observed, so the probability is at most about 10^{-4}.

$m < a$ merged classes. In our case $m = 3$. In step **3** of Box 18.4 we compute the sum of squares of groups for the m groups and divide it by the total sum of squares of the data set. The ratio between these two quantities is our test statistic, E^2. The distribution of E^2 is a complicated function and has not been tabled, except for the case of a few small equal-sized samples. Gaines and Rice (1990) suggest that one determine the probability by a sampling experiment as described in the box. Note that the outcome of the isotonic regression test is highly significant at $P < 0.0001$, since none of the 10,000 random trials resulted in an E^2-value as large as that for the observed sample. Thus, although we cannot claim a linear regression for the means, we can at least claim an isotonic decrease in the means in response to increases in density. A randomization test to determine the probability of the observed E^2-value would also be possible. We would perform a randomization test if we were concerned about the assumption of normality and homoscedasticity for these data (but such a test would require access to the original data, which were not available in this case).

Another test in which sampled randomization techniques are used is the **Mantel test** (Mantel, 1967; Sokal, 1979), which applies to data expressed as dissimilarities. In evolutionary biology and ecology, dissimilarity coefficients are frequently used to measure the degree of difference between individuals, populations, species, or communities. We have not yet encountered such coefficients in this text. Sometimes distance coefficients are used (actual geographical distances or multidimensional distances). The correlations between species of bees, discussed earlier in this section, can easily be turned into dissimilarities by subtracting them from 1. Then instead of indicating greater similarity, higher values of the coefficient would indicate greater dissimilarity. Dissimilarities can be computed from paired strands of DNA in different species, from gene frequencies that describe different population samples, or from morphological characters measured for individuals within a sample or among an array of samples. For purposes of learning the Mantel test, we will not be concerned with different methods for computing dissimilarities between a set of objects (= individuals, populations, species, communities), but we will assume these as givens. Readers specializing in the fields employing dissimilarities will know how to compute them. If dissimilarities are computed between all pairs of n entities, these values can be arranged in an $n \times n$ dissimilarity matrix, which, because it is symmetric, can be displayed as a half-matrix of $n(n-1)/2$ elements.

A Mantel test is used to estimate the association between two independent dissimilarity matrices describing the same set of entities and to test whether the association is stronger than one would expect from chance. For example, we may be interested in learning whether genetic distances between populations are related to their morphological dissimilarity. We would construct a genetic-distance matrix based on sequence similarity of nuclear or mitochondrial DNA, or on gene frequencies for various loci estimated for these populations. We would also construct a dissimilarity matrix based on morphological measurements. We would then ask whether larger genetic distances are associated with larger morphological dissimilarities. The Mantel test posits the null hypothesis that there is no association between the elements in one matrix and those in the other. Alternative scenarios include tests of whether differences in song patterns, measured spectrographically in local populations of birds, are related to genetic differences between these populations; whether language differences in human populations are related to genetic distances between them; and whether distances based on species composition of different habitats are related to microclimatic differences between these habitats. As we will see, the Mantel test is a very powerful procedure that can test other biological questions that, at first sight, do not appear to relate to dissimilarities but can be formulated to make them amenable to analysis by this test.

The example we will use to illustrate the Mantel test concerns genetic distances and geographic distances (in kilometers) between all pairs of 10 human populations. Are the genetic distances associated with the geographic distances? This is a question of spatial differentiation. If these populations had differentiated in situ, it would stand to reason that geographically closer populations would also be closer genetically. If, on the other hand, the populations had differentiated elsewhere, and then migrated into their current locations, no such association need occur. The data for this example, shown in Box 18.5, are from a study (Sokal, Smouse, and Neel, 1986) of the Yanomama Indians who live in a border area between Brazil and Venezuela and who have been studied very

Box 18.5 MANTEL TEST OF ASSOCIATION BETWEEN A GENETIC AND A GEOGRAPHIC DISTANCE MATRIX.

Genetic and geographic distances computed for 10 villages of the Yanomama Amerindians.

Genetic distances ($\times 10^5$):

Village	3G	8L-11P	8F	8ABC	11YZ	11D	11X	3RS	8O	15L
3G	0000									
8L-11P	2040	0000								
8F	4143	3944	0000							
8ABC	1213	2549	5285	0000						
11YZ	1246	1362	4473	1682	0000					
11D	2868	3908	4109	2195	4362	0000				
11X	3393	3347	3313	4774	4928	4567	0000			
3RS	3023	2816	8473	2804	2605	5730	5661	0000		
8O	3784	3229	3276	2445	2767	2986	4330	4805	0000	
15L	5698	6308	7491	3722	3377	6550	10242	5331	3661	0000

Geographic distances (in km):

Village	3G	8L-11P	8F	8ABC	11YZ	11D	11X	3RS	8O	15L
3G	0.0									
8L-11P	38.7	0.0								
8F	254.4	220.5	0.0							
8ABC	43.8	81.3	286.3	0.0						
11YZ	116.2	147.6	368.1	102.3	0.0					
11D	86.2	88.9	216.0	89.0	190.5	0.0				
11X	86.4	57.1	168.5	118.8	201.9	72.4	0.0			
3RS	174.2	203.3	349.6	136.1	203.6	141.0	211.1	0.0		
8O	118.9	83.3	137.3	155.3	230.8	109.0	39.0	249.3	0.0	
15L	303.4	298.3	249.2	299.9	401.6	217.2	252.9	250.5	263.8	0.0

SOURCE: Data extracted from Sokal, Smouse, and Neel (1986).

Computation

1. The Mantel coefficient Z is computed using Expression (18.1):

$$(0.02040 \times 38.7) + (0.04143 \times 254.4) + \cdots + (0.03661 \times 263.8) = 365.03379$$

Its significance is evaluated by random permutation of rows and columns of the geographic-distance matrix (see Table 18.2), followed by a recomputation of the

Box 18.5 Continued

Mantel coefficient. This procedure is repeated many times, until a reference distribution of Z-values has been established. The observed Z-value (in this case 365.03379) is then compared with this reference distribution.

2. If the number of replicated randomizations is N, the cumulative short tail of the distribution is evaluated as $(n_T + 1)/(N + 1)$, where n_T is the number of randomized Z-values equal to or above (or equal to or below) the observed value. In our case $n_T = 1$, $N = 249$, and the right-hand tail of the distribution is $(1 + 1)/(249 + 1) = 0.008$.

3. Alternatively, the correlation between the two matrices can be computed: $r = 0.51840$. The significance of this correlation can also be tested by random permutation of rows and columns of the geographic -distance matrix and recalculation of the correlation coefficient. If the same rows and columns are permuted as in step **1,** the numerical result will be identical. Otherwise, the results are likely to differ slightly.

intensively by J. V. Neel and his associates. These Amerindians are relatively untouched by Western civilization. All together, 50 populations were studied, but for this example we selected at random one sample from each of 8 village clusters and 2 samples from the Parima village cluster, making a total of 10 samples. We matched their genetic distances (based on 15 independent allele frequencies) with geographic distances between any two villages computed from their geographical coordinates.

As we show in Box 18.5, the Mantel statistic Z is computed as follows:

$$Z = \sum_{i=1}^{n-1} \sum_{j=i+1}^{n} X_{ij} Y_{ij} \tag{18.1}$$

where X_{ij} and Y_{ij} are elements of the two distance matrices $\mathbf{X}$ and $\mathbf{Y}$, respectively. This is called the Hadamard product of two matrices. It is the sum of the products of corresponding off-diagonal elements in the two matrices. Thus, the computation proceeds as shown in step **1** of Box 18.5, yielding 365.03379. If large genetic distances X_{ij} match large geographic distances for corresponding elements Y_{ij}, then the value of Z will be larger than average. Conversely, if there is a negative association—that is, if large values of X_{ij} match small values of Y_{ij} (and vice versa)—then the quantity Z will be small. To test whether the Z is an unusually deviant (statistically significant) value, we can carry out a randomization test. Whenever application of the Mantel test involves a substantial number of entities (more than seven), we carry out a sampled permutation test in which the elements of one matrix are randomly rearranged. We could permute the elements of one of the matrices (it does not matter which one) randomly, each time calculating Z for, say, 499 randomized values, making the observed Z the 500th. Then we could evaluate the probability of the observed value of Z by noting its position in the distribution of randomized outcomes. We must permute the n rows (and columns) of the matrix. Therefore, if we interchange entities 1 and 3, for example, these two rows and their corresponding columns in the matrix must be interchanged in their entirety. We show such an interchange for a 5×5 matrix in Table 18.2. Since an 8×8 matrix will have $8! = 40{,}320$ permutations, you can easily see why we take a sample of these permutations, rather than enumerating them completely. Usually a few hundred or a few thousand permutations are considered sufficient.

The Z-statistic is expressed in arbitrary units, and its implications are hard to understand without a significance test. For this reason, workers often employ the standardized Mantel coefficient, which is the product–moment correlation between the elements of the two dissimilarity matrices. However, one cannot use conventional significance levels for correlation coefficients, since the elements of the matrices are not independent bivariate observations, as is assumed by correlation theory. Instead we must enumerate the many permutations of rows and columns of one of the matrices, computing a correlation coefficient each time, and compare the observed coefficient against the reference distribution. A product–moment correlation coefficient can be computed automatically by a

Table 18.2 RANDOM PERMUTATION OF ROWS AND COLUMNS OF A DISTANCE MATRIX.

A schematic 5×5 distance matrix is shown here, followed by a random permutation of its rows and columns.

Distances

	1	2	3	4	5
1	0				
2	20	0			
3	41	39	0		
4	12	25	53	0	
5	13	14	45	17	0

Random permutation of rows and columns

	2	1	5	4	3
2	0				
1	20	0			
5	14	13	0		
4	25	12	17	0	
3	39	41	45	53	0

Mantel program if the matrices are first transformed by a suitable linear transformation. In consequence, the significance level of a Mantel coefficient will be exactly the same as that for the standardized coefficient. You can see that the application of the Mantel test requires a high-speed computer.

For the example in Box 18.5 we obtained a standardized Mantel statistic of $r = 0.51840$ for the association between genetic and geographic distances. The probability of this observed value was estimated as $P = 0.008$. (We carried out 249 random permutations of the rows and columns of the geographic-distance matrix. When the resulting correlation coefficients were arrayed in a frequency distribution, the observed value, the 250th item in the distribution, ranked second from the right tail, yielding a probability of $P = 2/250 = 0.008$). Spatial distance clearly is correlated with genetic distance.

When the size of the distance matrix is substantial (say, of dimension greater than 50×50), the computer time required for computation of a sampled permutation test starts to become prohibitive. Mantel (1967) worked out an approximate standard error for Z, which is asymptotically normally distributed. We find that this standard error gives good approximations to the sampled randomization probability values, even when matrices as small as 25×25 are analyzed. The formula for the standard error is complex and is not given here, but obtaining the significance of a Mantel test by an asymptotic approximation is an option in most computer programs for the Mantel test. When the asymptotic approximation was applied to our 10 Yanomama samples, we obtained $t_{[\infty]} = 2.21862$, which yields a probability of $P = 0.01326$. Although not as small as that obtained with the random permutation test, the result is still significant by conventional criteria.

We alluded in the preceding discussion to the wide applicability of the Mantel test. We have found that it can be used as an excellent nonparametric equivalent of analysis of variance as follows. Visualize a simple single-classification anova, with $a = 3$ groups and $n = 5$ replicates per group. Create a dissimilarity matrix between all pairs of 15 variates. The elements can be as simple as absolute differences between the variates. If the dissimilarity matrix is arranged as shown in Figure 18.4A, you will notice that all within-group dissimilarities are found in triangular submatrices, whereas all the between-group dissimilarities are shown as square submatrices. (They would be rectangular if sample sizes in the groups were unequal.) Next we construct a design matrix that matches the structure of the dissimilarity matrix. We place 0's in the triangular within-group submatrices and 1's in the square between-group submatrices (Figure 18.4B). We now subject these two matrices to a Mantel procedure. If differences between groups are greater than those within groups (rejection of the null hypothesis in an anovalike design), then the 1's in the design matrix will be associated with the largest differences. When we randomly permute rows and columns of one of these matrices, we find that the 1's no longer are necessarily associated with differences between groups, and the randomized Z-values are lower than the observed ones. Sokal et al. (1993) have shown that the power of such a test is nearly the

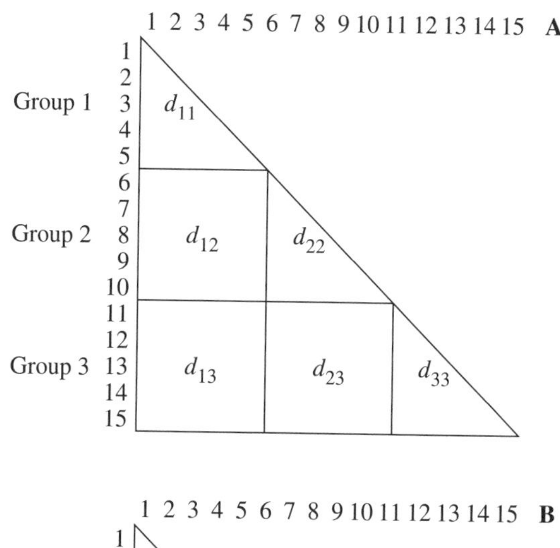

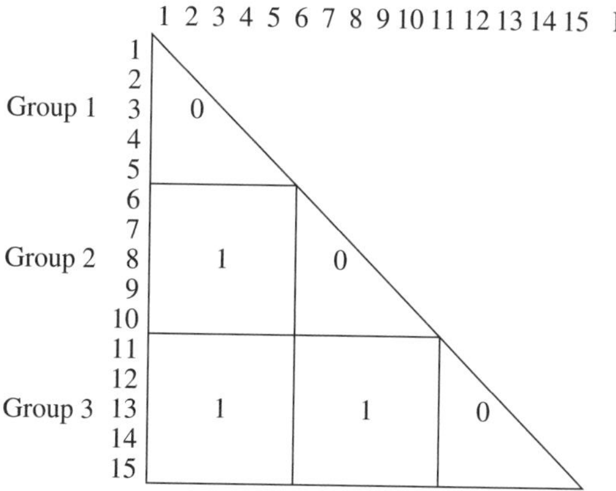

FIGURE 18.4 *Schematic distance (**A**) and design (**B**) matrices for carrying out a nonparametric analysis of variance by means of a Mantel test. The data consist of 15 observations grouped into 3 groups of 5 each. The* d_{ii} *symbolize distances within groups, the* d_{ij} *distances between groups. The design matrix replaces within-group distances by zeros, between-group distances by ones.*

same as that of an anova. By constructing various types of design matrices, a variety of hypotheses can be tested. For an illustration of the construction of such design matrices, see Livshits, Sokal, and Kobyliansky (1991).

Pairwise association between matrices has been extended to include three or more matrices by Smouse, Long, and Sokal (1986); see also Oden and Sokal (1992).

18.4 The Jackknife and the Bootstrap

This section introduces two other general-purpose techniques that are useful for analyzing either a novel statistic for which the mathematical distribution has not been fully worked out or a more ordinary statistic for which the distributional assumptions may not be true. In the previous section we obtained generality at the cost of reduced power by using a nonparametric test based on the assumption that all observed variates are equally likely. The **jackknife** technique is a parametric procedure that allows us to reduce the bias in our estimate of the population value for a statistic and provides a standard error for the statistic. The standard error allows us to make *t*-tests and to compute confidence intervals for the statistic (assuming the statistic is at least approximately normally distributed). Note that we are assuming normality for the jackknifed statistic, not for the individual variates themselves.

The jackknife procedure was developed by Tukey as a rough-and-ready statistical tool. Although there usually are more efficient specialized statistical methods for a given problem, the jackknife procedure can be a handy statistical tool despite its crudeness in some applications.

The idea is to split the observed data into groups (usually of size 1) and to compute values of the desired statistic, each time ignoring a different one of the groups of observations. The average of these estimates is used to reduce the bias in the statistic, and the variability among these values is used to estimate its standard error. An outline of the procedure is given in the first part of Box 18.6. For statistics that are bounded in range (such as a correlation coefficient, $-1 \leq r \leq 1$, or a variance, $s^2 \geq 0$), the jackknife procedure usually works more satisfactorily if the statistics are first transformed. For example, the jackknife procedure is usually applied to Fisher's *z*-transformations of the correlation coefficients and to the logarithm of a variance or functions of variances.

An important application of the jackknife procedure is in testing whether a correlation coefficient equals a specified nonzero value or in testing the equality of two correlation coefficients. The usual procedure using the *z*-transformation (see Box 15.4) is sensitive to a lack of normality. In order to jackknife r, we consider the sample of n observations to consist of n pairs of variates (Y_1, Y_2). The correlations are then computed with each such pair of variates ignored in turn. Thus r_{-i} is the correlation based on all $(n - 1)$ pairs except (Y_{1i}, Y_{2i}). This correlation is transformed to z's, z_{-i}, and then to "pseudovalues," ϕ_i, which are defined in step 3 of Box 18.6. The average of the ϕ's is our jackknifed estimate of z, which can be transformed back into a correlation coefficient as shown in Box 18.6. In this particular example the jackknife estimate of ρ is similar to the simple estimate based on all the data. The tests of significance and confidence limits are also similar to those that would be obtained using the methods described in Box 15.4.

Another application of this technique is in testing functions of sample variances. For example, we can jackknife the statistic $\mu = \ln(s_1/s_2)$ to test for the

Box 18.6 TUKEY'S JACKKNIFE METHOD.

Computation

1. Compute the desired sample statistic, St, based upon the complete sample (of size n).
2. Compute the corresponding statistics St_{-i} based upon the sample data with each of the observations i ignored in turn. If the sample size is very large, one may compute the statistics corresponding to the deletion of groups of k observations and then substitute the number of such groups for n in the following equations.
3. Compute the so-called pseudovalues, ϕ_i, as follows:

$$\phi_i = nSt - (n - 1)St_{-i}$$

4. The jackknifed estimate of the statistic is then simply

$$\widehat{St} = \frac{\sum \phi_i}{n} = \bar{\phi}$$

5. The approximate standard error of $\widehat{St}$ is

$$s_{\widehat{St}} = \sqrt{\frac{s_\phi^2}{n}}$$

$$= \sqrt{\frac{\sum (\phi_i - \bar{\phi})^2}{n(n - 1)}}$$

We may now assign approximate confidence limits to the statistic and perform approximate t-tests. Usually $s_{\widehat{St}}$ is considered to have $n - 1$ degrees of freedom. If the statistic is such that the ϕ_i can take on only a distinct values, however, the degrees of freedom are taken as $a - 1$. For example, if one were to jackknife the range, there would be only three different pseudovalues possible, generated by dropping the smallest, the largest, or any intermediate value. In such a case $a = 3$, and the degrees of freedom are 2.

Example

Test for the significance of the correlation coefficient and set 95% confidence limits to ρ using the data from Box 15.2. The jackknife procedure is applied to the z-transformation of r, which gives better results than r does.

1. The observed r based on $n = 12$ is 0.865186. The z-transformation ($\tanh^{-1} r$) yields 1.313,608,9.
2. The correlations r_{-i} are then computed. For example r_{-1} is computed using only observations 2 through 12, r_{-2} is computed using observations 1 and 3 through 12, and so forth. The z-transformations of r_{-i} are shown in the following table, along with their pseudovalues, $\phi_i = nz - (n - 1)z_{-i}$. For example $\phi_i = 12(1.313,608,9) - 11(1.415,141,2) = 0.196,752,7$.

BOX 18.6 CONTINUED

i	r_{-i}	z_{-i}	ϕ_i
1	0.888,580,9	1.415,141,2	0.196,752,7
2	0.884,185,4	1.394,630,3	0.422,372,9
3	0.891,958,7	1.431,427,1	0.017,608,2
4	0.830,244,4	1.188,922,5	2.685,159,2
5	0.811,793,3	1.132,265,8	3.308,382,4
6	0.863,850,4	1.308,322,8	1.371,755,2
7	0.875,486,9	1.356,106,5	0.846,134,8
8	0.872,939,5	1.345,300,7	0.964,998,4
9	0.845,792,2	1.241,181,3	2.110,312,5
10	0.867,189,7	1.321,634,0	1.225,332,2
11	0.858,372,4	1.287,127,8	1.604,900,6
12	0.877,860,9	1.366,363,8	0.733,304,4

3. The jackknifed $\hat{z}$ is then simply $\bar{\phi}$, which is 1.290,584,5. The inverse z-transformation yields an $\hat{r}$ value of 0.859,279,5, which is very similar to our original estimate.
4. The estimated standard error of $\hat{z}$ is $\sqrt{s_\phi^2/n} = 0.288,449,0$ with $n - 1 = 11$ degrees of freedom.
5. Test of $H_0: \rho = 0$:

$$t_s = (\hat{z} - 0)/s_{\hat{z}}$$
$$= 1.290,584,5/0.288,449,0 = 4.474$$

which is to be compared to $t_{\alpha[n-1]}$. Since $t_{.001[11]} = 4.437$, our $\hat{r}$ is significant at the $P < 0.001$ level, as was found in Box 15.4.

6. The final step is to set confidence limits. First we set 95% confidence limits to $\hat{\zeta}$:

$$L_1 = \hat{z} - t_{.05[11]}s_{\hat{z}}$$
$$= 1.290,584,5 - 2.201(0.288,449,0)$$
$$= 0.655,708,2$$

$$L_2 = \hat{z} + t_{.05[11]}s_{\hat{z}}$$
$$= 1.290,584,5 + 2.201(0.288,449,0)$$
$$= 1.925461$$

Transformed back to correlation coefficient scale, the limits for ρ are $L_1 = 0.57550$ $L_2 = 0.95836$. These values happen to be very similar to the limits we would have obtained if we had used the methods in Box 15.4 (which are not considered very reliable for such small sample sizes).

equality of two variances. The jackknife procedure correctly estimates the variability of μ for all underlying distributions, whereas the F-test, Bartlett's test, and related tests can be very inaccurate when the distributions are not normal (Miller, 1968). One can also jackknife ln (MS_{among}/MS_{within}) in a single-classification anova. In this case the pseudovalues would be based on the deletion of groups rather than of individual variates. Arvesen and Schmitz (1970) found the jackknife procedure to be more robust than the usual F-test for nonnormal distributions and to be equivalent to the F-test when the distributions are normal.

The jackknife does not always work. For example, the distribution of the jackknifed largest sample variate can be degenerate or nonnormal. The jackknife is also not useful for correcting for the effects of outliers, since they seem to affect the jackknifed estimate as much as they affect the original estimates. Hinkley (1978) warns that the jackknifed z-transformation of the correlation coefficient is quite sensitive to outliers. In such cases the jackknifed estimate of the standard error is a severe underestimate. By examining the pseudovalues, however, one can ascertain the effect of deleting variates. If the deletion of a particular variate yields a pseudovalue very different from the rest, that variate should be tested to see if it is an outlier. The quantity $\overline{\phi} - \phi_i$, which is the **sample influence function** of Devlin, Gnanadesikan, and Kettenrıng (1975), is a convenient measure of the influence that a given observation has on the statistic being studied.

Miller (1974) reviews the statistical properties of the jackknife procedure and provides an extensive bibliography. Manly (1977) gives an example of the calculation of approximate variances for estimates of population parameters using the jackknife procedure.

The **bootstrap** is a related technique for obtaining standard errors and confidence limits of various statistics. The basic idea of the bootstrap is quite simple. We take the sample for which we hope to estimate a parameter π using a sample statistic p. This statistic can be as simple as the arithmetic mean (for which, of course, analytical formulas are readily available, as we saw in Chapter 7, making the bootstrap unnecessary) or as complicated as the statistics that we discussed earlier in this section for the jackknife method. We compute p for the observed sample. If the sample is of size n, we then carry out a random resampling procedure by sampling with replacement using the n items from the observed sample as a parent population. For each of these samples we estimate the desired parametric statistic π. Because we are sampling with replacement, most samples will contain two or more replicates of the same variate in the observed sample and consequently will lack some of the variates from the observed sample. It has been shown that the mean of estimated statistics from these bootstrapped samples approximates the mean of the population, and that the standard deviation of such an estimate approximates the standard error of the statistic as if we had repeatedly sampled from the unknown population without replacement. This is an extremely important result because it permits us to calculate standard errors for all but a few refractory statistics (such as the sample median).

We can apply the bootstrap to the jackknife problem in Box 18.6. Because it involves heavy computation we will not illustrate this example in a box, but we describe it here. We take 1000 bivariate samples of size 12, sampled with replacement from the observed 12 bivariate pairs of Box 15.2. We compute the product–moment correlation coefficient for each of these samples and transform each to its z-transform, obtaining an array of 1000 z-values. Next we compute the mean of these 1000 z-values, as well as the standard deviation, which is the standard error of our estimate. We then set confidence limits by making use of the fact that the estimates should approximately follow the normal distribution. Table 18.3 compares estimates of the mean and standard error, as well as confidence limits, for the asymptotic standard errors of z, the jackknife estimate from Box 18.6, and the bootstrap estimates. Note that it has been shown that the jackknife and the bootstrap estimates approach each other asymptotically for large samples (Efron and Gong, 1983). In this example the estimates are rather different—probably because the sample is small ($n = 12$).

The bootstrap, although computationally very intensive, is a simple method for obtaining standard errors of complicated functions. Considered to be an improvement over the jackknife procedure, the bootstrap has been used extensively in biostatistical applications. For example, it has become quite popular in numerical taxonomic and phylogenetic research, where it was introduced by Felsenstein (1985). In phylogenetic estimation, the branching sequence of t species is estimated from a set of n characters, which vary in their states among species. The estimates are based, for example, on constructing the shortest tree, in terms of the amount of implied evolutionary change for a given data set. As a result of constructing such a tree, the systematist may conclude that species A

Table 18.3 COMPARISON OF ASYMPTOTIC, JACKKNIFE, AND BOOTSTRAP ESTIMATES, STANDARD ERRORS, AND CONFIDENCE LIMITS.

Data from Box 15.2.

	Method of estimation		
Statistic	Asymptotic	Jackknife	Bootstrap
$\hat{z}$	1.31363	1.29058	1.33394
s_z	0.33333	0.28845	0.28755
z L_1	0.98030	0.65571	0.77034
L_2	1.64696	1.92546	1.89754
$\hat{r}$	0.86519	0.85928	0.87021
r L_1	0.75320	0.57550	0.64713
L_2	0.92844	0.95836	0.95603

and B are closer to each other than they are to other species—that is, that they form a monophyletic taxon. How reliable is this taxon? Felsenstein suggested sampling n characters with replacement from the original data set to create new data sets from which new minimum-length trees are constructed. The result would be a number, say, $m = 100$ trees. If 95% or more of the trees contain the taxon $\{A, B\}$, then this branch of the tree is considered well substantiated. If, by contrast, only 30% of the bootstrapped trees show that set, little reliance can be placed on that portion of the taxonomic structure of the tree.

The BIOM-pc computer program carries out jackknife and bootstrap computations for selected statistics.

18.5 The Future of Biometry: Data Analysis

Although we indicated in Chapter 1 that biometry is a rapidly changing field and that it would be risky to try to anticipate its future development, we will nevertheless offer at least a glimpse of one of the developments on the horizon. Our discussion of randomization tests, the jackknife method, and the bootstrap method in this chapter leads us naturally into an area of great promise known as **data analysis.** Data analysis is defined as the systematic search through a set of data to reveal information and relationships of interest. The procedures of data analysis, although essentially numerical, are also experimental. In the previous two sections, we were, in a manner of speaking, carrying out experiments with a set of data. Data analysis consists of such techniques as transformations, to see what effect they have on the distribution of a sample or on the relationships between two variables in a sample; robust estimation, by minimizing the effects of a certain percentage of the extreme observations; the examination of residuals and influence functions; and especially graphic techniques, to suggest interesting patterns in the data. In essence these techniques are nothing new. They have been carried out by perceptive research workers and statisticians for many years. The development of high-speed computation—and especially of interactive data processing—however, permits an almost instantaneous reply to any query an investigator wishes to ask of a set of data. Therefore the investigator can ask many more questions, in the hope that some of them will yield insights through results that can then be followed up with additional questions or with subsequent experiments to elaborate and confirm new results.

Why should there be this emphasis on analyzing the actual research data instead of computing statistics that are abstractions or summaries of these data? More and more data are being obtained because of the enormous expansion of scientific activity and the development of automatic data-gathering devices. Research workers in many disciplines are confronted with masses of data the like of which have never been reckoned with before. Digestion of these data presents a serious challenge to the meaningful interpretation of underlying factors and regularities. In addition, there is an ever-increasing tendency toward quantification in all fields of human knowledge, as numerical methods enter fields previously untouched by quantification. Most important, however, is the revolution in the speed and capabilities of computing devices and in visual and graphic display units for the results of computation. At a computer terminal or workstation, it is now easy to truncate a sample and automatically recompute means and standard deviations or to transform a variable and study its new distribution curve, with all of these computations being carried out essentially instantaneously. Many attempts at data analysis, which previous researchers might wistfully have thought of undertaking but abandoned because of the tremendous amount of labor involved, can now be accomplished. Clearly, much of this analysis will be useless and wasted, but enough new information should emerge to give the overall approach great promise.

We can view data analysis as the systematic and largely automated scrutiny of sets of data to yield both summaries of and fresh insights into the relationships in a given study. Tukey (1977) places special emphasis on the analysis of residuals when carrying out an anova or a regression analysis. In data analysis, as much attention is paid to the residual deviations as to the main effects and interactions. An unusually large or small deviation is examined critically to discover reasons for such a deviation. Only with automatic computational devices is it feasible to inspect, routinely and systematically, all residual deviations. Since these deviations represent the remaining variability, they are, of course, the object of inquiry for further scientific research. General rules and procedures for data analysis cannot yet be given, inasmuch as the hardware and software necessary to carry out such work are not yet sufficiently standardized. It is quite clear, however, that data analysis will be a major aspect of future biometric work. One of the problems encountered in data analysis is that the longer one spends adjusting the data and checking various models, the less reliable the final statistical inferences become for those data. Some workers unthinkingly apply various statistical tests to the same data and then choose the test results that suit them best. Such a procedure is analogous to selecting data to fit models and then presenting the spectacularly accurate regression equation as if it represented a genuine random sample.

EXERCISES 18

Part I. Exercises 18.1 through 18.5 relate to the subject matter of Chapter 18.

18.1 An investigator believes that an increase in temperature speeds up development of the immature insects she is studying. This is a one-tailed hypothesis; the null hypothesis is that high and low temperatures produce equal developmental rate; the alternative hypothesis is that high temperatures produce more rapid rates. In the first experiment, 25 insects in individual containers are placed in each of two temperature chambers at low and high temperatures, respectively. The mean developmental time at the low temperature is 11.6 days, at the high temperature 7.4

days. Analysis of variance results in a value of $F_s = 2.94$. In a second experiment the investigator constructs an ingenious device in which the rank order of pupation in two experiments is recorded without the individuals being inspected twice a day, as had been necessary in the first experiment. With these rank orders of 50 pupae in each of two temperature conditions, she is able to carry out a Mann–Whitney U-test of the observations, which suggests again that the individuals at the higher temperature pupated earlier. The test statistic $t_s = 1.55$. Finally, the investigator wished to test so many individuals that neither of the previous two experimental setups would have been practical. She therefore arbitrarily started 500 larvae at low temperature and 495 larvae at high temperature. Seven days later 203 individuals in the low-temperature chamber had pupated, and 241 individuals of the high-temperature culture had pupated.

Do the individual tests each show significant acceleration of development at higher temperatures? Do the three tests jointly show this? *Answer:* $-2\Sigma \ln P = 16.738$, 6 *df*, $P < 0.025$ (using G with Williams's correction for the third experiment).

18.2 Compute the jackknife estimate of the biased estimate of the variance, $\Sigma y^2/n$, for the following data from Experiment 6.1 (see Table 6.1):

41 46 54 44 42

Compare your result with the usual unbiased estimate s^2. Which variate has the greatest influence on your sample estimate?

18.3 Compute the jackknife estimate of the coefficient of variation and its standard deviation for the data of Exercise 4.2. *Answer:* $\hat{V} = 92.1462$, $s_{\hat{V}} = 57.5$.

18.4 Test the fit of the linear, quadratic, and cubic regression analyses performed in Exercise 16.3 using a runs test on the signs of the deviations from regression.

18.5 Test for a trend in the flow variable F_7 in Exercise 16.4 using the runs up and down test and the runs above and below the median test. Comment on why they yield different results for these data. *Answer:* 7 runs up and down; 2 runs above and below median.

Part II. Exercises 18.6 through 18.41 range over the entire subject matter of the book. These problems do not require any computation, but serve as review problems for the student, who is required to indicate the appropriate method for a solution. The following steps should be taken for *each* of the exercises that follow:

(a) State the statistical method appropriate for providing the solution, and why.
(b) Outline the setup of the data.
(c) Give formulas for computations.
(d) State degrees of freedom; if none applicable, state "none."
(e) Draw attention to any requirements to which the data must conform to be tested by the method you have chosen, as well as to any corrections that might need to be applied.
(f) Enumerate the specific answers that could emerge as the result of your method.
(g) List any alternative method applicable.

18.6 A normal probability curve has been fitted to certain data. There are 15 classes. Are the deviations such as could be expected from accidents of sampling?

18.7 The amount of inorganic phosphorus in the blood plasma of 10 unrelated men is determined from specimens taken every morning for two weeks. Are there differences among individuals and/or does the amount vary from morning to morning? Would you expect the interaction to be negligible? If not, how would you design the experiment to demonstrate interaction?

18.8 An ecologist claims that there is a narrow optimum size for a burrowing animal, above or below which the animal is at a disadvantage. How could you investigate this hypothesis in a frequency distribution of 300 body diameters of this species?

18.9 You have data on the physical characteristics of 1221 sets of 4 siblings each. Among other traits, you know whether each individual's eye color is blue or not. Without studying any pedigrees how could you demonstrate that blue eyes are controlled wholly or in part by genetic factors? (If we can prove that sibs are more like each other than are randomly chosen individuals, we will consider this sufficient evidence, although theoretically sibs could be more alike because of common environment as well.)

18.10 We are measuring tarsal claw length of samples of male lice of species P taken from 8 different individuals of host species X. Is there an added variance due to differences between hosts? Remember that we are unlikely to obtain the same number of lice from each host individual.

18.11 Counts are made of the number of mites found on blades of grass species G in the sunlight and in the shade. Twenty blades of grass in the sun and 31 in the shade were examined. The number of mites per blade varied from 3 to 32. Are there more mites per blade in the shade?

18.12 An animal behaviorist is studying mating behavior in animal species X. Females of equal ages are introduced singly into a mating cage, where they are exposed to several males of known and demonstrable aggressiveness. In species X, however, the female can accept or reject copulation at will. Thus the number of copulations per unit time is an index of the female's submissiveness or sexual appetite. The investigator tests 100 females, and they range from 0 to 11 matings per unit time, with a mean of 3.2 matings. His hypothesis is that the females can be subdivided into at least two races or groups differing in submissiveness. How can he support his findings with an analysis of this experiment? Another way of looking at this analysis is to ask what results the scientist would expect if copulations were independent of the genetics of the female—that is, if they were merely a randomly varying variable.

18.13 A study of respiration under the influence of drugs in salamanders yields data on 7 animals for each of 5 drugs and 7 control animals. The variates are volume of oxygen consumed, and a preliminary investigation shows that the variances for each drug are anything but homogeneous. Before the investigator attempts to equalize variances by transformation, how could she test for differences among the drugs? If such differences are not present, she might not think it worthwhile to spend time looking for a suitable transformation.

18.14 An investigator studying weight gain in rats on different diets is comparing two groups of 20 rats each. Not only are the variances of the two samples different, but the variates are clearly not distributed normally. No transformation known to the investigator appears able to take care of the difficulty. Suggest a method of testing for differences in weight gain between the two samples.

18.15 Samples of moths were collected from four localities. The percentage of melanistic individuals varied from sample to sample. Can the four samples be considered as coming from a homogeneous population with regard to incidence of melanistic forms? (You possess the original data from which the percentages were calculated.) If not, are there any homogeneous subsets of localities?

18.16 An investigator wishes to demonstrate normality in a great number of frequency distributions. Computer facilities are not available. None of the frequency distributions has more than 20 items.

18.17 The lengths of 50 skulls of a certain species of field mouse from each of two localities are measured. Is there a significant difference in average length? (Variances of the two samples do not appear to differ greatly.)

18.18 You want to determine whether variations in body weight and amount of bile pigment in urine are due to common causes. You have data on 1243 male white Americans, ages 45 to 50.

18.19 The number of nests of bird species A per square mile of a certain forested area has been determined on the basis of a study of 60 square miles. This bird is known to be territorial (that is, nonrandomly distributed). How can you show this from your data, and would data based on square mile units necessarily show the phenomenon?

18.20 The mean number of dorsocentral bristles in dichaete *Drosophila melanogaster* is determined for flies emerging on each of five two-day intervals. Separate readings are obtained for each of 12 replicate bottles. Can the whole set of data be treated as if from a homogeneous population?

18.21 A biologist collects 357 specimens of a species of lizard in one locality. She measures tail length on all of them. How many must she measure in the future from any other locality so that she can pick up a significant difference equal to 5% of the present mean tail length of her lizards?

18.22 The variation "banded" is found in the dragonfly species L. Both males and females show this variation. During an afternoon's intensive collection in a rice paddy in the South, students in a field zoology class collect 117 mating pairs (in copula). As each pair is collected, the students classify it into one of the following four categories:
(a) both sexes common (unbanded) type
(b) both sexes banded
(c) ♀♀ common, ♂♂ banded
(d) ♀♀ banded, ♂♂ common
We wish to test whether mating was at random or whether these dragonflies prefer to mate with their own type.

18.23 Sixty-seven field mice and 23 pocket gophers were collected from a certain area. Femur length is measured for both species. Is length of the femur more variable in the mice?

18.24 Seven doses of chemical A (1, 2, 4, 8, 16, 32, and 64 mg) are administered to three-day-old chicks, 15 chicks per dose. All 105 chicks are caged together and distinguished by headstains denoting the dose. At the end of three weeks the chicks are killed, and the glycogen content of the liver is determined separately for each chick. Did changes in the dose have effects on glycogen content, and can this relation be quantified? Is a linear equation adequate to describe the effect of a dose of chemical A on glycogen content?

18.25 A botanist collects five samples of herbs of species B from different localities. The sample sizes vary. The correlation between height of plant and number of flowers has been calculated for all the samples and is significant for every one. Could the correlation between height and flower number be the same for the five localities? If so, what estimate can you give of ρ?

18.26 You would like to predict the total length of a cat from the length of its femur. What are the confidence limits of your predicted value for a given femur length?

18.27 An investigator wishes to establish a correlation between number of ovarioles and fecundity in drosophila. He first raises the female under optimal conditions, counting her egg output over a two-week period. Then he removes her ovaries by dissection, sections them, and counts the ovarioles. He has counted fecundity on 100 individuals but has sectioned only 25 ovaries to date. Before proceeding with his work he would like to have reassurance that correlation occurs in his data. What preliminary test would you suggest on the 25 paired readings amassed to date?

18.28 An ecologist watches a bird nest to record the food brought in by the female for her young. He can distinguish two size classes of insect prey, big and small. Reviewing a sequence of 30 consecutive observations taken in the course of one afternoon and containing 20 big and 10 small insects, he wonders whether the sequence of size classes is random.

18.29 To test one aspect of a new theory, an ecologist wished to compare two populations for an unconventional statistic that was unlikely to be distributed normally and for which the standard error was unknown.

18.30 The progeny of a cross was expected to yield dominants, heterozygotes, and recessives in a 1 : 2 : 1 ratio. Separate replicates were reared under five environmental conditions. Do the data fit the expected 1 : 2 : 1 ratio? Is there any evidence of differential survivorship in the different environments?

18.31 A pharmacologist wishes to compare the effects of two drugs on the longevity of hamsters, mainly to determine which drug has the greatest effect in shortening the life span. The variable recorded (length of life in days) is not distributed normally.

The investigator wishes to complete the experiment in as short a time as possible, but a few animals of each group live a very long time.

18.32 An immunologist compared the strength of two antigens by two separate tests. The first test induced an all-or-nothing skin reaction and was analyzed by a 2×2 test for independence. The second was a precipitin test, and the difference in turbidity was analyzed by a t-test (after the variates had been transformed). Both tests were inconclusive. How could an overall test of significance for both tests be constructed?

18.33 A physiologist wishes to study the effect of different levels of a new drug on mice. The response of the mice (gain in weight after two weeks) is expected to be influenced somewhat by their initial weights.

18.34 An ecologist would like to determine which of a series of environmental variables are the best predictors of the abundance of a species of plant.

18.35 We want to fit a linear model to a set of data in which it is assumed that variables A and B determine (cause) a variable C, which in turn directly affects another variable, D, which is also affected by variable A. How important is the indirect effect on B on variable D? The correlations between all pairs of variables have been computed.

18.36 You have a sample of 35 measurements of the height of a plant. You have concocted an unusual statistic describing the bimodality of the distribution. You would like to assign confidence limits to this statistic but do not know how to construct a formula for a standard error. How would you proceed?

18.37 Refer to Exercise 18.22. Assume that you have shown that mating pairs are preferentially of the same type. The data were obtained at the height of the season. Suppose you also have similar data for the early and the late season. Are the outcomes homogeneous, and if so, how can we estimate the preference jointly from the three data tables?

18.38 An investigator has studied mating behavior in an insect species. This behavior consists of a complex repertoire, which the ethologist has decomposed into some 20 steps. Samples from populations at 10 localities show differences in these steps, and the investigator is able to devise a distance measure indicating the differences in the repertoires of individuals from all pairs of the 10 localities. She would like to know whether these behavioral distances are related to the geographic distances of these populations.

18.39 Thirty patients with disease X were given a course of injections of drug D. Another sample of 25 with the same disease were given a placebo injection. The numbers of patients cured were recorded for both groups. How much more likely is a patient to be cured by drug D than by the placebo? Set confidence limits to this estimate.

18.40 Injection of sex hormone T brings on premature development of the reproductive organs in an animal species. An investigator would like to investigate the relation between dose of hormone and magnitude of effect (expressed as length of the reproductive organ) in samples of 20 animals per dose. He employs 10 doses ranging from low to high. It is obvious, after the results are inspected, that the function is not linear, nor is the regression significant. Attempts to transform the variables are not successful. The investigator would be satisfied with a simpler demonstration. Can it be shown that organ size is a nondecreasing function of dose?

18.41 An insecticide is administered to 8 batches of 50 insects for 7 increasing doses and one control. Calculate a regression equation for proportion mortality on dose and set confidence limits to your slope estimate.

Appendix: Mathematical Proofs

A.1 Demonstration that the sum of the deviations from the mean is equal to zero.

We must learn two common rules of statistical algebra. The first is that we can open a pair of parentheses with a Σ sign in front of them by treating the Σ as though it were a common factor:

$$\sum_{i=1}^{n} (A_i + B_i) = (A_1 + B_1) + (A_2 + B_2) + \cdots + (A_n + B_n)$$
$$= (A_1 + A_2 + \cdots + A_n) + (B_1 + B_2 + \cdots + B_n)$$

Therefore

$$\sum_{i=1}^{n} (A_i + B_i) = \sum_{i=1}^{n} A_i + \sum_{i=1}^{n} B_i$$

The second rule is that $\Sigma_{i=1}^{n} C$ developed during an algebraic operation, where C is a constant, can be computed as follows:

$$\sum_{i=1}^{n} C = C + C + \cdots + C \qquad (n \text{ terms})$$
$$= nC$$

Since in a given problem a mean is a constant value, $\Sigma^n \bar{Y} = n\bar{Y}$. You can check these rules using simple numbers. In the subsequent demonstration and others to follow, whenever all summations are over n items, we have simplified the notation by dropping subscripts for variables and superscripts above summation signs.

We wish to prove that $\Sigma y = 0$. By definition,

$$\begin{aligned}\sum y &= \sum (Y - \bar{Y}) \\ &= \sum Y - n\bar{Y} \\ &= \sum Y - \frac{n \sum Y}{n} \qquad \left(\text{since } \bar{Y} = \frac{\sum Y}{n}\right) \\ &= \sum Y - \sum Y\end{aligned}$$

Therefore $\Sigma y = 0$.

A.2 Demonstration of the effects of additive, multiplicative, and combination coding on means, variances, and standard deviations.

For this proof we must learn one more convention of statistical algebra. In Section A.1 we saw that $\Sigma C = nC$. When the Σ precedes both a constant and a variable, however, as in ΣCY, the constant can be placed before the Σ, since

$$\sum_{i=1}^{n} CY_i = CY_1 + CY_2 + \cdots + CY_n$$
$$= C(Y_1 + Y_2 + \cdots + Y_n)$$
$$= C\left(\sum_{i=1}^{n} Y_i\right)$$

Therefore

$$\sum_{i=1}^{n} CY_i = C\sum_{i=1}^{n} Y_i$$

Thus $\Sigma CY = C\Sigma Y$, $\Sigma C^2y^2 = C^2\Sigma y^2$, and $\Sigma 2\bar{Y}Y = 2\bar{Y}\Sigma Y$ (because both 2 and $\bar{Y}$ are constants).

MEANS OF CODED DATA

ADDITIVE CODING. The variable is coded $Y_c = Y + C$, where C is a constant, the additive code. Therefore

$$\sum Y_c = \sum (Y + C) = \sum Y + nC$$

and

$$\bar{Y}_c = \frac{\sum Y_c}{n} = \frac{\sum Y}{n} + \frac{nC}{n} = \bar{Y} + C$$

To decode $\bar{Y}_c$, subtract C from it and you will obtain $\bar{Y}$; that is, $\bar{Y} = \bar{Y}_c - C$.

MULTIPLICATIVE CODING. The variable is coded $Y_c = DY$, where D is a constant, the multiplicative code. Therefore

$$\sum Y_c = D\sum Y$$

and

$$\bar{Y}_c = \frac{\sum Y_c}{n} = D\frac{\sum Y}{n} = D\bar{Y}$$

To decode $\bar{Y}_c$, divide it by D and you will obtain $\bar{Y}$; that is, $\bar{Y} = \bar{Y}_c/D$.

COMBINATION CODING. The variable is coded $Y_c = D(Y + C)$, where C and D are constants, the additive and multiplicative codes, respectively. Therefore

$$\sum Y_c = D\sum(Y + C) = D\sum Y + nDC$$

and

$$\overline{Y}_c = \frac{\sum Y_c}{n} = D\frac{\sum Y}{n} + \frac{nDC}{n} = D\overline{Y} + DC$$

To decode $\overline{Y}_c$, divide it by D, then subtract C and you will obtain $\overline{Y}$; that is,

$$\overline{Y} = \frac{\overline{Y}_c}{D} - C$$

VARIANCES AND STANDARD DEVIATIONS OF CODED DATA

ADDITIVE CODING. The variable is coded $Y_c = Y + C$, where C is a constant, the additive code. By definition, $y = Y - \overline{Y}$, and

$$\begin{aligned} y_c &= Y_c - \overline{Y}_c \\ &= [(Y + C) - (\overline{Y} + C)] \qquad \text{(as shown for means above)} \\ &= [Y + C - \overline{Y} - C] \\ &= [Y - \overline{Y}] \\ &= y \end{aligned}$$

Therefore $\Sigma y_c^2 = \Sigma y^2$, and $\Sigma y_c^2/(n - 1) = \Sigma y^2/(n - 1)$.

Thus additive coding has no effect on sums of squares, variances, or standard deviations.

MULTIPLICATIVE CODING. The variable is coded $Y_c = DY$, where D is a constant, the multiplicative code. By definition, $y = Y - \overline{Y}$, and

$$\begin{aligned} y_c &= Y_c - \overline{Y}_c \\ &= DY - D\overline{Y} \qquad \text{(as shown for means above)} \\ &= D(Y - \overline{Y}) \\ &= Dy \end{aligned}$$

Therefore $y_c^2 = D^2y^2$, $\Sigma y_c^2 = D^2\Sigma y^2$, $\Sigma y_c^2/(n - 1) = D^2[\Sigma y^2/(n - 1)]$, and $s_c^2 = D^2s^2$.

Thus, when data have been subjected to multiplicative coding, a sum of squares or variance can be decoded by dividing it by the *square* of the multiplicative code; a standard deviation can be decoded by dividing it by the code itself, that is, $s^2 = s_c^2/D^2$ and $s = s_c/D$.

COMBINATION CODING. The variable is coded $Y_c = D(Y + C)$, where C and D are constants, the additive and multiplicative codes, respectively. By definition, $y = Y - \bar{Y}$, and

$$\begin{aligned} y_c &= Y_c - \bar{Y}_c \\ &= [D(Y + C) - (D\bar{Y} + DC)] \qquad \text{(as shown for means above)} \\ &= [DY + DC - D\bar{Y} - DC] \\ &= D[Y - \bar{Y}] \end{aligned}$$

Therefore $y_c = Dy$, as before.

Thus, in combination coding only the multiplicative code needs to be considered when decoding sums of squares, variances, or standard deviations.

A.3 Demonstration that the sum of squares within groups and the sum of squares among groups add to the total sum of squares in an analysis of variance (see Section 8.5).

By definition,

$$SS_{\text{total}} = \sum_{i=1}^{a} \sum_{j=1}^{n_i} (Y_{ij} - \bar{\bar{Y}})^2$$

Add and subtract $\bar{Y}_i$ within the parentheses:

$$= \sum^{a} \sum^{n_i} [(Y_{ij} - \bar{Y}_j) + (\bar{Y}_i - \bar{\bar{Y}})]^2$$

Square and then remove outer parentheses:

$$\begin{aligned} &= \sum^{a} \sum^{n_i} (Y_{ij} - \bar{Y}_i)^2 + 2 \sum^{a} \sum^{n_i} (Y - \bar{Y})(\bar{Y} - \bar{\bar{Y}}) + \sum^{a} n_i(\bar{Y}_i - \bar{\bar{Y}})^2 \\ &= SS_{\text{within}} + 2 \sum^{a} \sum^{n_i} (Y_{ij} - \bar{Y}_i)(\bar{Y}_i - \bar{\bar{Y}}) + SS_{\text{among}} \end{aligned}$$

The middle term is equal to zero, since $\Sigma^{n_i}(Y_{ij} - \bar{Y}_i)$ is zero within each group. Therefore, $SS_{\text{total}} = SS_{\text{within}} + SS_{\text{among}}$.

A.4 Derivation of simplified formulas for the standard error of the difference between two means.

The standard error squared, from Expression (9.2), is

$$\left[\frac{(n_1 - 1)s_1^2 + (n_2 - 1)s_2^2}{n_1 + n_2 - 2}\right]\left(\frac{n_1 + n_2}{n_1 n_2}\right)$$

When $n_1 = n_2 = n$, this expression simplifies to

$$\left[\frac{(n-1)s_1^2 + (n-1)s_2^2}{2n-2}\right]\left(\frac{2n}{n^2}\right) = \left[\frac{(n-1)(s_1^2 + s_2^2)(2)}{2(n-1)(n)}\right] = \frac{1}{n}(s_1^2 + s_2^2)$$

which is the square of the standard error of Expression (9.3).

When $n_1 \neq n_2$, but each is large enough that $(n_1 - 1) \approx n_1$ and $(n_2 - 1) \approx n_2$, the square of the standard error of Expression (9.2) simplifies to

$$\left[\frac{n_1 s_1^2 + n_2 s_2^2}{n_1 + n_2}\right]\left(\frac{n_1 + n_2}{n_1 n_2}\right) = \left[\frac{n_1 s_1^2}{n_1 n_2} + \frac{n_2 s_2^2}{n_1 n_2}\right] = \frac{s_1^2}{n_2} + \frac{s_2^2}{n_1}$$

which is the squared standard error of Expression (9.4).

A.5 Demonstration that t_s^2 obtained from a test of significance of the difference between two means (as in Box 9.6) is identical to the F_s-value obtained in a single-classification anova. This derivation is for the simple case of two equal-sized groups (as in Box 9.5).

$$t_s = \frac{\bar{Y}_1 - \bar{Y}_2}{\sqrt{\frac{1}{n(n-1)}\left(\sum^n y_1^2 + \sum^n y_2^2\right)}} \qquad \text{(from Box 9.6)}$$

$$t_s^2 = \frac{(\bar{Y}_1 - \bar{Y}_2)^2}{\frac{1}{n(n-1)}\left(\sum^n y_1^2 + \sum^n y_2^2\right)} = \frac{n(n-1)(\bar{Y}_1 - \bar{Y}_2)^2}{\sum^n y_1^2 + \sum^n y_2^2}$$

In the two-sample anova,

$$\begin{aligned}
MS_{\text{means}} &= \frac{1}{2-1}\sum^2 (\bar{Y}_i - \bar{\bar{Y}})^2 \\
&= (\bar{Y}_1 - \bar{\bar{Y}})^2 + (\bar{Y}_2 - \bar{\bar{Y}})^2 \\
&= \left(\bar{Y}_1 - \frac{\bar{Y}_1 + \bar{Y}_2}{2}\right)^2 + \left(\bar{Y}_2 - \frac{\bar{Y}_1 + \bar{Y}_2}{2}\right)^2 \quad \text{(since } \bar{\bar{Y}} = (\bar{Y}_1 + \bar{Y}_2)/2\text{)} \\
&= \left(\frac{\bar{Y}_1 - \bar{Y}_2}{2}\right)^2 + \left(\frac{\bar{Y}_2 - \bar{Y}_1}{2}\right)^2 \\
&= \tfrac{1}{2}(\bar{Y}_1 - \bar{Y}_2)^2 \quad \text{(since the squares of the numerators are identical)}
\end{aligned}$$

Then

$$MS_{\text{among}} = n \times MS_{\text{means}} = n[\tfrac{1}{2}(\bar{Y}_1 - \bar{Y}_2)^2]$$
$$= \frac{n}{2}(\bar{Y}_1 - \bar{Y}_2)^2$$

$$MS_{\text{within}} = \frac{\sum^n y_1^2 + \sum^n y_2^2}{2(n-1)}$$

$$F_s = \frac{MS_{\text{among}}}{MS_{\text{within}}}$$
$$= \frac{\frac{n}{2}(\bar{Y}_1 - \bar{Y}_2)^2}{\left(\sum^n y_1^2 + \sum^n y_2^2\right) \Big/ [2(n-1)]}$$
$$= \frac{n(n-1)(\bar{Y}_1 - \bar{Y}_2)^2}{\sum^n y_1^2 + \sum^n y_2^2}$$
$$= t_s^2$$

A.6 Derivation of Expression (11.3) for the expected error mean square of a completely randomized design estimated from the mean squares of a randomized-complete-blocks design.

Randomized-complete-blocks design

Source of variation	*df*	*MS*	Expected *MS*
Blocks	$b - 1$	MS_B	$\sigma^2_{(RB)} + a\sigma^2_B$
Treatments	$a - 1$	$MS_{A(RB)}$	$\sigma^2_{(RB)} + \frac{b}{a-1}\sum \alpha^2$
Error	$(a-1)(b-1)$	$MS_{E(RB)}$	$\sigma^2_{(RB)}$
Total	$ab - 1$		

Completely randomized design

Source of variation	*df*	*MS*	Expected *MS*
Treatments	$a - 1$	$MS_{A(CR)}$	$\sigma^2_{(CR)} + \frac{b}{a-1}\sum \alpha^2$
Error	$a(b-1)$	$MS_{E(CR)}$	$\sigma^2_{(CR)}$
Total	$ab - 1$		

If we assume that the total SS of the two designs is the same, we can write the following identity where each side represents the total SS:

$$(b-1)MS_B + (a-1)MS_{A(RB)} + (a-1)(b-1)MS_{E(RB)} = (a-1)MS_{A(CR)} + a(b-1)MS_{E(CR)}$$

Rewriting the identity in terms of the variance components of the expected means squares, we obtain

$$(b-1)(\sigma^2_{(RB)} + a\sigma^2_B) + (a-1)\left(\sigma^2_{(RB)} + \frac{b}{a-1}\sum\alpha^2\right) + (a-1)(b-1)\sigma^2_{(RB)} = (a-1)\left(\sigma^2_{(CR)} + \frac{b}{a-1}\sum\alpha^2\right) + a(b-1)\sigma^2_{(CR)}$$

$$[(b-1) + (a-1) + (a-1)(b-1)]\sigma^2_{(RB)} + a(b-1)\sigma^2_B + b\sum\alpha^2 = [(a-1) + a(b-1)]\sigma^2_{(CR)} + b\sum\alpha^2$$

$$(ab-1)\sigma^2_{(RB)} + a(b-1)\sigma^2_B = (ab-1)\sigma^2_{(CR)}$$

$$\sigma^2_{(CR)} = \sigma^2_{(RB)} + \frac{a(b-1)}{ab-1}\sigma^2_B$$

We rewrite this formula for $\sigma^2_{(CR)}$ in terms of the mean squares. Since $\sigma^2_{(RB)} = MS_{E(RB)}$ and $\sigma^2_B = (MS_B - MS_{E(RB)})/a$, we obtain

$$\begin{aligned} MS_{E(CR)} &= MS_{E(RB)} + a(b-1)\frac{MS_B - MS_{E(RB)}}{a(ab-1)} \\ &= MS_{E(RB)} + (b-1)\frac{MS_B}{ab-1} - (b-1)\frac{MS_{E(RB)}}{ab-1} \\ &= [(ab-1) - (b-1)]\frac{MS_{E(RB)}}{ab-1} + (b-1)\frac{MS_B}{ab-1} \\ &= \frac{b(a-1)MS_{E(RB)} + (b-1)MS_B}{ab-1} \end{aligned}$$

This is Expression (11.3).

A.7 Demonstration that t_s^2 obtained from a paired-comparisons significance test (as in Box 11.5) is identical to the F_s-value for treatments obtained in a two-way anova without replication (same box).

We will use the simpler $\Sigma^b (Y_1 - Y_2)$ in place of $\Sigma^b_{i=1} (Y_{i1} - Y_{i2})$. Also, remember that in such an anova $a = 2$.

$$t_s = \frac{\overline{D} - (\mu_1 - \mu_2)}{s_{\overline{D}}} \qquad \text{(from Box 11.5)}$$

where $\overline{D} = \overline{Y}_1 - \overline{Y}_2$, $\mu_1 - \mu_2$ is hypothesized to equal zero, and

$$s_{\overline{D}} = \sqrt{\frac{\sum^{b} [(Y_1 - Y_2) - (\overline{Y}_1 - \overline{Y}_2)]^2}{(b - 1)b}}$$

By squaring and rearranging a little, we can rewrite the expression for t_s as

$$t_s^2 = \frac{b(b - 1)(\overline{Y}_1 - \overline{Y}_2)^2}{\sum^{b} [(Y_1 - Y_2) - (\overline{Y}_1 - \overline{Y}_2)]^2}$$

In a randomized-complete-blocks anova with $a = 2$ groups

$$F_s = \frac{MS_{\text{among}}}{MS_{AB}}$$

In our case, since we have only two groups,

$$MS_{\text{among}} = b\sum^{a} (\overline{Y} - \overline{\overline{Y}})^2 = \frac{(\sum \text{Y}_1 - \sum Y_2)^2}{2b} = \frac{b}{2}(\overline{Y}_1 - \overline{Y}_2)^2$$

$$MS_{AB} = \frac{\sum_{i}^{a}\sum_{j}^{b} (Y_{ij} - \overline{Y}_i - \overline{Y}_j + \overline{\overline{Y}})^2}{b - 1}$$

Here $\overline{\overline{Y}} = (\overline{Y}_1 + \overline{Y}_2)/2$ because of equal sample size, and likewise $\overline{Y}_j = (Y_{1j} + Y_{2j})/2$. Since there are only two groups, we may drop the summation sign over a and write out the two squared terms for each pair of observations:

$$MS_{AB} = \frac{1}{b - 1}\sum_{j}^{b} \left\{ \left(Y_{1j} - \overline{Y}_1 - \frac{Y_{1j} + Y_{2j}}{2} + \frac{\overline{Y}_1 + \overline{Y}_2}{2} \right)^2 + \left(Y_{2j} - \overline{Y}_2 - \frac{Y_{1j} + Y_{2j}}{2} + \frac{\overline{Y}_1 + \overline{Y}_2}{2} \right)^2 \right\}$$

With a little rearrangement we find that

$$\begin{aligned} MS_{AB} &= \frac{1}{b - 1}\sum_{j}^{b} \left\{ \left[(Y_{1j} - \tfrac{1}{2}Y_{1j} - \overline{Y}_1 + \tfrac{1}{2}\overline{Y}_1) + \left(-\frac{Y_{2j}}{2} + \frac{\overline{Y}_2}{2} \right) \right]^2 \right. \\ &\quad \left. + \left[\left(Y_{2j} - \tfrac{1}{2}Y_{2j} - \overline{Y}_2 + \frac{\overline{Y}_2}{2} \right) + \left(-\frac{Y_{1j}}{2} + \frac{\overline{Y}_1}{2} \right) \right]^2 \right\} \\ &= \frac{1}{b - 1}\sum_{j}^{b} \{ [\tfrac{1}{2}(Y_{1j} - \overline{Y}_1) - \tfrac{1}{2}(Y_{2j} - \overline{Y}_2)]^2 \\ &\quad + [-\tfrac{1}{2}(Y_{1j} - \overline{Y}_1) + \tfrac{1}{2}(Y_{2j} - \overline{Y}_2)]^2 \} \end{aligned}$$

The two terms in square brackets above are identical except for the sign, which is not important since the terms are squared. The constant $\frac{1}{2}$ may also now be factored out (becoming $\frac{1}{4}$). We can therefore simplify this expression as follows:

$$MS_{AB} = \frac{1}{b-1}\frac{1}{4}\sum_{j}^{b} 2[(Y_{1j} - \overline{Y}_1) - (Y_{2j} - \overline{Y}_2)]^2$$
$$= \frac{1}{2(b-1)}\sum_{j}^{b}[(Y_{1j} - Y_{2j}) - (\overline{Y}_1 - \overline{Y}_2)]^2$$

F_s may now be written as follows:

$$F_s = \frac{\left[\frac{b}{2}(\overline{Y}_1 - \overline{Y}_2)^2\right]}{\frac{1}{2(b-1)}\sum_{j}^{b}[(Y_{1j} - Y_{2j}) - (\overline{Y}_1 - \overline{Y}_2)]^2}$$

After canceling the two 2's and moving $(b - 1)$ into the numerator, the expression for F_s is identical to our expression for t_s^2 given above.

A.8 Derivation of the formulas for b and a in regression. This derivation requires a knowledge of calculus.

We define

$$\sum d^2_{Y \cdot X} = \sum (Y - \hat{Y})^2$$
$$= \sum [Y - (a + bX)]^2$$
$$= \sum (Y - a - bX)^2$$

To minimize $\Sigma d^2_{Y \cdot X}$ with respect to the two parameters (a and b), we must first take partial derivatives with respect to each:

$$\frac{\partial \sum d^2_{Y \cdot X}}{\partial a} = 2\sum (Y - a - bX)(-1)$$
$$= -2\left[\sum Y - na - b\sum X\right]$$
$$\frac{\partial \sum d^2_{Y \cdot X}}{\partial b} = 2\sum (Y - a - bX)(-X)$$
$$= -2\left[\sum XY - a\sum X - b\sum X^2\right]$$

Next we set each partial derivative equal to zero and solve the pair of simultaneous linear equations for a and b.

$$\begin{cases} 0 = \sum Y - na - b\sum X \\ 0 = \sum XY - a\sum X - b\sum X^2 \end{cases}$$

If we multiply the first equation by $\Sigma X/n$ and then subtract it from the second equation, we cancel out the term involving a:

$$0 = \sum XY - \frac{\sum X \sum Y}{n} - b\sum X^2 + \frac{b\,(\sum X)^2}{n}$$

Thus

$$b = \frac{\sum XY - \dfrac{\sum X \sum Y}{n}}{\sum X^2 - \dfrac{(\sum X)^2}{n}}$$

The numerator of this expression

$$\sum XY - \frac{\sum X \sum Y}{n} = \sum XY - \overline{X}\sum Y \quad \text{or} \quad \sum XY - \overline{Y}\sum X$$

Add and subtract $\overline{Y}\Sigma X$:

$$= \sum XY - \overline{X}\sum Y - \overline{Y}\sum X + \overline{Y}\sum X$$

Factor out the summation signs and substitute $\Sigma\overline{X}\overline{Y}$ for $\overline{Y}\Sigma X$:

$$\begin{aligned} &= \sum [(XY - \overline{X}Y) - (X\overline{Y} - \overline{Y}\overline{X})] \\ &= \sum [(X - \overline{X})Y - (X - \overline{X})\overline{Y}] \\ &= \sum (X - \overline{X})(Y - \overline{Y}) \end{aligned}$$

Since $x = X - \overline{X}$ and $y = Y - \overline{Y}$,

$$\sum (X - \overline{X})(Y - \overline{Y}) = \sum xy$$

By analogous derivation, the denominator of this expression, $\Sigma XY - (\Sigma X)^2/n$, can be shown to equal $\Sigma(X - \overline{X})(X - \overline{X})$ which equals Σx^2. Therefore,

$$b = \frac{\sum xy}{\sum x^2}$$

Given b we can solve for a from the first equation:

$$\begin{aligned} a &= \frac{\sum Y}{n} - \frac{b\sum X}{n} \\ &= \overline{Y} - b\overline{X} \end{aligned}$$

A.9 Derivation of the formulas for b and a in regression. This derivation requires algebra only.

We need to minimize $\Sigma d^2_{Y \cdot X}$, the unexplained sum of squares

$$\begin{aligned}\sum d^2_{Y \cdot X} &= \sum (Y - \hat{Y})^2 \\ &= \sum (Y - a - bX)^2 \qquad (\text{since } \hat{Y} = a + bX) \\ &= \sum Y^2 + na^2 + b^2 \sum X^2 - 2a \sum Y - 2b \sum XY + 2ab \sum X\end{aligned}$$

Add and subtract the following three terms from this expression (not changing its value in the process):

$$\frac{(\sum Y)^2}{n}, \quad \frac{2b \sum X \sum Y}{n}, \text{ and } \frac{b^2(\sum X)^2}{n}$$

Rearranging the terms, we obtain

$$\begin{aligned}\sum d^2_{Y \cdot X} = &\left[\sum Y^2 - \frac{(\sum Y)^2}{n}\right] - 2b\left[\sum XY - \frac{\sum X \sum Y}{n}\right] \\ &+ b^2\left[\sum X^2 - \frac{(\sum X)^2}{n}\right] + \frac{(\sum Y)^2}{n} - \frac{2b \sum X \sum Y}{n} \\ &+ \frac{b^2(\sum X)^2}{n} + na^2 - 2a \sum Y + 2ab \sum X\end{aligned}$$

The first three terms can be rewritten as sums of squares (see Section A.3) or sums of products (see Section 14.4). For the remaining six terms, we rewrite ΣX and ΣY in terms of $\overline{X}$ and $\overline{Y}$, respectively. We obtain

$$\begin{aligned}\sum d^2_{Y \cdot X} = &\sum y^2 - 2b \sum xy + b^2 \sum x^2 + n\overline{Y}^2 - 2bn\overline{X}\overline{Y} + b^2 n\overline{X}^2 \\ &+ na^2 - 2an\overline{Y} + 2abn\overline{X}\end{aligned}$$

Factoring out n for the last six terms, we discover that they can be collected as

$$n(\overline{Y} - a - b\overline{X})^2$$

Therefore

$$\sum d^2_{Y \cdot X} = \sum y^2 - 2b \sum xy + b^2 \sum x^2 + n(\overline{Y} - a - b\overline{X})^2$$

By adding and subtracting $(\Sigma xy)^2/\Sigma x^2$ we obtain

$$\sum d_{Y\cdot X}^2 = \sum y^2 - \frac{(\sum xy)^2}{\sum x^2} + \frac{(\sum xy)^2}{\sum x^2} - 2b\sum xy + b^2\sum x^2 + n(\overline{Y} - a - b\overline{X})^2$$

$$= \sum y^2 - \frac{(\sum xy)^2}{\sum x^2} + \sum x^2\left[\frac{\sum xy}{\sum x^2} - b\right]^2 + n(\overline{Y} - a - b\overline{X})^2$$

If we wish to minimize $\Sigma d_{Y\cdot X}^2$ (barring trivial solutions), the two terms involving b and a must equal zero. Hence, when $\Sigma d_{Y\cdot X}^2$ is minimal, $b = \Sigma xy/\Sigma x^2$ and $a = \overline{Y} - b\overline{X}$.

Also since the last two terms must equal zero for minimal $\Sigma d_{Y\cdot X}^2$ we have obtained as a by-product a derivation of an alternative formula:

$$\sum d_{Y\cdot X}^2 = \sum y^2 - \frac{(\sum xy)^2}{\sum x^2} \qquad \text{[Expression (14.5)]}$$

A.10 Demonstration that the sum of squares of the dependent variable in regression can be partitioned exactly into explained and unexplained sums of squares, the cross products canceling out.

By definition (see Section 14.4),

$$y = \hat{y} + d_{Y\cdot X}$$

$$\sum y^2 = \sum(\hat{y} + d_{Y\cdot X})^2 = \sum \hat{y}^2 + \sum d_{Y\cdot X}^2 + 2\sum \hat{y}d_{Y\cdot X}$$

If we can show that $\Sigma\hat{y}d_{Y\cdot X} = 0$, then we have demonstrated the required identity. We have

$$\sum \hat{y}d_{Y\cdot X} = \sum bx(y - bx) \qquad \text{[since } \hat{y} = bx \text{ from Expression (14.3) and } d_{Y\cdot X} = Y - \hat{Y} = (Y - \overline{Y}) - (\hat{Y} - \overline{Y}) = y - \hat{y} = y - bx]$$

$$= b\sum xy - b^2\sum x^2$$

$$= b\sum xy - b\frac{\sum xy}{\sum x^2}\sum x^2 \qquad (\text{since } b = \textstyle\sum xy/\sum x^2)$$

$$= b\sum xy - b\sum xy$$

$$= 0$$

Therefore $\Sigma y^2 = \Sigma \hat{y}^2 + \Sigma d^2_{Y \cdot X}$ or, written in terms of variates,

$$\sum (Y - \bar{Y})^2 = \sum (\hat{Y} - \bar{Y})^2 + \sum (Y - \hat{Y})^2$$

A.11 Derivation of the standard error of a regression coefficient.

$$b = \frac{\sum xy}{\sum x^2}$$

By factoring out y, we express the regression coefficient as a weighted sum of the deviates $Y - \bar{Y}$:

$$\begin{aligned} b &= \sum \left(\frac{x}{\sum x^2} \right) y \\ &= \sum w_i y \qquad \left(\text{where } w_i = \frac{x_i}{\sum x^2} \right) \end{aligned}$$

Then (assuming that the deviations from regression are uncorrelated, and with the same variance, $s^2_{Y \cdot X}$) we obtain from Expression (7.1)

$$\begin{aligned} s^2_b &= \sum w_i^2 s^2_{Y \cdot X} \\ &= \left[\frac{1}{(\sum x^2)^2} \sum x_i^2 \right] s^2_{Y \cdot X} \\ &= \frac{s^2_{Y \cdot X}}{\sum x^2} \qquad \text{(as shown in Box 14.2)} \end{aligned}$$

A.12 Derivation of the variance of $\hat{Y}$, the estimated value of Y for a given value X.

In the equations below, an X without a subscript signifies the given value of X for which we wish to obtain $\hat{Y}$, the predicted value of Y. Values with subscripts of X and of x represent the observed variates and deviates of the sample.

First we must prove the identity $\Sigma xy = \Sigma xY$. By substitution,

$$\begin{aligned} \sum (X - \bar{X})(Y - \bar{Y}) &= \sum (XY - \bar{X}Y - X\bar{Y} - \overline{XY}) \\ &= \sum [(X - \bar{X})Y - \bar{Y}(X - \bar{X})] \\ &= \sum xY - \bar{Y} \sum (X - \bar{X}) \quad \left(\text{but since } \sum (X - \bar{X}) = 0\right) \\ &= \sum xY \end{aligned}$$

Then

$$\begin{aligned}
\hat{Y} &= a + bX \\
&= \bar{Y} - b\bar{X} + bX && (\text{since } a = \bar{Y} - b\bar{X}) \\
&= \bar{Y} + b(X - \bar{X}) \\
&= \frac{\sum Y}{n} + \frac{\sum xy}{\sum x^2}(X - \bar{X}) && \left(\text{since } \bar{Y} = \frac{\sum Y}{n} \text{ and } b = \frac{\sum xy}{\sum x^2}\right) \\
&= \sum\left(\frac{1}{n} + (X - \bar{X})\frac{x_i}{\sum x^2}\right)Y_i && \left(\text{since } \sum xy = \sum xY\right) \\
&= \sum w_i Y_i && \left(\text{where } w_i = \frac{1}{n} + (X - \bar{X})\frac{x_i}{\sum x^2}\right)
\end{aligned}$$

Assuming that the deviations from regression are uncorrelated, and with the same variance $s^2_{Y \cdot X}$, we obtain from Expression (7.1)

$$\begin{aligned}
s^2_{\hat{Y}} &= \sum w_i^2 s^2_{Y \cdot X} \\
&= \sum\left(\frac{1}{n} + (X - \bar{X})\frac{x_i}{\sum x^2}\right)^2 s^2_{Y \cdot X} \\
&= \sum\left[\frac{1}{n^2} + \frac{2(X - \bar{X})x_i}{n\sum x^2} + (X - \bar{X})^2\frac{x_i^2}{(\sum x^2)^2}\right]s^2_{Y \cdot X} \\
&= \left[\frac{1}{n} + \frac{2(X - \bar{X})}{n\sum x^2}\sum x_i + \frac{(X - \bar{X})^2}{\sum x^2}\right]s^2_{Y \cdot X} && \left(\text{since } \sum x_i = 0, \text{ the middle term in brackets vanishes}\right) \\
&= \left(\frac{1}{n} + \frac{(X - \bar{X})^2}{\sum x^2}\right)s^2_{Y \cdot X} && (\text{as shown in Box 14.2})
\end{aligned}$$

A.13 Proof that the variance of the sum of two variables is

$$s^2_{(Y_1+Y_2)} = s_1^2 + s_2^2 + 2r_{12}s_1s_2$$

where s_1 and s_2 are standard deviations of Y_1 and Y_2, respectively, and r_{12} is the parametric correlation coefficient between Y_1 and Y_2.

If $Z = Y_1 + Y_2$, then

$$s_Z^2 = \frac{1}{n}\sum(Z - \bar{Z})^2 = \frac{1}{n}\sum\left[(Y_1 + Y_2) - \frac{1}{n}\sum(Y_1 + Y_2)\right]^2$$

$$= \frac{1}{n}\sum\left[(Y_1 + Y_2) - \frac{1}{n}\sum Y_1 - \frac{1}{n}\sum Y_2\right]^2 = \frac{1}{n}\sum[(Y_1 + Y_2) - \bar{Y}_1 - \bar{Y}_2]^2$$

$$= \frac{1}{n}\sum[(Y_1 - \bar{Y}_1) + (Y_2 - \bar{Y}_2)]^2 = \frac{1}{n}\sum[y_1 + y_2]^2$$

$$= \frac{1}{n}\sum[y_1^2 + y_2^2 + 2y_1y_2] = \frac{1}{n}\sum y_1^2 + \frac{1}{n}\sum y_2^2 + \frac{2}{n}\sum y_1y_2$$

$$= s_1^2 + s_2^2 + 2s_{12}$$

Since $r_{12} = s_{12}/s_1s_2$, however, we have

$$s_{12} = r_{12}s_1s_2$$

Therefore,

$$s_Z^2 = s_{(Y_1+Y_2)}^2 = s_1^2 + s_2^2 + 2r_{12}s_1s_2 \quad \text{[Expression (15.9)]}$$

Similarly,

$$s_D^2 = s_{(Y_1-Y_2)}^2 = s_1^2 + s_2^2 - 2r_{12}s_1s_2 \quad \text{[Expression (15.9a)]}$$

A.14 Proof that the general expression for the G-test can be simplified to Expression (17.4).

In general, G is twice the natural logarithm of the ratio of the probability of the sample (with all parameters estimated from the data) and the probability of the sample (assuming the null hypothesis is true). Assuming a multinomial distribution, this ratio is

$$L = \frac{\dfrac{n!}{f_1!f_2!\cdots f_a!}p_1^{f_1}p_2^{f_2}\cdots p_a^{f_a}}{\dfrac{n!}{f_1!f_2!\cdots f_a!}\hat{p}_1^{f_1}\hat{p}_2^{f_2}\cdots \hat{p}_a^{f_a}}$$

$$= \prod_{i=1}^{a}\left(\frac{p_i}{\hat{p}_i}\right)^{f_i}$$

where f_i is the observed frequency, p_i is the observed proportion, and $\hat{p}_i$ the expected proportion of class i, while n is sample size, the sum of the observed frequencies over the a classes.

$$G = 2 \ln L$$
$$= 2\sum^{a} f_i \ln\left(\frac{p_i}{\hat{p}_i}\right)$$

Since $f_i = np_i$ and $\hat{f}_i = n\hat{p}_i$,

$$G = 2\sum^{a} f_i \ln\left(\frac{f_i}{\hat{f}_i}\right) \qquad \text{[Expression (17.4)]}$$

A.15 Proof that in a goodness-of-fit test with only two classes the deviations are always equal in magnitude but opposite in sign.

Let f_1 and f_2 be the two observed frequencies ($f_1 + f_2 = n$), and let $\hat{p}$ be the expected probability of the event that has an observed frequency of f_1. The expected frequencies $\hat{f}_1$ and $\hat{f}_2$ would then be

$$\hat{f}_1 = n\hat{p}$$
$$\hat{f}_2 = n(1 - \hat{p}) = n - n\hat{p}$$

The deviations from expectation are

$$f_1 - \hat{f}_1 = f_1 - n\hat{p}$$

and

$$\begin{aligned} f_2 - \hat{f}_2 &= f_2 - n + n\hat{p} \\ &= (n - f_1) - n + n\hat{p} \qquad (\text{since } f_1 + f_2 = n,\ f_2 = n - f_1) \\ &= -f_1 + n\hat{p} \\ &= -(f_1 - n\hat{p}) \qquad (\text{and since } n\hat{p} = \hat{f}_1) \\ &= -(f_1 - \hat{f}_1) \end{aligned}$$

A.16 Proof that Expression (17.5), $\Sigma^2(f_i - \hat{f}_i)^2/\hat{f}_i$, equals Expression (17.6), $(f_1 - \hat{p}n)^2/\hat{p}\hat{q}n$, or Expression (17.7), $(f_2 - \hat{q}n)^2/\hat{p}\hat{q}n$.

We have

$$X^2 = \frac{\sum^{2} (f_i - \hat{f}_i)^2}{\hat{f}_i} = \frac{(f_1 - \hat{f}_1)^2}{\hat{f}_1} + \frac{(f_2 - \hat{f}_2)^2}{\hat{f}_2}$$

In Section A.15 we showed that $(f_2 - \hat{f}_2) = -(f_1 - \hat{f}_1)$. The squares of these terms are therefore equal, and we can write

$$\frac{(f_1 - \hat{f}_1)^2}{\hat{f}_1} + \frac{(f_2 - \hat{f}_2)^2}{\hat{f}_2} = \frac{(f_1 - \hat{f}_1)^2}{\hat{f}_1} + \frac{(f_1 - \hat{f}_1)^2}{\hat{f}_2}$$

$$= \frac{(f_1 - \hat{f}_1)^2(\hat{f}_1 + \hat{f}_2)}{\hat{f}_1\hat{f}_2}$$

$$= \frac{(f_1 - \hat{f}_1)^2 n}{\hat{f}_1(n - \hat{f}_1)} \qquad (\text{since } \hat{f}_1 + \hat{f}_2 = n, \hat{f}_2 = n - \hat{f}_1)$$

$$= \frac{(f_1 - \hat{p}n)^2 n}{\hat{p}n(n - \hat{p}n)} \qquad (\text{since } \hat{f}_1 = \hat{p}n)$$

$$= \frac{(f_1 - \hat{p}n)^2}{\hat{p}(1 - \hat{p})n} \qquad (\text{by definition, } \hat{q} = 1 - \hat{p})$$

$$= \frac{(f_1 - \hat{p}n)^2}{\hat{p}\hat{q}n} \tag{17.6}$$

By replacing $(f_1 - \hat{f}_1)^2$ with $(f_2 - \hat{f}_2)^2 = (f_2 - \hat{q}n)^2$ in the numerator of the right-hand side of the derivation, we obtain

$$X^2 = \frac{(f_2 - \hat{q}n)^2}{\hat{p}\hat{q}n} \tag{17.7}$$

BIBLIOGRAPHY

Agresti, A. 1990. *Categorical Data Analysis.* Wiley, New York. 558 pp.

Allee, W. C., and E. Bowen. 1932. Studies in animal aggregations: Mass protection against colloidal silver among goldfishes. *J. Exp. Zool.* **61**:185–207.

Allee, W. C., E. S. Bowen, J. C. Welty, and R. Oesting. 1934. The effect of homotypic conditioning of water on the growth of fishes, and chemical studies of the factors involved. *J. Exp. Zool.* **68**:183–213.

Altman, D. G. 1978. Plotting probability ellipses. *Appl. Stat.* **27**:347–349.

Anderberg, M. R. 1973. *Cluster Analysis for Applications.* Academic Press, New York. 359 pp.

Anderson, V. L., and R. A. McLean. 1974a. *Design of Experiments.* Marcel Dekker, New York. 418 pp.

Anderson, V. L., and R. A. McLean. 1974b. Restriction errors: Another dimension in teaching experimental statistics. *Amer. Stat.* **28**:145–152.

Anscombe, F. J. 1948. The transformation of Poisson, binomial and negative binomial data. *Biometrika* **35**:246–254.

Archibald, E. E. A. 1950. Plant populations. II. The estimation of the number of individuals per unit area of species in heterogeneous plant populations. *Ann. Bot.,* N.S. **14**:7–21.

Arvesen, J. N., and T. H. Schmitz. 1970. Robust procedures for variance component problems using the jackknife. *Biometrics* **26**:677–686.

Asmundson, V. S. 1931. The formation of the hen's egg. *Sci. Agric.* **11**(9), 50 pp.

Balandra, K. P., and H. L. MacGillivray. 1988. Kurtosis: A critical review. *Amer. Stat.* **42**:111–119.

Bancroft, T. A. 1964. Analysis and inference for incompletely specified models involving the use of preliminary test(s) of significance. *Biometrics* **20**:427–442.

Band, H. T., and P. T. Ives. 1963. Genetic structure of populations. I. On the nature of the genetic load in the South Amherst population of *Drosophila melanogaster. Evolution* **17**:198–215.

Banta, A. M. 1939. *Studies on the Physiology, Genetics, and Evolution of Some Cladocera.* Carnegie Institution of Washington, Dept. Genetics, Paper 39. 285 pp.

Barnes, H., and F. A. Stanbury. 1951. A statistical study of plant distribution during the colonization and early development of vegetation on china clay residues. *J. Ecol.* **39**:171–181.

Barnett, V., and T. Lewis. 1978. *Outliers in Statistical Data.* Wiley, Chichester, UK.

Bartels, R. H., S. D. Horn, A. M. Liebetrau, and W. L. Harris. 1977. *A Computational Investigation of Conover's Kolmogorov–Smirnov Test for Discrete Distributions.* Johns Hopkins Univ. Dept. Math. Sci. Tech. Report No. 260. 14 pp.

Benjamini, Y. 1988. Opening the box of a box plot. *Amer. Stat.* **42**:257–262.

Bhalla, S. C., and R. R. Sokal. 1964. Competition among genotypes in the housefly at varying densities and proportions (the *green* strain). *Evolution* **18**:312–330.

Bishop, Y. M. M., S. E. Fienberg, and P. W. Holland. 1975. *Discrete Multivariate Analysis: Theory and Practice.* MIT Press, Cambridge, Mass. 557 pp.

Blakeslee, A. F. 1921. The globe mutant in the jimson weed (*Datura stramonium*). *Genetics* **6**:241–264.

Bliss, C. I. 1970. *Statistics in Biology.* Vol. 2. McGraw-Hill, New York. 639 pp.

Bliss, C. I., and D. W. Calhoun. 1954. *An Outline of Biometry.* Yale Co-op. Corp., New Haven, Conn. 272 pp.

Bliss, C. I., and R. A. Fisher. 1953. Fitting the negative binomial distribution to biological data and note on the efficient fitting of the negative binomial. *Biometrics* **9**:176–200.

Block, B. C. 1966. The relation of temperature to the chirp-rate of male snowy tree crickets, *Oecanthus fultoni* (Orthoptera: Gryllidae). *Ann. Entomol. Soc. Amer.* **59**:56–59.

Boardman, T. J. 1974. Confidence intervals for variance components—a comparative Monte Carlo study. *Biometrics* **30**:251–262.

Bortkiewicz, L. von. 1898. *Das Gesetz der kleinen Zahlen.* Teubner, Leipzig. 52 pp.

Box, G. E. P., and D. R. Cox. 1964. An analysis of transformations. *J. R. Stat. Soc.,* Ser. B **26**:211–243.

Bradley, J. V. 1968. *Distribution-free Statistical Tests.* Prentice-Hall, Englewood Cliffs, N.J. 388 pp.

Bradley, R. A., and S. S. Srivastava. 1979. Coding in polynomial regression. *Amer. Stat.* **33**:11–14.

Breiman, L., and J. H. Friedman. 1985. Estimating optimal transformations for multiple regression and correlation. *J. Amer. Stat. Assn.* **80**:580–598.

Brian, M. V., and J. Hibble. 1964. Studies of caste differentiation in *Myrmica rubra* L. 7. Caste bias, queen age and influence. *Insectes Sociaux* **11**:223–228.

Brower, L. P. 1959. Speciation in butterflies of the *Papilio glaucus* group. I. Morphological relationships and hybridization. *Evolution* **13**:40–63.

Brown, B. E., and A. W. A. Brown. 1956. The effects of insecticidal poisoning on the level of cytochrome oxidase in the American cockroach. *J. Econ. Entomol.* **49**:675–679.

Brown, F. M., and W. P. Comstock. 1952. Some biometrics of *Heliconius charitonius* (Linnaeus) (Lepidoptera, Nymphalidae). *Amer. Mus. Novitates* **1574.** 53 pp.

Buley, H. M. 1936. Consumption of diatoms and dinoflagellates by the mussel. *Bull. Scripps Inst. Oceanography,* Tech. Ser. **4**:19–27.

Burr, E. J. 1960. The distribution of Kendall's score S for a pair of tied rankings. *Biometrika* **47**:151–171.

Burrell, P. C., and T. W. Whitaker. 1939. The effect of indol-acetic acid on fruit-setting in muskmelons. *Amer. Soc. Hort. Sci.* **37**:829–830.

Burstein, H. 1973. Close approximation of binomial confidence limits at 10 confidence levels. *Commun. Stat.* **2**:185–187.

Carter, G. R., and C. A. Mitchell. 1958. Methods for adapting the virus of rinderpest to rabbits. *Science* **128**:252–253.

Casagrande, J. T., M. C. Pike and P. G. Smith. 1978. An improved approximate formula for calculating sample sizes for comparing two binomial distributions. *Biometrics* **34**:483–486.

Castillo, J. V. 1938. Determination of the optimum amount of fish meal in rations for growing ducklings. *Philippine Agric.* **27**:18–30.

Cattell, R. B. 1978. *The Scientific Use of Factor Analysis in Behavioral and Life Sciences.* Plenum Press, New York. 618 pp.

Clark, F. H. 1941. Correlation and body proportion in mature mice of the genus *Peromyscus. Genetics* **26**:283–300.

Clifford, H. T., and W. Stephenson. 1975. *An Introduction to Numerical Classification.* Academic Press, New York. 229 pp.

Clifford, P., S. Richardson, and D. Hémon. 1989. Assessing the significance of the correlation between two spatial processes. *Biometrics* **45**:123–134.

Cochran, W. G., and G. M. Cox. 1957. *Experimental Designs.* 2d ed. Wiley, New York. 611 pp.

Cohen, A. C., Jr. 1960. Estimating the parameter in a conditional Poisson distribution. *Biometrics* **16**:203–211.

Cohen, A. I. 1954. Studies on glycolysis during early development of the *Rana pipiens* embryo. *Physiol. Zool.* **27**:128–140.

Cohen, J. 1977. *Statistical Power Analysis for the Behavioral Sciences.* Rev. ed. Academic Press, New York. 474 pp.

Cohen, L. A., M. E. Kendall, E. Zang, C. Meschter, and D. P. Rose. 1991. Modulation of *N*-nitrosomethylurea-induced mammary tumor promotion by dietary fiber and fat. *J. Nat. Cancer Inst.* **83**:496–501.

Conahan, M. A. 1970. The comparative accuracy of the likelihood ratio and *X*-squared as approximations to the exact multinomial test. Ed.D. diss., Lehigh Univ. 64 pp.

Conover, W. J. 1972. A Kolmogorov goodness of fit test for discontinuous distributions. *J. Amer. Stat. Assn.* **67**:591–596.

Cowan, I. M., and P. A. Johnston. 1962. Blood serum protein variations at the species and subspecies level in deer of the genus *Odocoileus. Syst. Zool.* **11**:131–138.

Cramer, E. M. 1972. Significance tests and tests of models in multiple regression. *Amer. Stat.* **26**(4):26–30.

Crow, E. L. 1956. Confidence intervals for a proportion. *Biometrika* **43**:423–435.

Crow, E. L., and R. S. Gardner. 1959. Confidence intervals for the expectation of a Poisson variable. *Biometrika* **46**:441–453.

D'Agostino, R. B., W. Chase, and A. Belanger. 1988. The appropriateness of some common procedures for testing the equality of two independent binomial populations. *Amer. Stat.* **42**:198–202.

D'Agostino, R. B., and G. E. Noether. 1973. On the evaluation of the Kolmogorov statistic. *Amer. Stat.* **27**:81–82.

D'Agostino, R. B., and E. S. Pearson. 1973. Tests for departure from normality. Empirical results for the distribution of b_2 and $\sqrt{b_1}$. *Biometrika* **60**:613–622.

D'Agostino, R. B., and G. L. Tietjen. 1973. Approaches to the null distribution of $\sqrt{b_1}$. *Biometrika* **60**:169–173.

Davis, E. A., Jr. 1955. Seasonal changes in the energy balance of the English sparrow. *Auk* **72**:385–411.

Devlin, S. J., R. Gnanadesikan, and J. R. Kettenring. 1975. Robust estimation and outlier detection with correlation coefficients. *Biometrika* **62**:531–545.

Dixon, W. J. 1950. Analysis of extreme values. *Ann. Math. Stat.* **21**:488–506.

Dixon, W. J., ed. 1985. *BMDP Statistical Software.* Univ. of Calif. Press, Berkeley. 734 pp.

Dixon, W. J., and F. J. Massey, Jr. 1969. *Introduction to Statistical Analysis.* 3d ed. McGraw-Hill, New York. 638 pp.

Draper, N. R., and H. Smith. 1981. *Applied Regression Analysis.* 2d ed. Wiley, New York. 709 pp.

Dunn, G., and B. S. Everitt. 1982. *An Introduction to Mathematical Taxonomy.* Cambridge Univ. Press, New York. 152 pp.

Dunn, O. J., and V. A. Clark. 1974. *Applied Statistics: Analysis of Variance and Regression.* Wiley, New York. 387 pp.

Dunnett, C. W. 1980. Pairwise multiple comparisons in the homogeneous variance, unequal sample size case. *J. Amer. Stat. Assn.* **75**:789–795.

DuPraw, E. J. 1965. Non-Linnean taxonomy and the systematics of honeybees. *Syst. Zool.* **14**:1–24.

Dwass, M. 1960. Some *k*-sample rank-order tests. In *Contributions to Probability and Statistics, Essays in Honor of Harold Hotelling,* ed. I. Olkin et al. Stanford Univ. Press, Stanford, pp. 198–202.

Efron, B., and G. Gong. 1983. A leisurely look at the bootstrap, the jackknife, and cross-validation. *Amer. Stat.* **37**:36–48.

Eisenhart, C. 1947. The assumptions underlying the analysis of variance. *Biometrics* **3**:1–21.

Everitt, B. S. 1980. *Cluster Analysis.* 2d ed. Halsted Press, New York. 136 pp.

Everitt, B. S., and G. Dunn. 1992. *Applied Multivariate Data Analysis.* Oxford Univ. Press, New York. 304 pp.

Falconer, D. S. 1981. *Introduction to Quantitative Genetics.* 2d ed. Longman Press, New York. 340 pp.

Fawcett, R. F. 1990. The alignment method for displaying and analyzing treatments in blocking designs. *Amer. Stat.* **44**:204–209.

Felsenstein, J. 1985. Confidence limits on phylogenies: An approach using bootstrap. *Evolution* **39**:783–791.

Feuer, E. J., and L. G. Kessler. 1989. Test statistic and sample size for a two-sample McNemar test. *Biometrics* **45**:629–636.

Finney, D. J. 1971. *Probit Analysis.* 3d ed. Cambridge Univ. Press, London. 333 pp.

Finney, D. J., R. Latscha, B. M. Bennett, and P. Hsu. 1963. *Tables for Testing Significance in a 2 × 2 Contingency Table.* Cambridge Univ. Press, London. 102 pp.

Fisher, R. A. 1954. *Statistical Methods for Research Workers.* 12th ed. Oliver & Boyd, Edinburgh, 356 pp.

Fisher, R. A., and F. Yates. 1963. *Statistical Tables for Biological, Agricultural and Medical Research.* 6th ed. Oliver & Boyd, Edinburgh, 138 pp.

Fleiss, J. L. 1979. Confidence intervals for the odds ratio in case-control studies: The state of the art. *J. Chronic Diseases* **32**:69–77.

Fligner, M. A., and G. E. Policello. 1981. Robust rank procedures for the Fisher–Behrens problem. *J. Amer. Stat. Assn.* **76**:162–168.

Freeman, M. F., and J. W. Tukey. 1950. Transformations related to the angular and the square root. *Ann. Math. Stat.* **21**:607–611.

French, A. R. 1976. Selection of high temperatures for hibernation by the pocket mouse, *Perignathus longimembris:* Ecological advantages and energetic consequences. *Ecology* **57**:185–191.

Fröhlich, F. W. 1921. *Grundzüge einer Lehre vom Licht- und Farbensinn. Ein Beitrag zur allgemeinen Physiologie der Sinne.* Fischer, Jena. 86 pp.

Gabriel, K. R. 1964. A procedure for testing the homogeneity of all sets of means in analysis of variance. *Biometrics* **20**:459–477.

Gabriel, K. R. 1978. A simple method of multiple comparisons of means. *J. Amer. Stat. Assn.* **73**:724–729.

Gaines, S. D., and W. R. Rice. 1990. Analysis of biological data when there are ordered expectations. *Amer. Nat.* **135**:310–317.

Galambo, J. 1984. *Introduction to Probability Theory.* Marcel Dekker, New York. 200 pp.

Gallo, J., and A. I. Khuri. 1990. Exact tests for the random and fixed effects in an unbalanced mixed two-way cross-classification model. *Biometrics* **46**:1087–1095.

Games, P. A., and J. F. Howell. 1976. Pairwise multiple comparison procedures with unequal *N*'s and/or variances: A Monte Carlo study. *J. Educ. Stat.* **1**:113–125.

Gans, D. L. 1991. Preliminary test on variances. *Amer. Stat.* **45**:258.

Gart, J. J. 1969. An exact test for comparing matched proportions in crossover designs. *Biometrika* **56**:75–80.

Gartler, S. M., I. L. Firschein, and T. Dobzhansky. 1956. Chromatographic investigation of urinary amino-acids in the great apes. *Amer. J. Phys. Anthropol.* **14**:41–57.

Gaylor, D. W., and F. N. Hopper. 1969. Estimating the degrees of freedom for linear combinations of mean squares by Satterthwaite's formula. *Technometrics* **11**:691–705.

Geissler, A. 1889. Beiträge zur Frage des Geschlechtsverhältnisses der Geborenen. *Z. K. Sächs. Stat. Bur.* **35**:1–24.

Gideon, R. A., and D. E. Mueller. 1978. Computation of two-sample Smirnov statistics. *Amer. Stat.* **32**:136–137.

Gill, J. L. 1978. *Design and Analysis of Experiments in the Animal and Medical Sciences.* 3 vols. Iowa State Univ. Press, Ames, Iowa. Vol. 1, 409 pp.; Vol. 2, 301 pp.; Vol. 3, 173 pp.

Goddard, M. J. 1991. Constructing some categorical anomalies. *Amer. Stat.* **45**:129–134.

Goldstein, A. 1964. *Biostatistics.* Macmillan, New York. 272 pp.

Gonzalez, T., S. Sahni, and W. R. Franta. 1977. An efficient algorithm for the Kolmogorov–Smirnov and Lilliefors tests. *ACM Trans. Math. Software* **3**:60–64.

Goodstein R. 1989. Power and sample size via MS/PC-DOS computers. *Amer. Stat.* **43**:253–260.

Gordon, A. D. 1981. *Classification: Methods for Exploratory Analysis of Multivariate Data.* Chapman and Hall, New York. 193 pp.

Gorsuch, R. L. 1983. *Factor Analysis.* 2d ed. Lawrence Erlbaum Assoc. Hillsdale, N.J. 425 pp.

Gould, S. J. 1966. Allometry and size in ontogeny and phylogeny. *Biol. Rev.* **41**:587–640.

Gould, S. J. 1975. Allometry in primates, with emphasis on scaling and the evolution of the brain. *Contrib. Primatol.* **5**:244–292.

Goulden, C. H. 1952. *Methods of Statistical Analysis.* 2d ed. Wiley, New York. 467 pp.

Green, B. F. 1977. A practical interactive program for randomization tests of location. *J. Amer. Stat. Assn.* **31**:37–39.

Green, P. E.; with contributions by J. D. Carroll. 1978. *Analyzing Multivariate Data.* Dryden Press, Hinsdale, Ill. 519 pp.

Green, P. E., and J. D. Carroll. 1978. *Mathematical Tools for Applied Multivariate Analysis.* Academic Press, New York. 376 pp.

Greenland, S. 1991. On the logical justification of conditional tests for two-by-two contingency tables. *Amer. Stat.* **45**:248–251.

Greenwood, M., and G. U. Yule. 1920. An inquiry into the nature of frequency-distributions of multiple happenings. *J. R. Stat. Soc.* **83**:255–279.

Greig-Smith, P. 1964. *Quantitative Plant Ecology.* 2d ed. Butterworths, Washington, D.C. 256 pp.

Grizzle, J. E. 1967. Continuity correction in the X^2-test for 2×2 tables. *Amer. Stat.* **21**(4):28–32.

Groggel, D. J., and J. H. Skillings. 1986. Distribution-free tests for main effects in multifactor designs. *Amer. Stat.* **40**:99–102.

Grubbs, F. E. 1969. Procedures for detecting outlying observations in samples. *Technometrics* **11**:1–21.

Gumbel, E. J. 1954. Statistical theory of extreme values and some practical applications. U.S. Govt. Printing Office, Washington, D.C. 51 pp.

Gurland, J., and R. C. Tripathi. 1971. A simple approximation for unbiased estimation of the standard deviation. *Amer. Stat.* **25**:30–32.

Haine, E. 1955. *Biologisch-ökologische Studien an* Rhopalosiphoninus latysiphon *D.* Landwirtschaftsverlag, Hiltrup bei Münster (Westf.) 58 pp.

Haltenorth, T. 1937. Die verwandtschaftliche Stellung der Grosskatzen zueinander. *Z. Säugetierk.* **12**:97–240.

Hamilton, D. 1987. Sometimes $R^2 > r^2_{yx_1} + r^2_{yx_2}$. Correlated variables are not always redundant. *Amer. Stat.* **41**:129–132.

Hand, D. J. 1981. *Discrimination and Classification.* Wiley, New York. 218 pp.

Hand, D. J., and C. C. Taylor. 1987. *Multivariate Analysis of Variance and Repeated Measures: A Practical Approach for Behavioural Scientists.* Chapman and Hall, New York. 262 pp.

Hanna, B. L. 1953. On the relative importance of genetic and environmental factors in the determination of human hair pigment concentration. *Amer. J. Hum. Genet.* **5**:293–321.

Harman, H. H. 1976. *Modern Factor Analysis.* 3d ed. Univ. of Chicago Press, Chicago. 487 pp.

Harris, R. E., E. A. Zang, J. I. Anderson, and E. L. Wynder. 1993. Lung cancer risk associated with cigarette smoking in black and white Americans. *Internat. J. Epidem.* **22**:592–599.

Harter, H. L., H. J. Khamis, and R. E. Lamb. 1984. Modified Kolmogorov–Smirnov tests for goodness of fit. *Comm. Stat. Simul. Comput.* **13**:293–323.

Hartigan, J. A. 1975. *Clustering Algorithms.* Wiley, New York. 351 pp.

Hartley, H. O. 1950. The maximum F-ratio as a short cut test for heterogeneity of variances. *Biometrika* **37**:308–312.

Hartley, H. O. 1962. Analysis of variance. In *Mathematical Methods for Digital Computers,* Vol. 1, ed. A. Ralston and H. S. Wilf. Wiley, New York, pp. 221–230.

Hasel, A. A. 1938. Sampling error in timber surveys. *J. Agric. Res.* **57**:713–737.

Hedges, L. B., and I. Olkin. 1985. *Statistical Methods in Meta-Analysis.* Academic Press, New York. 369 pp.

Herr, D. G., and J. Gaebelin. 1978. Nonorthogonal two-way analysis of variance. *Psych. Bull.* **85**:207–216.

Hinkley, D. V. 1978. Improving the jackknife with special reference to correlation estimation. *Biometrika* **65**:13–21.

Hoaglin, D. C., and R. E. Welsch. 1978. The hat matrix in regression and ANOVA. *Amer. Stat.* **32**:17–22.

Hochberg, Y. 1974. Some generalizations of the *T*-method in simultaneous inference. *J. Multivar. Anal.* **4**:224–234.

Hochberg, Y. 1976. A modification of the *T*-method of multiple comparisons for one-way layout with unequal variances. *J. Amer. Stat. Assn.* **71**:200–203.

Hochberg, Y., and A. C. Tamhane. 1983. Multiple comparisons in a mixed model. *Amer. Stat.* **37**:305–307.

Hocking, R. R. 1976. The analysis and selection of variables in linear regression. *Biometrics* **32**:1–49.

Holm, S. 1979. A simple sequentially rejective multiple test procedure. *Scand. J. Stat.* **6**:65–70.

Hosmer, D. W., and Lemeshow, S. 1989. *Applied Logistic Regression.* Wiley, New York. 307 pp.

Hotelling, H. 1953. New light on the correlation coefficient and its transforms. *J. R. Stat. Soc.,* Ser. B **15**:193–232.

Hunt, G. L., and M. W. Hunt. 1976. Gull chick survival: The significance of growth rates, timing of breeding and territory size. *Ecology* **57**:62–75.

Hunter, P. E. 1959. Selection of *Drosophila melanogaster* for length of larval period. *Z. Vererbungsl.* **90**:7–28.

Hurvich, C. M., and C.-L. Tsai. 1990. The impact of model selection on inference in linear regression. *Amer. Stat.* **44**:214–217.

Jackson, J. E. 1991. *A User's Guide to Principal Components.* Wiley, New York. 569 pp.

Johnson, N. K. 1966. Bill size and the question of competition in allopatric and sympatric populations of dusky and gray flycatchers. *Syst. Zool.* **15**:70–87.

Johnson, N. L., and S. Kotz. 1969. *Distributions in Statistics: Discrete Distributions.* Vol. 1. Houghton Mifflin, Boston. 328 pp.

Jolicoeur, P. 1968. Interval estimation of the slope of the major axis of a bivariate normal distribution in the case of a small sample. *Biometrics* **24**:679–682.

Jolicoeur, P. 1975. Linear regressions in fishery research: Some comments. *J. Fish. Res. Board Can.* **32**:1491–1494.

Karten, I. 1965. Genetic differences and conditioning in *Tribolium castaneum. Physiol. Zool.* **38**:69–79.

Kempthorne, O. 1975. Fixed and mixed models in the analysis of variance. *Biometrics* **31**:473–486.

Kendall, M. G., and J. D. Gibbons. 1990. *Rank Correlation Methods.* 5th ed. Edward Arnold, London.

Kendall, M. G., and A. Stuart. 1961. *The Advanced Theory of Statistics.* Vol. 2, Hafner, New York. 676 pp.

Kermack, K. A., and J. B. S. Haldane. 1950. Organic correlation and allometry. *Biometrika* **37**:30–41.

Khamis, H. J. 1990. The δ-corrected Kolmogorov–Smirnov test for goodness of fit. *J. Stat. Planning Inference* **24**:317–335.

Khamis, H. J. 1992. The δ-corrected Kolmogorov–Smirnov test with estimated parameters. *Nonparametric Stat.* **2**:17–27.

Khuri, A. I., and R. C. Littell. 1987. Exact tests for the main effects variance components in an unbalanced random two-way model. *Biometrics* **43**:545–560.

Kim, J.-O., and C. W. Mueller. 1978. *Factor Analysis: Statistical Methods and Practical Issues.* Sage, Beverly Hills, Calif. 88 pp.

King, J. A., D. Maas, and R. G. Weisman. 1964. Geographic variation in nest size among species of *Peromyscus. Evolution* **18**:230–234.

Koehn, R. K. 1978. Physiology and biochemistry of enzyme variation: The interface of ecology and population genetics. In *Ecological Genetics: The Interface,* ed. P. F. Brussard. Springer-Verlag, New York, pp. 51–72.

Kotz, S., and D. F. Stroup. 1983. *Educated Guessing: How to Cope in an Uncertain World.* Marcel Dekker, New York. 187 pp.

Kouskolekas, C. A., and G. C. Decker. 1966. The effect of temperature on the rate of development of the potato leafhopper. *Empoasca fabae.* (Homoptera: Cicadellidae). *Ann. Entomol. Soc. Amer.* **59**:292–298.

Krebs, C. J. 1989. *Ecological Methodology.* Harper and Row, New York. 654 pp.

Kruskal, W. 1987. Relative importance by averaging over orderings. *Amer. Stat.* **41**:6–10.

Kruskal, W. H., and W. A. Wallis. 1952. Use of ranks in one-criterion variance analysis. *J. Amer. Stat. Assn.* **47**:583–621.

Krzanowski, W. J. 1988. *Principles of Multivariate Analysis.* Clarendon Press, Oxford. 563 pp.

Ku, H. H., and S. Kullback. 1974. Loglinear models in contingency table analysis. *Amer. Stat.* **28**:115–122.

Kuhry, B., and L. F. Marcus. 1977. Bivariate linear models in biometry. *Syst. Zool.* **26**:201–209.

Kullback, S. 1959. *Information Theory and Statistics.* Wiley, New York. 395 pp.

Kvålseth, T. O. 1985. Cautionary note about R^2. *Amer. Stat.* **39**:279–285.

Lande, R. 1977. On comparing coefficients of variation. *Syst. Zool.* **26**:214–217.

Larntz, K. 1978. Small-sample comparisons of exact levels for chi-squared goodness-of-fit statistics. *J. Amer. Stat. Assn.* **73**:253–263.

Leggatt, C. W. 1935. Contributions to the study of the statistics of seed testing. *Proc. Internat. Seed Testing Assn.* **5**:27–37.

Lewontin, R. C. 1966. On the measurement of relative variability. *Syst. Zool.* **15**:141–142.

Lewontin, R. C. 1974. The analysis of variance and the analysis of causes. *Amer. J. Hum. Genet.* **26**:400–411.

Lewontin, R. C., and J. Felsenstein. 1965. The robustness of homogeneity tests in $2 \times n$ tables. *Biometrics* **21**:19–33.

Lewontin, R. C., and M. J. D. White. 1960. Interaction between inversion polymorphisms of two chromosome pairs in the grasshopper, *Moraba scurra. Evolution* **14**:116–129.

Li, C. C. 1975. *Path Analysis—A Primer.* Boxwood Press, Pacific Grove, Calif. 346 pp.

Li, C. C. 1976. *First Course in Population Genetics.* Boxwood Press, Pacific Grove, Calif. 631 pp.

Lilliefors, H. W. 1967. The Kolmogorov–Smirnov test for normality with mean and variance unknown. *J. Amer. Stat. Assn.* **62**:399–402.

Lindeman, H. R. 1974. *Analysis of Variance in Complex Experimental Designs.* W. H. Freeman, San Francisco. 352 pp.

Ling, R. F. 1978. A study of the accuracy of some approximations for t, χ^2 and F tail probabilities. *J. Amer. Stat. Assn.* **73**:274–283.

Little, R. J. A. 1989. Testing the equality of two independent proportions. *Amer. Stat.* **43**:283–288.

Littlejohn, M. J. 1965. Premating isolation in the *Hyla ewingi* complex. *Evolution* **19**:234–243.

Livshits, G., R. R. Sokal, and E. Kobyliansky. 1991. Genetic affinities of Jewish populations. *Amer. J. Hum. Genet.* **49**:131–146.

Lunt, H. A. 1947. The response of hybrid poplar and other forest tree species to fertilizer and lime treatment in concrete soil frames. *J. Agric. Res.* **74**:113–132.

MacArthur, J. W. 1931. Linkage studies with the tomato. III. Fifteen factors in six groups. *Trans. R. Can. Inst.* **18**:1–19.

Mandel, J. 1964. *The Statistical Analysis of Experimental Data.* Wiley, New York. 410 pp.

Manly, B. F. J. 1977. A further note on Kiritani and Nakusuji's model for stage frequency data including comments on the use of Tukey's jackknife technique for estimating variances. *Res. Pop. Ecol.* **18**:177–186.

Manly, B. F. J. 1991. *Randmonization and Monte Carlo Methods in Biology.* Chapman and Hall, New York. 281 pp.

Mantel, N. 1967. The detection of disease clustering and a generalized regression approach. *Cancer Res.* **27**:209–220.

Mantel, N., and W. Haenszel. 1959. Statistical aspects of the analysis of data from retrospective studies of disease. *J. Nat. Cancer Inst.* **22**:719–748.

Mark, L. C., J. J. Burns, L. Brand, C. I. Campomanes, N. Trousof, E. M. Papper, and B. B. Brodie. 1958. The passage of thiobarbiturates and their oxygen analogs into brain. *J. Pharmacol. Exp. Therap.* **123**:70–73.

Markowski, C. A., and E. P. Markowski. 1990. Conditions for the effectiveness of a preliminary test of variance. *Amer. Stat.* **44**:322–326.

Marquardt, D. W. 1970. Generalized inverses, ridge regression, biased linear estimation and non-linear estimation. *Technometrics* **12**:591–612.

Marquardt, D. W., and R. D. Snee. 1975. Ridge regression in practice. *Amer. Stat.* **29**:3–20.

Martin, C. G., and P. A. Games. 1977. Anova tests for homogeneity of variance: Nonnormality and unequal samples. *J. Educ. Stat.* **2**:187–206.

Maxwell, A. E. 1970. Comparing the classification of subjects by two independent judges. *Brit. J. Psych.* **116**:651–655.

Mayer, L. S., and M. S. Younger. 1976. Estimation of standardized regression coefficients. *J. Amer. Stat. Assn.* **71**:154–157.

Mayr, E., and P. D. Ashlock. 1991. *Principles of Systematic Zoology.* 2d ed. McGraw-Hill, New York. 475 pp.

McGill, R., J. W. Tukey, and W. A. Larsen. 1978. Variations of box plots. *Amer. Stat.* **32**:12–16.

McNeil, D. R. 1977. *Interactive Data Analysis.* Wiley, New York. 186 pp.

Miller, R. 1968. Jackknifing variances. *Ann. Math. Stat.* **39**:567–582.

Miller, R. G. 1974. The jackknife—a review. *Biometrika* **61**:1–15.

Millis, J., and Y. P. Seng. 1954. The effect of age and parity of the mother on birth weight of the offspring. *Ann. Hum. Genet.* **19**:58–73.

Mittler, T. E., and R. H. Dadd. 1966. Food and wing determination in *Myzus persicae* (Homoptera: Aphidae). *Ann. Entomol. Soc. Amer.* **59**:1162–1166.

Montgomery, D. C. 1966. *Design and Analysis of Experiments.* Wiley, New York. 418 pp.

Morrison, D. F. 1990. *Multivariate Statistical Methods.* 3d ed. McGraw-Hill, New York. 495 pp.

Mulaik, S. A. 1972. *The Foundations of Factor Analysis.* McGraw-Hill, New York. 453 pp.

Mullett, G. M. 1972. Graphical illustration of simple (total) and partial regression. *Amer. Stat.* **26**(5):25–27.

Nelson, V. E. 1964. The effects of starvation and humidity on water content in *Tribolium confusum* Duval (Coleoptera). Ph.D. diss., Univ. of Colorado. 111 pp.

Newman, K. J., and H. V. Meredith. 1956. Individual growth in skeletal bigonial diameter during the childhood period from 5 to 11 years of age. *Amer. J. Anat.* **99**:157–187.

Noether, G. E. 1985. Elementary estimates: An introduction to nonparametrics. *J. Educ. Stat.* **10**:211–222.

O'Connell, C. P. 1953. *The Life History of the Cabezon Scorpaenichthys marmoratus (Ayres).* Calif. Div. Fish Game, Fish Bull. 93. 76 pp.

Oden, N. L., and R. R. Sokal. 1992. An investigation of three-matrix permutation tests. *J. Classif.* **9**:275–290.

Ogawa, J. 1974. *Statistical Theory in the Analysis of Experimental Design.* Marcel Dekker, New York. 465 pp.

Olson, E. C., and R. L. Miller. 1958. *Morphological Integration.* Univ. of Chicago Press, Chicago. 317 pp.

Ostle, B. 1963. *Statistics in Research.* 2d ed. Iowa State Univ. Press, Ames, Iowa. 585 pp.

Owen, D. B. 1962. *Handbook of Statistical Tables.* Addison-Wesley, Reading, Mass. 580 pp.

Page, E. B. 1963. Ordered hypotheses for multiple treatments: A significance test for linear ranks. *J. Amer. Stat. Assn.* **58**:216–230.

Park, W. H., A. W. Williams, and C. Krumwiede. 1924. *Pathogenic Microörganisms.* Lea & Febiger, Philadelphia. 811 pp.

Pearce, S. C. 1965. *Biological Statistics.* McGraw-Hill, New York. 212 pp.

Pearson, E. S., and H. O. Hartley. 1958. *Biometrika Tables for Statisticians.* Vol. 1. 2d ed. Cambridge Univ. Press, London. 240 pp.

Pettit, A. N., and M. A. Stephens. 1977. The Kolmogorov–Smirnov goodness-of-fit statistic with discrete and grouped data. *Technometrics* **19**:205–210.

Phillips, E. F. 1929. *Variation and Correlation in the Appendages of the Honeybee.* Cornell Univ. Agric. Exp. Sta. Mem. 121. 52 pp.

Phillips, J. R., and L. D. Newsom. 1966. Diapause in *Heliothis zea* and *Heliothis virescens* (Lepidoptera: Noctuidae). *Ann. Entomol. Soc. Amer.* **59**:154–159.

Pielou, E. C. 1977. *Mathematical Ecology.* Wiley, New York. 385 pp.

Pielou, E. C. 1984. *The Interpretation of Ecological Data: A Classification and Ordination.* Wiley, New York. 263 pp.

Powick, W. C. 1925. Inactivation of vitamin A by rancid fat. *J. Agric. Res.* **31**:1017–1027.

Prange, H. D., J. F. Anderson, and H. Rahn. 1979. Scaling of skeletal to body mass in birds and mammals. *Amer. Nat.* **113**:103–122.

Price, R. D. 1954. The survival of *Bacterium tularense* in lice and louse feces. *Amer. J. Trop. Med. Hyg.* **3**:179–186.

Quenouille, M. H. 1952. *Associated Measurements.* Academic Press, New York. 242 pp.

Radi, M. H., and D. C. Warren. 1938. Studies on the physiology and inheritance of feathering in the growing chick. *J. Agric. Res.* **56**:679–707.

Raktoe, B. L., A. Hedayat, and W. T. Federer. 1981. *Factorial Designs.* Wiley, New York. 209 pp.

Rao, P. 1971. Some notes on misspecification in multiple regressions. *Amer. Stat.* **25**(5):37–39.

Rayner, J. M. V. 1985. Linear relations in biomechanics: The statistics of scaling functions. *J. Zool.* (London), Ser. A **206**:415–439.

Reyment, R. A. 1991. *Multidimensional Paleobiology.* Pergamon Press, New York. 377 pp.

Ricker, W. E. 1954. Stock and recruitment. *J. Fish. Res. Board Can.* **11**:559–623.

Ricker, W. E. 1973. Linear regression in fishery research. *J. Fish. Res. Board Can.* **30**:409–434.

Ross, S. M. 1985. *Introduction to Probability Models.* 3d ed. Academic Press, Orlando, Fla. 502 pp.

SAS Institute. 1988. *SAS/STAT™ User's Guide.* Release 6.03 ed. SAS Institute, Cary, N.C. 1028 pp.

Scheirer, C. J., W. S. Roy, and N. Hare. 1976. The analysis of ranked data derived from completely randomized factorial designs. *Biometrics* **32**:429–434.

Scott, A., and C. Wild. 1991. Transformations and R^2. *Amer. Stat.* **45**:127–129.

Searle, S. R. 1989. Statistical computing packages: Some words of caution. *Amer. Stat.* **43**:189–190.

Seber, G. A. F. 1977. *Linear Regression Analysis.* Wiley, New York. 465 pp.

Seber, G. A. F., and C. J. Wild. 1989. *Nonlinear Regression.* Wiley, New York. 768 pp.

Shapovalov, L., and H. C. Taft. 1954. *The Life Histories of the Steelhead Rainbow Trout* (Salmo gairdneri gairdneri) *and Silver Salmon* (Oncorhynchus kisutch) *with Special Reference to Waddell Creek, California, and Recommendations Regarding Their Management.* Calif. Div. Fish Game, Fish Bull. 98. 375 pp.

Siegel, S. 1956. *Nonparametric Statistics for the Behavioral Sciences.* McGraw-Hill, New York. 312 pp.

Siegel, S., and N. J. Castellan, Jr. 1988. *Nonparametric Statistics for the Behavioral Sciences.* 2nd Ed. McGraw-Hill, New York.

Simpson, G. G., A. Roe, and R. C. Lewontin. 1960. *Quantitative Zoology.* Rev. ed. Harcourt, Brace, New York. 440 pp.

Sinnott, E. W., and D. Hammond. 1935. Factorial balance in the determination of fruit shape in *Cucurbita. Amer. Nat.* **64**:509–524.

Skinner, J. J., and F. E. Allison. 1923. Influence of fertilizers containing borax on growth and fruiting of cotton. *J. Agric. Res.* **23**:433–445.

Smith, F. L. 1939. A genetic analysis of red seed-coat color in *Phaseolus vulgaris. Hilgardia* **12**:553–621.

Smouse, P. E., J. C. Long, and R. R. Sokal. 1986. Multiple regression and correlation extensions of the Mantel test of matrix correspondence. *Syst. Zool.* **35**:627–632.

Sneath, P. H. A., and R. R. Sokal. 1973. *Numerical Taxonomy.* W. H. Freeman, San Francisco. 573 pp.

Snedecor, G. W., and W. G. Cochran. 1967. *Statistical Methods.* 6th ed. Iowa State Univ. Press, Ames, Iowa. 593 pp.

Snedecor, G. W., and W. G. Cochran. 1989. *Statistical Methods.* 8th ed. Iowa State Univ. Press, Ames, Iowa. 593 pp.

Snee, R. D. 1973. Some aspects of nonorthogonal data analysis. Part I. Developing prediction equations. *J. Quality Tech.* **5**:67–79.

Snee, R. D. 1974. Graphical display of two-way contingency tables. *Amer. Stat.* **28**:9–12.

Snee, R. D. 1977. Validation of regression models: Methods and examples. *Technometrics* **19**:415–428.

Sokal, R. R. 1952. Variation in a local population of *Pemphigus. Evolution* **3**:296–315.

Sokal, R. R. 1958. Quantification of systematic relationships and of phylogenetic trends. *Proc. Tenth Internat. Congr. Entomol.* **1**:409–415.

Sokal, R. R. 1962. Variation and covariation of characters of alate *Pemphigus populitransversus* in eastern North America. *Evolution* **16**:227–245.

Sokal, R. R. 1965. Statistical methods in systematics. *Biol. Rev.* (Cambridge) **40**:337–391.

Sokal, R. R. 1966. Pupation site differences in *Drosophila melanogaster. Univ. Kansas Sci. Bull.* **46**:697–715.

Sokal, R. R. 1967. A comparison of fitness characters and their responses to density in stock and selected cultures of wild type and black *Tribolium castaneum. Tribolium Inf. Bull.* **10**:142–147.

Sokal, R. R. 1979. Testing statistical significance of geographic variation patterns. *Syst. Zool.* **28**:227–231.

Sokal, R. R., and C. A. Braumann. 1980. Significance tests for coefficients of variation and variability profiles. *Syst. Zool.* **29**:50–66.

Sokal, R. R., and P. E. Hunter. 1955. A morphometric analysis of DDT-resistant and nonresistant housefly strains. *Ann. Entomol. Soc. Amer.* **48**:499–507.

Sokal, R. R., and I. Karten. 1964. Competition among genotypes in *Tribolium castaneum* at varying densities and gene frequencies (the black locus). *Genetics* **49**:195–211.

Sokal, R. R., and N. L. Oden. 1978a. Spatial autocorrelation in biology. 1. Methodology. *Biol. J. Linn. Soc.* **10**:199–228.

Sokal, R. R., and N. L. Oden. 1978b. Spatial autocorrelation in biology. 2. Some biological implications and four applications of evolutionary and ecological interest. *Biol. J. Linn. Soc.* **10**:229–249.

Sokal, R. R., N. L. Oden, B. A. Thomson, and J. Kim. 1993. Testing for regional differences in means: Distinguishing inherent from spurious spatial autocorrelation by restricted randomization. *Geogr. Anal.* **25**:199–210.

Sokal, R. R., and R. C. Rinkel. 1963. Geographic variation of alate *Pemphigus populitransversus* in eastern North America. *Univ. Kansas Sci. Bull.* **44**:467–507.

Sokal, R. R., and F. J. Rohlf. 1969. *Biometry.* 1st ed. W. H. Freeman, San Francisco. 776 pp.

Sokal, R. R., F. J. Rohlf, E. Zang, and W. Osness. 1980. Reification in factor analysis: A plasmode based on human physiology-of-exercise variables. *Mult. Beh. Res.* **15**:181–202.

Sokal, R. R., P. E. Smouse, and J. V. Neel. 1986. The genetic structure of a tribal population, the Yanomama Indians. XV. Patterns inferred by autocorrelation analysis. *Genetics* **114**:259–287.

Sokal, R. R., and F. J. Sonleitner. 1968. The ecology of selection in hybrid populations of *Tribolium castaneum. Ecol. Monogr.* **38**:345–379.

Sokal, R. R., and P. A. Thomas. 1965. Geographic variation of *Pemphigus populitransversus* in eastern North America: Stem mothers and new data on alates. *Univ. Kansas Sci. Bull.* **46**:201–252.

Sokoloff, A. 1955. Competition between sibling species of the *Pseudoobscura* subgroup of *Drosophila. Ecol. Monogr.* **25**:387–409.

Sokoloff, A. 1966. Morphological variation in natural and experimental populations of *Drosophila pseudoobscura* and *Drosophila persimilis. Evolution* **20**:49–71.

Späth, H. 1975. *Cluster-Analyse-Algorithmen.* R. Oldenbourg Verlag, München. 217 pp.

Spjøtvoll, E., and M. R. Stoline. 1973. An extension of the *T*-method of multiple comparison to include the cases with unequal sample sizes. *J. Amer. Stat. Assn.* **68**:975–978.

SPSS. 1990. *SPSS Reference Guide.* SPSS. Chicago. 949 pp.

StatXact. 1991. Version 2. Cytel Software Corp. Cambridge, Mass.

Steel, R. G. D., and J. H. Torrie. 1980. *Principles and Procedures of Statistics.* 2d ed. McGraw-Hill, New York. 633 pp.

Student (W. S. Gossett). 1907. On the error of counting with a haemacytometer. *Biometrika* **5**:351–360.

Sullivan, R. L., and R. R. Sokal. 1963. The effects of larval density on several strains of the housefly. *Ecology* **44**:120–130.

Sullivan, R. L., and R. R. Sokal. 1965. Further experiments on competition between strains of houseflies. *Ecology* **46**:172–182.

Swanson, C. O., W. L. Latshaw, and E. L. Tague. 1921. Relation of the calcium content of some Kansas soils to soil reaction by the electrometric titration. *J. Agric. Res.* **20**:855–868.

Swofford, D. L., and G. J. Olsen. 1990. Phylogeny reconstruction. In *Molecular Systematics,* ed. D. M. Hillis and C. Moritz. Sinauer Assoc., Sunderland, Mass., pp. 411–501.

Tate, R. F., and G. W. Klett. 1959. Optimal confidence intervals for the variance of a normal distribution. *J. Amer. Stat. Assn.* **54**:674–682.

Tatsuoka, M. M.; with contributions by P. R. Lohnes. 1988. *Multivariate Analysis: Techniques for Educational and Psychological Research.* 2d ed. Macmillan, New York. 479 pp.

Teissier, G. 1948. La relation d'allometrie: Sa signification statistique et biologique. *Biometrics* **4**:14–48.

Teissier, G. 1960. Relative growth. In *The Physiology of Crustacea,* Vol. 1, ed. T. H. Waterman. Academic Press, New York, pp. 537–560.

Thomson, G. H. 1951. *The Factorial Analysis of Human Ability.* 5th ed. Houghton Mifflin, New York. 383 pp.

Tsutakawa, R. K., and J. E. Hewett. 1978. Comparison of two regression lines over a finite interval. *Biometrics* **34**:391–398.

Tukey, J. W. 1949. One degree of freedom for non-additivity. *Biometrics* **5**:232–242.

Tukey, J. W. 1977. *Exploratory Data Analysis.* Addison-Wesley, Reading, Mass. 688 pp.

Ury, H. K. 1976. A comparison of four procedures for multiple comparisons among means (pairwise contrasts) for arbitrary sample sizes. *Technometrics* **18**:89–97.

Utida, S. 1943. Studies on experimental population of the Azuki bean weevil. *Callosobruchus chinensis* (L.) VIII. Statistical analysis of the frequency distribution of the emerging weevils on beans. *Mem. Coll. Agric. Kyoto Imp. Univ.* **54**:1–22.

Verner, J. 1964. Evolution of polygamy in the long-billed marsh wren. *Evolution* **18**:252–261.

Vollenweider, R. A., and M. Frei. 1953. Vertikale und zeitliche Verteilung der Leitfähigkeit in einem eutrophen Gewässer während der Sommerstagnation. *Schweiz. Z. Hydrol.* **15**:158–167.

von Neumann, J., R. H. Kent, H. R. Bellinson, and B. I. Hart. 1941. The mean square successive difference. *Ann. Math. Stat.* **12**:153–162.

Wainer, H. 1974. The suspended rootogram and other visual displays: An empirical validation. *Amer. Stat.* **28**:143–145.

Wallis, J. R., N. C. Matalas, and J. R. Slack. 1974. Just a moment! *Water Resources Res.* **10**:3–11.

Watson, R. A., and D. B. Tang. 1980. The predictive value of prostatic acid phosphatase as a screening test for prostatic cancer. *New England J. Med.* **303**:497–499.

Webber, L. G. 1955. The relationship between larval and adult size of the Australian sheep blowfly *Lucilia cuprina* (Wied.) *Austral. J. Zool.* **3**:346–353.

Webster, J. T., R. F. Gunst, and R. L. Mason. 1974. Latent root regression analysis. *Technometrics* **16**:513–522.

Weir, B. S. 1990. *Genetic Data Analysis: Methods for Discrete Population Genetic Data.* Sinauer Assoc. Sunderland, Mass. 377 pp.

Weir, J. A. 1949. Blood pH as a factor in genetic resistance to mouse typhoid. *J. Inf. Dis.* **84**:252–274.

Welsch, R. E. 1977. Stepwise multiple comparison procedures. *J. Amer. Stat. Assn.* **72**:566–575.

Whitaker, T. W. 1944. The inheritance of chlorophyll deficiencies in cultivated lettuce. *J. Hered.* **35**:317–320.

Whittaker, J., and M. Aitkin. 1978. A flexible strategy for fitting complex log-linear models. *Biometrics* **34**:487–495.

Whittaker, R. H. 1952. A study of summer foliage insect communities in the Great Smoky Mountains. *Ecol. Monogr.* **22**:1–44.

Wilkinson, L. 1989. *SYSTAT: The System for Statistics.* SYSTAT, Evanston, Ill. 638 pp.

Williams, C. B. 1964. *Patterns in the Balance of Nature.* Academic Press, London. 324 pp.

Williams, D. A. 1976. Improved likelihood ratio tests for complete contingency tables. *Biometrika* **63**:33–37.

Willis, E. R., and N. Lewis. 1957. The longevity of starved cockroaches. *J. Econ. Entomol.* **50**:438–440.

Woodson, R. E., Jr. 1964. The geography of flower color in butterflyweed. *Evolution* **18**:143–163.

Wright, S. 1968. *Evolution and the Genetics of Populations.* Vol. 1, *Genetic and Biometric Foundations.* Univ. of Chicago Press, Chicago. 469 pp.

Wright, S. 1969. *Evolution and the Genetics of Populations.* Vol. 2, *The Theory of Gene Frequencies.* Univ. of Chicago Press, Chicago. 511 pp.

Wuhrmann, K., and H. Woker. 1953. Über die Giftwirkungen von Ammoniak und Zyanidlösungen mit verschiedener Sauerstoffspannung und Temperatur auf Fische. *Schweiz. Z. Hydrol.* **15**:235–260.

Yamauchi, A., and H. Kimizuka. 1971. Study of bi-ionic potentials. *J. Theor. Biol.* **30**:285–295.

Young, L. C. 1941. On randomness in ordered sequences. *Ann. Math. Stat.* **12**:293–300.

Zang, E. A., and E. L. Wynder. 1992. Cumulative tar exposure: A new index for estimating lung cancer risk among cigarette smokers. *Cancer* **70**:69–76.

Zar, J. H. 1984. *Biostatistical Analysis.* 2d ed. Prentice-Hall, Englewood Cliffs, N. J. 718 pp.

Author Index

Agresti, A., 778
Aitkin, M., 757
Allee, W. C., 448, 450
Allison, F. E., 390
Altman, D. G., 592
Anderberg, M. R., 680
Anderson, V. L., 348, 387
Anscombe, F. J., 415
Archibald, E. E. A., 24
Arvesen, J. N., 823
Asnlock, P. D., 229
Asmundson, V. S., 683

Balandra, K. P., 114
Bancroft, T. A., 284
Band, H. T., 602
Banta, A. M., 220
Barnes, H., 88
Barnett, V., 229
Bartels, R. H., 714
Bayes, T., 70
Belanger, A., 740
Benjamini, Y., 152
Bernoulli, J., 3
Bhalla, S. C., 270
Bishop, Y. M. M., 744, 746, 755, 760
Blakeslee, A. F., 365
Bliss, C. I., 95, 519
Block, B. C., 534
Boardman, T. J., 213, 300
Bottini, E., 774
Bowen, E. S., 448
Box, G. E. P., 417
Bradley, J. V., 600
Bradley, R. A., 54
Braumann, C. A., 59
Breiman, L., 538
Brian, M. V., 790
Brower, L. P., 603
Brown, A. W. A., 267
Brown, B. E., 267
Brown, F. M., 581
Buley, H. M., 607
Burr, E. J., 597, 598
Burrell, P. C., 791
Burstein, H., 445

Calaprice, J., 554, 812
Calhoun, D. W., 95
Carroll, J. D., 622
Carter, G. R., 549
Casagrande, J. T., 767
Castellan, N. J., 426
Castillo, J. V., 554
Cattell, R. B., 681
Chase, W., 741
Clark, F. H., 606
Clark, V. A., 385, 386
Clifford, H. T., 602, 681
Clifford, P., 602
Cochran, W. G., 350, 360, 363, 387, 519, 678
Cohen, A. C., Jr., 93
Cohen, A. I., 361
Cohen, J., 578
Cohen, L. A., 792
Comstock, W. P., 581
Conahan, M. A., 702
Conover, W. J., 714
Cowan, I. M., 270
Cox, D. R., 417
Cox, G. M., 350
Cramer, E. M., 664
Crossley, D. A., 428
Crovello, T. J., 606
Crow, E. L., 157

Dadd, R. H., 789
D'Agostino, R. B., 175, 714, 741
Darwin, C., 4
Davis, E. A., Jr., 550
De Fermat, P., 3

De Moivre, A., 3
Decker, G. C., 551
Devlin, S. J., 823
Dixon, W. J., 229, 388, 406, 533
Draper, N. R., 663, 678
Dunn, G., 680, 681
Dunn, O. J., 385, 386
Dunnett, C. W., 252, 260
DuPraw, E. J., 269
Dwass, M., 431

Efron, B., 824
Ehrlich, P. R., 787
Eisenhart, C., 201
Everitt, B. S., 680, 681

Falconer, D. S., 274
Fawcett, R. F., 427
Felsenstein, J., 810, 824
Feuer, E. J., 786
Fienberg, S. E., 744, 746, 755, 760
Finney, D. J., 733
Fisher, R. A., 4, 95, 179, 185, 406, 522, 537, 575, 671, 678, 794
Fleiss, J. L., 763
Fligner, M. A., 431
Franta, W. R., 711
Freeman, M. F., 415, 422
Frei, M., 343
French, A. R., 366
Friedman, J. H., 538
Fröhlich, F. W., 534

Gabriel, K. R., 242, 245, 248, 257, 426, 431, 517
Gaebelin, J., 360
Gaines, S. D., 811, 813
Galambo, J., 70
Gallo, J., 360
Galton, F., 4, 39, 455
Games, P. A., 397, 403, 406
Gans, D. L., 405
Gardner, R. S., 157
Gart, J. J., 785
Gartler, S. M., 59, 205, 404
Gauss, K. F., 3
Gaylor, D. W., 300
Geissler, A., 80
Gibbons, J. D., 539
Gideon, R. A., 436
Gill, J. L., 260, 349, 387
Gnanadesikan, R., 823
Goddard, M. J., 778
Goldstein, A., 538
Gong, G., 824
Gonzalez, T., 711
Goodstein, R., 265
Gordon, A. D., 681
Gorsuch, R. L., 681
Gossett, W. S., 143
Gould, S. J., 535
Goulden, C. H., 538
Graunt, J., 3
Green, B. F., 809
Green, P. E., 622, 680
Greenland, S., 741
Greenwood, M., 97
Greig-Smith, P., 95
Grizzle, J. E., 731
Groggel, D. J., 447
Grubbs, F. E., 407
Gumbel, E. J., 229
Gunst, R. F., 663
Gurland, J., 53

Haenszel, W., 766
Haine, E., 367
Haldane, J. B. S., 544
Haltenorth, T., 17
Hamilton, D., 643
Hammond, D. H., 17
Hand, D. J., 680
Hanna, B. L., 320
Harman, H. H., 681
Harris, R. E., 792
Harter, H. L., 708
Hartigan, J. A., 680
Hartley, H. O., 102, 118, 140, 145, 154, 265, 387, 388, 397, 537, 671, 733
Hasel, A. A., 318
Hedges, L. B., 797
Hémon, D., 602
Herr, D. G., 360
Hewett, J. E., 518
Hibble, J., 790
Hinkley, D. V., 823
Hoaglin, D. C., 531, 532
Hochberg, Y., 244, 342, 406
Hocking, R. R., 663
Holland, P. W., 744, 746, 755, 760
Holm, S., 240
Hopper, F. N., 300
Hosmer, D. W., 778
Hotelling, H., 578, 579
Howell, J. F., 403, 406
Hunt, G. L., 681
Hunt, M. W., 681
Hunter, P. E., 104, 268, 287, 288, 527, 528
Hurvich, C. M., 669

Ives, P. T., 602

Jackson, J. E., 10, 680
Janzen, D. H., 782
Johnson, N. K., 177
Johnson, N. L., 95, 422

Johnston, P. A., 270
Jolicoeur, P., 544, 588

Karten, I., 337, 351
Kempthorne, O., 336
Kendall, M. G., 539, 560
Kermack, K. A., 544
Kessler, L. G., 786
Kettenring, J. R., 823
Khamis, H. J., 708, 711
Khuri, A. I., 360
Kim, J., 818
Kim, J.-O., 681
Kimizuka, H., 502, 503, 504
King, J. A., 365
Klett, G. W., 157
Kobyliansky, E., 819
Koehn, R. K., 667
Kotz, S., 70, 71, 95, 422
Kouskolekas, C. A., 551
Krebs, C. J., 95
Kruskal, W. H., 426, 662
Krzanowski, W. J., 10, 680
Ku, H. H., 760
Kuhry, B., 544, 545
Kullback, S., 690, 760
Kvålseth, T. O., 475, 538

Lamb, R. E., 708
Lande, R., 59
Laplace, P. S., 3
Larntz, K., 702
Leggatt, C. W., 90
Lemeshow, S., 778
Levinton, J., 682
Lewis, N., 189
Lewis, T., 229
Lewontin, R. C., 59, 179, 788, 810
Li, C. C., 642
Lilliefors, H. W., 711
Lindeman, H. R., 387
Ling, R. F., 797
Littell, R. C., 360
Little, R. J. A., 741
Littlejohn, M. J., 177
Livshits, G., 819
Long, J. C., 819
Lunt, H. A., 390

MacArthur, J. W., 692
MacGillivray, H. L., 114
Mandel, J., 542
Manley, B. F. J., 803, 823
Mantel, N., 766, 813, 818
Marcus, L. F., 544, 545
Mark, L. C., 683
Markowski, C. A., 397, 405
Markowski, E. P., 397, 405
Marquardt, D. W., 632, 663
Martin, C. G., 397
Mason, R. L., 663
Massey, F. J., Jr., 229
Maxwell, A. E., 786
Mayer, L. S., 618
Mayer, R., 6
Mayr, E., 229
McGill, R., 151
McLean, R. A., 348, 387
McNeal, D. R., 29
Meredith, H. V., 353
Miller, L., 571
Miller, R. G., 823
Miller, R. L., 31, 267, 398
Millis, J., 31, 45, 56, 266
Mitchell, C. A., 549
Mittler, T. E., 789
Montgomery, D. C., 387
Morrison, D. F., 680
Mueller, C. W., 681
Mueller, D. E., 436
Mulaik, S. A., 681
Mullett, G. M., 620

Neel, J. V., 814, 815, 816
Nelson, V. E., 459, 780
Newman, K. J., 353
Newsom, L. D., 551
Nightingale, F., 4
Noether, G. E., 539, 714

O'Connell, C. P., 535
Oden, N. L., 602, 818, 819
Ogawa, J., 387
Olkin, J., 797
Olsen, D., 449
Olsen, G. J., 681
Olson, E. C., 31, 267, 398
Ordway, K., 220
Ostle, B., 729
Owen, D. B., 118, 140, 437

Page, E. B., 541
Park, W. H., 448
Pascal, B., 3
Pearce, S. C., 387
Pearson, E. S., 102, 118, 140, 145, 154, 175, 265, 537, 671, 733
Pearson, K., 4, 559
Pettit, A. N., 714
Petty, W., 3
Phillips, E. F., 572
Phillips, J. R., 551
Pielou, E. C., 95, 680
Pike, M. C., 767
Poisson, S. D., 83
Policello, G. E., 431

Powick, W. C., 325
Prange, H. D., 553
Price, R. D., 389
Purves, W., 218, 674

Quenouille, M. H., 539
Quetelet, A., 3

Radi, M. H., 358
Raktoe, B. L., 308, 349, 385, 387
Rao, P., 662
Raynor, J. M. V., 545
Reyment, R. A., 680
Rice, W. R., 811, 813
Richardson, S., 602
Ricker, W. E., 535, 542, 544, 545
Rinkel, R. C., 259
Rohlf, F. J., 89, 228, 260, 275, 276, 332
Ross, S. M., 108

Sahni, S., 711
Scheirer, C. J., 445
Schmitz, T. H., 823
Scott, A., 538
Searle, S. R., 38
Seber, G. A. F., 493, 533, 678
Seng, Y. P., 31, 45, 56, 266
Shapovalov, L., 414
Siegel, S., 426, 600, 733
Simpson, G. G., 59, 535
Sinnott, E. W., 17
Skillings, J. H., 447
Skinner, J. J., 390
Smart, J. S., 597
Smith, F. L., 699
Smith, H., 663, 678
Smith, P. G., 767
Smouse, P. E., 814, 815, 819
Sneath, P. H. A., 680, 681
Snedecor, G. W., 360, 363, 386, 519, 678
Snee, R. D., 632, 663, 664, 740
Sokal, R. R., 18, 19, 27, 59, 95, 104, 150, 215, 237, 259, 260, 266, 270, 287, 288, 315, 319, 351, 366, 417, 421, 477, 496, 527, 536, 594, 601, 602, 649, 680, 681, 746, 801, 802, 804, 806, 813, 814, 815, 818, 819
Sokoloff, A., 550, 582
Sonleitner, F. J., 496, 804
Späth, H., 680
Spearman, C. E., 598
Spjøtvoll, E., 244
Srivastava, S. S., 54
Stanbury, F. A., 88
Steel, R. G. D., 360, 363, 385, 386, 387, 518
Stephens, M. A., 714
Stephenson, W., 681
Stoline, M. R., 244
Stroup, D. F., 70, 71
Stuart, A., 560
Student (W. S. Gossett), 84, 143
Sullivan, R. L., 150, 366, 714
Swanson, C. O., 363
Swofford, D. L., 681

Taft, H. C., 414
Tamhane, A. C., 342
Tang, D. B., 70, 71
Tate, R. F., 157
Tatsuoka, M. M., 680
Taylor, C. C., 680
Teissier, G., 535, 544
Thomas, P. A., 210, 259, 601
Thomson, B. A., 818
Thomson, G. H., 262
Tietjen, G. L., 175
Torrie, J. H., 360, 363, 385, 386, 387, 518
Tripathi, R. C., 53
Tsai, C.-L., 669
Tsutakawa, R. K., 518
Tukey, J. W., 28, 123, 151, 180, 213, 409, 415, 422, 820, 826

Ury, H. K., 239
Utida, S., 92

Verner, J., 788
Vollenweider, R. A., 343
von Bortkiewicz, L., 93
von Helmholtz, H., 6
von Neumann, J., 394

Wainer, H., 123
Wallis, J. R., 115
Wallis, W. A., 426
Warren, D. C., 358
Watson, R. A., 70, 71
Webber, L. G., 551
Webster, J. T., 663
Weir, B. S., 274
Weir, J. A., 317, 319, 339
Weldon, W. F. R., 4
Welsch, R. E., 252, 531, 532
Whitaker, T. W., 791
White, M. J. D., 788
Whittaker, J., 757
Whittaker, R. H., 23
Wild, C. J., 533, 539, 674
Wilkinson, L., 388
Williams, C. B., 95
Williams, D. A., 698, 702, 750
Williams, J. S., 213
Willis, E. R., 189
Willis, H. L., 247, 738
Woker, H., 372
Woodson, R. E., Jr., 447

Wright, S., 205, 443, 641, 642
Wuhrmann, K., 372
Wynder, E. L., 793

Yamauchi, A., 502, 503, 504
Yates, F., 406, 537, 671
Young, L. C., 394
Younger, M. S., 618
Yule, G. U., 97

Zang, E., 793
Zar, J. H., 176

Subject Index

a, number of groups, 180
a, Y-intercept, 455
A_i, random group effect, 196
α, type I error, 160
α_i, treatment effect, 191
$(\alpha\beta)_{ij}$, interaction effect of ith group of factor A and jth group of factor B, 329
AD, average deviation, 51
Absolute expected frequencies, 74
Acceptance region, 160
Accuracy of data, 13–16
ACE algorithm, 538
Added component due to treatment effects, 194
Added variance component among groups, 197
  computation of confidence limits, *Box 9.3,* 215
  estimation of, 213
Additive coding, 54, 834–836
Additivity assumption in analysis of variance, 407–409
Adjusted Y-values, 488
Alienation, coefficient of, 565
Allocation of resources, optimal, 309–317
Allometric growth curve, 533
Alternative hypothesis, 159–169
  conceptual, 223
  fundamental, 223
  operational, 223
Analysis
  cluster, 680
  data, 825–826
  discriminant function, 678
  factor, 680
  multivariate, 592, 680
  path, 634–649
  principal component, 593
  probit, 538
  stepwise multiple regression, 656–664
Analysis of covariance, 494, 499–521
  assumptions, 518
  computation, *Box 14.9,* 504
  table, 520
  testing differences among adjusted means, 516
  uses, 518–519
Analysis of variance
  assumptions of, 392–409
  by computer, 387–388
  designs expressed as factorial anova without replication, 387
  factorial
    computation, *Box 12.1,* 372
    design, 369–370
    expected mean squares, 381–385
    higher-order, 381–385
    higher-order interaction sums of squares, 385
    three-way, 370–381
  hierarchic, *see* Analysis of variance, nested
  introduction to, 179–205
  model I, 196, 201–203
    comparison of means, *see* Planned and Unplanned comparisons
  model II, 196, 203–205
  multivariate, 680
  multiway, 369–388
  nested, 272–317
    computation for
      three levels, *Box 10.5,* 289
      three levels with unequal sample sizes, *Box 10.7,* 302
      two levels, *Box 10.1,* 276
      two levels mixed model, *Box 10.4,* 286
      two levels with unequal sample sizes, *Box 10.6,* 294
    design, 272–275
    mixed model, 273
    relations to two-way anova, 340
    three-level, 288
    three-level with unequal sample sizes, 301
    two-level, 275
    two-level (mixed model), 283

Analysis of variance, nested *(continued)*
with unequal sample sizes, 292–309
partitioning of total sum of squares and degrees of freedom, 197–200
rules for pooling mean squares, *Box 10.3,* 284
single classification, 207–265
computation for
equal sample sizes, *Box 9.4,* 218
two groups with equal sample sizes, *Box 9.5,* 220
unequal sample sizes, *Box 9.1,* 210
table, 198
three-way, computation without replication, *Box 12.1,* 372
two-way, 321–363
computation
with replication, *Box 11.1,* 324, *Box 11.2,* 332
without replication, *Box 11.3,* 343
design, 321–323
disproportional and unequal subclass sizes, 360
mixed model, 322
model II with replication, 339
nonparametric methods in lieu of, 440–447
relations to nested anova, 340–342
with replication, 323–331
without replication, 342–352
missing observation, computation, *Box 11.7,* 361
significance testing, 331–342
unequal but proportional subclass sizes, 357
computation, *Box 11.6,* 358
unequal subclass sizes, 357–360
Ancova. *See* Analysis of covariance
Angular transformation, 419–422
Anova. *See* Analysis of variance
Antimode, 47
A posteriori comparisons, 230
Approximate t-test for equality of two means assuming unequal variances, 405
computation, *Box 13.4,* 404
Approximate test of equality of means assuming unequal variances, 403
computation, *Box 13.3,* 402
A priori comparisons, 230
Arcsine transformation, 419–422
Areas of normal curve, 102–103
Area under curve, 100
Arithmetic mean, 40–43
Array, 21
Association, 741
degree of, 557
nonparametric tests for, 593–601
Olmstead and Tukey's corner test, 601
outline, *Box 15.8,* 599
Assumptions in analysis of variance, 392–409
additivity, 407–409
homogeneity of variances, 396–406
independence of errors, 393–396
normality, 406–407
randomness, 393
Attributes, 12
Autocorrelation, 394
Average
deviation, 51
unweighted, 43
variance within groups, 182
weighted, 42, 133

b, regression coefficient, 455
$b_{Yj\cdot}$, partial regression coefficient, 612
$b'_{Yj\cdot}$, standard partial regression coefficient, 613
$b_{Y\cdot X}$, regression coefficient of variable Y on variable X, 455
β, parametric value for regression coefficient, 470
β, type II error, 162
β_j, fixed treatment effect of jth group of factor B, 328
Bar diagram, 30
Bartlett's test of homogeneity of variances, 397
computation, *Box 13.2,* 398
Bartlett's three-group method for model II regression, 544–545
computation, *Box 14.12,* 546
Batch mode, 36
Bayes' theorem, 70
Behrens-Fisher test, 406
Belt
confidence, 473
prediction, 475
Berkson case, 542
Berkson's fallacy, 778
Biased estimator, 52
Bimodal distribution, 47, 114
Binomial distribution, 71–81, 687
clumping, 76
general formula, 72
repulsion, 76
Bioassay, 538
Biological statistics, 2
Biometry, 1
history of, 3–5
BIOM-pc program, 36
Biostatistics, 2
Bivariate
mean, confidence ellipses for, 586–593
normal distribution, 560
scattergram
computation of equal frequency ellipse for, *Box 15.6,* 589
computation of principal axes for, *Box 15.6,* 589

Blocks design
  incomplete, 349, 387
  randomized complete, 346–352
    computation, *Box 11.4,* 350
Bonferroni method, 240, 702–703
  sequential tests, computation, *Box 9.9,* 241
  single-classification goodness of fit, 703
Bootstrap, 823–825
Box-Cox transformation, 417–419, 536
Box plots, 151

c_{ij}, coefficient of linear comparison, 233, 523, 530, 672
C_n, correction for bias in s, 53, 58, 143
χ^2, chi-square, 152
$\chi^2_{\alpha[\nu]}$, critical chi-square at probability level α, and degrees of freedom ν, 154
CD, coefficient of dispersion, 87
CV, coefficient of variation, 58
Calculator, electronic, 34
Causation, 486
Central limit theorem, 132
Central tendency, measure of, 39
Character, 9
Characteristic roots and vectors, 588
Chi-square distribution, 152–154, 175–176
Chi-square tables, 154
Chi-square test, 695, 704
  for goodness of fit, 696, 704
    with more than two classes, 697
    single classification, expected frequencies based on extrinsic hypothesis,
    computation, *Box 17.1,* 699
  for heterogeneity, 715
  of independence, 736
    degrees of freedom, 740
  interaction, 715
Class
  interval, 24
  limits, implied, 13, 25
  mark, 24
Classes in anova, 180
Classes, grouping of, 24
Clumped distribution, 76, 78, 83, 88, 114
Clumping
  departure from binomial distribution, 76
  departure from Poisson distribution, 88, 90, 92
Cluster analysis, 680
Cochran's *Q*-test, 786
  computation, *Box 17.16,* 780
Coding of data, 53–54, 834–836
  additive, 54, 807, 808
  combination, 54, 808
  multiplicative, 54, 807, 808
Coefficient of
  alienation, 565
  association, 741
  concordance, Kendall's, 600
  correlation, *see* Correlation coefficient
  determination, 564
  disarray, 598
  dispersion, 87
  intraclass correlation, 213
    computation, *Box 9.2,* 214
  leverage, 531
  linear comparison, 233, 523, 530, 672
  multiple correlation, 653
  multiple determination, 623, 653
  multiple nondetermination, 624
  nondetermination, 565
  partial correlation, 649
  partial regression, 612
  path, 636
  rank correlation
    Kendall's, 593
      computation, *Box 15.7,* 594
    Spearman's, 598
  regression, 455, 565–566
  standard partial regression, 613
  variation, 57–59
    standard error of, 138, 148
Combination coding, 54, 835, 836
Combining probabilities from independent tests of significance, 794–797
  computation, *Box 18.1,* 795
Comparison intervals, 246, 249, 517–518
Comparison limits, 245, 248–249, 257, 517–518
  computation, equal sample sizes, *Box 9.10,* 245
  contrasts, *Box 9.13,* 257
  unequal sample sizes, *Box 9.11,* 248
Comparison of
  a single observation with the mean of a sample, computation, *Box 9.7,* 248–249
  dependent variates in regression, 488
  means
    planned, 229–240
    unplanned, 240–260
  regression lines, 493–499
Comparisons
  paired, 352–356, 567–569
    computation, *Box 11.5,* 353
  tests, multiple, 240–260
Compilers, 36
Computed variables, 16
Computer, 34–35
  anova by, 387–388
  language, 34
  programs, 35–37
Conceptual alternative hypothesis, 223
Concordance, Kendall's coefficient of, 600
Confidence
  belt, 473
  ellipses for the bivariate mean, 586–593
  interval, 140

Confidence *(continued)*
limits, 139–143, 146–152
multiple, 530–531
limits for
added variance components, computation, *Box 9.3,* 215
correlation coefficients, *Box 15.4,* 576
cumulative frequency distribution, 714
difference between two means, computation, *Box 9.6,* 225
estimated means, 483
μ, computation, *Box 7.2,* 147
normally distributed statistic, 146–152
Poisson parameter, *Box 7.4,* 156
proportions, 444
regression coefficients, 472, 483
regression statistics
computation for more than one value of Y per X, *Box 14.5,* 484
computation, one value of Y per X, *Box 14.3,* 471
variance components, 213–217
variances, 154–157
computation by shortest unbiased confidence intervals, *Box 7.3,* 156
regions for a bivariate scattergram, computation, *Box 15.6,* 589
Conditional probability, 69
Conservative test, 168
Consumer's risk, 163
Contagious distribution, 76, 83, 94–95
Contingency tables, 726
Continuous variables, 11
frequency distributions of, 98–100
Contrasts, unplanned among means, 252–259
Control, statistical, 489
Corner test for association, Olmstead and Tukey's, 601
outline, *Box 15.8,* 599
Correction
for bias in s, 53
for continuity, 703, 704
Williams', 698, 730, 737
Yates', for continuity, 730, 737
Correlated proportions, testing of, 779
computation, *Box 17.16,* 780
Correlation
applications of, 583–586
illusory, 585
matrix, 574, 680
between means and ranges, 401
between means and variances, 401
nonsense, 584, 585
part-whole, 570
and regression, contrasted, 556–558
serial, 396
significance tests in, 574–583
spurious, 573
Correlation coefficient
computation of product–moment, 569–574
multiple, 653
partial, 649
product–moment, 559–566
formula for, 559–560
computation for frequency distribution, *Box 15.3,* 572
computation for small and medium samples, *Box 15.2,* 571
rank, 593–601
relation with paired comparisons test, 567–569
test of homogeneity among, *Box 15.5,* 581
test of significance for, computation, *Box 15.4,* 576
Cost function, 311
Covariance, 194, 462, 559
analysis of, 363, 494, 499–521
Covariate, 499
Criterion variable, 634
Critical
number of runs, 798
region, 160
value, 145
value of rank sum in Wilcoxon's signed ranks test, 444
values of Mann-Whitney statistic, 427
Cumulative
distribution function, 102
normal curve, 102–103, 116–117, 537
Curve
allometric growth, 533
area under, 100
cumulative normal, 102–103, 116–117, 537
dosage-mortality, 537
empirically fitted, 487
power, 167
Curvilinear regression, 665–678
testing for, *Box 16.6,* 672

$d_{Y \cdot X}$, unexplained deviation in regression, 463, 467
df, degrees of freedom, 144
δ, true difference, 262
δ-corrected Kolmogorov–Smirnov test, 711
Data, 2
accuracy of, 13–16
analysis of, 825–826
coding of, 53–54
handling of, 33
processing of
economy in, 37–38
efficiency in, 37–38
Deciles, 46
Degree of association, 557

Degrees of freedom, 53
Density, 98
Dependent variable, 453
Dependent variates, comparison of, 488
Derivative of a function, 455
Derived variables, 16–19
Description of scientific laws, 487
Descriptive statistics, 39–59
Design
  completely randomized, compared with randomized complete blocks, 350–352, 838–839
  experimental, 385
  incomplete blocks, 349, 387
  Latin square, 385
  nested anova, 272–275
  paired comparisons, 352–356, 440
  randomized complete blocks, 346–352
    computation, *Box 11.3,* 343
  split plot, 386
Determination, coefficient of, 564
Deviate, 50
  Freeman-Tukey, 750
  normal equivalent, 116–117, 537
  ranked normal, 118
  standard, 105
  standard normal, 105
Deviation
  average, 51
  mean, 51
  from regression, mean square for, 481
  standard, 51, 54–57, 137
Deviations from the mean, sum of, 833
Dichotomized data, runs test for, computation, *Box 18.2,* 799
Difference
  between two means, 219–227
    confidence limits of, 226
      computation, *Box 9.6,* 225
    significance of, 221
    simplified formulas for standard error, 838
    t-test for, computation, *Box 9.6,* 225
    t_s^2, equal to F_s, 839
  between two variances, testing significance of, computation, *Box 8.1,* 189
  honestly significant, 244
  least significant, 243
  minimum significant, 244
  significant, 163
  true, 262
Differences among regression coefficients, unplanned tests, computation, *Box 14.8,* 500
Disarray, coefficient of, 598
Discrepance, 345
Discrete variables, 11
Discriminant functions, 678
Dispersion
  coefficient of, 87
  statistics of, 39, 48–59
Distribution
  bimodal, 47, 114
  binomial, 71–81, 687
  bivariate normal, 560
  chi-square, 152–154
  clumped, 76, 78, 83, 88, 114
  contagious, 76, 83, 94–95
  F-, 184–189
  frequency, *see* Frequency distribution
  function, cumulative and normal, 102
  hypergeometric, 94
  leptokurtic, 114
  logarithmic, 95
  lognormal, 125
  of means, 128–136
  multimodal, 47
  multinomial, 729
  negative binomial, 95
  normal, 21, 101–123
    applications, 109–111
    model for, 106–109
    ordinate, 101, 111
  overdispersed, 77
  platykurtic, 114
  Poisson, 81–93
  preparation of frequency, *Box 2.1,* 26
  probability, 62, 73
  repulsed, 76, 78, 83, 92, 114
  standard normal, 123
  Student's *t*-, 143–146
  truncated Poisson, 93
  underdispersed, 77
Distribution-free methods. *See* Nonparametric methods
Dosage-mortality curves, 537
Dosages, 537
Dunn-Šidák method, 239, 702–703

ϵ_{ij}, random deviation of the *j*th individual of group *i*, 202
η, serial independence statistic, 394
Economy in data processing, 37–38
Effects
  main, 327
  random group, 196
  treatment, 191
Efficiency
  in data processing, 37–38
  relative, 311
Eigenvalues, 588
Eigenvectors, 588
Electronic calculators, 34
Empirically fitted curves, 487
Equal frequency ellipses, 592

Equal tails method of finding confidence limits for sample variances, 155
Equality of
  two means assuming unequal variances, 405
    approximate t-test, 405
      computation, *Box 13.4,* 404
  two proportions, 726
  two variances, test, 189–190
  variances, assumption in anova, 396–406
Equation
  multiple regression, 610–613
  normal, 614, 615, 622
  regression, 455, 460
Error
  rate, experimentwise, 239
  restriction, 348
  standard, *see* Standard error
  type I, 158–169
  type II, 159–169
Errors, test for independence of, 394–396
  computation, *Box 13.1,* 395
Estimate
  of added variance components, Model II anova, 213
    computation, *Box 9.2,* 214
  less than fully efficient
    mean, 55, 120
    standard deviation, 55, 120
  of mean in regression
    confidence limits for, 483
    standard error of, 483
  of value of Y, 460
  of X from Y, 491–493
    computation, *Box 14.6,* 492
  of ζ, unbiased, 583
Estimators, unbiased, 52, 139, 560, 583
Events, independence of, 67
  simple, 64
Exact randomization test, 803–808
Exact test
  Fisher's, 730–736
    computation, *Box 17.7,* 734
  randomization, 803
  single classification, 702
Expected
  binomial frequencies, computation, *Box 5.1,* 82
  frequencies
    absolute, 74
    relative, 74
  mean squares
    factorial anova, 381–385
    four-factor factorial anova, 382
  normal frequencies, computation, *Box 6.1,* 112
  Poisson frequencies, computation, *Box 5.2,* 84
  value, 133
    for Y, given X, 456
Experimental design, 385
Experimentwise error rate, 239
  Dunn-Šidák method, 239
Extrinsic hypothesis, 694, 698

f, observed frequency, 74
$\hat{f}$, absolute expected frequencies, 74
$\hat{f}_{rel}$, relative expected frequency, 74
F, variance ratio, 185
F_s, sample statistics of F-distribution, 185
F_{max}, maximum variance ratio, 399
F_{max}-test, 397
  computation, *Box 13.2,* 399
F-distribution, 184–189
F-test
  one-tailed, 186
  two-tailed, 187
Factor analysis, 586, 680
Factorial anova
  computation, *Box 12.1,* 372
  design, 369–370
  fractional, 387
  higher-order, 381–385
  higher-order interaction sums of squares, 385
  three-way, 370–381
  without replication, to represent other designs, 387–388
Factorial, mathematical operation, 72
Figures, significant, 14
Fisher's exact test, 730–736
  computation, *Box 17.7,* 734
Fitting a normal distribution to observed data, 111–113
  computation, *Box 6.1,* 112
Fractional factorial anova, 387
Freedom, degrees of, 53
Freeman-Tukey deviates, 750
Frequencies
  absolute expected, 74
  observed, 74
  relative expected, 74
Frequency data, randomized blocks for, 778–787
  computation, *Box 17.16,* 780
Frequency distribution
  computation of g_1, and g_2, *Box 6.2,* 115
  computation of median, *Box 4.1,* 45
  of continuous variables, 98–100
  graphic test for normality, *Box 6.3,* 118
  L-shaped, 21
  meristic, 22
  normal, 21, 101–123
    conditions for, 109
  preparation of, *Box 2.1,* 26
  qualitative, 22
  quantitative, 22, 23
  two-way, 573
  U-shaped, 21

Frequency distributions, 19–31, 61–62
Frequency polygon, 31
Friedman's method for randomized blocks, 440
  computation, *Box 13.10,* 441
Function, 452
  derivative of, 455
  discriminant, 678
  probability density, 98
  slope of, 453–455
Functional relationships, 452
Fundamental alternative hypothesis, 223

g_1, measure of skewness, 114
g_2, measure of kurtosis, 114
g_1 and g_2
  computation in a frequency distribution, *Box 6.2,* 115
G, sample statistic of log likelihood ratio test, 689, 690, 692
G_H, heterogeneity *G*, 717, 718, 721
G_P, pooled *G*, 717, 721
G_T, total *G*, 717, 721
GM_Y, geometric mean, 43
γ_1, parametric measure of skewness, 114
γ_2, parametric measure of kurtosis, 114
G-test, 690
  general expression, 847–848
  for goodness of fit, single classification, expected frequencies
    based on extrinsic hypothesis, computation, *Box 17.1,* 699
    based on intrinsic hypothesis, computation, *Box 17.2,* 706
  heterogeneity, 715
    pooled and total, meaning of, 720
  of independence, 729
    degrees of freedom, 740
  interaction, 715, 718
  with Yates' correction, 730
Gaussian multipliers, 622, 631
General structural relation, 545
Geometric mean, 43
  regression, 544
Goodness of fit tests
  by chi-square, 697, 704–705
    for more than two classes, 697
  for two classes, 704, 848
  by *G*-test, 689–695
  introduction, 686–697
  Kolmogorov-Smirnov test, *Box 17.3,* 712
  replicated, 715–724
    computation, *Box 17.4,* 716
    unplanned tests, computation, *Box 17.5,* 722
  single classification, 697–715
Graph paper
  probit, 538
Graphic methods
  estimate of mean and standard deviation, *Box 6.3,* 118
  test for normality, 116–123
    by normal quantile plots, *Box 6.3,* 118
Grouping of classes, 24
  procedure illustrated, *Box 2.1,* 26
Groups
  average variance within, 182
  in anova, 180
  variance among, 183
  variance within, 182
GT2-method, 244, 261, 499
  computation, *Box 9.11,* 247, *Box 14.8,* 500

h_i, leverage coefficients, 531
H, statistic for Kruskal-Wallis test, 423
H_0, null hypothesis, 158
H_1, alternative hypothesis, 159
H_Y, harmonic mean, 43
Hanging histogram, 123
Hanging rootogram, 123
Harmonic mean, 43
Heterogeneity
  chi-square test for, 715
  G-test for, 715
Heteroscedasticity, 397
Hierarchic analysis of variance. *See* Analysis of variance, nested
Histogram, 30, 100
  hanging, 123
History of biometry, 3–5
Homogeneity
  among correlation coefficients, test of, computation, *Box 15.5,* 581
  of replicates tested for goodness of fit, unplanned tests, computation, *Box 17.5,* 722
  of variances
    assumption in analysis of variance, 396–406
    tests, computation, *Box 13.2,* 398
Homoscedasticity, 397
Honestly significant difference, 244
Hyperellipsoids, 592
Hypergeometric distribution, 94
  mean of, 94
Hypothesis
  alternative, 159–169
    conceptual, 223
    fundamental, 223
    operational, 223
  extrinsic, 694, 698
  intrinsic, 694, 705
  null, 158–169
  testing, 157–169

I, information, 690

Illusory correlations, 585
Implied class limits, 13, 25
Incomplete blocks design, 349, 387
Independence
  assumption in analysis of variance, 393–396
  of events, 67
  exact test of, *see* Fisher's exact test
  serial, 394–396
    computation, *Box 13.1,* 395
  test of
    chi-square, 736
    by G, 729
    in multiway tables, 743–760
    $R \times C$, 737
      computation, *Box 17.8,* 738
    in three-way tables, computation, *Box 17.10,* 746
    in two-way tables, 724–743
Independent variable, 453
Index, 17
Individual degree of freedom comparisons of means in analysis of variance, *Box 14.10,* 525
  nonorthogonal, 235
  orthogonal, 231, 522
Individual mean square, 199
Individual observations, 8
Influence function, 823
Information, 690
Interaction, 326–331
  chi-square test for, 713
  G-test for, 718
  partitioning of, in a two-way anova, computation, *Box 16.7,* 674
  second-order, 370, 379
  sum of squares, 326
  third-order, 370
Interactive mode, 36
Interdependence, 557
Interference, 328
Interpreters, 36
Interquartile range, 151
Intersection, 64
Interval
  class, 24
  comparison, 246, 249, 517–518
  confidence, 140
Intraclass correlation, coefficient of, 213
  computation, *Box 9.2,* 214
Intragroup mean square, 199
Intrinsic hypothesis, 694, 705
Inverse prediction, 493
Isotonic regression, 811–813
  computation, *Box 18.4,* 812
Item, 9

Jackknife, 820–823
  computation, *Box 18.6,* 821

k, sample size of a single binomial sample, 71
Kendall's coefficient of concordance, 600
Kendall's coefficient of rank correlation, 593
  computation, *Box 15.7,* 594
Kendall's robust line-fit method, 539
  computation, 540
Kolmogorov-Smirnov test for goodness of fit, 708–715
  computation, *Box 17.3,* 712
  δ-corrected, 711
Kolmogorov-Smirnov two-sample test, 434–439
  computation, *Box 13.9,* 435
Kruskal-Wallis test, 423–427
  computation, *Box 13.6,* 424
  Scheirer–Ray–Hare extension, 445–447
    computation, *Box 13.12,* 446
Kurtosis, 114

L, likelihood ratio, 689
L_1, lower confidence limit, 140
L_2, upper confidence limit, 140
LD_{50}, median lethal dose, 47, 493
LSD, least significant difference, 243
Latent roots and vectors, 588
Latin square design, 385
Least significant difference, 243
Least squares, 460–461
Leptokurtic curve, 114
Level, significance, 160, 163
Leverage coefficients, 531
Liberal test, 168
Likelihood, 689
Likelihood ratio test, 689
Limits
  comparison, *see* Comparison limits
  confidence, *see* Confidence limits
  implied class, 13, 25
  prediction, 475, 483
Linear comparison
  in anova, 521–531
  coefficients of, 233, 522
Linear regression. *See* Regression
Location, statistics of, 39, 40–47
Log-anova test for homogeneity of variances, 397
  computation, *Box 13.2,* 399
Logarithmic
  distribution, 95
  transformation, 413–415, 533
Logistic equation, 538
Logistic regression, 767–778
  continuous independent variable, *Box 17.15,* 774
  multistate categorical independent variable, *Box 17.14,* 772
Logit, 538, 761, 769
  empirical, 762

Log likelihood ratio test, 689–692
Log-linear models, 743–744
Lognormal distribution, 125
Log-odds, 759

$m_{\alpha[k^*,\nu]}$, studentized maximum modulus, 244, 517
M, median, 44–47
MS, mean square, 199
$MS_{\hat{Y}}$, mean square for regression, 481
$MS_{Y \cdot X}$, mean square for deviations from regression, 481
MSD, minimum significant difference, 244
MSR, minimum significant range, 244
μ, parametric mean, 52
  confidence limits for, computation, *Box 7.2,* 147
μ_Y, expected value for variable *Y* for any given value of *X*, 456
Main effects, 327
Major axis
  reduced, 544
  standard, 544
Mann-Whitney *U*-test, 427–431
  computation, *Box 13.7,* 428
Manova, multivariate analysis of variance, 680
Mantel test, 813–819
  computation, *Box 18.5,* 815
Mantel-Haenszel
  estimate, 766
  test, *Box 17.12,* 764
Mark-recapture studies, 94
Mathematical model, structural, 487
Matrix, correlation, 574, 680
Maximum likelihood method, 769
McNemar test for significance of changes, 779
  computation, *Box 17.16,* 780
Mean
  approximate estimates of, 55
  arithmetic, 40–43
  deviation, 51
  geometric, 43
  graphic estimate of, *Box 6.3,* 118
  harmonic, 43
  of hypergeometric distribution, 94
  methods for computing, 54–57
  parametric, 52
  of Poisson distribution, 87
  range, 55
  of a sample, comparison of a single observation with, computation, *Box 9.7,* 228
  square, 51, 199
    among groups, 199
    for deviations from regression, 481
    error, 199
    explained, 467
    individual, 199
    intragroup, 199
    due to linear regression, 467, 481
    total, 199
    unexplained, 467
  sum of the deviations from, 833
Means
  in anova, individual degree of freedom comparisons, computation, *Box 14.10,* 525
  approximate test of equality assuming unequal variances, 403
    two means, computation, *Box 13.4,* 404
    two or more means, computation, *Box 13.3,* 402
  distribution of, 128–136
  and ranges, correlation between, 401
  t-test of the difference between two, 223–227
    computation, *Box 9.6,* 225
  variance among, 128–136, 182
  and variances, correlation between, 401
Measure of central tendency, 39
Measurement variables, 11
Median, 44–47
  computation in a frequency distribution, *Box 4.1,* 45
  effective dose, 47
  lethal dose, 47
  runs test above and below, 800
  test for equality, *Box 17.9,* 742
Meristic
  frequency distribution, 22
  variables, 11
Meta-analysis, 797
Midrange, 55
Minimum significant difference, 244
Minimum significant range, 244
Missing observation in two-way anova without replication, 361–363
  computation, *Box 11.7,* 361
Mixed model
  nested anova, 273
  two-way anova, 322
Mode, 47–48
Model I
  anova, 196, 201–203
  regression
    assumptions, 455–457
    one *Y* per *X*, 457–476
    several *Y*'s per *X*, 476–486
  two-way frequency table, 727, 740
Model II
  anova, 196, 203–205
    estimation of added variance component, 213
      computation, *Box 9.2,* 214
    two-way with replication, 338–342
  nested anova, 273
  regression, 457, 541–549, 557–558

Model II, regression *(continued)*
computation, *Box 14.12,* 546
two-way frequency table, 727–728, 740
Model III, two-way frequency table, 728, 740
Model, hierarchical, 757–759
Model, structural mathematical, 487
Moment statistics, 114
Monte Carlo methods, 810
Moriguti-Bulmer procedure. *See* Confidence limits for added variance components
Multimodal distributions, 47
Multinomial distribution, 730
Multiple comparisons tests. *See* Unplanned comparisons
Multiple confidence limits, 530
Multiple correlation coefficient, 653
Multiple determination, coefficient of, 623, 653
Multiple nondetermination, coefficient of, 624
Multiple regression, 609, 610–634
backward elimination procedure, 655
computation, *Box 16.1,* 616
equation, 610–613
forward selection procedure, 655
overall significance test, 624
significance tests, computation, *Box 16.2,* 626
standard errors, *Box 16.3,* 633
stepwise procedure, 656–664
Multiplicative coding, 54, 834, 835
Multivariate analysis, 592, 680
of variance, 680
Multiway analysis of variance, 369–385
Multiway tables, tests of independence in, 743–760

n, sample size, 41
n_0, average sample size in analysis of variance, 212
NED, normal equivalent deviate, 117
ν, degrees of freedom, 144
Negative binomial distribution, 95
Nested anova. *See* Analysis of variance, nested
Neyman's example, 778
Nominal data, 12
Nonadditivity: Tukey's test for, 409
computation, *Box 13.5,* 410
Nondetermination, coefficient of, 565
Nonorthogonal comparisons among means, 235–240, 261
Nonparametric methods
in lieu of single classification anova, 423–439
in lieu of two-way anova, 440–447
power of, 168
simultaneous test procedure, 431–434
computation, *Box 13.8,* 432
tests for association, 593–601
tests for regression, 539–541
unplanned test, 431–434
Nonsense correlations, 584, 585
Normal
curve
areas of, 102–103
cumulative, 102–103, 116–117, 537
height of ordinate, 101
deviates
ranked, 118
standard, 105
distribution, 21, 101–123
applications, 109–111
bivariate, 560
computation of expected frequencies, *Box 6.1,* 112
conditions for, 109
function, 102
model for, 106–109
ordinate of, 101, 111
properties of, 101–105
standard, 123
equations, 614, 615, 622
equivalent deviate, 116–117, 537
probability
density function, 101–105
quantile plots, 118
Normality
assumption of, in analysis of variance, 406–407
of a frequency distribution
graphic test by normal quantile plots, *Box 6.3,* 118
testing departures from, *Box 6.3,* 118, 708–714
Null hypothesis, 158–169
Number of replicates needed to detect given true difference between means, computation, *Box 9.14,* 263
Numerical taxonomy, 681, 806

ω_{MH}, Mantel–Haenszel estimate of joint odds ratio, 766
Object-oriented programming, 36
Observations, individual, 8
Observed data, fitting a normal distribution to, 111–113
Observed frequencies, 74
Odds ratio, 761
joint, 766
test of homogeneity, *Box 17.12,* 764
Mantel–Haenszel, 766
Olmstead and Tukey's corner test for association, 601
outline, *Box 15.8,* 599
One-tailed tests, 79, 168
Operational alternative hypothesis, 223
Optimal allocation of resources, 309–317
in multi-level sampling, computation, *Box 10.8,* 315

Ordinate for a normal distribution, 101, 111
Orthogonal
  comparisons, 231, 261
  polynomials, 671–678
  single degree of freedom comparisons, 522
Outlier detection, 406
Overdispersed distributions, 77

p, binomial probability, 71
$\hat{p}$, parametric binomial probability, 78
P, probability, 63, 172, 262
ϕ, phi coefficient, 741–743
Paired comparisons, 352–356, 567–569
  computation, *Box 11.5,* 353
  nonparametric tests, 440–445
  t-test for, 356
    computation, *Box 11.5,* 354
    related to correlation, 567–569
    t_s^2, identical to F_s, 837–838
Parameters, 52
Parametric
  mean, 52
  product–moment correlation coefficient, 560
  regression coefficient, 456, 470
  variance, 52
Part-whole correlation, 570
Partial correlation coefficient, 649
Partial regression coefficient, 612
Partitioning
  among groups sum of squares, 235, 521–531
  chi-square, 715
  G, 715–721
  interaction in a two-way anova, computation, *Box 16.7,* 674
  total sum of squares and degrees of freedom, in anova, 197–200
Path analysis, 634–649
Path coefficient, 636
Path diagram, 635
Pearson statistic, X^2, 696
Percentages, 17
  drawbacks of, 17
  testing equality of two, 741
    sample size required, *Box 17.13,* 768
Percentiles, 46
Planned comparisons, 230
  among means, 229–240
    in single classification Model I anova, computation, *Box 9.8,* 237, *Box 14.10,* 525
Platykurtic curve, 114
Poisson distribution, 81–93
  calculation of expected frequencies, *Box 5.2,* 84
  mean, 87
  truncated, 93
  variance, 87
Polynomials, orthogonal, 671–678
Pooling mean squares in anova, rules for, *Box 10.3,* 284
Population, 9
  statistics, 52
Positive predictive value, 70
Power curve, 167
Power of an anova, 264
Power of a test, 167–168
Power transformation, 417
Precision of data, 13–16
Predicted mean in regression, standard error for, 483
Predicted sample mean based on k items (in regression), standard error of, 475
Prediction, 487, 557–558
  belts, 475
  inverse, 493
  limits, 475, 483
    for the sample mean, 475
Predictor variable(s), 634
  choosing, 654
Prevalence, 69
Principal axes, 586–593
  for a bivariate scattergram, computation, *Box 15.6,* 589
Principal component analysis, 593
Probability, 63
  conditional, 69
  density function, 98
    normal, 101–105
      parameters, 101
  distribution, 62, 73
  graph paper, *see* Normal probability
  space, 65
Probit
  analysis, 538
  graph paper, 538
  transformation, 537
Probits, 537
Producers' risk, 163
Product–moment correlation coefficient, 559–566
  computation of, 569–574
  computation for frequency distribution, *Box 15.3,* 572
  computation for small and medium samples, *Box 15.2,* 571
  formula for, 559–560
  parametric, 560
Products, sum of, 462
Proportions, testing of correlated, 779
  computation, *Box 17.16,* 780
Pseudovalue, 820

q, binomial probability, 71
$\hat{q}$, parametric binomial probability, 78
$Q_{\alpha[k,\nu]}$, studentized range, 244
$Q'_{\alpha[k,\nu]}$, studentized augmented range, 244

$Q-Q$ plots, 118
Q-test, Cochran's, 786
  computation, *Box 17.16,* 780
Qualitative frequency distribution, 22
Quantile–quantile plots, 118
Quantiles, 46
Quantitative frequency distribution, 22, 23
Quartiles, 46
Quintiles, 46

r, number of sample items in a hypergeometric distribution, 94
r_I, coefficient of intraclass correlation, 213
r_s, Spearman's coefficient of rank correlation, 598
r_{jk}, product–moment correlation coefficient, 560
ρ_{jk}, parameter of product–moment correlation coefficient, 560
$R \times C$ test of independence, 737
  computation, *Box 17.8,* 738
RE, relative efficiency, 311
Random group effect, 196
Randomization, tests, 803–819
  exact, 803–808
  sampled, 808–819
Randomized complete blocks, 346–352
  design, 346–348
    compared with completely randomized design, 838–839
    computation, *Box 11.3,* 343
    missing values in, 361–363
  for frequency data, 778–787
    computation, *Box 17.16,* 780
  Friedman's method for, 440
    computation, *Box 13.10,* 441
Randomness, tests for, 797–803
Random numbers, 74, 104
Random sampling, 393
Range, 48–49
  mean, 55
  minimum significant, 244
  simultaneous test procedure, 244
  studentized, 244
    augmented, 244
Rank correlation, 593–601
  Kendall's coefficient of, 593
    computation, *Box 15.7,* 594
  Spearman's coefficient of, 598
Ranked normal deviates, 118
Ranked variable, 12
Ranking tests, 423
Rankits, 118
  used to test for normality and estimate mean and standard deviation, *Box 6.3,* 121
Rates, 17
Ratios, 16
  drawbacks of, 17
Reciprocal transformation, 536
Reduced major axis, 544
Region
  acceptance, 160
  critical, 160
  rejection, 160
Regression
  analysis, stepwise multiple, 656–664
  assumptions, Model I, 456–457
  coefficient, 455, 565–566, 841
    confidence limits for, 472, 483
    parametric value for, 456, 470
    partial, 612
    standard error of, 468, 470, 483
    standard partial, 613
    test of significance of, 470
  coefficients, unplanned comparisons among, computation, *Box 14.8,* 500
  contrasted with correlation, 556–558
  curvilinear, 665–678
  equation
    intrinsically linear, 677
    linear, 455, 457–466
    multiple, 610–613
  estimation of X from Y, 491–493
    computation, *Box 14.6,* 492
  geometric mean, 544
  introduction to, 452–455
  isotonic, 811–813
    computation, *Box 18.4,* 812
  line, linear, 461
  linear, mean square due to, 470
  lines, comparison of, 493–499
    test of equality of slopes, computation, *Box 14.7,* 495
  logistic, 767–778
    computation, *Box 17.15,* 774
  model I, several Y's per X, 476–486
  model II, 457, 541–549
    computation, *Box 14.12,* 546
  more than one value of Y per X, 476–486
    unequal sample sizes, computation, *Box 14.4,* 477
  multiple, *see* Multiple regression
  nonlinear, 677
  nonparametric tests, 539–541
  through the origin, 474
  of regression coefficient on a criterion variable, 499
  significance tests in, 466–476
  single Y for each X, computation, *Box 14.1,* 465
  standard error
    of observed sample mean, 473
    of $\hat{Y}$, 473
  standard errors, *Box 14.2,* 468

statistics
derivation of, 843
standard errors and confidence limits
more than one Y per X, computation, *Box 14.5,* 484
single Y for each X, computation, *Box 14.3,* 472
transformations in, 532–539
uses of, 486–491
Rejection region, 160
Relation d'allométrie, 544
Relationships, functional, 452
Relative efficiency, 311
Relative expected frequencies, 74
Remainder sum of squares, 345
Repeated testing of the same individuals, 345, 779
computation, *Box 17.16,* 780
Replacement, in sampling, 67
Replicated goodness of fit tests, 715–724
computation, *Box 17.4,* 716
Repulsed distribution, 76, 78, 83, 92, 114
Repulsed, spatially or temporally uniform distribution, 83
Repulsion, 76, 92
Residuals, examination of, 531–532
standardized, 531
Restriction error, 348
Risk
consumers', 163
producers', 163
relative, 761
Robust line-fit method, Kendall's, 539
Roots
characteristic, 588
latent, 588
Rounding, 15
Runs, 797
critical number of, 798
test, 797–803
above and below the median, 800
for dichotomized data, computation, *Box 18.2,* 799
for trend data, computation, *Box 18.3,* 802
up and down test, 801
computation, *Box 18.3,* 802

s, sample standard deviation, 52
s^2, sample variance, 52
s_{12}, covariance of Y_1 and Y_2, 559
s_A^2, sample estimate of added variance component among groups, 197
s_b, standard error of regression coefficient, 468
$s_{\bar{Y}}$, standard error of the mean, 137
$s_{\hat{Y}}^2$, mean square due to linear regression, 467
$s_{\hat{Y}}$, standard error of predicted Y_i for a given value of X_i, 468
$\hat{s}_{\bar{Y}}$, standard error of predicted mean, 468
$s_{Y \cdot X}^2$, unexplained mean square, 467
SE, See Standard error
SEM, standard error of the mean, 149
SS, sum of squares, 51, 199
SS-STP, sum of squares simultaneous test procedure, 254
σ, parametric standard deviation, 53
σ^2, parametric variance, 52
σ_A^2, parametric value of added variance component, 197
$\sigma_{B \subset A}^2$, parametric value of added variance component of B in A, 274
Sample, 8
influence function, 823
mean
prediction limits for, 475
standard error in regression, 483
standard error of, 137–138
size, 41
n required for a test, 260–265
required to detect a given difference between
observed and expected correlation, computation, *Box 15.4,* 578
two percentages, computation, *Box 17.13,* 768
space, 64
statistics, 52
variance, 52
Sampled randomization test, 808–819
Sampling, random, 393
Scattergram, bivariate, computation of equal-frequency ellipse for, *Box 15.6,* 589
Scheffé-Box test. *See* Log-anova test for homogeneity of variances
Scheffé method, 252, 261
Scheirer–Ray–Hare extension of the Kruskal–Wallis test, 445–447
computation, *Box 13.12,* 446
Scientific laws, description of, 487
Second-order interaction, 370
Serial independence, 394–396
computation, *Box 13.1,* 395
Sensitivity, 69
Shortest unbiased confidence intervals for proportion and Poisson parameter, 157
computation, *Box 7.4,* 156
variance, 155
computation, *Box 7.3,* 156
Šidák. *See* Dunn-Šidák method.
Sign test, 444–445
Signed-ranks test, Wilcoxon's, 440–444
computation, *Box 13.11,* 443
Significance
of changes, McNemar's test for, 779
computation, *Box 17.16,* 780
of correlation coefficients, computation, *Box 15.4,* 576

Significance *(continued)*
  of the difference between two means, 221
  levels, 160,163
  statistical, 163
  tests
    combining probabilities from, 794–797
      computation, *Box 18.1,* 795
    in correlation, 574–583
      computation, *Box 15.4,* 576, *Box 15.5,* 581
    of the deviation of a statistic from its parameter, 173–175
      computation, *Box 7.5,* 174
    of regression statistics, computation, *Box 14.3,* 471, *Box 14.5,* 484
    of a sample variance from a parametric variance, 175–177
    of a single variate against a parametric mean, 169–170
    of a statistic, computation, *Box 7.5,* 174
Significant
  difference, least, 243
  figures, 14
Simple event, 64
Simpson's paradox, 778
Simultaneous test procedure
  for homogeneity of replicates tested for goodness of fit, 724
    computation, *Box 17.5,* 722
  nonparametric, 431–434
    computation, *Box 13.8,* 432
  sum of squares, 254–259
Single classification
  anova, *see* Analysis of variance, single classification
  goodness of fit test, *see* Goodness of fit tests, single classification
Single degree of freedom comparisons. *See* Individual degree of freedom comparisons
Single observation compared to mean of a sample, 227–229
  computation of, *Box 9.7,* 228
Size, sample, 41
Skewness, 111
Slope of a function, 453–455
Software, 35–37
Spatially uniform distribution, 83
Spearman's coefficient of rank correlation, 598
Specificity, 69
Split plot design, 386
Spurious correlations, 573
Square, mean, 51, 199
Square root transformation, 415–417
Squares
  least, 460–461
  sum of, 51, 199
    interaction, 326
    remainder, 345
    subgroups within groups, 275
    total, 198
Standard
  deviate, 105
  deviation, 51, 54–57, 137
    of a binomial distribution, 78
    computation
      frequency distribution, *Box 4.3,* 56
      unordered data, *Box 4.2,* 56
  error, 137
    of coefficient of variation, 138, 148
    for common statistics, *Box 7.1,* 138
    of difference between two means, 223–224, 836–837
    of estimated mean in regression, 468, 483
    of estimated value $\hat{Y}$, along regression line, 468, 473, 633, 634
    of mean, 133–136, 138
    of multiple regression statistics, *Box 16.3,* 633
    of observed sample mean in regression, 468, 473, 483, 633, 634
    of partial regression coefficient, 632, 633
    of predicted mean in regression, 468, 483, 633, 634
    of predicted sample mean based on k items, 468, 475, 483
    of regression coefficient, 468, 470, 483
      derivation, 845
    of regression statistics, *Box 14.2,* 468
    of standard partial regression coefficient, 632, 633
  major axis, 544
  normal deviate, 105
  normal distribution, 123
  partial regression coefficient, 613
Standardization, 110
Standardized deviate, 105
Standardized residual, 531
Statistic, 3
  testing significance, *Box 7.5,* 174
Statistical
  control, 489
  significance, 163
    conventional statement of, 172
  thinking, 5–7
Statistics, 2
  biological, 2
  descriptive, 39–59
  of dispersion, 39, 48–59
  of location, 39, 40–47
  moment, 114
  population, 52
  sample, 52
Stem-and-leaf display, 28–29
  back-to-back, 29
Stepwise multiple regression analysis, 656–664

computation, *Box 16.4,* 657
Stepwise test procedures, multiple comparisons tests, 252
Structural mathematical model, 487
Structural relation, 545
Student-Newman-Keuls procedure, 260
Student's *t*-distribution, 143–146
Studentized augmented range, 244
Studentized maximum modulus, 244, 517
Studentized range, 244
Substitution of variables, 491
Sum of products, 462
Sum of squares, 51, 199
among groups, computational rule for, 208
computational formula, 54
of dependent variable, partitioning of, 467, 844
explained, 464
interaction, 326
simultaneous test procedure, 254–259
total, 198
unexplained, 464
Sum of two variables, variance of, 567, 846
Synergism, 328

$t_{\alpha[\nu]}$, critical values of Student's distribution for ν degrees of freedom, 146
t_s, sample statistic of *t*-distribution, 171
T, critical value of rank sum of Wilcoxon's signed ranks test, 444
τ, Kendall's coefficient of rank correlation, 593
t-distribution, Student's, 143–146
t-tables, 145
t-test
for difference between a single specimen and a sample, 227–229
computation, *Box 9.7,* 228
for difference between two means, 223–227
computation, *Box 9.6,* 225
for equality of two means, approximate, 405
computation, *Box 13.4,* 404
for paired comparisons, 356
computation, *Box 11.5,* 353
2*I*, sample statistic of log likelihood ratio test, 690
T-method, 244, 261
computation, *Box 9.10,* 245, *Box 14.8,* 500
T'-method, 244, 261, 499
computation, *Box 9.11,* 247, *Box 14.8,* 500
Table
contingency, 726
two-by-two frequency, 726
two-way frequency, 726
Taxonomy, numerical, 681, 806
Testing, hypothesis, 157–169
Test(s)
of association, 741
nonparametric, 593–601
Behrens-Fisher, 406
chi-square, 695, 704
conservative, 168
for correlated proportions, 779
computation, *Box 17.16,* 780
for curvilinear regression, *Box 16.6,* 672
of departures from normality, 116, 708–714
of difference between two variances, 189–190
computation, *Box 8.1,* 189
of equality
of means of two samples whose variances are unequal, computation, *Box 13.4,* 404
of medians, *Box 17.9,* 742
of odds ratios, *Box 17.12,* 764
of slopes of several regression lines, computation, *Box 14.7,* 495
of two percentages, 741
exact randomization, 803–808
Fisher's exact, 730–736
computation, *Box 17.7,* 734
for goodness of fit, introduction, 686–697
Kolmogorov-Smirnov, computation, *Box 17.3,* 712
replicated, 715–724
computation, *Box 17.4,* 716
single classification, 697–715
heterogeneity chi-square, 715
of homogeneity
among correlation coefficients, computation, *Box 15.5,* 581
of variances, computation, *Box 13.2,* 398
Bartlett's, 398
F_{max}, 399
log-anova, 399
Scheffé-Box, 399
of independence
chi-square, 736
multiway tables, 743–760
$R \times C$, computation, *Box 17.8,* 738
2×2, computation, *Box 17.6,* 731
three-way table, computation, *Box 17.10,* 746
two-way tables, 724–743
Wald, 770
interaction chi-square, 715
Kolmogorov-Smirnov goodness of fit, 708–715
computation, *Box 17.3,* 712
δ-corrected, 711
Kolmogorov-Smirnov two-sample, 434–439
computation, *Box 13.9,* 435
Kruskal-Wallis, 423–427
computation, *Box 13.6,* 424
liberal, 168
likelihood ratio, 689

Test(s) *(continued)*
log likelihood ratio, 689–692
Mann-Whitney U, computation, *Box 13.7,* 428
Mantel, 813–819
computation, *Box 18.5,* 815
McNemar, 779
nonadditivity, Tukey's, 409
computation, *Box 13.5,* 410
one-tailed, 79, 168–169
ordering, 539
Page's L for ordered alternatives, 541
power of a, 167–168
randomization, 803–819
for randomness, 797–803
ranking, 423
repeated, of same individuals, 345, 779
runs, 797–803
above and below the median, 800
dichotomized data, computation, *Box 18.2,* 799
up and down, 801
computation, *Box 18.3,* 802
sampled randomization, 803–819
of serial independence, *Box 13.1,* 395
sign, 444–445
of significance
combining probabilities from, 794–797
for correlation coefficients, computation, *Box 15.4,* 576
in correlation, 574–583
of the regression coefficient, 470
of a statistic, computation, *Box 7.5,* 174
two-tailed, 79, 158
Wilcoxon's signed-ranks, 440–444
computation, *Box 13.11,* 443
Wilcoxon's two-sample, computation, *Box 13.7,* 429
Third-order interaction, 370
Three-group method, Bartlett's, for Model II regression, 544–545
computation, *Box 14.12,* 546
Three-level nested anova, 288–292
computation, *Box 10.5,* 289
with unequal sample sizes, 301–308
computation, *Box 10.7,* 302
Three-way factorial anova, 370–381
Three-way table, analysis by log-linear models, 743–760
computation, *Box 17.10,* 746
Total mean square, 199
Total sum of squares, 198
Transformation
angular, 419–422
arcsine, 419–422
Box-Cox, 417–419
logarithmic, 413–415, 533
logit, 538
power, 417
probit, 537
reciprocal, 536
square root, 415–417
Transformations, 409–422
in regression, 532–539
Treatment effects, 191
added component due to, 194
True difference, 262
Truncated Poisson distributions, 93
Tukey-Kramer method, 252, 261, 499
computation, *Box 9.11,* 251, *Box 14.8,* 500
Tukey's test for nonadditivity, 409
computation, *Box 13.5,* 410
Two-by-two table, 726
Two-by-two tests of independence, 726–737
computation, 731, *Box 17.7,* 734
Two-sample test, Wilcoxon's, 427–431
computation, *Box 13.7,* 429
Two-tailed F-test, 187
Two-tailed test, 79, 158
Two-way
analysis of variance, *see* Analysis of variance, two-way
frequency distributions, 573
frequency table, 726
graphic representation, 740
Model I, 727
Model II, 727–728
Model III, 728
test of independence, 724–743
Type I error, 158–169
Type II error, 159–169

U_s, Mann-Whitney sample statistic, 427
U-shaped frequency distributions, 21
U-test, Mann-Whitney, 427–431
computation, *Box 13.7,* 428
Unbiased estimate of ζ, 583
Unbiased estimators, 52, 139, 560, 583
Underdispersed distributions, 77
Unexplained mean square, 467
Unexplained sum of squares, 464
Union, 65
Unordered data, computation of $\bar{Y}$, and s, from, *Box 4.2,* 56
Unplanned comparisons among means, 230, 240–260
diagrammatic representation, 246, 249, 259–260
GT2-method, 244, 261, 499
computation, *Box 9.11,* 247, *Box 14.8,* 500
T-method, 244, 261
computation, *Box 9.10,* 245, *Box 14.8,* 500
T'-method, 244, 261, 499
computation, *Box 9.11,* 247, *Box 14.8,* 500

Tukey-Kramer method, 252, 261, 499
computation, *Box 9.11,* 247, *Box 14.8,* 500
Welsch step-up procedure, 252, 261
computation, *Box 9.12,* 253
Unplanned contrasts among means, 252–259
computation, *Box 9.13,* 255
GT2-method, 254, 261
Scheffé method, 252, 261
Sum of squares simultaneous test procedure, 254, 261
T-method, 252, 261
Unplanned tests
of homogeneity of replicates tested for goodness of fit, 721–722
computation, *Box 17.5,* 722
of k regression coefficients, 499
computation, *Box 14.8,* 500
nonparametric, 431–434
computation, *Box 13.8,* 432
Unweighted average, 43
Up and down runs test, 801

V, coefficient of variation, 58
V^*, corrected coefficient of variation, 58
Value, expected, 133
Variable, 9, 11
computed, 16
continuous, 11
criterion, 634
dependent, 453
derived, 16–19
discrete, 11
independent, 453
measurement, 11
meristic, 11
predictor, 634
ranked, 12
Variables, substitution of, 491
Variance, 51–53
among groups, 183
among means, 182
analysis of, *see* Analysis of variance
components
added among groups, 197
confidence intervals of, computation, *Box 9.3,* 215
estimation of, *Box 9.2,* 214
confidence limits, 154–157
computation by shortest unbiased confidence intervals, *Box 7.3,* 156
of a hypergeometric distribution, 94
of means, 128–136, 182
of a Poisson, 87
parametric, 52
sample, 52
of sum of two variables, 567–569
of $\hat{Y}$, derivation, 845
within groups, 182
Variance inflation factors, 632
Variances
Bartlett's test for homogeneity of, 397
equality of, 396–406
tests of homogeneity of, computation, *Box 13.2,* 398
Variate, 12
Variation, coefficient of, 57–59

w_i, weighting factor, 42
W, Kendall's coefficient of concordance, 600
Wald test, 770
Weber-Fechner law, 533
Weighted average, 42
Welsch step-up procedure, 252, 261
computation, *Box 9.12,* 253
Wilcoxon's signed-ranks test, 440–444
computation, *Box 13.11,* 443
Wilcoxon's two-sample test, 427–431
computation, *Box 13.7,* 429
Winsorizing, 407

X^2, sample statistic of chi-square distribution, 176, 397, 580, 696–697
X^2-test. *See* Chi-square test

y, deviation from the mean, 50
$\hat{y}$, explained deviation from regression, 463
$\hat{Y}_i$, estimated value of Y_i, 460
$\bar{Y}$, arithmetic mean, 41
$\bar{\bar{Y}}$, mean of means, 182
Y-intercept, 455
Y-values, adjusted, 488
Yates' correction for continuity, 730, 737

z, transformation for r, 575
z_u, unbiased estimate of ζ, 583
z^*, Hotelling's transformation for r, 579
Z, height of ordinate of normal curve, 101, 111
Z, Mantel statistic, 816
ζ, parametric value of z, 575

KEY TO STATISTICAL METHODS

	Design or Purpose	Measurement Variables	Ranked Variables	Attributes
1 variable 1 sample	Examination of a single sample	Procedure for grouping a frequency distribution, Box 2.1; stem-and-leaf display, Section 2.5; testing for outliers, Section 13.4; boxplots, Section 7.5 Test for serial independence, Box 13.1 Computing median of frequency distribution, Box 4.1 Computing arithmetic mean: unordered sample, Box 4.2; frequency distribution, Box 4.3 Computing standard deviation: unordered sample, Box 4.2; frequency distribution, Box 4.3 Setting confidence limits: mean, Box 7.2; variance, Box 7.3 Computing g_1 and g_2, Box 6.2		Confidence limits for a percentage, Section 17.1 Confidence limits for binomial and Poisson parameters, Box 7.4 Runs test for randomness in dichotomized data, Box 18.2
	Comparison of a single sample with an expected frequency distribution	Normal expected frequencies, Box 6.1 Goodness of fit tests: parameters from an extrinsic hypothesis, Box 17.1; from an intrinsic hypothesis, Box 17.2 Kolmogorov–Smirnov test of goodness of fit, Box 17.3 Graphic "tests" for normality: normal quantile plots and rankits, Box 6.3 Test of sample statistic against expected value, Box 7.5		Bionomial expected frequencies, Box 5.1 Poisson expected frequencies, Box 5.2 Goodness of fit tests: parameters from an extrinsic hypothesis, Box 17.1; from an intrinsic hypothesis, Box 17.2
1 variable ≥ 2 samples	Single classification	Single classification anova: unequal sample sizes, Box 9.1; equal sample sizes, Box 9.4 Planned comparison of means in anova, Box 9.8; single degree of freedom comparisons of means, Box 14.10; Dunn–Šidák and Bonferroni methods, Section 9.6; sequential Bonferroni tests, Box 9.9 Unplanned comparison of means: T-method, equal sample sizes, Box 9.10; T′, GT2, and Tukey–Kramer, unequal sample sizes, Box 9.11; Welsh step-up, Box 9.12; STP test, Section 9.7; contrasts using Scheffé, T, and GT2, Box 9.13; multiple confidence limits, Section 14.10 Estimate variance components: unequal sample sizes, Box 9.2; equal sample sizes, Box 9.3 Setting confidence limits to a variance component, Box 9.3 Tests of homogeneity of variances, Box 13.2 Tests of equality of means when variances are heterogeneous, Box 13.3 Required sample size and power, Box 9.14 Nonparametric test of equality of medians, Box 17.9	Kruskal–Wallis test, Box 13.6 Unplanned comparison of means by a nonparametric STP, Box 13.8	*G*-test for homogeneity of percentages, Boxes 17.5 and 17.8 Comparison of several samples with an expected frequency distribution, Box 17.4; unplanned analysis of replicated tests of goodness of fit, Box 17.5
	Nested classification	Two-level nested anova: equal sample sizes, Box 10.1; unequal sample sizes, Box 10.6 Three-level nested anova: equal sample sizes, Box 10.5; unequal sample sizes, Box 10.7		
	Two-way or multi-way classification	Two-way anova: with replication, Box 11.2; without replication, Box 11.3; unequal but proportional subclass sizes, Box 11.6; with a single missing observation, Box 11.7 Three-way anova, Box 12.1 More-than-three-way classification, Section 12.3 and Box 12.2 Test for nonadditivity in a two-way anova, Box 13.5	Friedman's method for randomized blocks, Box 13.10 Two-way replicated anova, Box 13.12	Three-way log-linear model, Box 17.9 Randomized blocks for frequency data (repeated testing of the same individuals), Box 17.16